Inhalation Toxicology

THIRD EDITION

Inhalation Toxicology

THIRD EDITION

Edited by

Harry Salem
Sidney A. Katz

CRC Press
Taylor & Francis Group
Boca Raton London New York

CRC Press is an imprint of the
Taylor & Francis Group, an **informa** business

CRC Press
Taylor & Francis Group
6000 Broken Sound Parkway NW, Suite 300
Boca Raton, FL 33487-2742

First issued in paperback 2016

© 2015 by Taylor & Francis Group, LLC
CRC Press is an imprint of Taylor & Francis Group, an Informa business

No claim to original U.S. Government works

Version Date: 20140620

ISBN 13: 978-1-138-03366-5 (pbk)
ISBN 13: 978-1-4665-5273-9 (hbk)

This book contains information obtained from authentic and highly regarded sources. While all reasonable efforts have been made to publish reliable data and information, neither the author[s] nor the publisher can accept any legal responsibility or liability for any errors or omissions that may be made. The publishers wish to make clear that any views or opinions expressed in this book by individual editors, authors or contributors are personal to them and do not necessarily reflect the views/opinions of the publishers. The information or guidance contained in this book is intended for use by medical, scientific or health-care professionals and is provided strictly as a supplement to the medical or other professional's own judgement, their knowledge of the patient's medical history, relevant manufacturer's instructions and the appropriate best practice guidelines. Because of the rapid advances in medical science, any information or advice on dosages, procedures or diagnoses should be independently veri-fied. The reader is strongly urge to consult the relevant national drug formulary and the drug companies' printed instructions, and their websites, before administering any of the drugs recommended in this book. This book does not indicate whether a particular treatment is appropriate or suitable for a particular individual. Ultimately it is the sole responsibility of the medical professional to make his or her own professional judgements, so as to advise and treat patients appropriately. The authors and publishers have also attempted to trace the copyright holders of all material reproduced in this publication and apologize to copyright holders if permission to publish in this form has not been obtained. If any copyright material has not been acknowl-edged please write and let us know so we may rectify in any future reprint.

Library of Congress Cataloging-in-Publication Data

Inhalation toxicology (Salem)
   Inhalation toxicology / editors, Harry Salem, Sidney A. Katz. -- Third edition.
   p. ; cm.
   Includes bibliographical references and index.
   ISBN 978-1-4665-5273-9 (hardback : alk. paper)
   I. Salem, Harry, 1929- editor. II. Katz, Sidney A., 1935- editor. III. Title.
   [DNLM: 1. Air Pollutants--toxicity. 2. Hazardous Substances--toxicity. 3. Inhalation--drug effects. 4. Inhalation Exposure. 5. Toxicology--methods. WA 754]
   RA1245
   615.9'1--dc23
                                       2014022612

Visit the Taylor & Francis Web site at
http://www.taylorandfrancis.com

and the CRC Press Web site at
http://www.crcpress.com

This book is dedicated to the memory of Florence Salem, who passed away on February 22, 2010. She was Harry Salem's greatest inspiration and most ardent supporter.

In addition, we dedicate the book to our colleagues, whose excellent contributions provided us with the material for Inhalation Toxicology, *as well as to our families—the Salem family (Jerry and Amy, Joel, Marshall, and Abby Rose) and the Katz family (Sheila, Craig and Ji, Kevin and Hana, Jeff and Wendy, and Syd and Daniel)—whose patience, support, and encouragement enabled us to bring together the many aspects of this finished product.*

Contents

Preface

The human body is subjected to numerous chemical exposures from the external environment. The atmosphere is the largest component of this environment, and, of the three portals for intoxication of the human body, the lungs are the largest surface exposed to this environment. Protecting the lungs from exposures to hazardous chemicals in the environment is often more difficult than protecting the human body from ingestion and dermal absorption hazards. Hazardous chemicals enter the body through the lungs in both the occupational and domestic environments. Among the substances frequently inhaled with air are an array of gases, aerosols, and particulates from natural and anthropogenic sources. In addition, some humans occasionally inhale therapeutic and/or abused drugs.

The objective of *Inhalation Toxicology* is to provide the practicing professional as well as the aspiring student with a pragmatic textbook. It includes contributions from scientists in the academic, commercial/industrial, and governmental sectors and focuses on the regulatory aspects of exposure and testing, testing equipment and procedures, respiratory allergy and irritation of the respiratory tract, risk assessment, toxicology theory and toxicology modeling, and toxic effects of some individual toxicants. The contributors are from China, Germany, Israel, Slovenia, Switzerland, and United States.

Inhalation Toxicology recognizes that the 100 m^2 surface of the lungs coupled with the 15 L/min respiration rate provide a significant opportunity for the entry of toxic and therapeutic chemicals into the human body. Transfer across the alveolar–gas interface is rapid and often enhanced by large partition coefficients.

This third edition of *Inhalation Toxicology* includes chapters on topics ranging from the collection and characterization of airborne particulate matter to the inhalation toxicology of asbestos fibers and nanoparticles to the development of *lung-on-a-chip technology* for predicting in vivo responses. To enhance its educational utility, some thought-provoking questions and answers have been added at the end of each chapter.

The compassionate friendship of Phyllis Meltzer is gratefully acknowledged and deeply appreciated. Her support and encouragement made it possible for Harry Salem to survive his loss and remain focused on life in this world.

Special acknowledgments and thanks to Donna M. Hoffman and the publisher, who helped us finalize the submission.

Harry Salem
Sidney A. Katz

Editors

Harry Salem, BSc, MA, PhD, is chief scientist at the US Army Edgewood Chemical Biological Center, Aberdeen Proving Ground, Maryland. He has been a visiting professor at Drexel University, Rutgers University, Temple University, and the University of Pennsylvania and has held a variety of positions in the pharmaceutical industry and in commercial toxicology laboratories. He has also served as the editor in chief of the *Journal of Applied Toxicology*. He was elected a Congressional Fellow of the Society of Toxicology in 2001 and is also a Fellow of the New York Academy of Sciences, the American College of Clinical Pharmacology, the American College of Toxicology, and the Academy of Toxicological Sciences.

Sidney A. Katz, PhD, is emeritus professor of chemistry at Rutgers University, Camden, New Jersey. During his long tenure there, he was invited for visiting professorships to Canada, England, Germany, Hungary, Slovenia, and South Africa. He has been a research fellow at the US Army Edgewood Chemical Biological Center and a Senior NATO Fellow at the University of London Reactor Centre. He has also received two NSF/AEC postdoctoral fellowships. Prior to his academic career, he held technical positions with E.I. DuPont de Nemours and Company, Inc., and with the R.M. Hollingshead Corporation. From 1997 to 2002, he was a New Jersey commissioner for hazardous waste facilities siting. His research activity is in environmental bioanalytical chemistry.

Contributors

Melvin E. Andersen
The Hamner Institutes for Health Sciences
Research Triangle Park, North Carolina

Bryan Ballantyne (deceased)
Consultant
Charleston, West Virginia

David M. Bernstein
Consultant
Geneva, Switzerland

Vincent Castranova
Health Effects Laboratory Division
National Institute for Occupational
    Safety and Health
Morgantown, West Virginia

Finis Cavender
Cavender Enterprises, Inc.
Highland Haven, Texas

Harvey J. Clewell III
Center for Human Health Assessment
The Hamner Institutes for Health Sciences
Research Triangle Park, North Carolina

Stephanie Cole
Edgewood Chemical Biological Center
US Army
Aberdeen, Maryland

Rhian B. Cope
Toxicology Excellence for Risk Assessment
Cincinnati, Ohio

Joseph L. Corriveau
Edgewood Chemical Biological Center
US Army
Aberdeen, Maryland

Russell Dorsey
Edgewood Chemical Biological Center
US Army
Aberdeen, Maryland

Mike Dourson
Toxicology Excellence for Risk Assessment
Cincinnati, Ohio

Arik Eisenkraft
NBC Protection Division
Office of the Assistant Minister of Defense
    for CBRN Defense
Ministry of Defense
Tel Aviv, Israel

Nabil M. Elsayed
Independent Consultant
Chatham, New Jersy

Avshalom Falk
NBC Protection Division
Israeli Ministry of Defense
Tel Aviv, Israel

Michael Feasel
Edgewood Chemical Biological Center
US Army
Aberdeen, Maryland

Shayne C. Gad
Gad Consulting Services
Cary, North Carolina

Brian E. Hawkins
Battelle
CBRNE Defense—Threat Assessment
Columbus, Ohio

Terry J. Henderson
Edgewood Chemical Biological Center
US Army
Aberdeen, Maryland

Tom Ingersoll
Joint CBRN Center of Excellence
Defense Threat Reduction Agency
Aberdeen, Maryland

Gregor Jereb
Department of Health Sciences
University of Ljubljana
Ljubljana, Slovenia

and

School for Environmental Sciences
University of Nova Gorica
Nova Gorica, Slovenia

Pius Joseph
Toxicology and Molecular Biology Branch
National Institute for Occupational
    Safety and Health
Morgantown, West Virginia

Sidney A. Katz
School for Environmental Sciences
University of Nova Gorica
Nova Gorica, Slovenia

and

Department of Chemistry
Rutgers University at Camden
Camden, New Jersey

Steven G. Kelsen
Division of Pulmonary, Critical Care,
    and Allergy Medicine
Department of Medicine
School of Medicine
Temple University
Philadelphia, Pennsylvania

Albert P. Li
In Vitro ADMET Laboratories, LLC
Columbia, Maryland

Wenli Li
Department of Toxicology
Fourth Military Medical University
Xi'an, Shaanxi, People's Republic of China

Paul J. Lioy
Robert Wood Johnson Medical School
Rutgers, The State University of New Jersey
and
Environmental and Occupational Health
    Sciences Institute
Piscataway, New Jersey

Janna Madren-Whalley
Edgewood Chemical Biological Center
US Army
Aberdeen, Maryland

Gediminas Mainelis
Department of Environmental Sciences
Rutgers, The State University of New Jersey
New Brunswick, New Jersey

Loyda B. Mendez
Department of Microbiology and Molecular
    Genetics
University of California at Irvine
Irvine, California

Glenn Milner
Principal Toxicologist
Center for Toxicology and Environmental
    Health
North Little Rock, Arkansas

Janet Moser
Chemical Security Analysis Center
US Department of Homeland Security
Gunpowder, Maryland

Patricia Nance
Toxicology Excellence for Risk Assessment
Cincinnati, Ohio

Yevgen Nazarenko
Department of Atmospheric
    and Oceanic Sciences
McGill University
Montréal, Québec, Canada

Michael J. Oldham
Regulatory Affairs
Altria Client Services
Virginia Commonwealth University
Richmond, Virginia

Jonathan Oyler
Medical Research Institute
    of Chemical Defense
US Army
Aberdeen, Maryland

Juergen Pauluhn
Department of Experimental
    Toxicology-Inhalation
Bayer Pharma AG
Wuppertal, Germany

Robert F. Phalen
Health Effects Laboratory Division
Centers for Disease Control
    and Prevention
University of California at Irvine
Irvine, California

Borut Poljšak
Departmant of Health Sciences
University of Ljubljana
Ljubljana, Slovenia

and

School for Environmental Sciences
University of Nova Gorica
Nova Gorica, Slovenia

Dale W. Porter
Health Effects Laboratory Division
National Institute for Occupational
    Safety and Health
Morgantown, West Virginia

Patricia Richter
Office on Smoking and Health
Centers for Disease Control and Prevention
Atlanta, Georgia

Harry Salem
Edgewood Chemical Biological Center
US Army
Aberdeen, Maryland

Jeffry D. Schroeter
Applied Research Associates, Inc.
Raleigh, North Carolina

Rajendran Sellamuthu
Department of Environmental Health
School of Public Health
Indiana University
Bloomington, Indiana

Douglas R. Sommerville
Edgewood Chemical Biological Center
US Army
Aberdeen, Maryland

Shuichi Takayama
Department of Biomedical Engineering
Biointerfaces Institute
University of Michigan
Ann Arbor, Michigan

Christina Umbright
Toxicology and Molecular Biology Branch
National Institute for Occupational
    Safety and Health
Morgantown, West Virginia

Joshua B. White
Department of Biomedical Engineering
Biointerfaces Institute
University of Michigan
Ann Arbor, Michigan

Ian C. Whittaker
Battelle
CBRNE Defense—Threat Assessment
Columbus, Ohio

Patrick H. Wilson
Battelle
CBRNE Defense—Threat Assessment
Columbus, Ohio

David J. Winkel
Battelle
CBRNE Defense—Threat Assessment
Columbus, Ohio

Chao Zhang
Department of Biomedical Engineering
Biointerfaces Institute
University of Michigan
Ann Arbor, Michigan

Introduction

At the conclusion to his review of the materials and methods for conducting inhalation studies, Dorato (1990) stated, "Inhalation studies have been conducted for well over 100 years. While the basic principles may have changed little, inhalation technology has been in a state of continuous evolution." His overview describes the significant advances made in the design of exposure chambers, the introduction of test materials and the determination of their concentrations, the assessment of toxicity, and the care of the subjects.

In addition to the advances described in *Overview of Inhalation Toxicology* (Dorato, 1990), the trend in alternatives to animals for toxicological testing has become a more recognized research vehicle. Normal human tracheal–bronchial epithelial cells have been cultured using serum-free media to form a three-dimensional, tissue-like structure that closely resembles the epithelial tissue of the respiratory tract. This material is commercially available as EpiAirways™ from Mat Tek Corporation (Ashland, Massachusetts). Among the many uses of EpiAirways™ for inhalation toxicology research are the detection of agents triggering asthma in the occupational sector (Shaesgreen et al., 1999), the evaluation of medication delivery efficiency by the nasal route (Quay et al., 2001), and the comparison with bovine nasal explants for nasal medication transport (Chemuturi et al., 2005). While EpiAirways™ has found extensive utilization in inhalation toxicology research, it has the shortcoming of being a static model for a dynamic system. The lung-on-a-chip model does not suffer from this handicap.

As pointed out in Chapter 3 by Takayama et al., recent advances in the field of microfluidics have advanced lung-on-a-chip technology. A lung on a chip is typically not a model of the entire lung or a device made with the intent of performing gas exchange functions for animals or humans, but rather for enhancing in vitro cell culture models for improving predictions about chemical toxicity and/or efficacy and to gain mechanistic insights into lung pathophysiology. One example of the utility of such devices is the reproduction of the drug toxicity–induced pulmonary edema observed in human cancer patients treated with interleukin-2 at similar doses and over the same time frame. These studies also led to the identification of potential new therapies including angiopoietin-1 and a new transient receptor potential vanilloid 4 ion channel inhibitor, which might prevent this life-threatening toxicity of interleukin-2 in the future (Huh et al., 2012). This achievement was heralded as a significant advance in imitating organs that can serve as testing grounds for novel medications to treat a wide variety of diseases. The lung on a chip not only modeled a serious respiratory ailment but also accurately predicted the toxicity of the compound causing it as well as the ability of a new medication to prevent it (Service, 2012). Future advances in lung-on-a-chip technology could reduce the need for slower, more expensive, and highly controversial animal testing and lead to more efficient drug development and patient treatment.

A new development in technology is sometimes associated with hazards as well as with benefits. The hazardous aspect of a material raises questions on the possibility of release into the environment and the subsequent exposure of humans and animals. Nazarenko et al. (Chapter 9) and Porter and Castranove (Chapter 10) have addressed these questions from different perspectives. Nanotechnology is a new science, and, consequently, the research on exposure is incomplete. The present risk assessments may be conservative (Rudd, 2008; Seaton et al., 2009; Goldstein, 2010; Sahu and Prusty, 2010). As the gaps in the research are filled with more information, risk assessment will be reevaluated in the future.

In Chapter 5, Hawkins et al. maintain the application of inhalation toxicology concepts to inform consequence mitigation strategies through the predictive modeling of health outcomes is an essential facet of twenty-first-century toxicology. This requires the specification of endpoints, the statistical estimation of dose–response data describing those endpoints based on toxicological information, and the rigorous mathematical application of dose–response relationships. Misapplication of toxicological concepts can lead to exaggerated estimates of consequences that are likely to confuse, or even mislead, decision makers. This error is frequently encountered in emergency response and dispersion modeling analyses that characterize consequences based on protective guidelines, such as acute exposure guidance levels (AEGLs). Such analyses generally significantly overestimate consequences. Multidisciplinary approaches merging mathematical, statistical, and toxicological techniques can ensure that predictive modeling provides the best possible guidance through a basis in sound toxicology concepts.

Inhalation risk assessments focusing on the derivation of acute emergency response guidance values, such as the ERPG (emergency response planning guidelines) or AEGL (acute exposure guidance levels), require a quantitative appreciation for the impact of concentration × time (C × t) on the point of departure (POD) for nonlethal effects. Experimental/mathematical procedures are available for the estimation of time-adjusted median lethal concentration (LCt_{50}) values for multiple exposure durations. However, these procedures often fail when applied to LCt_{01} and LCt_{05}. Estimations of these values have had to rely upon subjective comparisons and multiple assumptions that vary between experts, with the result that much controversy surrounds this activity. Consequently, toxicologists have been somewhat reticent to provide these estimates in the absence of state-of-the-art concentration × time–response data and an understanding of the toxicity endpoint's principal mode of action. Relationships found valid in rodent bioassays may not be valid for humans if the species-specific variables of *inhaled dose* and associated metric of dose are insufficiently characterized (Chapter 6).

Yu et al. (2010) have expressed concerns about the lack of generally accepted inhalation toxicology methodologies for evaluating the health effects of nanoparticles. They suggested a battery of tests for monitoring the inhalation toxicity testing chamber, including a differential mobility analyzing system (DMAS) to measure the number of particles, their size distribution and surface area, and the estimated mass dose, as well as transmission electron microscopy (TEM) or scanning electron microscopy (SEM) and energy dispersive x-ray analysis (EDXA) to determine the chemical composition for assessing the risks of inhalation exposure to nanoparticles. These are the techniques employed by Jereb et al. in Chapter 1. In their investigation on the origin of airborne particulate matter deposited in and around the marine terminal they describe in Chapter 1.

In their article, "A short history of the toxicology of inhaled particles," Donaldson and Seaton (2012) stated, "Particle toxicology arose in order to understand the mechanisms of adverse effects of 3 major particle types that had historically the greatest toll of human health–quartz, coal and asbestos." In Chapter 18, Joseph et al. present their research on the relationships between silica-induced toxicity and changes in global gene expression. Barlow et al. (2013) have expressed the opinion that the findings from studies on the effects of fiber type in mesotheliomal cells were flawed because they lacked dose–response relationships. In Chapter 14, Bernstein carefully and comprehensively compares and contrasts the physical characteristics and physiological effects of serpentine and amphiboli asbestos.

The mechanisms for the acute toxicology and pathophysiology of phosgene have been described in detail by Li and Pauluhn (in Chapter 19). Haber's rule was among the toxicological considerations they describe in Chapter 19. In the simulations comparing causality numbers using the Haber model and the ten Berg model they describe in Chapter 7, Ingersoll et al. found both predictive models tend to overestimate causality numbers and considerations of the toxic load result in more accurate predictions of casualty numbers. These models were based on exposure to toxic substances. The physiologically based kinetic models Clewell et al. discuss in Chapter 8 are more focused on the metabolic events following inhalation of volatile compounds. In Chapter 4, Cope et al. consider the influences of size on the penetration of particulate matter into the respiratory system as well as Haber's law with the ten Berg modification in their presentation on the risk assessment of inhaled materials. Salem et al. apply these principles to the inhalation toxicology of riot control agents, and Corriveau and Feasel make similar applications for the toxicology of incapacitating agents in Chapters 11 and 12, respectively. Cavender and Milner consolidate and review the data on the lethal and nonlethal effects of ammonia exposure, and they describe several incidents of accidental ammonia releases in Chapter 13.

Inhalation need not be limited to volatile compounds. In their review, Hiraiwa and van Eden (2013) proposed, "Air pollutants can be classified by their source, chemical composition, size, mode of release (gaseous or particulate), and space (indoor or outdoor)." The size distribution of airborne particulate matter was among the topics Phalen et al. include in their review of nose-only exposure systems in Chapter 2. Another of their topics included in this chapter is aerosol dispersion for controlled exposure of experimental animals. Kelson's review clearly demonstrates how animal models have greatly expanded our knowledge on the pathogenesis of chronic obstructive pulmonary disease and the mechanisms involved in emphysema and their connection with cigarette smoke and altered gene expression in Chapter 16. In Chapter 15, Li et al. describe an experimental system for measuring the cytotoxicity of eight cigarette smoke condensates and nicotine to three pulmonary cell types.

In Chapter 17, Eisenkraft and Falk describe medical responses and interventions for toxic inhalation exposures to chemicals released in industrial accidents and terroristic actions. Details for the treatment of exposures to chlorine, phosgene, and smoke are also included in this chapter. In addition to the inhalation of toxic materials, therapeutic agents have been administered by the inhalation route. Gad has reviewed the safety assessments for such administrations in Chapter 22.

The scope and depth of the material presented in the current edition of *Inhalation Toxicology* contribute significantly to resolving some of the issues raised in preceding works.

Harry Salem
US Army Edgewood Chemical Biological Center
Aberdeen Proving Ground, Maryland

Sidney A. Katz
Rutgers University
Camden, New Jersey

REFERENCES

Barlow, C.A., Lievense, L., Gross, S., Ronk, C.J., and Paustenbach, D.J. (2013) The role of genotoxicity in asbestos-induced mesothelioma: An explanation for the differences in carcinogenic potential among fiber types, *Inhal. Toxicol.*, 25(9), 553–567.

Chemuturi, N.V., Hayden, P., Klausner, M., and Dononvan, M.D. (2005) Comparison of Human Trachael/ Bronchial Epithelial Cell Culture and Bovine Nasal Respiratory Explants for Nasal Drug Transport Studies, *J. Pharm. Sci.*, 94(9), 1976–1985.

Donaldson, K. and Seaton, A. (2012) A short history of the toxicology of inhaled particles, *Particle Fiber Toxicol*, 9(13). http://www.particleandfibertoxicology.com/conten/t9/1/13. doi: 10.1186/1743-8977-9-13, accessed April 6, 2014.

Dorato, M.A. (1990) Overview of inhalation toxicology, *Environ Health Perspect*, 85, 163–170.

Goldstein, B.D. (2010) The Scientific Basis for the Regulation of Nanoparticles: Challenging Paracelsus and Paré, *J. Environ. Law*, 28(7), 7–28.

Hiraaiwa, K. and van Eden, S.F. (2013) Contribution of lung macrophages to the inflammatory response induced by exposure to air pollutants, *Mediat Inflam.* http://dx.doi.org/1155.org/1155/2013/619523, doi:10.1155/2013/619523.

Huh, D., Leslie, D.C., Matthews, B.D., Fraser, J.P., Jurek, S., Hamilton, G.A., Thorneloe, K.S., MacAlexander, M.A., and Ingber, D.E. (2012) A Human Disease Model of Drug Toxicity–Induced Pulmonary Edema in the Lung-on–a-Chip Microdevice, *Sci. Trasnl. Med.*, 159ra, 147.

Quay, S.C., deMeirelel, J., Wormuth, D., Vangalla, S., and Biswas, M., (2001) Effects of enhancers on macromolecules formation on membrane penetration, cell viability and resistance using Epiairway™ tissue model, *AAPS Pharm. Sci.*, 3(3), S–57).

Rudd, J. (2008) Regulating the Impacts of Engineered Nanoparticles under TSCA: Shifting Authority from Industry to Government, Columbia *J. Environ. Law*, 33(2), 215–282.

Sahu, S.K. and Prusty, A.K. (2010) Toxicological and Regulatory Considerations of Pharmacologically Important Nanoparticles, *J. Current Pharm. Res.*, 3(1), 8–12.

Seaton, A., Tran, L., Aitken, R., and Donaldson, K. (2009) Nanoparticles, human health hazard and regulation, *J. Royal Soc. Interface*, 7(Suppl. 1), S 119–129. doi: 10.1098/rsif.2009.0252.focus.

Service, R.F., (2012), Lung–on–a–Chip Breaths New Life into Drug Discovery, *Science*, 338(6108), 731.

Shaesgreen, J., Klausner, M., Kubilus, J., and Ogle, P. (1999) Reconstructed Differentiated Airway Epithelial Cultures to Detect Occupational Asthma Causing Agents, *Toxicologist*, 48(1–S), 126.

Yu, I.J., Ji, J.H., and Ahn, K.H. (2010) Development of International Standards for Nanotechnology and Risk Assessment, IEEE–NANO, 10th IEEE Conference on Nanotechnology, August 17–20.

1 Collection and Characterization of Particulate Matter Deposition

Gregor Jereb, Sidney A. Katz, and Borut Poljšak

CONTENTS

1.1 INTRODUCTION

1.1.1 Particulate Matter Deposition

Atmospheric dust (large or coarse particles, also known as dust or grit) has traditionally been considered mainly as a cause of nuisance rather than a health hazard. However, particulate matter deposition not only represents a major concern due to nuisance for people but can also be a cause of health and other problems (Hall et al., 1993), either from direct inhalation or ingestion or through secondary pathways after particles have deposited on vegetables, fruits, or drinking water supplies.

According to the process of deposition, particles can be removed from the atmosphere and deposited at the Earth's surface in two ways, dry deposition or wet deposition, depending on the phase in which it strikes the Earth's surface (Finlayson Pitts and Pitts, 2000). Dry deposition is characterized by direct transfer of gas phase and particulates from air to ground, vegetation, water bodies, and other Earth surfaces. This transfer can occur by sedimentation, impaction, and diffusion to surfaces or, in case of plants, by physiological uptake. In the absence of precipitation, dry deposition plays the major role in removing pollutants from the atmosphere (Finlayson Pitts and Pitts, 2000). Wet deposition includes rain, snow, fog, cloud, dew formation, and washout from the atmosphere (Godish, 2004; Vallero, 2008).

Particulate matter deposition is usually measured using simple passive devices. However, these methods of sample collection have some deficiencies due to the aerodynamic characteristics of the sampling gauge and also the direct influence of wind (wind flow around the gauge either prevents the entry of particles into the gauge opening or removes particles after deposition due to a wind-generated circulation inside the gauge) (Hall et al., 1993). Due to exposure to various disturbances, such measurements are less reliable. Nevertheless, with some improvements of the measuring device, results are indicative and therefore useful. In the work presented here, some improvements of particulate matter deposition sampling are described as well as different analyzing techniques used for the identification and characterization of particulate deposition in order to correlate it with the source.

1.1.2 Port of Koper

The Port of Koper is located in the Republic of Slovenia at the top of the Adriatic Sea across from Venice and just below Trieste. The activity of the port has shown annual increases during the last 20 years following the independence of the Republic of Slovenia. During the year 2010, throughput at the port was 15,372,047 tons. Almost a half million 10-ton containers and a quarter million automobiles came through the port. The greatest mass of material handled at the port, however, was dry bulk cargo, 6,363,557 tons (Luka Koper 2012). Coal constitutes nearly two-thirds of this dry bulk cargo. Iron ore is another significant part of this dry bulk cargo. Coal and iron ore are exported to both Italy and Austria (Carinska uprava Republike Slovenije 2012). Coal is imported into Slovenia from the Far East and from South Africa, and it is stored temporarily at the Port of Koper. Open storage covering an area of 100,000 m² can accommodate 500,000 tons of coal and 300,000 tons of iron ore. The open coal storage piles can be seen in Figure 1.1.

The coal storage piles are now surrounded by an enclosure 11 m high, and a water spray system has been installed to reduce dusting during loading operations. A part of this dust mitigation/containment system is shown in Figure 1.2. Wet cleaning of the transportation roads around the European Energy Terminal (EET) has been initiated as an additional dust control measure.

Some of this coal is used domestically, but a large part of it is exported to neighboring European states. ENEL, Italy's largest power company, imports much of the coal from the Port of Koper. A Siwertell bulk coal loader supplied by Cargotec with a capacity of 2000 tons/h (for coal with density between 0.8 and 1.2 tons/m³) is now in operation at the Port of Koper (MacGregor 2012).

FIGURE 1.1 Port of Koper and open coal stockpiles, as indicated by the arrow. (Courtesy of Public Relations Office at the Port of Koper, Koper, Slovenia.)

FIGURE 1.2 Part of the dust mitigation/containment system at the Port of Koper. (Courtesy of Public Relations Office at the Port of Koper, Koper, Slovenia.)

1.1.3 FUGITIVE DUST EMISSIONS

The inhabitants of Ankaran and Rožnik, residential districts within 2 km to the northeast of the coal storage depot, have voiced many complaints about black dust emissions alleged to originate from the EET in the Port of Koper. The closest residence is 1000 m from the storage sites. Particulate matter depositions were visible in yards, gardens, terraces, and freshly washed laundry. Evidence for this deposition following a period of strong westerly wind during October of 2006 is shown in Figure 1.3. According to a 2005 investigation of environment pollution in Ankaran (ARSO, 2005) made by the Environmental Agency of the Republic of Slovenia (ARSO), the 24 h limit value for

(a)

(b)

FIGURE 1.3 Photographs taken after strong winds in October 2006 showing a clean gauze (a) and the same gauze (b) after wiping a terrace floor at a residence in Rožnik approximately 1000 m from the coal stockpile. (From Jereb, G. et al., *Int. J. Sanit. Eng. Res.,* 3(1), 4, 2009. With permission.)

particles PM_{10} was exceeded. On the basis of annual data collected from 2003 to 2008, Cepak and Marzi (2009) reported, "The legally prescribed limit (40 µg/m³) has never been exceeded." However, data presented in this report show that the legally prescribed limits for total dust mass had been exceeded at some sampling sites in and around the Port of Koper.

1.1.4 PUBLIC HEALTH IMPLICATIONS

Several epidemiological studies (Clancy et al., 2002; Hoek et al., 2002; Brook et al., 2004; Dominici et al., 2006; Miller et al., 2007; Pope III et al., 2008) have shown correlations between increased concentrations of particulate matter in air and increased incidences of respiratory and cardiovascular diseases.

The adverse effects of coal dust inhalation among coal miners are well known as the black lung disease (Wouters et al., 1994; Cohen et al., 2008). Evidence and concern for adverse respiratory effects among children residing in the vicinity of coal mining operations and coal transportation routes are increasing (Bradin et al., 1994; Pless-Mulloli et al., 2001). Schins and Borm (1999) have reviewed the inhalation toxicology of coal dust. Their emphasis is on the interactions of macrophages and neutrophils, epithelial cells, and fibroblasts with reactive oxygen species, cytokines, growth factors, and other macromolecules in relation to tissue damage in the respiratory tract. Kania et al. (2011) have reported on the controlled inhalation exposure of rats to coal dust and to coal dust plus cigarette smoke. The results of these experiments indicate that such exposures initiate systemic oxidative stress. Animals exposed to 12.5 mg/m³ of PM_{10} coal dust for 5 min showed statistically significant increases in malondialdehyde levels in blood and statistically significant decreases in superoxide dismutase activities in serum compared to controls. These differences were more pronounced following 5 min exposures to cigarette smoke after the coal dust exposures.

1.1.5 PURPOSE OF WORK

The purpose of the work described in this chapter was to collect and characterize dust deposition in and around of the Municipality of Koper in order to determine what, if any, impact the outdoor coal and iron ore storage depots at the port may have on the deposition of this material on the residential areas. Of particular concern were the spatial distributions of the dust depositions around the coal storage depots and characteristics of the dust depositions collected from various locations around the coal storage depot. Until now, no study dealing with the contributions of particulate matter emissions from the Port of Koper to the surrounding area has been conducted. The work described here is based on two 1-year studies. The focus of the first study was the development of methodologies for sample collection and characterization. The second study refined these techniques and applied them to assessing the impact of the coal storage depot on dust depositions in the residential areas of Ankaran and Rožnik.

1.2 MATERIALS AND METHODS

Standard and modified methods for the collection of particulate matter deposition were employed. Some modifications were made in order to optimize collection. Deposition was determined gravimetrically. In addition, further analyses of metal contents in the samples, examinations by electron microscopy, and measurements of stable carbon isotopes were performed with the aim to detect and characterize particles in the deposition and possibly correlate these particles with activities at the EET at the Port of Koper. In addition, a new technique for quickly estimating the direction and quantity of airborne particulate matter was developed in the course of this study.

1.2.1 SAMPLING SITES AND SAMPLING METHODS

Sampling sites 1–6 were used in the 2005–2006 phase of the study. Their locations relative to the coal storage depot are shown in Figure 1.4. In the 2007–2008 study, four more sites were added. Sampling sites 6 and 7 were used as control sites. Sampling site 6 was located in the residential area of Hrvatini. Sampling site 7 was located on Debeli Rtič, several kilometers north of the Port of Koper.

The sampling sites used for the collection of particulate matter deposition and for the estimation of any influence of the Port of Koper on the total dust deposition in surrounding areas during the preliminary study 2005/2006 were located northeast of the port as shown in Figure 1.4. Some of the sampling sites were placed directly in the residential area. Locating a sampling site on the micro level was based on the anticipated influence on emissions by the immediate area. Micro level, in this sense, means the location of a measurement site and the influences the immediate surroundings

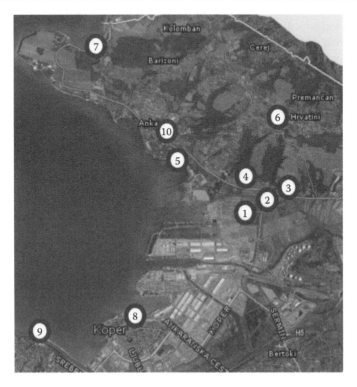

FIGURE 1.4 Location of sampling sites: sites 1–6 were used in the 2005–2006 part of the study. Sites 7–10 were added for the 2007–2008 part of the study.

(distance between the nearby buildings, trees, etc.) may have on this location. The sampling gauges were placed 1½–2 m above ground level (the height of the breathing zone for an average adult). Care was taken to insure no obstacles that could have influence on the airflow passing the sampling gauge were in close proximity of the sampling sites.

During the first year of the study, from October 2005 to October 2006, six sampling sites in the area of Ankaran and its surrounding were established. Their locations are shown in Figure 1.4. The first five sampling sites were placed in the prevailing direction of the wind or in the direction toward the residential area. Sites 1, 2, and 3 were located downwind from the coal storage depot in order to estimate the influence of distance on dust sedimentation. The sixth sampling site was located in the residential area of Hrvatini. Due to its distance from the Port of Koper and its height above sea level, a negligible influence was expected from the coal storage depot at the EET in the Port of Koper. Site 6 was used as a control sample to measure the background deposition.

The standard Bergerhoff sedimentators were used for the collection of particulate matter deposition in accord with the method 2119 part 2 from the Association of German Engineers, Bestimmung des Staubniederschlags mit Auffanggefassen aus Glas (Bergerhoff-Verfahren) oder Kunststoff (Determination of Dust Precipitation with Glass [Bergerhoff method] or Plastic Collection Receptacles) (VDI, 1996). The same method is specified by the Association of Clean Air Societies of Yugoslavia number 201 (SDČVJ, 1987). This method is used as the standard method for the measurement of particulate matter deposition in Slovenia. In the first year of the study, three Bergerhoff sedimentators were placed on each of the sampling sites according to the Verein Deutscher Ingenieure/The Association of German engineers (VDI) standard (1996).

The method for the collection of particulate matter deposition described in VDI 2119 part 2 (1996) is deficient in one respect. It is subject to potential wind-driven losses of particulate matter. When a wind blows through the sample container, some particles may be blown away and not be counted as a part of the deposited material. These particles may precipitate at another location when the wind abates or when an obstacle is encountered. For this reason, a metal screen oriented toward the storage depot was added as such an obstacle. To improve the collection method, distilled water was added to prevent the wind from blowing away the particulate matter deposition from the containers. The purpose of the modifications was to establish and quantify the fraction of wind-borne particulate matter from the direction of the storage depot. The modified (Model B) and unmodified (Model A) samplers are shown schematically in Figure 1.5 and actually in Figure 1.6.

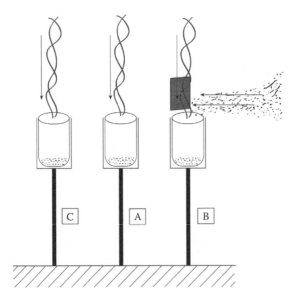

FIGURE 1.5   Schematic of samplers. (From Jereb, G. et al., *Int. J. Sanit. Eng. Res.*, 3(1), 4, 2009. With permission.)

FIGURE 1.6 Photograph of samplers. (From Jereb, G. et al., *Int. J. Sanit. Eng. Res.,* 3(1), 4, 2009. With permission.)

During the first year of the study, three sample collectors were placed at each sampling site. The mass of material deposited in the first container, container A, was determined monthly. The mass of material deposited in the second container, container C, was meant to be determined annually. The mass of material deposited in the third container, container B, representing both the vertical and the horizontal contributions to deposition, was determined monthly. The yearly sample from container C was unusable due to the presence of decayed insects and plant leaves and contamination with algae. The yearly value of particulate matter deposition was calculated from the monthly values. Therefore, during the second year of the study, from December 2007 through December 2008, the third collector (C) was omitted. During the second year, four more sampling sites were added, increasing their number from 6 to 10. The locations of the second-year sampling sites are shown in Figure 1.4. Those numbered 1 through 6 were retained from the 2005–2006 study. Sampling sites identified as numbers 7–10 are the sites added for the 2007–2008 sampling year.

1.2.2 Gravimetric Analysis

During first study, the samples were collected for gravimetric analysis on the 15th day of each month for a 1-year period from October 2005 to October 2006. During the second study, the sample period was shortened to 15 days, and the samples were collected for gravimetric analysis twice per month on the 1st and 15th of each month from December 2007 through November 2008.

The masses of the particulate matter deposited at each sampling site were determined by gravimetric analysis. The analytical scheme allowed the separation and determination of particles larger than and smaller than 3 μm. The glass collecting bottles from each of the samplers were taken to the laboratory. Insects or parts of insects, bits of leaf, etc., when found, were removed prior to initiating the gravimetric analysis.

Particulate matter was washed from the collecting bottles with distilled water. The washings were filtered through a previously dried (105°C ± 5°C), cooled, and weighed cellulose nitrate membrane filter of pore size 3 μm using a vacuum filtering device. A modification of procedure ISO 11923 (1997) was used. Three blank filters were included in the procedures used for processing the samples.

FIGURE 1.7   Filters used to filter samples from the control site (left) and from the site closest to the coal depot (right).

Shown in Figure 1.7 are the filters used to filter the sample from control sampling site, site 6, on the left and from the sampling site closest to the coal depot, site 1, on the right. Both samples were collected during the period February 15–March 15, 2006. The filters containing particles larger than 3 μm were dried for 60 min at 105°C ± 5°C and weighed using an analytical balance, capable of weighing with a precision of at least ± 0.1 mg.

The volumes of the filtrates containing particles smaller than 3 μm were measured, and 250 mL aliquots of each of the filtrates were transferred to individual previously dried (105°C ± 5°C), cooled, and weighed evaporating dishes. The liquids were evaporated to dryness at a temperature of 80°C ± 5°C. After that, the evaporating dishes were dried at 105°C ± 5°C for 1 h, cooled, and reweighed. The masses of particulate matter larger and smaller than 3 μm were calculated and presented as mg/m²·day. The particulate matter was retained for examination by atomic spectrometry, gamma spectrometry, $^{13}C/^{12}C$ mass spectrometry, and electron microscopy.

1.2.3   METAL DETERMINATIONS

During the 2005–2006 part of the study, the determination of metals was made by atomic absorption spectrometry for each sampling location on the particulate matter deposition collected during the entire year. These measurements were made in the Chemical Laboratory at the Koper Public Health Institute.

Samples collected during the 2007–2008 part of the study were delivered to the Laboratory for Sanitary Chemistry at the Public Health Institute Celje for the determination of aluminum, cadmium, chromium, copper, iron, lead, nickel, and zinc. After weighing the filters containing particulate matter larger than 3 μm, half of each filter was prepared by microwave digestion with 10 mL HNO_3 followed by dilution up to a final volume of 500 mL. The concentrations of the metals were determined by inductively coupled plasma mass spectrometry (ICP-MS), and the results for each metal were calculated as μg/m²·day. For various reasons (vandalism, mechanical destruction of the sample, organic pollution), some of the filters were discarded without making the analyses.

1.2.4   GAMMA-RAY SPECTROMETRY

Because radionuclides are frequently present in coal as impurities (Van Hook, 1979; Dowdall et al., 2004; Ward and Suárez-Ruiz, 2008), samples of coal and iron ore were measured for the emission of gamma radiations. When the gamma-ray activity were found to be above background levels, further analyses of the samples of particulate matter deposition would be performed. The gamma-ray emitter's activity in the sample of coal and iron ore was measured using high-resolution gamma-ray spectrometry (HRG) following LMR-DN-10 (Korun and Glavič Cindro, 2003) procedure in the Jožef Stefan Institute. The reported uncertainties were calculated in accordance with GUM guidelines (ISO/IEC, 1995). The results of the measurement of samples with gamma-ray spectrometry are the activities of all gamma-ray emitters present in the sample in concentrations exceeding their minimum detectable activities. The measured activity of ^{40}K makes possible a comparison with the results of a chemical analysis, offering a way for an independent check of activity measurements.

The specific activity of ^{40}K measured in the sample of coal was 25 ± 4 Bq/kg corresponding to 780 ± 120 ppm of potassium. The specific activity of ^{40}K measured in the sample of iron ore was 35 ± 6 Bq/kg corresponding to 1100 ± 190 ppm of potassium. The activity of ^{226}Ra is determined from activities of short-lived radon progenies (^{214}Pb and ^{214}Bi). The factor describing the equilibrium between radium and radon progeny was calculated from the emanation rate of radon from the sample and the time interval between sample preparation and measurement. The activity of uranium is determined assuming that ^{238}U is in equilibrium with progenies ^{234}Th and ^{234m}Pa and that the ratio of concentrations of ^{235}U and ^{238}U equals its natural value.

Radionuclide analyses were performed only during the 2005–2006 study on samples of coal and iron ore collected at the respective stockpiles at the Port of Koper. Specific activities of gamma-ray emitters were calculated back to the date August 25, 2006. A subsample of the coal and a 1160 g of iron ore were analyzed. Additionally, a sample of soil from an uncultivated area in Slovenia was analyzed as a reference for the normal amount of radionuclides in the soil.

1.2.5   SCANNING ELECTRON MICROSCOPY–X-RAY FLUORESCENCE SPECTROMETRY

Problems are frequently encountered when attempting to visually identify and quantify particles of coal in environmental samples containing plant fragments, insect fragments, pollen, etc. For this reason, scanning electron microscopy–energy-dispersive x-ray fluorescence spectrometry (SEM–EDXS) was employed for determining the morphological structure of particles in particulate matter deposition as well as measuring some of its metallic components. The SEM–EDXS measurements were performed in the Department for Nanostructured Materials at Jožef Stefan Institute.

From the preliminary study (2005–2006), only samples collected between February 15, 2006, and March 15, 2006, were evaluated by SEM–EDXS because extremely high particulate matter depositions occurred during this time. For the SEM–EDXS analysis, only small fractions of filters were taken from original filters. In order to determine coal and iron ore in the samples of particulate matter deposition originating from the EET depot at the Port of Koper, shape, size, and elemental chemical composition of collected particulate matter were determined.

For electron microscopy during the 2007–2008 part of the study, portions from the individual sampling collected over the entire year at each sampling site were combined. A part of each original filter was selected for further processing. Samples were analyzed qualitatively and quantitatively.

The cellulose nitrate membrane filters (pore size of 3 μm) had overall diameters of 47 mm and effective diameters of 41 mm. A circle having a diameter of 6 mm (1/61 the surface of the sample) was cut from each filter. Due to various reasons (deliberate contaminated sample [sand, firecracker, etc.], organic pollution of the sample [leaves and the droppings of birds], the error in the process [spilled filter with alcohol, the occurrence of copper oxide, and the occurrence of salts in the sample]), it was necessary to eliminate from further analysis a few samples from individual sampling sites throughout the year. Cut sections of the filter (2r = 6 mm) were combined (in annual terms) and inserted in a test tube having a volume of 50 mL. The particulate matter was resuspended in 50 mL of distilled water after several cycles of ultrasonic homogenization.

From each homogenate, 10 mL was taken out using automatic pipette and transferred to the Macherey Nagel polycarbonate membrane filter (pore size 0.4 μm, 2r = 47 mm) and filtered using a vacuum filtering device. Specimens prepared in this way were transferred to the Department for Nanostructured Materials at Jožef Stefan Institute, where the scanning electron microscopic examinations were conducted.

Approximately 5 × 5 mm specimens were cut out in the center of pre-prepared polycarbonate filters (pore size 0.4 μm) and prepared for processing using SEM. Pieces of filters were fixed on the sample holder using double-sided self-adhesive conductive tape. A thin layer of carbon was deposited on filter samples with Balzers carbon coater.

The samples were investigated at 500× magnification. On each sample, up to 12 areas were investigated. The number of particles in each analyzed area was calculated per cm^2 of the filter. The average number of particles has been multiplied by the number of 2551.02 (normalization to 1 cm^2). The result was given as the average number of particles per cm^2 of the filter.

Qualitative elemental analysis, morphological appearance, and size measurements of the individual particles were performed. Also quantitative stereological analysis was performed—the number of the particles was counted in an area of known size. Results were given as number of particles on the filter surface.

1.2.6 MASS SPECTROMETRY

Stable isotopes of light biogenic elements (including carbon) are ideal for monitoring the natural flux of biogeochemical and transport processes in different environments. Using ^{13}C, the circulation of carbon in the environment can be monitored. In addition, the origin of organic carbon can be determined. Different sources of carbon in the environment have specific ratios of the carbon isotopes ^{13}C and ^{12}C. Analyses of the $^{13}C/^{12}C$ isotopic ratio in samples of the particulate matter deposition were used to estimate the proportion of carbon in the samples that originated from the coal in the stockpile at the EET.

The ratios of stable carbon isotopes $^{13}C/^{12}C$ in samples of the particulate matter depositions were determined using mass spectrometry. The isotopic composition or the ratio between the heavier and lighter isotopes of carbon is expressed as a δ-value. The δ-value represents the relative difference of isotopic composition, and it is expressed in ‰ using the following equation:

$$\delta^{13}C = \frac{R_{se} - R_{ss}}{R_{ss}} \times 1000 \tag{1.1}$$

where
  R_{se} is the $^{13}C/^{12}C$ isotope ratio in the sample
  R_{ss} is the $^{13}C/^{12}C$ isotope ratio in standard

International standards for isotopic measurements are available from the International Atomic Energy Agency (IAEA) in Vienna. Standards are selected according to the similarity of the average prevalence of certain isotopes in nature. For carbon, a carbonate standard V-PDB (Vienna Pee Dee Belemnite) is used with R_{ss} = 0.0112372 (Craig, 1957).

The ratio of carbon isotopes $^{13}C/^{12}C$ was measured only in samples from the 2007 to 2008 study. For analyses, one-quarter of each of the two filters from the same month was combined. In this way, 12 monthly samples were obtained for each sampling site. From each quarter of the filter, the deposited particulate matter was removed by scraping with a surgical scalpel and placed in a test tube (volume 10 mL). Further analysis of the samples was completed at the Department of Environmental Sciences of Jožef Stefan Institute. Each sample was inserted into the silver capsule for analysis.

Inorganic carbon was removed by adding HCl to the silver capsule containing the sample and heating the capsule and its contents at 60°C until the sample was completely dry. Measurements of carbon isotopic composition were performed by mass spectrometry. Accuracy and precision of the measurements was monitored with the use of laboratory standards. For carbon, urea standard with $\delta^{13}C$ value of –30.6 ± 0.2‰ was used. In addition, the quality of the measurements was monitored using reference standards IAEA-NBS (oil), IAEA-CH-7, and IAEA-CH-6 with $\delta^{13}C$ values of 29.7 ± 0.2‰, 31.8 ± 0.2‰, and 10.4 ± 0.2‰, respectively. The precision of measurement was usually ± 0.2‰.

The control sample (blank cellulose nitrate filter) and a sample of coal from the EET stockpile were analyzed under the same conditions as the standards. On the basis of the ratios of stable carbon isotopes $^{13}C/^{12}C$ in the monthly samples, in the control and in the coal from the EET depot, the coal content of the samples was calculated according to the following isotopic mass balance equation:

$$\delta^{13}C_{sample} = F_{coal} \times \delta^{13}C_{coal} + F_{control} \times \delta^{13}C_{control}$$

$$1 = F_{coal} + F_{control}$$

(1.2)

where

$\delta^{13}C_{sample}$ is the isotopic composition of the sample
$\delta^{13}C_{coal}$ is the isotopic composition of the coal
$\delta^{13}C_{control}$ is the isotopic composition of the control
F_{coal} is the fraction of coal in the sample
$F_{control}$ is the fraction of control in the sample

1.2.7 DIRECTIONAL SAMPLE COLLECTION

For a quick estimation of the direction from which the particulate matter was deposited and the quantity of particulate matter deposited, both of which were necessary for determining the main sources of emissions of dust particles, a simple alternative measurement device based on deposition and adhesion was developed. The basic idea was to collect the particulate matter from air on adhesive material (medical petrolatum). The measurement device enabled collection of particles in both the horizontal and vertical directions. For the measuring device, plastic balls covered with medical petrolatum were used. A hole at the bottom was fitted with a wooden stick serving as a carrier. This device is described in detail elsewhere (Goličnik et al., 2008). The diameter of the ball is 20 cm (0.20 m); the calculated area of the ball (the trap surface) is 1256 cm² (0.1256 m²). Sampling devices were placed 1.5–1.7 m above ground level (the height of breathing of an adult person) and located around the coal and iron ore depot in the Port of Koper.

This directional sampling device was placed at sampling sites 1–5 and 10 for different 1-month sample collection intervals during the 2007–2008 study year. These sites are among those marked in Figure 1.4. After the exposure time, the balls were analyzed in the laboratory. Particles were counted manually using magnifier lenses on each major point of the compass (N, S, E, and W) and in the direction oriented toward the EET ore depot. On each side of the ball, particles were counted five times on the surface of 1 cm² using a special template model.

In addition to manual particle counting, a digital photograph of each ball was taken with single-lens reflex camera (Canon EOS 350D) for further analyses. Photographs taken between October 15 and November 15, 2008, were processed and analyzed, as opposed to manual counting of particles, where six 1-month average samples were analyzed.

Computerized particle counting with software (IT 3.0, Image Tool 3) was used to analyze the digital photographs. This program is an open code program for analysis and image processing. The program was developed by a group of experts Don Wilcox, Brent Dove, Doss McDavid, and David Greer from the University of Texas Health Science Center in San Antonio (UTHSCSA). This program enables viewing, editing, analyzing, compressing, storing, and printing images. The program may process different types of image formats, including BMP, PCX, IF, GIF, and JPEG. The program allows automatic particle counting (black dots) in pre-prepared pictures. Results of counting were presented as number of particles per cm^2.

Additionally, SEM–EDXS was used to examine the morphology structure of particles trapped on the sampling device as well as to determine their chemical composition. The SEM–EDXS measurements were made in the Department for Nanostructured Materials at Jožef Stefan Institute. The surface of 1 cm^2 was scraped and transferred onto the Macherey Nagel polycarbonate membrane filter (pore size 0.4 μm, 2r = 47 mm) and analyzed by electron microscopy.

1.3 RESULTS AND DISCUSSION

1.3.1 LEGISLATION

In Slovenia, the particulate matter deposition was previously regulated by the decree on limiting values, alert thresholds, and critical emission values for substances in the atmosphere (1994). The maximum value for the monthly (350 mg/m^2·day) and annual (200 mg/m^2·day) deposition was enforced. On July 2007, this decree was annulled. Consequently, these dust deposition limiting values have been used only as recommended values since 2007 in Slovenia. No statutory or official air quality criterion for dust annoyance (dust deposition) has been set by UK, European, or World Health Organization (WHO) agencies; however, an unofficial guideline has been used widely in environmental assessments in several countries throughout the world (Environment Agency, 2004):

- In England and Wales, a *custom and practice* limit of 200 mg/m^2·day is used for measurements with dust deposition gauges.
- In the United States, Washington has set a state standard of 187 mg/m^2·day for residential areas.
- The German TA Luft criteria for *possible nuisance* and *very likely nuisance* are 350 and 650 mg/m^2·day, respectively.
- Western Australia also sets a two-stage standard, with *loss of amenity first perceived* at 133 mg/m^2·day and *unacceptable reduction in air quality* at 333 mg/m^2·day.
- The Swedish limits promoted by the Stockholm Environment Institute range from 140 mg/m^2·day for rural areas to 260 mg/m^2·day for town centers.

1.3.2 GRAVIMETRIC ANALYSIS

Two collection methods were employed. Method A collected vertical deposition, and the particulate matter deposited both vertically and horizontally was collected by Method B. The samplers for Methods A and B are shown in Figures 1.5 and 1.6. The measured monthly vertical depositions of particulate matter using the standard method (Method A) of collection at each sampling site (in mg/m^2·day) are presented in Figures 1.8 and 1.9. Figures 1.10 and 1.11 show that more dust was collected when the sampling device was modified by adding a 20 cm × 30 cm metal screen (Figures 1.5 and 1.6) oriented toward the iron ore and coal stockpiles (Method B) than was collected using the standard method (Method A). With the metal screen, the horizontal transport of the dust by wind could be collected also. Also with the addition of distilled water, the loss of dry dust sample from the glass by the action of strong wind was prevented. It is important to emphasize that the limit (recommended) value applies only for the collection of particulate matter deposition using a

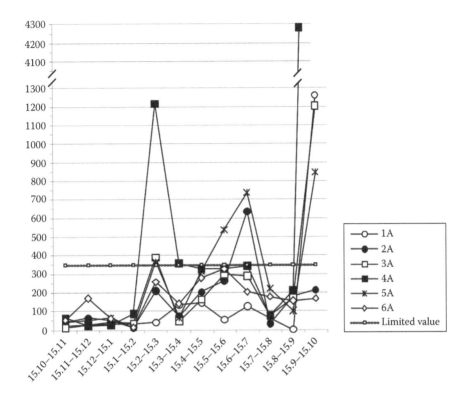

FIGURE 1.8 Particulate matter deposition during 2005–2006. Method A—limit value 350 mg/m²·day. (From Jereb, G. et al., *Int. J. Sanit. Eng. Res.*, 3(1), 4, 2009. With permission.)

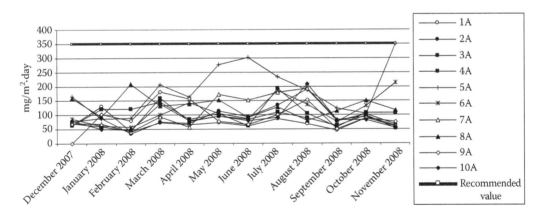

FIGURE 1.9 Particulate matter deposition during 2007–2008. Method A—recommended value 350 mg/m²·day.

standard measuring device (Method A). The limit value of 350 mg/m²·day is included in Figures 1.8 through 1.11 (for the second part of the study—2007/2008) for reference purposes only. During the sampling period from October 2005 to October 2006, the total amount of particulate matter deposition at all of the sampling sites in the area of the municipality of Ankaran and its surroundings exceeded the limit value of 350 mg/m² · day 11 times, corresponding to 15% of all measurements.

The highest monthly values were recorded in the period from September 15, 2006, to October 15, 2006. During this period, the monthly emission limit value was exceeded at nearly all of the

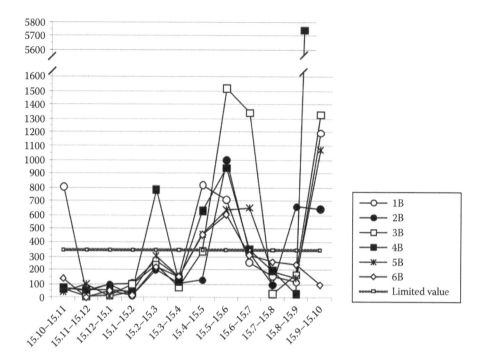

FIGURE 1.10 Particulate matter deposition during 2005–2006. Method B—limit value 350 mg/m²·day. (From Jereb, G. et al., *Int. J. Sanit. Eng. Res.*, 3(1), 4, 2009. With permission.)

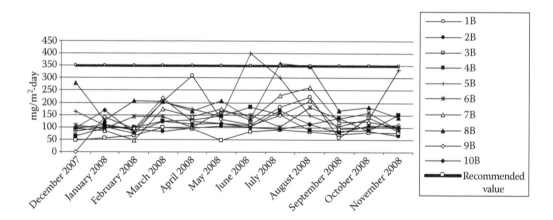

FIGURE 1.11 Particulate matter deposition during 2007–2008. Method B—recommended value 350 mg/m²·day.

sampling sites. A similar increase in particulate matter deposition values was observed in the February 15 to March 15, 2006, and June 15 to July 15, 2006, time periods. The greatest monthly value for particulate matter deposition was measured at sampling site 4 (Rožnik residential area) during the September 15 to October 15, 2006, time period (4283 mg/m²·day). At the same sampling site, extremely high values were observed also during the February 15 to March 15, 2006, time period (1216 mg/m²·day). At sampling sites 1, 3 and 5, very high values were observed between September 15 and October 15, 2006. These high values could be attributed to strong wind, which blew for a short time (a few hours) from the direction of the sea across the coal and iron ore storage piles toward the residential area. The consequences were reported also by the local residents as dust

TABLE 1.1

Particulate Matter Deposition: Method A (Particles <3 μm during 2005–2006 [mg/m²·Day])

	1A	2A	3A	4A	5A	6A
October 15–November 15, 2005	18	28	5	54	37	31
November 15–December 15, 2005	15	48	13	5	30	97
December 15, 2005–January 15, 2006	28	25	39	23	49	52
January 15–February 15, 2006	13	5	24	39	5	5
February 15–March 15, 2006	35	198	365	329	341	245
March 15–April 15, 2006	115	53	25	125	42	110
April 15–May 15, 2006	129	178	122	146	247	176
May 15–June 15, 2006	45	229	291	301	498	279
June 15–July 15, 2006	115	625	265	311	712	149
July 15–August 15, 2006	50	23	56	34	152	148
August 15–September 15, 2006	[a]	68	149	180	79	53
September 15–October 15, 2006	1228	207	1094	4157	828	148

[a] Sample was not analyzed (sample was destroyed).

deposition seen on yards, gardens, and facades. The photographs shown in Figure 1.3 were taken in the proximity of sampling site 4 in October 2006, during the period when the values of particulate matter deposition were extremely high.

Initiation of corrective measures at the Port of Koper in 2007 apparently reduced the fugitive dust emissions. Installation of water spray at the iron ore and coal stockpiles and construction of a dust curtain around these stockpiles were completed in October of 2006 and became operational shortly thereafter.

The monthly depositions of particles smaller than 3 μm collected by Method A and Method B during the 2005–2006 and during the 2007–2008 phases of the investigation are listed on a site-by-site basis in Tables 1.1 through 1.4. The results show considerable differences between individual months by both sampling methods, A and B.

At almost all sampling sites, higher values of particulate matter deposition were collected in the Model B sampler compared to the Model A sampler. Spearman's correlation coefficient (ρ) for collections using Models A and B is relatively high. The only exception is sampling site 1, where no correlations were observed. For sampling site 2, $\rho = 0.84$ ($p = 0.001$); for sampling site 3, $\rho = 0.797$ ($p = 0.002$); for sampling site 4, $\rho = 0.699$ ($p = 0.011$); for sampling site 5, $\rho = 0.902$ ($p = 0.000$); and for sampling site 6, $\rho = 0.783$ ($p = 0.003$) for Models A and B, respectively. In the case of windblown dust, the particles are *released* when they encounter the metal screen on the Model B sampler. Consequently, the amount of particulate matter deposition in the Model B sampler is higher than that deposited in the Model A sampler. Natural obstacles such as buildings, trees, and other vegetations could cause similar effects. In addition, it seems that the quantity of particulate matter deposition is dependent on weather-related factors, wind and rain, more than on the manipulation of the coal and the quantity of coal in the stockpiles at the EET depot.

1.3.2.1 Meteorological Data

The data on wind direction and velocity were organized to match the sampling intervals, and the percentage of the wind blowing across the EET storage depots for the coal and the iron ore toward the sampling sites was shown in Table 1.5. Wind velocity, relative humidity, and ambient temperature are factors associated with dust deposition. However, since the time interval for sampling the

TABLE 1.2
Particulate Matter Deposition: Method A (Particles <3 µm during 2007–2008 [mg/m²·Day])

	1A	2A	3A	4A	5A	6A	7A	8A	9A	10A
December 1–December 15, 2007	50	58	39	56	152	61	60	110	—[a]	62
December 15, 2007–January 1, 2008	25	40	58	25	111	40	28	—[a]	—[a]	38
January 1–January 15, 2008	160	59	49	85	67	50	43	60	84	40
January 15–February 1, 2008	60	69	46	99	67	56	65	84	74	49
February 1–February 15, 2008	28	21	34	23	59	29	27	242	90	20
February 15–March 1, 2008	40	37	39	190	84	58	37	91	54	47
March 1–March 15, 2008	29	44	43	46	133	55	35	—[a]	94	95
March 15–April 1, 2008	94	86	228	193	194	108	115	119	—[a]	63
April 1–April 15, 2008	26	29	54	43	100	127	14	83	199	28
April 15–May 1, 2008	49	53	42	63	165	51	53	82	55	56
May 1–May 15, 2008	16	28	25	20	103	49	41	64	16	23
May 15–June 1, 2008	78	107	82	103	379	89	229	130	77	74
June 1–June 15, 2008	52	50	71	43	243	60	98	76	58	47
June 15–July 1, 2008	—[a]	84	48	49	215	67	108	70	57	59
July 1–July 15, 2008	49	110	57	67	252	226	170	123	94	66
July 15–August 1, 2008	92	95	111	249	117	43	127	78	60	80
August 1–August 15, 2008	44	221	110	133	—[c]	177	211	34	—[b]	337
August 15–September 1, 2008	61	61	37	35	138	46	144	89	141	51
September 1–September 15, 2008	62	70	67	72	172	82	111	82	—[a]	113
September 15–October 1, 2008	13	43	11	8	26	34	20	—[a]	48	7
October 1–October 15, 2008	49	80	85	76	102	85	44	102	49	52
October 15–November 1, 2008	93	97	86	108	—[c]	138	118	166	92	76
November 1–November 15, 2008	32	46	34	133	—[c]	33	40	101	46	26
November 15–December 1, 2008	51	47	29	33	305	287	43	72	—[b]	49

[a] Sample was not analyzed (sample was destroyed).
[b] High values due to presence of NaCl or Cu compounds.
[c] High values due to presence of organic pollution of sample (decay parts of plants or insects, algae).

TABLE 1.3
Particulate Matter Deposition: Method B (Particles <3 µm during 2005–2006 [mg/m²·Day])

	1B	2B	3B	4B	5B	6B
October 15–November 15, 2005	88	57	32	54	18	117
November 15–December 15, 2005	23	15	5	45	46	[a]
December 15, 2005–January 15, 2006	64	81	12	47	5	16
January 15–February 15, 2006	78	8	77	19	24	5
February 15–March 15, 2006	165	171	203	419	281	139
March 15–April 15, 2006	92	63	49	70	113	85
April 15–May 15, 2006	687	79	265	217	256	231
May 15–June 15, 2006	613	916	1305	846	591	599
June 15–July 15, 2006	222	290	1286	309	605	203
July 15–August 15, 2006	108	75	20	160	140	234
August 15–September 15, 2006	64	628	111	17	100	146
September 15–October 15, 2006	1145	625	1317	5657	918	58

[a] Sample was not analyzed (sample was destroyed).

TABLE 1.4

Particulate Matter Deposition: Method B (Particles <3 µm during 2007–2008 [mg/m²·Day])

	1B	2B	3B	4B	5B	6B	7B	8B	9B	10B
December 1–December 15, 2007	90	80	56	49	262	92	56	167	—[a]	79
December 15, 2007–January 1, 2008	23	52	−5	26	37	69	61	—[a]	—[a]	49
January 1–January 15, 2008	111	86	36	53	90	71	53	73	48	197
January 15–February 1, 2008	58	73	48	86	80	78	64	107	190	80
February 1–February 15, 2008	15	126	52	56	59	82	28	115	—[b]	34
February 15–March 1, 2008	90	35	49	97	81	98	46	—[a]	88	70
March 1–March 15, 2008	43	26	34	62	115	64	99	—[a]	—[b]	82
March 15–April 1, 2008	249	108	113	97	190	92	144	138	213	70
April 1–April 15, 2008	117	53	52	126	—[c]	68	72	107	150	50
April 15–May 1, 2008	217	53	52	59	165	90	79	89	29	70
May 1–May 15, 2008	31	29	25	48	95	80	69	85	23	21
May 15–June 1, 2008	151	100	—[a]	118	—[c]	140	133	173	115	101
June 1–June 15, 2008	59	78	69	248	342	101	109	101	108	67
June 15–July 1, 2008	111	66	49	49	—[c]	98	98	95	75	66
July 1–July 15, 2008	87	—[a]	70	151	370	61	289	—[a]	164	74
July 15–August 1, 2008	221	138	63	84	126	45	124	321	101	46
August 1–August 15, 2008	138	64	92	—[a]	—[c]	218	223	—[b]	—[b]	118
August 15–September 1, 2008	140	89	29	52	104	41	—[a]	270	183	34
September 1–September 15, 2008	66	70	69	59	134	77	157	85	—[a]	66
September 15–October 1, 2008	8	49	3	26	20	57	18	—[a]	61	18
October 1–October 15, 2008	115	69	70	72	123	143	64	198	59	75
October 15–November 1, 2008	144	74	75	73	—[c]	110	160	148	118	106
November 1–November 15, 2008	47	35	61	76	126	42	55	88	81	27
November 15–December 1, 2008	67	52	36	138	460	73	84	—[a]	—[b]	112

[a] Sample was not analyzed (sample was destroyed).
[b] High values due to presence of NaCl or Cu compounds.
[c] High values due to presence of organic pollution of sample (decay parts of plants or insects, algae).

particulate matter deposition is too long, no correlation between wind and dust deposition was obtained. Meteorological data on temperature, precipitation, wind direction, and wind strength during the second phase of the study were obtained from the ARSO and from the Port of Koper as measured by the Primorska Institute for Natural Sciences and Technology (PINT).

The data on wind direction and strength included in the study were measured at the meteorological stations located on the roof of the administration building at the Port of Koper. Additionally, data on temperature and precipitation as well as data on wind speed and direction were collected also at the meteorological station of Koper located on Markovec Hill near Koper. Both meteorological stations are managed by ARSO. Information on the speed and direction of wind was obtained also from the Port of Koper Department of Environmental and Occupational Health. These measurements were performed by PINT at the meteorological station in the Port of Koper. Data are not recorded for some days and only daily values are provided on others. There are inconsistencies among the data for the wind direction and speed measured at the meteorological stations within the Port of Koper (ARSO and PINT) and the data collected at the meteorological station on Markovec Hill (ARSO). Although the distance between these three measuring locations is less than 4 km, significant differences can be observed (Figures 1.12 through 1.14), most probably due to the topology of the terrain.

TABLE 1.5
Percent of Wind Blowing from the Storage Depots toward Sampling Sites

	% of Wind				
	1 (SW)	2 (SW)	3 (SW)	4 (SW)	5 (S)
October 15–November 15, 2005	1.4	1.4	1.4	1.4	8
November 15–December 15, 2005	2.4	2.4	2.4	2.4	12.6
December 15, 2005–January 15, 2006	a	a	a	a	a
January 15–February 15, 2006	1.3	1.3	1.3	1.3	9.7
February 15–March 15, 2006	2.8	2.8	2.8	2.8	8.7
March 15–April 15, 2006	3.2	3.2	3.2	3.2	5.8
April 15–May 15, 2006	3.2	3.2	3.2	3.2	5.8
May 15–June 15, 2006	2.8	2.8	2.8	2.8	6.4
June 15–July 15, 2006	2.5	2.5	2.5	2.5	4.5
July 15–August 15, 2006	2.1	2.1	2.1	2.1	6.8
August 15–September 15, 2006	2.6	2.6	2.6	2.6	8.4
September 15–October 15, 2006	a	a	a	a	a

a Data were not available.

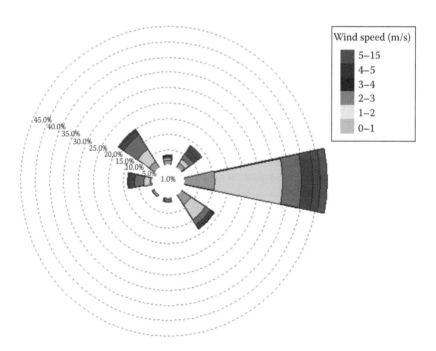

FIGURE 1.12 (**See color insert.**) Wind rose diagrams for December 2007 to November 2008 (station EET–ARSO).

The data on wind direction were divided into the eight points of the compass: N, NE, E, SE, S, SW, W, and NW. The data for wind speed were divided into six classes: from 0 to 1 m/s, from 1 to 2 m/s, from 2 to 3 m/s, from 3 to 4 m/s, from 4 to 5 m/s, and greater than 5 m/s. The wind rose diagrams (United States Department of Agriculture 2012) presented in Figures 1.12 through 1.14 show how measurements of meteorological data collected at different locations in the same area can yield different results. The data from the automated monitoring station at Markovec Hill

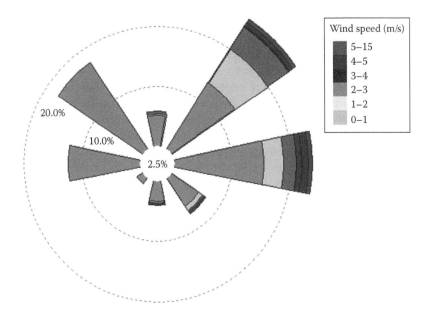

FIGURE 1.13 **(See color insert.)** Wind rose diagrams for December 2007 to November 2008 (inside of the Port of Koper–measured by PINT).

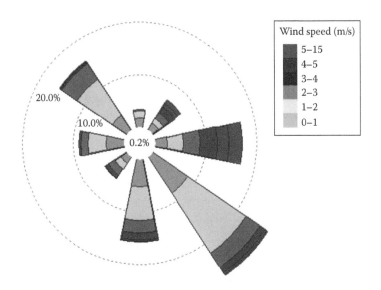

FIGURE 1.14 **(See color insert.)** Wind rose diagrams for December 2007 to November 2008 (meteorological station Markovec, Koper—ARSO).

(ARSO) were recorded at half-hour intervals (ARSO), the measuring station on the roof of the administration building (ARSO) provided hourly data, and the measurement stations within the Port of Koper (PINT) provided only once-a-day values for wind speeds and directions. The interpretation of meteorological results requires additional care because, on the basis of the acquired data from all three measuring sites, it could well be that the measured values differ and probably depend on the micro location and the morphology of the terrain at the site. As an extreme example, the measuring station on the roof of the administration building recorded a maximum wind speed of 10.8 m/s (38.9 km/h), the measuring station at Markovec Hill recorded a higher maximum value of 12.4 m/s (44.64 km/h), and the measuring station inside the Port of Koper reported a maximum value of 7.3 m/s (26.3 km/h) during the same period of time (2007–2008).

Additional wind rose diagrams (United States Department of Agriculture 2012) based on the wind direction and wind velocity data for each bimonthly sampling period during the second phase of the study were prepared. Data that were available represented wind directions and wind speeds expressed as percent of frequency. From the results of these wind rose diagrams, it could be seen that the east wind dominates in both frequency and strength. From April to June, the wind is usually from the northwest, and from September to November, the wind is frequently from the southeast. As already mentioned, occasionally the strong mistral blows from the west to the northwest for a very short period of time. After such events, the inhabitants of Rožnik report heavy depositions of black particles in their living environments. Due to the short duration of these strong, westerly winds, there were no records of extreme weather events such as the sudden emergence of a strong wind blowing from stockpiles at the port toward the sampling sites.

The maximum temperature over the entire time period was 34.6°C measured on June 26, 2008, at 15:30 h. Maximum rainfall of 35.5 L/m² was recorded on July 18, 2008. Daily fluctuation of temperatures and amount of precipitations are presented in Figure 1.15. During the entire time of the study, 910 mm of precipitation fell (December 1, 2007 to November 30, 2008). Most of the recorded rainfall during the year of the study on the coastal area is observed in the summer. Almost 60% (526 mm of 910 mm) of precipitation fell during the summer months from April to August.

Typical correlations between particulate deposition and wind frequency on a month-by-month basis are shown in Figures 1.16 and 1.17. Sampling sites 1, 2, and 3 were along a line (Figure 1.4) in the predominant wind direction (NE–SW) to evaluate the impact of distance from the EET depot for coal and iron ore on the quantity of particulate matter deposition. A sketch of the topography of terrain is shown in Figure 1.18.

Information on the quantities of coal and iron ore off- and on-loaded at the EET in the Port of Koper was provided by the Port of Koper Department of Environmental and Occupational Health. The time intervals for the movement of the coal and the iron ore were adjusted to the time intervals for particulate matter deposition and the speed and direction of the wind. The movement of these materials could not be correlated with the quantities of particulate matter deposited in the vicinity. The highest Spearman's rho correlation is 0.44. Typical relationships between particulate matter deposition and coal and iron ore movements at the EET are presented in Figures 1.19 and 1.20.

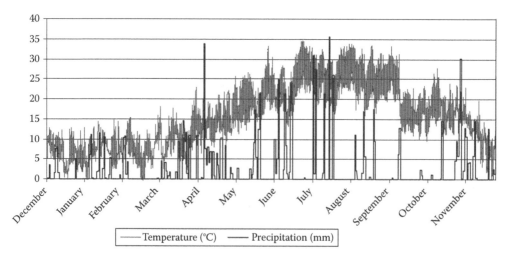

FIGURE 1.15   Daily temperature and precipitation from December 1, 2007, to November 30, 2008, recorded at the Automatic Measuring Station Markovec—ARSO.

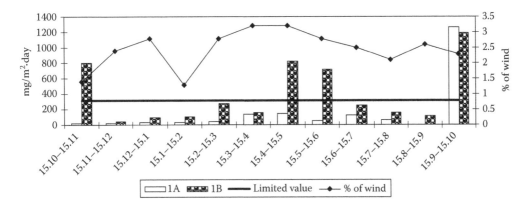

FIGURE 1.16 Correlations between particulate matter deposition and wind velocity at sampling site 1.

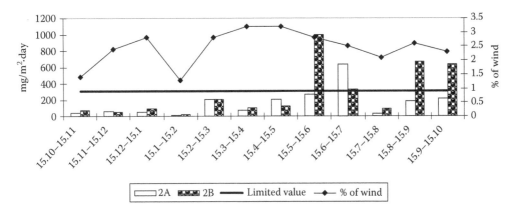

FIGURE 1.17 Correlations between particulate matter deposition and wind velocity at sampling site 2.

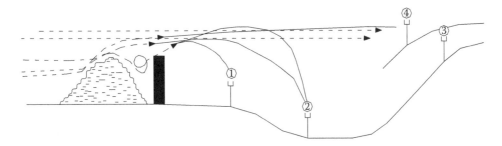

FIGURE 1.18 The potential transport of suspended particles from the EET depot toward sampling sites 1, 2, 3, and 4.

1.3.2.2 Comparison of Methods A and B

The standard method for the collection of particulate matter deposition using the Bergerhoff sedimentators according to VDI and Deutsches Institut für Normung/German Institute for Standardization (DIN) guidelines (VDI, 1996) may be flawed because it does not capture airborne particulate matter carried by the wind in the horizontal direction. Consequently, the method for sampling airborne particulate matter deposition was modified by adding metal screens oriented toward the EET depot to the collectors. Results comparing the modification, Method B, to the standard method, Method A, presented in Figure 1.21 show that the horizontal contribution by the wind

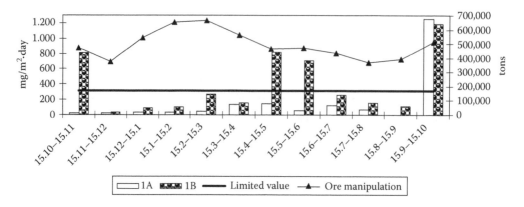

FIGURE 1.19 Comparison among particulate matter deposition and ore manipulation at the EET—sampling sites 1A and 1B.

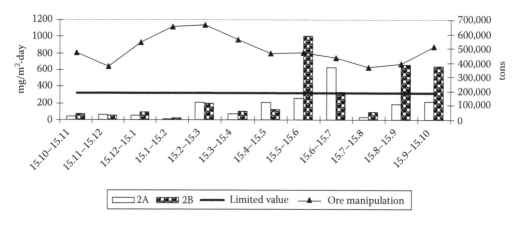

FIGURE 1.20 Comparison among particulate matter deposition and ore manipulation at the EET—sampling sites 2A and 2B.

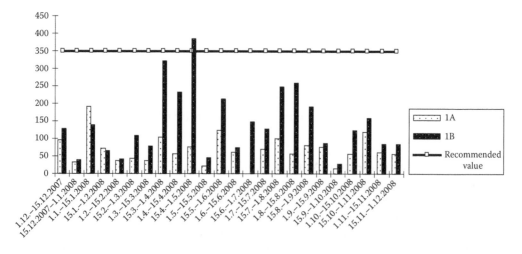

FIGURE 1.21 Comparison between the vertical deposition and horizontal contribution to particulate matter deposition (sampling site 1).

added significantly to the quantities of dust collected at sampling site 1, which is the nearest to the stockpile. In 22 out of 24 times, the mass of particulate matter deposition collected by Method B was greater than the mass of particulate matter collected by Method A. The Spearman's correlation coefficient (ρ) for sampling sites 1A and 1B is almost 0.7 with p value of 0.000. Similar observations were made at other sampling sites. In at least 15 out of 24 instances, the masses of particulate matter deposition collected by Method B were significantly greater than those collected by Method A.

1.3.3 DETERMINATION OF METALS

Several metals were determined in the coal and iron ore at the EET stockpiles and in the soil from an adjacent area. These determinations were made in the Chemical Laboratory of the Public Health Institute Koper and in the Laboratory for Sanitary Chemistry at the Public Health Institute Celje. These results, presented in Table 1.6, show none of the metals were found in iron ore or in the coal at levels comparable to those in the soil sample. A high value of iron in the sample of soil reflects the fact that the surrounding terra rosa is rich in iron.

The results presented in Table 1.7 for the determination of aluminum, cadmium, chromium, copper, iron, nickel, and zinc were obtained for the 1-year samples of particulate matter depositions collected

TABLE 1.6

Content of Metals in Coal, Iron Ore, and Soil

	Coal[a] (mg/kg d.w.)	Soil[a] (mg/kg d.w.)	Iron Ore[a] (mg/kg d.w.)	Iron Ore[b] (mg/kg d.w.)
Fe	4,800	29,600	649,000	558,240
Al	9,800	13,500	3,200	3,150
Cu	8.9	46.9	41.1	40
Zn	15.1	116.8	18.7	16
Cd	<1	<1	<1	0.1
Pb	2.5	25.0	3.6	10
Cr	5.9	66.7	6.5	9
Ni	4.5	94.8	1.6	5

[a] Public Health Institute, Koper.
[b] Public Health Institute, Celje.

TABLE 1.7

Values of Metals in the Yearly Sample (2005/2006) of Particulate Matter Deposition ($\mu g/m^2 \cdot Day$)

Sampling Site	Fe	Al	Cu	Zn	Cd	Pb	Cr	Ni
1A	3,054	2,584	66	146	<0.3	25	27	14
2A	7,740	7,263	93	239	<0.3	35	39	26
3A	14,433	16,805	93	278	<0.3	40	90	35
4A	29,450	31,942	371	861	1.2	87	98	50
5A	6,892	6,110	80	93	<0.3	24	28	14
6A	13,081	8,748	172	324	<0.3	60	46	42
Limited value	—	—	—	400	2	100	—	—

Source: Jereb, G. et al., *Int. J. Sanit. Eng. Res.*, 3(1), 4, 2009. With permission.

Note: —, Limited value is not determined.

TABLE 1.8
Ignition Residues from Particles Smaller than 3 μm, %

	1A	1B	2A	2B	3A	3B	4A	4B	5A	5B	6A	6B
May 15–June 15, 2006	a	13	a	4	a	1	a	7	a	2	a	1
June 15–July 15, 2006	1	7	1	4	2	4	1	4	10	9	5	7
July 15–August 15, 2006	a	a	a	a	a	a	<1	4	4	12	<1	13
August 15–September 15, 2006	a	5	3	8	13	18	26	30	8	0	11	6
September 15–October 15, 2006	10	10	3	4	4	11	10	12	12	10	6	1

[a] No data available.

at all six of the original sampling sites during the 2005–2006 preliminary study. The limit value for zinc (Decree, 1994) was exceeded only at sampling site 4A. The regulations do not specify limits for amounts of aluminum, chromium, copper, iron, and nickel. The results also show increased values of the other metals, iron, aluminum, copper, zinc, and lead, at sampling site 4. The next highest values of metals in the deposition are present at one of the control locations (sampling site 6A).

As might be expected from the apparent reduction in fugitive dust emissions accompanying the initiation of mitigation/control measures mentioned earlier, the metal deposition during the 2007–2008 phase of the investigation was significantly diminished relative to those of the previous year. Lead deposition was a few $\mu g/m^2 \cdot day$, well below the recommended value of 100 $\mu g/m^2 \cdot day$. Deposition of cadmium was 0.1 $\mu g/m^2 \cdot day$ or less and that of arsenic was 0.2 $\mu g/m^2 \cdot day$ or less. Zinc deposition was reduced by approximately 10-fold, but copper deposition increased to a few hundred $\mu g/m^2 \cdot day$. Contamination from a copper wire added to the sampling bottle to inhibit the growth of algae may be responsible for this increase. Iron and aluminum deposition was reduced by approximately tenfold during the second phase of the investigation.

It was not possible to distinguish the coal dust suspected of originating at the Port of Koper from other organic sources of carbon (burning of fossil fuel, traffic, pollen, etc.) on the basis of the results from the metal determinations, nor could the ore stockpile at the Port of Koper be distinguished from iron compounds in the soil on the basis of the results from the metal determinations.

Many of the dust particles collected as particulate matter deposition have organic origins, including pollen, plant parts, algae, and insect parts, as well as carbon from coal particles and soot. The particulate matter appears to be largely organic in nature. The percentages remaining after ignition are listed in Table 1.8.

The amount of iron in PM_{10} reported in the ARSO study, "Measurements of air pollution in Ankaran from 28 June to 11 September 2005" (ARSO, 2005), varied between 6.6 and 13.9 mg/kg and showed a high correlation between the amounts of iron and the amounts of PM_{10}. Žitnik et al. (2005) reported a correlation between the iron ore movement in the EET and the iron concentration in PM_{10}. Topič (2009) found that the surface of the iron ore stockpile developed a film that prevented dusting as long as the stockpile was not disturbed. Consequently, some studies report correlation between iron ore manipulation and iron concentrations in air particulates. Conversely to the situation with the iron ore, coal storage and manipulation represent continuous sources of dusting because no such film develops on the coal surface. The coal must be continuously compressed in order to prevent spontaneous ignition. Therefore, the coal storage stockpile represents a source of airborne particulate matter particularly in the event of a strong wind.

1.3.4 DETERMINATION OF GAMMA-EMITTING NUCLIDES

Many naturally occurring radionuclides are present as impurities in coal and in iron ore (Van Hook, 1979; Dowdall et al., 2004; Ward and Suárez-Ruiz, 2008). Representative samples of coal and iron

TABLE 1.9

Specific Activity of Gamma Emitters in Coal Sample

Nuclide	Specific Activity (Bq/kg)
[134]Cs	<0.04
[137]Cs	<0.04
[210]Pb	19 ± 11
[214]Bì	21 ± 2
[226]Ra	21 ± 2
[228]Ra	6.1 ± 0.3
[228]Th	6.1 ± 0.2
[230]Th	70 ± 34
[238]U	14 ± 3

TABLE 1.10

Specific Activity of Gamma Emitters in Iron Ore Sample

Nuclide	Specific Activity (Bq/kg)
[134]Cs	<0.07
[137]Cs	<0.25
[210]Pb	8.7 ± 2.8
[214]Bì	10.2 ± 1.1
[226]Ra	9.8 ± 1.1
[228]Ra	7.7 ± 0.6
[228]Th	7.9 ± 0.5
[238]U	7.6 ± 2.1

ore from the EET depot in the Port of Koper were assayed for gamma-ray activity. The results of these measurements are listed in Tables 1.9 and 1.10. Specific activities of these gamma-ray emitters were calculated back to August 15, 2006. The number following the ± symbol is the numerical value of the combined standard uncertainty in the specific activity and corresponds to the 68% confidence interval. The value following the < symbol represents the minimum detectable activity for a given radionuclide and also corresponds to the 68% confidence interval. In addition to the coal and iron ore samples, the gamma radiations from soil samples were measured as a reference for the radionuclides normally present in the area. These measurements were made at the Jožef Stefan Institute. The results of these measurements are presented in Table 1.11. The presence of ^{137}Cs is the major difference between results of gamma-ray emitter activity in coal and iron ore samples and results of gamma-ray emitter activity in the soil sample. The presence of the ^{137}Cs in the soil sample is attributed to the Chernobyl accident. Janković et al. (2011) reported the presence of [226]Ra, [232]Th, [40]K, [235]U, [238]U, ^{210}Pb, and ^{137}Cs in coal from Serbia, and they too attributed to the presence of ^{137}Cs to anthropogenic sources. Because the activity of the gamma-ray emitters from sample of coal and iron ore was not significantly increased, further evaluations of radioactivity in the particulate matter were discontinued. Akkurt et al. (2009) determined [238]U, [232]Th, and [40]K in coal samples by sodium iodide gamma-ray spectrometry for purposes of assessing the radiological hazard on the basis of the radium equivalent. Their results showed that the exposure doses were below the recommended limits. Saraevic et al. (2009) also employed gamma spectrometry to measure the radioactivity of [238]U and its progeny, [232]Th and its progeny, and [40]K in coal as well as in foods grown and raised in the vicinity of coal mines in Bosnia

TABLE 1.11
Specific Activity of Gamma Emitters
in Soil Sample

Nuclide	Specific Activity (Bq/kg)
^{7}Be	2.1 ± 1.2
^{40}K	364 ± 36
^{137}Cs	32.4 ± 1.6
^{210}Pb	63 ± 40
^{214}Bi	39.0 ± 3.9
^{226}Ra	39.0 ± 2.7
^{228}Ra	30.1 ± 1.3
^{228}Th	29.0 ± 1.5
^{230}Th	<90.6
^{238}U	19.6 ± 5.3

and Herzegovina. In addition to the laboratory measurements, the dose rates mentioned earlier were measured to evaluate the risks associated with radioactivity from windblown coal dust into domestic and agricultural areas. Outside of the area immediately surrounding the coal mine, the absorbed dose rate was found to be relatively close to the background value for Bosnia and Herzegovina, 111 nGy/h.

1.3.5 ELECTRON MICROSCOPY

Evidence for the presence of coal and iron ore in the particulate matter deposition collected around the EET depot at the Port of Koper was obtained by SEM. The presence of both coal and iron ore particles linked the dust deposition with the coal and iron ore stockpiles at the port. The presence of coal was confirmed with morphological and chemical analysis by electron microscopy of the filters. A typical electron micrograph and a typical x-ray fluorescence spectrum are shown in Figures 1.22 through 1.24, respectively.

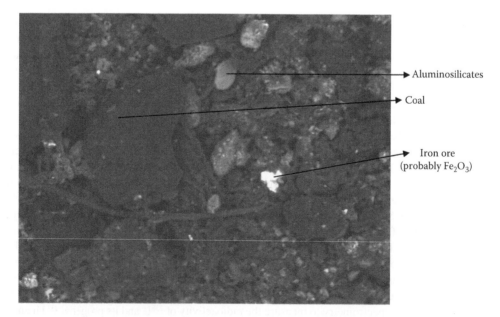

FIGURE 1.22   Sample 4A (330× magnification, image width 400 μm).

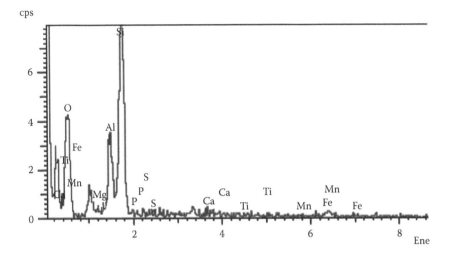

FIGURE 1.23   Typical EDXS spectrum of aluminosilicate.

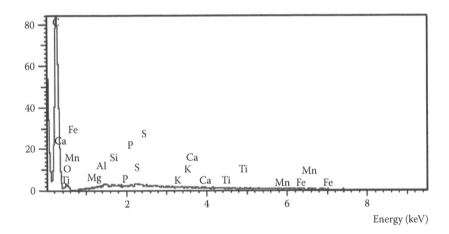

FIGURE 1.24   Typical EDXS spectrum of coal particle.

Iron oxide and coal particles were detected in samples of particulate matter deposition collected from February 15, 2006, to March 15, 2006, at sampling sites 1–5. In the sample collected at sampling site 6, the control site, iron was also present, not as Fe_2O_3 but always in the presence of silica and alumina. Therefore, it could be speculated that the origin of such iron is probably the soil from the surrounding environment, the so-called terra rosa. Coal particles in the sample of particulate matter collected at sampling site 6 were not detected. Also, morphological characteristics of the rust (scale), which could contaminate the samples, were not detected. Coal and iron ore particles were also not detected at the additional reference sampling sites in the Beli Križ area near Portorož (approximately 15 km from the Port of Koper). However, silica, calcium carbonate, and magnesium carbonate were detected.

The results of the SEM are strong indications that the movement of iron ore and coal at the Port of Koper contributes to the deposition of particulate matter in the surroundings. The presence of iron ore and coal in this particulate matter was confirmed on the basis of morphology and chemical composition. Using a volumetric collection method and SEM and ICP-MS for characterization, Anaja et al. (2012) found that the national limit for PM_{10} was exceeded in 10 of

12 samples from a site adjacent to one dwelling and in 6 of 12 samples from a site adjacent to another in the vicinity of an active surface coal mine in the community of Roda in rural Virginia (United States). Among the similar applications of this technique for characterizing airborne particulate matter are those of Stizmann et al. (1999) in the United Kingdom and Ault et al. (2012) and Xie et al. (2009) in China.

1.3.6 Mass Spectrometry

Measurements of the ratios of carbon isotopes $^{13}C/^{12}C$ were employed as an attempt to determine the presence of coal in samples of particulate matter deposition. Ohuchi et al. (1995) have employed the significantly different intrinsic $^{13}C/^{12}C$ ratios of coal and of bitumen as tracers for quantifying the amount of coal incorporated into distillate fractions. Coal from one source has a more or less a constant ratio of carbon isotopes, and, consequently, a more or less constant $\delta^{13}C$ value. Analysis of coal from the EET depot showed a $\delta^{13}C$ of $-28.2‰$. The $\delta^{13}C$ for a blank filter was $-25.0‰$. The results of isotope ratios of carbon isotopes $^{13}C/^{12}C$ from different sampling sites are presented in Table 1.12. From the results of the carbon isotopic ratio for each monthly sample and the results from the coal and control, the content of the carbon in samples that originated from coal was estimated. These results indicate that the largest amounts of carbon

TABLE 1.12
$\delta^{13}C$ Values Determined in Particulate Matter Deposition, ‰

	1A	1B	2A	2B	3A	3B	4A	4B	5A	5B
December	−26.4	−26.6	−26.0	−26.4	−25.5	−26.5	−25.7	−25.9	−26.7	−25.4
January	−26.2	−26.3	−26.5	−25.8	−25.3	−25.3	−25.9	−26.2	−26.2	−25.2
February	−26.6	−27.1	−28.1	−27.4	−26.3	−27.1	**−31.0**	−26.7	−28.2	−27.6
March	−25.5	**−28.4**	—	−27.2	−25.1	−25.8	−25.7	−25.5	−25.2	−25.5
April	**−28.9**	−27.3	−27.6	−25.4	**−28.2**	**−28.6**	**−34.8**	**−27.7**	**−29.9**	**−29.4**
May	−25.1	−24.5	−26.0	−25.5	−26.9	−25.2	−22.5	−25.3	−25.9	−25.6
June	—	−25.6	−26.0	−25.4	−26.0	−25.0	—	−23.9	−25.7	−25.5
July	—	—	−25.4	−25.2	−24.4	−25.2	−24.6	−25.3	−25.5	−26.0
August	−25.1	−26.0	−26.0	−25.2	−24.8	−25.2	−24.5	−24.4	−25.9	−25.4
September	−25.8	−25.7	−26.2	−26.2	−26.6	−23.5	−26.6	−26.3	−26.4	−26.2
October	−25.1	−24.7	−25.3	−24.8	−25.6	−23.6	−24.1	−24.7	−23.1	−24.7
November	−25.0	−26.6	−26.4	−25.8	−27.2	−26.2	−24.4	−26.0	−25.5	−25.4

	6A	6B	7A	7B	8A	8B	9A	9B	10A	10B
December	*−26.4*	*−26.2*	*−25.9*	*−26.5*	−25.8	−26.7	—	—	−26.1	−25.8
January	*−25.3*	***−29.4***	*−25.8*	*−25.4*	−25.9	−26.4	−26.1	−26.2	−26.0	−25.3
February	*−27.5*	*−28.0*	*−26.9*	*−26.0*	−27.7	−26.7	**−28.7**	−25.9	−26.2	−26.1
March	—	*−25.4*	*−25.4*	*−25.9*	—	−25.5	−25.1	—	—	−26.1
April	*−25.9*	*−26.6*	***−29.1***	***−29.7***	**−31.7**	**−27.6**	**−26.5**	**−28.1**	−21.5	−27.2
May	*−25.0*	*−23.5*	*−26.7*	*−26.9*	−25.8	−25.0	−25.7	−25.9	−26.7	−25.4
June	—	*−26.0*	*−26.1*	*−26.1*	−24.7	−25.4	—	—	−24.8	−26.4
July	*−24.5*	*−26.3*	*−25.4*	*−25.9*	−25.6	−25.5	−25.1	−25.5	−24.8	−25.5
August	*−24.2*	*−24.7*	*−25.0*	*−25.0*	−24.4	−25.9	−24.9	−25.1	−25.8	−25.0
September	*−26.3*	*−26.9*	*−25.0*	*−26.3*	−25.3	−26.9	−26.1	−26.9	−25.7	−26.1
October	*−24.1*	*−24.8*	*−23.9*	*−25.2*	−26.4	−24.1	−24.8	−25.1	−24.9	−25.7
November	*−25.5*	*−26.4*	*−25.9*	*−26.6*	—	−26.9	—	**−29.0**	**−27.7**	**−28.7**

Note: Bold indicates values of $^{13}C/^{12}C$ isotope ratio comparable to values in the coal. Italics indicate control sampling sites (6 and 7).

in samples that could be attributed to coal appeared in the February to March and April to May 2008 sampling periods in almost all sites. This is especially true in the February to March samples. However, it is difficult to correlate these results with the influence of wind or other weather parameters. During the February to March 2008 sampling period, the prevailing wind was the Bora from the N and NE. During this time, the temperatures were relatively low. This raises the questions of the potential impact of the heating dwellings using fossil fuels and the possible association with the lower $\delta^{13}C$ values from this source rather than indicating that carbon could originate from coal dust from the stockpile at the port. Temperatures were practically the same during the entire winter and were not lower during February. Therefore, it is difficult to correlate the results of the mass spectrometric measurements with the amounts of particulate matter deposition. The amount of dust deposition was relatively low during February. Greater amounts were obtained during hot, dry summer months. However, the amount of particulate matter deposition represents the total suspended particles that settle to the ground as opposed to the results for the isotopic composition of carbon, which comes from coal. The carbon contents of the samples collected by Method B (vertical plus horizontal deposition) were greater than those collected by Method A, suggesting the coal stockpile as their origin because all of the modified sampling devices were oriented toward the EET and the Port of Koper. It is necessary, however, to stress that the use of carbon isotopes $^{13}C/^{12}C$ for analyses represents a new approach to identifying the source of carbon in the dust deposition samples. The possibility of error must be highlighted because the samples are not homogeneous in composition and structure. They contain numerous particles from the environment. Among the possible sources of carbon are indeed particles of coal but also fragments of insects, pollen, plant parts, algae, soot, and others. The isotopic value ($\delta^{13}C$) is specific for each of these sources, and in case of measurements on nonhomogeneous samples, the $\delta^{13}C$ values could represent the average value of these various sources, which may indicate the false conclusion about the origin of the carbon. In addition, the estimate with which the proportion of carbon originating from coal was made was given as the percent of carbon in the sample. Actually, the amount of carbon in the sample was not determined. Therefore, the correlation between the carbon in the sample, which probably originated from the coal at each of the sampling sites, and the amount of particulate matter deposition (gravimetric analysis) cannot be derived from the data collected thus far. For further investigation, therefore, quantification of the total carbon content of the samples as well improving the method would be necessary.

1.3.7 CORRELATION OF PARTICULATE MATTER DEPOSITION WITH ORE AND COAL MOVEMENTS

Data for coal and iron ore manipulation at the EET were obtained from the Department of Environmental and Occupational Health of the Port of Koper. In Figures 1.25 and 1.26, correlations between particulate matter deposition and iron ore and coal manipulation at the EET in the Port of Koper are presented. On the basis of the results from the first phase of the study, confirmation of a correlation was not expected because the particulate matter deposition sample collection time was not sufficiently frequent. As seen from Figures 1.25 and 1.26, correlations between coal and iron ore manipulation and quantities of particulate matter deposition collected at the sampling sites in the port's surroundings during the second phase of the investigation could not be confirmed.

The proportion of particulate matter deposition increased during the dry summer months. The reason for this is most likely dryness of the ground, which contributed to the dustiness in the surroundings. In addition, during the summer, the wind blows predominantly from the sea across the stockpiles toward the land. This is reflected by the increased quantities of particulate matter deposition during the summer months. The amount of ore off- and on-loading is rather constant during the entire year with slightly increased quantities of coal off-loading during winter. For this reason, the hypothesis that the emission values of dust particles at the port are significantly affected by the process of off- or on-loading of coal and iron ore cannot be confirmed.

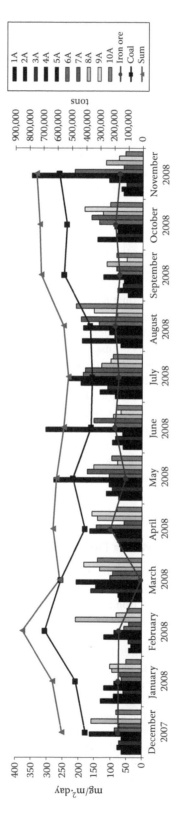

FIGURE 1.25 (See color insert.) Correlation among particulate matter deposition and coal and iron ore manipulation at the EET—sampling site A.

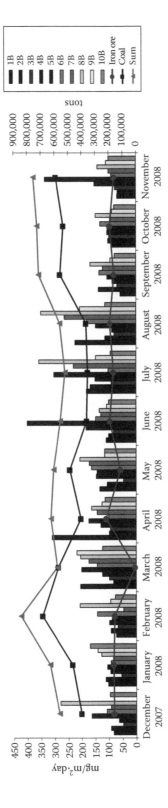

FIGURE 1.26 (See color insert.) Correlation among particulate matter deposition and coal and iron ore manipulation at the EET—sampling site B.

1.3.7.1 Quantification of Particles Collected in the Annual Sample

The presence of coal in the particulate matter deposition was demonstrated in the SEM examination of samples collected during the first part of the study. During the second part of the study, samples from the entire year were analyzed after combining samples from each sampling period on a site-by-site basis. For these analyses, a part of each filter was selected for further processing. Since particles were examined not only qualitatively (size, chemical deposition, and morphology of individual particles) but also quantitatively (number of particles from individual sampling sites), dilution of the samples was necessary. Because particulate matter deposition was rather low during the entire sampling period compared to the preliminary results, a lower number of particles were expected, and consequently, lower numbers of coal and iron ore particles were anticipated. This prediction was confirmed by SEM examination. More particles of aluminosilicates originating from the surrounding soil were observed than were particles of coal and iron ore. The inhabitants of nearby residential area, Rožnik, did not report disturbing emissions of black dust during the second part of the study (2007–2008).

The number of particles per cm^2 of filter is calculated as the number of particles on the filter after dilution. For the conversion to the number of particles in the entire dust deposition, results should be multiplied additionally by a factor of 3079.5. But since all the samples were processed using the same procedure, the values are given as the number of particles per cm^2 of treated filter. Results given according to the size and chemical composition of a particles grouped by individual size classes are shown in Table 1.13.

The particles present in the greatest numbers were identified as aluminosilicates (compounds having an empirical formula of $KAlSiO_4$). Their sizes ranged from 1 to 10 μm. The majority of these particles were assumed to have natural origins. In addition to the aluminosilicate particles, many titanium dioxide (TiO_2) particles were found, all of which were present in the smaller size fractions (less than 1 up to 5 μm). Coal particles, majority of which were larger than 5 μm, were detected also. None of the coal particles was smaller than 1 μm. Coal particles were found at the control sampling sites, 6 and 7, too. A possible explanation for the source of these coal particles at control sites 6 and 7 could be the boats transporting coal from the Port of Koper to their destinations in Trieste. A few particles of iron oxide were identified. The majority of them were rather small (from 1 to 5 μm). Sampling site 1, which is nearest to the depot, had the largest number of iron oxide particles. The relative amounts of these particles at each sampling site are shown in Figure 1.27.

1.3.8 Directional Sample Collection

A simple, inexpensive alternative device was developed for the rapid estimation of direction and magnitude of airborne particulate matter. This device, based on deposition/impaction and adhesion on the surface of a plastic sphere, enabled the collection of particles in both the horizontal and vertical directions.

1.3.8.1 Visual Perception

Directional sampling devices were placed in the residential neighborhoods around the coal and iron ore stockpiles at the Port of Koper. Visual inspection of the loading of particles on the adhesive surface of the balls allowed estimations of both direction and intensity of the particulate deposition and subsequent speculations on their sources. Upon visual assessment, uneven deposition of particles was observed on the surface of the balls. These could be associated with the direction from which the dust particles came. As shown in Figure 1.28, higher amounts of particles were obtained on the side of the sampler oriented toward the Port of Koper and the EET depot (b) than on the same ball on the side oriented away from the EET ore depot (a).

TABLE 1.13

Average Number of Particles per Square Centimeter of Filter Surface Area

Size (μm) Sample	Compounds with Al, Si, K (Silicates, Aluminosilicates, etc.)				Titanium Dioxide				Iron Oxides (FeO, Fe$_2$O$_3$, Fe$_3$O$_4$)				Particles Containing Carbon (Coal)				Sum
	<1	1–5	5–10	>10	<1	1–5	5–10	>10	<1	1–5	5–10	>10	<1	1–5	5–10	>10	
1A	0	20,918	4,082	1,020	2,041	4,082	0	0	0	1,531	0	0	0	510	0	0	34,184
1B	0	7,289	1,822	364	1,093	3,644	0	0	1,093	2,551	364	364	0	1,822	364	364	21,137
2A	0	5,102	1,701	850	9,354	14,456	0	0	0	850	0	0	0	0	0	0	32,313
2B	0	5,952	6,803	0	11,054	9,354	850	0	0	850	0	0	0	0	7,653	2,551	45,068
3A	0	0	0	0	638	0	0	0	0	0	0	0	0	1,913	638	0	3,189
3B	1,020	1,276	255	0	765	1,020	0	0	255	0	0	0	0	510	0	255	5,357
4A	283	1,134	0	0	850	567	0	0	0	283	0	0	0	283	283	0	3,685
4B	1,786	2,041	0	0	255	255	0	0	255	0	0	0	0	255	255	255	5,357
5A	0	425	0	0	0	0	0	0	0	0	0	0	0	0	0	0	425
5B	0	1,160	696	0	4,638	0	0	0	0	232	0	0	0	696	928	232	8,581
6A	0	2,551	5,102	0	1,701	17,857	0	0	0	0	0	0	0	2,551	4,252	0	34,014
6B	0	5,102	638	1,913	7,015	6,378	638	0	0	0	0	0	0	0	3,189	2,551	27,423
7A	0	1,701	0	0	5,952	0	0	0	0	0	0	0	0	3,401	850	0	11,905
7B	6,633	14,286	510	0	7,143	4,592	0	0	0	0	0	0	0	0	2,551	1,020	36,735
8A	3,189	9,566	0	0	4,464	5,102	0	0	0	0	0	0	0	638	1,276	0	24,235
8B	1,913	5,740	0	638	5,102	0	0	0	638	1,276	0	0	0	0	0	0	15,306
9A	1,913	7,653	638	638	3,827	4,464	0	0	0	0	0	0	0	0	1,276	0	20,408
9B	3,189	16,582	5,102	0	1,913	8,929	0	0	0	0	0	0	0	1,276	5,102	0	42,092
10A	18,707	52,721	2,551	850	3,401	20,408	0	0	0	0	0	0	0	0	0	2,551	101,190
10B	12,755	38,265	51,020	2,551	58,673	114,796	0	0	0	0	0	0	0	0	0	7,653	285,714

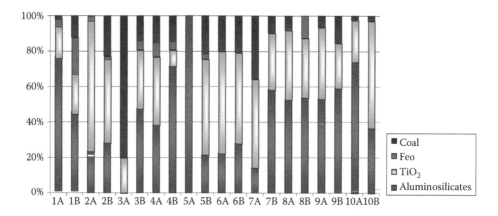

FIGURE 1.27 Percentages of different particles at individual sampling site.

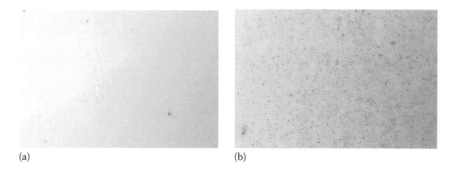

(a)   (b)

FIGURE 1.28 Surface of sampling device after 1 month of exposure oriented away (a) and toward (b) the Port of Koper and coal and iron ore depot.

1.3.8.2 Manual Particle Counting

To quantify deposition, the number of particles per unit area was counted manually. Results show that particles on the surface of the alternative sampler are not evenly distributed. Differences of up to 10 times, and sometimes even more, were observed in the number of particles at different locations on the surface of the sampler. The largest numbers of particles were found on the south and southwest side of the sampler, the side oriented toward the Port of Koper and the EET depot.

The time intervals April 1 to May 1, May 1 to June 1, August 1 to September 1, September 15 to October 15 and October 15 to November 15 were selected randomly for directional sampling at sites 1–5 and 10. Samples from all sampling sites and time intervals were analyzed by manual counting. Counting particles using computer software was performed only on samples from time interval October 15 to November 15. The results of manual particle counting are listed in Table 1.14.

1.3.8.3 Computerized Particle Counting

The numbers of particles per cm^2 counted on the alternative sampling device (ball) using computer software (IT 3,0) are presented in Table 1.15. Particles were counted only for the sample collected during the October 15 to November 15, 2008, time period. The maximum number of particles was observed on parts of the sampling devices oriented toward the EET depot.

1.3.8.4 Comparison of Manual and Computerized Particle Counting

Figure 1.29 compares the number of particles per cm^2 counted manually with the number of particles counted using the computer software at each point of the compass for each of the five sampling sites. These results show very good correlation between manual and computerized particle counting.

TABLE 1.14
Number of Particles on 1 cm² — Sampling Device 1 (the Ball)

Location / Period		Sampling Site 1					Sampling Site 2					Sampling Site 3					Sampling Site 4					Sampling Site 5					Sampling Site 10				
Direction		N	S	E	W	SWᵃ	N	S	E	W	SWᵃ	N	S	E	W	SWᵃ	N	S	E	W	S,SWᵃ	N	S	E	W	Sᵃ	N	S	E	W	Sᵃ
1.4. do 1.5.2008		15	57	32	146	134	21	12	17	130	108	41	71	12	32	134	13	11	13	101	85	11	200	15	39	200	20	85	41	25	85
		21	64	24	149	184	25	15	26	124	78	33	98	18	34	184	8	12	8	154	103	21	151	8	51	151	26	84	44	16	84
		7	94	31	178	148	20	16	10	149	42	28	98	26	26	148	13	21	13	95	128	19	209	15	17	209	21	88	40	20	88
																						17	204	8	31	204	25	63	23	20	63
																						23	187	17	67	187	13	79	36	16	79
No. of particles/cm²																															
Average no. of particles/cm²		14,3	71,7	29	157,7	155,3	22	14,3	17,7	134,3	76	34	89	18,7	30,7	155,3	11,3	14,7	11,3	116,7	105,3	18	190	13	41	190	21	80	37	19	80
1.5. do 1.6.2008		41	84	11	186	309																2	17	3	16	17	19	26	36	26	26
		28	53	19	232	301																14	25	2	17	25	13	28	38	29	28
		34	63	25	211	292																7	17	4	19	17	8	34	24	20	34
		39	86	36	141	248																4	10	3	6	10	5	37	39	18	37
		28	45	35	238	224																8	20	1	10	20	12	33	42	21	33
No. of particles/cm²																						7									
Average no. of particles/cm²		34	66,2	25,2	201,6	274,8																7	17,8	2,6	14	17,8	11	32,2	36	23	32
1.8. do 1.9.2008		28	99	46	85	153	16	47	18	50	102	68	25	29	60	16	29	25	29	69	52										
		31	53	50	117	152	15	39	21	56	93	60	14	41	39	19	31	14	41	91	62										
		31	79	36	95	136	17	30	16	53	96	59	22	25	42	14	32	22	25	87	47										
		30	78	36	64	150	7	34	19	39	69	59	26	21	41	14	9	26	21	61	27										
		32	45	44	116	148	14	31	8	47	66	35	17	22	48	7	13	17	22	58	45										
No. of particles/cm²																															
Average no. of particles/cm²		30,4	70,8	42,4	95,4	147,8	14	36,2	16,4	49	85	56	20,8	27,6	46	14	23	20,8	27,6	73,2	46,6										

(continued)

TABLE 1.14 (continued)

Number of Particles on 1 cm² — Sampling Device 1 (the Ball)

15.9. do 15.10.2008

Location / Direction	Sampling Site 1					Sampling Site 2					Sampling Site 3					Sampling Site 4				
	N	S	E	W	SWᵃ	N	S	E	W	SWᵃ	N	S	E	W	SWᵃ	N	S	E	W	S,SWᵃ
	41	35	74	171	/											53	57	36	28	/
	49	29	109	184	/											29	60	63	36	/
	27	37	89	153	/											53	49	49	17	/
	14	33	62	131	/											48	39	31	16	/
No. of particles/cm²	11	34	89	159	/											32	40	28	13	/
Average no. of particles/cm²	28,4	33,6	84,6	159,6	/											43	49	41,4	22	/

15.10. do 15.11.2008

Location / Direction	Sampling Site 1					Sampling Site 2					Sampling Site 3					Sampling Site 4					Sampling Site 5					Sampling Site 10				
	N	S	E	W	SWᵃ	N	S	E	W	SWᵃ	N	S	E	W	SWᵃ	N	S	E	W	S,SWᵃ	N	S	E	W	Sᵃ	N	S	E	W	Sᵃ
	40	173	55	68	332	12	65	39	82	177	48	29	23	101	153	32	68	33	67	87	48	146	15	23	146	34	109	49	14	109
	26	111	82	117	324	12	42	49	103	174	53	35	32	128	92	32	39	39	77	81	47	159	18	41	159	35	129	48	20	129
	22	181	83	108	283	10	60	33	99	158	51	37	21	120	101	31	54	27	52	85	29	124	17	21	124	27	105	39	21	105
	26	186	71	42	228	11	77	38	80	194	14	31	18	108	115	36	30	21	57	86	13	91	24	25	91	31	122	40	12	122
No. of particles/cm²	24	111	84	101	256	6	57	42	113	131	25	45	22	123	73	15	32	49	64	77	27	105	18	16	105	37	116	49	13	116
Average no. of particles/cm²	27,6	152	75	87,2	284,6	10	60,2	40,2	95,4	167	38	35	23,2	116	106,8	29	44,6	33,8	63,4	83,2	33	125	18	25	125	33	116	45	16	116

ᵃ direction towards EET depot.

TABLE 1.15

Number of Particles on 1 cm² — Counting with Software IT 3,0

Location	Sampling Site 1					Sampling Site 2					Sampling Site 3					Sampling Site 4					Sampling Site 5					Sampling Site 10				
Direction	N	S	E	W	SWᵃ	N	S	E	W	SWᵃ	N	S	E	W	SWᵃ	N	S	E	W	S.SWᵃ	N	S	E	W	Sᵃ	N	S	E	W	Sᵃ
No. of particles/cm²	30	185	64	46	450	14	257	30	45	122	79	52	35	119	125	53	55	45	50	104	8	141	15	40	141	13	128	55	6	128
	96	95	58	107	475	9	269	31	33	142	111	48	35	123	120	55	52	87	63	92	12	140	18	38	140	20	150	52	18	150
	43	154	62	61	429	14	209	58	57	194	69	40	48	134	108	17	34	27	47	80	14	163	14	39	163	26	161	30	19	161
	26	89	57	84	408	7	103	51	70	179	94	60	55	158	143	25	36	61	78	82	25	124	41	16	124	24	178	28	7	178
	33	103	65	43	445	16	88	14	78	173	73	74	48	142	90	44	33	38	66	85	37	112	42	28	112	27	179	45	15	179
Average no. of particles/cm²	45	125	61	68	441	12	185	37	57	162	85	55	44	135	117	39	42	52	61	89	19	136	26	32	136	22	159	42	13	159

ᵃ Direction towards EET depot.

15.10.do 15.11.2008

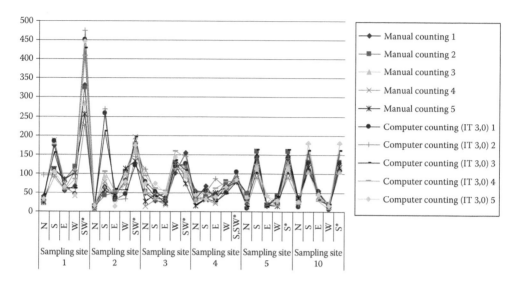

FIGURE 1.29 **(See color insert.)** Comparison between manual and computer counting.

The Spearman's correlation coefficient (ρ) for computerized counting vs. manual counting is relatively high—$\rho = 0.874$ ($p = 0.000$). However, in spots where very large numbers of particles were deposited on 1 cm^2, computer counting gave a higher number of particles. In such cases, more accurate results are probably obtained using computer software, because errors are more likely when a large number of particles appear in a small area. It is difficult to focus on each individual particle in cases of high saturation. An example of such an instance is the results for sampling site 1 where 35% more particles were counted using IT 3,0 compared to manual counting.

1.3.8.5 Morphology and Chemical Composition of Particulate Matter Deposited on Alternative Sampling Device Using SEM–EDXS

Dust particles collected using alternative collection device were evaluated also by electron microscopy and EDXS. Particles of coal, iron oxide, and titanium oxide, as well as particles of organic origin from the natural environment, were observed as shown in Figure 1.30. The particles of coal as well the particles of iron oxides could originate from the depot of coal and iron ore at the EET in the Port of Koper. These particles are morphologically similar to the particles analyzed in particulate matter deposition described in Section 1.3.4.

1.4 SUMMARY AND CONCLUSIONS

A variety of approaches were employed to investigate the possible relationships between the fugitive dust emissions associated with the dry cargo handling operations at the Port of Koper and the occasional contamination of the nearby residential neighborhoods with heavy deposits of black particulate matter. Modification of the traditional Bergerhoff sedimentator allowed differentiation between the horizontal and vertical deposition of airborne particulate matter, and the development of a directional deposition/adhesion sampling device enabled the identification of potential sources of airborne particulate matter. Coupling digital photography and computerized particle counting to the directional deposition/adhesion sampling device provided a means for the rapid estimation of the origin and intensity of airborne particulate matter deposition. The correlation of airborne particulate matter deposition with the prevailing weather conditions demonstrated the importance of considering meteorological data. Success of ^{13}C/^{12}C mass spectrometry, elemental chemical analysis, and gamma spectrometry on matching the coal and iron ore stockpiles at the Port of Koper with the airborne particulate matter depositions in the residential areas of Ankaran

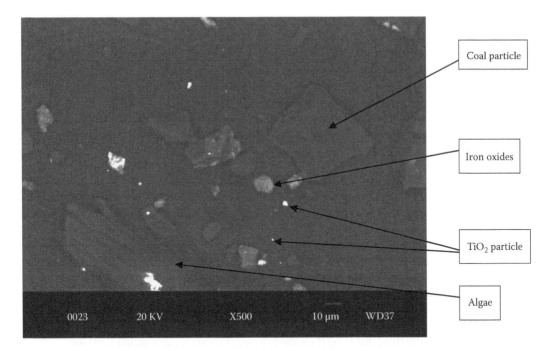

FIGURE 1.30 Sample from simple measuring device (ball) analyzed using electron microscopy.

and Rožnik was minimal. SEM, on the other hand, was of great utility in identifying coal and iron ore in these airborne particulate matter depositions.

During the first phase of the investigation, 2005–2006, extreme values of particulate matter deposition in the neighborhood northeast of the port were detected. Values exceeded the limit values during individual months. Also, yearly limit values were exceeded. During the second phase of the investigation, 2007–2008, elevated levels of suspended particles were not detected. The reason for low values of dust pollution could be attributed to the remedial actions taken at the Port of Koper to reduce the fugitive dust emissions associated with activities involving the movement of coal and iron ore. These actions are an effective approach to reduce or limit emissions of dust particles to the adjacent residential neighborhood.

The new sampling device is suitable for quick estimation of direction of major airborne particle sources. Since the cost of sampling device is relatively low, many such devices could be used for quick screening of the terrain in order to find the most representative locations for later placement of more sophisticated, accurate, and expensive devices for air monitoring.

QUESTIONS

The particulate matter collected at sampling site 2B over a 15-day period in a Bergerhoff sedimentator having a circular collection surface with an active diameter of 72 mm was washed out of the glass collection vessel onto a cellulose nitrate filter membrane of 3 μm pore size and 41 mm active diameter with 500 mL of distilled water and filtered. The initial mass of the dry filter membrane was 71.5 mg, and the mass of the membrane with the particulate matter was 78.5 mg after drying. The residue obtained by evaporating 250 mL of filtrate had a mass of 7.1 mg.

1. Calculate the deposition of >3 μm particulate matter at sampling site 2B in mg/m²·day.
   Answer: $(78.5 - 71.5)/[15(72/2000)^2\pi] = 114.5$ mg/m²·day
2. Calculate the deposition of <3 μm particulate matter at sampling site 2B in mg/m²·day.
   Answer: $2(7.1)/[15(72/2000)^2\pi] = 235.5$ mg/m²·day

3. What is the recommended limit value for particulate matter deposition in mg/m²·day in the Republic of Slovenia?

   Answer: 350 mg/m²·day

4. What is the difference in mg/m²·day between the total particulate matter deposition at sampling site 2B and the recommended limit value?

   Answer: (114.5 + 235.5) − 350 = 0 mg/m²·day

5. Why measure particulate matter deposition if there is no regulated limiting value?

   Answer: Particulate matter deposition represents a great deal of annoyance for inhabitants. Although the majority of particles are large particles (with diameter larger than 10 µ), among larger particles, smaller ones (PM10 or smaller) are also present (they could have a direct impact on health). Larger, coarse particles have, besides annoyance, also an influence on the upper respiratory tract, so they can also be a cause of health and other problems, either from direct inhalation or ingestion or through secondary pathways after particles have deposited on vegetables, fruits, or drinking water supply. Therefore, chemical composition and also morphology and physical structure of the particles are of great importance to correlate particulate matter deposition with health impacts. Measuring particulate matter deposition is also a simple, inexpensive method for which no power supply on the field is required, so we could, with a relatively easy method, continue monitoring changes in the environment regarding particulate matter deposition over the years. Additionally, results could be compared with results from recent history, since in the vicinity of major power plants and other rather dusty industries, usually there are some data about particulate matter deposition available.

ACKNOWLEDGMENTS

This research would not have been possible without the active participation and support of the Port of Koper and experts from their department for environmental protection Marzi Boris, MSc, and Franka Cepak, MSc. For this we gratefully acknowledge not only the support and participation but also the continued encouragement throughout the course of this work. We also acknowledge the assistance of Mateja Dominko at the Port of Koper Public Relations Office in arranging for permission to use the photographs in Figures 1.1 and 1.2. Special thanks are given to Professor Dr. Goran Dražić from the Department for Nanostructured Materials and to Assistant Professor Dr. Nives Ogrinc from the Department of Environmental Sciences at Jožef Stefan Institute; to Tanja Bolte, MSc, from ARSO; and to the students (now graduates) in the BSc study program of Sanitary Engineering at the University of Ljubljana Blaž Goličnik and Bojana Trivunčević.

REFERENCES

Akkurt, I., Mavi, B., Akyıldırım, H., and Günoglu, K. (2009) Natural radioactivity of coals and its risk assessment, *Intern. J. Phys. Sci.*, 4(7), 403–406.

Anaja, V.P., Isherwood, A., and Morgan, P. (2012) Characterization of particulate matter (PM10) related to surface coal mining operations in Appalachia, *Atmosph. Environ.*, 54, 496–501.

ARSO (Environmental Agency of the Republic of Slovenia) (2005) Air pollution measurement in Ankaran city from 28th of June to 11th of September 2005—Study report. (Meritve onesnaženosti zraka v Ankaranu od 28. junija do 11. septembra 2005). Ljubljana, Slovenia.

Ault, A.P., Peters, T.M., Sawvelt, G.S., Casuccioŝ, G.S., Willis, R.D., Norris, G.A., and Grasslan, V.H. (2012) Single—Particle SEM—EDX analysis of iron—Containing coarse particulate matter in an urban environment: Sources and distribution of iron within Cleveland, Ohio, *Environ. Sci. Technol.*, 46(8), 4331–4339.

Bradin, B., Smith, M., Millgan, P., Benjamin, C., Dunne, E., and Pearson, M. (1994) Respiratory morbidity in Merseyside schoolchildren exposed to coal dust and air pollution, *Arch. Dis. Child.*, 70, 305–312.

Brook, R.D., Franklin, B., Cascio, W., Hong, Y., Howard, G., Lipsett, M., Luepker, R. et al. (2004) Air pollution and cardiovascular disease: A statement for healthcare professionals from the expert panel on population and prevention science of the American Heart Association, *Circulation*, 109, 2655–2671.

Carinska uprava Republike Slovenije/Customs Administration of the Republic of Slovenia. http://www.carina. gov.si/fileadmin/curs.gov.si/internet/Za_medije/Sporocila_za_medije/Port_of_Koper_Zornada_Vrabec. pdf (accessed August 20, 2012).

Cepak, F. and Marzi, B. (2009) Environmental impact of the Port of Koper, *Varstvo Narave*, 22, 97–116.

Clancy, L., Goodman, P., Sinclair, H., and Dockery, D.W. (2002) Effect of air-pollution control on death rates in Dublin, Ireland: An intervention study, *Lancet*, 360(9341), 1210–1214.

Cohen, R.A., Patel, A., and Green, F.H. (2008) Lung disease caused by exposure to coal mine and silica dust, *Semin. Respir. Crit. Care Med.*, 29(6), 651–661.

Craig, H. (1957) Isotopic standards for carbon and oxygen and correction factors for mass spectrometric analysis for carbon dioxide, *Geochim. Cosmoschim. Acta*, 12, 133.

Decree on limit values, alert thresholds and critical emission values for substances into the atmosphere (1994) (Uredba o mejnih, opozorilnih in kritičnih imisijskih vrednostih snovi v zraku). Ur. L. RS št.73/1994.

Dominici, F., Peng, R.D., Bell, M.L., Pham, L., McDermott, A., Zeger, S.L., and Samet, J.M. (2006) Fine particulate air pollution and hospital admission for cardiovascular and respiratory diseases, *JAMA*, 295(10), 1127.

Dowdall, M., Vicat, K., Frearson, I., Gerland, S., Lind, B., and Shaw, G. (2004) Assessment of the radiological impacts of historical coal mining operations on the environment of Ny-A Lesund, Svalbard, *J. Environ. Radioact.*, 71, 101–114.

Environment Agency (2004) Monitoring of particulate matter in ambient air around waste facilities, Technical Guidance Document (Monitoring) M17. UK Environment Agency, Bristol, U.K.

Finlayson Pitts, B.J. and Pitts, J.N. (2000) *Chemistry of the Upper and Lower Atmosphere. Theory, Experiments and Applications.* Academic Press, San Diego, CA, pp. 15–43, 349–440.

Godish, T. (2004) *Air Quality*, 4th edn. Lewis, Boca Raton, FL.

Goličnik, B., Jereb, G., Poljšak, B., Katz, S.A., and Bizjak, M. (2008) Alternate method for quick estimation of direction and quantity of particulate matter deposition, *The European Aerosol Conference 2008*, Thessaloniki, Greece, Abstract T04A033P.

Hall, D.J., Upton, S.L., and Marsland, G.W. (1993) Improvements in dust gauge design. In: *Measurement of Airborne Pollutants*, Couling, S. (ed.). Butterworth-Heinemann Ltd, Oxford, U.K., 301pp.

Hoek, G., Brunekreef, B., Goldbohm, S., Fischer, P., and van den Brandt, P.A. (2002) Association between mortality and indicators of traffic-related air pollution in the Netherlands: A cohort study, *Lancet*, 360(9341), 1203–1209.

ISO Standard 11923 (1997) Water quality. Determination of suspended solids by filtration through glass-fiber filters. International Organization for Standardization—ISO/TC 147/SC 2; 6.

ISO/IEC Guide (1995) Guide to the expression of uncertainty in measurement (GUM). International Organization for Standardization, Geneva, Switzerland, 98pp.

Janković, M.M., Todorović, D.J., and Nikolić, J.D. (2011) Analysis of natural radionuclides in coal, slag and ash in coal—Fired power plants in Serbia, *J. Min. Metall. B*, 47, 149–155.

Jereb, G., Marzi, B., Cepak, F., Katz, S.A., and Poljšak, B. (2009) Collection and analysis of particulate matter deposition around the Port of Koper. Sanitarno inženirstvo, *Int. J. Sanit. Eng. Res.*, 3(1), 4–16.

Kania, N., Setiawan, B., and Chandra Kusuma, H.M.S. (2011) Oxidative stress in rats caused by coal dust plus cigarette smoke, *Universa Medicina*, 30(2), 80–87.

Korun, M. and Glavič Cindro, D. (2003) Visokoločljivostna spektrometrija gama v laboratoriju. Delovno navodilo LMR-DN-10. Institut Jožef Stefan, Ljubljana, Slovenia.

Luka Koper/Port of Koper, Terminals and Cargo, http://www.luka-kp.si/eng/terminals-and-cargo (accessed August 20, 2012).

MacGregor, Cargotech equipment handles increased coal and container capacity at Koper, http://www. macgregor-group.com (accessed August 20, 2012).

Miller, K.A., Siscovick, D.S., Sheppard, L., Shepherd, K., Sullivan, J.H., Anderson, G.L., and Kaufman, J.D. (2007) Long-term exposure to air pollution and incidence of cardiovascular events in women, *N. Engl. J. Med.*, 356(5), 447–458.

Ohuchi, T., Muehlenbachs, K., Steer, J., and Chambers, A. (1995) Insights into coal solubilization during coal—Bitumen coprocessing as monitored by 13C/12C ratios, *Fuel Technol.*, 41(2), 101–124.

Pless-Mulloli, T., Howel, D., and Prince, H. (2001) Prevalence of asthma and other respiratory symptoms in children living near and away from opencast coal mining sites, *Int. J. Epidemiol.*, 30, 556–563.

Pope III, C.A., Renlund, D.G., Kfoury, A.G., May, H.T., and Horne, B.D. (2008) Relation of heart failure hospitalization to exposure to fine particulate air pollution, *Am. J. Cardiol.*, 102(9), 1230–1234.

Saracevic, L., Samek, D., Mihalj, A., and Gradascevic, N. (2009) The natural radioactivity in vicinity of the brown coal mine Tusnica—Livno, BiH, *Radioprotection*, 44(5), 315–320.

SDČVJ (1987) Određivanje taložne tvari (sediment)—Measurement of deposited matter. Smjernica SDČVJ 201—prijedlog. Savez Društava za čistoću Vazduha Jugoslavije, Sarajevo, Jugoslavija.

Schins, R.P.E. and Borm, P.J.A. (1999) Mechanisms and mediators in coal dust induced toxicity: A review, *Ann. Occup. Hyg.*, 43(1), 7–33.

Sitzmann, B., Kendall, M., Watts, J.F., and Williams, I.D. (1999) Characterization of airborne particulates in London by computer—Controlled scanning electron microscopy, *Sci. Total Environ.*, 241(1–3), 63–73.

Topič, N. (2009) An analysis of fugitive dust emission from coal stockpile in the Port of Koper. Master thesis. Jožef Stefan International Postgraduate School, Ljubljana, Slovenia, June 2009.

United States Department of Agriculture, Natural Resources Conservation Service, Water and Climate Center, Wind Rose Data, http://www.wwc.nrcs.usda.gov/climate/windrose.html (accessed August 20, 2012).

Vallero, D. (2008) *Fundamentals of Air Pollution*, 4th edn. Elsevier/Academic Press, Oxford, U.K., pp. 64–65, 355–313, 396–359.

Van Hook, R.I. (1979) Potential health and environmental effects of trace elements and radionuclides from increased coal utilization, *Environ. Health Perspect.*, 33, 227–247.

VDI 2119 Blatt 2 (1996) Bestimmung des Staubniederschlags mit Auffanggefassen aus Glass (Bergerhoff-Verfahren) oder Kunststoff. Verein Deutscher Ingenieure, Dusseldorf, Germany.

Ward, C.R. and Suárez-Ruiz, I. (2008) Chapter 1: Introduction to applied coal petrology. In: *Applied Coal Petrology: The Role of Petrology in Coal Utilization*, Suárez-Ruiz, I. and Crelling, J.C. (eds.). Elsevier Ltd., Amsterdam, the Netherlands, pp. 1–18.

Wouters, E.F., Joma, T.H., and Westenend, M. (1994) Respiratory effects of coal dust exposure: Clinical effects and diagnosis, *Exp. Lung Res.*, 20(5), 385–394.

Xie, R.K., Seip, H.M., Liu, L., and Zhang, D.S. (2009) Characterization of individual airborne particles in Taiyuan City, China, *Air Qual. Atmos. Health*, 2(3), 123–131.

Žitnik, M., Jakomin, M., Pelicon, P., Rupnik, Z., Simi, J., Budnar, M., Grlj, N., and Marzi, B. (2005) Port of Koper—Elemental concentrations in aerosols by PIXE, *X-Ray Spectrom.*, 34(4), 330–334.

2 Nose-Only Aerosol Exposure Systems
Design, Operation, and Performance

Robert F. Phalen, Loyda B. Mendez, and Michael J. Oldham

CONTENTS

2.1 INTRODUCTION

Several methods are available for exposing laboratory animal models to aerosols, each having its advantages and disadvantages (Table 2.1). Whole-body exposure systems (chambers) have a long history in inhalation toxicology, mainly due to their animal loading capacity and adaptability to several species. Chambers are often unsuitable for use with radioactive, infectious, and other potentially hazardous exposure materials, due to the need for large amounts of exposure material, inefficient delivery of aerosols to animals, multiple routes of exposure (e.g., oral, skin, and eyes), and the external contamination of exposed subjects. Also, chambers are costly, require considerable laboratory space, require expensive air supplies, and are difficult to decontaminate.

Nose-only systems have emerged as particularly useful for exposing laboratory animals to hazardous airborne materials for several reasons including the following: (1) they can be made compact enough to fit into secondary enclosures; (2) they are well tolerated by animal subjects when designed with animal comfort in mind (and the animals undergo preliminary training sessions); (3) respiratory rates, volumes, flows, and other physiological parameters can be acquired during exposures; (4) they provide reliable, uniform, and efficient aerosol delivery; (5) in most circumstances, they are acceptable simulations of natural aerosol exposures; (6) they are relatively easy to decontaminate; and (7) several functional designs are well documented in the literature (e.g., Cheng and Moss, 1995; Phalen, 1997; Pauluhn, 2003; Jaeger et al., 2005; Wong, 2007; Stone et al., 2012).

TABLE 2.1

Comparison of Inhalation Exposure Methods

Method	Advantages	Disadvantages	Issues to Consider
Whole-body (chambers)	Accommodate a large number of subjects	Multiple routes of exposure: skin, eyes, oral	Cleaning throughput air
	Efficient for chronic studies	Variability of doses	Construction materials
	Do not require restraint	Cannot pulse exposure	Losses of study material
	Have good environmental control	Poor investigator contact with subjects	Noise, vibration, air temperature, RH[a]
		Expensive investment	Cleaning exhausted air volume
		Animal by-product air contaminants (dander, ammonia, etc.)	
Head-only	Good for repeated exposures	Stress to subjects	Pressure fluctuations
	Limits routes of exposure	Seal around neck	Losses of study material
	More efficient use of study material than chambers	Labor in handling subjects	Air temperature, RH
		Rebreathing of exhaled air	Subject comfort
			Subject restraint
Nose-/mouth-only (including masks)	Exposure mostly to respiratory tract	Stress to some species	Exposure tube design
	Uses less exposure material	Seal about face/nose	Body temperature
	Has good containment	Labor in handling subjects	Subject comfort
		Exposure times limited to hours	Losses in system plumbing
Lung-only or partial lung-only	Precision of dose	Technically difficult	Air temperature and RH
	Uses less exposure material	Invasive	Stress to subjects
	Can use unexposed control tissue from the same animal	Artifacts in dose and response	Physiological support may be required
	Bypasses the nose	Bypasses the nose	Surgery or intubation is required

[a] RH = Relative humidity.

2.2 NOSE-ONLY EXPOSURES

Typical nose-only exposure systems, including secondary containment and an aerosol generation, are depicted in Figures 2.1 and 2.2. The exposure portion has a central plenum (or exposure manifold) into which an aerosol enters after generation and conditioning (e.g., by dilution, drying, and electrical discharging). Confinement (exposure) tubes with one end protruding into the exposure aerosol can be arranged about the plenum in various ways to permit simultaneous exposure of several animals. If the exposure tubes are well designed, animals will enter, move forward, and stick their noses out through a hole at the front of the tube. Animals may breathe from a static (i.e., no airflow) plenum or have flow past, or toward (i.e., directed flow), their noses. Mixed exhaled and throughput air is exhausted from the system, where it may be cleaned free of aerosols and exposure gases prior to discharge into the environment. For nose-only systems inside a secondary containment (e.g., a chamber, biosafety hood, or a glove box), it may be necessary to use an insulated enclosure for the aerosol-generating equipment to isolate exposed animals from heat, noise, and vibration. Such a system using several glove boxes for exposing rodents is described by Hoskins et al. (1997).

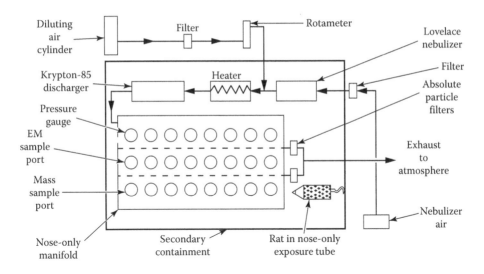

FIGURE 2.1 A nose-only exposure system with secondary containment (stainless steel chamber) suitable for radioactive aerosol exposures. (Courtesy of the Air Pollution Health Effects Laboratory, University of California, Irvine, CA.)

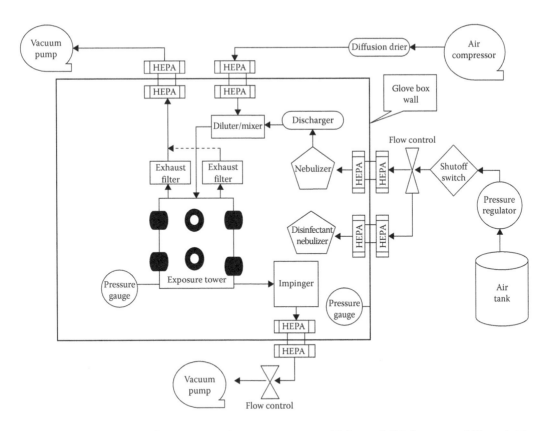

FIGURE 2.2 A high-containment nose-only exposure system with internal disinfectant capability suitable for infectious aerosol exposures. Double HEPA filters prevent contamination of the laboratory. (Courtesy of the Air Pollution Health Effects Laboratory, University of California, Irvine, CA.)

In order to prevent contamination of the laboratory during exposure, the nose-only system can be maintained at a pressure negative with respect to the secondary containment, and the secondary containment held at a pressure negative with respect to the exposure room. If environmental contamination is a concern, the entire laboratory building can have a negative pressure relative to the outside environment.

Nose-only exposures can be labor intensive and stressful to the animal subjects if the system is not well designed. Training animals to accept restraint in nose-only tubes is recommended to reduce stress and produce normal physiological states. Also, thermoregulation of the subjects must be considered both in tube design and by cooling and/or isolation from heat sources (e.g., pumps and heaters used for drying aerosols).

Rodents are particularly adaptable to close confinement in nose-only exposure tubes. Rabbits, ferrets, and most other small species will also accept such exposures, but they may require additional training. Some species, for example, sheep, pigs, rabbits, dogs, and equines, can be trained to accept nose-only exposure muzzle masks, but they may require sedation prior to attaching them to exposure systems and may require restraint during exposures. Nonhuman primates are usually exposed in head-only systems (using a full helmet-like enclosure of the head), but sedation may be required (Dabisch et al., 2010).

2.3 HISTORY

Nose-only exposure systems have a long history of use. Early exposure studies and systems include those reported by Barnes (1947), Henderson (1952), Bair et al. (1961), Casarett (1964), Thomas (1969), Johnson and Ziemer (1971), and Raabe et al. (1973), to name a few. When the first nose-only exposure was used in an inhalation study is unknown, and perhaps unknowable, as it seems reasonable that the method would be used by any investigator who wanted to minimize the amount of exposure material required and/or to avoid contamination of the fur (as by infectious, toxic, or radioactive materials) during an inhalation study. MacFarland (1983), Cheng and Moss (1995), Phalen (1997), Pauluhn (2003), Jaeger et al. (2005), and Wong (2007) have reviewed nose-only exposure systems and included some historical information. Several systems are in current use based on designs by Raabe et al. (1973) and Cannon et al. (1983). New designs have emerged, often for specific applications (e.g., Yeh et al., 1987; Pauluhn, 1994; Jaeger et al., 2005; Oldham et al., 2009; Werley et al., 2009; Stone et al., 2012). Several of these systems are commercially available.

2.4 MINIMIZING ANIMAL STRESS

Unless careful attention is paid to the entire nose-only system design, animals may experience stressful conditions during aerosol exposures. Foremost is the requirement for close fitting yet comfortable exposure tubes that invite the animals to enter and move forward where the nose enters the exposure atmosphere. The body enclosure may be snug, but it must be comfortable, especially for nonsedated animals. The portion of the tube enclosing the head should be designed to painlessly fit the head of the species used. If the tubes are poorly designed, awake animals may not enter freely, may struggle or turn around, and even suffocate in their attempts to escape. Poorly designed exposure tubes can produce abnormal physiological states and thereby distort exposure-related data.

Maintaining physiologically tolerable environmental conditions during nose-only exposures is another essential design requirement. Thermal stress in exposure tubes can lead to animal fatalities, so thermally conductive or vented construction materials are sometimes used. Heat sources, such as pumps and heated aerosol drying devices, should be isolated from the animals, and environmental temperatures monitored and controlled within tolerable limits. Rats in particular regulate body temperatures by adjusting blood flow through their hairless tails, so close tail confinement without provision for cooling should be avoided if possible, unless exposure durations are short. Provision of cool air flowing over the exposure tubes may be required for any species. Rodents require frequent access to drinking water, which limits exposure durations to a few (e.g., up to 6) hours. Inhumane exposure durations are species dependent and must be avoided.

Stress may also be reduced by progressive subject training that involves an initial brief sham exposure followed by confinement periods that will actually be used in the study. Such training for rats and mice may require up to 14 days for respiratory and cardiac functions to completely normalize (Narciso et al., 2003). Properly trained animals will usually enter the tubes freely and relax during exposures without attempting to escape. When animals struggle during confinement, the reason should be determined and the problem remedied. Sedation is seldom required to obtain low-stress exposures, but it can be used if necessary. Exposures that produce significant acute respiratory tract irritation should be avoided unless sedation or pain relief is provided.

Vibration, noise, and unfamiliar odors also stress rodents and other species. As many rodents are active at night, daytime exposures may be preferable. If elevated respiratory rates are required, CO_2 can be added to the exposure atmosphere. In any case, it is wise to measure the physiological status of the animals under exposure conditions. Control groups simultaneously exposed to a sham atmosphere are strongly recommended, as cage controls will not experience the handling, confinement, and environmental conditions associated with the exposure.

2.5 ACQUIRING BREATHING DATA

Breathing data, especially frequencies and tidal volumes, are needed to estimate inhalation doses, and they are also useful measures of the effects of exposures. Mauderly (1990) reviewed several methods for acquiring such data for laboratory animals during inhalation exposures. One method, plethysmography, is readily adaptable to nose-only exposures. Plethysmography is any of several techniques for measuring and recording volume changes in a subject or a portion of the subject. Whole-body, head-out, and limb- or digit-only are descriptions of plethysmographic techniques. When the enclosed portion of the subject changes size (e.g., due to respiration or changes in blood flow), the pressure in the plethysmograph enclosure varies. Measurement of the pressure change or airflow in and out of the enclosure is recorded. If the device is calibrated (e.g., using a syringe or ventilator pump), the recording can be converted to provide accurate volume changes. Such a technique has been described for unsedated guinea pigs by Murphy and Ulrich (1964) and for unsedated rats by Mautz and Buffalino (1989). The apparatus used by Mautz (1997) is shown in Figure 2.3. Similar methods have been described by Mauderly (1986, 1990). For masked animals, flow meters or spirometers (in series or parallel with the animal) can be used to acquire breathing data. Boecker et al. (1964) described such a system for dogs, which could be adapted to other species.

2.6 INHALABILITY

Inhalability has had several definitions in inhalation toxicology, including exposure efficiency of the deep lung and that portion of the exposure material that enters the trachea. The modern concept of *aerosol particle inhalability* evolved from sampler efficiency studies (e.g., Ogden and Birkett, 1978). Standards for inhalable particle sampling have been developed for application to adult workers (Soderholm, 1989). The aerosol inhalability (I, AI, or IF) is defined as the fraction of aerosol in the immediate breathing zone that is actually inhaled into the nose and/or mouth. Inhalability is also referred to as the aspiration efficiency or the sampling efficiency of the exposed subject. It is analogous to the sampling efficiency of an aerosol monitor or collector. For workers, the sampling efficiency depends on the external wind speed, ventilation rate, orientation of the face to the wind, and most importantly the aerosol particle aerodynamic diameter. Table 2.2 shows the American Conference of Governmental Industrial Hygienists (ACGIH®, 2012) inhalability values for particles up to 100 μm aerodynamic diameter for low wind speeds averaged over all orientations

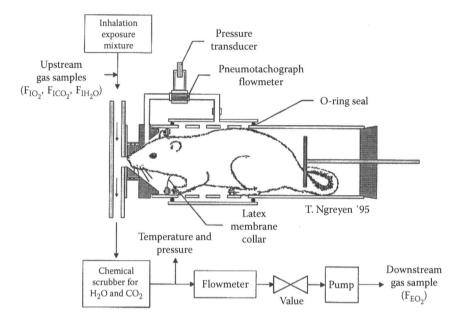

FIGURE 2.3 A plethysmograph/nose-only aerosol exposure system for rats. (From Mautz, W.J., Animal monitoring, in *Methods in Inhalation Toxicology*, Phalen, R.F., ed., CRC Press, Boca Raton, FL, 1997, Chapter 6, pp. 85–99; Courtesy of the Air Pollution Health Effects Laboratory, University of California, Irvine, CA.)

to the wind. These widely used values apply only to the *average* adult worker. Body size, level of exertion, high or zero wind speeds, and other factors modify aerosol inhalability.

Inhalability concepts have also been applied to the efficiency of aerosol delivery to animals, for example, while exposed in nose-only systems (Menache et al., 1995). In a well-designed nose-only exposure system, an unbiased sample from the breathing zone (e.g., from an unused exposure port) can be used to determine the sampling efficiency of the subject. Total deposition in the subject

TABLE 2.2
Inhalability versus Particle Aerodynamic Diameter for Workers

% Sampling Efficiency (Inhalability)	D_{ae}[a] (µm)
100	0
97	1
94	2
87	5
77	10
65	20
58	30
54.5	40
52.5	50
50	100

Source: Data from ACGIH®, *American Conference of Governmental Hygienists*, Cincinnati, OH, 2012.
[a] D_{ae} = Aerosol aerodynamic diameter.

TABLE 2.3

Percent Inhalability (I) for Rats versus Particle Aerodynamic Diameter for Miller (2000) Using Menache et al. (1995) Calculations, and Asgharian et al. (2003) as Measured during Nose-Only Exposures[a]

I, Miller (2000)	I, Asgharian et al. (2003)	Aerodynamic Diameter of Particle (μm)
97	97	0.5
95	85	0.8
93	82	1
85	68	2
77	48	3
71	37	4
65	30	5
55	21	7
44	15	10
Not done	7	20

[a] Data are averages estimated from graphs for three breathing rates; slow, normal, and fast.

divided by an accurate breathing-zone sample yields the inhalability. As shown in Table 2.3, Miller (2000), using the equations of Menache et al. (1995), published inhalability values for rats. Also shown in Table 2.3 are the lower values for rats measured in a nose-only system that were reported by Asgharian et al. (2003). Animal exposures are more complex than those of workers who typically experience orientation-averaged low wind speeds. Air velocities directed at the noses of animals in exposure tubes can be very high, which not only alters the aerosol inhalability but may also produce avoidance behavior, a sign of stress to the exposed subjects.

The respiration rate and tidal volume may be abnormal in confined subjects. These parameters should be measured under actual exposure conditions. An important design requirement is to supply breathing air in excess of the animals' requirement (e.g., 2–10 times the minute ventilation). Lower supply rates can produce rebreathing of exhaled air, which complicates determinations of the dose. Rates that are too high can produce avoidance behaviors and unknown inhaled doses and even harm the animals. As breathing and inhalability parameters depend on the animal characteristics and the specific system used for exposure, literature values are seldom adequate, so actual characterization is recommended.

2.7 AEROSOL DEPOSITION

2.7.1 MEASURING AEROSOL DEPOSITION

An important aspect of an inhalation study is to determine the dose delivered to the animals. Inhaled aerosol dose determinations are more complex than those associated with other common dosing techniques (e.g., oral, dermal, or injection). Although essentially all nose-only inhalation systems allow for the measurement of the aerosol concentration during the exposure, the actual aerosol deposition within the animals' respiratory tracts should be measured or at least closely estimated. Otherwise, the delivered dose will be unknown. Measuring *total* (anywhere in the respiratory tract) aerosol deposition is usually easier than measuring *regional* (e.g., nose, tracheo-bronchial tree, and alveoli) deposition. Counting particle concentrations in inhaled and exhaled air can be used to measure total deposition and using radioactive aerosols or sacrifice techniques can be used to measure regional deposition. Clearance curves for radioactive particles that can be acquired with external detectors have been used to measure the regional deposition of inhaled particles by performing initial counts (for total) and analyzing the clearance at 20–24 h to separate out regional deposition. A common assumption is that poorly soluble material retained in the animal more than 24 h after exposure was deposited distal to the ciliated airways, which may not always be true.

2.7.2 EXPERIMENTAL AEROSOL DEPOSITION

The most complete data to date on measuring aerosol deposition after delivery with a nose-only system were published in a series of studies by Raabe et al. (1977, 1988), in which total and regional deposition of monodisperse radioactive aerosols in several species (i.e., mice rats, hamsters, guinea pigs, and rabbits) of unsedated small animals was reported. In the 1988 paper, ventilation values based on formulae published by Guyton (1947) in mice, rats, hamsters, guinea pigs, and rabbits were used. The near monodisperse particles ranged from about 0.2 to 10 μm in diameter. Tables 2.4 and 2.5 show deposition data for mice and rats. Because nose-only exposures to tracer particles lasted up to 45 min, early clearance would have occurred. Material detected in the gastrointestinal tract was assigned to the *Head and Larynx* region in Tables 2.4 and 2.5. In addition to Raabe and colleagues, other groups have reported deposition fractions for rodents using additional (both smaller and larger) particle sizes (Alessandrini et al., 2008; Thomas et al., 2008; Oldham et al., 2009) to characterize their nose-only inhalation systems and the aerosol delivery in their studies. The data imply that even for small animal models, large diameter particles have some, albeit small pulmonary deposition (as is shown in Tables 2.4 and 2.5).

TABLE 2.4
Percent of Presented Aerosols Depositing in Airways of CF$_1$ Mice

D_{ae}^a (μm)	Head and Larynx	Trachea and Bronchi	Pulmonary	Total Respiratory Tract
0.27	10.5	14	45	70
1.09	43	6.1	9.7	59
3.45	90	2.7	0.9	94
4.49	99	1.1	0.23	100
5.98	98	1.4	0.4	100
9.65	99	0.66	0.04	100

Source: Data adapted from Raabe, O.G. et al., *Ann. Occup. Hyg.*, 32(S1), 53, 1988.
[a] D_{ae} = Aerosol aerodynamic diameter.

TABLE 2.5
Percent of Presented Aerosols Depositing in Airways of Fischer 344 Rats

D_{ae}^a (μm)	Head and Larynx	Trachea and Bronchi	Pulmonary	Total Respiratory Tract
0.29	4.9	8.4	13	27
1.02	7.4	4.9	6.5	19
1.03	10	8.3	11	29
3.11	88	4.4	6.6	99
4.26	80	15	4.8	100

Source: Data adapted from Raabe, O.G. et al., *Ann. Occup. Hyg.*, 32(S1), 53, 1988.
[a] D_{ae} = Aerosol aerodynamic diameter.

2.7.3 PREDICTED AEROSOL DEPOSITION

Traditional mechanistic respiratory tract aerosol deposition models (ICRP, 1994; NCRP, 1997; MPPDep, 2002) as well as computational fluid dynamic (CFD) techniques (Jeon et al., 2012) have been used to design and interpret results from experiments conducted in nose-only inhalation systems. Traditional mechanistic models apply the three main deposition mechanisms (i.e., inertial impaction, gravitational sedimentation, and Brownian diffusion) to particles flowing through simplified airways based on morphometric measurements. Inhaled particle deposition calculations may be regional (extrathoracic, tracheobronchial, and pulmonary), or tracheobronchial airway generation number by airway generation number. By utilizing species-specific and strain-specific anatomy in these mechanistic models, predictions of aerosol deposition can be used to guide the target exposure concentrations and/or duration to get a desired deposition dose for a study as well as to extrapolate and/or interpret results.

CFD techniques have also been used to predict regional deposition such as in the nasal cavities of rodents (Fischer 344 and Sprague Dawley rat), New Zealand white rabbit, and Rhesus monkey (Morgan et al., 1991; Kimbell et al., 1993, 1997a; Godo et al., 1995; Corley et al., 2009, 2012; Jiang and Zhao, 2010). Although these predictions mainly included gaseous toxicants (formaldehyde, acrylic acid, acetone, etc.) (Cohen Hubal et al., 1997; Kimbell et al., 1997b; Bush et al., 1998; Kimbell and Subramaniam, 2001), CFD techniques have also been used to predict aerosol deposition in the nasal cavities of rats (Garcia and Kimbell, 2009). Because of the computational requirements of CFD techniques and the complexity of tracheobronchial airways, there are few predictions for aerosol deposition in these airways (Corley et al., 2012). Currently, only a few species and strains have appropriate anatomical respiratory data suitable for use in aerosol deposition computational models.

2.7.4 EXPERIMENTAL VERSUS PREDICTED AEROSOL DEPOSITION

A few studies have attempted to validate or verify aerosol deposition predictions with empirical data (Nadithe et al., 2003; Wichers et al., 2006; Oldham et al., 2009). In these studies, experimental aerosol deposition measurements appear to agree with predicted values of aerosol deposition. CFD techniques have also been used to design nose-only inhalation apparatus specific for nano-aerosols (Jeon et al., 2012). Dosimetry prediction differences from the measured deposition ranged from 10% up to about 40% depending on the particle diameter (Wichers et al., 2006; Oldham et al., 2009; Jeon et al., 2012). These differences in deposition efficiencies can be due to strain-specific respiratory tract anatomy, ventilation parameters (e.g., measured versus estimated), as well as the design of the nose-only inhalation systems (Oldham et al., 2009).

2.8 OTHER APPLICATIONS FOR NOSE-ONLY INHALATION SYSTEMS

A recent review (Cathcart et al., 2011) highlighted the use of exhaled breath and exhaled breath condensate in toxicology studies of large animals. Much less work has been done with rats and mice, most likely due to the challenge of collection of exhaled breath from enough animals for the reliable detection of potential markers. A nose-only inhalation system is well suited to this challenge. Nose-only exposure systems with a low internal volume provide good detection sensitivity. de Broucker et al. (2012) collected exhaled breath from a nose-only exposure system and determined that statistically significantly elevated levels of NO_x and H_2O_2 resulted from lipopolysaccharide-induced acute lung injury in rats. Among the many applications of animal nose-only exposure systems, Stone et al. (2012) designed a system for testing the effectiveness of various filter media for preventing airborne transmission of infections.

2.9   SUMMARY AND DISCUSSION

Nose-only aerosol exposure systems have a long and successful history. As not all investigators have had success with this exposure modality, it is important to consider how some of the problems can be corrected. Early nose-only systems used glass (e.g., soda bottles) or plastic (e.g., centrifuge tubes) for holding animals during exposures. Although inexpensive and convenient, such tubes seldom provide good fits to the animal's bodies or heads. Poor fit can not only cause animals to resist placement in the tubes but also cause significant abrasion damage to rodent's eyes (which may be in direct contact with the tapered end of the tube). Attempts to confine animals in the exposure tubes have included plungers that are advanced against the rear of the animal with significant force. Such force may crush the animal or at least distort its posture enough to produce significant stress. Also, the use of rubber stoppers at the rear of glass or plastic tubes can lead to overheating and, in extended confinements, even deaths. The use of well-designed animal exposure tubes that are neither too snug nor too loose and that are shaped to accommodate the animals' heads is essential for successful exposures. Animals will generally enter properly sized and shaped tubes and move forward without being forced. Even same-aged inbred rodents will vary in body size, so a selection of tube sizes should be available.

In some unsuccessful designs, thermal stress has led to fatalities. Aerosol-generating and conditioning equipment can generate significant heat, so it is imperative that such heat sources should not be placed near the exposure tubes if exposures last for more than a few minutes. Even when there are no nearby heat sources, chilled air may be needed to cool the exposure tubes in order to overcome metabolic heat production.

Rodents and other small animal species use their extensive olfactory apparatus to detect predators so unfamiliar odors can produce agitation and escape behaviors. Therefore, odor control is important. Exposure tubes should be thoroughly washed (e.g., in a high-temperature dishwasher) before each use. Technicians that handle animals should wear fresh, clean lab coats and not use perfumes or other scented personal care products on exposure days. Odor control also applies to carts used to transport animals to the exposure area and to the components of the exposure system. Lack of attention to such details can cause a nose-only exposure experiment to be stressful to subjects (and experimenters) and even fail.

Gentle handling of animals is also essential to successful exposures. Technicians should be trained in proper handling techniques. The goal is to obtain the confidence and thus cooperation of the animal subjects. It should be clear that good system design, maintenance, and operation are all required for conducting nose-only exposures.

Nose-only exposure systems just are one of the several useful techniques for exposing laboratory animals to aerosols. This exposure modality is particularly effective for use with potentially hazardous or expensive exposure materials that require control of doses and unwanted contamination. Although many designs are available, investigators should select or build systems that are appropriate for their species and study design. Animal comfort is an important consideration. When properly designed and operated, nose-only exposure systems can be considered as the gold standard for measuring particle-size-dependent deposition efficiencies in small animals.

QUESTIONS

A. Multiple choice questions: Select the best answer.

1. What type of exposure system would be preferred for a chronic study of a relatively nontoxic aerosol using rodents? Exposures are 24 h per day for 9 months.
   a. Nose-only
   b. Partial-lung-only
   c. Whole-body
   d. All of the above are suitable
   Answer: c

2. What type of exposure system would be recommended for a single 30 min ferret exposure to anthrax spores? The exposure concentrations are 1 and 10 spores per cubic centimeter of air, and the group size is 10 animals.
   a. Nose-only
   b. Partial-lung-only
   c. Whole-body
   d. All of the above are suitable

   Answer: a

3. You are designing a rat study to examine the toxicity to deep-lung macrophage cells of plutonium particles produced by an explosion. The particles are 20 μm in aerodynamic diameter, and the exposure is very brief (e.g., a few seconds). What exposure method would you select?
   a. Nose-only to prevent contamination of the fur
   b. Whole-body as it is most realistic
   c. Lung-only in order to bypass filtration in the rats' noses
   d. Substitute external gamma ray exposure to mimic the radiation effect

   Answer: c

B. Write a brief essay to answer each question.

1. You are designing a nose-only exposure system for use with mice, rats, and rabbits using aerosols that will be nebulized from a water suspension and then heat-treated to dry and fuse tiny aerosol particles together. How would you prevent thermal stresses to the animals?

   Answer for full credit: Provide isolation of the heating source from the exposure location; and use exposure tubes that conduct body heat away from the animals, and are comfortable.

2. What factors in a nose-only exposure might influence the inhalability of the exposure aerosol?

   Answer for full credit includes three of the following: the aerosol characteristics, the air velocities in the breathing zone of the animals, the exposure tube designs, and the ventilation characteristics of the animals.

ACKNOWLEDGMENTS

Support for Dr. Phalen is provided by the University of California Irvine Center for Occupational and Environmental Health and by the Charles S. Stocking Fund. Dr. Mendez is supported by the California Air Resources Board Contract No. ARB-08-30, awarded to Dr. Michael Kleinman of the University of California, Irvine. Leslie Kimura and Robin Ferguson provided word-processing and administrative support.

REFERENCES

ACGIH® 2012 TLVs® and BEIs®, ACGIH®, *American Conference of Governmental Hygienists*, Cincinnati, OH, 2012.

Alessandrini, F., Semmler-Behnke, M., Jakob, T., Schulz, H., Behrendt, H., and Kreyling, W., Total and regional deposition of ultrafine particles in a mouse model of allergic inflammation of the lung, *Inhal. Toxicol.*, 20: 585–593, 2008.

Asgharian, B., Kelly, J.T., and Tewksbury, E.W., Respiratory deposition and inhalability of monodisperse aerosols in Long-Evans rats, *Toxicol. Sci.*, 71: 104–111, 2003.

Bair, W.J., Willard, D.H., and Temple, L.A., Plutonium inhalation studies-I. The retention and translocation of inhaled Pu23902 particles in mice, *Health Phys.*, 7: 54–60, 1961.

Barnes, J.M., The development of anthrax following the administration of spores by inhalation, *Br. J. Exp. Pathol.*, 28: 385–394, 1947.

Boecker, B.B., Agular, F.L., and Mercer, T.T., A canine inhalation exposure apparatus utilizing a whole-body plethsymograph, *Health Phys.*, 10: 1077–1089, 1964.

Bush, M.L., Frederick, C.L., Kimbell, J.S., and Ultman, J.S., A CFD-PBPK hybrid model for simulating gas and vapor uptake in the rat nose, *Toxicol. Appl. Pharmacol.*, 150: 133–145, 1998.

Cannon, W.C., Blanton, E.F., and McDonald, K.E., The flow-past chamber: An improved nose-only exposure system for rodents, *Am. Ind. Hyg. Assoc. J.*, 44: 923–928, 1983.

Casarett, L.J., Distribution and excretion of Polonium-210, *Radiat. Res. Suppl.*, 5: 148–165, 1964.

Cathcart, M.P., Love, S., and Hughes, K.J., The application of exhaled breath gas and exhaled breathe condensate analysis in the investigation of the lower respiratory tract in veterinary medicine: A review, *Vet. J.*, 191: 282–291, 2011.

Cheng, Y.S. and Moss, O.R., Inhalation exposure systems, *Toxicol. Methods*, 5: 161–197, 1995.

Cohen Hubal, E.A., Schlosser, P.M., Conolly, R.B., and Kimbell, J.S., Comparison of inhaled formaldehyde dosimetry predictions with DNA-protein cross-link measurements in the rat nasal passages, *Toxicol. Appl. Pharmacol.*, 143: 47–55, 1997.

Corley, R.A., Kabilan, S., Kuprat, A., Carson, J., Minard, K., Jacob, R., Timchalk, C. et al., Comparative computational modeling of airflows and vapor dosimetry in the respiratory tracts of a rat, monkey and human, *Toxicol. Sci.*, 128: 500–510, 2012.

Corley, R.A., Minard, K.R., Kabilan, S., Einstein, D.R., Kuprat, A.P., Harkema, J.R., Kimbell, J.S., Gargas, M.L., and Kinzell, J.H., Magnetic resonance imaging and computational fluid dynamics (CFD) simulations of rabbit nasal airflows for the development of hybrid CFD/PBPK models, *Inhal. Toxicol.*, 21: 512–518, 2009.

Dabisch, P.A., Kline, J., Lewis, C., Yeager, J., and Pitt, M.L.M., Characterization of a head-only aerosol exposure system for nonhuman primates, *Inhal. Toxicol.*, 22: 224–233, 2010.

de Broucker, V., Hassoun, S.M., Hulo, S., Cherot-Kornobis, N., Neviere, R., Matran, R., Sobaszek, A., and Edme, J.-L., Non-invasive collection of exhaled breath condensate in rats: Evaluation of pH, H_2O_2 and NO_x in lipopolysaccharide-induced acute lung injury, *Vet. J.*, 194: 222–228, 2012.

Garcia, G.J. and Kimbell, J.S., Deposition of inhaled nanoparticles in the rat nasal passages: Dose to the olfactory region, *Inhal. Toxicol.*, 14: 1165–1175, 2009.

Godo, M.N., Morgan, K.T., Richardson, R.B., and Kimbell, J.S., Reconstruction of complex passageways for simulations of transport phenomena: Development of a graphical user-interface for biological applications, *Comput. Methods Programs Biomed.*, 47: 97–112, 1995.

Guyton, A.C., Analysis of respiratory patterns in laboratory animals, *Am. J. Physiol.*, 150: 78–83, 1947.

Henderson, D.W., An apparatus for the study of airborne infection, *J. Hyg.*, 50: 53–58, 1952.

Hoskins, J.A., Brown, R.C., Cain, K., Clouter, A., Houghton, C.E., Bowskill, C.A., and Hibbs, L.R., The construction and validation of a high containment nose-only rodent inhalation facility, *Ann. Occup. Hyg.*, 41: 51–61, 1997.

ICRP, *Human Respiratory Tract Model for Radiological Protection*, International Commission on Radiological Protection Publication 66, Pergamon Press, New York, 1994.

Jaeger, R.J., Shami, S.G., and Tsenova, L., Directed-flow aerosol inhalation exposure systems: Application to pathogens and highly toxic agents. In *Inhalation Toxicology*, 2nd edn., Salem, H. and Katz, S.A., eds., Taylor & Francis, Boca Raton, FL, 2005, Chapter 4, pp. 73–90.

Jeon, K., Yu, I.J., and Ahn, K.H., Evaluation of newly developed nose-only inhalation exposure chamber for nanoparticles, *Inhal. Toxicol.*, 24: 550–556, 2012.

Jiang, J.B. and Zhao, K., Airflow and nanoparticle deposition in rat nose under various breathing and sniffing conditions: A computational evaluation of the unsteady and turbulent effect, *J. Aerosol Sci.*, 41: 1030–1043, 2010.

Johnson, R.F., Jr. and Ziemer, P.L., The deposition and retention of inhaled 152–154 Europium oxide in the rat, *Health Phys.*, 20: 187–193, 1971.

Kimbell, J.S., Godo, M.N., Gross, E.A., Joyner, D.R., Richardson, R.B., and Morgan, K.T., Computer simulation of inspiratory airflow in all regions of the F344 rat nasal passages, *Toxicol. Appl. Pharmacol.*, 145: 388–398, 1997a.

Kimbell, J.S., Gross, E.A., Joyner, D.R., Godo, M.N., and Morgan, K.T., Application of computational fluid dynamics to regional dosimetry of inhaled chemicals in the upper respiratory tract of the rat, *Toxicol. Appl. Pharmacol.*, 121: 253–263, 1993.

Kimbell, J.S., Gross, E.A., Richardson, R.B., Conolly, R.B., and Morgan, K.T., Correlation of regional formaldehyde flux predictions with the distribution of formaldehyde-induced squamous metaplasia in F344 rat nasal passages, *Mutat. Res.*, 380: 143–154, 1997b.

Kimbell, J.S. and Subramaniam, R.P., Use of computational fluid dynamics model for dosimetry of inhaled gases in the nasal passages, *Inhal. Toxicol.*, 13: 325–334, 2001.

MacFarland, H., Designs and operational characteristics of inhalation exposure equipment—A review, *Fundam. Appl. Toxicol.*, 3: 603–613, 1983.

Mauderly, J.L., Respiration of F344 rats in nose-only inhalation exposure tubes, *J. Appl. Toxicol.*, 6: 25–30, 1986.

Mauderly, J.L., Measurement of respiration and respiratory responses during inhalation exposures, *J. Am. Coll. Toxicol.*, 9: 397–405, 1990.

Mautz, W.J., Animal monitoring. In *Methods in Inhalation Toxicology*, Phalen, R.F., ed., CRC Press, Boca Raton, FL, 1997, Chapter 6, pp. 85–99.

Mautz, W.J. and Buffalino, C., Breathing patterns and metabolic rate responses of rats exposed to ozone, *Resp. Physiol.*, 76: 69–78, 1989.

Menache, M.G., Miller, F.J., and Raabe, O.G., Particle inhalability curves for humans and small laboratory-animals, *Ann. Occup. Hyg.*, 39: 317–328, 1995.

Miller, F., Dosimetry of particles: Critical factors having risk assessment implications, *Inhal. Toxicol.*, 12(S3): 389–395, 2000.

Morgan, K.T., Kimbell, J.S., Monticello, T.M., Patra, A.L., and Fleishman, A., Studies of inspiratory airflow patterns in the nasal passages of the F344 rat and rhesus monkey using nasal molds: Relevance to form-aldehyde toxicity, *Toxicol. Appl. Pharmacol.*, 110: 223–240, 1991.

MPPDep, Multiple path particle Deposition model V. 1.11, RIVM Report 650010030, National Institute of Public Health, Bilthoven, the Netherlands and the Environment and Centers for Health Research (CIIT), Research Triangle Park, NC, 2002. Currently version 2.11 available at http://www.ara.com/products/ MPPD.html, accessed April 8, 2012.

Murphy, S.D. and Ulrich, C.E., Multi-animal test system for measuring effects of irritant gases and vapors on respiratory function in guinea pigs, *Am. Ind. Hyg. Assoc. J.*, 25: 28–36, 1964.

Nadithe, V., Rahamatalla, M., Finlay, W.H., Mercer, J.R., and Samuel, J., Evaluation of nose-only aerosol inhalation chamber and comparison of experimental results with mathematical simulation of aerosol deposition in mouse lungs, *J. Pharm. Sci.*, 92: 1066–1076, 2003.

Narciso, S.P., Nadziejko, E., Chen, L.C., Gordon, T., and Nadziejko, C., Adaptation to stress induced by restraining rats and mice in nose-only inhalation holders, *Inhal. Toxicol.*, 15: 1133–1143, 2003.

NCRP, Report 125—Deposition, retention and dosimetry of inhaled radioactive substances, National Council on Radiological Protection and Measurements, Bethesda, MD, 1997.

Ogden, T.L. and Birkett, J.L., Inhalable dust sampler for measuring hazard from total airborne particulate, *Ann. Occup. Hyg.*, 21: 41–50, 1978.

Oldham, M.J., Phalen, R.F., and Budiman, T., Comparison of predicted and experimentally measured aerosol deposi-tion efficiency in BALB/C mice in a new nose-only exposure system, *Aerosol Sci. Technol.*, 43: 970–977, 2009.

Pauluhn, J., Validation of an improved nose-only exposure system for rodents, *Appl. Toxicol.*, 14: 55–62, 1994.

Pauluhn, J., Overview of testing methods used in inhalation toxicity: From facts to artifacts, *Toxicol. Lett.*, 140(SI): 183–193, 2003.

Phalen, R.F., Inhalation exposure methods. In *Methods in Inhalation Toxicology*, Phalen, R.F., ed., CRC Press, Boca Raton, FL, 1997, Chapter 5, pp. 69–84.

Raabe, O.G., Al-Bayati, M.A., Teague, S.V., and Rasolt, A., Regional deposition of inhaled monodisperse coarse and fine aerosol particles in small laboratory animals, *Ann. Occup. Hyg.*, 32(S1): 53–63, 1988.

Raabe, O.G., Bennick, J.E., Light, M.E., Hobbs, C.H., Thomas, R.L., and Tillery, M.I., An improved apparatus for acute inhalation exposure of rodents to radioactive aerosols, *Toxicol. Appl. Pharmacol.*, 26: 264–273, 1973.

Raabe, O.G., Yeh, H.C., Newton, G.J., Phalen, R.F., and Velasquez, D.J., Deposition of inhaled monodisperse aerosols in small rodents. In *Inhaled Particles IV*, Part I, Walton, W.H., ed., Pergamon Press, Oxford, U.K., 1977, pp. 1–21.

Soderholm, S.C., Proposed international conventions for particle size-selective sampling, *Ann. Occup. Hyg.*, 33: 301–320, 1989.

Stone, B.R., Heimbuch, B.K., Wu, C-Y., and Wander, J.D., Design, construction and validation of a nose-only inhalation exposure system to measure infectivity of filtered bioaerosols in mice, *J. Appl. Microbiol.*, 113: 757–766, 2012.

Thomas, R.J., Webber, D., Sellors, W., Collinge, A., Frost, A., Stagg, A.J., Bailey, S.C. et al., Characterization and deposition of respirable large- and small-particle bioaerosols, *Appl. Environ. Microbiol.*, 74: 6437–6443, 2008.

Thomas, R.L., Deposition and initial translocation of inhaled particles in small laboratory animals, *Health Phys.*, 16: 417–428, 1969.

Werley, M.S., Lee, K.M., and Lemus, R., Evaluation of a novel inhalation exposure system to determine acute respiratory responses to tobacco and polymer pyrolysate mixtures in Swiss-Webster mice, *Inhal. Toxicol.*, 21: 719–729, 2009.

Wichers, L.B., Rowan, W.H., Nolan, J.P., Ledbetter, A.D., McGee, J.K., Costa, D.L., and Watkinson, W.P., Particle deposition in spontaneously hypertensive rats exposed via whole-body inhalation: Measured and estimated dose, *Toxicol. Sci.*, 93: 400–410, 2006.

Wong, B.A., Inhalation exposure systems: Design, methods and operation, *Toxicol. Pathol.*, 35: 3–14, 2007.

Yeh, H.C., Snipes, M.B., and Brodbeck, R.D., Nose-only exposure system for inhalation exposures of rodents to large particles, *Am. Ind. Hyg. Assoc. J.*, 48: 247–251, 1987.

3 Lungs-on-a-Chip

Shuichi Takayama, Joshua B. White, and Chao Zhang

CONTENTS

3.1 INTRODUCTION

Many drug and toxicology tests cannot be performed directly on humans. Therefore, animal tests are routinely used to predict human patient responses; however, anatomical and physiological differences between animals and humans make it difficult to extrapolate results from animals to humans. Furthermore, for reasons of ethics and cost, as well as to better predict human, rather than animal, responses, there is an increasing interest in the construction of more physiological in vitro cell culture models. Recent advances in the field of microfluidics, which is the technology to manipulate small volumes of fluid, have advanced lung-on-a-chip technology to address these needs. It is noted that lungs-on-a-chip are typically not models of the entire lung or devices made with the intent of performing gas exchange functions for animals or humans, but rather for enhancing in vitro cell culture models for improving predictions about chemical toxicity and/or efficacy and to gain mechanistic insights into lung pathophysiology. This chapter will introduce the readers to basic concepts and representative examples of this field of lungs-on-a-chip. We will start with a more in-depth discussion of why one may want to go through the extra effort of creating these micro-engineered cell cultures. Then we will explain the basic methods of microfabrication used to construct these systems. The heart of this chapter will be two sections that describe microfluidic models of the small airways and the alveoli. We will finally end with conclusions and a discussion of future directions.

3.2   WHY CREATE LUNGS-ON-A-CHIP?

Advances in cell biology now allow in vitro culture of many types of cells, including primary human lung epithelial cells. Cultures of such cells in conventional culture dishes and microwells are well established, are readily performed using off-the-shelf culture tools, and have proven beneficial in many cell biological studies. What might microfluidic lung cell culture provide that conventional in vitro cultures cannot? A major difference between the environment of the lung in vivo and that of the culture dish is that the former is dynamic and accompanies movement with every breath, while the dish is rigid and static. Biochemical advances now allow easy exposure of cells to relevant biomolecules by adding them to cell culture media. Cell responses, however, are also modulated significantly by mechanical stimulation. Typical cell culture platforms do not re-create physiological mechanical stimulations critical for physiological cell function. Other differences between typical dish culture and the in vivo lung environment are that epithelial cells in the lung are exposed to air on their apical side and interact with other cell types, such as endothelial cells and fibroblasts, on their basal side. Microfluidic lung-on-a-chip platforms aim to fill these gaps between typical dish cultures and the physiological lung environment (Figure 3.1). This is enabled through the creation of culture platforms made of flexible material that stretches (Figure 3.1a), through the culture of cells in micro-channels where liquid plugs can be generated (Figure 3.1b), and through the growth of cells on porous membranes (Figure 3.1c and d).

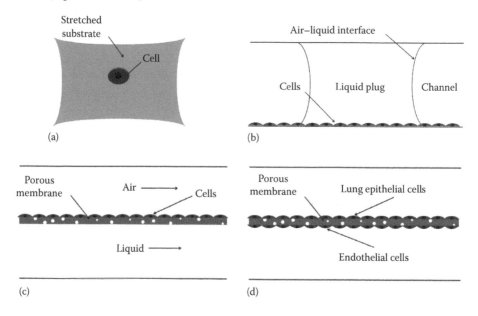

FIGURE 3.1   The physiological lung environment exposes lung epithelial cells to (a) stretch and compression; (b) fluid mechanical stresses, particularly when fluid accumulates in the lung due to various diseases, such as congestive heart failure, and the fluid propagates over the surfaces of cells; (c) air–liquid interfaces; and (d) interactions with other types of cells, such as endothelial cells.

3.3 COMMONLY USED MICROFABRICATION TECHNIQUES

Lung-on-a-chip construction requires microstructures that mimic the small airways and alveoli. The basic underlying technology used to fabricate these microstructures originated in photolithography, a technology developed for the microelectronics industry. Because of the cost-intensive nature and material constraints (mainly limited to silicon substrates) of photolithography, an alternative microfabrication technique, commonly referred to as soft lithography, has been developed. The next paragraphs provide a brief overview of these two technologies.

3.3.1 PHOTOLITHOGRAPHY

Photolithography is a multistep process originally developed to create structures in silicon. A schematic of the process is shown in Figure 3.2. The procedure is typically performed in a highly specialized clean room facility that facilitates the complex fabrication steps required to create the necessary microstructures. The highly optimized process yields very precise structures but is expensive and not readily accessible to many laboratories because of the specialized equipment required.

3.3.2 SOFT LITHOGRAPHY

To overcome some of the limitations of photolithography, such as high cost and silicon-based materials, an alternative microfabrication technique has been developed. In this method called soft lithography (Whitesides et al. 2001), photolithography is still required to make one microstructured master mold. A polymer replica of the master mold is then made by curing prepolymer poured over the master mold (Figure 3.3). Although a variety of different polymers can be used, the most commonly used material is a silicone elastomer, poly(dimethylsiloxane), commonly referred to as PDMS for brevity. The replica structure created in PDMS is bonded to a flat substrate to create the final microchannel device.

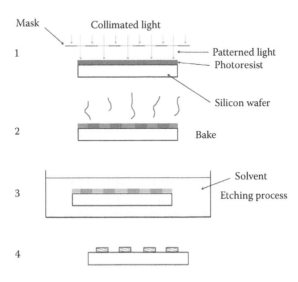

FIGURE 3.2 Basic steps of photolithography. Coating of silicon wafer with photoresist (not shown). Exposure of photoresist to patterned light (step 1). Baking (step 2). Use of solvent to region-specifically dissolve photoresist (step 3). Further baking and drying gives a silicon substrate with micropatterned photoresist structures that can be used as a master mold in a rapid prototyping technique called soft lithography (step 4; also see Figure 3.3 for details of soft lithography).

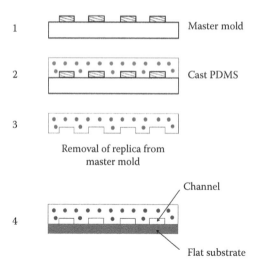

FIGURE 3.3 Basic steps of soft lithography. The starting master mold is the patterned photoresist structure obtained from photolithography (step 1). Poly(dimethylsiloxane) prepolymer is poured over the master mold and cured by heating (step 2). The cured polymer replica is peeled off of the master mold (step 3). This replica is sealed against a flat substrate to give closed microchannels (step 4).

The soft lithography microfabrication technique has become widespread, particularly for biomedical research applications such as fabrication of lungs-on-a-chip for several reasons:

1. Accessibility and low cost. While the master mold requires special facilities of typical photolithography, the subsequent polymer casting is simple, inexpensive, and readily performed in any lab with minimal capital and time investment. Multiple microchannel devices can be fabricated from one master mold that is used repeatedly.
2. Easy bonding. PDMS pieces can be bonded together with a variety of materials including other PDMS pieces by plasma activation of the surface or by using PDMS prepolymer as glue.
3. Biocompatibility. PDMS is nontoxic and gas permeable. Long-term cell culture in small-volume microchannels benefits from this gas permeability.
4. Optical transparency. PDMS is optically transparent and nonfluorescent at most wavelengths. This makes PDMS microchannels convenient for microscopy.
5. Flexibility. PDMS is an elastomer that can be stretched and bent. This flexibility is useful for re-creating the stretch environment of the lung. The material properties, such as Young's modulus, can also be adjusted to match that of typical lung tissue.

There are, however, some limitations to soft lithography. Because of its hydrophobic nature, PDMS can absorb hydrophobic chemicals, including many drugs, thereby limiting its application to the testing of some drugs (Toepke and Beebe 2006; Wang et al. 2012). Large-scale manufacturing with PDMS is difficult due to the inability to use PDMS for injection molding and the difficulty of aligned bonding due to the inherent flexibility of the material. Regardless, many of the lung-on-a-chip devices published to date utilize soft lithography and PDMS. Some representative examples are described in the following text.

3.4 MICROFLUIDIC MODELS OF THE SMALL AIRWAYS

The walls of small airways are lined with a thin film of liquid, and under many pathologic and some normal conditions, the liquid lining can become unstable and form liquid plugs (Halpern and Grotberg 2003). Inspiration and expiration can force the movement and subsequent rupture of such liquid plugs, which exposes airway epithelial cells to significant fluid mechanical forces (Fujioka and Grotberg 2004; Tai et al. 2011). In some diseases, the airway itself can collapse, exerting further mechanical forces on the airway epithelial cells (Gaver et al. 1996; Heil 1998). The next few paragraphs describe some representative microfluidic models of the airway that re-create some of these physiological microenvironments in vitro.

3.4.1 Parallel-Plate Model

The pioneering experimental work that established the importance of fluid mechanical effects (in the absence of solid mechanical effects) of airway reopening on cellular damage was performed in the Gaver Lab (Bilek et al. 2003; Kay et al. 2004; Jacob and Gaver 2005). This section describes these early models of the airway reopening.

3.4.1.1 Device Construction

This in vitro model system of airway reopening utilizes a millimeter-sized parallel-plate flow chamber. In this model, one plate has lung epithelial cells cultured on it. Another plate is placed parallel to this cell-cultured surface with a small gap between the two surfaces that is on the order of the diameter of small airways. Although the width of the channels is much wider than that found in actual airways, the fluid mechanical parameters of the airways are closely mimicked by the parallel-plate model, and thus, the fluid mechanical forces are physiologically relevant.

3.4.1.2 Cell Culture

In this system, cells are cultured under conventional in vitro cell culture conditions. That is, cells are grown submerged in culture media and without exposure to air on the apical side for any prolonged period. Fluid mechanical stress injury experiments, as described in the next paragraph, are performed on these cell monolayers.

3.4.1.3 Air Fingers Generate Fluid Mechanical Stress

In this model of the airway, a finger of air is propagated through the liquid-filled chamber lined with pulmonary epithelial cells by blowing air into the device. Cellular injury is evaluated by staining cells with live/dead stains, which provide information about cell viability based on membrane permeability, and by analyzing cellular detachment. By propagating the air–liquid meniscus of the air finger at different velocities, as well as by using liquids with different viscosities, this model demonstrated that a steep pressure gradient near the advancing meniscus of the air finger primarily determines the extent of epithelial cell damage. Furthermore, the use of surfactants, which decrease surface tension of the liquid, decreases tissue injury significantly. This study, relevant to a newborn's first breath, gives insights into the understanding of lung injury in pulmonary diseases such as neonatal respiratory distress syndrome, but does not address the effect of the movement and rupture of a liquid plug with a finite length (e.g., Figure 3.4a) on cellular damage. Furthermore, the extent of cellular injury and cellular detachment observed in this system may be more severe than what occurs physiologically because the epithelial cells not exposed to air–liquid culture conditions (see next section for more detail on air–liquid interface culture) are often less robust.

3.4.2 Microchannel Model

Although semi-infinite bubbles, as utilized in the previous parallel-plate model of the airway, are relatively straightforward to create and control, they lack many of the features of airway closure and reopening caused by liquid plugs (Figure 3.4a). Physiologically, liquid plugs form stochastically by liquid film instabilities that occur randomly. This makes systematic studies of the effects of liquid plugs difficult to perform in vivo because plug formation, propagation, and rupture

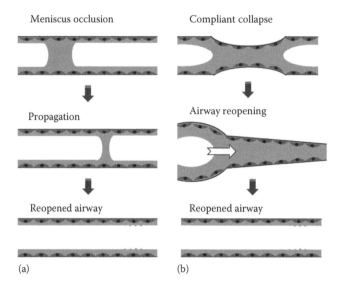

FIGURE 3.4 Two major modes of airway closure. (a) Meniscus occlusion and (b) compliant collapse. Airway reopening is initiated by the propagation of a liquid plug and is completed when the plug loses a sufficient volume of liquid and ruptures.

cannot be well controlled. Conventional in vitro cultures in dishes cannot re-create such liquid plugs. Even the parallel-plate models described earlier have difficulties in controlled formation and propagation of nanoliter- and picoliter-sized liquid plugs reliably. This led to the construction of a microchannel model of a small airway that incorporates a specialized liquid plug generator (Huh et al. 2007; Tavana et al. 2010, 2011). Furthermore, this system is designed to culture lung epithelial cells at an air–liquid interface, which induces cellular differentiation into a more physiologically functional monolayer and makes the cells more resistant to applied fluid mechanical stresses.

3.4.2.1 Device Construction
The microchannel models of the small airway are constructed using variations of soft lithography. Master molds of the required channel features are generated by the first four steps of photolithography (Figure 3.2); these molds are then used to create negative replicas in PDMS using soft lithography (Figure 3.3). In these devices, there are two different master molds that are created. One mold provides channel features of the upper channel, and another provides the channel features of the lower channel (Figure 3.4a). These two channels are aligned and made to sandwich a porous polyester membrane with 400 nm pores on which cells are subsequently cultured. The pore size is small enough that cells cannot pass through but are large enough that fluids and nutrients can pass. The two layers and membrane are bonded together with PDMS prepolymer that acts as mortar/glue to fill gaps and hold all pieces together (Chueh et al. 2007).

3.4.2.2 Cell Culture
In this system, lung epithelial cells such as primary human small airway epithelial cells (Huh et al. 2007) or human lung epithelial cell lines (Tavana et al. 2011) are cultured on the porous membrane sandwiched between the upper and lower channels (Figure 3.5). Once the cells are seeded, attached, and grown to confluence, the cell culture medium in the upper channel can be removed and replaced with air. The epithelial cells will thus have their apical side exposed to air while they can be fed with culture media from below through the porous membrane. In the

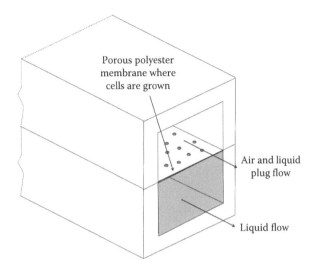

Porous polyester
membrane where
cells are grown

Air and liquid
plug flow

Liquid flow

FIGURE 3.5 Microchannel model of the small airway. Lung epithelial cells are cultured in the upper channel on a porous membrane. The apical side of cells is exposed to air. The cells are fed with cell culture medium from the lower channel. (Adapted from Huh, D. et al., *Proc. Natl. Acad. Sci. USA*, 104, 18886, 2007.)

case of primary human small airway cells, physiological structural and functional differentiation can be observed after several weeks as confirmed by staining for tight junction and production of specialized proteins (Huh et al. 2007). As described more in the next section, the air–liquid interface cultured epithelial cells are more robust against fluid mechanical injury as measured by cell membrane damage and detachment.

3.4.2.3 Liquid Plug Propagation and Rupture Generates Fluid Mechanical Stresses

A key feature of the microchannel model of the small airway is the ability to generate small liquid plugs in a controlled manner (Huh et al. 2007). The steps used to generate liquid plugs are depicted in Figure 3.6. The K-shaped microfluidic channel has two inlets and two outlets. Air and liquid enter from the two inlets, and under the appropriate flow conditions, liquid and air exit out the waste outlet, while only air flows into the channel that leads to the cell culture chamber. When a liquid plug needs to be sent to the cell culture chamber, air flow is briefly stopped, and liquid fills the entire width of the channel. When air flow is resumed, a small liquid plug is generated and propagated down the length of the cell culture chamber. Control of pressure differences and duration of air flow interruption regulate the size of the plugs and the velocity with which the plugs move through the channel (Tavana et al. 2010). Under the appropriate operation conditions, the plugs can be made to rupture in the cell culture region. Such plug rupture events mimic analogous events that occur in the lung and are identified as crackle sounds upon auscultation of the chest by a stethoscope (Huh et al. 2007).

Such plug propagation and rupture events generate significant fluid mechanical stresses that can injure lung epithelial cells (Huh et al. 2007). The controlled environment of these microfluidic airway models enabled correlation of increased cellular injury with increasing number of plug propagation and rupture events. It was also observed that cellular-level injury increased as the liquid plugs became shorter and propagated fast over the cells. The microfluidic small airway injury model was also used to demonstrate the protective effects of clinically used surfactant therapies against such liquid plug propagation events (Tavana et al. 2011). Cellular injuries caused by liquid plugs of phosphate-buffered saline (high surface tension) were mitigated upon incorporation of clinically used concentrations of the surfactant Survanta™ (which decreases surface tension).

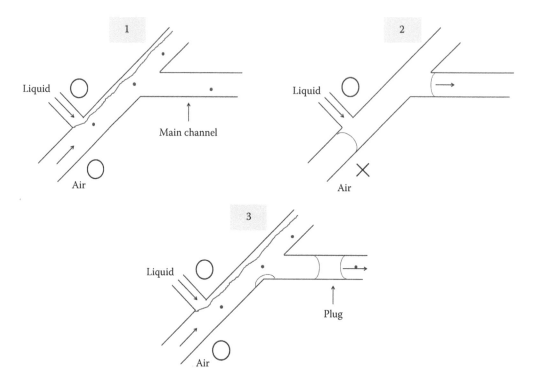

FIGURE 3.6 Microfluidic mechanism for the formation of nanoliter liquid plugs in microchannels. In the base state, air and liquid flow in parallel. The cell culture channel only receives flow of air (state 1). When air flow is briefly stopped, the channel fills with liquid (state 2). As the flow of air is restarted, a small liquid plug is sent down the cell culture channel (state 3). (Adapted from Huh, D. et al., *Proc. Natl. Acad. Sci. USA*, 104, 18886, 2007.)

3.5 MICROFLUIDIC MODELS OF THE ALVEOLI

With every breath, alveoli stretch and expand, then relax and contract. Under some conditions, the alveoli can collapse or become filled with liquid, which exposes alveolar epithelial cells to solid and fluid mechanical forces. This section describes two representative microfluidic models of the alveoli that culture lung epithelial cells under such mechanically stimulated conditions.

3.5.1 MICROCHANNEL MODEL

An elegant model of an alveolus that incorporates both lung epithelial cells and endothelial cells exposes the endothelial cells to fluid flow similar to the flow found in pulmonary vasculature and exposes both sets of cells to dynamic stretching under conditions of stretching and with blood vessel side fluid flow was developed by the Ingber Lab (Huh et al. 2010).

3.5.1.1 Device Construction

The device construction is similar to the two-layer channel system used for the microchannel-based small airway model (Figure 3.5), but incorporates porous membranes fabricated from PDMS rather than polyester membranes. The device also has two separate side channels that allow the PDMS membrane to be stretched upon the application of a vacuum (Figure 3.7) (Huh et al. 2010). Fabrication of this system also starts with fabrication of the two PDMS slabs with channel features of the upper and lower channels. Next, a thin and porous PDMS membrane is fabricated

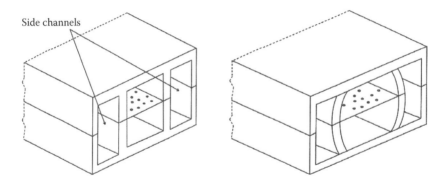

Side channels

FIGURE 3.7 An alveolus-on-a-chip model that incorporates flexible PDMS membranes and vacuum-actuated side channels that expose cells cultured on the membrane to uniaxial stretch. (Adapted from Huh, D. et al., *Science*, 328, 1662, 2010.)

by casting PDMS prepolymer against a microfabricated pillar array structure to create an array of through holes. These three PDMS pieces comprising the upper PDMS channel structure, porous PDMS membrane, and lower channel pieces are then bonded together. Finally, a solvent-based PDMS etching technique (Takayama et al. 2001) is used to selectively etch away the PDMS membranes in the side channels.

3.5.1.2 Cell Culture

A key feature of this alveolus-on-a-chip model is the use of cocultures of lung epithelial cells and endothelial cells. The two cell types are seeded on opposite sides of the PDMS porous membrane but can interact through the pores. Similar to the microfluidic airway model, the cells can be cultured with the apical side exposed to air with feeding from the lower channel. Interestingly, neutrophils flowed through the endothelial channel can extravasate through the endothelial cell layer, through the pores of the PDMS membrane, and across the lung epithelial cell layer when bacteria or certain types of nanoparticles are present on the epithelial cell side (Huh et al. 2010). This alveolus-on-a-chip system represents one of the most sophisticated multicell in vitro microfluidic cell culture models reported to date.

3.5.1.3 Linear Stretching Exacerbates Nanoparticle and Drug Toxicity

In addition to having a physiological epithelial–endothelial coculture, this alveolus-on-a-chip system has the unique feature of being able to expose the cells to stretch while also performing high-resolution optical microscopy on the cell layers. Confocal microscopy, for example, showed enhanced uptake of nanoparticles by the alveolar epithelium and underlying endothelial cell layer upon stretch compared to static conditions (Huh et al. 2010). Interestingly, inflammatory responses, such as reactive oxygen species production and neutrophil recruitment in response to exposure to silica nanoparticles or carboxylated quantum dots, are exacerbated by physiological cell stretching. Importantly, these in vitro microfluidic alveoli results were confirmed to be physiologically relevant through animal model studies. This nanoparticle toxicity study may be the most relevant example of inhalation toxicity in microfluidic models since nanoparticle toxicity is often triggered by the inhalation of nanoparticular aerosols.

In another study, the microfluidic alveolus model was used to re-create interleukin-2 (IL-2) toxicity-induced pulmonary edema (Huh et al. 2012). Again, physiological stretch was shown to play an important role in increased permeability associated with pathologic toxicity. Additionally, in this chapter, the authors demonstrated the ability to treat the pathological symptoms-on-a-chip with therapeutics. For the case of IL-2 and stretch-induced pulmonary edema, angiopoietin-1 (Ang-1) and a transient receptor potential vanilloid 4 (TRPV4) ion channel inhibitor (GSK2193874) were demonstrated to reduce the symptoms (Huh et al. 2012).

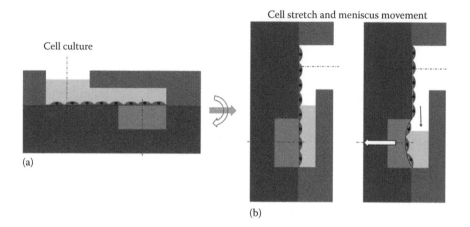

FIGURE 3.8 A terminal sac alveolus-on-a-chip model that can simultaneously expose alveolar epithelial cells to both stretch and fluid mechanical stress. (a) The device in cell culture mode all filled with cell culture medium. (b) The actuation state where the device is half filled with air and turned sideways to allow the meniscus to move up and down over the cultured cells as they are exposed to stretch. (Adapted from Douville, N.J. et al., *Lab Chip*, 11, 609, 2011.)

3.5.2 MICROFLUIDIC TERMINAL SAC MODEL

A model of an alveolus that aims to study the fluid mechanical effects together with effects of stretching was developed in the Takayama Lab (Douville et al. 2011).

3.5.2.1 Device Construction

This device also has a three-layer PDMS structure similar to the device in the previous section (Figure 3.8). The middle PDMS membrane on which cells are cultured, however, is not porous. The cells and membrane are exposed to stretching by the application of a negative fluidic pressure on the back side of the membrane opposite where alveolar epithelial cells are cultured. A key feature of this device design, which has a dead-end channel, is the ability to not only stretch the alveolar epithelial cells but also expose them to the movement of an air–liquid meniscus when the channel is partly filled with culture medium or other liquid. This enables the evaluation of simultaneous solid and fluid mechanical stimulation on cells.

3.5.2.2 Cell Culture

Alveolar epithelial cells are cultured on the PDMS membrane under submerged culture conditions. Because the PDMS membrane is not porous, long-term culture with the apical side exposed to air is not possible. Short-term (several hours) exposure to air, however, is possible and allows the evaluation of the combined effects of different solid and fluid mechanical effects on cell injury. Both a human alveolar cell line (A549 cells) and primary murine alveolar epithelial cells have been successfully cultured in this system.

3.5.2.3 Combined Solid and Fluid Mechanical Stress Exacerbates Cell Injury

The main novel capability of this microfluidic alveolar model is the ability to simultaneously expose alveolar epithelial cells to both stretch and fluid mechanical stresses exerted on the cells by the air–liquid interface of a meniscus moving over the cells. The alveolus model was evaluated for cell injury operating under three conditions: (1) healthy conditions where the alveolus is filled with air, (2) fetal conditions where the entire lung is filled with liquid, and (3) edematous conditions where the alveolus is partially filled with liquid and an air–liquid meniscus moves over cells upon breathing-associated volume changes in the alveolus. While the fully air- or liquid-filled conditions of (1) and (2) resulted in minimal injury to cells, conditions of (3) with a partially liquid-filled alveolus resulted in severe cell injury and even cell-layer delamination. While physiological or even pathological levels of stretch do not injure cells immediately, when the stretch is combined with steep pressure gradients generated at the leading edge of a moving meniscus, the cellular injury is quick and extensive (Douville et al. 2011). Importantly, many of these pathological symptoms-on-a-chip are mitigated when liquid is enriched with therapeutic surfactants. This is another example of how advanced microfluidic in vitro cell culture models can be useful in determining injury mechanisms and testing therapeutics in ways not possible with conventional cell cultures.

3.6 CONCLUSIONS

This chapter provides a brief overview of the state of the art in microfluidic lung cell culture models commonly referred to as lungs-on-a-chip. The field is still in its infancy but is already demonstrating the ability to test therapeutics and understand disease physiology and toxicology in ways that are otherwise impossible. Key differences between conventional dish culture and these microfluidic models of the lung include incorporation of lung dynamics and mechanical stimulation. While the incorporation of cell stretching is possible with nonmicrofluidic models, exposure of lung cells to fluid mechanical stresses associated with a propagating liquid plug or moving meniscus is only possible using microchannel devices. In fact, systematic studies of the role of fluid mechanical stresses in the lung may only be possible using microfluidic models since manipulation of nanoliter volumes of fluids in the lung in vivo is currently impossible.

There are of course many limitations and areas that need significant improvement. Many diseases involve interactions between many types of cells, tissues, and organs. A stand-alone lung-on-a-chip will have limitations in this regard for understanding toxicology. With ongoing efforts to create many types of organs-on-a-chip and efforts to integrate them together, this is an area where significant improvements can be expected in the coming years. Another area of need is analysis. While optical microscopy and fluorescent cell counting and analysis are convenient and informative, the readouts are often simple (such as cell numbers and live/dead ratios). Such readouts may represent certain extreme disease conditions but are difficult to relate to actual patients and human outcomes. Thus, more in-depth biomarker analysis is an area of great need. While fluid can be sampled from microfluidic devices, the small volumes of fluids in microfluidic cell culture models make fluid manipulations nontrivial and require development of novel microscale analysis methods. Finally, correlating the miniscule microfluidic organ models to life-sized human physiology requires mathematical modeling and scaling analysis to be most useful.

In summary, the field of lungs-on-a-chip is an exciting interdisciplinary field with many challenges and opportunities. Although in its infancy, the field has achieved promising preliminary successes; it will continue to expand rapidly and make significant findings with increased collaborations between micro-device engineers, toxicologists, and computational modelers.

QUESTIONS

1. What are advantages and disadvantages of using the techniques of soft lithography to create microchannels for lungs-on-a-chip?

   Answer: Advantages include ease of fabrication through the use of molding, easy bonding of structures using plasma activation or polymer curing, and mechanical flexibility that makes it possible to create stretchable substrates for lung cell cultures to mimic breeding actions. Disadvantages include the hydrophobic nature of the material that absorbs hydrophobic chemicals and drugs. The flexibility of the material can also be a challenge for precise fabrication and scale-up of manufacturing.

2. What specific dynamic mechanical forces are lung cells exposed to in the small airways, and how can microfluidic devices re-create an aspect of these forces for in vitro cell culture?

3. What specific dynamic mechanical forces are lung cells exposed to in the alveoli, and how can microfluidic devices re-create an aspect of these forces for in vitro cell culture?

Answer for 2 and 3: Lung cells can experience both solid and fluid mechanical stimulation. For example, breathing motions exert stretching of the cells. When the airways and alveoli are liquid filled, movement of the air–liquid meniscus interface can also exert substantial fluid mechanical stresses especially under conditions of surfactant dysfunction. Microfluidic devices have been created that can generate defined liquid plugs to re-create fluid mechanical forces within the lung. Using flexible materials, stretching motions of the lung can also be re-created and exerted onto cells.

4. What are future challenges and opportunities for the field of lungs-on-a-chip?

Answer: Some of the challenges include sourcing of appropriate cells and materials, incorporating multiple cell types, and even development of systems that link with other organ systems to better re-create human physiology. The complexity of the devices also presents challenges for scale-up and manufacturing for high-throughput applications such as drug discovery. Another challenge is the development of convenient and reliable readouts to correlate microfluidic lung-on-a-chip function with actual human physiology and to be able to direct efficacy and toxicity of drugs.

REFERENCES

Bilek, A. M., Dee, K. C., Gaver, D. P. III. 2003. Mechanisms of surface-tension-induced epithelial cell damage in a model of pulmonary airway reopening. *J. Appl. Physiol.* 94: 770–783.

Chueh, B. H., Huh, D., Kyrtsos, C. R., Houssin, T., Futai, N., Takayama, S. 2007. Leakage-free bonding of porous membranes into layered microfluidic array systems. *Anal. Chem.* 79: 3504–3508.

Douville, N. J., Zamankhan, P., Tung, Y.-C., Li, R., Vaughan, B. L., White, J., Grotberg, J. B., Takayama, S. 2011. Combination fluid and solid mechanical stresses contribute to cell death and detachment in a microfluidic alveolar model. *Lab Chip* 11, 609–619.

Fujioka, H., Grotberg, J. B. 2004. Steady propagation of a liquid plug in a two-dimensional channel. *J. Biomech. Eng. Trans. ASME* 126: 567–577.

Gaver, D. P. III, Halpern, D., Jensen, O. E., Grotberg, J. B. 1996. The steady motion of a semiinfinite bubble through a flexible-walled channel. *J. Fluid Mech.* 319: 25–65.

Halpern, D., Grotberg, J. B. 2003. Nonlinear saturation of the Rayleigh instability due to oscillatory flow in a liquid-lined tube. *J. Fluid Mech.* 492: 251–270.

Heil, M. 1998. Airway closure: Occluding liquid bridges in strongly buckled elastic tubes. *J. Biomech. Eng.* 121: 487–493.

Huh, D., Fujioka, H., Tung, Y.-C., Futai, N., Paine, R., Grotberg, J. B., Takayama, S. 2007. Acoustically detectable cellular-level lung injury induced by fluid mechanical stresses in microfluidic airway systems. *Proc. Natl. Acad. Sci. USA* 104: 18886–18891.

Huh, D., Leslie, D. C., Matthews, B. D., Fraser, J. P., Jurek, S., Hamilton, G. A., Thorneloe, K. S., McAlexander, M. A., Ingber, D. E. 2012. A human disease model of drug toxicity-induced pulmonary edema in a lung-on-a-chip microdevice. *Sci. Transl. Med.* 4: 159ra147.

Huh, D., Matthews, B. D., Mammoto, A., Montoya-Zavala, M., Hsin, H. Y., Ingber, D. E. 2010. Reconstituting organ-level lung functions on a chip. *Science* 328: 1662–1668.

Jacob, A. M., Gaver, D. P. III. 2005. An investigation of the influence of cell topography on epithelial mechanical stresses during pulmonary airway reopening. *Phys. Fluids* 17: 031502.

Kay, S. S., Bilek, A. M., Dee, K. C., Gaver, D. P. III. 2004. Pressure gradient, not exposure duration, determines the extent of epithelial cell damage in a model of pulmonary airway reopening. *J. Appl. Physiol.* 97: 269–276.

Tai, C.-F., Bian, S., Halpern, D., Zheng, Y., Filoche, M., Grotberg, J. B. 2011. Numerical study of flow fields in an airway closure model. *J. Fluid Mech.* 677: 483–502.

Takayama, S., Ostuni, E., Qian, X., McDonald, J. C., Jiang, X., LeDuc, P., Wu, M.-H., Ingber, D. E., Whitesides, G. M. 2001. Topographical micropatterning of poly(dimethylsiloxane) using laminar flows of liquids in capillaries. *Adv. Mater.* 13: 570–574.

Tavana, H., Kuo, C.-H., Lee, Q. Y., Mosadegh, B., Huh, D., Christensen, P. J., Grotberg, J. B., Takayama, S. 2010. Dynamics of liquid plugs of buffer and surfactant solutions in a micro-engineered pulmonary airway model. *Langmuir* 26: 3744–3752.

Tavana, H., Zamankhan, P., Christensen, P. J., Takayama, S., Grotberg, J. B. 2011. Epithelium damage and protection during reopening of occluded airways in a physiologic microfluidic pulmonary airway model. *Biomed. Microdevices* 13: 731–742.

Toepke, M. W., Beebe, D. J. 2006. PDMS absorption of small molecules and consequences in microfluidic applications. *Lab Chip* 6: 1484–1486.

Wang, J. D., Douville, N. J., Takayama, S., El-Sayed, M. 2012. Quantitative analysis of molecular absorption into PDMS microfluidic channels. *Ann. Biomed. Eng.* 40: 1862–1873.

Whitesides, G. M., Ostuni, E., Takayama, S., Jiang, X. Ingber, D. E. 2001. Soft lithography in biology and biochemistry. *Ann. Rev. Biomed. Eng.* 3: 335–373.

4 Human Health Risk Assessment of Inhaled Materials

Rhian B. Cope, Patricia Nance, and Mike Dourson

CONTENTS

4.1 BASIC PRINCIPLES AND DEFINITIONS

4.1.1 RISK AND HAZARD ASSESSMENT PARADIGMS

A human health risk assessment is a method used to define the nature and the probability of adverse health effects in humans who may be exposed (now or in the future) to chemicals in contaminated environmental media. The objectives of the process are to (1) define the context(s), purpose, scope, and technical approaches to be used in the assessment process (planning and scoping); (2) identify and characterize the spectrum of recognized adverse health effects (or threats) associated with the exposure of humans to a specific material or mixture (hazard identification); (3) document the relationship between exposure (or dose) and the individual hazards and/or adverse effects (dose–response assessment); (4) assess the magnitude of exposure (or dose) of the affected populations (exposure assessment); and (5) determine an overall qualitative or quantitative conclusion regarding the risk(s) pertaining to a particular set of circumstances associated with the specific material or mixture. In practice, this last step may involve the setting of an exposure (or dose) threshold that is associated with an acceptable level of risk (e.g., a respiratory reference concentration) or determining a margin of exposure or cancer inhalation unit risk (cancer slope factor (SF)) for a particular population or subpopulation.

4.1.2 FUNDAMENTAL ASSUMPTIONS THAT UNDERPIN THE RISK ASSESSMENT PARADIGM

Some of the critical assumptions of the current human health risk assessment paradigm are as follows:

- The relationship between the exposure and the adverse effect is cause and effect and not just association. Overall, this means that the hazard is a direct consequence of the relevant exposure; that is, the exposure is sufficient and/or necessary and/or contributory to the observed adverse effect. This assumption implies that the Bradford-Hill criteria for causation have been met. If the relationship between the exposure and the adverse event is not directly causal, then changing the level of exposure may not affect the level of risk.
- The hazard can be identified; that is, the adverse event can be expressed in some form of identifiable manner.
- The magnitude of the hazard, adverse effect, or event can be measured.
- The level of exposure (or dose) can be detected and measured.
- The degree of risk is related somehow to the exposure (or dose); that is, within some definable range on the dose–response curve, reducing the dose will reduce the risk and increasing the dose will increase the risk. The dose–response relationship can be a threshold or be continuous. This is really a component of the cause and effect assumption.
- Given that risk assessments are deliberately and consciously developed within the context of a particular set of stated circumstances, the results of the analysis are therefore only directly relevant provided that the relevant frame of reference has remained the same. Particular care is needed when extrapolating the results of a risk assessment beyond its stated frame of reference and/or context. For example, the results of a risk assessment developed for a physically fit and young worker subpopulation may not be directly extrapolated to a population of people living in a nursing home.

4.1.3 SOME BASIC DEFINITIONS

Within the human health risk paradigm, the term *hazard* means a potential source of harm or adverse health effect on a person or persons. This definition illustrates a critical, but often unstated underlying theme in human health risk assessment: the concern with, and importance of, the welfare of individuals over their entire lifespan within a population. A second unstated underlying theme and assumption follows from this: If individuals are either not affected or are only minimally affected by a hazard, then the overall population will survive and thrive. This concern for the individual is somewhat different than the focus of other types of chemical risk assessment where the concern is the *big picture* survival of populations, and the fate of individuals is of a lesser concern. The concept of *hazard* also incorporates some estimate (often qualitative or implied) of the seriousness of the harm or adverse health effect; for example, death is judged to be a more *harmful* hazard than respiratory sensory irritation.

The term *risk* refers to the likelihood (chance or probability) that a person may be harmed or suffers an adverse health effect if exposed to a hazard. Following on from this definition, a *risk assessment* is effectively asking the following question: Under a particular defined set of circumstances, how frequently are the effects of a particular hazard likely to occur? It further follows that *risk assessments* are actually quantitative, semiquantitative, or qualitative probability estimates that will follow the principles of probability theory; that is, they rely upon the concepts of probability distributions and mathematical central limit theory.

The term *gas* refers to a state of matter consisting of particles that have neither a defined volume nor a defined shape. Within toxicology, the term *gas* is often restricted to materials that meet this definition at normal temperatures and pressures; that is, the T-critical and P-critical for the chemical are outside of the range encountered under normal atmospheric circumstances.

The term *vapor* refers to a condensable gas, that is, a state of matter that can be converted to another physical form by alterations in temperature and/or pressure. The difficulty with this definition is that for all practical purposes, there are no significant chemical or physical differences between a vapor and a gas. The main difference between the terms is one of connotations. The term *vapor* is more commonly used when (1) there is a gaseous phase that is in equilibrium with a corresponding liquid or solid; (2) there is an emphasis on the gaseous phase of a chemical material; or (3) by the practice of common usage; for example, water in the gaseous phase in the atmosphere is usually referred to as *water vapor* rather than *water gas* even though both are technically correct.

The term *particle* refers to small localized objects that can be ascribed volume and mass. The term is a very general and imprecise one and is dependent on the scale of the context: An object with a size that is small or negligible relative to the size of the body can be considered a particle. The term *particulate* refers to a suspension of unconnected particles within a medium (usually air). The *unconnected* part of the definition is toxicologically important—it means that the movement of each individual particle is essentially independent of the movement of any other particle.

The separation of gases/vapors from particles is very important within the inhalation risk assessment paradigm because their dosimetry is treated very differently. The dosimetry and the evaluation of dose responses, toxicokinetics, and toxicodynamics of materials that lie on the boundary between the gas and particle physical forms and of substances that change physical form within the respiratory tract can be challenging. An unstated assumption is that gases/vapors are inherently inhalable; that is, 100% of the mass fraction of the gas/vapor in air can enter into the respiratory system by either the mouth or the nose.

The term *aerosol* refers to a suspension of solid particles or liquid in air. It has been recommended that the term *aerosol* be restricted to suspensions of particles in air. However, there remains much mixed usage of the term. Irrespective of whether or not an aerosol is composed of solid particles or liquid droplets, it is treated as particulate in terms of dosimetry and dose–response analysis.

The term *dust* is commonly used in a generic manner to mean particles suspended in air and/or aerosols. The term *mist* is commonly used to refer to a suspension of liquid droplets in air. The size of dust or mist particles (or fibers) is related to the amount of energy involved in creation: The higher the energy, the smaller the particle created, and the lower the energy, the larger the particle created.

The size of particles (and aerosols) in inhalation toxicology is usually defined by their mass median aerodynamic diameter (MMAD or D50) and aerodynamic equivalent diameter (AED). MMAD is usually expressed in micrometers and should always be accompanied by geometric standard deviation value (g or σg). The MMAD is the median of the distribution of airborne particle mass with respect to the aerodynamic diameter, that is, the average particle diameter by mass.

The AED of a particle is the diameter of a unit density sphere (i.e., a perfectly spherical droplet of pure water) that would have the identical settling velocity as the equivalent particle of water. The use of AED takes into account the factors that affect particle-settling velocity. In effect, AED is a form of statistical normalization of ratings that takes into account the true size, specific gravity, shape factor, surface properties, and slip factor of a particle.

The σg characterizes the variability of the geometric size distribution of the particle population. The σg is critical in particle dosimetry since it allows for the calculation of the fraction of an overall particle concentration or dose that can be inhaled and that deposits in particular parts of the respiratory tract.

The term *inhalable fraction* refers to the mass fraction of particles capable of entering into the respiratory system. Particles within the inhalable fraction can deposit anywhere within respiratory tract; that is, the inhalable fraction encompasses the extrathoracic, thoracic, and respirable particulate fractions. In humans, the inhalable particulate fraction 50% cutoff is generally regarded as AED 100 μm. However, the actual cutoff is significantly influenced by wind direction and speed. The fractional respiratory penetration of inspirable dusts is summarized in Table 4.1.

The extrathoracic fraction is the fraction of inhalable particles that deposit anywhere in the area of the respiratory tract lying between the nostrils/mouth and the distal end of the larynx. In humans, inhalable particles with AED of >25 μm will generally fall into the extrathoracic fraction.

The thoracic fraction is the fraction of inhalable particles that can penetrate the head airways and enter the airways of the lung. The thoracic fraction includes the respirable particle fraction. In humans, 98% of particles with AED ≤ 25 μm fall into the thoracic fraction category, and the MMAD for the category is 10 μm (i.e., 50% of particles with an AED of 10 μm will penetrate into the lung airways). The fractional penetration of particles for this category is summarized in Table 4.2.

TABLE 4.1

Aerodynamic Equivalent Diameter versus % Penetration into the Respiratory System

Aerodynamic Equivalent Diameter (μm)	% Penetration
0	100
5	87
10	77
40	54.5
100	50

TABLE 4.2

Aerodynamic Equivalent Diameter versus % Penetration into the Thoracic Region

Aerodynamic Equivalent Diameter (μm)	% Penetration
0	100
2	94
4	89
10	50
14	23
18	9.5
25	2

TABLE 4.3
Aerodynamic Equivalent Diameter versus %
Penetration into the Gas-Exchange Areas of the Lung

Aerodynamic Equivalent Diameter (μm)	% Penetration
0	100
2	91
4	50
6	17
8	5
10	1

The respirable fraction is the mass fraction of particles capable of penetrating the respiratory tract to the level of the on-ciliated airways and gas-exchange regions of the lungs (i.e., the structures distal to the terminal bronchioles: the respiratory bronchioles, the alveolar ducts, and the alveoli). In humans, 99% of particles with AED ≤10 μm fall into the respirable category, and the MMAD for the category is 4 μm. The fractional penetration of particles for this category is summarized in Table 4.3.

4.2 RISK ASSESSMENT PROCESS

The risk assessment process consists of five approximately sequential steps:

1. Planning and scoping
2. Hazard identification
3. Dose–response assessment
4. Exposure assessment
5. Risk characterization

4.2.1 PLANNING AND SCOPING

The objective of the planning and scoping phase is to carefully frame the risk assessment problem being addressed, as discussed by a number of organizations (e.g., Dourson et al., 2013). To this end, risk assessors typically attempt to define the answers to a series of questions in order to provide a context and frame of reference for the risk assessment process:

- Who is at risk? The answers to this question could include individuals, the general population, specific subpopulations such as workers, specific life-stages, and population subgroups.
- What is the specific agent of concern? The answers to this question could include a specific chemical, a complex mixture, a combination of effects, and so forth.
- What is the source of the risk? This could include workplaces, point sources, nonpoint sources, natural sources, and so forth.
- How does the exposure occur? Within the context of this text, the route of exposure is by inhalation. However, it should be remembered that injury to the respiratory system can occur via noninhalation routes. This question also addresses the more general issue of exposure pathways: inhaled materials can derive from the air, from surface water, from groundwater, from soil, from solid wastes, and so forth.
- What does the human body do to the substance? In other words, what are the key toxicokinetic features (absorption, distribution, metabolism, and excretion) of the substance?

- What are the key health effects? In other words, what does the substance do to the human body (including toxicodynamic effects)? Within the context of inhalation toxicology, key questions within this area will include the following: (a) Are the adverse effects primarily occurring within the respiratory tract (e.g., site of first contact effects)? (b) Are the adverse effects dependent on systemic absorption and distribution to a target organ or tissue that is distal to the respiratory tract? (c) Does a combination of respiratory tract and systemic effects occur? (d) Are the effects reversible (via repair or other mechanisms) and/or does tolerance or adaptation occur over time?
- What is the time scale involved in the generation of the adverse effects? In other words, how long does it take to produce the effect and do the adverse effects depend upon *critical windows* of exposure or exposure during critical life-stages. For example, teratogenic effects are mostly associated with the period of organogenesis during fetal development, and almost all teratogenic effects are dependent on a *critical window* exposure period during development.
- Is the available body of knowledge sufficient enough and reliable enough to proceed? Typically this and exploratory process that involves searching for all relevant data, collating the data, a data gap analysis, and a preliminary assessment of quality and reliability of the available data set.

4.2.2   Hazard Identification

The overall objectives of the hazard identification process are to determine (1) whether or not exposure to a substance can cause an increase in the incidence of specific adverse health effects and (2) whether the adverse health effect is likely to occur in humans. These objectives can be broken down into several specific aims: (1) identification of the types of adverse health effects that can be caused by exposure substance; (2) an objective assessment of the human relevance of the MOAs that result in the putative adverse health effects; and (3) an objective assessment of the quality and the weight of evidence supporting the putative adverse health effects. This includes an objective assessment of the suitability of the data for risk assessment purposes.

For noncancer toxicity, hazard identification seeks to identify the target organ or critical effect of exposure (i.e., the first adverse effect or its known and immediate precursor that occurs as the dose rate increases, US EPA, 2002). However, the body may exhibit a spectrum of responses, not all of which are adverse. The risk assessor must discern adaptive effects, in which the body's ability to withstand challenge is enhanced, or compensatory effects, in which the body makes adjustments to maintain overall function without further detriment, from the true critical effect. The choice of critical effect is used as a basis for the dose–response assessment.

For cancer, hazard identification also seeks to identify the target organs and precursor events for the cancer development process, and both cancer and noncancer hazard identification involve the evaluation of incidence of the critical effect or tumors in humans (epidemiology) and animals, mechanistic or mode of action (MOA) data, chemical and physical properties, structure–activity relationships (SARs), and metabolic and pharmacokinetic properties to culminate in a judgment of overall weight of evidence (e.g., US EPA, 2005). Cancer evaluations often also consider the results of mutagenicity tests, since one of the likely precursors to tumor development is the formation of mutations. In this case, an important question is whether the mutations play an early causal role in the formation of the tumor, or whether the mutations play a secondary role, due to either the amplification of preexisting mutations or the late occurrence of mutations.

4.2.2.1   Sources of Data for Hazard Identification

Sources of data for hazard identification include direct human experimental studies (rarely available for many substances because of ethical considerations), epidemiological studies of human (and possibly other species) populations, experimental studies in animals, and in vitro studies.

Epidemiology is the study of the distribution and determinants of disease or health status in human populations. Well-conducted epidemiologic studies that demonstrate a positive correlation between a substance and a disease are considered strongly convincing evidence of human risk. Critically, it is difficult for epidemiology studies to demonstrate true cause and effect relationships. For the most part, epidemiology studies demonstrate associations rather than direct cause and effect relationships. However, human data adequate for showing such a relationship are rare. Occupational and accidental exposures to an agent are difficult to translate into meaningful risk estimates as exposure levels are not quantified, exposure often occurs via multiple routes, and the duration of exposure is often unknown. Epidemiological data are very informative, but corroborating evidence from other data sources often is needed to inform our understanding of an agent's risk.

Clinical studies are controlled experiments designed to assess particular effects in humans of exposure to a substance. Clinical studies are able to control dose, duration, and route of exposure, as well as the age, sex, and health and behavioral characteristics (e.g., smokers vs. nonsmokers) of the subjects.

Experimental studies conducted in animals (typically mammals) can, in many cases, be used as a surrogate for human data. These studies can be designed to control many pertinent variables (dose, route of exposure, exposure duration, age, etc.). Of course, doses in experimental animals must be converted to relevant human doses: a process known as interspecies extrapolation. This can be done in a number of ways, for example, by division of the experimental animal dose by an uncertainty factor (UF), allometric scaling, by making dosimetric adjustments that take into account structural and physiological differences between humans and the test species, or developing and using a physiologically based pharmacokinetic (PBPK) model.

In vitro tests are those conducted outside of a living organism. Compared with in vivo tests that are conducted in living organisms, in vitro tests are generally cheaper and faster to conduct. In vitro tests may be conducted in bacteria or cell cultures, or in animal tissue.

Read-across is a technique that is largely used to fill data gaps. The principle underlying the read-across technique is that the members of a structurally related chemical family commonly have similar toxicological properties; thus if specific data are lacking for a specific chemical, the data gap can be filled by inference from the known hazardous properties of members of the same chemical family. Two basic approaches are used in read-across: the analog approach and the category approach. The analog approach is based upon grouping of a small number of chemicals, whereas the category approach relies upon data from large groups or families of related chemicals. Detailed guidance on the read-across method can be found in the European Center for Ecotoxicology and Toxicology of Chemicals (ECETOC) Technical Report 116.

In the absence of toxicity data, SAR analysis and physicochemical data can be used to predict health effects, based on our understanding of structurally similar chemicals and specific responses in test systems. Modern SAR systems make extensive use of computer modeling and fall within the concept of in silico (i.e., *in silicon*) analysis (derived from the in silico phase of pharmaceutical development). When extrapolating from analogs, the focus is on key reactive chemical structural groups that are thought to be the most likely characteristic of the chemical to determine toxicity. Quantitative structure–activity relationship models (QSAR models) are regression or classification models based on these principles. Current QSAR systems are based on the Cramer Descriptors (synonym: Cramer Rules) or a related variant of this system.

The identification of adverse effects also typically involves the process of hazard classification. The current internationally accepted guideline for hazard classification for chemicals is the United Nations Globally Harmonized System (GHS) of Classification and Labeling. GHS has been incorporated into European law as the Classification, Labeling and Packaging (CLP) Regulation. The primary purpose of GHS is to create one universal standard for hazard classification and hazard communication.

4.2.2.2 Bradford-Hill Criteria

The Bradford-Hill criteria (synonym: Hill's criteria) are a group of minimal conditions that are necessary to fulfill in order to support a causal relationship between an incidence and a consequence, particularly within the context of epidemiology studies. However, the Bradford-Hill criteria are commonly extrapolated beyond epidemiology and are applied in a broader sense in the study of the etiology of disease (including toxicological disease). The criteria are as follows:

- Strength: A small association does rule out a causal effect; however, the stronger or larger the association, the more likely the relationship is causal.
- Consistency: Consistency across different laboratories, different studies, and study designs and across different populations increases the likelihood of a cause and effect relationship.
- Specificity: Causation is likely if a very specific population at a specific site and disease with no other likely explanation.
- Temporality: The effect has to occur after the exposure. This is the only absolute requirement within the criteria.
- Biological gradient (dose response): In general, as dose increases, so should the effect. Alternatively, there may be a threshold-type dose response. This criterion is often misinterpreted to suggest that the same agent must cause differing effects that have the same dose response: this is not correct and is not the correct way to apply this criterion. Occasionally, the effect relationship may be a protective one; that is, as the dose increases, the effect decreases, or the dose response may display hormesis.
- Biological plausibility: Although not absolutely essential, a biologically plausible MOA is helpful in establishing a cause and effect relationship. However, the understanding of potential MOAs may be limited by current knowledge. It is also possible that something entirely new to science has been observed (i.e., a paradigm shift).
- Coherence: Coherence between epidemiological studies and laboratory-based studies increases the likelihood that the relationship is cause and effect. However, the lack of coherency does not nullify the fact that epidemiology has shown an association (however, association is not cause and effect).
- Experiment: Experimental evidence may support a cause and effect relationship.
- Analogy: The effect of similar factors may be considered.

4.2.2.3 Assessment of Modes of Action

The phrase *mode of action* is defined as a sequence of key events and processes, starting with the interaction of an agent with its target organ, tissue, or cell, proceeding through pathophysiological and anatomical changes, and resulting in an adverse health effect. A key event is a critical, detectable, and measurable precursor step that is essential for the MOA to occur (or a biologically based marker for such an element). Multiple MOAs may apply for a single chemical, even within the same target organ, tissue, or cell.

Some examples of common MOAs include the following:

- Direct DNA reactivity leading to mutations
- Indirect DNA reactivity (e.g., interaction with spindle protein)
- Cell proliferation (e.g., cytoxicants and mitogens)
- Receptor binding (e.g., TCDD and peroxisome proliferators)
- Hormone disruption (e.g., thyroid follicular cell carcinogenesis)

A mechanism of action is distinguished from an MOA by the presence of a more detailed and intricate description of the process leading to the adverse health effect (often at the molecular level).

4.2.2.4 Human Relevance Framework for Modes of Action

The International Program for Chemical Safety (IPCS) initially developed a human relevance framework for cancer MOAs. This framework has now been adopted and applied by other investigators

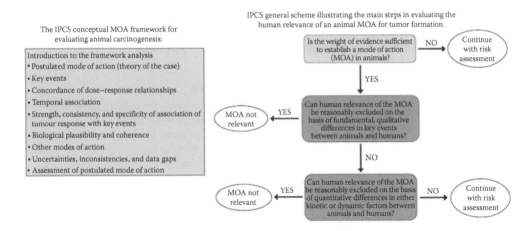

FIGURE 4.1 The International Program on Chemical Safety (IPCS) Human Cancer Relevance Framework. A number of general points and conclusions follow from the development of this framework: (1) Prior to embarking on a framework analysis, there needs to be careful evaluation of the weight of evidence for a carcinogenic response in experimental animals; (2) Peer involvement and independent review are essential prerequisites for the general acceptance and scientific defensibility of a new MOA; (3) The framework is applicable to all MOAs for carcinogens, including DNA reactivity; (4) Although human relevance is likely to be assumed for most DNA reactive carcinogens, the human relevance analysis is a valuable approach to enhance understanding, improve characterization of the hazard and risk, and identify exceptions; (5) When dealing with a chemical that may operate through a novel MOA, the analysis is focused on the chemical and entails a detailed evaluation via the HRF. However, when a specific chemical produces a tumour response consistent with an already established and peer-reviewed MOA through which other chemicals have been shown to operate, the analysis is then focused on the established MOA and a determination of whether the chemical produces its carcinogenic effect via the same key events established for the pathway; (6) When evaluating the human relevance of a tumor response found in experimental animals, the concordance analysis of key events is for the MOA and is not necessarily a chemical-specific evaluation. Chemical-specific and generic information relevant to the carcinogenic process can be valuable in the analysis. As knowledge advances, MOAs will become less chemical-specific and based even more on the key biological processes involved, allowing greater generalization of human relevance from one compound to another; (7) The biological understanding and significance of the key events can inform the approach to dose–response extrapolation for cancer risk, and thus, understanding of the MOA can have a profound effect on the hazard and risk characterization, particularly when non-linear processes or dose transitions are inherent in the relevant biology; (8) It is recommended that a database of generally accepted MOAs and informative case studies be established and maintained. It should provide examples that add to the existing case studied developed by ILSI/RSI and IPCS, and that are instructive in the application of the framework analysis. This database is particularly important as experience continues to evolve in the development of modes of action of carcinogens; (9) It is important to consider potentially susceptible subgroups and different life stages in the analysis.

and organizations to noncancer risk assessment as well. The IPCS framework is essentially a stepwise process (Figure 4.1) that is based upon the following critical questions:

- Is the weight of evidence sufficient to establish the MOA in animals?
- Are key events in the animal MOA plausible in humans?
- Taking into account kinetic and dynamic factors, are key events in the animal MOA plausible in humans?

4.2.2.5 Assessment of Quality, Weight of Evidence, and Suitability of Data for Risk Assessment

A commonly used system for assessing the reliability of animal toxicology studies for regulatory purposes is the Klimisch score (Table 4.4). The ECETOC Technical Report No. 104 (2009) also provides an internationally accepted scheme for categorizing the relevance of human data for use in

TABLE 4.4

Klimisch Scoring: A Method of Assessing the Reliability of Toxicology Studies for Regulatory Purposes

Score	Description	Details
1	Reliable without restriction	This includes studies or data from the literature or reports that were carried out or generated according to generally valid and/or internationally accepted testing guidelines (preferably performed according to GLP) or in which the test parameters documented are based on a specific (national) testing guideline (preferably performed according to GLP) or in which all parameters described are closely related/comparable to a guideline method.
2	Reliable with restriction	This includes studies or data from the literature, reports (mostly not performed according to GLP), in which the test parameters documented do not totally comply with the specific testing guideline, but are sufficient to accept the data or in which investigations are described that cannot be subsumed under a testing guideline, but that are nevertheless well documented and scientifically acceptable.
3	Not reliable	This includes studies or data from the literature/reports in which there are interferences between the measuring system and the test substance or in which organisms/test systems were used that are not relevant in relation to the exposure (e.g., unphysiological pathways of application) or that were carried out or generated according to a method that is not acceptable, the documentation of which is not sufficient for an assessment and which is not convincing for an expert judgment.
4	Not assignable	This includes studies or data from the literature, which do not give sufficient experimental details and are only listed in short abstracts or secondary literature (books, reviews, etc.).

Source: Klimisch, H.J. et al., *Regul. Toxicol. Pharmacol.*, 25, 1, 1997.

risk assessment (Figure 4.2), a matrix scoring system for assessing the quality, reliability, and suitability of animal data for human health risk assessment that incorporates the key concepts from the International Program on Chemical Safety Human Relevance Framework (Figure 4.3) and a clear system for determining whether the available human or animal data should take precedence in hazard and risk assessment processes. The ECETOC matrix system is driven by the inherent strengths and weaknesses of each data source rather than the study outcomes. The great advantage of the ECETOC matrix approach is that it provides a clear matrix for the integration of human and animal data based on the assessment of data quality for human health risk assessment purposes (Figure 4.4).

4.2.2.6 Types of Animal Studies Commonly Available for Hazard Identification

Animal inhalation toxicology studies can be divided into two major groups based upon the method of exposure: whole-body exposure and nose-only exposure. The whole-body exposure method involves placing the whole animal (unrestrained) within an inhalation chamber. The claimed advantages of such a system are the ability to expose multiple animals at the same time and within the same chamber (thus reducing some of the logistical and dosimetry challenges), the use of minimal restraint (which becomes increasingly important in long-duration repeat-dose studies), reduced labor requirement (with the possibility of automation), and the particular suitability of the technique for chronic, near-lifetime exposure studies. However, this methodology involves the requirement of large amounts of the test article, multiple potential routes of exposure (oral vial grooming, dermal via skin absorption, and ocular via the ocular mucous membranes), the difficulty in obtaining uniform concentrations and distribution of the test article within the test chamber, control of the chamber environment, greater potential exposure of personnel to the test article and animals filtering their exposure through their hair coats.

The alternative methods are head-only and nose-only exposures. The advantages of these systems are that only the respiratory tract is exposed, better containment of highly toxic test articles, and the use of the test article is relatively efficient. The big disadvantage of these systems is that they are more difficult to apply to long-duration repeat-exposure studies because of the

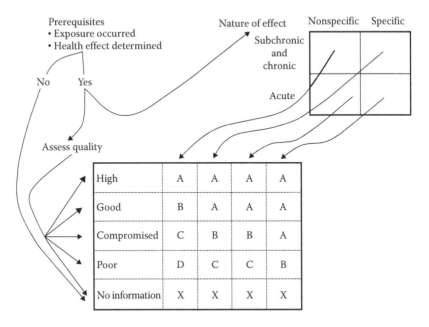

FIGURE 4.2 The ECETOC System for Scoring Human Data with Regard to Its Quality for Risk Assessment Purposes. The scheme involves (1) data prerequisites to be met: exposure to the substance in question must be demonstrated, the exposure must precede the claimed adverse effects, the adverse health effect should be determined on the basis of published criteria that are generally accepted as valid disease endpoints or symptom definitions, and other causal factors that have a large influence on the occurrence of the adverse effect or disease should be taken into account; (2) if the prerequisites are met, the data are assessed in detail and assigned to a quality category (high, good, compromised, poor, or no information); and (3) the nature of the health effect is then considered, for example, specific or nonspecific, acute or subchronic. This is then combined with the category to produce a quality score on a scale of A (highest quality) to D (lowest quality) or X (no information). High-quality human data are characterized by several studies of a commonly recognized epidemiological design with >2/3 of the studies displaying consistent results with one another and consistency with other biological evidence. High-quality human data also have quantified exposure data that are linked to individuals, clear measurement and control of confounding factors or evidence, and/or convincing argumentation that the effect of potentially confounding factors does not affect the interpretation of the study. Good-quality human data are characterized when most of the criteria for high-quality data are met, and no more than three of the following limitations apply: (1) Consistency between studies is not high, yet still suggestive (>50% of the studies are concordant); (2) Exposure data are not always quantifiable or linked to individuals; (3) There is no good biological understanding to underpin the results; (4) Not all strong confounding factors can be ruled out in the majority of the studies; (5) Health outcome measurements are not well validated; (6) Confidence intervals are more than an order of magnitude; and (7) A monotonic dose–response relationship (or lack thereof for null data) exists in the majority of the studies. Case reports and raw incidence/prevalence data or studies with major methodological flaws fall into the category of poor-quality data for risk assessment purposes. (From ECETOC, Framework for the integration of human and animal data in chemical risk assessment, Technical Report 104, ISSN-0773-8071-104, European Center for Ecotoxicology and Toxicology of Chemicals, Brussels, Belgium, 2009.)

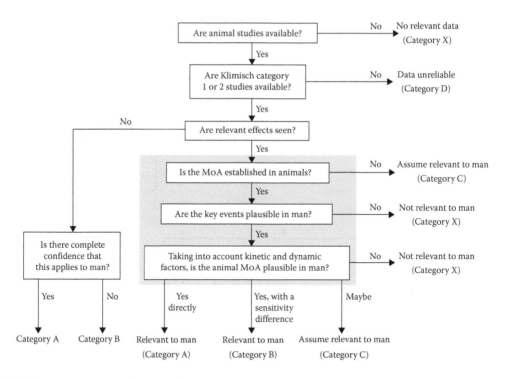

FIGURE 4.3 The ECETOC System for Categorizing the Relevance of Animal Data for Use in Human Risk Assessment. MOA: mode of action. *Note*: The shaded area represents the IPCS human relevance framework. (From ECETOC, Framework for the integration of human and animal data in chemical risk assessment, Technical Report 104, ISSN-0773-8071-104, European Center for Ecotoxicology and Toxicology of Chemicals, Brussels, Belgium, 2009.)

potential induction of restraint stress in the animals, the higher labor requirements, and difficulties in maintaining adequate seals around the head and nose.

A third group of more specialized study techniques have been used on occasion, including lung-only exposure and partial lung exposure. These studies generally require anesthesia plus physiological support (e.g., a respirator) and are thus unsuitable for longer repeat-exposure studies. Furthermore, such studies are open to criticism regarding *unphysiological* exposure conditions, the logistical limitations on the number of animals per exposure group, and the formation of centralized deposits of the test article (whereas other techniques result in a more even distribution in the lung). Specialized lung-only and partial lung studies are more commonly used to study lung toxicokinetics.

Inhalation studies can also be divided based upon the exposure system used. Older studies have used static exposure systems. A static exposure system is where a defined quantity of the test article is introduced into a closed system (i.e., there is no airflow or air exchange) and allowed to mix with the trapped air. These systems are efficient in terms of the utilization of the test article; however, their use is limited by oxygen depletion, accumulation of wastes, and difficulties with accurate dosimetry due to loss of the test agent. Static exposure systems are generally not suitable for the current standards of toxicological testing.

Dynamic exposure systems have become the expected norm for regulatory toxicology studies. Dynamic exposure systems are characterized by continuous replacement of air and the test article within the exposure chamber. Under these conditions, the concentration of the test article within the chamber will reach a theoretical equilibrium value based upon the formula:

$$t_{99} = \frac{(4.605 \times \text{chamber volume})}{\text{Chamber airflow}}$$

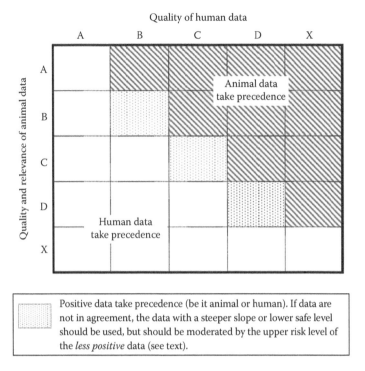

Quality of human data

FIGURE 4.4 The ECETOC Matrix for Integrating Human and Animal Data Based on Assessment of Data Quality for the Purposes of Human Health Risk Assessment. Category A human data (high-quality data) always take precedence as the basis for human health risk assessment irrespective of the quality of the animal data. Where human and animal data have equivalent data quality, it may be possible to use both data sources, particularly when the data from the different animal species are essentially concordant. However, when deriving a human health toxicological threshold, the data that require the fewer adjustment, assessment, or UFs are generally preferred. In most cases, this will be the human data. If the human and animal data are not concordant, then the following approach is suggested: (1) "Hazard assessment: Where human and animal data scores are identical (i.e., B/B, C/C, or D/D), the data that suggest a hazard should generally take precedence," and (2) "Dose–response assessment: Where the animal data suggest a lower safe level (i.e., they are 'more positive') than the human data, or vice versa, and the scores are the same (i.e., B/B, C/C, or D/D), the data resulting in the 'most positive' (i.e., lower) safe level should take precedence; the other data source should be used as the upper limit. In this way, a range that is consistent with both data sets is obtained, while allowing the most positive data to be used in a protective manner." (From ECETOC, Framework for the integration of human and animal data in chemical risk assessment, Technical Report 104, ISSN-0773-8071-104, European Center for Ecotoxicology and Toxicology of Chemicals, Brussels, Belgium, 2009.)

Provided that chamber equilibration is sufficient, accurate dosimetry is possible. It is critical that sufficient information regarding chamber equilibration and concentration of the test article over time is recorded. These data must be presented in the study report.

Irrespective of the method of exposure and the type of exposure system used, it is absolutely critical that the control (including vehicle-only controls) animals are treated exactly the same as the test-article-exposed groups. This is accomplished using *sham* exposures, scheduling, and randomization. Environmental factors such as temperature, humidity, atmospheric pressure, noise, vibration, airflow, and air quality can all affect the outcomes of the study. It is absolutely critical that all inhalation chambers used in a study be maintained within a given set of specifications and that both the specifications and the actual achieved readings for each relevant parameter are recorded in the study report.

It should also be noted that the testing of highly reactive materials (e.g., reaction potential with moisture and oxygen) involves additional challenges. Under such circumstances, it may be critical to determine what humans are actually exposed to. It may be more relevant that degradation products are used as the test article rather than the parent chemical if humans are more likely to be exposed to the degradation products. It is also common for highly reactive materials to produce site of first contact injury to tissues outside of the respiratory tract resulting in a significant animal welfare challenge as well as substantial confounding factors when trying to interpret the results of such studies.

An interesting challenge may also present itself resulting from the common usage of dehumidified *dry* air. Air humidity may significantly affect toxicological events within the respiratory tract. Thus, as a general rule, it is critical to assess the conditions under which exposure actually occurs in humans and to replicate the relevant toxicologically significant aspects within the exposure system.

As with most other types of toxicological studies, inhalation studies can also be classified on the basis of the duration of exposure. Classically, acute inhalation toxicology studies (e.g., OECD Test Guidelines 403 and 436, used in conjunction with Series on Testing and Assessment Number 39 Guidance Document on Acute Inhalation Toxicity Testing) are mostly performed in rats and mice, use either whole-body or nose-only exposure, and typically consist of a single exposure lasting between 1 and 4 h. Some general design aspects of acute inhalation toxicity tests are summarized in Table 4.5. The overall aims of acute inhalation toxicity testing are to establish an index of relative toxicity for classification and labeling purposes, to justify exposure concentrations for longer-duration studies, and to aid in the development of immediately dangerous to life and health exposure limits and short-term exposure limits.

Many acute studies will utilize a limit test design, particularly when the test article is known to have low inhalation toxicity. For particles and aerosols, the exposure limit is usually defined as the maximum attainable concentration (MAC) with MMAD of 1–4 μm of 5 mg/L (based on the GHS of Classification and Labeling). If the lethality at the MAC is less than 50%, no further testing is required. It should be noted that the GHS MAC of 5 mg/L is usually technically impossible to achieve and is well above human real-world exposures. In particular, as aerosol/particle concentration increases, there is a greater tendency for particle aggregation or droplet coalescence. This results in an increase in the MMAD of the aerosol, thus affecting its respirability and deposition within the respiratory tract. Thus, at high concentrations, it is quite common to observe a spurious decrease in toxicity.

High concentrations of reactive polymers and dry powder aerosols present other specialized challenges. At high concentrations, these materials tend to form conglomerates in the upper respiratory tract that results in the physical obstruction of the upper airways (dust loading). Given that rodents are obligate nose-only breathers (except for agonal respiration near death), these effects can be spuriously interpreted as a toxicological effect.

TABLE 4.5

Some Typical Design Parameters for Acute Inhalation Toxicity Studies

Group size of five animals per sex

Up to five exposure levels (including no exposure and vehicle-only exposure controls as needed)

Two species used (typically mice and rats)

Whole-body or nose-/head-only exposure

Single 1–4 h exposure

14-day postexposure observation period

Necropsy and gross anatomic pathology required (histopathology recommended)

As a general rule of thumb, concentrations of 2 g/L can be achieved without problems of aggregation/coalescence/conglomeration for most aerosol and particle test articles. Testing at concentrations above 2 g/L is usually technically challenging.

For gases, the limit test is usually performed at 20,000 ppm and for vapors, at 20 mg/L. In cases where these concentrations cannot be achieved, the MAC is used. In theory, the MAC for vapors should be the same as the *headspace* saturated vapor concentration, CS. The CS can be estimated using the following formula:

$$ppm = \frac{\text{Vapor pressure (mmHg)}}{760 \text{ mmHg}} \times 10^6$$

As a matter of practicality, it is often technically difficult to generate concentrations equivalent to the CS under the experimental circumstances used in inhalation toxicology studies. As a rough rule of thumb, the MAC for vapors under these circumstances is often about one-third of the CS.

Given that causing a laboratory explosion is generally an inadvisable practice for acute inhalation toxicology studies, attention needs to be paid to the upper and lower explosive limits and the flash point for the relevant test article, particularly when performing limit tests. The upper explosive limit is the concentration above which there is insufficient oxygen for an explosion to occur. The lower explosive limit is the concentration below which there is insufficient fuel to cause an explosion. The flash point is the lowest temperature at which a liquid gives off enough vapor to form an ignitable mixture.

A final design consideration may be important in acute inhalation toxicity studies. There are genuine animal welfare and scientific concerns surrounding the use of high concentrations of irritants and corrosives in acute inhalation toxicity testing. The maximum dose used in these studies needs to be limited to those below the level known to produce significant skin/eye irritancy and corrosion. The implications of this are that the skin/eye irritancy characteristics of the test article need to be known before conducting acute inhalation toxicity testing.

There are currently OECD guidelines for subacute, subchronic, and chronic repeat-inhalation-exposure toxicity studies (Test Guidelines 412, 413, and 452/453, respectively). Subacute studies have a typical duration of 28 days with 6 h per day exposure periods. Subchronic studies typically have 90-day duration, whereas chronic studies typically have durations of at least 12 months. Whole-body exposure systems are more likely to be used in these study designs for reasons of practicality and reduction of animal stressors. Other typical design parameters are summarized in Tables 4.6 through 4.8. Satellite study groups are often included in these protocols in order to test for reversibility, allow for interim sacrifices and specialized techniques such as bronchoalveolar lavage, additional clinical pathology, and additional histopathology evaluations. With the advent of

---

TABLE 4.6

Some Typical Design Parameters for Subacute (28-Day) Inhalation Toxicity Studies

Study design may incorporate a preliminary dose range–finding study.

Group size of five animals per sex.

Three or more exposure levels (including no exposure and vehicle-only exposure controls as needed).

Satellite groups for the study of reversibility or specialized investigations (e.g., bronchoalveolar lavage) can be included.

Two species used (typically mice and rats).

Whole-body or nose/head-only exposure.

Exposure for 6 h/day, 5 days/week (can be up to 7 days/week).

Gross and microscopic anatomic pathology required.

TABLE 4.7

Some Typical Design Parameters for Subchronic (90-Day or 13-Week) Inhalation Toxicity Studies

Group size of 10 animals per sex.

Three or more exposure levels (including no exposure and vehicle-only exposure controls as needed).

Satellite groups for the study of reversibility or specialized investigations (e.g., bronchoalveolar lavage) can be included.

Two species used (typically mice and rats).

Whole-body or nose/head-only exposure.

Exposure for 6 h/day, 5 days/week (can be up to 7 days/week).

Hematology is typically included.

Clinical chemistry is typically included.

Urinalysis is typically included.

Gross and microscopic anatomic pathology required.

TABLE 4.8

Some Typical Design Parameters for Chronic (12-Month or Longer) Inhalation Toxicity Studies

Group size of 20 animals per sex.

Three or more exposure levels (including no exposure and vehicle-only exposure controls as needed).

Satellite groups for the study of reversibility or specialized investigations (e.g., bronchoalveolar lavage) can be included.

Two species used (typically mice and rats).

Whole-body or nose/head-only exposure.

Exposure for 6 h/day, 5 days/week (can be up to 7 days/week).

Hematology is typically included.

Clinical chemistry is typically included.

Urinalysis is typically included.

Gross and microscopic anatomic pathology required.

relatively cheap specialized diagnostic imaging equipment for rodents, magnetic resonance imaging, computer-aided tomography, and positron-emission tomography studies have become important procedures in these types of studies.

4.2.2.7 Globally Harmonized System of Classification and Labeling

The United Nations GHS classification system has been progressively adopted as the internationally accepted method of classifying chemical hazards. Within the European Union, the system is largely integrated into the CLP Regulation. Other countries have made similar legislative adjustments. Tables 4.9 and 4.10 summarize some of the GHS classification category cutoffs that are relevant to inhalation toxicology hazard assessment.

4.2.2.8 Adjusting for the Duration of Exposure: Haber's Law and the Ten Berge Modification

Ernest Warren and Fritz Haber noted during their studies of the acute (exposure durations between 1 and 120 min) lethality effects of poison gases (specifically gas weapons in the case of Fritz Haber, notably phosgene, methylchloroformate, cyanide gas, chloroacetone, xylylbromide, and chlorine) that exposure to a low concentration of a poisonous gas for a long time often had the same effect (death) as exposure to a high concentration for a short time. Generally, a simple mathematical relationship applied: $C \times t = k$, where C is the concentration of the poisonous gas, t is the time of exposure, and k is a constant or *toxic load*. If different concentrations and times of exposure are used, this implies

TABLE 4.9
Globally Harmonized System Classification LC_{50} Criteria for Acute (4 h Exposure) Toxicity Effects of Inhaled Substances

Form of Substance	Category 1	Category 2	Category 3	Category 4
Gases (ppmV)	100	500	2500	20,000
Vapors (mg/L)	0.5	2.0	10	20
Dusts and Mists (mg/L)	0.05	0.5	1.0	5

If the acute toxic class method is used, the following conversions are applied:

Form of Substance	Classification Category or Experimentally Obtained Acute Toxicity Range Estimate	Converted Acute Toxicity Point Estimate
Gases (ppmV)	0 < Category 1 ≤ 100	10
	100 < Category 2 ≤ 500	100
	500 < Category 3 ≤ 2,500	700
	2,500 < Category 4 ≤ 20,000	4500
Vapors (mg/L)	0 < Category 1 ≤ 0.5	0.05
	0.5 < Category 2 ≤ 2.0	0.5
	2.0 < Category 3 ≤ 10.0	3
	10 < Category 4 ≤ 20.0	11
Dusts/Mists (mg/L)	0 < Category 1 ≤ 0.05	0.005
	0.05 < Category 2 ≤ 0.5	0.05
	0.5 < Category 3 ≤ 1.0	0.5
	1.0 < Category 4 ≤ 5.0	1.5

Notes:
- The acute toxicity estimate (ATE) is calculated by dividing LD_{50} by the LC_{50}.
- Saturated vapor concentration may be used in some regulatory systems to provide for specific health and safety protection (e.g., UN Recommendations for the Transportation of Dangerous Goods).
- Inhalation cutoff values are based on 4 h testing exposures. Conversion of existing inhalation toxicity data that have been generated according to 1 h exposures should be by dividing by a factor of 2 for gases and vapors and 4 for dusts and mists.
- The units for inhalation toxicity are a function of the form of the inhaled materials. Values for dusts and mists are expressed in mg/L. Values for gases are expressed in ppmV. Acknowledging the difficulties in testing vapors, some of which consist of mixtures of liquid and vapor phases, the table provides values in units of mg/L. However, for those vapors that are near the gaseous phase, classification should be based on ppmV.
- For some substances, the test atmosphere will not just be a vapor but will consist of a mixture of liquid and vapor phases. For other substances, the test atmosphere may consist of a vapor that is near the gaseous phase. In these latter cases, classification should be based on ppmV as follows: Category 1 (100 ppmV), Category 2 (500 ppmV), Category 3 (2,500 ppmV), Category 4 (20,000 ppmV).
- Category 5 classifications are intended to allow for the identification of substances with low acute toxicity hazard but may still present a danger to vulnerable populations. The requirements for Category 5 classification are as follows:
  - LD_{50} in the 2000–5000 mg/kg range with equivalent doses for inhalation
  - Toxicologically significant in effects in humans
  - Any mortality is observed when tested up to Category 4 values
  - Expert judgment confirms significant clinical signs of toxicity when tested up to Category 4 values except for diarrhea, piloerection, or an ungroomed appearance
  - Where expert judgment confirms reliable information indicating the potential for significant acute effects from other animal studies
- Of particular importance is the use of well-articulated values in the highest hazard categories for dusts and mists. Inhaled particles between 1 and 4 μm MMAD will deposit in all regions of the rat respiratory tract. This particle size range corresponds to a maximum dose of about 2 mg/L. In order to achieve applicability of animal experiments to human exposure, dusts and mists would ideally be tested in this range in rats.
- In addition to the classification for inhalation toxicity, if data are available that indicates that the mechanism of toxicity was corrosivity of the substance, certain authorities may choose to label it as corrosive to the respiratory tract. Corrosion of the respiratory tract is defined by the destruction of the respiratory tract tissue after a single limited period of exposure analogous to skin corrosion; this includes destruction of the mucosa.

TABLE 4.10

Globally Harmonized System Classification for Repeat-Exposure Target Organ Toxicity Following Inhalation Exposure

Form of Substance (Species)	Category 1 Guidance Value	Category 2 Guidance Value
Gas (rat)	$\leq$50 ppmV/6 h/day	50 < Category 2 $\leq$ 250 ppmV/6 h/day
Vapor (rat)	$\leq$0.2 mg/L/6 h/day	0.2 < Category 2 $\leq$ 1.0 mg/L/6 h/day
Dust/mist/fume	$\leq$0.02 mg/L/6 h/day	0.02 < Category 2 $\leq$ 0.2 mg/L/6 h/day

that $C_1 \times t = C_2 \times t$. This relationship, in theory, can also be used to extrapolate concentration values between short-term and long-term exposures:

$$C_2 = \frac{C_1 \times t_1}{t_2}$$

In modern risk assessment, the ten Berge modification of Haber's law is commonly used: $C_n \times t = k$. The exponential n is a regression coefficient for the exposure concentration–exposure duration relationships for the relevant effect. In general, the value of n lies between 1 and 3. If suitable data are not available to derive n, a default value of n = 1 is used for extrapolating from shorter to longer exposure durations, and a default value of n = 3 is used for extrapolating from longer to shorter exposure durations. Using the ten Berg modification, the Haber's law equation becomes

$$C_2 = \frac{C_1^n \times t_1}{t_2}$$

However, note that there are many cases where Haber's law and the ten Berge modified Haber's law do not accurately describe the dose–time relationships for the toxicological effects of gases. The use of these simple relationships may seriously over- or underestimate the degree of toxicological effects, particularly when there are large extrapolations in terms of the time of exposure. High-quality data for the specific duration of exposure of interest are preferable to the use of Haber's law or the ten Berge modification.

4.2.3 Dose–Response Assessment

The usual primary aims of a dose–response assessment are to (1) define the type and shape of the dose–response curve, the area of greatest interest being the left-hand side of the curve (i.e., the shape at lower doses); and (2) determine a toxicological threshold that has acceptable safety properties for the population at risk or an acceptable defined risk. The term *acceptable safety properties* is generally accepted to mean adequate protection even for susceptible subpopulations. The determination of a human threshold of this nature typically involves extrapolation to account for different durations of exposure, extrapolation from animal data, potential adjustments, and accounting for uncertainties. Within this context, the term *uncertainty* means accounting for the unknown (we don't know what we don't know). If a reasonable quantitative estimate can be made for a factor known to affect a particular endpoint, then the factor is no longer an *uncertainty* although adjustment of the toxicological threshold may still be required.

As can be surmised from this, the setting of human toxicological thresholds with acceptable safety properties is more often than not a semiquantitative best estimate. This has at least two clear implications: (1) a human toxicological threshold set by these methods may have an uncertainty of perhaps an order of magnitude (i.e., ± 100.5) and (2) human toxicological thresholds are not set in stone. They evolve over time within the contexts of increased knowledge and *postmarket* surveillance. Thus, it is entirely possible, although hopefully rare, that a correctly calculated human toxicological threshold will not provide adequate protection for some hitherto-unknown susceptible population. Within this context, it is critical that human toxicological thresholds are not perceived to have greater accuracy, precision, and predictive power than what they are actually intended to indicate.

The derivation of human toxicological thresholds (human equivalent concentrations (HECs) or derived no effect levels, DNEL) following respiratory exposure generally consists of a number of sequential steps: (1) determination of relevant toxicological thresholds from the left side of a dose–response curve; (2) modification of the dose–response threshold to the correct starting point; (3) the application of UFs; and (4) a holistic overview of the resulting HEC/DNEL to ensure that it makes biological and practical sense.

4.2.3.1 Types of Effects Used for the Determination of Toxicological Thresholds

In general, toxicological effects are divided into two main groups: threshold and nonthreshold effects. A threshold effect implies that there is a clearly definable dose threshold below which the adverse effect no longer occurs. This implies, at least in theory, that the risk associated with exposures below the threshold is zero. However, in reality, the *threshold* is often determined by the ability to detect an effect above background in a toxicology study (the study sensitivity or signal-to-noise ratio). The ability to detect an effect is substantially affected by practical experimental design considerations (e.g., animal group size) that limit experimental statistical power. In reality, a typical regulatory toxicology study has the statistical power to detect a change in a biological response of about 5%–10% above background (occasionally 1% under exceptional circumstances). Thus, a no observed adverse effect level (NOAEL) in a typical toxicology study may actually be equivalent to a 5%–10% risk (i.e., a detectable effect in 5–10 individuals out of 100). In general, the concept of a threshold toxicological effect is usually superimposed upon a biological variable that is continuous (has continuous probability distributions); that is, within the limits of the variable's range, any value is possible. Most biological phenomena fall under the purview of continuous variables.

The alternative type of data encountered in toxicological threshold setting is nonthreshold dose responses. For nonthreshold effects, it is impossible (at least theoretically) to determine a dose below which there is no measurable adverse effect. Classically, nonthreshold dose–response relationships are applied to classical directly DNA-reactive mutagenesis and classical mutagenic carcinogenesis. Often this type of data is discrete (count data), that is, one that cannot take on all values within the limits of the variable. The nonthreshold dose–response relationship implies that there can never be a level of exposure where the risk is zero. In these circumstances, risk assessors typically determine a dose or exposure that is associated with an acceptable level of risk and how the risk changes for a given change in dose or exposure. Typically, the acceptable risk for carcinogenesis is in the 1 additional case per 100,000 population (10^{-5}) to 1 additional case per 1,000,000 population (10^{-6}). Classically, the linear low-dose extrapolation model has been used to derive these parameters, particularly when the exact carcinogenic MOA has not been established or when dose extrapolation using a toxicodynamic model is not yet possible.

4.2.3.2 Determination of Relevant Toxicological Thresholds from the Left Side of a Dose–Response Curve

This step typically involves the use of animal data and may involve the determination of multiple thresholds due to the presence of multiple effects or key events. It is critical at this stage to determine thresholds for all the potential adverse effects (and key steps) and not just the most apparently sensitive endpoint since different UFs may be applied to the different effects. Because of the different application of uncertainty effects, it is possible in some rare situations that the most sensitive adverse effect does not result in the lowest HEC or DNEL.

In risk assessment, dose–response curves are generally assumed to be monotonic; that is, the magnitude of the effect increases (or decreases) in some form of proportionate relationship as dose changes. However, there are examples of nonmonotonic dose–response curves (e.g., hormesis and dose–response curves with inflections) that have been documented. Vitamins, essential nutrients, hormonal effects, and endocrine disruption are known examples of nonmonotonic dose–response curves. Nonmonotonic dose–response relationships are more likely to be present at the low-dose left-hand side of the curve: exactly the area of greatest interest and concern when setting toxico-logical thresholds. This presents a substantial challenge to classical techniques for the setting of toxicological thresholds since these techniques generally involve large extrapolations below the actual experimental doses used and rely on the assumption (often unproven and with little evidence to support its validity) that effects at low to very low doses can be reasonably predicted by the effects observed at higher (experimentally tested) doses.

The fact that risk assessors are most interested in the low-dose left-hand side of the dose–response curve also has substantial implications for the design of toxicological studies. The first implication is on dose-level selection for studies. Current recommendations include at least three dose levels that span the range from the NOAEL through to either a detectable level of toxicity or until other logistical/regulatory limits occur (e.g., the MAC and regulatory limit dose). In terms of defining toxicological thresholds, it is especially helpful if there are at least two dose levels that display intermediate levels of toxicity and that these dose levels are relatively closely spaced above the NOAEL. This improves the ability to accurately define the characteristics of the left side of the dose–response curve.

Classically, there have been five general criteria for defining the high dose in a toxicology study: (1) the maximum tolerated dose (MTD); (2) regulatory limit doses; (3) top dose based on the saturation of exposure (derived from toxicokinetic/toxicodynamic studies); (4) maximum feasible/practical dose (e.g., MAC or CS); or (5) a dose providing a 50-fold margin of exposure in relation to human exposures. The MTD is defined as the highest dose that will be tolerated for the study duration. The primary purpose of deriving an MTD is to identify a dose where target organ toxicity of interest is observable, but the dose is not so high that the study integrity is not jeopardized by morbidity, mortality, or other significant confounding abnormal biology.

The difficulty in setting the maximum dose is that there is no clear consensus or international standardization on what constitutes the MTD. One rule of thumb indicates that the MTD is the dose that results in up to 10% loss of body weight, no mortality, and no signs of severe toxicity. Other recent guidance on setting MTDs has defined this dose level as one that produces moderate toxicity as evidenced by one or more of the following clinical signs:

- Weight loss of up to 20%
- Food and water consumption less than 40% of normal for 72 h
- Staring coat—marked piloerection
- Subdued animal that shows subdued behavior patterns even when provoked
- Little peer interaction
- Intermittent hunched posture
- Oculonasal discharge present
- Intermittent abnormal breathing patter
- Intermittent tremors
- Intermittent convulsions
- Transient prostration (less than 1 h)
- No self-mutilation

For threshold adverse effects, two basic approaches are currently in use: the classical NOAEL/lowest observable adverse effect level (LOAEL) technique and the benchmark dose technique. The NOAEL/LOAEL technique is the oldest and simplest of these techniques. The NOAEL is defined as the highest dose level that does not produce a significant increase in adverse effects in comparison to the control group. It should be noted that the emphasis is on biological significance. Statistical significance is not an absolute requirement for the setting of an NOAEL. The NOAEL is not the same as a no observed effect level since a test article may produce effects that are not necessarily biologically adverse or evidence of nonadverse biological adaptation may be present. There are three basic types of findings in toxicology studies that are used to determine the NOAEL: (1) overt toxicity (i.e., clinical signs and anatomic pathology findings); (2) surrogate markers of adverse toxicity (i.e., changes in clinical chemistry parameters); and (3) exaggerated pharmacodynamic effects. If the study dose range does not incorporate an NOAEL, then it is possible to use the LOAEL. However, this will increase the level of uncertainty associated with the toxicological threshold.

The limitations regarding the NOAEL/LOAEL techniques are as follows: (1) the NOAEL/LOAEL values can be substantially affected by experimental design and are heavily dependent upon the dose range and dose spacing used in the study; (2) the NOAEL/LOAEL is heavily dependent upon the sensitivity of the study, that is, the statistical power to detect an effect above background. This, in turn, is heavily dependent upon the data variance, the number of animals per treatment group, the number of dose levels used, and the dose-level spacing. In general, studies involving fewer animals tend to produce higher NOAELs; (3) since no two toxicological studies have the same statistical power, different NOAEL/LOAEL values will define different levels of actual statistical risk. Thus, it is not possible to quantitatively compare different NOAEL/LOAEL values across a chemical class, chemical family, or even across different adverse effects observed in the same toxicological study. Such comparisons are qualitative at best; (4) determination of the NOAEL/LOAEL is often based upon best scientific judgment and is a frequent source of controversy; (5) the slope of the dose–response curve plays little or no role in determining the toxicological threshold; (6) the NOAEL does not define the size of the risk for a given level of exposure, nor does it document how the risk might change as the exposure level changes. Analysis of existing published values suggests that the level of risk associated with an NOAEL is about 10^{-4} (1 per 10,000); however, there is considerable variation around this value; (7) the NOAEL is heavily dependent on the population mean; for example, the NOAEL might be a dose that produces a mean response of 10% within the studied population; however, the actual individual responses may range from 5% (resistant individuals) to 15% (susceptible individuals) at this dose. Thus, the NOAEL does not inherently account for susceptible subpopulations. Substantial toxicity may be present in susceptible subpopulations exposed to the NOAEL dose; however, these effects are *averaged out* by the lack of adverse effects in the average and resistant individuals; (8) by definition, the NOAEL must be one of the doses tested in the study; (9) the NOAEL/LOAEL technique does not incorporate a measure of data variance and uncertainty; (10) an LOAEL cannot be used to derive an NOAEL that has not been detected in a study. The current approach is to apply an additional UF of 10 to the LOAEL and hope that this results in a number that approximates the true NOAEL; (11) the NOAEL/LOAEL technique rewards poor experimental designs that inherently favor higher values; and (12) due to the logistical limitations on the statistical power of toxicology studies, the NOAEL is actually a dose that defines a 5%–20% biological response, not a 0% biological response; that is, the risk at the NOAEL is not zero. The main advantages of the NOAEL/LOAEL approach are that it is simple and easy to understand, it is not dependent on an assumed dose–response model, and it can be used for continuous or discrete data.

A benchmark dose (often referred to as the benchmark dose low or BMDL) is a statistical lower confidence limit (typically the lower 95% confidence interval, but the lower 99% confidence interval may be used in some circumstances) for a dose that produces a predetermined change in response rate (or a predetermined level of risk) of an adverse effect (the benchmark response, BMR) compared with the background (Figure 4.5). The BMDL approach involves fitting a mathematical model to the available dose–response data. The BMR is typically set near the lower level of responses that can be accurately measured given the statistical power of the study (typically, a 5%–10% response level, although a 1% response level may be justified for particularly good studies).

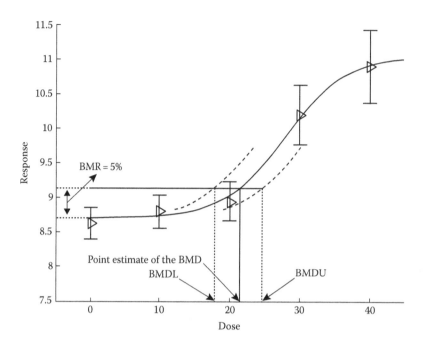

FIGURE 4.5   Key concepts of the BMD approach.

BMDL determination is usually performed using relevant software packages. The advantages of the BMDL approach are as follows: (1) It is applicable to all toxicological effects (including threshold and nonthreshold effects); (2) it makes use of all of the available dose–response data to estimate the shape of the dose–response curve for a particular endpoint; (3) it is less influenced by experimental design; (4) it incorporates a statistical measure of variability and uncertainty of the experimental data; (5) it accounts for the possibility of susceptible populations by using the lower confidence interval of the dose associated with a defined BMR; (6) the BMDL approach provides a formal quantitative evaluation of data quality by taking into account all aspects of the specific data; (7) even in the face of poor-quality data, the meaning of the BMDL value remains well defined: it reflects a dose level where the associated size of the effect is unlikely to be larger than the defined BMR; (8) the BMDL approach can interpolate between different dose levels with a defined level of statistical accuracy. A BMDL is always associated with a predefined effect size for which the corresponding dose has been calculated (which differs from a NOAEL where the corresponding effect size is mostly not calculated); (9) the BMDL approach allows for much more quantitative comparisons of potency across chemical classes and across different adverse endpoints. The development of quantitative relative potency estimates is possible using the BMDL technique; (10) the effects of covariates on the dose–response relationship (e.g., sex, exposure duration, and chemical coexposures) can be analyzed using the BMDL approach; (11) the BMDL approach provides a higher level of confidence in derived toxicological thresholds because data quality and all the available data are taken into account in a better manner than the NOAEL/LOAEL technique; and (12) the BMDL approach rewards better experimental designs, particularly those designs that increase statistical power and reduce data variance. The risk associated with the BMDL technique is that the results will become model driven rather than data driven. This can be avoided by careful application of the relevant techniques involved. Like any technique (including the NOAEL/LOAEL method), there is always the problem of *garbage in, garbage out*. The BMDL approach will not resurrect studies that are scientifically hopeless. It is not a panacea for poor science. The only solution in this case is to discard the study and look for better data. The other problem associated with the BMD is that it is more time-consuming, it requires appropriate computer systems (some BMD platforms are

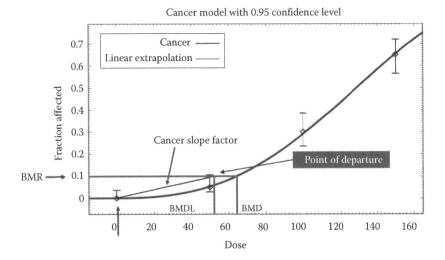

FIGURE 4.6 The cancer linear low-dose extrapolation technique.

notoriously incompatible with some secure network types), and it requires training and expertise to perform the technique and to review the results.

It is important to remember that the NOAEL/LOAEL and BMD techniques are not mutually incompatible. It is common for both techniques to be used on the same data set as a form of *ballpark reality checking* of the outcomes.

Dose–response analysis for classical DNA-reactive mutagens and classical mutagenic (*genotoxic*) carcinogenesis has been traditionally performed using the linear low-dose extrapolation (LLDE) technique for each specific tumor type whenever possible (Figure 4.6). For this procedure, a straight line is extended from a point on the dose–response curve (point of departure, POD) to the origin. The lowest POD that is adequately supported by the available data is used. In most circumstances, this is the lower 95% confidence interval limit on the lowest dose that can be supported for modeling by the data. Often the POD is the BMDL (95% confidence interval) for a 1%–10% cumulative tumor incidence BMR (often termed the lower effective dose or LED10 for a BMR of 10%). The slope of the extrapolated line is called the SF. The SF is effectively an upperbound estimate or risk per dose increment. The SF in mg/kg body weight/day can thus be used to estimate extra risk probabilities for a given exposure level (risk-specific dose (RSD)), or it can be used to calculate an exposure level. Assuming a 10% BMR (or 10% risk for the LED10), the formula for SF is as follows:

$$SF = \frac{\Delta Risk}{\Delta Dose} = \frac{BMR}{BMD_L} = \frac{0.1}{LED_{10}} = \mu g/m^3$$

In the particular case of inhalation, the SF is the same as unit risk, and no unit conversion is needed. Notably, unit risk estimates will often assume a standardized daily respiratory volume (i.e., m^3/day air) and body weight (kg). These values may need to be adjusted to match the population of interest (e.g., the standardized daily respiratory volume for a physically active worker will not be the same as for a person in a sedentary occupation).

It should be noted that the LLDE technique is really a default technique that is used when there are insufficient data regarding the shape of the low-dose area of the dose–response curve. Available high-statistical power studies have demonstrated that supralinear and sublinear cancer dose–response curves can occur in different tissues and with different cancer types in the same animal at the same time! Sublinear responses are less of a concern because the SF will tend to overestimate the risk and thus provide an additional margin of safety regarding any regulatory toxicological thresholds that are set from the analysis. However, supralinear responses are of significant concern because the

LLDE technique may seriously underestimate the risks. For these reasons, toxicodynamic models for low-dose extrapolation of cancer risk are currently preferred. However, sufficient mechanistic information and data for these techniques are lacking for a wide range of substances.

4.2.3.3   Regional Gas/Vapor Dose Rate Adjustments

For the purposes of risk assessment, it may be necessary to define what fraction of an inhaled gas or vapor acts in specific parts of the respiratory tract (or systemically) in humans versus an animal model. For this reason, the US EPA has developed three separate regional gas dose rate (RGDR) equations based upon the site of action and toxicokinetic characteristics of the gas type. The general formula for the use of the RGDR to adjust a relevant toxicological threshold to account for a specific site of action is as follows:

$$\text{Region-specific human equivalent toxicological threshold} =$$
$$\text{Toxicological threshold (e.g., NOAEC or BMD}_L) \times \text{RGDR}$$

Category 1 gases are highly water soluble and/or irreversibly reactive in the surface liquids and tissues of the extrathoracic and tracheobronchial regions of the respiratory tract. For this reason, they tend to primarily act as the site of first contact toxicants provided that the respiratory *sink* is not overwhelmed by extreme exposures. Examples of gases/vapors within this category are hydrogen fluoride, chlorine, and formaldehyde.

$$\text{RGDR}_{\text{Extrathoracic region}} = \frac{\left(\dfrac{V_E}{SA_{ET}}\right)\text{Animal}}{\left(\dfrac{V_E}{SA_{ET}}\right)\text{Human}}$$

$$\text{RGDR}_{\text{Tracheobronchial region}} = \frac{\left(\dfrac{V_E}{SA_{TB}}\right)\text{Animal}}{\left(\dfrac{V_E}{SA_{TB}}\right)\text{Human}}$$

$$\text{RGDR}_{\text{Pulmonary region}} = \frac{\left(\dfrac{Q_{ALV}}{SA_{PU}}\right)\text{Animal}}{\left(\dfrac{Q_{ALV}}{SA_{PU}}\right)\text{Human}}$$

V_E = Respiratory minute volume in mL/min

SA_{ET} = Surface area of the extrathoracic region in cm^2

SA_{TB} = Surface area of the tracheobronchial region in cm^2

SA_{PU} = Surface area of the pulmonary region in cm^2

Q_{ALV} = Aveolar ventilation rate in mL/min which can be approximated by $0.6 \times V_E$

Category 3 gases/vapors are those that are relatively insoluble in water and are not reactive in the extrathoracic and tracheobronchial regions of the respiratory tract. These gases/vapors are not *scrubbed out* in the upper respiratory tract and conducting airways. They therefore penetrate into the deep pulmonary areas where they become available for absorption into the systemic circulation. Classical examples are styrene and most of the common anesthetic gases and vapors. The RGDR formulation for these gases is applicable for systemic effects that occur outside of the respiratory tract:

$$RGDR_{Systemic\ Category\ 3} = \frac{H_{b/g}Animal}{H_{b/g}\ Human}$$

$H_{b/g}$ = The blood:gas partition coefficient for the gas/vapor

Category 2 gases are intermediate between Category 1 and Category 3 gases. They are moderately water soluble, rapidly and reversibly reactive, and/or moderately to slowly irreversibly metabolized within respiratory tissues. These intermediate gases have the potential for both sites of first contact and systemic toxic effects. A classical example is sulfur dioxide. The earlier RGDR formulae for Category 1 and Category 3 gases can be used according to what site of action is of greatest interest.

4.2.3.4 Regional Deposited Dose Ratios for Aerosols, Mists, and Dusts

Regional deposited dose ratios (RDDRs) are a dosimetric adjustment factor that essentially perform the same role as RGDRs except that they are used for inhaled particle exposures. An RDDR is the ratio of the regional deposited dose calculated for a given exposure in the animal species of interest to the regional deposited dose of the same exposure in a human. This ratio is used to adjust the exposure effect level for interspecies dosimetric differences to derive an HEC for particles. RDDRs are best calculated using various currently available software modeling systems. In general, the determination of an RDDR will require data regarding the MMAD and σg.

4.2.3.5 Accounting for *We Don't Know What We Don't Know*: The Use of UFs

Within the earlier context, it is critical to separate the concepts of variability and uncertainty. Variability occurs due to heterogeneity of response in time and space and is inherent to the particular population under examination. Variability is often derived from bona fide toxicokinetic and toxicodynamic differences between individuals and across populations. The collection of additional data may help to clarify the extent of variability and the mechanisms that cause it, but it will not reduce the amount of variability present. On the other hand, uncertainty is essentially driven by a lack of knowledge. Uncertainty can be reduced by the collection of additional data or by performing a better scientific experiment (particularly in the area of recognizing and controlling confounding factors or by using better techniques).

- UFs are used to account for unknown factors relating to the following:
- Dose–response differences between average humans and sensitive subpopulations (UFH)
- Extrapolation of toxicology data from animals to humans (UFA)
- Extrapolation from subchronic to long-term (chronic) exposure durations (UFS)
- Extrapolation from an LOAEL to account for the lack of an experimentally determined NOAEL (UFL)
- Database deficiencies and data quality uncertainties (UFD)
- A food quality protection factor (FQPA) is used in the United States to account for the perceived higher sensitivity of children to the effects of pesticides

Importantly, different countries and organizations use different values for particular types of UFs (Table 4.11). Typically, a default value of 10 is used for each UF, and different UFs are multiplied together, that is,

$$UF_{Total} = UF_H \times UF_A \times UF_S \times UF_L \times UF_D \times FQPA$$

TABLE 4.11

Comparison of Uncertainty Factors Used by Different Countries and Organizations

UFs	Health Canada	IPCS	RIVM	ATSDR	EPA
Interindividual (H)	$10 (3.16 \times 3.16)$	$10 (3.16 \times 3.16)$	10	10	$10 (3.16 \times 3.16)$
Interspecies (A)	$10 (2.5 \times 4.0)$	$10 (2.5 \times 4.0)$	10	10	$\leq 10 (3.16 \times 3.16)$
Subchronic to chronic (S)	1–100	1–100	10	NA	≤ 10
LOAEL to NOAEL (L)			10	10	≤ 10
Incomplete database (D)			NA	NA	≤ 10
Modifying factor (MF)	1–10	1–10	NA	NA	0 to ≤ 10 (discontinued)

The default value of 10 for each UF type was largely derived from engineering practices rather than biological reality. However, empirical observation over a long period of time has generally supported the default value, although a value of 10 may over- or underestimate the amount of uncertainty present in particular circumstances.

The two most commonly used UFs are the animal to human extrapolation UF and the intrahuman variability UF. These UFs can each be subdivided into toxicokinetic (i.e., what the body does to the chemical) and toxicodynamic (what the chemical does to the body) components. Thus,

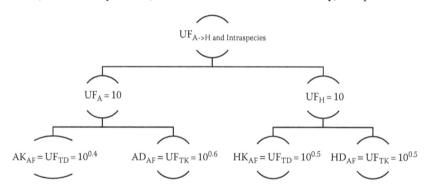

$$UF_{A \to H \text{ and Intraspecies}} = UF_A \times UF_H = AK_{AF} \times AD_{AF} \times HK_{AF} \times HD_{AF}$$

TK = toxicokinetic
TD = toxicodynamic
A = animal
H = human
AF = adjustment factor

This subdivision of the UFA and the UFH factors allows for the development of chemical-specific adjustment factors (CSAFs) that utilize any available interspecies comparative toxicokinetic and toxicodynamic data. It is critical to note that CSAFs can be developed only if there are data pertaining to the ultimate toxicological form (ultimate toxicant) of the substance in question. Toxicokinetic and toxicodynamic information on an absorbed metabolic precursor is not sufficient for the derivation of a CSAF.

Assuming that both human and animal toxicokinetic data are available for the ultimate toxicological form of the substance (i.e., the active chemical species), there are two further steps involved in the derivation of a toxicokinetic interspecies CSAF (i.e., the AKAF): (1) the choice of metric and (2) the examination of the experimental data for suitability. The decisions pertaining to the choice of metric are based around the question of "what toxicokinetic parameter is the best predictor of the target organ dose and response?" Effects at the target organ may be related to the total amount of the toxicant delivered to the site of action or to their maximum concentration. The peak plasma concentration (C_{Max}) and area under the time versus tissue (plasma) concentration curve (AUC) usually reflect concentrations of the toxicant at the site of action (unless there is significant protein binding). Whole-body clearance (CL) is inversely related to the AUC. Notably, AUC, C_{Max}, and CL are all normalized to body mass. The toxicant half-life (T½) is not normalized to body mass and cannot be used for derivation:

$$\text{For subchronic or chronic repeat exposures}: \frac{\left(\dfrac{AUC_{Human}}{Dose}\right)}{\left(\dfrac{AUC_{Animal}}{Dose}\right)}$$

$$\text{For subchronic or chronic repeat exposures}: \frac{CL_{Human}}{CL_{Animal}}$$

$$\text{For single acute exposures}: \frac{\left(\dfrac{C_{Max\ Human}}{Dose}\right)}{\left(\dfrac{C_{Max\ Animal}}{Dose}\right)}$$

It should be noted that the equations using AUC or CL ratios are usually more health protective. Other types of interspecies toxicokinetic metrics are possible if they are supported from data. If the AUC ratio calculation is to be used, it is important that the AUCs are determined after toxicokinetic steady-state conditions have been established. If chemical- and/or species-specific AUC, C_{Max}, or CL data are not available, it may be possible to use a rate-limiting chemical-related parameter such as glomerular filtration rate or renal blood flow. PBPK modeling generally takes into account the AKAF, and separate values for this parameter are usually not required.

Evaluation of the relevance of the population sources of the data is a key step in the development of CSAFs. A CSAF is based upon a comparison of mean animal data with mean human data. This means that the human toxicokinetic or toxicodynamic study chosen needs to be sufficient representative of the at-risk population. Differences in life-stages examined between the animal and human studies must also be taken into account. In general, data derived from sensitive subgroups of the human population should not be used for the derivation of interspecies CSAFs since differences due to variation in the susceptibility of the human population fall under the UFH and not the UFA.

Toxicokinetic studies in animals and humans that are used for the determination of the AKAF should be performed using the same route of exposure as the adverse effect of interest; otherwise, route-to-route extrapolation may need to be applied to the data. Ideally, data that are used to derive an AKAF should be based on animals exposed to doses that are equivalent or similar to the relevant BMDL or NOAEL and, under dosage conditions, similar to those in the toxicity studies on which the BMDL or NOAEL is based. If this is not possible, the implications of the differences should be carefully evaluated, particularly the potential for nonlinear kinetics. Ideally, the doses used in the human studies that support the derivation of an AKAF should be similar to the estimated or predicted real-world human exposure level. The sample or experimental group size in the studies used to support the derivation of an AKAF should be sufficiently large to allow for a reliable estimate of central tendency (i.e., determination of a reliable mean or median).

The default equation used for the determination of an interspecies toxicodynamic CSAF (ADAF) is

$$AD_{AF} = \frac{\text{Effective dose}_{Animal\ 10\%\ response}}{\text{Effective dose}_{Human\ 10\%\ response}}$$

This equation is applicable irrespective of whether or not the human and animal dose–response curves are parallel (i.e., have the same mathematical model). In special cases where the dose–response curves are parallel, the effective dose for any level of response between 10% and 90% can be used. As with the AKAF, only data on the penultimate toxicant/active chemical species can be used. The issues pertaining to the relevance of the populations used for comparison, relevance of the routes of exposure and the adequacy of the experimental group size, and the determination of the AKAF are also directly relevant to the derivation of ADAF values.

The HKAF can be calculated using the following formulae:

$$HK_{AF} = \frac{AUC_{95\% \text{ Confidence Interval}}}{AUC_{Mean}} \quad OR \quad \frac{C_{Max \text{ 95\% Confidence Interval}}}{C_{Max \text{ Mean}}} \quad OR \quad \frac{CL_{Mean}}{CL_{5\% \text{ Confidence Interval}}}$$

For single acute exposures, the C_{Max} ratio is the usual default method. The AUC ratio and the CL ratios are generally the default methods for subchronic to chronic repeat exposures. The previous discussions regarding the relevance of the populations used for comparison, relevance of the routes of exposure, and the adequacy of the experimental group size are also pertinent to the derivation of the HKAF. HKAF values are commonly determined on the basis of PBPK modeling.

Most chemicals encountered will not have sufficient human data to allow for the derivation of an HDAF. However, if such data are available, the following equation can be used:

$$HD_{AF} = \frac{\text{Mean effective dose}_{10\% \text{ response}}}{5\% \text{ Confidence interval of the effective dose}_{10\% \text{ response}}}$$

4.2.3.6   Allometric Scaling

The principle of allometric scaling is used as an alternative to the CSAF methods for deriving part of an interspecies UF. Allometry is the study of the relationship between body size to shape, anatomy, and physiology. In toxicology, toxicokinetic parameters and some toxicodynamic parameters scale reasonably accurately between species based upon the differences in body surface area or caloric demand or basal metabolic rate. Allometric scaling is commonly used for estimating the maximum safe exposure in initial clinical trials of pharmaceuticals in healthy adult humans. Some regulatory authorities may also use allometric scaling in chemical risk assessment since it takes into account some toxicokinetic and toxicodynamic components of the UFA.

The fundamental equation used for allometric scaling in toxicology is

$$\text{Dosimetric adjustment factor} \left(DAF \right) = \left(\frac{\text{Body weight}_{Animal}}{\text{Body weight}_{Human}} \right)^{0.25}$$

This equation is based on the premise that basal metabolic rate, caloric demand, glomerular filtration, glucose turnover, food consumption, and water consumption all scale in proportion to the body weight 0.734 in mammalian species ranging in size from mice to elephants. The exponential value of 0.734 is commonly rounded off to 0.75. It should be noted that exponents between 0.6 and 0.8 have been reported for metabolic rate, and different regulatory authorities may use different exponentials within this range. Other toxicologically pertinent mechanisms are known to allometrically scale in proportion to body surface area (an exponential of 0.67).

In terms of using allometric scaling to account for interspecies toxicokinetic differences (i.e., AKAF), it is important to note the following:

- Respiratory parameters such as surface areas within the respiratory tract and respiratory volumes generally scale allometrically in relation to body weight.
- As a general rule, absorption across biological membranes (i.e., transcellular absorption) is largely determined by the physicochemical properties of the toxicant and not by the species. There are some notable exceptions to this general rule; for example, dogs are more adept at absorbing lipophilic materials from the gut than other species.
- Toxicant distribution is largely determined by blood flow, rates of diffusion, and rates of active transport. Blood flow across species scales allometrically. However, there are substantial interspecies differences in the capacity of toxicants to bind to protein within the circulation. Protein binding essentially limits the volume of distribution of the bound toxicant fraction to the plasma volume, thus limiting the amount of toxicant that is available to diffuse out of the circulation to a target tissue. Accordingly, allometric scaling is generally considered as being unsuitable for toxicants that have significant binding to plasma proteins. Provided no plasma protein binding occurs, it is notable that the volume of distribution of a toxicant scales allometrically with exponentials ranging from 0.8 to 1.1. Although data are lacking, active transport mechanisms are dependent on metabolic rate and are thus expected to scale allometrically.
- As a general rule, there are often substantial and complex interspecies differences in the metabolism/biotransformation of xenobiotics. From an overall holistic perspective, drug biotransformation systems are not related to body size, and they may not be amenable to allometric scaling. This implies that allometric scaling may not apply to toxicants that undergo extensive metabolism/biotransformation.
- The current evidence supports the notion that allometric scaling is relevant to toxicants that are largely excreted by renal filtration. However, allometric scaling is not predictive of excretion via the biliary route.
- Studies of pharmaceuticals have demonstrated that about 25% of drugs will scale allometrically. Allometric scaling is most predictive of drugs that have minimal to no metabolism/biotransformation and are largely excreted by renal glomerular filtration.
- There are often large interspecies pharmacodynamic differences, particularly if there is a receptor or second-messenger-based MOA. Toxicants that are known to display large interspecies differences in their spectrum of effects and in the shape of the dose–response curves generally do not scale allometrically.
- There are often substantial toxicokinetic and toxicodynamic differences between different life-stages of the same species. Allometric scaling between different life-stages based on body weight or surface area is generally not recommended.
- As a general rule, the effects of acute single exposures often do not allometrically scale with body weight or skin surface area.

Given this discussion, it is usually preferable to use the CSAF methodology rather than allometric scaling if possible. However, allometric scaling may be useful for toxicants that do not display marked toxicodynamic differences between species, are not bound to protein in the circulation, do not display marked species differences in metabolism/biotransformation, and are largely excreted in urine.

Given that allometric scaling will account for only part of the interspecies UF, it is prudent that it is used in conjunction with a factor that compensates for any remaining interspecies toxicokinetic and toxicodynamic differences, that is,

$$UF_A = \text{Allometric scaling factor} \times UF_{\text{Remaining differences}}$$

The difficulty is that there is no internationally accepted consensus on the value for the UFRemaining Differences parameter. As a general rule of thumb, the UFRemaining Differences parameter should be at ≥ 100.5 if allometric scaling is used.

The European Chemicals Agency (ECHA) REACh method for allometric scaling

The current ECHA REACh method of allometric scaling basically incorporates a version of Haber's law combined with allometrically scaled standard respiratory volumes in order to adjust a rat 6 h/day, 5 days/week exposure NOAEC or BMDL:

$$\text{Worker NOAEC}_{\text{Corrected 8 h/day, 5 days/week}} = \text{NOAEC} \times \frac{\text{Exposure conditions}_{\text{Rat}}}{\text{Exposure conditions}_{\text{Human}}}$$

$$= \text{NOAEC} \times \frac{\text{Exposure duration}_{\text{Rat}}}{\text{Exposure duration}_{\text{Human}}}$$

$$\times \frac{\text{Standard respiratory volume}_{8\,\text{h Human sedentary}}}{\text{Standard respiratory volume}_{8\,\text{h Human light activity}}}$$

$$= \text{NOAEC} \times \frac{6\,\text{h}}{8\,\text{h}} \times \frac{6.7\,\text{m}^3}{10\,\text{m}^3}$$

To derive a corrected NOAEC for the general human population with 24 h of exposure 7 days/week,

$$\text{General population NOAEC}_{\text{Corrected 24 h/day, 7 days/week}} = \text{NOAEC} \times \frac{5\,\text{days exposure per week}}{7\,\text{days/week}}$$

$$\times \frac{6\,\text{h/day exposure}}{24\,\text{h/day exposure}}$$

The human standard respiratory volumes used are based on those for a 70 kg average human. Sex differences are not taken into account.

4.2.3.7 Putting It All Together

The overall final objective of the whole process of dose–response assessment is to determine an inhalation toxicological threshold that is suitable for the human population (inclusive of susceptible subpopulations). The overall equation for this process is

$$\text{Human-derived toxicological threshold}\left(\text{e.g., RfC or DNEL}\right) = \frac{\text{Adjusted toxicological threshold}}{\text{UF}}$$

For threshold toxicological endpoints, the relevant steps in this process can be summarized as follows:

1. Determine the effects of interest (hazard assessment).
2. Determine if the effects of interest are adverse or adaptive (hazard assessment).
3. Perform a dose-response analysis to determine an NOAEC or BMD$_L$.
4. Adjust the NOAEC or BMD$_L$ to the desired exposure duration using Haber's law or the ten Berge modification.
5. Adjust the NOAEC or BMD$_L$ to account for regional effects in the human respiratory tract or for systemic effects using the RGDR or RDDR equations.

6. Some systems like the REACh DNEL system will adjust the NOAEC or BMD_L based on the Haber's law plus standard respiratory volumes per unit time. This technique is really a variant of allometric scaling.

7. Derive the total UF either by using the default values (if there are insufficient data to do anything else) or by incorporating the CSAF and/or allometric scaling techniques as appropriate.

8. Divide the adjusted NOAEC or BMC_L by the total UF.

9. Check that the derived RfC or DNEL makes biological and toxicological sense.

10. Avoid the *analysis paralysis* phenomenon, that is, avoid a state of overanalyzing (or overthinking) a situation to the point that a definitive decision or action is never taken.

11. *Extinct by instinct* is just as fatal as *analysis paralysis*. *Extinct by instinct* means that poor or fatal decisions or conclusions are reached based on hasty or incomplete assessment of the available data or *gut reactions*.

For nonthreshold endpoints, the procedure is much the same except that the toxicological threshold is derived from the cancer SF based upon a predetermined level of acceptable risk, that is, an RSD:

$$RSD = \frac{\text{Target risk (e.g., } 10^{-6})}{\text{Unit risk}}$$

4.2.4 Exposure Assessment

The US EPA defines exposure as "contact between an agent and the visible exterior of a person (e.g., skin and openings into the body)." Thus, exposure assessment is the process of quantitatively assessing the magnitude, frequency, and duration of human exposure to an environmental agent (inhaled in this case). Exposure assessment also makes predictions or estimates of future potential exposures to an agent that has not yet been released. Inevitably, exposure assessment defines the size, nature, and types of human populations that are exposed to the toxicant in question. It also includes information on the level of uncertainty of the exposure scenarios and conclusions.

In some cases, exposure is still measured directly. However, in modern risk assessment practice, potential inhalation exposures are more commonly determined by modeling. There are numerous published computer-based models for the determination of exposure to inhaled materials under different circumstances and conditions (e.g., US EPA APEX/TRIM.Expo$_{Inhalation}$, US EPA HAPEM4, US EPA pNEM, and ECETOC TRA). Different countries and regulatory regions commonly have their own set of preferred models or customized systems. As with all modeling, particular care must be taken to ensure that the results generated by models are reality checked with real-world data. The big risk associated with the modeling approach is that the various systems are commonly extrapolated or applied beyond their original design parameters or design frame of reference. This may result in a time-consuming, all-be-it pleasant journey through a toxicological fantasyland. Modeling is also highly susceptible to the *garbage in, garbage out* phenomenon meaning that modeling is only as good as the fundamental underlying data and assumptions on which it is based. It is critical that the results of modeling systems be checked to make sure that they make basic physical and biological sense.

The other concern about inhalation exposure modeling is that many of the current systems are designed to be highly conservative in order to minimize the chances of the results being underprotective. While in itself this is perfectly valid, the combination of conservative exposure modeling with the inherent conservatism of other aspects of risk assessment may result in an extremely (possibly overly) conservative outcome. Extreme and overconservative outcomes may be just as damaging as underprotective outcomes.

4.2.5 RISK CHARACTERIZATION

The final step of a risk assessment is risk characterization. This step integrates exposure data and dose response (informed by hazard characterization) to obtain risk estimates and provides risk managers with information regarding the probable nature and distribution of health risks. It has both quantitative and qualitative components and clearly delineates uncertainty and data gaps. The key components of risk characterization are transparency, clarity, consistency, and reasonableness.

One tool used to communicate the actual likelihood of risk to exposed populations is the hazard quotient. Individual risk is estimated by comparing the daily exposure with the RfC for each chemical of concern as expressed by the following equation:

$$\text{Hazard quotient} = \frac{\text{Exposure}}{\text{RfC (or RSD)}}$$

When the hazard quotient is greater than one, exposure exceeds the RfC, and the exposed populations might be at risk. The actual risk is related to the degree to which exposure exceeds the RfC as well as characteristics of the exposed population. Therefore, using an RfC aims to describe a protective exposure limit rather than predicting risk for a given level of chemical exposure.

The ultimate goal of a risk assessment is to provide the risk manager with sufficient information, clearly communicated in comprehensible manner, to give them a grasp of what is known and unknown regarding the risk of a specific situation.

QUESTIONS

1. Which of the following is not a typical objective for a human health risk assessment?
   a. To define the context(s), purpose, scope, and technical approaches to be used in the assessment process
   b. To identify and characterize the spectrum of recognized adverse health effects (or threats) associated with the exposure of humans to a specific material or mixture
   c. To document the relationship between exposure (or dose) and the individual hazards and/or adverse effects (dose–response assessment)
   d. To assess the magnitude of exposure (or dose) of the affected populations (exposure assessment)
   e. To assess the mass fractional distribution within different environmental media
   Answer: e
2. Which of the following are fundamental assumptions associated with risk assessments?
   a. The relationship between the exposure and the adverse effect is cause and effect and not just association
   b. The hazard can be identified; that is, the adverse event can be expressed in some form of identifiable manner
   c. The magnitude of the hazard, adverse effect, or event can be measured
   d. A and B
   e. All of the above
   Answer: e
3. Which of the following are not absolute requirements for a risk assessment?
   a. The level of exposure (or dose) can be detected and measured.
   b. That the degree of risk does not necessarily have to be somehow related to the level of exposure.
   c. The relationship between exposure and effect must be causal.
   d. The magnitude of the hazard, adverse effect, or event can be measured.
   e. The hazard can be identified.
   Answer: b

4. Which of the following Bradford-Hill Criteria absolutely must be met in order to demonstrate a causal relationship?
   a. Strength of association
   b. Temporal relationship
   c. Consistency
   d. A and B
   e. All of the above
   Answer: b

5. Which of the following is not a feature of hazard assessments?
   a. *Hazard* means a potential source of harm or adverse health effect on a person or persons.
   b. Hazard assessments include an objective probability estimate of how often a particular hazard is likely to result in actual harm.
   c. A hazard assessment often includes a descriptor of the magnitude or severity of the hazard.
   d. A and B.
   e. All of the above.
   Answer: b

6. Which of the following would not be a high-reliability source of data for a hazard and risk assessment?
   a. A well-reported OECD chronic toxicity study that was conducted according to GLP
   b. A well-reported multisite large-scale prospective human epidemiology study conducted according to good clinical practices
   c. A patient's medical chart and clinical notes
   d. A well-designed and reported large-scale meta-analysis study
   e. A well-conducted and well-reported laboratory study that did not use an OECD test method
   Answer: c

7. Which of the following are important features of an inhalation toxicology study?
   a. It is critical that sufficient information regarding chamber equilibration and concentration of the test article over time is recorded.
   b. Irrespective of the method of exposure and type of exposure system used, it is absolutely critical that the control (including vehicle-only controls) animals are treated *exactly* the same as the test-article-exposed groups.
   c. It is absolutely critical that all inhalation chambers used in a study be maintained within a given set of specifications and that both the specifications *and the actual achieved readings for each relevant parameter* are recorded in the study report.
   d. A and B only.
   e. All of the above.
   Answer: e

8. A dose threshold for which of the following xenobiotics is most likely to scale allometrically with reasonable accuracy?
   a. An extensively metabolized xenobiotic
   b. A xenobiotic that is largely excreted in bile
   c. A xenobiotic that is predominantly excreted unchanged (i.e., not metabolized) by renal filtration
   d. A xenobiotic that is heavily bound to plasma proteins
   e. A xenobiotic whose dose threshold is directly proportional to the surface area of the gas-exchange regions of the lung
   Answer: c

REFERENCES

Dourson M., Becker R.A., Haber LT., Pottenger LH., Bredfeldt T., Fenner-Crisp P., (2013). Advancing human health risk assessment: Integrating recent advisory committee recommendations. *Crit Rev Toxicol* 43(6): 467–492.

Klimisch H.J., Andreae M., Tillmann U. (1997). A systematic approach for evaluating the quality of experimental toxicological and ecotoxicological data. *Regul Toxicol Pharmacol* 25: 1–5.

BIBLIOGRAPHY

Bogen KT., (1990). Of apples, alcohol, and unacceptable risk. *Risk Anal* 10(2): 199–200.

Bolt HM, Kappus H, Buchter A, Bolt HM (1976). Disposition of [1,2-14C] vinyl chloride in the rat. *Arch Toxicol* 35: 153–162.

Clewell HJ, Covington TR, Crump KS (1995b). The application of a physiologically based pharmacokinetic model for vinyl chloride in a noncancer risk assessment. Prepared by ICF Kaiser/Clement Associates for the National Center for Environmental Assessment, U.S. Environmental Protection Agency, Washington, DC, under EPA contract number 68-D2-0129.

Clewell HJ, Gentry PR, Gearhart JM, Allen BC, Covington TR, Andersen ME (1995a). The development and validation of a physiologically based pharmacokinetic model for vinyl chloride and its application in a carcinogenic risk assessment for vinyl chloride. Prepared by ICF Kaiser for the Office of Health and Environmental Assessment, U.S. Environmental Protection Agency, and the Directorate of Health and Standards Programs, Occupational Safety and Health Administration, Washington, DC.

ECETOC (2009) Framework for the integration of human and animal data in chemical risk assessment. Technical Report 104. ISSN-0773-8071-104. European Center for Ecotoxicology and Toxicology of Chemicals, Brussels, Belgium.

Farrar D, Allen B, Crump K, Shipp A (1989). Evaluation of uncertainty in input parameters to pharmacokinetic models and the resulting uncertainty in output. *Toxicol Lett* 49(2–3): 371–385.

Hill AB (1965). The environment and disease: Association or causation? *Proc R Soc Med* 58: 295–300.

IPCS (International Programme on Chemical Safety) (2005). Chemical-specific adjustment factors for interspecies differences and human variability: Guidance document for use of data in dose/concentration–response assessment. Available at: http://whqlibdoc.who.int/publications/2005/9241546786_eng.pdf.

Meek ME, Boobis AR, Crofton KM, Heinemeyer G, Raaji MV, Vickers C (2011). Risk assessment of combined exposure to multiple chemicals: A WHO/IPCS framework. *Regul Toxicol Pharmacol* 60(2): S1–S14.

NAS (National Academy of Science) (1983). Risk assessment in the federal government: Managing the process. Committee on the Institutional Means for Assessment of Risks to Public Health, National Academy Press, Washington, DC.

NAS (National Academy of Science) (2009). Science and decisions: Advancing risk assessment. National Research Council, National Academies Press, Washington, DC. AKA, "Silverbook".

US EPA (United States Environmental Protection Agency) (October 1994). Methods for derivation of inhalation reference concentrations and application of inhalation dosimetry. EPA/600/8-90/066F.

US EPA (United States Environmental Protection Agency) (2000). Science policy council handbook: Risk characterization. EPA 100-B-00-002. Offices of Science Policy & Research and Development, Washington, DC.

US EPA (Environmental Protection Agency) (December 2002a). A review of the Reference Dose (RfD) and Reference Concentration (RfC) processes. Risk Assessment Forum. EPA/630/P-02/002F. Environmental Protection Agency, Washington, DC.

US EPA (United States Environmental Protection Agency) (2002b). Determination of the appropriate FQPA safety factor(s) in tolerance assessment. US Environmental Protection Agency, Office of Pesticide Programs, Washington, DC. Available at: http://www.epa.gov/pesticides/trac/science/determ.pdf.

US EPA (United States Environmental Protection Agency) (March 2005). Guidelines for carcinogen risk assessment. EPA/630/P-03/001B. Available at: http://www.epa.gov/ncea/iris/backgr-d.htm.

US EPA (United States Environmental Protection Agency) (June 2012). Benchmark dose technical guidance document. Final Draft. EPA/100/R-12/001. Available at: http://www.epa.gov/raf/publications/benchmarkdose.htm.

5 Application of Inhalation Toxicology Concepts to Risk and Consequence Assessments

Brian E. Hawkins, David J. Winkel,
Patrick H. Wilson, and Ian C. Whittaker

CONTENTS

5.1 APPLICATION OF INHALATION TOXICOLOGY TO OCCUPATIONAL HEALTH

Prior to 1984, much effort in toxicology focused on limiting occupational exposures to hazardous materials. These early efforts produced well-known safety limits such as the permissible exposure limit (PEL), the recommended exposure limit (REL), and the threshold limit value (TLV). These limits, conservative by design, were established to protect workers from short- and long-term exposures over the course of their working lifetimes (OSHA, 2006; ACGIH, 2012; NIOSH, 2012).

- TLV: American Conference of Governmental Industrial Hygienists (ACGIH) guidelines for airborne concentrations of substances to which nearly all workers may be repeatedly exposed, day after day, without adverse health effects. TLVs can be represented by the following (ACGIH, 2012):
  - TWA: Time-weighted average exposure concentration for a conventional 8 h workday and 40 h work week for a working lifetime.
  - STEL: Short-term exposure limit represented by a 15 min average concentration, which should not result in irritation, chronic or irreversible tissue damage, or narcosis. The STEL supplements the TWA.
  - C: Ceiling exposure concentration that should not be exceeded during any part of the working lifetime.

- PEL: Occupational Safety and Health Administration (OSHA) regulation limits on the amount or concentration of a substance in the air. They may also contain a skin designation for dermal hazards. PELs can be represented by the following (OSHA, 2006):
  - TWA: Time-weighted average exposure concentration over a conventional 8 h workday for a working lifetime.
  - C: Ceiling exposure concentration that should not be exceeded.
- REL: National Institute for Occupational Safety and Health (NIOSH) recommended maximum concentrations of a chemical to which a worker could be safely exposed. RELs are also represented by the following (NIOSH, 2012):
  - TWA: Time-weighted average exposure concentration for up to a 10 h workday during a 40 h workweek.
  - STEL: Short-term exposure limit represented by a 15 min TWA concentration that should not be exceeded at any time during a workday.
  - C: Ceiling exposure concentration that should not be exceeded at any time.

Industrial hygiene limits have built-in safety factors aimed at minimizing the likelihood of adverse health effects. Such considerations are absolutely appropriate and necessary in the workplace, where low levels of exposure to hazardous chemicals are real and present safety concerns. Thus, the guidelines are set quite low to avoid any injury to the working population. Such levels could be considered similar to no effects level (NOEL) values or the effective dose resulting in a zero probability of injury to the (presumably healthy adult) working population.

5.2 APPLICATION OF INHALATION TOXICOLOGY TO EMERGENCY RESPONSE

On December 3, 1984, a leak at a pesticide plant in Bhopal, India, resulted in the release of more than 40 tons of methyl isocyanate. Approximately 3,800 people in the surrounding area died shortly after the release, and an estimated 20,000 more deaths over the next 20 years were linked to the leak (Broughton, 2005). Following the incident, the field of inhalation toxicology began to expand from industrial hygiene applications to include emergency planning and the protection of the general population. In 1986, the National Advisory Committee for the Development of Acute Exposure Guideline Levels (AEGLs) for Hazardous Substances (AEGL Committee) was formed to develop and recommend AEGLs for chemical emergencies (EPA, 2012a). In 1988, the Emergency Response Planning (ERP) Committee, originally the Organizational Resources Counselors, Inc., began producing Emergency Response Planning Guidelines (ERPGs) (AIHA, 2006). Following extensive effort in developing appropriate methods for defining exposure levels, the first AEGL Chemical Priority List was published in 1997 (EPA, 2012b). In addition to AEGLs and ERPGs, Temporary Emergency Exposure Limits (TEELs), defined by the US Department of Energy Subcommittee on Consequence Assessment and Protective Actions (SCAPA), also provided emergency planning levels. The definition and levels for each guideline are presented in the following text.

- AEGLs: AEGLs were developed for five different short-term exposure periods (10 min, 30 min, 1 h, 4 h, and 8 h). The following AEGL levels specify the airborne concentration of a substance above which it is predicted the general population, including susceptible individuals, experience the following effects (NRC, 2001):
  - AEGL-1: Notable discomfort, irritation, or certain asymptomatic, nonsensory effects. These effects, however, are not disabling and are transient and reversible upon the cessation of exposure.
  - AEGL-2: Irreversible or other serious, long-lasting adverse health effects or an impaired ability to escape.
  - AEGL-3: Life-threatening adverse health effects or death.

- ERPGs: ERPGs specify the maximum airborne concentrations below which it is believed that nearly all individuals could be exposed for up to 1 h without experiencing or developing the following effects (AIHA, 2006):
  - ERPG-1: Mild, transient adverse health effects or the perception of a clearly defined objectionable odor.
  - ERPG-2: Irreversible or other serious health effects or symptoms that could impair the ability to take protective action.
  - ERPG-3: Life-threatening health effects.
- TEELs: Exposure limits developed by the US Department of Energy SCAPA. TEELs specify the maximum airborne concentration below which it is believed that nearly all individuals could be exposed without experiencing or developing the following effects (DOE, 2008):
  - TEEL-1: Mild, transient adverse health effects or the perception of a clearly defined objectionable odor.
  - TEEL-2: Irreversible or other serious health effects or symptoms that could impair the ability to take protective action.
  - TEEL-3: Life-threatening health effects.

There are a few differences between these three general population limits that must be understood.

- AEGLs and TEELs are defined as exposure levels that result in injury, while the ERPGs are defined as exposure levels that result in no injury.
- TEELs are temporary values intended for use until AEGLs or ERPGs are adopted.
- The exposure limits are based on different time intervals. The ERPGs and (effectively) the TEELs are based on 1 h of exposure, while AEGLs are generated for multiple exposure durations including 10 min, 30 min, 60 min, 4 h, and 8 h.
- Since exposure duration is limited to 1 h for TEELs and ERPGs, they do not reflect the toxic load (TL) of the chemical if applied to other exposure durations (see Chapter 6 of this book). However, the TL is considered in the determination of the five time-dependent AEGLs.

The chemical industry has widely adopted these planning guidelines and used them to develop emergency response plans based on hazard zones indicating exceedance of these emergency planning levels. The use of these emergency planning levels to determine the risk to the population within the hazard zones or appropriate response plans following a hazardous release can be misleading. Emergency planning levels incorporate safety and uncertainty factors applied to experimental data to protect the general population (Bogen, 2005). As with the TLVs, PELs, and RELs discussed in the previous section, the concentration limits presented here can also be considered similar to a NOEL, as the concentrations represent the highest estimated concentration to not result in the adverse health effect. Furthermore, a correction factor is applied to some emergency planning guidelines to further protect susceptible individuals.

5.3   APPLICATION OF INHALATION TOXICOLOGY TO RISK AND CONSEQUENCE ASSESSMENTS

In the twenty-first century, risk and consequence assessments gained acceptance as approaches to prioritize limited resources against a broad spectrum of potential hazards. Current computing technology and predictive modeling are able to generate finely resolved concentration estimates as functions of both space and time for scenarios of interest. These concentration data define inhalation doses for potential victims and can be converted to the probability of likely injuries specific to a location and time. The knowledge of probable injuries can then be used to determine where aid is most useful following an emergency. Therefore, as the demand for performing risk and consequence assessments increases, so does the demand on toxicologists to identify injury endpoints of concern; statistical estimation of dose–response data for defined injury endpoints based on the interpretation of available toxicological information; and rigorous mathematical application of dose–response considerations. The use of occupational health or emergency planning levels fails to accurately assess the potential consequences to the population within the hazard zones.

5.3.1   IDENTIFICATION OF APPROPRIATE INJURY ENDPOINTS

The first step in assessing consequences of a scenario is to ensure that the injury endpoints suitably inform mitigation decisions. Injury endpoints are categories of injuries selected for the purposes of the assessment based on a combination of symptoms, necessary health care, or any other information that provides criteria useful for grouping injuries. The choice of injury endpoints can facilitate estimating the potential consequences and subsequent burden on the medical response system, quantifying the benefit of the medical response, or comparing outcomes across various hazard scenarios.

One example of identifying appropriate injury endpoints is the medical mitigation modeling thrust of the 2012 Department of Homeland Security Chemical Terrorism Risk Assessment (CTRA). The injury endpoints identified for the CTRA are listed as follows (Good et al., 2013):

- Life-threatening: A life-threatening injury is considered a direct threat to the individual's life; these injuries cause death if not sufficiently treated.
- Severe: A severe injury is a nonlethal injury; these victims have injuries that cause performance degradation or otherwise affect the abilities of the individual. These victims would seek care and be admitted for care under normal conditions.
- Mild-to-moderatate: A mild-to-moderate injury is a nonlethal injury; these victims have injuries of sufficient severity such that most of them would seek care under normal conditions.

The symptoms relating to the each injury endpoint for chlorine are listed as examples in the following text:

- Life-threatening: Pulmonary edema, bronchitis, and chemical pneumonia.
- Severe: Debilitating cough, bronchospasm, drooling (difficulty swallowing), and dyspnea.
- Mild-to-moderatate: Nondebilitating cough, bronchospasm, and dyspnea.

5.3.2   MATHEMATICALLY INTERPRETING DOSE-PROBIT

As discussed in the preceding section, safety thresholds (TLVs, PELs, etc.) are not applicable for estimating consequences of exposure events through modeling. The underlying assumptions of such guidelines are quite conservative (rightfully so in an occupational environment in which worker exposures should be minimized) and can lead to greatly inflated estimates of consequences. Of greater utility to predictive modeling are dose–response techniques that estimate outcomes of an exposure, specifically the probability of injuries of each level resulting from an exposure. The following sections explain in detail the application of dose–response data to predict outcomes of an exposure. Specifically, a step-by-step probit calculation to translate effective dosage to the likelihood of injury will be completed.

5.3.2.1   Probit Transform

Dose–response data, typically plotted on a log scale in the field of toxicology, is sigmoidal in shape. As explained in Chapter 6 of this book, the concepts introduced by Bliss (1934) and Finney (1947) resulted in the probit method for transforming a nonlinear dose–response curve into a straight line. Figure 5.1 illustrates the probit transformation using a hypothetical, idealized data set.

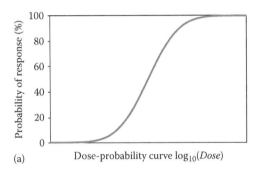

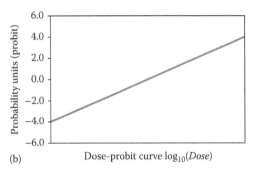

FIGURE 5.1   Illustration of the probit transformation of dose–response data. (a) Dose-probability curve and (b) dose-probit curve.

As seen in Figure 5.1, transforming the y-axis from probability of response into probability units (or *probit*) yields a linear curve. A discussion on the construction of dose-probit curves is included in Figure 5.2.

The construction of a dose-probit curve, such as the one shown in Figure 5.1, is the first step in using dose–response data for predictive modeling purposes. A linear regression of the dose-probit curve provides a line of the following form:

$$Y^{-1}(P) = b \cdot \log_{10}(Dose) + a \tag{5.1}$$

where

$$a = -b \cdot \log_{10}(Dose_{50}) \tag{5.2}$$

And $Y^{-1}(P)$ denotes the probit function, or the probit value corresponding to the probability of response P given the effective dosage denoted by $Dose$, b represents the probit slope, a represents the probit intercept parameter, and $Dose_{50}$ denotes the effective dose that corresponds to a 50% probability of response.

Although the mathematics behind the probit transform are somewhat complex, construction of the dose–probit curve in Figure 5.1 is quite simple given the right tools. For example, Microsoft® Excel allows for the simple calculation of any probit value from the corresponding probability value using the "NORMSINV" or "NORMINV" functions. Each of these functions returns the value of the probit for a given probability (assuming, as in this case, that the distribution of responses is represented by a normal distribution). The only difference between the two functions is that the NORMSINV function automatically utilizes the parameters of the standard normal distribution (i.e., a mean of 0 and a standard deviation of 1), and the NORMINV requires specification of the mean and standard deviation. A simple probit calculation for one instance of dose–response data using Excel is shown in this figure. In addition to using Excel, probit values for the z-distribution are often tabulated in statistics books, and such tables can also be found on the Internet.

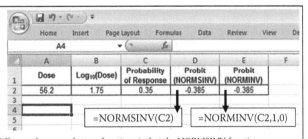

The probability associated with a given probit can also be used to calculate the probability of response. In this case, the "NORMSDIST" or "NORMDIST" functions are used. As with the functions introduced earlier, the main difference between these two functions is that the NORMSDIST function assumes the parameters of the standard normal distribution (i.e., a mean of 0 and a standard deviation of 1), and the NORMDIST requires specification of the mean and standard deviation. The NORMDIST function also has a fourth argument that must be set to "TRUE" to return the appropriate value.

FIGURE 5.2   Calculation of probit values using Microsoft® Excel.

Recall that the dose–response data were transformed under the assumption that the data can be represented by the standard normal distribution. This transform has the effect of concentrating probability density near the origin (i.e., near $Y^{-1}(P) = 0$). Substituting Equation 5.2 into Equation 5.1, with $Y^{-1}(P) = 0$ (and knowing that this must correspond to a dosage of $Dose_{50}$), allows for the calculation of an equation for the probit intercept, a, as provided in Equation 5.2.

When used in predictive modeling, alternative interpretations of the methodology in Equations 5.1 and 5.2 include P being denoted by the percentage of an exposed population that is injured by the effective dosage being denoted by $Dose$ and $Dose_{50}$ denoting the effective dose required to injure half (50%) of the exposed population.

5.3.2.2 Applying Probit Methodologies to Predictive Modeling

Assuming the probit parameters (the slope and the intercept) are known quantities, the application of a dose-probit curve for predictive modeling is straightforward. For example, consider the probit parameters representing a life-threatening injury caused by inhalation exposure to chlorine (Table 5.1). Note that for this example, the $Dose$ parameter can be approximated using the ten Berge TL model (discussed in detail in Chapter 6 of this book) by Juergen Pauluhn to estimate inhalation dosages with the consideration of both the duration of the exposure and the severity of the exposure through the use of the TL exponent (ten Berge and van Heemst, 1983; ten Berge et al., 1986).

Using the data in Table 5.1, Equation 5.1 is rewritten as

$$Y^{-1}(P_{LT}) = b_{LT} \cdot \log_{10}(Dose) + a_{LT} = 6.0 \cdot \log_{10}(Dose) - 62.5 \tag{5.3}$$

The LT subscripts in the terms in Equation 5.3 denote life-threatening injuries. Equation 5.3 can also be written in the following form, which has been solved for P_{LT}.

$$P_{LT} = Y(b_{LT} \cdot \log_{10}(Dose) + a_{LT}) = Y(6.0 \cdot \log_{10}(Dose) - 62.5) \tag{5.4}$$

In Equation 5.4, the Y operator indicates that the subsequent value in parentheses is a z value (not a probability value, P). Consider a hypothetical constant inhalation exposure to chlorine of 900 mg/m^3 for a duration of 60 min, yielding an effective $Dose$ of $(900 \text{ mg/m}^3)^n \times 60$ min. Given the value of the TL exponent ($N = 2.8$), the $Dose$ is calculated to be 1.1×10^{10} mg-min/m^3 (i.e., a $\log_{10}(Dose)$ of 10). The dose-probit curve for chlorine is plotted in Figure 5.3, and the example $\log_{10}(Dose)$ is indicated by a dotted line.

The probability of an individual sustaining a life-threatening injury from the $Dose$ can then be calculated as 1.4% using Equation 5.4, reproduced with specific numeric values in Equation 5.5:

$$P_{LT} = Y\left(6.0 \cdot \log_{10}\left((900 \text{ mg/m}^3)^{2.8} \cdot 60 \text{ min}\right) - 62.5\right) = Y(-2.19) = 1.4\% \tag{5.5}$$

TABLE 5.1

Probit Parameters for Inhalation Exposure to Chlorine Resulting in Life-Threatening Injury

Parameter	Value
Probit slope ($\log_{10}$), b_{LT}	6.0
$Dose50_{LT}$ [(mg/m^3)^N(min)][a]	2.6×10^{10}
Probit intercept, a_{LT}	−62.5

[a] The exponent N represents the TL exponent, assumed to equal 2.8.

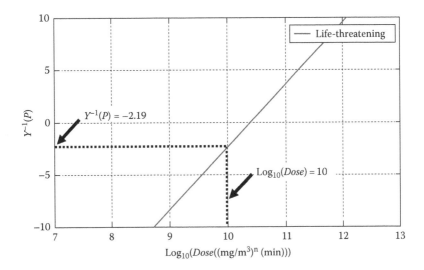

FIGURE 5.3   Graphical illustration of chlorine probit parameters (single probit).

When applied to predictive modeling, the probability of life-threatening injury calculated in Equation 5.5 can be applied to the entire exposed population (assuming a homogeneous population). For example, if 1000 people were exposed in this hypothetical scenario, 14 of the people can be assumed to have suffered life-threatening injuries.

5.3.2.2.1   Extension of Probit Methodologies to Multiple Injurious Levels

A large area hazard will result in a variety of injuries to the exposed population. Not all injuries, for example, will be of the life-threatening type. Some will be rather mild, and others will be debilitating but not life threatening. The performance of multiple probit calculations to account for these additional injury types is possible and requires additional mathematical and logical considerations beyond the simple probit relationship displayed previously (in addition to the identification of probit parameters for each injury type). The probit relationship in the preceding equations provides the probability of an injury at least as serious as the stated injury type. Therefore, to avoid multiple counting of victims, the probability of any particular injury level must be adjusted by subtracting the probability of all of the more serious injuries at that particular dose.

Ideally, probit parameters would be known for every distinct or desired injury level or endpoint. In reality, only a limited number of unique dose-probit curves can be generated empirically for a given chemical due to the difficulty associated with estimating these values. The following analysis incorporates chlorine dose-probit data for three distinct injury types: life-threatening, severe, and mild-to-moderate (MM) injuries. Equations 5.6 through 5.8 show the necessary adjustments to the previously introduced equations in this chapter for the handling of multiple probits. Equation 5.6, a reproduction of Equation 5.4, allows for the calculation of the probability of life-threatening injury and is unchanged. Equations 5.7 and 5.8 allow for the calculation of the probability of severe (SEV) and mild-to-moderate injuries, respectively, by subtracting the probability of the more serious injury level(s).

$$P_{LT} = Y(b_{LT} \cdot \log_{10}(Dose) + a_{LT})$$ (5.6)

$$P_{SEV} = Y(b_{SEV} \cdot \log_{10}(Dose) + a_{SEV}) - P_{LT}$$ (5.7)

TABLE 5.2

Probit Parameters for Inhalation Exposure to Chlorine Resulting in Life-Threatening, Severe, and Mild-to-Moderate Injuries

	Injury Type–Specific Value		
Parameter	Life-Threatening (LT)	Severe (SEV)	Mild-to-Moderate (MM)
Probit slope ($\log_{10}$), b	6.0	6.0	6.0
$Dose50$ [$(mg/m^3)^N(min)$][a]	2.6×10^{10}	4.1×10^9	1.1×10^9
Probit intercept, a	−62.5	−57.7	−54.2

[a] The exponent N represents the TL exponent, assumed to equal 2.8 for all injury types. Note that although it is identical for all injury types in this calculation, it can be injury level dependent for other chemicals.

$$P_{MM} = Y(b_{MM} \cdot \log_{10}(Dose) + a_{MM}) - P_{LT} - P_{SEV} \qquad (5.8)$$

The probit parameters representing the different injury levels caused by inhalation exposure to chlorine from the 2012 CTRA are shown in Table 5.2.

When compared to thresholds or guidelines such as AEGLs or ERPGs, injury endpoint dose-probit data provide a better estimate of injuries resulting from a large-scale exposure. Figure 5.4 denotes threshold limits and median probit values for multiple injury endpoints for a short-duration exposure. The injury endpoints require concentrations about three orders of magnitude higher than several widely accepted threshold/guideline values. Use of the threshold values, therefore, predicts significantly more injuries (likely to be orders of magnitude higher than the number of injuries predicted by dose-probit data).

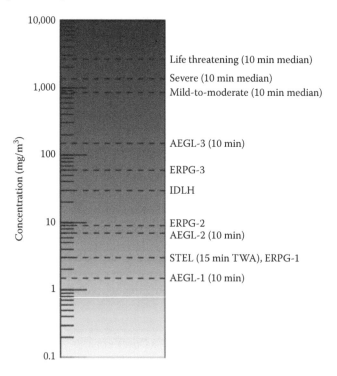

FIGURE 5.4 Short exposure time thresholds and median probit values for chlorine.

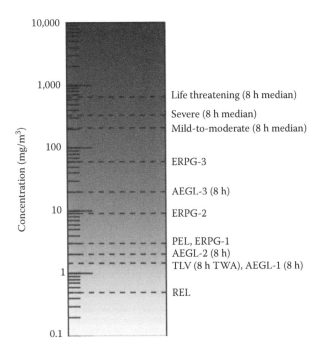

FIGURE 5.5   Long exposure time thresholds and median probit values for chlorine.

For long exposure times (8 h), it is still true that using threshold values to model injuries will provide inflated consequence results. For the example of chlorine, Figure 5.5 depicts injury endpoint concentrations that are multiple orders of magnitude higher than their occupational guideline and emergency response counterparts (e.g., severe injury concentrations are roughly two to three orders of magnitude higher than the ERPG-2 and AEGL-2).

The example calculation provided earlier for the probability of life-threatening injuries can now be extended to the consideration of additional injury types. As shown earlier, the probability of an individual sustaining a life-threatening injury can be calculated as 1.4% (Equation 5.9). The probability of an individual sustaining a severe injury can then be calculated as 98.2% (Equation 5.10). The probability of an individual sustaining a mild-to-moderate injury can then be calculated as 0.4% (Equation 5.11). Although the probabilities in this example sum to 100%, this result is merely due to the nature of the probit curves. They are quite far apart from one another, and the selected dosage term was chosen to ensure that injuries for each type were realized in this example. These results are shown graphically in Figure 5.6.

$$P_{LT} = Y\left(6.0 \cdot \log_{10}\left((900 \text{ mg/m}^3)^{2.8} \cdot 60 \text{ min}\right) - 62.5\right) = Y(-2.19) = 1.4\% \qquad (5.9)$$

$$P_{SEV} = Y\left(6.0 \cdot \log_{10}\left((900 \text{ mg/m}^3)^{2.8} \cdot 60 \text{ min}\right) - 57.7\right) - 1.4\%$$

$$= Y(2.62) - 1.4\% = 99.6\% - 1.4\% = 98.2\% \qquad (5.10)$$

$$P_{MM} = Y\left(6.0 \cdot \log_{10}\left((900 \text{ mg/m}^3)^{2.8} \cdot 60 \text{ min}\right) - 54.2\right) - 1.4\% - 98.2$$

$$= Y(6.05) - 99.6\% = 100.0\% - 99.6\% = 0.4\% \qquad (5.11)$$

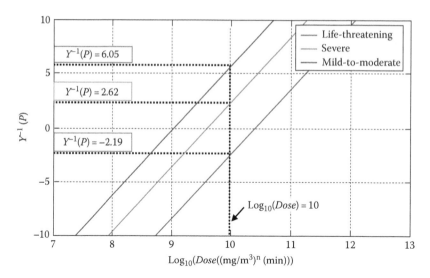

FIGURE 5.6 Graphical illustration of multiple chlorine probit parameters and corresponding injury probabilities (multiple probits).

Different injury levels can be the result of a chemical or other hazard acting on different organ systems or through different modes of actions. The probit parameters may have different slopes, TL exponents, etc., to reflect this variability. As a result, it is possible for the probit curves to intersect in a given region of the parameter space, or for a lesser injury curve to fall completely below a more serious injury curve. In either case, application of Equations 5.7 and 5.8 would produce a negative probability of injury, which of course is mathematically impossible. For this reason, Equations 5.7 and 5.8 can be modified such that the probability of injury is never less than zero. This modification is reflected in Equations 5.12 and 5.13, where a *Max* operator is used to implement a zero probability of injury if the injury probability resulting from the probit calculation was negative. Effectively, this additional logic forces lesser injury probit curves to be equal to or greater than their more serious injury counterparts. Note that no adjustment for the intersection of probit curves is needed for the most serious injury level (life-threatening injury).

$$P_{SEV} = Max\left[Y\left(b_{SEV} \cdot \log_{10}(Dose) + a_{SEV}\right) - P_{LT}, 0\right] \qquad (5.12)$$

$$P_{MM} = Max\left[Y(b_{MM} \cdot \log_{10}(Dose) + a_{MM}) - P_{LT} - P_{SEV}, 0\right] \qquad (5.13)$$

Suppose that the previously introduced mild-to-moderate probit parameters for chlorine were altered, as shown in Figure 5.7, to allow for increased probability of such injuries at low effective doses.

Clearly, for the same *Dose* considered in previous example calculations, the mild-to-moderate probit curve is now below the severe probit curve. While the probability of life-threatening and severe injuries is the same as previously shown, the probability of an mild-to-moderate injury is now 0% as shown in Equation 5.14.

$$P_{MM} = Max\left[Y(1.25) - 1.4\% - 98.2\%, 0\right] = Max\left[89.4\% - 99.6\%, 0\right] = 0\% \qquad (5.14)$$

5.3.3 INCORPORATING STATISTICAL UNCERTAINTY

Typically, the treatment of uncertainty in inhalation toxicology involves the application of order-of-magnitude uncertainty factors to translate data from experimental animal studies to protective

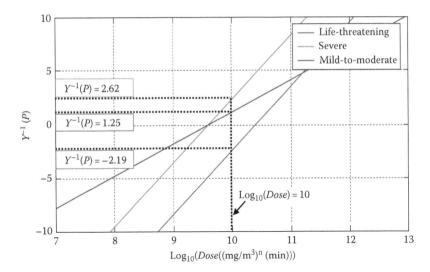

FIGURE 5.7   Graphical illustration of adjusted probit parameters and corresponding injury probabilities (multiple intersecting probits).

guidelines or industrial hygiene exposure standards for humans. As a result, the term uncertainty in toxicology is generally associated with categorical reductions in the dose required to incur injury or effect. While such reductions are appropriate for establishing intentionally conservative threshold values corresponding to the potential onset of injuries, incorporation of uncertainty in dose–response parameters used for predictive modeling requires a different approach.

Uncertainty is more appropriately incorporated into modeling efforts through the estimation of statistical confidence bounds on dose–response parameters. For example, Figure 5.8 shows the results of a probit-based analysis of experimental dose–response data using standard commercial software, such as Microsoft® Excel or SYSTAT® SigmaPlot, in which the uncertainty in the median dose (Ct_{50}) and probit slope have been incorporated via statistical regression.

Although there are more rigorous methods employing weighting schemes, the Regression Data Analysis capability within Microsoft® Excel provides a readily available method to estimate the confidence intervals of dose–response data. The effect of incorporating uncertainty in this fashion transforms a single dose-probit curve into a density of dose-probit curves, which can be visualized

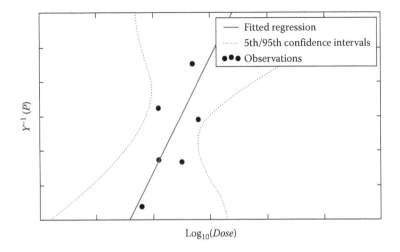

FIGURE 5.8   Example of probit-based regression incorporating uncertainty.

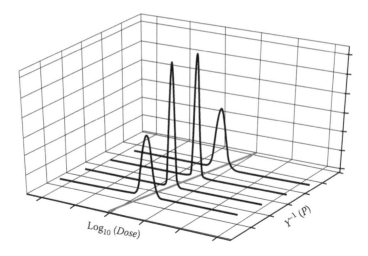

FIGURE 5.9   Three-dimensional visualization of uncertainty in probit analysis.

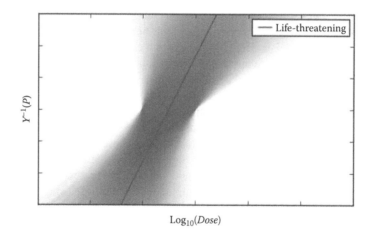

FIGURE 5.10   Color density visualization of uncertainty in probit analysis.

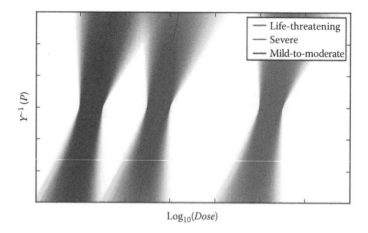

FIGURE 5.11   Visualization of dose–response uncertainty for multiple probit analyses.

using a third dimension (Figure 5.9) or color density scheme (Figure 5.10). Note that, in contrast to uncertainty considerations in protective guidelines that offset the estimated values to lower doses, incorporation of uncertainty in dose–response efforts for consequence assessment creates a Gaussian spread of uncertainty around the experimental observations. This concept can further be applied to multiple probits as illustrated in Figure 5.11. Subsequent modeling efforts can then sample from the resulting distributions of probit parameters using Monte Carlo techniques to provide the best possible estimate of consequences consistent with the constraints of the available data.

5.4 SUMMARY

The application of inhalation toxicology concepts to inform consequence mitigation strategies through predictive modeling of health outcomes is an essential facet of twenty-first century toxicology. This requires specification of endpoints of concern; statistical estimation of dose–response data describing those endpoints based on toxicological information; and rigorous mathematical application of dose–response relationships. Misapplication of toxicological concepts can lead to exaggerated estimates of consequences that are likely to confuse or even mislead decision makers. This error is frequently encountered in emergency response and dispersion modeling analyses that characterize consequences based on protective guidelines, such as AEGLs. Such analyses significantly overestimate consequences. Decisions made based on this type of misuse of exposure guidelines can actually exacerbate consequences by driving critical resources to be focused on less impactful activities. Only multidisciplinary approaches that merge mathematical, statistical, and toxicological techniques can ensure that predictive modeling provides the best possible guidance through a basis in sound toxicology concepts.

QUESTIONS

1. Using the chlorine probit information found in Table 5.2, determine the probability of having a life-threatening injury when exposed to a chlorine concentration of 1500 mg/m³ for 30 min.
   Answer: 39% probability of life-threatening injury
2. If the probability of having a life-threatening injury is 80%, what is the concentration of chlorine that an individual has been exposed to for a period of 90 min?
   Answer: 1180 mg/m³ concentration of chlorine
3. What are the probabilities of life-threatening, severe, and mild-to-moderate injuries resulting from an exposure to a chlorine concentration of 2000 mg/m³ for 10 min.
   Answer: 0.01% probability of mild-to-moderate injury, 85% probability of severe injury, and 15% probability of life-threatening injury
4. When establishing occupational guidelines, a factor of 10 is often used when data are sparse or when extrapolating between species.
   (a) What would the probabilities of injury be for question 3 if the median exposures *Dose 50s* were a factor of 10 higher (less toxic)?
   Answer: 87% probability of mild-to-moderate injury, 1.3% probability of severe injury, and 0% probability of life-threatening injury if all three D50s are a factor of 10 higher (less toxic).
   (b) What if they were a factor of 10 lower (more toxic)?
   Answer: 0% probability of mild-to-moderate injury, 0% probability of severe injury, and 100% probability of life-threatening injury if all three D50s are a factor of 10 lower (more toxic).
5. Because they are estimated from limited dose-response data, probit slopes often represent a significant source of uncertainty.
   (a) What would the probabilities of injury be for question 3 if the probit slopes were a factor of 2 higher?
   Answer: 0% probability of mild-to-moderate injury, 98% probability of severe injury, and 2% probability of life-threatening injury if all three probit slopes are a factor of 2 higher.

(b) What if they were a factor of 2 lower?
Answer: 2.9% probability of mild-to-moderate injury, 66.8% probability of severe injury, and 30.3% probability of life-threatening injury if all three probit slopes are a factor of 2 lower.

6. Answer the following questions based on the figure below:
   a. Assuming that severity of injury increases with dose, identify the curves that correspond to the most and least severe injury types.
      Answer: Curve 1 represents the least severe injury type. Curve 3 represents the most severe injury type.
   b. Assume that Curves 1, 2, and 3 correspond to mild-to-moderate, severe, and life-threatening injury types, respectively. Given the shallow slope of Curve 1 compared to the slope of Curves 2 and 3, what can be concluded about the toxic mechanism and/or target organs of a mild-to-moderate injury compared to a severe or life-threatening injury?
      Answer: The different slope indicates a different toxic mechanism and/or target organ for mild-to-moderate injury and creates a high capacity for these injuries to occur.
   c. Based on the relative proximity/spacing of Curves 2 and 3, what can be concluded about the potential for the injury type represented by Curve 2?
      Answer: Relatively low capacity/window for Curve 2 injury type (Severe).

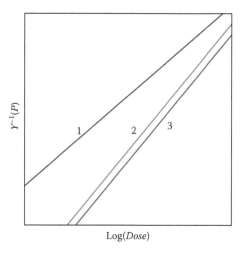

Log(*Dose*)

REFERENCES

ACGIH (American Conference of Governmental Industrial Hygienists). 2012. Products—TLV® chemical substances introduction. https://www.acgih.org/Products/tlvintro.htm (accessed March 26, 2013).

AIHA (American Industrial Hygiene Association). 2006. AIHA ERP committee procedures and responsibilities. http://www.aiha.org/insideaiha/GuidelineDevelopment/ERPG/Documents/ERP-SOPs2006.pdf (accessed March 26, 2013).

Bliss, C.I. 1934. The method of probits. *Science* 79:38–39.

Bogen, K.T. 2005. Risk analysis for environmental health triage. *Risk Anal* 25:1085–1095.

Broughton, E. 2005. The Bhopal disaster and its aftermath: A review. *Environ Health* 4:6.

Finney, D.J. 1947. *Probit Analysis: A Statistical Treatment of the Sigmoid Response Curve.* Cambridge, U.K.: University Press.

Good, K., D. Winkel, M. VonNiederhausern, B. Hawkins, J. Cox, R. Gooding, and M. Whitmire. 2013. Medical mitigation model: Quantifying the benefits of the public health response to a chemical terrorism attack. *J Med Toxicol* 9(2):125–132.

NIOSH (National Institute for Occupational Safety and Health). 2012. NIOSH pocket guide to chemical hazards, introduction. http://www.cdc.gov/niosh/npg/pgintrod.html (accessed March 26, 2013).

NRC (National Research Council) Subcommittee on Acute Exposure Guideline Levels. 2001. Standing operating procedures for developing acute exposure guideline levels for hazardous chemicals (executive summary). Washington, DC: The National Academies Press.

OSHA (Occupational Safety and Health Administration). 2006. Occupational safety and health standards, standard 1910 subpart Z—Toxic and hazardous substances, 1910.1000—Air contaminants. http://www.osha.gov/pls/oshaweb/owadisp.show_document?p_table=STANDARDS&p_id=9991 (accessed March 26, 2013).

ten Berge, W.F. and M.V. van Heemst. 1983. Validity and accuracy of a commonly used toxicity-assessment model in risk analysis. In *Fourth International Symposium on Loss Prevention and Safety Promotion in the Process Industries*, Vol. 1, Institute of Chemical Engineers, Rugby, U.K.

ten Berge, W.F., A. Zwart, and L.M. Appelman. 1986. Concentration-time mortality response relationship of irritant and systemically acting vapours and gases. *J Hazard Mater* 13:310–309.

US DOE (Department of Energy). 2008. Temporary emergency exposure limits for chemicals: Methods and practice. DOE-HDBK-1046-2008. http://www.hss.doe.gov/nuclearsafety/techstds/docs/handbook/DOE-HDBK-1046-2008.pdf (accessed March 26, 2013).

US EPA (Environmental Protection Agency). 2012a. History of the AEGL program. http://www.epa.gov/opptintr/aegl/pubs/history.htm (accessed March 26, 2013).

US EPA (Environmental Protection Agency). 2012b. Chemical priority lists. http://www.epa.gov/opptintr/aegl/pubs/priority.htm (accessed March 26, 2013).

6 Time Scaling of Dose and Time to Response
The Toxic Load Exponent

Juergen Pauluhn

CONTENTS

6.1 INTRODUCTION

The concept that the product of the concentration (C) of a substance and the length of time (t) animals are exposed to it produces a fixed level of effect for a given endpoint has been ascribed to Ferdinand Flury and Fritz Haber, who described the behavior of war gases (Witschi, 1999) in the early 1900s. While it was recognized that $C \times t = k$ was applicable only under certain conditions, many toxicologists have used this rule to analyze experimental data whether or not the respective chemicals, biological endpoints, bioassay, and exposure scenarios were suitable candidates for the rule. In the early days of toxicology, the mathematical solutions were straightforward using a log–log plot of C and t composed of two or more rectilinear segments, for when the log–log plot was curvilinear, and for when the slope of the dosage–mortality curve was a function of C. Experimental data from this relationship suggest that C and t are interchangeable within the short time limits given, but that under specific conditions, concentration becomes more important than time in predicting toxicity. Such *dose-rate-dependent variables* may come into play when the *inhaled dose* is not any longer proportional to C because the rate of *intake* is affected by time-related changes in ventilation. Likewise, *intake* does not necessarily mean *uptake*, which, for gases, is retention dependent. Retention changes with time due to the t-dependent saturation of the immediate compartment of *uptake*. Hence, with increasing exposure duration, the tissues of the respiratory tract become increasingly saturated, and inhaled gases, depending on the chemical reactivity and water solubility, may be exhaled without being retained. At steady state, the retention of an inhaled gas decreases toward a plateau, and ventilation becomes less important than perfusion. This picture might further be complicated by time-dependent changes in wash-in/wash-out equilibria. Accordingly, $C \times t$ relationships have to be analyzed and interpreted with caution and need to be distinguished for exposure durations representing the early yet non-steady-state condition and that occurring at steady state. When substituting the toxic load exponent n = a/b in $C^a \times t^b$, this equation is simplified to $C^n \times t = constant\ effect\ outcome$, which appears to embrace most of the variables affecting the dose rate of uptake.

The toxic load exponent can hardly be considered chemical-specific as n appears to be contingent on species-specific changes in ventilation and uptake. Also the time window the empirical data are determined is important and whether t is sufficiently long to attain steady state or not. Thus, the toxic load exponent "n" cannot be solely attributed to C but rather is a composite derivation of power functions depending on both C and t (see also Miller et al., 2000). These authors also noted that quantification of the key factor, inhaled dose, which is responsible for the effect occurring, remains a complex endeavor in inhalation toxicology.

Especially for respiratory tract irritants, this picture is further complicated by intensity–time *plateauing* with time during which physiological effects adapt within a relatively short period of inhalation exposure. Depending on whether the test substance is volatile, an aerosol, chemically reactive, and/or sufficiently water soluble, a C-dependent anterior–posterior gradient of site-specific dosing and injury within the respiratory tract may occur. Due to location-specific differences in the relative abundance of *irritant receptors*, including the site-specific inhaled dose, the hypoventilation commonly observed in rodents may invert to hyperventilation in humans. The implications of these variables affecting the temporal dynamics of portal-of-entry dosing within a heterogeneous system such as the respiratory tract have incompletely been explored and elucidated to date. The objective of this chapter is to analyze the extent to which C- and t-dependent physiological changes in rats and mice occur when exposed to sensory irritants and whether they can be considered to be a major cause for deviations from Haber's rule.

6.2 BACKGROUND ON REQUIREMENTS FOR TIME SCALING

Toxicity is a function of exposure, which depends on concentration and noninterrupted exposure duration (time). In cases when an inhaled substance is efficiently eliminated or is instantly reacted away at the initial site of deposition within the respiratory tract, the integrated injury (i.e., effect) accumulates over time and is directly proportional to time and dose rate. This condition may not be fulfilled for intermittent exposures when detoxifying or adaptive pathways come into play. The complexity associated with time scaling increases precipitously when addressing intermittent repeated exposure patterns because there must be an accounting for kinetic factors. Thus, when elimination, repair, and reconstitution can take place within the time period of interest, the accumulated dose may not be directly proportional to the ultimate effect or the duration of the total exposure time. Figure 6.1 delineates that an intermittent high-level exposure over 1–4 weeks may result in C × t–proportional lung burdens and toxic effects accordingly. At that point in time where the fractional amount cleared between each exposure interval matches the fractional inhaled dose, the steady state has been attained. Then the body or lung burden becomes independent of the duration of exposure.

It is far more difficult to characterize an inhaled dose than a dose from any other route of administration. With oral or parenteral routes, a discrete amount of a chemical is commonly given in a bolus. Conversely, the inhaled dose depends on the substance's physicochemical properties, exposure concentration, ventilation pattern, and additional factors that determine the deposition, retention, and clearance of inhaled substances within the respiratory tract. Deposition patterns within the various regions of the respiratory tract are also important and may present themselves in a concentration × time-dependent manner.

Despite similar C × t relationships, the actually inhaled dose may differ in either a C- or a t-dependent manner. Changes in ventilation often occur C-dependently due to the stimulation of afferent nociceptive reflexes. In small laboratory rodents, such stimulation goes along with hypoventilation, whereas in humans the opposite response is often observed. The extent of change depends on whether these afferents are triggered within the upper and/or lower respiratory tract. Similarly, also the depth of penetration of vapors or gases into the tract is concentration dependent,

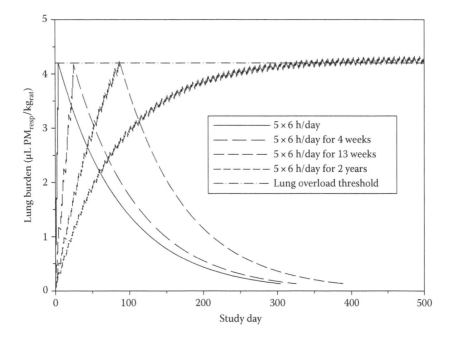

FIGURE 6.1 Modeling of cumulative lung burdens of rats exposed to respirable particle (PM_{resp}) volume concentrations of 0.86, 0.24, 0.105, and 0.069 $\mu L/m^3$ 5 days/week for 1, 4, 13, or 104 weeks, respectively. The volumetric overload threshold is defined as 4.2 μL PM_{resp}/kg_{rat}. (From Pauluhn, J., *Toxicology*, 279, 176, 2011a.)

especially in small laboratory rodents where reactive and water-soluble gases are scrubbed to an appreciable extent within the upper airways. Time-dependent variables have to be accounted for when retention changes within the time period of interest. For instance, ammonia, a highly water-soluble gas, is predominantly retained in the upper airways. In context with human inhalation studies with ammonia, it was concluded that when the rise in concentration of ammonia in the absorbing lining fluids/surface tissues of the upper respiratory tract was becoming more saturated with this gas, the retained percentage decreased to 20% (Silverman et al., 1949). Consequently, the deduction that inspired ammonia would go progressively deeper into the respiratory tract was not borne out by the symptoms observed. Presumably, the inspired ammonia continued to be removed by the upper passages, which then returned progressively larger amounts to expired air as it passed over these surfaces, consistent with a wash-in/wash-out phenomenon. Opposite to marked hypoventilation in small rodents (see Figure 6.2, upper panel), humans exposed to 500 ppm of ammonia elaborated a substantial hyperventilation.

An inhalation risk assessment needs to address multiple exposure scenarios, but not all of them are supported by empirical data that exactly mirror the exposure pattern of interest. These exposure scenarios may range from brief acute exposures, as occur in accidental situations, to continuous exposures to environmental contaminants, as well as recurrent intermittent exposures at the workplace. The objective of time scaling in inhalation toxicity assessment actually attempts to adjust *accumulated inhaled dose* at different *continuous acute* exposure patterns.

Especially for inhalation risk assessments that focus on the derivation of acute Emergency Response Guidance Values, such as the Emergency Response Planning Guidelines (ERPGs) or Acute Exposure Guidance Levels (AEGLs), a quantitative appreciation is required for the impact of concentration × time (C × t) on the point of departure (POD) for nonlethal effects (e.g., degree of transient pulmonary inflammation or edema) and lethal effects (e.g., statistically derived lethal threshold concentration at 1% or 5%, which is LC_{01} or LC_{05}). Experimental/mathematical

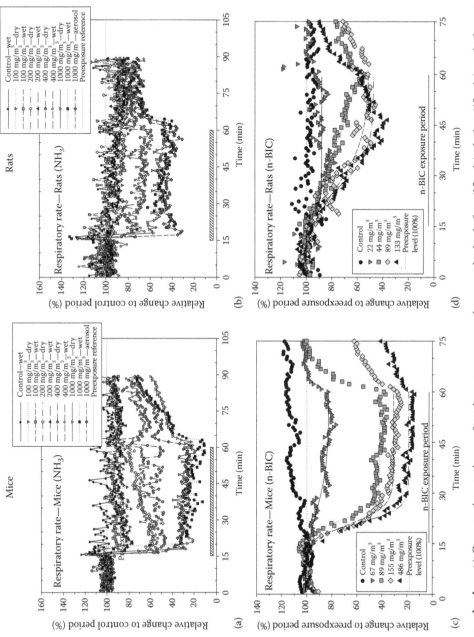

FIGURE 6.2 **(See color insert.)** Comparative respiratory function measurements in rats and mice before (15 min), during (45 min), and after exposure to (a and b) the upper respiratory tract water-soluble irritant ammonia (NH₃) gas and (c and d) the respiratory tract moderately water-soluble irritant n-BIC (vapor). Data represent means of four animals/group.

procedures are available for the estimation of time-adjusted median lethal concentration (LCt_{50}) values for multiple exposure durations. However, these procedures often fail when applied to LCt_{01} and LCt_{05}. Estimations of these values have had to rely upon subjective comparisons and multiple assumptions that vary between experts, with the result that much controversy surrounds this activity. Consequently, toxicologists have been somewhat reticent to provide these estimates in the absence of state-of-the-art concentration × time–response data and an understanding of the toxicity endpoint's principal mode of action. Relationships found valid in rodent bioassays may not be valid for humans if the species-specific variables of *inhaled dose* and associated metric of dose are insufficiently characterized.

6.3 TIME DEPENDENCE OF TOXICOLOGICAL OUTCOMES

Rodent inhalation studies that use nose-only exposure systems make it possible to monitor time-course changes of key physiological endpoints contemporaneous with exposure. The primary endpoints of interest are changes in body temperature and cardiopulmonary functions. Particularly in rodent inhalation studies, it is important to recognize potential biases in toxic outcome that may be caused or modulated by changes in body temperature. Unlike humans, small laboratory rodents can reflexively undergo a hypothermic response when exposed to respiratory tract irritants due to their small thermal inertia (Figure 6.3). This irritation-related hypothermia often coincides with a reduction in ventilation (Figure 6.2), heart rate (see Chapter 19 of this book on Phosgene),

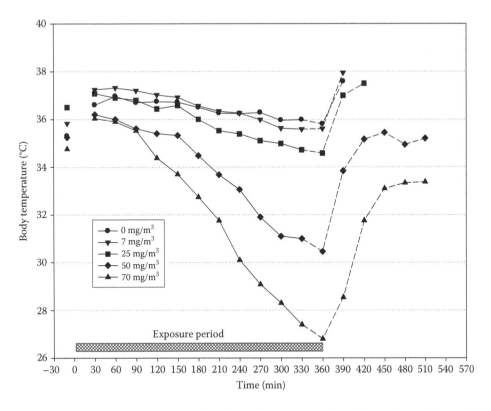

FIGURE 6.3 Nose-only 6 h exposure of mice to increasing concentrations of a respiratory irritant aliphatic monoisocyanate. Body temperatures were measured before exposure (–30 to 0 min.) and during exposure (>0 to 360 min) in nose-only restrainer tubes (hashed bar). After cessation of the irritant stimulus and normal housing body temperatures increased again. Body temperatures were measured with subcutaneously implanted transponders, which provide manipulation-free data collection.

and metabolism. Each endpoint follows its own species-specific inertia. These responses, collectively described as reflex bradypnea, may confer a higher resistance to irritant toxicants. Due to their small body weight, mice are markedly more thermolabile than rats (Gordon et al., 2008).

Rats exposed to vapors of the aliphatic monoisocyanate n-butylisocyanate (n-BIC) or the aromatic toluene diisocyanate (TDI) showed a concentration-dependent decreased respiratory rate of as much as 70% of the preexposure level (Figure 6.4). For respiratory tract irritants, the tidal volume decreases at low concentrations and reverses to an increase at concentrations high enough to gain access to the alveolus (Figure 6.4). This appears to be a typical reflection of upper respiratory tract sensory irritation and afferents originating from trigeminal nerves (Kraetschmar reflex) that depress ventilation. At high concentrations, lower respiratory tract irritation (Paintal reflex) occurs and is manifested by a shallow, high-frequent breathing pattern. The product of both commonly is a decreased respiratory minute volume (MV; Figure 6.5). Despite the somewhat maximum depression in the respiratory MV, the depth of penetration may vary from one concentration to another and cannot reliably be estimated from the measurements shown. Thus, adjustments of the inhaled dose in relation to the exposure concentration and time, as well as the resulting integrated physiological response, are far from trivial.

These species-specific differences need to be appreciated when using small-rodent animal models; otherwise such small-rodent-specific changes secondary to hypothermia may be inappropriately considered adverse. As evidenced in Figures 6.2 and 6.5, depressed respiration occurs almost instantly with the onset of exposure to sensory irritant concentrations but becomes more time dependent at lower concentrations, especially for less water-soluble vapors such as n-BIC (Figure 6.2, lower panel). Mice attain stable breathing patterns somewhat faster and more pronounced than rats due to their two-and-a-half-times higher respiratory rate (Figure 6.2). For the highly water-soluble ammonia gas, the time to attain the maximum effect appears to be more dependent on the water solubility of the irritant rather than the species. The increase in concentration of water-soluble gases and vapors can trigger J-receptor reflexes within lower respiratory tract commonly evidenced by an increased tidal volume (Figure 6.4). As shown for the reactive but poorly water-soluble gas phosgene (Figure 6.5, lower panel), the reflexively induced depression in the ventilation of rats may wane shortly after the onset of exposure. This makes short exposure durations dosimetrically more variable and less precise with a general tendency toward overestimation of the inhaled dose (and underestimation of toxicity). Despite the marked changes in ventilation, only minimal changes in the respiratory rate occurred. This suggests that the variations in ventilation are attributed to changes in tidal volume.

The concentration-dependent changes in tidal volume and respiratory rate suggest that each C × t may have its preferential site of injury as the *depth of penetration* of vapor or gas depends more on concentration than on exposure duration. Consequently, the principal site of injury and mode of action is concentration-dependent as illustrated for the onset of mortality of n-BIC-exposed rats (see Figure 6.6). Exposure durations up to 1 h utilized concentrations of n-BIC high enough to penetrate the lower airways with resultant acute lung edema as the cause of death within 1-day postexposure. With lower concentrations at proportionally longer exposure durations of 4 and 6 h, this mortality pattern changed entirely toward late-onset patterns on postexposure days 8–14 in the absence of any edema-related early mortality. This delayed type of mortality is typical for irritation-related airway injury and ensuing obliterating inflammation (*bronchiolitis obliterans*) and mucus production, which eventually leads to airway plugging and a lethal mismatch of the alveolar ventilation: perfusion relationship.

These examples demonstrate that C × t relationships need to be analyzed thoughtfully while bearing in mind that the same endpoint (mortality) may involve several critical site-of-dosing/injury-dependent mode of actions. Such dependence seems to be restricted to reactive, water-soluble gases and vapors. Conversely, nonreactive lipophilic gases, vapors, and aerosols do not show similar penetration-dependent *scrubbing effects* within the airways so the critical dosing site is that with the

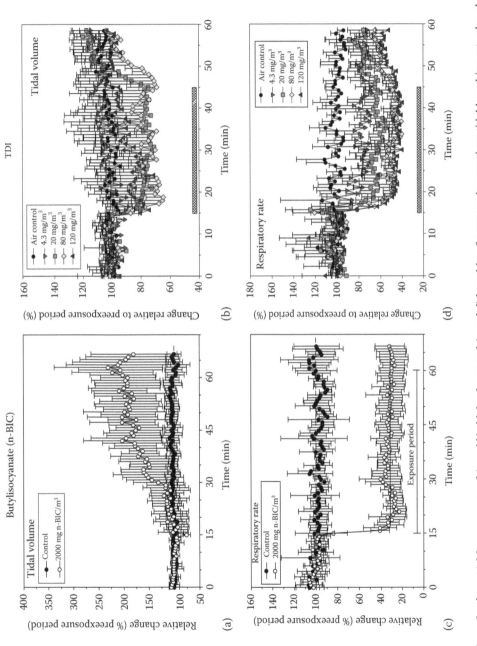

FIGURE 6.4 (See color insert.) Measurements of (a and b) tidal volume and (c and d) breathing frequency in conscious tidal breathing rats placed into volume displacement plethysmographs before, during, and after exposure to the vapor phase of aliphatic n-butyl(monoisocyanate (n-BIC) and aromatic toluene diisocyanate (TDI). The n-BIC- and TDI-exposed animals were exposed to the sensory irritant for 45 and 30 min, respectively. Data represent averages ±SD of at least four rats/group.

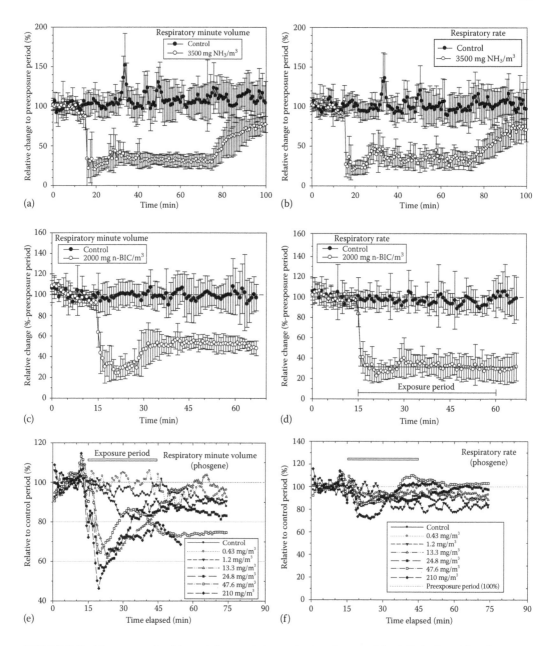

FIGURE 6.5 Measurements of (a, c, e) the respiratory minute volume and (b, d, f) the breathing frequency of conscious tidal breathing rats in volume displacement plethysmographs before, during, and after exposure to the upper respiratory tract irritant ammonia (gas), the respiratory tract irritant n-BIC (vapor), and lower respiratory tract irritant phosgene (gas). Data represent means±SD of four animals/group (±SD omitted in the lower panel).

highest surface area and vulnerability to injury—the alveoli. Of note is these substances show close adherence to Haber's rule since the absence of upper respiratory tract sensory irritation minimizes C × t-dependent changes in ventilation and retention. With these physiological variables in mind, mathematical models for C × t analyses need to be applied thoughtfully and with caution prior to any default extrapolation from rodents to humans.

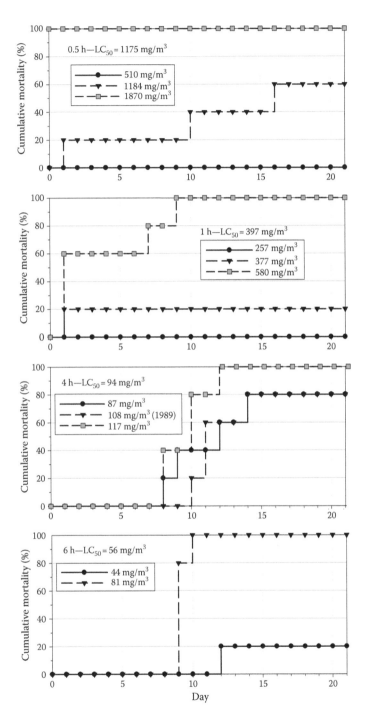

FIGURE 6.6 C × t–dependence of the occurrence of mortality during a 3-week postexposure period in rats exposed for 0.5, 1, 4, or 6 h to the vapor phase of the aliphatic n-butyl(mono)isocyanate. The median lethal concentrations (LC_{50}) were calculated by probit analysis.

6.4   HABER'S RULE AND THE TOXIC LOAD EXPONENT

As alluded to earlier text, Haber's rule (Haber, 1924) is based on the early work by Flury (1921) and is commonly understood in inhalation toxicology as a constant product of the two factors *exposure concentration in the inspired air* (C) and the *duration* (t) during which this concentration is inhaled to yield an identical intensity of biological response (Witschi, 1999). In the original work from Flury (1921), a third factor had been considered, namely, the actual volume inhaled by the animal during the exposure period (t). Retention-related attributes of inhaled dose at non-steady-state and steady-state conditions were not considered at that time. The negligence of these test system–specific variables implies that the inhaled volume of a specified concentration of a hazardous substance in air is constant across exposure groups and need not be considered any further. This simplification is subject to challenge, especially when analyzing C × t relationships of irritant gases in small laboratory rodents that are known to instantly depress their ventilation concentration dependently while opposite changes may occur in humans. Historically, Haber's rule has been used for concentration × time extrapolations assuming that each unit of damage is irreversible, that no repair takes place during the exposure period and, therefore, that each unit of exposure is 100% cumulative. However, this is generally not the case for acutely toxic responses to short-term exposures.

Haber's rule was further refined and validated using an array of structurally diverse chemicals by ten Berge et al. (1986, 1989) and Zwart et al. (1988, 1990) by introducing the toxic load exponent (n) to combine the exposure concentration (C) and exposure time (t) using the following equation: $C^n \times t = k$, where k is a numerical constant. Exponential weighing factors are incorporated to better mathematically describe the relative contribution of C and t (Miller et al., 2000); however, for practicability, the toxic load "n" is a combination of both. This empirical relationship must be derived from experimental values and is therefore limited by the experimental data available. This equation can be generalized using a two-variate surface plot relating toxicity (mortality) and time as follows: $y = b_0 + b_1 \log(C) + b_2 \log(t)$, where $n = b_1/b_2$. The "n" is considered to be chemical-specific and toxic endpoint-specific exponent. An assessment by ten Berge et al. (1986) of LC_{50} data for an array of chemicals demonstrated chemical-specific relationships between exposure concentration and exposure duration that were often exponential.

The *dependent variable* "y" representing the effect commonly is probit transformed and b_0, b_1, and b_2 are empirically derived constants. The definite advantage of this approach is that the entire matrix of empirical data can be used to calculate the LCt_{50} and any fraction thereof. Nonetheless, it does not resolve the issue whether any deviation from Haber's rule (which requires n = 1) is primarily related to C, to inhaled cumulated dose-related factors, or to truly t-dependent factors. So far, limited published data are available to better understand the ventilation-related *variables of dosing* relative to the LCt_{50} and toxic load exponent.

The initial experimental validation of Haber's rule involved the endpoint of pulmonary edema and mortality in experimental animals exposed to phosgene gas, although this endpoint has subsequently been modeled for chlorine and other pulmonary irritants (ten Berge, 1986; Zwart and Woutersen, 1988). Similarly, more recent experimental studies on rats demonstrate that for pulmonary irritant vapors (e.g., phosgene, Pauluhn, 2002, 2006a,b; Pauluhn et al., 2007) and pulmonary irritant aerosols (polymeric methylenediphenyl-4,4′-diisocyanate, Pauluhn, 2002), Haber's rule is followed for both lethal and nonlethal endpoints. These findings appear to support the notion that pulmonary irritants, as long as the exposure time is long enough to *neutralize* the initially observed transient depression in ventilation (e.g., see Chapter 19 of this book on Phosgene and Figure 6.5), follow the $C^n \times t$ paradigm with n = 1 because the inhaled factional exposure concentration $C_I = C \times$ respiratory MV is constant over time and the critical sites of initial injury are independent of C. In contrast, more water-soluble and/or reactive airway irritants have, in addition to the airways of upper respiratory tract, at least two additional C-dependent sites of injury that have to be considered, namely, the bronchial airways and the alveoli. This goes along with an additional C-dependence on the occurrence of mortality as detailed in Figure 6.6. These findings call for a modified approach that takes into account the C × t-interrelationship of the specific endpoint or POD chosen and the magnitude of irritation-related changes of the inhaled dose observing critically whether the selected concentration is high enough to penetrate the distal airways C- or C × t-dependently.

6.4.1 Haber's Rule, Irritation, and Variables That Affect Dosimetry

The relationships of C-dependent changes in ventilation due to sensory irritation have to be put into the context of the respective POD derived from the experimental data. In case a maximum decrease in ventilation has already been attained below the C-range of interest, as shown for ammonia (Figure 6.5), ventilation-dependent deviations from Haber's rule might not be occurring to any appreciable extent. However, in case this prerequisite is not fulfilled, test species–specific modifiers of uptake of the inhaled substance need to be considered. This is usually the case for moderately to less water-soluble gases and vapors, such as n-BIC and phosgene (Figure 6.5). For ammonia, the LC_{01} on rats at 10, 30, 60, and 240 min exposure duration was 19,727, 11,442, 8,113, and 4,080 mg/m^3, respectively. These concentrations are higher than the RD_{50} of 900 mg/m^3 with little differences whether the rats were exposed to the dry gas or ammonia dissolved in an aqueous aerosol. There was a general tendency that ammonia in dry air was a slightly more potent sensory irritant. Accordingly, maximum depression of ventilation can be expected at all lethal exposure levels. Therefore, the deviation from n = 1 is likely to be related to any time-related diminution of retention as well as C-dependent changes in retention sites with differing susceptibilities rather than changes in respiratory MV.

Applying this rationale to the logic of the ERPGs results in the following ERPG-2 (the maximum concentration in air below which it is believed that nearly all individuals could be exposed for up to 1 h without experiencing or developing irreversible or other serious health effects or symptoms that could impair their abilities to take protective action) and ERPG-3 (the maximum concentration in air below which it is believed that nearly all individuals could be exposed for up to 1 h without experiencing life-threatening health effects) values for ammonia. The ERPG-3 is derived as follows:

$$\text{ERPG-3} = \text{POD}_A \times \frac{\text{MV}_A}{\text{MV}_H} \times \frac{r_A}{r_H} \times \frac{1}{\text{AF}} = 10{,}067 \frac{\text{mg}}{\text{m}^3} \times \frac{0.8 \times 0.3}{0.25 \times 3} \times \frac{1}{3}$$

$$= \frac{3{,}221}{3} \frac{\text{mg}}{\text{m}^3} \approx \frac{4{,}500}{3} \text{ppm} = 1{,}500 \text{ ppm}$$

where POD_A is based on this rat inhalation study (1-h LC_{01} = 10,067 mg/m^3), MV_A/MV_H is the ratio of the minute ventilation (kg body weight–based) of rats (A) and humans (H) adjusted to the respective maximum hypo-(A) and hyperventilation (H) expected to occur. In rats, the maximum depression of ventilation is 30% of normal (Figure 6.5), while that of humans is 300% of normal (Silverman et al., 1949) of the normal ventilation. In rats and humans, the latter is defined as 0.8 and 0.25 L/kg-min, respectively (Bide et al., 2000; Pauluhn and Thiel, 2007). Steady state was reported to occur in humans within 10–30 min (at 500 ppm), and a similar range is also expected to occur in rats. Accordingly, species-specific differences in retention (r_A/r_H) cancel out. The site of major gas retention at 1500 ppm is still confined to the upper respiratory tract. Therefore, no adjustment factor (AF) is deemed necessary for any *breakthrough* into the posterior airways and preexisting disease present at this location. Due to the very steep concentration–mortality relationship of ammonia and the fact that laryngospasm and glottis edema have been reported to occur in humans in the range of or exceeding 1700 ppm (DFG, 1986), an AF = 3 of the above effect threshold (4500/3 ppm = 1500 ppm) is considered to be defensible as an ERPG-3. This estimated threshold appears to be not at variance with the reported lethal range of 2500–6500 ppm (exposure duration 30 min) in humans (DFG, 1986). The respective AEGL-3 equivalent is 1100 ppm (AEGL, 2007). These similarities give credence to the approach taken.

Similar analyses were made for the respiratory tract irritant n-BIC and lower respiratory tract phosgene. While n-BIC-exposed rats displayed a stable plateau of decreased respiration similar to ammonia, phosgene elicited just a transient response (Figure 6.5). Each change decreases the inhaled dose with an impact on "n." For the respiratory tract irritant n-BIC, the 4 h LC_{50}: RD_{50} ratio in rats is 60 mg/m^3: 40 mg/m^3 = 1.5. Based on the depression in MV shown for n-BIC (Figures 6.2 and 6.5, the "n" of the LC^nt_{50} is likely to be affected by C-dependent changes in ventilation. When adjusting the actually measured exposure concentration "C" by the proportional decrease in respiratory minute ventilation ($C' = C^n \times (1 - f_{MVd})$, with f_{MVd} being fractionally decreased respiratory minute ventilation relative to normal baseline, the scaling to the actually inhaled C' results in n = 1.04 instead of n = 0.73 as delineated in Figure 6.7. This outcome corroborates the contention that "n" represents more a correction factor for the "*inhaled dose* rather than any chemical-specific exponent.

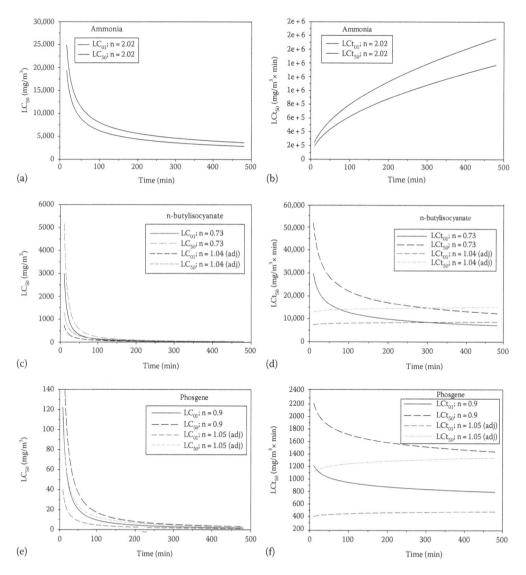

FIGURE 6.7   **(See color insert.)** $C^n \times$ t–dependence of the median lethal concentration (LC_{50}) and LCt_{50} products as well as the respective nonlethal threshold concentrations (LC_{01} and LCt_{01}) of rats exposed to (a and b) the upper respiratory tract irritant ammonia (gas) (see Figure 6.5, a and b), (c and d) the respiratory tract irritant n-BIC (vapor) (see Figures 6.5, c and d), and (e and f) lower respiratory tract irritant phosgene (gas) (see Figure 6.5, e and f). The actual exposure concentrations were analyzed as measured or were adjusted for the time-dependent depression in ventilation (adj.).

When transient changes in respiration occur as observed in phosgene-exposed rats, the "n" seems to be independent of the total duration of exposure as long as the exposure time is long enough to compensate for transient changes occurring at the start of exposure (see Figure 6.5). Accordingly, the pulmonary toxicity is likely to be underestimated using high and short $C^n \times t$ products. Under such conditions, the outcome is biased to yield n < 1. Likewise, at longer exposure durations, breathing patterns become normal with "n" approaching 1.

This demonstrates that under exposure conditions that trigger rodent-specific protective reflexes, rodents may receive an appreciably lower dose than equally exposed larger species. Another factor that influences "n" is that higher concentrations are likely to travel deeper into the lung. The $C \times t$–related local injury of the bronchial airways and alveoli often results in biphasic mortality patterns. Similar to the time-course patterns of survival shown in Figure 6.6, alveolar injury is reflected by early onset of mortality at exposure concentrations higher than 117 mg/m^3, whereas at lower concentrations, the toxic outcome is predominated by airway injury. These examples show that the toxic load exponent "n" is dependent on the experimental regimen chosen, the extent as to which the key determinant of the inhaled dose, the ventilation, is affected, and additionally whether the experimental window chosen changes from nonsteady state to steady state with decrease in retention. In sum, this analysis supports the hypothesis that the toxic load exponent "n" is contingent on the sensory irritant potency of the inhaled substance and associated rodent-specific depression in ventilation. If not accounted for, the toxicity of inhaled substances with sensory irritant potential is subject to underestimation at very short durations of exposure.

6.4.2 HABER'S RULE, EXPOSURE REGIMEN, AND SPECIES DIFFERENCES

The examples shown demonstrate that rodent-specific factors may affect inhalation dosimetry that can only be overcome in using larger nonrodent species. This is common praxis for inhalation pharmaceuticals but the exception for chemical substances. The extrapolation of PODs from small laboratory rodent bioassays using a high C and short t exposure regimen is increasingly complex due to the rodent-specific nociceptive responses described earlier. It is commonly believed that human exposure patterns likely to occur under accidental chemical release conditions can at least partially be modeled in bioassays using larger animal species. However, this often requires individually engineered solutions to address the specific features of both the test substance and the test species. Comparison of different species is sometimes difficult to perform since the mode of exposure used for small rodents and for larger laboratory animals, such as dogs or monkeys, is not necessarily identical. Animal exposure units must fit properly to prevent passive exposure of technical staff. Larger animals usually identify any physical weakness of the exposure system and will abuse it. These factors mandate the construction of tight-fitting face masks and/or effective restraint of larger animals, which might induce distress, alter the breathing behavior, and make the required analytical characterization of the animals' atmosphere within the breathing zone difficult or even impossible. Further problems are salivation into the tubing system of the face mask (dogs) and the dead space of the tubing system in relation to the species-specific tidal volume. These considerations demonstrate that weight-of-evidence approaches should not necessarily go with the weight of test species used to evaluate the $C \times t$–dependence of inhaled substances rather than the robustness of the animal model chosen. Therefore, before applying data from any specialized experimental approach, the strengths and weaknesses of each course taken must be fully understood and implemented.

6.5 IMPLICATIONS FOR HAZARD IDENTIFICATION AND RISK ASSESSMENT

This analysis shows that the most robust empirical inhalation data are generated when employing minimally irritating concentrations of a sufficient duration of exposure. Under such conditions, concentration-dependent physiological changes that would decrease the inhaled dose are minor, and the extrapolation across species is associated with least uncertainties. At irritant levels of exposure, rodent-specific secondary changes cannot be ruled out with an impact on dosimetry and/or metabolism. Such variables add an additional dimension of complexity to the extrapolation of toxicological data from small rodents to humans.

The thermal instability observed in mice and rats has rarely any corollary in humans due to their markedly higher body weight thermal inertia. The nociceptive reflex systems in obligate nasal breathing rodents, which protect the lower respiratory tract from inhaling noxious substances, cannot directly be translated to oronasally breathing humans. The kinetics of *loading* also is highly species dependent, taking into account that rats breathe about 100–150 breaths/min, while humans breathe at a markedly lower frequency. Nociceptive reflexes are triggered by afferent sensory neurons that then carry nerve impulses from the sense organ toward the central nervous system. Although afferent reflexes are also operative in larger species, the resultant efferent response to nociception is often translated and phenotypically manifested differently and species specifically. Sensory irritation phenomena have been observed in humans (Shusterman et al., 2006), but they appear to be more confined to mere *sensations* and the associated psychological and/or psychophysical sequelae. The reflexively mediated ostensive protective changes in ventilation are often opposite to those observed in rodents.

The experimental data shown in Figures 6.5 and 6.7 support the conclusion that extrapolation of data from acute, short high-concentration exposure regimens is biased to deliver outcomes that underestimate toxicity due to rodent-specific depressed ventilation. Consequently, time adjustments based on acute long exposure durations and lower concentrations appear to be inherently more conservative than regimens using the inverse approach. Likewise, for longer-term exposures at which steady state is attained, the exposure time per se is increasingly less important relative to the fractional dose inhaled and retained per unit of time (Figure 6.1). However, despite an equal inhaled dose, the retention decreases as well when approaching the steady state. Most regulatory approaches, in the absence of other supporting data to determine the reasonableness of any alternative time adjustment, recommend that a toxic load exponent of $n = 1$ is to be used when extrapolating from shorter to longer exposure durations, and a value of $n = 3$ is to be used when extrapolating from longer to shorter durations (NRC, 2001). This analysis is subject to challenge as long as the attributes *ventilation vs. retention*, which cause the deviation from $n = 1$, are yet incompletely characterized and understood.

6.6 CONCLUSION

To date, neither Haber's rule nor other more generalized power law models have been systematically evaluated with respect to their predictions across species for reactive water-soluble and nonreactive/less soluble airway or pulmonary irritants under varying $C \times t$ conditions at equal exposure intensities. Indeed, the $C^n \times t = k$ concept produces a fixed level of effect for a given endpoint assuming all dose-related variables remain constant across different concentrations and species. Especially for respiratory tract irritants, the rodent-specific depression in ventilation must be accounted for by overproportionally increased experimental concentrations at short exposure durations to attain the desired *equivalence of effect*. This then produces experimental conditions differing increasingly from Haber's rule. Therefore, it appears to be timely to reconsider Flury's early call for ventilation and dosimetry adjustments in $C \times t$ analyses as articulated one century ago.

In summary, broadening of models to take into account the full range of information available regarding the physical chemistry, dosimetry, and biological impact of specific airborne toxicants will likely produce more robust successors to Haber's rule for the explanation of concentration–time relationships in experimental inhalation toxicology. Thus, physiology-based pharmacokinetic modeling may become an increasingly important tool for toxicologists to better understand and execute time adjustments across species.

QUESTIONS

1. Haber's rule is fulfilled when $C^n \times t$ = constant for n = 1. What is meant by *constant*?
   (a) Constant endpoint
   (b) Constant effect
   (c) The inhaled dose
   Answer: b

2. The toxic load exponent "n" of the equation $C^n \times t$ = constant is believed to better match the $C^1 \times t$ = constant relationship. What is the role of the exponent "n"?
   (a) A mathematical correction factor to better fit the data.
   (b) A substance-specific potency factor for toxicity.
   (c) A substance-specific potency factor for irritation-induced changes in ventilation.
   Answer: c

3. You want to reproduce brief high-level accidental exposures to an irritant substance in an animal inhalation bioassay. Which of the following animal species and exposure regimen would you consider most appropriate?
   (a) Mouse—exposure duration 0.5, 1, and 4 h.
   (b) Monkey—exposure duration 1, 2, and 5 min.
   (c) Rat—exposure duration 4 h.
   Answer: c

4. You have to anticipate the toxic load exponent "n" of an (I) irritant lipophilic vapor or gas with a (II) highly reactive nonvolatile irritant highly respirable aerosol. What appears to be the most common outcome?
   (a) n = 1 for both I and II
   (b) n < 1 for I and n = 1 for II
   (c) n # 1 for both I and II
   Answer: a

5. The toxic load exponent "n" in the equation $C^n \times t$ = constant was defined for lethality in rats.
   (a) The same "n" can be applied for any other endpoint in any species.
   (b) The same "n" can be applied for any other endpoint but cannot be extrapolated across species.
   (c) The "n" is endpoint specific and can only be extrapolated across species when the principal mode of action is the same and appreciable differences in dosimetry do not occur.
   Answer: c

REFERENCES

AEGL. 2007. Acute exposure guideline levels for selected airborne chemicals, Vol. 6, Ammonia. Prepared by the National Research Council of the National Academies. The National Academies Press, Washington, DC, pp. 58–114.

Bide, R.W., Armour, S.J., and Yee, E. 2000. Allometric respiration/body mass data for animals to be used for estimates of inhalation toxicity to young adult humans. *J. Appl. Toxicol.* 20:273–290.

DFG. 1986. The MAK collection for occupational health and safety: MAK value documentation for ammonia updated 1973, 1986, 1993, 1996, 1999, and 2000. Available at: Wiley Online Library (http://onlinelibrary.wiley.com/book/10.1002/3527600418/topics). Accessed April 8, 2014.

Haber, F. 1924. Zur Geschichte des Gaskrieges (On the history of gas warfare). In: *Fünf Vorträge aus den Jahren 1920–1923 (Five Lectures from the Years 1920–1923)*, Springer, Berlin, pp. 76–92.

Flury, F. 1921. Über Kampfgasvergiftungen. I: Über Reizgase. *Z. Ges. Exp. Med.* 13:1–15.

Gordon, C.J., Spencer, P.J., Hotchkiss J., Miller, D.B., Hinderliter, P.M., and Pauluhn, J. 2008. Thermoregulation and its influence on toxicity assessment. *Toxicology* 244:87–97.

Miller, F.J., Schlosser, P.M., and Janszen, D.B. 2000. Haber's rule: A special case in a family of curves relating concentration and duration of exposure to a fixed level of response for a given endpoint. *Toxicology* 149:21–34.

NRC (National Research Council). 2001. *Standing Operating Procedures for Developing Acute Exposure Guideline Levels for Hazardous Chemicals*. National Academy Press, Washington, DC.

Pauluhn, J. 2000a. Acute inhalation toxicity of polymeric diphenyl-methane-4,4'-diisocyanate (MDI) in rats: Time course of changes in bronchoalveolar lavage. *Arch. Toxicol.* 74:257–269.

Pauluhn, J. 2000b. Short-term inhalation toxicity of polyisocyanates aerosols in rats: Comparative assessment of irritant-threshold concentrations by bronchoalveolar lavage. *Inhal. Toxicol.* 14:287–301.

Pauluhn, J. 2006a. Acute nose-only exposure of rats to phosgene. Part I: Concentration × time dependence of LC_{50}s and non-lethal-threshold concentrations and analysis of breathing patterns. *Inhal. Toxicol.* 18:423–435.

Pauluhn, J. 2006b. Acute nose-only exposure of rats to phosgene. Part II: Concentration × time dependence of changes in bronchoalveolar lavage during a follow-up period of 3 months. *Inhal. Toxicol.* 18:595–607.

Pauluhn, J. 2011a. Poorly soluble particulates: Searching for a unifying denominator of nanoparticles and fine particles for DNEL estimation. *Toxicology* 279:176–188.

Pauluhn, J., Carson, A., Costa, D.L., Gordon, T., Kodavanti, U., Last, J.A., Matthay, M. A., Pinkerton, K.E., and Sciuto, A.M. 2007. Workshop summary—Phosgene-induced pulmonary toxicity revisited: Appraisal of early and late markers of pulmonary injury from animals models with emphasis on human significance. *Inhal. Toxicol.* 19:789–810.

Pauluhn, J., Eben, A., and Kimmerle, G. 1990. Functional, biochemical, and histopathological evidence of airway obstruction in rats following a four-hour acute inhalation exposure to n-butyl isocyanate. *Exp. Pathol.* 40:197–203.

Shusterman, D., Matovinovic, E., and Salmon, A. 2006. Does Haber's law apply to human sensory irritation? *Inhal. Toxicol.* 18:457–471.

Silverman, L., Whittenberger, J.L., and Muller, J. 1949. Physiological response of man to ammonia in low concentrations. *J. Ind. Hyg. Toxicol.* 31:74–78.

ten Berge, W.F. and Zwart, A. 1989. More efficient use of animals in acute inhalation toxicity testing. *J. Hazard Mater.* 21: 65–71.

Witschi, H. 1999. Some notes on the history of Haber's rule. *Toxicol. Sci.* 50:164–168.

Zwart, A., Arts, J.H.E., Klokman-Houweling, J.M., and Schoen, E.D. 1990. Determination of concentration-time-mortality relationships to replace LC50 values. *Inhal. Toxicol.* 2:105–117.

Zwart, A. and Woutersen, R.A. 1988. Acute inhalation toxicity of chlorine in rats and mice: Time–concentration–mortality relationships and effects on respiration. *J. Hazard Mater.* 19:195–208.

7 Improper Use of Haber's Law Results in Erroneous Fatality Estimation from Predictive Models*

Tom Ingersoll, Janet Moser, Douglas R. Sommerville, and Harry Salem

CONTENTS

7.1 INTRODUCTION

The dose of chemical toxicant to which a poisoned victim is exposed over time is a function of the specific toxicant, its concentration, route of exposure (e.g., inhalation, ingestion, and skin absorption), duration of exposure, and the physiological state of the victim (Eaton and Gilbert 2013). Physiological state can include many factors such as age, sex, genetic makeup, health, and nutritional status. Outstanding among physiological attributes is the victim's species. For inhalation exposure within a species, the important predictor variables are the concentration of the toxicant and exposure duration (Haber 1924; Prentiss 1937). There are two common measures for inhalation exposure to a vapor/gas toxicant: the product of concentration and time and the toxic load.

The historical and most often used estimator for inhalation exposure is the product of concentration and time.

$$K = C \times T \qquad (7.1)$$

where
K is a constant representing toxic load
C is the concentration of chemical to which an individual is exposed
T is the duration of exposure

---

* The views and opinions expressed in this chapter are those of the author and should not be construed as an official Department of the Army position, policy, or decision, unless so designated by other official documentation. Citation of trade names in this chapter does not constitute an official Department of the Army endorsement or approval of the use of such commercial items.

Equation 7.1 is called Haber's law (Haber 1924) and is often used to predict the probability of death following exposure. Haber's law was developed in the early 1920s. However, it was soon determined that Ct was inadequate to account for changes in toxicity as a function of exposure duration. Thus, the toxic load was introduced and subsequently was popularized in the 1980s (ten Berge et al. 1986). The ten Berge model for toxic load is expressed by the following formula:

$$K = C^n \times T \tag{7.2}$$

The value n is the toxic load exponent and is a property that differs for each toxicant. The toxic load exponent determines the relative weighting of concentration vs. duration when determining the toxic load. When n is less than 1, increased duration of exposure has a stronger effect on toxic load than increased concentration. When n is greater than 1, increased concentration has a greater effect than increased exposure duration. A special case occurs when n is equal to 1, where Equation 7.2 is reduced to the formula for its predecessor, Haber's law. The value of n can be determined from experimental data using maximum-likelihood estimation (Sommerville et al. 2006).

Most historic animal model toxicity studies use variable concentration at fixed exposure duration, often 10 min or less. Predictive models based on such data will lose precision when exposures diverge from the special case of fixed duration. Studies with variable exposure durations must be performed in order to estimate toxic load exponents and therefore increase the reliability of prediction for more general cases. A well-executed example of a toxic load study with variable exposure duration, providing detailed explanation of parameter estimation including the toxic load exponent, is described by Mioduszewski et al. (2001). A general description of the mathematics of toxic load is provided by Sommerville et al. (2006).

One shortcoming of both the ten Berge and Haber methods is that they assume constant exposure concentration across time. This may be approximately the case when toxic gases or aerosols are well mixed and released into a confined space such as a test chamber. However, when fixed amounts of such chemicals are released into the open, exposure concentration drops with time as the chemical disperses over an increasingly larger area. In this case, concentration for some small time interval may be approximated as a function of time $C(t)$, and overall toxic load becomes the sum of toxic load across all time intervals. Toxic load for arbitrarily small time intervals may then be represented by the following expression: (Sommerville et al. 2006)

$$K = \int_0^T C(t)^n dt \tag{7.3; Sommerville et al. 2006}$$

where the toxic load is integrated across all time intervals from $t = 0$ through $t = T$ (other terms remain the same as in Equation 7.1). When a fixed amount of aerosol disperses over a spatially uniform population, it encounters a larger population with each time step; meanwhile, the concentration of the chemical decreases as the volume it occupies increases. The result is an exposure concentration that changes over time and unequal exposure within the population, where those individuals closest to the release location receive to higher toxic loads than those further away. Those individuals located a short distance from the release point may be exposed to lethal concentrations of a toxicant, whereas at great distance, toxic loads may be small enough to be sublethal.

This unequal exposure within a population affects differences in fatality estimation between Haber and ten Berge-based models. When a toxicant has a toxic load exponent greater than 1, low toxicant concentrations at a distance from the point of release tend to make lethality estimates from Haber models larger than those from ten Berge model. Small initial releases dispersing over large populations generally result in overestimates when using Haber's law, as most known toxic load exponents are greater than 1. However, Haber models may give underestimates in special cases when large releases occur over concentrated populations, depending on the initial amount of toxicant released, its rate of dispersal over the population, its toxic load exponent, the spatial arrangement of the population, and the duration of exposure.

A relationship between toxic load and expected number of fatalities (or casualties, in general) can be established experimentally using laboratory animals (Boyland et al. 1946; Mioduszewski et al. 2001) or derived from historic data of accidental exposure in humans (Nelson 2006). Data required are counts of deaths vs. survival, chemical concentrations, and exposure duration. Probability of mortality is then modeled using binomial generalized-linear models with a probit link function and maximum-likelihood parameter estimation. The ten Berge formulation for this model is

$$\text{Probit}(\Pr(m)) = \beta_0 + \beta_1 C^n T \tag{7.4}$$

where
Pr(m) is the probability of mortality
β_0 and β_1 are scalar coefficients

A model version based on Haber's law sets the exponent n to a value of 1. An important consequence of fitting this model to data is that the intercept value, β_0, may be unrealistically estimated at a nonzero value. This can adversely affect fatality estimates at low toxic loads. Linear models, such as those based on Haber's law, may be more likely to erroneously estimate the intercept value than models where curvature is more flexible, such as with the ten Berge formulation.

In typical practice, and in deference to traditional dose–response methods, independent variables C and T are often log-transformed before fitting the model:

$$\text{Probit}(\Pr(m)) = \beta_0 + n \times \log(C) + \log(T) \tag{7.5}$$

Note, however, that this transformation changes the value of the intercept β_0 and confounds estimation of the coefficients β_0 and β_1 for back-transformation into Equation 7.4. Unconfounded estimation of the intercept β_0 is important as it measures background mortality, where toxicant concentration is sublethal. For this reason, the untransformed formulation given by Equation 7.4 may be preferred for fatality predictions. Nonfixed time interval inhalation toxicity studies are rare enough that standard formulation is not well established in literature.

Akaike's information criterion (AIC; Akaike 1973) is an objective method for comparing and selecting between models. Calculating AIC scores compares the relative likelihood of models, while also penalizing model complexity. AIC is used to select the model that provides the best fit and explanation for the results, without adding an excess of model terms. Unlike the more traditional tests of hypothesis, such as likelihood-ratio tests or analysis of variance, AIC may be used to select between models when terms are not nested (Burnham and Anderson 2002). AIC may be more appropriate than tests of hypothesis when the analytic goal is to estimate parameters, such as the toxic load exponent, rather than establish the confidence level for a null hypothesis. As such, AIC is a suitable tool for selecting between Haber's law and ten Berge's toxic load model.

When the toxic load exponent does not equal 1, fatalities are erroneously estimated when the toxic load is modeled on Haber's law because time is unrealistically weighted against concentration. For the many chemicals with toxic load exponents greater than 1, the Haber model can overestimate fatalities at low concentrations or long durations and underestimate fatalities at high concentrations or low durations (Sommerville et al. 2006). This relationship should reverse for those few chemicals with toxic load exponents known to be less than 1. In all cases, fatalities at low toxic loads may be erroneously estimated when model intercept values are nonzero or otherwise diverge from background mortality.

Chlorine provides an example where casualty predictions for accidental releases differ from what has been observed historically. Although it is acknowledged that Haber's law does not apply to chlorine, many of the toxicity estimates used with currently popular atmospheric transport and dispersion models are based upon Haber's law. Utilizing the toxic load model, a new human estimate for chlorine inhalation lethality was derived, and the impact of the new estimate was evaluated through a series of transport and dispersion modeling runs (Sommerville et al. 2010). Using the new estimate, the predicted downwind hazard distances were consistent with what has been witnessed historically.

7.2 METHODS

We used published data from Mioduszewski et al. (2001) to estimate lethality in response to toxic load. In this study, rats are used as the animal model for determining the probability of mortality in response to toxic loads following exposure to the chemical warfare agent, sarin. The data describe rat fatalities at exposure concentrations ranging from 2.1 to 54.4 mg/m^3 and exposure durations from 5 to 360 min. Time and concentration are scaled so that longer durations occur at lower concentrations. The study uses 10 rats at 34 different concentration/time levels, for a total of 340 rats.

We used a maximum-likelihood method to select the exponent for a toxic load model that produced the highest model likelihood for data given by Mioduszewski et al. (2001). Assuming that rat mortality is a binomially distributed response, we calculated the binomial probability for each observed outcome, death or survival, across a range of toxic load exponent values. The likelihood was calculated as the product of these probabilities for all the rats in the study. The toxic load exponent value that gave the highest likelihood was its maximum-likelihood estimate.

We then fit two models to the rat data: a Haber model with an exponent (n) of 1 and a toxic load model with the exponent we estimated via maximum likelihood. We also compared information content of the two models using AIC. Parameter values were extracted from these models to estimate the probability of mortality in relation to toxic load and for use in lethality estimation in simulated populations (see later).

First, we simulated a spatially uniformly distributed population of rats in the open, occupying a 200 m circle, with 10 rats per m^2. At time = 0, we simulated a release of 3 g of sarin in the center of the circle. We used a simple model of uniform radial diffusion, where the sarin formed cylinders of fixed height and well-mixed uniform concentration. The cylinders increased in diameter with each time step, so that per meter concentration diminished as the cylinder increased in volume with time, and the additional number of rats exposed to toxicant increased with each time step. We used 20 intervals of 1 min each and diffusion of 1 m per minute. (Note that this was not meant to represent realistic diffusion rates but provided the simplest diffusion model since diffusion was not our parameter of interest.) We calculated total lethality predicted by the toxic load and Haber models and graphed these as time series for comparison.

Second, we simulated the release of 3 g of sarin down a 10 m wide tunnel over a spatially uniformly distributed rat population of 10 rats per m^2, at uniform 1 m per minute dispersal rate, dispersing down a single linear axis, at a well-mixed uniform concentration, and for a 20 min duration. This scenario allowed the exposed area to add the same number of new rats exposed with each time step, with the concentration dropping at a constant rate as the affected area increased.

Third, we increased the initial release to 5 g of sarin over the same tunnel population, for a duration of 10 min, calculating lethality predicted by the two models. Finally, we simulated initial release over a range of 1–6 g and exposure durations from 1 to 20 min. We calculated the difference, between Haber and toxic load model predictions across these ranges, and graphed the difference as time series.

For each scenario, we calculated two space-by-time matrices that represented sarin exposure in the rats for each simulated population. One matrix used a toxic load exponent of 1 (Haber's law), and one matrix used a toxic load exponent of 1.6 (toxic load). Matrix values were the sum of accumulated dose at each distance from the release point and at each time point. For simplicity, we assumed that no amount of chemical was eliminated by the rats.

We used toxic load and Haber formulations of the binomial generalized-linear models to predict lethality across their respective exposure matrices. We then compared lethality predictions between the two models, assuming that the model selected by AIC provided the most realistic predictions.

7.3 RESULTS

Maximum-likelihood estimation indicated that a toxic load exponent (n) of 1.6 fit the highest likelihood for the observed data. AIC values for a ten Berge model with an exponent (n) of 1.6, compared to a Haber model with an exponent (n) of 1, are presented in Table 7.1 and indicate that the Haber model is virtually uninformative compared to the ten Berge model.

Graphs of lethality probabilities extracted from the models indicated nonzero intercept values for both the ten Berge and Haber models (Figure 7.1). Thus, both models unrealistically predicted some fatalities even when toxic loads were zero, resulting in fatality overestimates at very low toxic loads. This was a consequence of fitting the model to all of the data, rather than fitting locally to the extreme value at the intercept. The more flexible ten Berge model produced an intercept closer to zero, so that low-dose fatality overestimates would be less than those from the Haber model (Figure 7.1).

Simulation of radial diffusion across a circle of rats resulted in predicted survival of 18.63% of the population using the Haber model and 69.90% using the ten Berge model (Figure 7.4a). In this scenario, the Haber model gave fatality estimates at 270% higher than those of the ten Berge model. Across time, the Haber model gave slight underestimates of fatalities until approximately minute 3, after which lethality was overestimated.

TABLE 7.1

AIC Table Comparing Models

Model	Model Terms	D.F.	Log Likelihood	AIC	ΔAIC	w_i
Toxic load	$K \sim C^{1.6}T$	2	−200.54	405.1	0	1
Haber	$K \sim CT$	2	−232.2527	468.5	63.4	1.7e−14

Note: Model weight (w_i) of 1.7e−14 indicates that the Haber model provides very little information when compared to the toxic load model ($w_i = 1$).

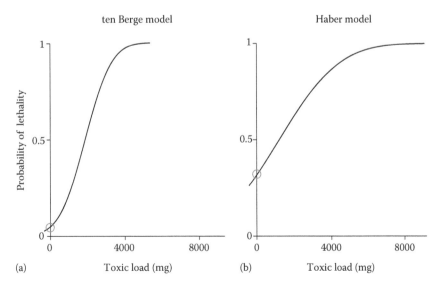

FIGURE 7.1 Comparison of toxic load (a) and Haber (b) toxic load models for sarin exposure to rats. Estimates are from data from Mioduszewski et al. (2001). Nonzero lethality at y-intercepts (circles), where toxic loads are zero, is unrealistic for both models, though closer to zero in the more flexible toxic load model.

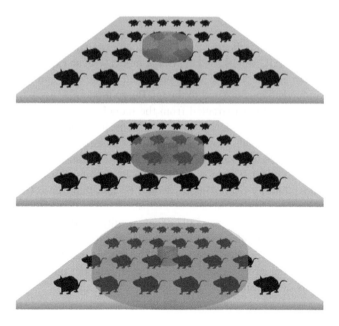

FIGURE 7.2 Schematic representation of radial dispersal over a spatially uniform population. With each time step, a larger number of new individuals are added to the exposed population (shaded area).

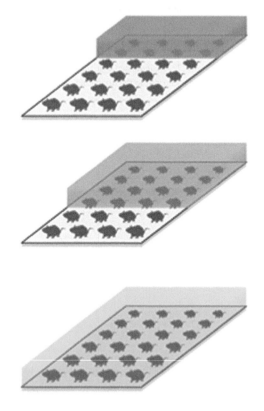

FIGURE 7.3 Schematic representation of linear dispersal over a spatially uniform population. With each time step, equal numbers of new individuals are added to the exposed population (shaded area).

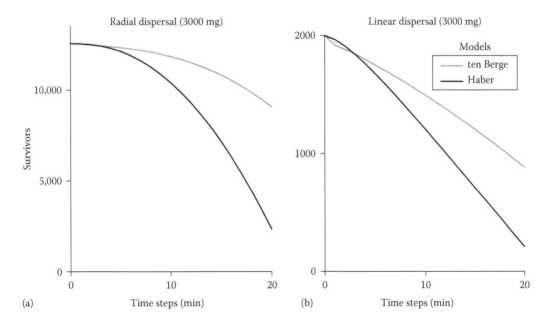

FIGURE 7.4 Predictive simulation for 20 min exposure time series of survival for radial (a) and linear (b) dispersal of 3 g sarin across spatially uniformly distributed rat populations. Radial dispersal differed from linear dispersal in that it encountered a larger population increase with each time step as the radius of the affected area increased.

Simulation of linear diffusion across a rectangle of rats resulted in predicted survival of 10.56% of the population using the Haber model and 44.12% using the ten Berge model (Figure 7.4b). In this scenario, the Haber model gave fatality estimates at 160% higher than those of the ten Berge model. Across time, the Haber model again gave underestimates of fatalities until approximately minute 3, after which lethality was overestimated.

Increasing the initial amount of sarin released to 5 g and decreasing overall time to 10 min produced a simulation of linear diffusion across a rectangle of rats with predicted survival of 19.74% of the population using the Haber model and 36.11% using the ten Berge model (Figure 7.5). In this scenario, the Haber model gave overall fatality estimates at 126% higher than those of the ten Berge model. Across time, the Haber model gave underestimates of fatalities until approximately minute 5, after which lethality was overestimated.

When time and quantity of release were varied, fatality estimates from the Haber model in most cases were high compared with the ten Berge model (Figure 7.6). The exception was when toxicant concentrations were high and exposure duration was low. In those special cases, the Haber model gave lower fatality estimates than the ten Berge model.

7.4   DISCUSSION

Inhalation toxicology studies are often bound by traditional methods found in the seminal literature for the discipline. Such methods are cautious, well established, and limited by the often narrow scope of available data. These methods frequently rely on animal models based on fixed and brief exposure duration at constant toxicant concentration. Historically, as data for more variable exposure duration became available, it became apparent that the original estimator for inhalation exposure, Haber's law, was realistic only for special cases and that its extension via the ten Berge toxic load model added realism for the more general case. However, reliance on Haber's law for casualty estimates in simulated scenarios remains persistent.

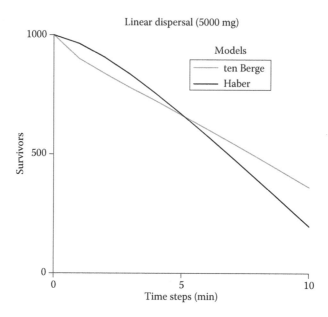

FIGURE 7.5 Predictive simulation for 10 min exposure time series of survival for linear dispersal of 5 g sarin across spatially uniformly distributed rat populations. High initial concentrations illustrated the interaction between concentration and time affecting the direction, under or over, of erroneous lethality estimates.

Traditional methods that rely on Haber's law assume a linear response in probit-transformed lethality probability across all exposure concentration levels, including low concentrations that might be expected to be sublethal. While an assumption of linear nonthreshold effects at low concentrations is based on simple theory and is convenient for modeling, it may not always be based on careful observation or experimental data (Calabrese 2013). Some nonlinear effect must occur in

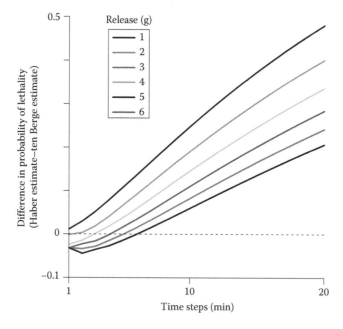

FIGURE 7.6 Difference between Haber and toxic load model predictions, using different initial release quantities. Values below zero (dashed line) indicate potential fatality underestimates by the Haber model at high concentrations and short durations. Values above zero indicate overestimates.

nature for the lowest concentrations to be sublethal. Linear models unrealistically account for this effect by placing the lethality intercept, depicted as the y-axis intercept in Figure 7.1, at values above or below zero. Intercept values above zero would imply that some of the lowest doses are lethal. Intercept values below zero would imply that small doses enhance survival above background levels, where no toxicant is applied. Because the toxic load model is exponential with respect to concentration, rather than linear, it has the flexibility to more closely approximate nonlinear sublethality at low concentrations than does the Haber model.

We used parameters developed from experimental data to populate predictive mechanistic models for simulated scenarios. We found important differences in predicted fatalities between Haber and ten Berge models, differences that could explain casualty overestimates encountered when applying Haber models to data from accidental release of toxicants into open spaces. When toxicants disperse freely over a spatially uniform population, most of that population will receive relatively low doses following exposure to low concentrations. This favors overestimates of lethality by the Haber model (Figure 7.6). However, we caution that the Haber model may underestimate casualties when high concentrations of toxicants are released into confined spaces.

As with all models, neither the toxic load nor the Haber model is absolutely correct. Both models show an unrealistic divergence from plausible lethality values when exposure is zero (Figure 7.1). However, the toxic load model can fit plausible low-dose values closer to zero than Haber models when dose–response deviates from linear trends. In addition, we expect the toxic load model fit by maximum likelihood to give more accurate predictions than the Haber model across the range of exposures when toxicants are released in the open; this would apply for sarin and for many other toxicants where toxic load exponents are determined experimentally to be not equal to 1. AIC model selection indicated a very strong preference for the toxic load over the Haber model when estimating parameters for rat data found in Mioduszewski et al. (2001). AIC model weights (w_i) of 1 (toxic load) vs. 1.7e−14 (Haber) indicate virtually no support for the Haber model compared to the toxic load model (Table 7.1). These AIC values indicated that the greater complexity of the toxic load model was justified, due to increases in model likelihood as applied to experimental data.

Erroneous casualty estimation may have far-reaching impacts. For example, to assess chemical threats and the need for mitigation countermeasures (i.e., medical countermeasures, decontamination, detection, and protection capabilities), many organizations are involved in predicting public health impacts of chemical spills and intentional chemical attacks against soldiers and civilians. The predictions depend upon accurate toxicity data and the dispersion models commonly used to predict numbers of casualties from releases of highly toxic chemicals such as chemical warfare agents, pesticides, toxic industrial chemicals, and toxic industrial materials. The common assumption is that the chemicals follow Haber's law. However, as noted, for many chemicals of concern (e.g., sarin and hydrogen cyanide), Haber's law does not apply, and numbers of casualties are overestimated. Casualty overestimation would likely result in inaccurate threat assessment and, subsequently, inappropriate development and placement of mitigating countermeasures.

We have demonstrated that the use of Haber's law to predict fatalities may lead to erroneous estimations of casualties from releases of many toxic compounds. Accurate predictions of casualty numbers in an exposed population can be greatly improved by employing a toxic load model, which takes into account concentration-dependent mechanisms and employs an exponent to increase the importance of the concentration as the concentration becomes the controlling factor of dose.

QUESTIONS

1. Define Haber's law and the toxic load model and write their mathematical expressions.
   Haber's law states that the product of toxicant concentration and time is a constant: $K = C \times T$.
   Answer: The toxic load model states that the product of concentration raised to the toxic load exponent and time is a constant: $K = C^n \times T$.

2. Which model, Haber's law or the toxic load model, is the correct model?
   Answer: Neither model is correct, but the toxic load model provides better lethality approximations in most cases.
3. What is meant by the intercept value for a toxic load model?
   Answer: The intercept is the lethality when dose, concentration, or duration of exposure is zero.
4. When the concentration of a toxicant and its toxic load exponent are high, will Haber's law overestimate casualties at short exposure durations?
   Answer: No, casualties tend to be underestimated under these conditions.

ACKNOWLEDGMENT

Our thanks to Steven Skurski of the Edgewood Chemical Biological Center Public Affairs Office, for providing Figures 7.2 and 7.3.

REFERENCES

Akaike, H. 1973. Information theory as an extension of the maximum likelihood principal. In *Second International Symposium on Information Theory*, eds. B. N. Petrov and F. Caski, pp. 267–281. Budapest, Hungary: Akademiai Kiado.

Boyland, E., F. F. McDonald, and M. J. Rumens. 1946. The variation in toxicity of phosgene for small animals with the duration of exposure. *British Journal of Pharmacology* 1: 81–89.

Burnham, K. P. and D. R. Anderson. 2002. *Model Selection and Multimodel Inference: A Practical Information Theoretic Approach*. New York: Springer.

Calabrese, E. J. 2013. How the US National Academy of Sciences misled the world community on cancer risk assessment: new findings challenge historical foundations of the linear dose response. *Arch Toxicol* 87:2063–2081.

Eaton, D. L. and S. G. Gilbert. 2013. Chapter 2: Principles of toxicology. In *Casarett & Doull's Toxicology: The Basic Science of Poisons*, 8th edn., ed. C. D. Klaassen, pp. 13–48. New York: McGraw-Hill Education.

Haber, F. R. 1924. *Zur Geschichie des gaskrieges, in Funf Vortage aus Jahren 1920–1923*. Berlin, Germany: Springer.

Mioduszewski, R. J., J. H. Manthel, R. A. Way et al. 2001. ECBC low level operational toxicology program: Phase 1—Inhalation toxicity of sarin vapor in rats as a function of exposure concentration and duration. Technical Report 183. Aberdeen Proving Ground, MD: Edgewood Chemical Biological Center.

Nelson, G. 2006. Effects of carbon monoxide in man: Exposure fatality studies. In *Carbon Monoxide and Human Lethality: Fire and Non-Fire Studies*, ed. M. M. Hinschler, pp. 3–62. New York: Taylor & Francis.

Prentiss, A. M. 1937. *Chemicals in War*. New York: McGraw Hill.

Sommerville, D. R., J. J. Bray, S. A. Reutter-Christy, R. E. Jablonski, and E. E. Shelly. 2010. Review and assessment of chlorine mammalian lethality data and the development of a human estimate. *Military Operations Research* 15: 59–86.

Sommerville, D. R., K. H. Park, M. O. Kierzewski, M. D. Dunkel, M. I. Hutton, and N. A. Pinto. 2006. Toxic load modeling. In *Inhalation Toxicology*, eds. H. Salem and S. A. Katz, pp. 137–158. New York: CRC Press.

ten Berge, W. F., A. Zwart, and L. M. Appelman. 1986. Concentration-time mortality response relationship of irritant and systemically acting vapours and gases. *Journal of Hazardous Materials* 13: 301–309.

8 Physiologically Based Kinetic Modeling of Inhalation

Harvey J. Clewell III, Jeffry D. Schroeter,
and Melvin E. Andersen

CONTENTS

8.1 INTRODUCTION

Pharmacokinetics is the quantitative study of factors that control the time course for absorption, distribution, metabolism, and excretion of chemicals within the body. Pharmacokinetic (PK) models provide sets of equations that simulate the time courses of chemicals and their metabolites in various tissues throughout the body. The interest in PK modeling in toxicology arose from the need to relate internal concentrations of active compounds at their target sites with the doses of chemical given to an animal or human subject. The reason, of course, is a fundamental tenet in toxicology that both beneficial and adverse responses to compounds are related to the concentrations of active chemicals reaching target tissues rather than the amounts of chemical at the site of absorption. The relationships between tissue dose and administered dose can be complex, especially in high dose toxicity testing studies, with multiple, repeated daily dosing, or when metabolism or toxicity at routes of entry alters uptake processes for various routes of exposure. PK models of all kinds are primarily a tool to assess chemical dosimetry at target tissues for a wide range of exposure situations.

In physiologically based pharmacokinetic (PBPK) modeling, compartments correspond to discrete tissues or to groupings of tissues with appropriate volumes, blood flows, and pathways for metabolism of test chemicals (Bischoff and Brown, 1966). These PBPK models include pertinent biochemical and physicochemical constants for metabolism and solubility in each compartment. Routes of dosing (routes of administration) are included in their proper relationship to the overall physiology. The equations that form the basis of the PBPK model also account for the time-sequence of dose input into test subjects and permit input by multiple routes if necessary for specific exposure situations. Each compartment in the model is described with a mass-balance differential equation

(MB-DE) whose terms mathematically represent biological processes. The set of equations is solved by numerical integration to simulate tissue time-course concentrations of chemicals and their metabolites.

Some PBPK models account for interactions of circulating compounds with specific receptors or the covalent interactions of chemicals with tissue constituents. Modeling these reversible and irreversible molecular interactions with cell constituents is the initial step in developing physiologically based pharmacokinetic/pharmacodynamic (PBPK/PD) models for effects of chemicals on biological processes. A number of reviews are available on PBPK modeling (Himmelstein and Lutz, 1979; Gerlowski and Jain, 1983; Leung, 1991), including a comprehensive review of published PBPK models (Reddy et al., 2005).

8.2   HISTORY OF PBPK MODELING OF INHALATION

Some of the first applications of PBPK modeling were by inhalation anesthesiologists interested in understanding the role of ventilation rates, blood flow rates, and tissue solubility on the uptake and distribution of volatile anesthetics to the central nervous system. In the 1920s, Haggard (1924a,b) quantitatively described the importance of physiological factors for the uptake of ethyl ether into the body during the first few breaths. Accomplishing this analysis required writing an equation for the relationship between inhaled ether and the concentration of ether in blood. Tools for solving this equation over time were not available, so the mathematical analysis was limited to the first few breaths when venous concentrations remained small. Henderson and Haggard (1943) provided the first detailed discussion of the toxicology of inhaled compounds in the context of the principles that control exposure, absorption, and physiological actions. It is the first articulation of a PBPK modeling strategy in occupational and environmental toxicology.

More complete PBPK models for inhalation were provided by Kety (1951), Mapleson (1963), and Riggs (1963). In these models, body tissues were lumped together based on blood perfusion rates, giving sets of tissues referred to as richly perfused or poorly perfused. Mapleson (1963) solved the set of equations using an analog computer to give solutions to the complete time course within the various tissue groups. These analog computer PBPK models for inhaled gases and vapors were extended by Fiserova-Bergerova and colleagues (1975, 1979, 1980) to focus on compounds in the occupational environment and to describe metabolism of these compounds in liver. The extension to include metabolism was particularly important for subsequent work in toxicology because most compounds of interest in occupational toxicology are metabolized and metabolites are often involved in toxic responses.

In the 1930s, Teorell (1937a,b) provided a set of equations for uptake, distribution, and elimination of drugs from the body. These papers are regarded as providing the first physiological model for drug distribution. However, computational methods were not available to solve the sets of equations at this time. Exact mathematical solutions for distribution of compounds in the body could only be obtained for simplified models in which the body was reduced to a small number of compartments that did not correspond directly with specific physiological entities. Over the next 30 years, PK modeling focused on these simpler compartmental descriptions rather than on developing models more concordant with the structure and content of the biological system itself. Two areas of concern that served as a challenge to compartmental PK modeling arose in the 1960s and early 1970s: (1) the saturation of elimination pathways and (2) the possibility that blood flow rather than metabolic capacity of an organ might limit clearance.

Compartmental models were brought to toxicology and risk assessment in a series of innovative studies designed to examine PK behavior where specific elimination pathways, both metabolic and excretory, become saturated at high doses (Gehring et al., 1976, 1977, 1978). These approaches were applied to a series of compounds of toxicological and commercial importance including herbicides (Sauerhoff et al., 1976, 1977), solvents (McKenna et al., 1982), plastic monomers (McKenna et al., 1978a,b), and hydrocarbons (Young et al., 1979; Ramsey et al., 1980). The final piece of technology needed to bring a full PBPK approach to studying factors that determine chemical disposition came with the rapid development of digital computation by the engineering community.

Scientists trained in chemical engineering and computational methods developed PBPK models for chemotherapeutic compounds, that is, chemicals used in cancer therapy (Bischoff and Brown, 1966). Many of these compounds are highly toxic and have therapeutic efficacy by being slightly more toxic to rapidly growing cells (the cancer cells) than to normal tissues. Initial successes with methotrexate (Bischoff et al., 1971) led to PBPK models for other compounds, including 5-fluorouracil (Collins et al., 1982) and cisplatin (Farris et al., 1988). These seminal contributions showed the ease with which realistic descriptions of physiology and relevant pathways of metabolism could be incorporated into PBPK models for chemical disposition and paved the way for more extensive use of PBPK modeling in toxicology and chemical risk assessment. These models took advantage of the increasing availability of digital computation on main frame computers for solving sets of MB-DEs.

Ramsey and Andersen (1984) applied a PBPK modeling approach to describe the disposition of styrene in rats and humans for a range of concentrations and for several routes of administration. Using scale-up methods common for engineering models (Dedrick, 1973), the interspecies PBPK model for styrene (Figure 8.1) was able to predict blood and exhaled air time-course curves for oral

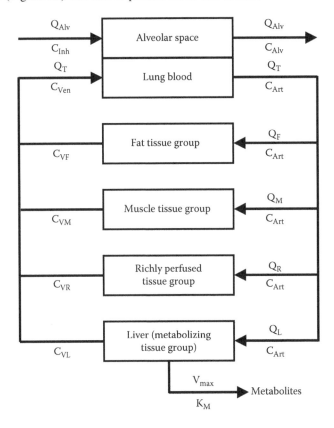

FIGURE 8.1 Diagram of a physiologically based pharmacokinetic model for styrene. In this description, groups of tissues are defined with respect to their volumes, blood flows (Q), and partition coefficients for the chemical. The uptake of vapor is determined by the alveolar ventilation (Q_{Alv}), cardiac output (Q_T), blood:air partition coefficient (P_B), and the concentration gradient between arterial and venous pulmonary blood (C_{Art} and C_{Ven}). The dashed line reflects the fact that the lung compartment is described by a steady-state equation, assuming that diffusion between the alveolar air and lung blood is fast compared to ventilation and perfusion. Metabolism is described in the liver with a saturable pathway defined by a maximum velocity (V_{max}) and affinity (K_M). The mathematical description assumes equilibration between arterial blood and alveolar air as well as between each of the tissues and the venous blood exiting from that tissue.

and intravenous dosing in the rat and for inhalation exposures in human volunteers. This ability to support extrapolation to untested (and sometimes untestable) conditions is an essential part of risk assessment and has made these PBPK models attractive tools in human health risk assessments of various kinds (Clewell and Andersen, 1985; National Research Council, 1987). In the styrene PBPK model, the liver was split off as a separate compartment (i.e., rather than embedded in a central compartment), metabolism in the liver was saturable (i.e., followed Michaelis–Menten kinetics), and styrene clearance from tissues was directly based on blood flow and metabolic characteristics of tissues.

Among the opportunities offered by PBPK approaches are (1) creating models from physiological, biochemical, and anatomical information, entirely separate from the collection of detailed concentration time-course curves; (2) evaluating mechanisms by which biological processes govern disposition of a wide range of compounds by comparison of PK results with model predictions; (3) using chemicals as probes of the biological processes to gain more general information on the way chemical characteristics govern the importance of various transport pathways in the body; (4) applying the models in risk assessments for setting exposure standards; and (5) using annotation of a modeling database as a repository of information on toxicity and kinetics of specific compounds.

The advent of biologically structured PBPK models had a dramatic influence on the nature of the experiments conducted to determine PK behavior and to estimate tissue dosimetry. In PBPK descriptions, time-course behavior is not an intrinsic property of the organism accessible only by direct experimentation. Instead, it is a composite behavior, governed by more fundamental physiological and biochemical processes. More importantly, these fundamental processes can be studied in simpler systems to obtain the necessary PBPK model parameters in experiments separate from collection of time-course concentration curves. Based on these parameters and an appropriate model structure, tissue time-course behaviors can be predicted by computer simulation with PBPK models and compared to data as a test of model performance (Figure 8.2).

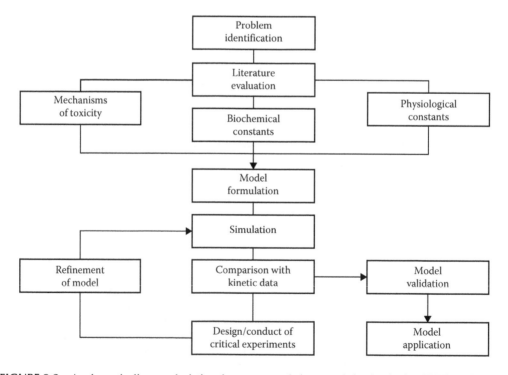

FIGURE 8.2 A schematic diagram depicting the recommended approach for developing PBPK models.

8.3 PBPK MODELING OF VOLATILE COMPOUNDS

Volatile chemicals have provided a good test bed for examining this approach to PBPK modeling. The disposition of volatiles in the body is related to breathing rates, tissue volumes, tissue blood flow rates, tissue partition coefficients, and kinetic constants for metabolism of the chemical in particular tissues. Physiological factors important in developing an appropriate and useful PBPK model have been discussed by Krishnan and Andersen (2001). Physiological parameters can be found in the bio-medical literature and were compiled by Brown et al. (1997). Partition coefficients can be measured by equilibrating tissue homogenates in a vial with an atmosphere containing the test chemical (Sato and Nakajima, 1979a; Gargas et al., 1989). Constants for metabolism (enzyme kinetic constants for saturable, first-order or second-order reactions) can be determined in vitro with tissue homogenates, microsomal preparations, liver slices, etc., by supplementing these preparations with reactants to pro-mote metabolic reactions (Sato and Nakajima, 1979b; Hilderbrand et al., 1981; Kedderis et al., 1993). Another method for assessing metabolic parameters in vivo relies on closed chamber inhalation tech-niques. Here, a small number of live animals are placed in a closed chamber to measure the rate of loss of chemical at a variety of chamber concentrations (Hefner et al., 1975; Filser and Bolt, 1979; Gargas et al., 1986). These in vitro and in vivo experiments can provide all the parameters necessary for constructing a PBPK model for the parent chemical, and time-course behavior is now *predictable*, based on results of these ancillary studies. Other approaches for developing predictive PBPK models include using structure–activity relationships (SARs) to estimate model parameters for classes of compounds (Poulin and Krishnan, 1996, 1999; Parham et al., 1997; Parham and Portier, 1998).

A variety of low molecular weight volatile compounds are known to cause lung tumors in mice and to cause damage to nasal tissues, especially to the olfactory epithelium. The initial PBPK models available for DCM or even more comprehensive models for styrene and its metabolites (Csanady et al., 1994), another compound that causes mouse lung tumors and rodent nasal tissue toxicity, were not developed to look at uptake of vapors directly from airway to the epithelial tissues. Airway equilibration of soluble hydrophilic compounds restricts the uptake of chemical into the deeper lung and into systemic blood, and PBPK models for equilibration with tissues have been developed for this phenomenon (Johanson, 1991; Kumagai and Matsunaga, 1995).

Recently, more complete PBPK models for the airway/tissue interface in the lungs were applied to coordinate a risk assessment with styrene for mouse lung tumors and mouse and rat nasal tissue toxicity (Sarangapani et al., 2002). Another styrene model that includes airway equilibration and local metabolism in epithelial tissues in the lungs appeared recently (Csanady et al., 2003). This model uti-lized attributes of previous work on airflow modeling in the nasal airway and equilibration processes in the tracheobronchial region to relate toxic responses with metabolized dose of styrene in tissues. A wide range of research undergirds the assignment of toxic responses with tissue exposure to specific epoxide metabolites derived from the oxidation of styrene by cytochrome P450 family enzyme com-plexes in epithelial tissues. In developing and applying these newer models for styrene, discrepancies in dose–response relationships for inhalation toxicity in the lung, liver, and kidney became clarified for styrene and for other volatile compounds, such as chloroform, by elaborating the description of the respiratory tract. For relatively soluble compounds that have high systemic clearance, responses appear in pulmonary sites at lower exposure concentrations than in the liver and kidney because of the direct equilibration of airway tissues with inhaled gases and vapors. The liver and kidney only equilibrate with chemical in blood. This exercise also showed that earlier models with metabolism in the lung occurring only from compound delivered to pulmonary tissues by arterial blood were not realistic.

8.4 CFD MODELING

Inhaled irritant gases frequently elicit responses at the initial site of contact in the body. Gases that are highly soluble or reactive are primarily absorbed in the upper respiratory tract upon inhalation, making the nasal passages the target tissue of interest. The nasal passages of laboratory animals and humans possess complex airway shapes and turbinate structures which promote the dosimetry of inhaled particles and gases. Studies of inspiratory airflow in nasal molds revealed that the distribution of nasal lesions was due in part to the locations of major airflow streams (Morgan et al., 1991). Other factors influencing vapor deposition include regional dose, tissue susceptibility, absorption properties of the nasal lining, and the physicochemical properties of the inhaled gas. The effects of nasal geometry and chemical properties such as air- and tissue-phase diffusivity, solubility, and reactivity on nasal uptake patterns can be investigated using computational fluid dynamic (CFD) models of the upper respiratory tract.

CFD modeling involves the numerical analysis of fluid flows and embedded materials. For inhalation toxicology applications, we are concerned with the transport and deposition of inhaled vapors in air. CFD uses numerical methods to solve the Navier–Stokes equations, which are a set of nonlinear partial differential equations that describe the motion of fluids. CFD modeling involves four major steps: model development, mesh generation, the solution procedure, and post-processing of results. Respiratory geometries have been derived from sequential cross-sections of MRI or CT scans and consist of anatomically accurate 3D representations of the respiratory space. Since the problem formulation typically involves a large number of nodes, resulting in large problems requiring hours of computing time, CFD solutions are often performed using third-party commercial software (e.g., Fluent, ANSYS, Inc.). Recent advances in computing power and image analysis tools have allowed CFD models to rapidly progress to include segments of the animal or human respiratory tract and the solution of airflow and material transport through these regions.

The first CFD model used in inhalation toxicology analyses was developed by Kimbell et al. (1993) and consisted of the anterior region of the F344 rat. The motivation for this model was to simulate uptake patterns of inhaled soluble, reactive gases (e.g., formaldehyde) and compare regions of high uptake with corresponding lesion distribution. This model was subsequently extended to include the entire F344 rat nasal passages, including the olfactory region (Kimbell et al., 1997a). The model was able to characterize the regional airflow streams throughout the rat nasal passages and the complex flow patterns present in the ethmoid region. Simulation results for formaldehyde uptake in the rat nose also supported the hypothesis that the distribution of formaldehyde-induced squamous metaplasia was correlated with the locations of high uptake (Kimbell et al., 1997b). The CFD rat nasal model was later used to predict tissue dose of inhaled ozone (Cohen Hubal et al., 1996) and hydrogen sulfide (Moulin et al., 2002), which are two other gases that elicit effects in the nasal cavities of rodents. CFD models have also been generated for the nasal passages of the Sprague Dawley rat to investigate odorant uptake in the olfactory region (Yang et al., 2007) and also to study dosimetry of reactive gases in the upper and lower respiratory airways of the Sprague Dawley rat (Corley et al., 2012).

CFD models were also developed for the monkey and human nasal passages. The monkey model was constructed from tracings of airway outlines of serial sections of tissue specimens from the right nasal passages and was used to predict inhaled airflow and gas uptake patterns (Kepler et al., 1998). Initial efforts in human nasal modeling focused on characterizing airflow patterns in human nasal models that were developed from MRI or CT data (Keyhani et al., 1995; Subramaniam et al., 1998). By today's standards, the imaging resolution and numerical meshes of these models were fairly coarse. However, recent advances in mesh generation software and imaging analysis tools have led to a rapid expansion in monkey and human nasal modeling efforts; 3D reconstructions of the human nose are now developed from high-resolution CT data that allow for rapid reconstruction of high-fidelity, anatomically accurate models that can be used to simulate vapor uptake and particle deposition (e.g., Wang et al., 2009). Corley et al. (2012) developed a CFD model of an infant rhesus monkey that was based on CT imaging of the upper airways and MR imaging of a silicone cast of lung airways.

Nasal CFD models of laboratory animals and humans can be used to compare estimates of regional nasal airflow, inhaled gas uptake, and particle deposition locations for interspecies extrapolation of risk estimates. The anatomically accurate representations allow for site-specific predictions of regional dose that can elucidate the dose–response behavior that may differ between species. In the work by Kimbell et al. (2001a,b), rat, monkey, and human nasal CFD models were used to compare localized predictions of formaldehyde flux between species. The boundary condition governing formaldehyde absorption on nasal airway walls was calibrated to nasal uptake measurements in rats; the same mass transfer coefficient was then used in all species. Given the high predicted nasal uptake in all species (90%, 67%, and 76% in the rat, monkey, and human, respectively), large anterior-to-posterior gradients in wall mass flux patterns were observed. The localized, site-specific nature of the CFD formaldehyde flux predictions was used to calculate average flux in specific regions of the nasal passages for subsequent use in models of formaldehyde carcinogenesis.

While earlier CFD modeling efforts focused on the development of the nasal geometries, study of inhaled airflow patterns, and uptake of highly reactive gases such as formaldehyde, more recent studies have focused on the incorporation of chemical-specific parameters into the boundary conditions governing gas uptake. The behavior of inhaled chemicals in the airspace depends on the airflow patterns and the chemical air-phase diffusivity and is therefore very similar for all gases. However, the tissue disposition of absorbed gases can vary widely depending on the solubility, tissue-phase diffusivity, and metabolic rates. These clearance processes in the nasal mucosa affect the nasal resistance of the gas and the rate of uptake from the airspace and must be accounted for in order to study the effects of these parameters on nasal absorption.

Cohen Hubal et al. (1996) incorporated a virtual nasal lining on the CFD model of the rat nasal passages to study the uptake of inhaled ozone. Tissue disposition of inhaled ozone was modeled according to the reaction-diffusion equation, with ozone uptake and tissue transport determined by its solubility, diffusivity, and a first-order clearance term representing ozone reactivity. Due to the linearity of all terms in the reaction-diffusion equation, an analytical solution could be obtained for tissue disposition and ozone flux from the airspace. This methodology was then used by Schroeter et al. (2006a) to study uptake of hydrogen sulfide (H_2S) gas in the rat nasal passages. Nasal uptake of H_2S was shown to exhibit a saturable behavior, with lower uptake fractions at higher exposure concentrations. In order to replicate this behavior, both first-order and saturable clearance terms were included in the reaction-diffusion equation governing H_2S tissue disposition and were numerically solved to compute flux from the airspace. The first-order term and the Michaelis–Menten parameters were calibrated to nasal uptake data in rats. The first-order term and maximal metabolic rate were then allometrically scaled and used in the human nasal CFD model to predict nasal uptake of inhaled H_2S in humans (Schroeter et al., 2006b). Flux results in the olfactory region from the rat and human nasal CFD models were subsequently used to estimate acceptable H_2S exposure limits in humans. A similar procedure was used to predict nasal uptake of inhaled acrolein in rats and humans due to its saturable nasal uptake (Schroeter et al., 2008). A detailed discussion of the quantitative risk assessment approach used for acrolein is described in Section 19.8.

8.5 HYBRID CFD–PBPK MODELING

The strengths of high-fidelity approaches such as CFD modeling lie in their abilities to calculate the influence of airflow patterns on site-specific uptake and wall mass flux in anatomically accurate representations of the respiratory system. However, CFD models of the respiratory tract rely on high-resolution imaging data, can be difficult to construct, and are computationally expensive. The objective of hybrid CFD–PBPK models is to utilize information on airflow allocation and air-phase mass transfer coefficients that can be derived from CFD models in a PBPK model of vapor uptake that is still representative of the respiratory anatomy, but is more flexible and adaptable for calculating tissue concentrations for use in risk assessment.

Hybrid CFD–PBPK models for vapor uptake in the nasal passages are based on the seminal work of Morris et al. (1993), who developed a PBPK model for nasal uptake of nonreactive vapors. In this work, the nasal cavity of the rat was divided into three compartments: one compartment representing the ventral lateral airstream receiving 88% of the inhaled airflow and two compartments (one each for respiratory and olfactory epithelium) in the dorsal medial airstream receiving the remaining 12% of airflow. Each tissue compartment was modeled as a stack of multiple layers representing the mucus, epithelium, and submucosal layers. Vapor transport and clearance in the epithelial compartments was modeled according to the vapor diffusivity, metabolism, and blood perfusion in the appropriate epithelial levels. Vapor concentrations in the mucus layer were assumed to equilibrate with the air concentration according to the mucus:air or blood:air partition coefficient. A similar model structure was used (Plowchalk et al., 1997; Bogdanffy et al., 1999) to describe nasal uptake and metabolism of vinyl acetate in the rat nasal cavity.

The equilibration of inhaled vapors between mucus and air is not necessarily a valid assumption for all gases, since it does not account for the effect of resistance to molecular transport in the air and tissue phases. To overcome this limitation, vapor concentrations at the air:mucus interface were coupled by a permeability coefficient that incorporated mass transfer coefficients in the air and mucus (Bush et al., 1998). The mucus-phase mass transfer coefficient was computed as a function of the mucus-phase diffusivity. The air-phase mass transfer coefficients were computed from CFD simulations in the rat nasal model. In the CFD model, a "C = 0" boundary condition was implemented on all nasal airways walls, meaning that there was no resistance to mass transfer in the mucus or tissue. Therefore, nasal uptake was only limited by the air-phase resistance of the gas, which is a function of the flow rate, airway geometry, and air-phase diffusivity. By calculating the uptake in different regions of the nasal cavity, the air-phase mass transfer coefficients could be computed for each compartment in the rat nasal PBPK model. Using this model structure, predictions of nasal uptake for acetone, acrylic acid, and isoamyl alcohol compared well with experimental measurements (Bush et al., 1998).

Frederick et al. (1998) extended this approach to the rat and human by designing compartmental structures in a CFD–PBPK model based on the nasal anatomy of each species. Inhaled air passed over the squamous epithelium located in the nasal vestibule and then split into dorsal and ventral streams before joining again in the nasopharynx. In the rat model, the ventral stream passed over two ventral respiratory compartments and the dorsal stream passed over a dorsal respiratory compartment followed by two olfactory compartments. The human model was similar and contained two ventral respiratory compartments but only one dorsal olfactory compartment since the human olfactory epithelium covers a much smaller surface area than the rat. Airflow allocations computed from the rat and human nasal CFD models (Kimbell et al., 1997a; Subramaniam et al., 1998) were used to designate the fraction of inhaled flow to each compartment. This hybrid CFD–PBPK model was used to compare interspecies tissue concentrations of inhaled acrylic acid. They found that the human olfactory region is exposed to twofold to threefold lower tissue concentrations of acrylic acid than the rat. Variations on this model structure were also used to estimate nasal uptake of acetaldehyde (Teeguarden et al., 2008) and diacetyl (Morris and Hubbs, 2009). These hybrid CFD–PBPK models are capable of estimating target tissue concentrations in distinct regions (e.g., the olfactory epithelium) of the nasal cavity for use in quantitative risk assessment calculations.

8.6 RISK ASSESSMENT APPLICATIONS

The main reason for the rapid expansion of PBPK modeling over the past 10–15 years has undoubtedly been the contributions of the technology to dose–response assessment and extrapolations in chemical risk assessments. The first application of a PBPK model in a risk assessment was with DCM. A PBPK model for DCM was first developed to explore causality between various measures of tissue dose (referred to as dose metrics) and carcinogenicity (Andersen et al., 1987). The PBPK model contained tissue clearance by oxidation and glutathione (GSH) conjugation in the liver and kidney, accounted for dosing by inhalation or drinking water, and allowed simulation of expected tissue dose metrics in mice and humans. With this PBPK model, it was possible to calculate expected tissue exposures to metabolites from the oxidative and conjugation pathways for different exposure conditions in the liver and lung for mice and humans.

For DCM, the dose metrics chosen for analyzing tumor responses were integrated intensity of tissue exposure to reactive intermediates, that is, the rate of metabolism through a specific pathway per volume of tissue per time. The carcinogenic responses for DCM correlated well with the GSH pathway metabolism but not with the oxidation pathway metabolism. The work with DCM was the first use of a PBPK model for low dose and interspecies extrapolation based on tissue dose metrics. This extrapolation used human-specific parameters for tissue volumes, breathing rates, and distribution of enzymes involved in oxidation and conjugation in the model structure. Risk extrapolation assumed that mouse and human tissues would be equally responsive to equivalent tissue exposures to the reactive GSH pathway intermediates. This PBPK model has been cited and used in risk assessments by Health Canada (Health Canada, 1993) and by the Occupational Safety and Health Administration (OSHA) in the USA (OSHA, 1997). The risk assessment exercise with DCM established a pattern for application of a PBPK model in risk assessment that is still closely followed today. In addition, the PBPK modeling spurred a variety of research to refine variability and uncertainty analysis for PBPK models (Clewell and Clewell, 2008) and to conduct targeted research to confirm the association of toxicity with GSH pathway metabolites. The refinement of the DCM PBPK model structure based on targeted mechanistic studies was particularly useful in establishing the GSH pathway mode of action for DCM carcinogenicity and increasing confidence in application of this PBPK model in risk assessment.

Since the first proposal to use a PBPK model in risk assessment with DCM in 1987, the field of PBPK modeling in relation to risk modeling has grown steadily. In more recent years, there has been increasing acceptance of the use of these PBPK dosimetry models in a variety of risk assessment applications, and criteria for their evaluation have been described (Clewell and Clewell, 2008; McLanahan et al., 2012). The US EPA's Reference Concentration (RfC) documentation explicitly includes routine application of interspecies differences in dosimetry in assessing RfCs (US EPA, 1994). The RfC documentation for vinyl chloride (VC) in the US EPA's Integrated Risk Information System, IRIS, specifically describes and uses a PBPK model for standard setting and dose route extrapolations (US EPA, 2000a). A risk assessment with acrylic acid (Andersen et al., 2000) applied a PBPK model linked to computational fluid dynamic (CFD) previously defined calculations of nasal airflow (Frederick et al., 1998) to derive an RfC. The Hazardous Air Pollutants (HAPS) Test Rule (Federal Register, 1996) invited increased efficiency/efficacy in testing by providing possible substitution of certain oral toxicity tests instead of requiring new inhalation studies, for those instances where a validated PBPK model is available to conduct extrapolations across dose routes.

8.7 PBPK MODELING EXAMPLE: VINYL CHLORIDE

The risk assessment for VC provides a useful example of the application of a PBPK model in risk assessment. The PBPK model in this case was initially developed to support a cancer risk assessment for VC(Clewell et al., 2001) but was actually applied by EPA for both their cancer and noncancer assessments (US EPA, 2000a). VC is a gas that is used primarily as a chemical precursor in the production of plastics and resins. It is a trans-species carcinogen, producing tumors in a variety of tissues, from both inhalation and oral exposures, across a number of species. In particular, exposure to VC has been associated with a rare tumor, liver angiosarcoma, in a large number of studies in mice, rats, and humans. The mode of action for the carcinogenicity of VC appears to be a relatively straightforward example of DNA adduct formation by a reactive metabolite, leading to mutation, mistranscription, and neoplasia.

The primary route of metabolism of VC is by the action of the mixed function oxidase (MFO) system, now referred to as cytochrome P450 or CYP, to form chloroethylene oxide (CEO). The metabolism of VC is a dose-dependent, saturable process. CEO is a highly reactive, short-lived epoxide that rapidly rearranges to form chloroacetaldehyde (CAA), a reactive α-halocarbonyl compound; this conversion can also be catalyzed by epoxide hydrolase. The main detoxification of these two metabolites is conjugation with glutathione. CAA has been associated with the cytotoxicity of VC; CEO is responsible for its carcinogenicity.

The carcinogenicity of VC is related to the production of reactive metabolic intermediates. The most appropriate pharmacokinetic dose metric for a reactive metabolite is the total amount of the metabolite generated divided by the volume of the tissue into which it is produced (Andersen et al., 1987). In the case of VC, a reasonable dose metrics for angiosarcoma would be the total amount of metabolite formation divided by the volume of the liver (RISK).

The PBPK model for VC is shown in Figure 8.3. For a poorly soluble, volatile chemical like VC, only four tissue compartments are required in the PBPK model: a richly perfused tissue compartment which includes all of the organs except the liver, a slowly perfused tissue compartment which includes all of the muscle and skin tissue, a fat compartment which includes all of the fatty tissues, and a liver compartment where metabolism occurs. All metabolisms were described in the liver, which is a good assumption in terms of the overall kinetics of VC, but which would have to be revised to include target-tissue-specific metabolism if a serious attempt were to be made to

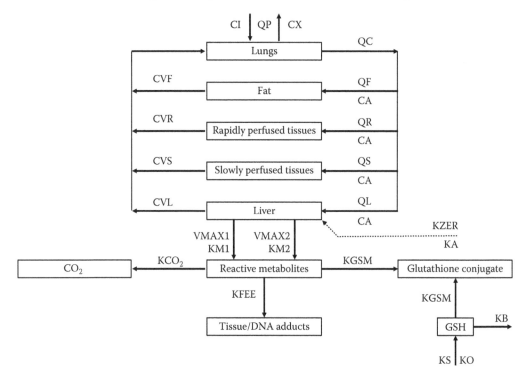

FIGURE 8.3 Diagram of the PBPK model used in a cancer risk assessment for vinyl chloride. *Abbreviations:* CA—arterial blood concentration; QP—alveolar ventilation; CI—inhaled concentration; CX—exhaled concentration; QC—cardiac output; QF, CVF—blood flow to, and venous concentration leaving, the fat; QR, CVR—blood flow to, and venous concentration leaving, the richly perfused tissues (most organs); QS, CVS—blood flow to, and venous concentration leaving, the slowly perfused tissues (e.g., muscle); QL, CVL—blood flow to, and venous concentration leaving, the liver; VMAX1, KM1—capacity and affinity for the high affinity oxidative pathway enzyme (CYP 2E1); VMAX2, KM2—capacity and affinity for the lower affinity oxidative pathway enzymes, (e.g., CYP 2C11/6); KZER—zero-order rate constant for uptake of VC from drinking water; KA—first-order rate constant for uptake of VC from corn oil; KCO₂—first-order rate constant for metabolism of VC to CO₂; KGSM—first-order rate constant for reaction of VC metabolites with GSH; KFEE—first-order rate constant for reaction of VC metabolites with other cellular materials, including DNA; KB—first-order rate constant for normal turnover of GSH; KO—zero-order rate constant for maximum production of GSH; KS—parameter controlling rate of recovery of GSH from depletion.

perform a VC risk assessment for a tissue other than the liver (Andersen et al., 1987). The model also assumes flow-limited kinetics, or venous equilibration, that is, that the transport of VC between blood and tissues is fast enough for steady state to be reached within the time it is transported through the tissues in the blood.

Metabolism of VC is modeled by two saturable pathways, one high affinity, low capacity (with parameters VMAX1C and KM1) and one low affinity, high capacity (with parameters VMAX2C and KM2). Subsequent metabolism is based on the VC metabolic scheme described earlier: the reactive metabolites (whether CEO, CAA, or other intermediates) may then either be metabolized further, leading to CO_2, react with GSH, or react with other cellular materials, including DNA. Because exposure to VC has been shown to deplete circulating levels of GSH, a simple description of GSH kinetics was also included in the model.

The parameters for the model are documented in Clewell et al. (2001). The physiological parameters are US EPA reference values (US EPA, 1988), except for the alveolar ventilation (QPC) in the human, which was calculated from the standard US EPA value for the ventilation rate in the human, 20 m^3/day, assuming a 33% pulmonary dead-space (US EPA, 1988). The value for the cardiac output (QCC) in the human was selected to correspond to the same workload as the standard US EPA ventilation using data from Astrand and Rodahl (1970). For modeling of the closed chamber studies with human subjects, more typical resting values of 15 for cardiac output (QCC) and 18 for alveolar ventilation (QPC) were used. In some cases, it was also necessary to slightly vary the alveolar ventilation and cardiac output of the animals in the closed chamber studies in order to obtain acceptable simulations of those experiments. The partition coefficients for Fisher-344 (F344) rats were taken from Gargas et al. (1989) and those for Sprague Dawley rats were taken from Barton et al. (1995). The Sprague Dawley values were also used for modeling of Wistar rats. Blood/air partition coefficients for the other species were obtained from Gargas et al. (1989), and the corresponding tissue/blood partition coefficients were estimated by dividing the Sprague Dawley rat tissue/air partition coefficients by the appropriate blood/air value.

The affinity for the 2E1 pathway (KM1) in the rat, mouse, and hamster was set to 0.1 on the basis of studies of the competitive interactions between CYP2E1 substrates in the rat (Barton et al., 1995). The affinity used for the non-2E1 pathway (KM2) in the mouse and rat was set during the iterative fitting of the rat total metabolism, glutathione depletion, and rate of metabolism data, described in the following text. The capacity parameters for the two oxidative pathways (VMAX1C and VMAX2C) in the mouse, rat, and hamster were estimated by fitting the model to data from closed chamber exposures with each of the species and strains of interest (Bolt et al., 1977; Gargas et al., 1990; Barton et al., 1995), holding all of the other model parameters fixed and requiring a single pair of values for VMAX1C and VMAX2C to be used for all of the data on a given sex/strain/species.

Initial estimates for the subsequent metabolism of the reactive metabolites and for the glutathione submodel in the rat were taken from the model for vinylidene chloride (D'Souza and Andersen, 1988). These parameter estimates, along with the estimates for VMAX2C and KM2, were then refined for the case of VC in the Sprague Dawley rat using an iterative fitting process which included the closed chamber data for the Sprague Dawley and Wistar rat (Bolt et al., 1977; Barton et al., 1995) along with data on glutathione depletion (Jedrychowski et al., 1985), total metabolism (Gehring et al., 1978), and CO_2 elimination (Watanabe and Gehring, 1976). The reactive metabolite/glutathione submodel parameters obtained for the rat were used for the other species with appropriate allometric scaling (e.g., body weight to the −1/4 for the first-order rate constants).

Parameterization of the P450 metabolism pathways in the human was accomplished as follows: There is no evidence of high capacity, low affinity P450 metabolism for chlorinated ethylenes in the human based on the results of occupational kinetic studies for another chemical, TCE; therefore, VMAX2C in the human was set to zero. The ratio of VMAX1C to KM1 could be estimated by fitting the model to data from closed chamber studies with two human subjects (Buchter et al., 1978), in a manner entirely analogous to the method used for the animal closed chamber analysis.

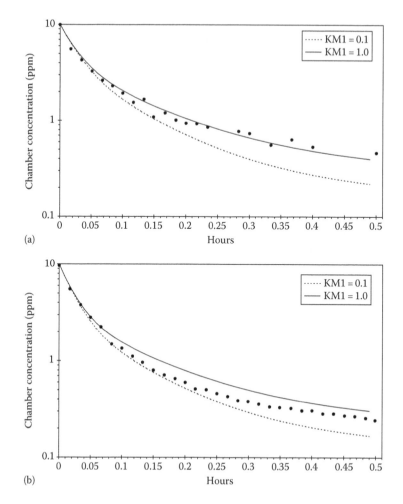

FIGURE 8.4 Model predictions (lines) and experimental data (symbols) for the chamber concentration during exposure of human subjects to VC in a closed, recirculated chamber (Data taken from Buchter et al. 1978). (a) Subject A—the lines show the model predictions for (top to bottom) KM1 = 1.0 and 0.1; (b) Subject B—the lines show the model predictions for (top to bottom) KM1 = 1.0 and 0.1.

The result of fitting the data on one of the two subjects is shown by the upper panel in Figure 8.4, and the model prediction using the value estimated from this subject is compared with the data from the second subject in the lower panel. The precision of the estimate of VMAX1C/KM1 can be evaluated by a comparison of the two model runs shown in these figures, for KM1 = 1.0 and KM1 = 0.1. It can be seen that the ratio of VMAX1C/KM1 varies between the two subjects. The wide variability in human CYP2E1 activity is an important consideration for estimating the potential difference between average population risk and individual risk in a human cancer risk assessment for materials like VC, whose carcinogenicity depends on metabolic activation.

In order to obtain separate estimates of VMAX1C and KM1 in the human, higher exposure concentration closer to metabolic saturation would be required. Fortunately, cross-species scaling of CYP2E1 between rodents and humans appears to follow allometric expectations for metabolism very closely; that is, the metabolic capacity scales approximately according to body weight raised to the 3/4 power (Andersen et al., 1987). Support for the application of this principle to VC can be obtained from data on the metabolism of VC in nonhuman primates (Buchter et al., 1980). Based on data for the dose-dependent metabolic elimination of VC in the rhesus monkey, the maximum capacity for metabolism can be estimated to be about 50 μmol/h/kg. This equates to a VMAX1C (the allometrically scaled constant used in the model) of approximately 4 mg/h for a 1 kg animal, a value which is in the same range as those estimated for the rodents from the closed chamber exposure data. The similarity of VMAX1C in humans and rats is also supported by an in vitro study which found the activity of human microsomes to be 84% of the activity of rat microsomes. Based on these comparisons, the human VMAX1C was set to the primate value, and KM1 was calculated using this value of VMAX1C and the ratio of VMAX1C/KM1 obtained from the closed chamber analysis. The ability of the resulting human model to reproduce constant concentration inhalation exposure data (Buchter et al., 1978) is shown in Figure 8.5. From a comparison of the model predictions for KM1 = 1.0 and KM1 = 0.1, it can be seen that the reproduction of parent chemical concentrations in a constant concentration inhalation exposure is not a particularly useful test of the accuracy of the metabolism parameters in a PBPK model of a volatile compound. Based on the results of the Monte Carlo analysis, the discrepancies or agreement between the model and the data are primarily due to details of the physiological description of the individual, such as fat content, ventilation rate, blood/air partition, etc., rather than the rate of metabolism.

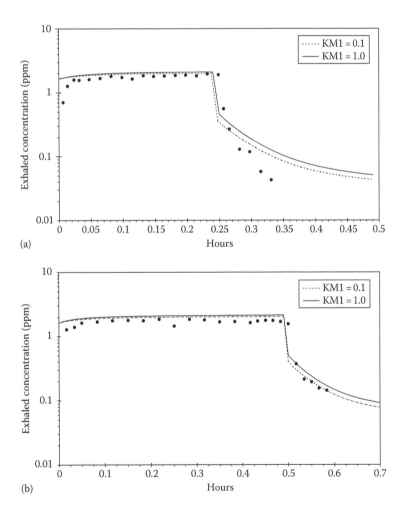

FIGURE 8.5 Model predictions (lines) and experimental data (symbols) for the exhaled air concentration during and following inhalation exposure of human subjects to a constant concentration of 2.5 ppm VC (Data taken from Buchter et al. 1978). (a) Subject A—the lines show the model predictions for (top to bottom) KM1 = 1.0 and 0.1; (b) Subject B—the lines show the model predictions for (top to bottom) KM1 = 1.0 and 0.1.

The model was used to calculate the pharmacokinetic dose metric for angiosarcoma (RISK) in the most informative of the animal bioassays, as well as for human inhalation exposure. These RISK dose metrics (mg metabolized per L liver per day, assuming the density of the liver is 1 kg/L) were then used to estimate the 95% upper confidence limit (UCL) on the human risk for lifetime exposure to 1 ppm VC from each of the sets of animal bioassay data. The resulting risk estimates for lifetime human exposure to 1 ppm VC, based on the RISK dose metric, range from 1.52×10^{-3} to 15.7×10^{-3}. Because saturation of metabolism occurs well above the 1 ppm concentration in the human, estimates of risk below 1 ppm can be adequately estimated by assuming linearity (e.g., the risk estimates for lifetime human exposure to 1 ppb of VC would range from approximately 1.52×10^{-6} to 15.7×10^{-6}). It should be noted that although the animal studies represent both inhalation and oral exposure, the risk predictions in each case are for human inhalation exposure.

There are no consistent differences between risk estimates based on male and female animals, with the female-based risks being higher than the male-based risks in some studies, and lower in others, but generally agreeing within a factor of two to three. The risk estimates based on inhalation studies with mice (0.5×10^{-3} to 4.3×10^{-3}/ppm) agree very well with those based on inhalation studies with rats (1.46×10^{-3} to 5.94×10^{-3}/ppm), demonstrating the ability of pharmacokinetics to integrate dose–response information across species.

The risks estimated from the dietary administration of VC (0.94×10^{-3} to 3.13×10^{-3}/ppm) are also in good agreement with those obtained from the inhalation bioassays, showing good route-to-route correspondence of potency based on the pharmacokinetic dose metric. However, the estimates based on oral gavage of VC in vegetable oil (6.58×10^{-3} to 16.3×10^{-3}/ppm) are about sixfold higher than either dietary or inhalation exposure. It has previously been noted in studies with chloroform that administration of the chemical in corn oil results in more marked hepatotoxic effects than observed when the same chemical is provided in an aqueous suspension (Bull et al., 1986). The toxicity and oxidative environment created in the liver by continual dosing with large volumes of vegetable oil could serve to potentiate the effects of genotoxic carcinogens in the liver. A similar phenomenon could be responsible for the apparently higher potency of VC when administered by oil gavage compared to incorporation in the diet.

In order to evaluate the plausibility of the risks predicted on the basis of the animal data, risk calculations were also performed on the basis of available epidemiological data (Clewell et al., 2001). The lifetime risk of liver cancer per ppm VC exposure estimated from the three studies only ranges over about one order of magnitude: from 0.4×10^{-3} to 4.2×10^{-3}. Moreover, these estimates are in excellent agreement with the estimates based on animal data. The agreement of the pharmacokinetic animal-based risk estimates with the pharmacokinetic human-based risk estimates provides strong support for the assumption used in this study: that cross-species scaling of lifetime cancer risk can be performed on a direct basis of lifetime average daily dose (without applying a body surface area adjustment) when the risks are based on biologically appropriate dose metrics calculated with a validated PBPK model. The consistency of this conclusion likely depends on the characteristics of the carcinogen, that is, trans-species carcinogens with common target organs across species.

The pharmacokinetic cancer risk assessment for VC demonstrates all of the attributes of an effective target tissue dose metric. First, the form of the metric (total daily metabolism divided by the volume of the liver) was consistent with the mode of action for the endpoint of concern (liver tumors), which involves DNA adduct formation by a highly reactive chloroethylene epoxide produced from the metabolism of vinyl chloride. Secondly, while the dose response for liver tumors vs. exposure concentration of VC is highly nonlinear, with a plateau at several hundred ppm, the dose response for liver tumors vs. the metabolized-dose metric is essentially linear from 1 to 6000 ppm. Finally, and most impressively, when the potency of VC liver carcinogenicity was expressed in terms of the metabolized-dose metric, essentially the same potency was calculated from both inhalation and oral studies in the mouse and rat, as well as from occupational inhalation exposures in the human. While it is rare to find a case where there is such consistency across widely diverse studies, a dose metric that adequately represents the biologically effective dose should generally have lower values under exposure conditions with no effect and higher values for toxic exposures, regardless of differences in exposure scenario, route, or species.

8.8 CFD MODELING EXAMPLE: ACROLEIN

The quantitative risk assessment of inhaled acrolein provides a good example of the use of CFD models in the estimation of exposure limits for inhaled irritant gases. Acrolein is a soluble and reactive aldehyde that is primarily used in the synthesis of other chemical compounds and is also formed from the combustion of organic materials. Human exposure to acrolein occurs mostly from inhalation of cigarette smoke, automobile exhaust, or being in close proximity to wood smoke (e.g., forest fires). Acrolein vapor has an acrid, pungent odor with an odor threshold of about 0.25 ppm (ATSDR, 2007). Exposure concentrations are typically less than 12 ppb in indoor air but can be much higher near cigarette smokers or other combustion sources.

Acrolein is a potent respiratory tract irritant that induces cytotoxicity at the site of contact. Inhaled acrolein has been shown to deplete rat nasal glutathione (Lam et al., 1985), stimulate cell proliferation in the rat respiratory tract (Roemer et al., 1993), and induce nasal pathology in rodents (Feron et al., 1978; Cassee et al., 1996; Dorman et al., 2008). Acrolein-induced nasal effects include olfactory neuronal loss, and inflammation, hyperplasia, and squamous metaplasia of the respiratory epithelium. Nasal lesions in rodents were used by the US EPA to establish an inhaled RfC for inhaled acrolein (US EPA, 1994). The US EPA inhalation risk assessment of acrolein is currently based on interspecies extrapolation of critical effects in the nasal passages of rodents, but relies on default assumptions of uptake and dosimetry between species.

Nasal uptake of acrolein was measured in the isolated upper respiratory tracts of anesthetized rats and was found to be flow rate and concentration dependent (Morris, 1996; Struve et al., 2008). Nasal uptake decreased as exposure concentration increased which is suggestive of saturable metabolic or reactive processes. In the upper respiratory tract of the rat, uptake ranged from as low as 28% at an exposure concentration of 9.1 ppm and a flow rate of 300 mL/min up to 98% at an exposure concentration of 0.6 ppm and a flow rate of 100 mL/min (Morris, 1996; Struve et al., 2008). Struve et al. (2008) also measured acrolein nasal uptake in rats that were preexposed to acrolein and reported an approximate 13%–38% increase in nasal uptake versus control animals that were preexposed to air.

Anatomically accurate CFD models of the rat and human nasal passages were developed and used to simulate steady-state inspiratory airflow and acrolein uptake to compare dosimetry among species (Figure 8.6) (Schroeter et al., 2008). A two-tiered epithelial structure of the nasal mucosa was used to include the effects of acrolein diffusivity, partitioning, first-order and saturable clearance terms, and blood perfusion on nasal uptake (Figure 8.7). The saturable component of acrolein clearance in the nasal mucosa was necessary to replicate the uptake behavior observed in the experimental studies. The physicochemical properties of acrolein were taken from the literature and the clearance terms, including the first-order term (k_f) and the Michaelis–Menten parameters (V_{max}, K_m), were calibrated to nasal uptake data in rats. Similar clearance properties were assumed in the human nasal passages, and the kinetic parameters were scaled accordingly to account for species differences in metabolic rates. A scaling factor based on the surface area of mucus-coated epithelium in the nasal passages was applied to the first-order clearance term, $k_f V_t$ (V_t is the nonsquamous tissue volume) and the maximal metabolic rate, V_{max}. The Michaelis–Menten parameter K_m was assumed to be species invariant.

Acrolein uptake simulations were conducted at exposure concentrations ranging from 0.1 to 10 ppm. Nasal uptake predictions in rodents showed good agreement with experimental data and accurately captured the flow rate and concentration dependence on uptake efficiency. Interspecies differences in acrolein nasal uptake were evaluated by conducting CFD simulations at twice the estimated minute volume for each species (434 mL/min in the rat and 13.8 L/min in the human). Predicted uptake in the human CFD model also demonstrated saturable uptake behavior, and uptake values were consistently

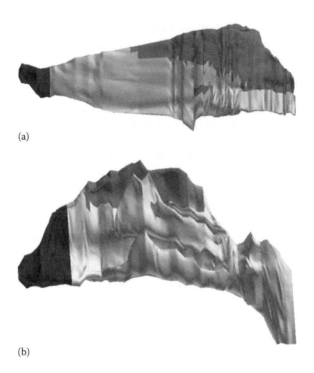

(a)

(b)

FIGURE 8.6 CFD models of (a) the rat and (b) human nasal passages used for acrolein nasal dosimetry calculations. The nasal surfaces are divided into squamous (black), olfactory (dark gray), and respiratory/transitional (light gray) epithelium.

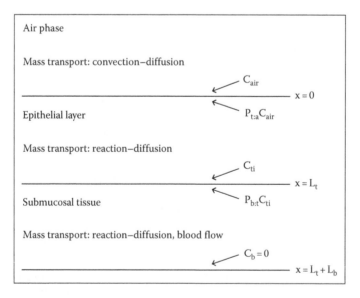

FIGURE 8.7 Schematic of the boundary condition governing the pharmacokinetics of mass transfer in the epithelial layer of the nasal CFD models for inhaled acrolein.

less than those in the rat nose. A sensitivity analysis on model parameters revealed that uptake predictions were most sensitive to the maximal metabolic rate, the air:tissue partition coefficient, and nasal tissue depth.

The risk assessment of inhaled acrolein relies on extrapolation of effects observed in the nasal passages of rodents. The current inhalation RfC, which is defined as an estimate of a continuous inhalation exposure to humans that is likely to be without appreciable risk of adverse effects over a person's lifetime (US EPA, 1994), is 2×10^{-5} mg/m^3 or about 9 ppt. The current RfC was based on nasal pathology in rats with a lowest observed adverse effect level (LOAEL) of 0.4 ppm (Feron et al., 1978). A minimal LOAEL was used in the absence of a no observed adverse effect level (NOAEL) for nasal pathology from the Feron et al. (1978) study. Inhalation RfC calculations by the US EPA rely on dosimetric adjustment factors to account for species differences in delivered dose. Once an NOAEL (or LOAEL) is identified from a critical effect, an adjusted NOAEL (NOAEL$_{[ADJ]}$) is computed by time-adjusting the exposure conditions from the study to continuous exposures (6–24 h/day, 5–7 days/week in this case). This value is then multiplied by the regional gas dose ratio (RGDR) to obtain an NOAEL human equivalent concentration (NOAEL$_{[HEC]}$). The RGDR is computed from the ratios of minute volume to upper respiratory tract surface areas between species and assumes complete absorption of the gas with a uniform dose throughout the nose. The nasal uptake studies (Morris, 1996; Struve et al., 2008) and the CFD modeling efforts (Schroeter et al., 2008) revealed that neither of these two assumptions was valid for acrolein. In other words, nasal uptake was shown to be <100% and the CFD results showed a nonuniform anterior-to-posterior gradient in wall absorption. The CFD models can be used to estimate species differences in nasal dosimetry of inhaled acrolein to replace the default assumptions used in the calculation of the RGDR.

The nasal toxicity of acrolein was further studied by Dorman et al. (2008). Lower exposure concentrations were used to establish an NOAEL for respiratory and olfactory effects. In their study, adult male F344 rats were exposed by inhalation to acrolein concentrations of 0, 0.02, 0.06, 0.2, 0.6, or 1.8 ppm for 6 h/day, 5 days/week for up to 90 days. Nasal histopathology was evaluated at transverse nasal sections corresponding to six standard levels (Morgan, 1991). Cross-sections located at Levels I–IV consisted of respiratory and olfactory epithelium; Levels V and VI passed through the nasopharyngeal duct. They identified an NOAEL of 0.2 ppm for respiratory epithelial hyperplasia, inflammation, and squamous metaplasia and an NOAEL of 0.6 ppm for olfactory neuronal loss. Even though the NOAEL for respiratory effects was lower than that for olfactory neuronal loss, the olfactory NOAEL was used in subsequent risk assessment calculations with the CFD models since the olfactory areas in the rat received lower tissue doses of inhaled acrolein.

For regional analysis of nasal histopathology, cross-sectional airway perimeters from Levels I to IV were divided into regions based on anatomical landmarks (Figure 8.8). Regions were of similar length and were numbered consecutively beginning on the ventral aspect of the septum. All of Level I and most of Levels II and III were comprised of respiratory epithelium. Olfactory epithelium was present in region 6 of Level II, regions 4–6 and 16 of Level III, and regions 1–15 of Level IV. Lesion incidence was determined in each region of each level for a detailed localized mapping of acrolein-induced nasal histopathology. Nasal lesions on respiratory epithelium were observed in Levels I and II following 0.6 or 1.8 ppm acrolein exposure. Respiratory lesions in Level III were only observed in the 1.8 ppm exposure case. Olfactory lesions were observed in Levels II–IV following exposure to 1.8 ppm acrolein.

Coronal cross-sections corresponding to Levels I–IV were identified in the rat nasal CFD model (Figure 8.8). The airway perimeters were divided into regions that were identical in scope to those used for the lesion incidence mapping. Average wall mass flux of inhaled acrolein was computed in each region of each level at the exposure concentrations used in the Dorman study. Acrolein air:tissue flux showed a gradual decrease in magnitude, progressing from Levels I to IV due to higher absorption rates in the anterior regions of the nose.

Average flux in each region predicted by the CFD model and lesion incidence were ranked, and the correlation between the two sets of data was tested using the Spearman rank correlation coefficient. A correlation between flux and lesion incidence could not be established in Levels I and II since most lesion incidence values were at or near 100%. Strong correlations were found between flux and lesion

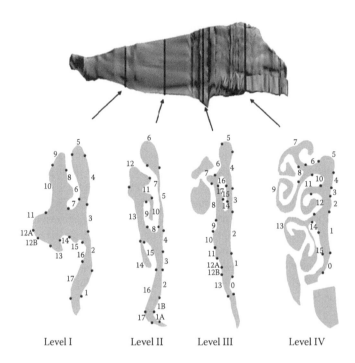

FIGURE 8.8 Septal view of the rat nasal CFD model displaying the locations of transverse cross-sections (Levels I–IV) where average acrolein flux predictions were compared with lesion incidence.

incidence on respiratory epithelium in Level III and on olfactory epithelium in Levels III and IV, indicating that high tissue doses of inhaled acrolein are suggestive of lesion formation (Figure 8.9). Using this knowledge, CFD results for acrolein dosimetry were used to estimate a tissue-dose-based point-of-departure to form the basis of a quantitative risk assessment for inhaled acrolein that does not rely on simplifying assumptions for interspecies extrapolation. In Level IV, the highest predicted flux at an exposure concentration of 0.6 ppm (the NOAEL) was 191 pg/cm^2-s, occurring in region 4. Relatively high flux values also occurred in multiple regions in Level IV at this exposure concentration (Figure 8.9). However, flux values <191 pg/cm^2-s occurred at an exposure concentration of 1.8 ppm (the LOAEL) in regions 7–9, 11, and 13 (Figure 8.9). In regions 7, 9, and 13, lesion incidence was not statistically elevated over control animals. Lesion incidence in regions 8 and 11 was significant (8/12 and 7/12, respectively). The lowest flux value in these two regions was 72 pg/cm^2-s. This value was considered a threshold flux value for olfactory lesions in the rat nose and, assuming that equal tissue doses will elicit similar responses in olfactory epithelium of rats and humans, was used as a tissue-dose-based point-of-departure in subsequent calculations of exposure limits for humans.

Olfactory flux values were calculated in the human CFD model over a wide range of acrolein exposure concentrations. At each concentration, olfactory flux values at each nodal point of the computational mesh were rank-ordered to determine the 99th percentile value (i.e., the flux value which was greater than 99% of all olfactory flux values). The 99th percentile flux value was used in place of the actual maximum flux to overcome possible inaccuracies in computing wall mass fluxes in the nasal passages. An exposure concentration of 45 ppb acrolein produced a 99th percentile olfactory flux value equal to the threshold flux value of 72 pg/cm^2-s from the rat. Subsequently adjusting for the noncontinuous exposure conditions used in the Dorman study, a tissue-dose based NOAEL$_{[HEC]}$ for acrolein was estimated to be approximately 8 ppb.

Uncertainty factors (UFs) are usually applied by the US EPA to the NOAEL[HEC] to arrive at the final RfC. In its acrolein assessment, the US EPA applied a UF of 3 for interspecies extrapolation, a UF of 10 for human variability, a UF of 10 for adjustment from subchronic to chronic, and a UF of 3 to account for the lack of an NAOEL in the Feron study. For the quantitative risk assessment

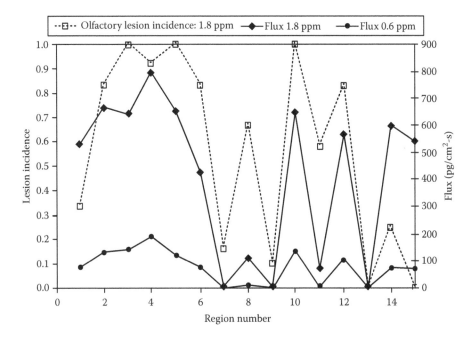

FIGURE 8.9 Acrolein-induced olfactory lesion incidence (1.8 ppm) and wall mass flux predictions (0.6 and 1.8 ppm) in Level IV of the rat nasal cavity.

approach using the CFD analyses, Schroeter et al. (2008) advocated using a composite UF of 30, which includes a factor of 10 for human variability and a factor of 3 for possible species differences in pharmacodynamics. The UF of 10 for human variability was not changed since the CFD analysis only considered a single adult human nasal model and did not examine effects of interhuman changes in nasal geometry on acrolein dosimetry. A UF of 3 for pharmacodynamics differences was also applied since the CFD models only looked at pharmacokinetic differences. The UF of 10 for subchronic to chronic extrapolation was deemed unnecessary for agents that target the olfactory neuron, since past studies with other agents have shown little to no progression of olfactory lesions after subchronic exposure. The UF of 3 for lack of an NOAEL was also not needed since the traditional NOAEL/LOAEL approach was abandoned in favor of a tissue-dose-based quantitative analysis. Applying the UFs of 30, a proposed inhalation RfC of 0.27 ppb was obtained.

Soluble and reactive gases such as acrolein display moderate to high uptake in the upper respiratory tract. Default assumptions used in the risk assessment process do not take into account interspecies differences in uptake or absorption rates, which can be vastly different between laboratory animals and humans. In this case study with acrolein, CFD models of the rat and human nasal passages were used to elucidate differences in uptake patterns between the two species and use highly localized predictions of acrolein uptake rates to perform a quantitative risk assessment for this chemical. Since human nasal extraction of inhaled acrolein is much lower than the rat, there is also concern about lung penetration. However, Corley et al. (2012) recently showed that lung flux rates of inhaled acrolein were lower than those found in the nasal passages using a CFD model of the combined upper and lower airways. However, acrolein vapor did penetrate further during oral breathing due to lower uptake rates in the mouth. This could be a concern when considering oronasal breathing of inhaled vapors.

8.9 CONCLUDING REMARKS

The PBPK modeling approaches discussed in this chapter represent an evolution from simpler, compartmental kinetic models toward more realistic, biological descriptions of the determinants that regulate the disposition of inhaled chemicals in the body. The development of these PBPK models was frequently met with skepticism because they required a large number of variables for metabolism, transport, binding, etc. The older classical PK models had a much more sparse set of variables. Concerns were raised about the ability to obtain values for all these parameters. However, many of these constants were identified from experiments separate from collection of time-course studies or were known from physiological or anatomical research. These independently measured parameters could be introduced into structured PBPK models and simulations (predictions) from the model compared to data to decide if the model structure were accurate or in need of reparameterization or more extensive reformulation. This iteration of proposed physiological structures for the models and testing them against data is simply a form of adherence to the scientific method: hypothesis generation and testing applied to problems of chemical inhalation and disposition.

In the disciplines of toxicology and chemical risk assessment, PBPK modeling has become a well-established tool for research and analysis. The technologies applied so successfully in developing PBPK models will naturally support their migration to applications in systems biology for describing cell signaling pathways, perturbations of these pathways by exogenous compounds, and adaptation to perturbations with prolonged exposures. The immediate future has many opportunities for the quantitative analysis of various pharmacokinetic and biological studies using computer modeling and systems approaches for kinetics and dynamics.

QUESTIONS

1. Describe at least three opportunities offered by PBPK approaches.
   Answer: Among the opportunities offered by PBPK approaches are (1) creating models from physiological, biochemical, and anatomical information, entirely separate from collection of detailed concentration time-course curves; (2) evaluating mechanisms by which biological processes govern disposition of a wide range of compounds by comparison of PK results with model predictions; (3) using chemicals as probes of the biological processes to gain more general information on the way chemical characteristics govern the importance of various transport pathways in the body; (4) applying the models in risk assessments for setting exposure standards; and (5) using annotation of a modeling database as a repository of information on toxicity and kinetics of specific compounds.

2. A vial equilibration technique has been used to determine the partitioning of a volatile compound in rat tissues. In this method, the chemical is introduced into the headspace of the vial and allowed to equilibrate with a tissue homogenate. The headspace is then sampled to estimate the tissue:air ratio. The blood:air, muscle:air, liver:air, and fat:air partitions determined for this compound are 2.4, 1.0, 1.7, and 24.0, respectively. What are the tissue:blood partition coefficients for this compound in the rat?
   Answer: Muscle:blood = 0.4, liver:blood = 0.7, and fat:blood = 10.0.

3. For the compound in question 2, the human blood:air is 1.2. What are the tissue:blood partition coefficients for this compound in the human?
   Answer: Muscle:blood = 0.8, liver:blood = 1.4, and fat:blood = 20.0.

4. What is the role of CFD modeling in hybrid CFD/PBPK models?
   Answer: Estimation of tissue uptake of the chemical from the upper respiratory tract airway.

5. For inhalation of relatively soluble compounds that are rapidly metabolized in the liver, why might toxicity be produced in the upper respiratory tract at lower exposure concentrations than in the liver or kidney?
   Answer: Because of the direct equilibration of upper airway tissues with inhaled chemical in the airway before it is absorbed in the alveolar region.

REFERENCES

Agency for Toxic Substances and Disease Registry. (2007). *Toxicological Profile for Acrolein.* Atlanta, GA: U.S. Public Health Service.

Andersen, M.E., Clewell, H.J., Gargas, M.L., Smith, F.A., and Reitz, R.H. (1987). Physiologically based pharmacokinetics and the risk assessment process for methylene chloride. *Toxicol. Appl. Pharmacol.* 87, 185–205.

Andersen, M.E., Sarangapani, R., Gentry, P.R., Clewell, H.J., III, Covington, T.R., and Frederick, C.B. (2000). Application of a hybrid CFD-PBPK nasal dosimetry in an inhalation risk assessment: An example with acrylic acid. *Toxicol. Sci.* 57, 312–325.

Astrand, P. and Rodahl, K. (1970). *Textbook of Work Physiology.* McGraw-Hill, New York.

Barton, H.A., Creech, J.A., Godin, C.S., Randall, G.M., and Seckel, C.S. (1995). Chloroethylene mixtures: Pharmacokinetic modeling and in vitro metabolism of vinyl chloride, trichloroethylene, and trans-1,2-dichloroethylene in the rat. *Toxicol. Appl. Pharmacol.* 130, 237–247.

Bischoff, K.B. and Brown, R.H. (1966). Drug distribution in mammals. *Chem. Eng. Prog. Symp. Series* 62, 33–45.

Bischoff, K.B., Dedrick, R.L., Zaharko, D.S., and Longstreth, J.A. (1971). Methotrexate pharmacokinetics. *J. Pharm. Sci.* 60, 1128–1133.

Bogdanffy, M.S., Sarangapani, R., Plowchalk, D.R., Jarabek, A., and Andersen, M.E. (1999). A biologically-based risk assessment for vinyl acetate-induced cancer and non-cancer inhalation toxicity. *Toxicol. Sci.* 51, 19–35.

Bolt, H.M., Laib, R.J., Kappus, H. et al. (1977). Pharmacokinetics of vinyl chloride in the rat. *Toxicology* 7, 179–188.

Brown, R.P., Delp, M.D., Lindstedt, S.L., Rhomberg, L.R., and Beliles, R.P. (1997). Physiological parameter values for physiologically based pharmacokinetic models. *Toxicol. Ind. Health* 13, 407–484.

Buchter, A., Bolt, H.M., Filser, J.G., Goergens, H.W., Laib, R.J., and Bolt, W. (1978). Pharmakokinetic und karzinogenese von vinylchlorid. Arbeitsmedizinische Risikobeurteilung. Verhandlungen der Deutchen Gesellschaft fuer Arbeitsmedizin, vol. 18, Gentner Verlag, Stuttgart, Germany, pp. 111–124.

Buchter, A., Filser, J.G., Peter, H., and Bolt, H.M. (1980). Pharmacokinetics of vinyl chloride in the Rhesus monkey. *Toxicol. Lett.* 6, 33–36.

Bull, R.J., Brown, J.M., Meierhenry, E.A. et al. (1986). Enhancement of the hepatotoxicity of chloroform in B6C3F1 mice by corn oil: Implications for chloroform carcinogenesis. *Environ. Health Perspect.* 69, 49–58.

Bush, M.L., Frederick, C.B., Kimbell, J.S., and Ultman, J.S. (1998). A CFD-PBPK hybrid model for simulating gas and vapor uptake in the rat nose. *Toxicol. Appl. Pharmacol.* 150, 133–145.

Cassee, F.R., Groton, J.P., and Feron, V.J. (1996). Changes in the nasal epithelium of rats exposed by inhalation to mixtures of formaldehyde, acetaldehyde, and acrolein. *Fundam. Appl. Toxicol.* 29, 208–218.

Clewell, H.J., III and Andersen, M.E. (1985). Risk assessment extrapolations and physiological modeling. *Toxicol. Ind. Health* 1, 111–131.

Clewell, H.J., Gentry, R., Gearhart, J.M., Allen, B.C., and Andersen, M.E. (2001). Comparison of cancer risk estimates for vinyl chloride using animal and human data with a PBPK model. *Sci. Total Environ.* 274, 37–66.

Clewell, R.A. and Clewell, H.J. 2008. Development and specification of physiologically based pharmacokinetic models for use in risk assessment. *Regul. Toxicol. Pharm.* 50, 129–143.

Cohen Hubal, E.A., Kimbell, J.S., and Fedkiw, P.S. (1996). Incorporation of nasal-lining mass-transfer resistance into a CFD model for prediction of ozone dosimetry in the upper respiratory tract. *Inhal. Toxicol.* 8, 831–857.

Collins, J.M., Dedrick, R.L., Flessner, M.F., and Guarino, A.M. (1982). Concentration dependent disappearance of fluorouracil from peritoneal fluid in the rat: Experimental observations and distributed modeling. *J. Pharm. Sci.* 71, 735–738.

Corley, R.A., Kabilan, S., Kuprat, A.P., Carson, J.P., Minard, K.R., Jacob, R.E., Timchalk, C. et al. (2012). Comparative computational modeling of airflows and vapor dosimetry in the respiratory tracts of rat, monkey, and human. *Toxicol. Sci.* 128, 500–516.

Csanady, G.A., Kessler, W., Hoffmann, H.D., and Filser, J.G. (2003). A toxicokinetic model for styrene and its metabolite styrene-7,8-oxide in mouse, rat and human with special emphasis on the lung. *Toxicol. Lett.* 138, 75–102.

Csanady, G.A., Mendrala, A.L., Nolan, R.J., and Filser, J.G. (1994). A physiologic pharmacokinetic model for styrene and 7,8-styrene oxide in mouse, rat, and man. *Arch. Toxicol.* 68, 143–157.

Dedrick, R.L. (1973). Animal scale-up. *J. Pharmacokinet. Biopharm.* 1, 435–461.

Dorman, D.C., Struve, M.F., Wong, B.A., Marshall, M.W., Gross, E.A., and Willson, G. (2008). Respiratory tract responses in male rats following subchronic acrolein inhalation. *Inhal. Toxicol.* 20, 205–216.

D'Souza, R.W. and Andersen, M.E. (1988) Physiologically based pharmacokinetic model for vinylidene chloride. *Toxicol. Appl. Pharmacol.* 95, 230–240.

Farris, F.F., Dedrick, R.L., and King, F.G. (1988). Cisplatin pharmacokinetics: Applications of a physiological model. *Toxicol. Lett.* 43, 117–137.

Federal Register. (June 26, 1996). Part II Environmental Protection Agency. 40 CFR 799. Proposed test rule for hazardous air pollutants; Proposed rule. *Fed. Regist.* 33178–33200.

Feron, V.J., Kruysse, A., Til, H.P., and Immel, H.R. (1978). Repeated exposure to acrolein vapor: Subacute studies in hamsters, rats and rabbits. *Toxicology* 9, 47–57.

Filser, J.G. and Bolt, H.M. (1979). Pharmacokinetics of halogenated ethylenes. *Archiv. Toxicol.* 42, 123–136.

Fiserova-Bergerova, V. (1975). Mathematical modeling of inhalation exposure. *J. Conibust. Toxicol.* 32, 201–210.

Fiserova-Bergerova, V. and Holaday, D.A. (1979). Uptake and clearance of inhalation anesthetics in man. *Drug Metab. Rev.* 9, 43–60.

Fiserova-Bergerova, V., Vlach, J., and Cassady, J.C. (1980). Predictable "individual differences" in uptake and excretion of gases and lipid soluble vapours simulation study. *Br. J. Ind. Med.* 37, 42–49.

Frederick, C., Bush, M.L., Subramaniam, R.P., Black, K.A., Finch, L., Kimbell, J.S., Morgan, K.T., Subramaniam, R.P., Morris, J.B., and Ultman, J.S. (1998). Application of a hybrid computational fluid dynamics and physiologically based inhalation model for interspecies dosimetry extrapolation of acidic vapors in the upper airways. *Toxicol. Appl. Pharmacol.* 152, 211–231.

Gargas, M.L., Andersen, M.E., and Clewell, H.J., III (1986). A physiologically based simulation approach for determining metabolic constants from gas uptake data. *Toxicol. Appl. Pharmacol.* 86, 341–352.

Gargas, M.L., Burgess, R.J., Voisard, D.E., Cason, G.H., and Andersen, M.E. (1989). Partition coefficients of low-molecular-weight volatile chemicals in various liquids and tissues. *Toxicol. Appl. Pharmacol.* 98, 87–99.

Gargas, M.L., Clewell, H.J., III, and Andersen, M.E. (1990). Gas uptake inhalation techniques and the rates of metabolism of chloromethanes, chloroethanes, and chloroethylenes in the rat. *Inhal. Toxicol.* 2, 285–309.

Gehring, P.J., Watanabe, P.G., and Blau, G.E. (1976). Pharmacokinetic studies in evaluation of the toxicological and environmental hazard of chemicals. In: *New Concepts in Safety Evaluation* (*Advances in Modern Toxicology*, vol. 1), M.A. Mehlman, R.E. Shapiro, and H.Blumenthal, eds., Hemisphere Publishing Corporation, New York, pp. 193–270.

Gehring, P.J., Watanabe, P.G., and Park, C.N. (1978). Resolution of dose-response toxicity data for chemicals requiring metabolic activation: Example—Vinyl chloride. *Toxicol. Appl. Pharmacol.* 44, 581–591.

Gehring, P.J., Watanabe, P.G., and Young, J.D. (1977). The relevance of dose-dependent pharmacokinetics in the assessment of carcinogenic hazard of chemicals. In: *Origins of Human Cancer, Book A: Incidence of Cancer in Humans* (*Cold Spring Harbor Conferences on Cell Proliferation*, vol. 4). H.H. Hiatt, J.D. Watson, and J.A. Winsten, eds., Cold Spring Harbor Laboratory, Cold Spring-Harbor, New York, pp. 187–203.

Gerlowski, L.E. and Jain, R.J. (1983). Physiologically based pharmacokinetic modeling: Principles and applications. *J. Pharm. Sci.* 72, 1103–1126.

Haggard, H.W. (1924a). The absorption, distribution, and elimination of ethyl ether. II. Analysis of the mechanism of the absorption and elimination of such a gas or vapor as ethyl ether. *J. Biol. Chem.* 59, 753–770.

Haggard, H.W. (1924b). The absorption, distribution, and elimination of ethyl ether. III. The relation of the concentration of ether, or any similar volatile substance, in the central nervous system to the concentration in the arterial blood, and the buffer action of the body. *J. Biol. Chem.* 59, 771–781.

Health Canada. (1993). Canadian environmental protection act, priority substances list assessment report. Dichloromethane. Canada Communications Group—Publishing, Ottawa, Ontario, Canada, K1A 0S.

Hefner, R.E., Watanabe, P.G., and Gehring, P.J. (1975). Preliminary studies of the fate of inhaled vinyl chloride monomer in rats. *Ann. N. Y. Acad. Sci.*, 246, 135–148.

Hendersen, Y. and Haggard, H.W. (1943). *Noxious Gases and the Principles of Respiration Influencing Their Action.* American Chemical Society Monograph Series, Reinhold Publishing Corporation, New York.

Hilderbrand, R.L., Andersen, M.E., and Jenkins, L.J. (1981). Prediction of in vivo kinetic constants for metabolism of inhaled vapors from kinetic constants measured in vitro. *Fundam. Appl. Toxicol.* 1, 403–409.

Himmelstein, K. and Lutz, R.J. (1979). A review of the application of physiologically based pharmacokinetic modeling. *J. Pharmacokinet. Biopharm.* 7, 127–137.

Jedrychowski, R.A., Sokal, J.A., and Chmielnicka, J. (1985). Comparison of the impact of continuous and intermittent exposure to vinyl chloride, including phenobarbital effects. *J. Hyg. Epidemiol. Microbiol. Immunol.* 28, 111–120.

Johanson, G. (1991). Modeling of respiratory exchange of polar-solvents. *Ann. Occup. Hyg.* 35, 323–339.

Kedderis, G.L., Carfagna, M.A., Held, S.D., Batra, R., Murphy, J.E., and Gargas, M.L. (1993). Kinetic analysis of furan biotransformation by F-344 rats in vivo and in vitro. *Toxicol. Appl. Pharmacol.* 123, 274–282.

Kepler, G.M., Richardson, R.B., Morgan, K.T., and Kimbell, J.S. (1998). Computer simulation of inspiratory nasal airflow and inhaled gas uptake in a rhesus monkey. *Toxicol. Appl. Pharmacol.* 150, 1–11.

Kety, S.S. (1951). The theory and applications of the exchange of inert gas at the lungs. *Pharmacol. Rev.* 3, 1–41.

Keyhani, K., Scherer, P.W., and Mozell, M.M. (1995). Numerical solution of airflow in the human nasal cavity. *J. Biomech. Eng.* 117, 429–441.

Kimbell, J.S., Godo, M.N., Gross, E.A., Joyner, D.R., Richardson, R.B., and Morgan, K.T. (1997a). Computer simulation of inspiratory airflow in all regions of the F344 rat nasal passages. *Toxicol. Appl. Pharmacol.* 145, 388–398.

Kimbell, J.S., Gross, E.A., Joyner, D.R., Godo, M.N., and Morgan, K.T. (1993). Application of computational fluid dynamics to regional dosimetry of inhaled chemicals in the upper respiratory tract of the rat. *Toxicol. Appl. Pharmacol.* 121, 253–263.

Kimbell, J.S., Gross, E.A., Richardson, R.B., Conolly, R.B., and Morgan, K.T. (1997b). Correlation of regional formaldehyde flux predictions with the distribution of formaldehyde-induced squamous metaplasia in F344 rat nasal passages. *Mutat. Res.* 380, 143–154.

Kimbell, J.S., Overton, J.H., Subramaniam, R.P., Schlosser, P.M., Morgan, K.T., Conolly, R.B., and Miller, F.J. (2001a). Dosimetry modeling of inhaled formaldehyde: Binning nasal flux predictions for quantitative risk assessment. *Toxicol. Sci.* 64, 111–121.

Kimbell, J.S., Subramaniam, R.P., Gross, E.A., Schlosser, P.M., and Morgan, K.T. (2001b). Dosimetry modeling of inhaled formaldehyde: Comparisons of local flux predictions in the rat, monkey, and human nasal passages. *Toxicol. Sci.* 64, 100–110.

Krishnan, K. and Andersen, M.E. (2001). Physiologically based pharmacokinetic modeling in toxicology. In: *Principles and Methods of Toxicology*, A.W. Hayes, ed., Taylor & Francis, Philadelphia, PA pp. 193–241.

Kumagai, S. and Matsunaga, I. (1995). Physiologically-based pharmacokinetic model for acetone. *Occup. Environ. Med.* 52, 344–352.

Lam, C.W., Casanova, M., and Heck, H.D. (1985). Depletion of nasal mucosal glutathione by acrolein and enhancement of formaldehyde-induced DNA-protein cross-linking by simultaneous exposure to acrolein. *Arch. Toxicol.* 58, 67–71.

Leung, H.W. (1991). Development and utilization of physiologically based pharmacokinetic models for toxicological applications. *J. Toxicol. Environ. Health* 32, 247–267.

Mapleson, W.W. (1963). An electric analog for uptake and exchange of inert gases and other agents. *J. Appl. Physiol.* 18, 197–204.

McKenna, M.J., Zempel, J.A., and Braun, W.H. (1982). The pharmacokinetics of inhaled methylene chloride in rats. *Toxicol. Appl. Pharmacol.* 65, 1–10.

McKenna, M.J., Zempel, J.A., Madrid, E.O., Braun, W.H., and Gehring, P.J. (1978a). Metabolism and pharmacokinetic profile of vinylidene chloride in rats following oral administration. *Toxicol. Appl. Pharmacol.* 45, 821–835.

McKenna, M.J., Zempel, J.A., Madrid, E.O., and Gehring, P.J. (1978b). The pharmacokinetics of [14C]vinylidene chloride in rats following inhalation exposure. *Toxicol. Appl. Pharmacol.* 45, 599–610.

McLanahan, E.D., El-Masri, H., Sweeney, L.M., Kopylev, L., Clewell, H.J., Wambaugh, J., and Schlosser, P.M. (2012). Physiologically based pharmacokinetic model use in risk assessment—Why being published is not enough. *Toxicol. Sci.* 126(1), 5–15.

Morgan, K.T. (1991). Approaches to the identification and recording of nasal lesions in toxicology studies. *Toxicol. Pathol.* 19, 337–351.

Morgan, K.T., Kimbell, J.S., Monticello, T.M., Patra, A.L., and Fleishman, A. (1991). Studies of inspiratory airflow patterns in the nasal passages of the F344 rat and rhesus monkey using nasal molds: Relevance to formaldehyde toxicity. *Toxicol. Appl. Pharmacol.* 110, 223–240.

Morris, J.B. (1996). Uptake of acrolein in the upper respiratory tract of the F344 rat. *Inhal. Toxicol.* 8, 387–403.

Morris, J.B., Hassett, D.N., and Blanchard, K.T. (1993). A physiologically based pharmacokinetic model for nasal uptake and metabolism of non-reactive vapors. *Toxicol. Appl. Pharmacol.* 123, 120–129.

Morris, J.B. and Hubbs, A.F. (2009). Inhalation dosimetry of diacetyl and butyric acid, two components of but-
ter flavoring vapors. *Toxicol. Sci.* 108, 173–183.

Moulin, F.J.M., Brenneman, K.A., Kiimbell, J.S., and Dorman, D.C. (2002). Predicted regional flux of hydrogen sulfide correlates with distribution of nasal olfactory lesions in rats. *Toxicol. Sci.* 66, 7–15.

National Research Council. (1987). *Pharmacokinetics in Risk Assessment in Drinking Water and Health*,
vol. 8. National Academy Press, Washington, DC, 487pp.

Occupational Safety and Health Administration (OSHA). (1997). Occupational exposure to methylene chloride; final rule. *Fed. Reg.* 62(7), 1493–1619.

Parham, F.M., Kohn, M.C., Matthews, H.B., DeRosa, C., and Portier, C.J. (1997). Using structural information
to create physiologically based pharmacokinetic models for all polychlorinated biphenyls. *Toxicol. Appl.
Pharmacol.* 144, 340–347.

Parham, F.M. and Portier, C.J. (1998). Using structural information to create physiologically based pharmacokinetic models for all polychlorinated biphenyls. II. Rates of metabolism. *Toxicol. Appl. Pharmacol.*
151, 110–116.

Plowchalk, D.R., Andersen, M.E., and Bogdanffy, M.S. (1997) Physiologically-based modeling of vinyl acetate uptake, metabolism and intracellular pH changes in the rat nasal cavity. *Toxicol. Appl. Pharmacol.*
142, 386–400.

Poulin, P. and Krishnan, K. (1996). Molecular structure-based prediction of the partition coefficients of organic
chemicals for physiological pharmacokinetic models. *Toxicol. Methods* 6, 117–137.

Poulin, P. and Krishnan, K. (1999). Molecular structure-based prediction of the toxicokinetics of inhaled vapors
in humans. *Int. J. Toxicol.* 18, 7–18.

Ramsey, J.C. and Andersen, M.E. (1984). A physiologically based description of the inhalation pharmacokinetics of styrene in rats and humans. *Toxicol. Appl. Pharmacol.* 73, 159–175.

Ramsey, J.C., Young, J.D., Karbowski, R., Chenoweth, M.B., McCarty, L.P., and Braun, W.H. (1980).
Pharmacokinetics of inhaled styrene in human volunteers. *Toxicol. Appl. Pharmacol.* 53, 54–63.

Reddy, M.B., Yang, R.S.H., Clewell, H.J., and Andersen, M.E. (2005). *Physiologically Based Pharmacokinetic
Modeling: Science and Applications*. John Wiley & Sons, Hoboken, NJ.

Riggs, D.S. (1963). *The Mathematical Approach to Physiological Problems: A Critical Primer*. MIT Press,
Cambridge, MA, 445pp.

Roemer, E., Anton, H.J., and Kindt, R. (1993). Cell proliferation in the respiratory tract of the rat after acute
inhalation of formaldehyde or acrolein. *J. Appl. Toxicol.* 13, 103–107.

Sarangapani, R., Teeguarden, J.G., Cruzan, G., Clewell, H.J., and Andersen, M.E. (2002). Physiologically
based pharmacokinetic modeling of styrene and styrene oxide respiratory-tract dosimetry in rodents and
humans. *Inhal. Toxicol.* 14, 789–834.

Sato, A. and Nakajima, T. (1979a). Partition coefficients of some aromatic hydrocarbons and ketones in water,
blood and oil. *Br. J. Ind. Med.* 36, 231–234.

Sato, A. and Nakajima, T. (1979b). A vial equilibration method to evaluate the drug metabolizing enzyme activity for volatile hydrocarbons. *Toxicol. Appl. Pharmacol.* 47, 41–46.

Sauerhoff, M.W., Braun, W.H., Blau, G.E., and Gehring, P.J. (1976). The dose dependent pharmacokinetic
profile of 2,4,5-trichlorophenoxy acetic acid following intravenous administration to rats. *Toxicol. Appl.
Pharmacol.* 36, 491–501.

Sauerhoff, M.W., Braun, W.H., and LeBeau, J.E. (1977). Dose dependent pharmacokinetic profile of silvex
following intravenous administration in rats. *J. Toxicol. Environ. Health* 2, 605–618.

Schroeter, J.D., Kimbell, J.S., Andersen, M.E., and Dorman, D.C. (2006b). Use of a pharmacokinetic-driven
computational fluid dynamics model to predict nasal extraction of hydrogen sulfide in rats and humans.
Toxicol. Sci. 94, 359–367.

Schroeter, J.D., Kimbell, J.S., Bonner, A.M., Roberts, K.C., Andersen, M.E., and Dorman, D.C. (2006a).
Incorporation of tissue reaction kinetics in a computational fluid dynamics model for nasal extraction of
inhaled hydrogen sulfide in rats. *Toxicol. Sci.* 90, 198–207.

Schroeter, J.D., Kimbell, J.S., Gross, E.A., Willson, G.A., Dorman, D.C., Tan, Y.M., and Clewell, H.J. III.
(2008). Application of physiological computational fluid dynamics models to predict interspecies nasal
dosimetry of inhaled acrolein. *Inhal. Toxicol.* 20, 227–243.

Struve, M.F., Wong, V.A., Marshall, M.W., Kimbell, J.S., Schroeter, J.D., and Dorman, D.C. (2008). Nasal
uptake of inhaled acrolein in rats. *Inhal. Toxicol.* 20, 217–225.

Subramaniam, R.P., Richardson, R.B., Morgan, K.T., Kimbell, J.S., and Guilmette, R.A. (1998). Computational
fluid dynamics simulations of inspiratory airflow in the human nose and nasopharynx. *Inhal. Toxicol.*
10, 91–120.

Teeguarden, J.G., Bogdanffy, M.S., Covington, T.R., Tan, C., and Jarabek, A.M. (2008). A PBPK model for evaluating the impact of aldehyde dehydrogenase polymorphisms on comparative rat and human nasal tissue acetaldehyde dosimetry. *Inhal. Toxicol.* 20, 375–390.

Teorell, T. (1937a). Kinetics of distribution of substances administered to the body. I. The extravascular modes of administration. *Arch. Int. Pharmacodyn.* 57, 205–225.

Teorell, T. (1937b). Kinetics of distribution of substances administered to the body. II. The intravascular mode of administration. *Arch. Int. Pharmacodyn.* 57, 226–240.

US Environmental Protection Agency (USEPA). (1988). Reference physiological parameters in pharmacokinetic modeling. EPA/600/6-88/004. Office of Health and Environmental Assessment, Washington, DC.

US Environmental Protection Agency. (1994). Methods for derivation of inhalation reference concentrations and application of inhalation dosimetry, EPA/600/8-90/066F. Office of Research and Development, Washington, DC.

US Environmental Protection Agency. (2000a). Toxicological review of vinyl chloride. Appendices A-D. EPA/635R-00/004. Washington, DC.

Wang, S.M., Inthavong, K., Wen, J., Tu, J.Y., and Xue, C.L. (2009). Comparison of micron- and nanoparticle deposition patterns in a realistic human nasal cavity. *Respir. Physiol. Neurobiol.* 166, 142–151.

Watanabe, P.G. and Gehring, P.J. (1976). Dose-dependent fate of vinyl chloride and its possible relationship to oncogenicity in rats. *Environ. Health Perspect.* 17, 145–152.

Yang, G.C., Scherer, P.W., Zhao, K., and Mozell, M.M. (2007). Numerical modeling of odorant uptake in the rat nasal cavity. *Chem. Senses* 32, 273–284.

Young, J.D., Ramsey, J.C., Blau, G.E., Karbowski, R.J., Nitschke, K.D., Slauter, R.W., and Braun, W.H. (1979). Pharmacokinetics of inhaled or intraperitoneally administered styrene in rats. *Toxicology and Occupational Medicine: Proceedings of the Tenth Inter-America Conference on Toxicology and Occupational Medicine*, Elsevier/North Holland, New York, pp. 297–310.

9 Nanomaterial Inhalation Exposure from Nanotechnology-Based Consumer Products

Yevgen Nazarenko, Paul J. Lioy, and Gediminas Mainelis

CONTENTS

9.1 INTRODUCTION

9.1.1 NANOTECHNOLOGY AND ITS DISTINCT POSITION IN RESEARCH

The US National Nanotechnology Initiative (NNI) defines nanotechnology as "the understanding and control of matter at dimensions between approximately 1 and 100 nanometers, where unique phenomena enable novel applications" (National Science and Technology Council 2007). It is necessary to note that alternative definitions of nanotechnology exist (Balogh 2010; Dionysios 2004; Romig Jr et al. 2007; Schummer 2007) and any specific dimensional boundaries of matter should not always be considered as strict limits since the effects attributed to the dimensional parameters between the atomic (approximately 0.2 nm) and "bulk" levels are also observed outside of the 1–100 nm range (Cedervall et al. 2007; Hu et al. 2006; Kim et al. 2004; Konan et al. 2002; Perrault and Chan 2009; Shaw 2011; Vayssieres 2003).

The dimensional characteristics of nanomaterials are better defined according to ISO/TS 27687 as cited by Iavicoli et al. (2010). There, the term "nano-object" is defined as material with one, two, or three external dimensions in the size range of approximately 1–100 nm. A "nanoplate" is a nano-object with one external dimension of 1–100 nm. A "nanofiber" is a nano-object with two external dimensions of 1–100 nm, with a "nanotube" being defined as a hollow nanofiber and a "nanorod" defined as a solid nanofiber. Finally, a "nanoparticle" is a nano-object with all three external dimensions in the range of 1–100 nm. "Nanomaterial" then refers to any kind of nano-objects in the pure form or incorporated into a larger matrix or substrate.

Despite the ongoing debate about the suitability of different definitions of nanotechnology and the manifestation of different size-related effects of materials in various size ranges, the size range that is receiving most attention from researchers working in the field of nanotechnology is between approximately 1 and 100 nm. Incidentally, below 10 nm, the optical, magnetic, and electronic properties of materials are influenced by quantum effects (Haglund Jr. 1998). These various dimension-related properties of nanomaterials can be translated into the desired properties of consumer products, which drives the development and/or marketing of nanotechnology-based products (Wardak et al. 2008). Although the primary size of nano-objects in consumer products is on a nanometer scale, they can agglomerate or aggregate* to sizes exceeding 1 μm and as large as 10 μm in diameter (Lioy et al. 2010; Nazarenko et al. 2011, 2012b). At the same time, issues surrounding human exposure to nanomaterials and the sites of potential contact and deposition in the human body depend on the final size and form of the nanomaterial at the time of contact (Lioy 2010).

9.1.2 NANOMATERIAL PRODUCTION AND USE IN CONSUMER PRODUCTS

The interest in developing technologies based on the unique behavior of nanoscale materials and structures has been steadily increasing. The research and industrial production of nanomaterials and nano-engineered structures and their introduction into common consumer products are rapidly growing (Baxter et al. 2009; Sarma 2008). The use of nanoscale products also elevates the risk of nanomaterial exposure and substantial releases into the environment. (Hansen et al. 2008; Majestic et al. 2010; Maynard and Aitken 2007; McCall 2011). Such releases are of great concern due to potential exposures (Lioy 2010) and negative health and environmental effects caused by nano-objects (Colvin 2003; Dionysios 2004; Drobne 2007; Gwinn and Vallyathan 2006; Hoet et al. 2004; Holgate 2010; McCall 2011; Nel et al. 2006; Nowack and Bucheli 2007; Riediker 2009; Schmid et al. 2009; Teow et al. 2011).

Deliberately added nanomaterial ingredients are already present in a wide array of products on the market including personal care and cosmetic sprays and powders, dietary supplements and medications, cleaning and disinfectant liquids and sprays, sports equipment, and clothing (Bradford et al. 2009; Maynard 2007; Woodrow Wilson International Center for Scholars 2011). Based upon the number of such products, exposures are occurring to members of the general population. It is hard to estimate the exact number of nanotechnology-based products in any local and worldwide markets as their registration and adequate labeling and marketing often range from nonexistent to limited (Chatterjee 2008; Fischer 2008; Gruère 2011; Michelson 2008). The best attempt to estimate the number of nanotechnology-based consumer products in the market and catalog them was done through the creation of an inventory of nanotechnology-based consumer products within The Project on Emerging Nanotechnologies that as of May 31, 2012, listed "1317 products, produced

---

* The term "agglomerate" is used to describe loosely attached particles and the term "aggregate"—to describe tightly attached particles that form after release as an aerosol. If "loose," the agglomerated particles can separate from each other under certain conditions, including after deposition in the respiratory system. If "tight," particles may not be separated and remain as larger aggregates (USEPA 2006).

by 587 companies, located in 30 countries" (Woodrow Wilson International Center for Scholars 2011). At the same time, an online search conducted for just one type of nanotechnology-based consumer products—nanosilver consumer sprays—using the same methodology as used by the Nanotechnology Consumer Products Inventory (Fauss 2008) revealed a much higher number of such products available for purchase by consumers than listed in the earlier-mentioned inventory (Nazarenko et al. 2011), thus leading to potentially widespread exposures. Moreover, the fact that many of the products still listed in this database are no longer available is an example of high fluidity of the nanoproduct market itself. It points to the difficulties regarding the construction and maintenance of any database cataloging consumer nanoproducts, as well as inherent inadequacies of such inventories.

At the same time, recent research identified the possibility of nanosized materials being present in consumer products not labeled as nanotechnology-based (Nazarenko et al. 2011, 2012b). The way some regular products are manufactured can lead to certain ingredients being dispersed in them at nanoscale, creating a possibility for nanomaterial inhalation exposure during manufacturing and use of such products. As mentioned earlier, nanosized particles present in the original product may agglomerate into larger particles during product shelf life or storage before and between its uses. Upon exposure to a person, this agglomeration phenomenon would lead to different sites of particle deposition and potentially different health end points (Lioy et al. 2010; Nazarenko et al. 2011, 2012a) compared to exposures to particles of initial size.

9.1.3 Potential Implications of Nanomaterial Exposure

The concern about exposure to nanomaterials is based on findings that the physical and chemical properties of nanosized materials, including their toxicity, biological and health effects, differ substantially from properties of the same materials in bulk (Maynard et al. 2006). Hazards associated with engineered nanomaterials are challenging conventional approaches to risk assessment and management (Maynard 2007; Wittmaack 2007). The physicochemical properties of nano-objects that may play a role in causing toxic effects include particle size and size distribution, agglomeration state, shape, crystalline structure, chemical composition, surface area, surface chemistry, surface charge, and porosity (Oberdörster et al. 2005). However, most often, health-related studies consider only one or two of these parameters, for example, only particle size and crystalline structure, or mean particle size and most abundant chemical substance comprising the particles. However, inhalation studies do not consider the form of the nano-objects inhaled by a human, that is, agglomerated versus nonagglomerated, which is a serious error (Lioy et al. 2010).

Along with the dermal (cutaneous) route, the inhalation route of nanomaterial exposure is one of the two major ways for nanomaterials from nanotechnology-based consumer products to come in contact with the human body. When certain types of nanotechnology-based consumer products, such as sprays or powders, are used, particles can easily be aerosolized. Once airborne, aging and dynamics of these particles, especially agglomeration and size, change due to evaporation and condensation (Hinds 1999) and complicate their characterization and evaluation of potential health effects, since the physicochemical properties of the nanomaterial-containing aerosol are likely to be modified between the point of aerosolization and the site(s) of particle contact and deposition in the respiratory system.

Once inhaled, some particles (including those containing nanomaterials) deposit in the oral and nasal cavities and the pharynx. The rest of the particles travel deeper into the tracheobronchial and alveolar regions of the respiratory system. There they can deposit due to particle inertia, diffusion, gravitational settling, electrostatic effects, and other mechanisms. Some particles that had deposited in the pharynx can be washed off with saliva and swallowed. Some particles that had deposited in the tracheobronchial region can also be swallowed, since they move back up into the pharynx via the mucociliary escalator, and can be swallowed. Thus, inhaled particles may end up in both the respiratory system and the digestive tract. From there, single and agglomerated nano-objects can translocate through the blood stream and nervous tissue into other organs, as observed in rodent studies, with translocation depending on the size of nano-objects (Borm et al. 2006; Kreyling et al. 2007, 2011; Oberdorster et al. 2004; Singh and Nalwa 2007). The presence of free nano-objects following nanomaterial inhalation exposure was observed in the cardiovascular system, liver, brain, testes, spleen, stomach, and kidneys (Bakand et al. 2012; El-Ansary and Al-Daihan 2009; Reijnders 2012). There is evidence that certain nano-objects pass through the placental barrier as well (Keelan 2011; Wick et al. 2009). Numerous mechanisms are suspected to be the cause of negative health effects of nanomaterials on the systemic and cellular levels. Most recently, mechanistic studies showed that inhalation exposure to nanomaterials that leads to inflammation in lung tissues causes formation of metabolic stressors and platelet-leukocyte aggregates (Plummer et al. 2011; Reijnders 2012; Xiong et al. 2012). The metabolic stressors and platelet-leukocyte aggregates in turn have been linked to chronic inflammation in other organs, cardiovascular disease, and arteriosclerosis, and they also negatively affect the development of fetus (Gomez-Mejiba et al. 2008; Jackson et al. 2012; Reijnders 2012; Tabuchi and Kuebler 2008; Tedgui and Mallat 2006). The effects of nano-object aggregation and agglomeration are thought to assist in particle uptake and translocation (Borm et al. 2006; Reijnders 2012).

Numerous studies have been completed on biological effects caused by specific pure nano-objects. For example, ^{13}C particles (36 nm count median diameter) generated by electric discharge from [^{13}C] graphite rods were found to translocate to the olfactory bulb of the rat central nervous system, following whole-body exposure (Oberdorster et al. 2004); so were manganese oxide nanoparticles (30 nm, exposure concentration of ~500 $\mu g/m^3$) with resulting inflammatory changes following whole-body exposure or intranasal instillation (Elder et al. 2006). Titanium dioxide aerosol particles of 22 nm count median diameter inhaled by rats were later (1 and 24 h) found on the luminal side of airways and alveoli, in all major lung tissue compartments and cells, and within capillaries (Geiser et al. 2005). When pulmonary macrophages and red blood cells were exposed to fluorescent polystyrene microspheres of three different sizes (1000, 200, and 78 nm), particle uptake occurring due to diffusion or adhesive interactions was greater for the nanosized particles compared with the other particles. Specifically, on average, 77% ± 15% (mean ± SD) of the macrophages contained 78 nm particles, while 21% ± 11% contained 200 nm particles, and 56% ± 30% contained 1000 nm particles (Geiser et al. 2005). However, only the difference between the 78 and 200 nm particles was statistically significant. Particle size and state of agglomeration of titanium dioxide nanoparticles administered through various exposure routes, including inhalation, have been shown to affect inflammatory response in various tissues of mice (Grassian et al. 2007a,b; Wang et al. 2007). Another study investigated pulmonary effects (pulmonary inflammation, cytotoxicity, and adverse lung tissue effects) following rodent exposure to chemically identical and similarly sized (~25 and ~100 nm primary particle size) nano-TiO_2 particles that had different crystalline structure: rutile, anatase, or their combination. The researchers concluded that different pulmonary responses could be related to different crystalline structures, inherent pH of the particles, or surface chemical reactivity (Warheit et al. 2007). Nanosized zero-valent iron (nZVI; $Fe^0_{[s]}$) toxicity to bronchial epithelial cells has been shown and its mechanisms investigated (Keenan et al. 2009). Penetration of maghemite (γ-Fe_2O_3) (Baroli et al. 2007) and zinc oxide (Cross et al. 2007) nano-objects into (although not through) the human skin has also been observed.

These and other studies clearly point to potential health effects due to exposures to pure nanomaterials. A summary of nanotoxicology studies indicated that based on animal and animal cell studies, nanomaterial exposure in humans could be associated with a number of acute and chronic health effects, including inflammation, exacerbation of asthma and metal fume fever, fibrosis and chronic inflammatory lung diseases, and carcinogenesis (Bakand et al. 2012). However, until the results of clinical observations or epidemiological studies based upon nano-object exposures become available, it is difficult to ascertain any particular health effect mechanism or define exposure–response relationships.

9.2   NANOMATERIALS IN CONSUMER PRODUCTS

As described earlier, biological effects due to exposure to pure nanomaterials, including inhalation exposure, have been explored to some degree in animal studies. However, the exposure and subsequent effects associated with nanomaterials due to the use of actual consumer products and/ or contact with the associated waste presents unknown health and environmental risks (Bradford et al. 2009; Keenan et al. 2009), which were virtually ignored through the first decade of the twenty-first century (Lioy et al. 2010). At the same time, a great variety of nanotechnology-based consumer products are available in the market (Bekker et al. 2013; Gruère 2012; Nowack et al. 2012). The understanding of the potential for exposure to nanomaterials from such products and the resulting health effects are critical for the development of any safety regulations and guidelines (Drobne 2007; Frater et al. 2006; Schmid et al. 2009; Segal 2004; VanCalster 2006; Warheit et al. 2007), in order to properly regulate nanoproducts. It is also likely that many de facto nanoproducts are not identified as "nano" due to inadequate labeling or vague definitions of what constitutes a nanotechnology-based product. Table 9.1 shows a list of some nanotechnology-based products, currently or previously available in the market, that were investigated by independent researchers, with results published in peer-reviewed scientific literature. Some of these studies are reviewed later in this chapter.

When nanomaterial(s) are incorporated into a product, their size, surface area, surface chemistry, solubility, and possibly shape may potentially affect their toxicology (Kanarek 2007; Maynard et al. 2006; Shrader-Frechette 2007; Warheit et al. 2007). In addition, the concentration, distribution, and size characteristics of nanomaterial(s) and other components in a product could also affect the potential for exposure and adverse health effects (Hansen et al. 2008). Agglomeration of nanoobjects present in a product is another important process, because particles composed of multiple nano-objects can produce different biological effects compared with homogeneous particles of similar size (Bermudez et al. 2004). Finally, if airborne particles are released during product use, the potential differences between the aerosol characteristics immediately after the particle release and the aerosol characteristics at the point of contact with a living tissue need to be characterized and used to design toxicological and epidemiological studies.

Use of consumer products where nano-objects that are not fixed within a solid material is of a special concern because easily dispersed nano materials have a higher potential to cause inhalation and dermal exposure (Hansen et al. 2008; Shimada et al. 2009). Human exposure to such free nanoobjects is most likely to occur through the use of commercially available nanomaterial-containing consumer products in the form of liquid or powder dispersions (Hansen et al. 2008; Nazarenko et al. 2011, 2012b; Oberdörster et al. 2005; Shimada et al. 2009; Wardak et al. 2008). The resulting exposure to nanomaterials as individual particles or agglomerates is likely to be elevated when such liquids or powders are used, causing the release of airborne particles near or in the breathing zone.

The way consumer liquids or powders containing nanomaterials are aerosolized can substantially affect the extent of inhalation exposure (Shimada et al. 2009; Wolff and Niven 1994). For example, the use of a manually activated nonpressurized pump sprayer compared to a pressurized gas propellant sprayer will result in very different aerosol particle size distributions, leading to different particle inhalability and deposition in various regions of the respiratory tract: head airways, tracheobronchial, and alveolar. One study using a nanosilver product showed no measurable release of nanosized particles when a pump sprayer was used, but a substantial release of silver nanoparticles was observed when a gas propellant sprayer was used for the same product (Hagendorfer et al. 2010). Hagendorfer et al. (2010), however, did not measure the release of nanosilver as agglomerates when a pump sprayer was used and, therefore, the absence of individual airborne silver nanoparticles in this case does not necessarily mean a complete absence of nanosilver in the produced aerosol. Another study of four commercial nanoproducts (Lorenz et al. 2011) assessed three propellant sprays and one spray product in a pump sprayer. Again, no nanosized aerosol was detected when a pump-sprayed product was used, even though silver nanoparticles were

TABLE 9.1

List of Spray and Powder Products, Labeled and/or Marketed as Nanotechnology Based, That Were Acquired from End-Consumer Market and Investigated by Independent Researchers, with Results Published in Peer-Reviewed Scientific Literature

Product	Reported Engineered Nanomaterial Component(s)	Purpose of Application	Reference
Bathroom (propellant spray)	Nano-TiO$_2$	Bathroom cleaner/sanitizer	Chen et al. (2010)
Nanosilver (pump spray)	Nano-Ag	Unknown	Hagendorfer et al. (2010)
Nanosilver (propellant spray)	Nano-Ag	Unknown	Hagendorfer et al. (2010)
"Nivea® Silver Protect" antiperspirant (propellant spray)	Unknown	Antiperspirant	Lorenz et al. (2011)
"Nano Schmutz Blocker" shoe impregnation (propellant spray)	Unknown	Shoe treatment	Lorenz et al. (2011)
"Nano Wet Bloc" Shoe impregnation (propellant spray)	Unknown	Shoe treatment	Lorenz et al. (2011)
"Nano-Argentum 10" plant-strengthening agent (pump spray)	Nano-Ag	Plant treatment	Lorenz et al. (2011)
Silver Nanospray (pump spray)	Nano-Ag	Antibacterial general purpose	Nazarenko et al. (2011)
Facial Nanospray (pump spray)	Copper, Calcium, Magnesium, Zinc	Topical cosmetic	Nazarenko et al. (2011)
Hair Nanospray (pump spray)	Unknown	Fixing spray for Nanofibre™ hair additions	Nazarenko et al. (2011)
Disinfectant Nanospray (pump spray)	Unknown	Surface disinfectant, cleaner, and deodorizer	Nazarenko et al. (2011)
Skin Hydrating Nanomist (pump spray)	Unknown	Topical cosmetic	Nazarenko et al. (2011)
Wheel Nanocleaner (pump spray)	Unknown	Vehicle wheel cleaner	Nazarenko et al. (2011)
Nanopowder M	Unknown	Moisturizer	Nazarenko et al. (2012a,b)
Nanopowder D	Unknown	Blusher	Nazarenko et al. (2012a,b)
Nanopowder K	Nano-TiO$_2$Nano-ZnO	Loose powder sunscreen	Nazarenko et al. (2012a,b)
Nanofilm (pump spray)	Likely none	Coating of nonabsorbing floor materials	Nørgaard et al. (2009)
Nanofilm (pump spray)	Likely none	Coating of ceramic tiles	Nørgaard et al. (2009)
Nanofilm (pump spray)	Nanoscale anatase TiO$_2$	Glass coating	Nørgaard et al. (2009)
Nanofilm (propellant spray)	Likely none	Multipurpose coating product	Nørgaard et al. (2009)
Magic-Nano (pump spray)	Likely none	Household surface treatment	Pauluhn et al. (2008)
Magic Nano Glass and Ceramic (propellant spray)	Likely none	Household surface treatment	Pauluhn et al. (2008)
Magic Nano Bath (propellant spray)	Likely none	Household surface treatment	Pauluhn et al. (2008)
"Silver Scent" Hunter (pump spray)	Nano-Ag	Antiodor spray for hunters	Quadros and Marr (2011)
"Wellness Colloidal Silver Throat Spray" (pump spray)	Nano-Ag	Throat treatment spray	Quadros and Marr (2011)

found in the original liquid product using transmission electron microscopy (TEM). The conclusion was that nanomaterials were aerosolized in agglomerated form as particles larger than 100 nm. Importantly, Lorenz's study also demonstrated that aerosol particles smaller than 100 nm can in some cases be formed during aerosol generation even when no nano-objects are present in the original spray liquid. Generation of aerosol containing a high number of individual nanoparticles from products not containing nanomaterials was observed by Nørgaard et al. (2009). All these studies indicate that aerosol size distribution produced during product use may substantially depend on the aerosolization method and might not be a reliable indicator of the presence or absence of engineered nanomaterials in the product. It is also evident that presence or absence of particles smaller than 100 nm in the aerosol resulting from the use of a consumer product does not necessarily indicate that engineered nanomaterials are present or absent in this product.

A story with the German bathroom cleaning sprays "Magic-Nano," "Magic Nano Glass & Ceramic," and "Magic Nano Bath" (Kleinmann GmbH; Sonnenbühl, Germany) illustrates several earlier-mentioned issues presented by nanotechnology-based consumer products: (1) ambiguous labeling and marketing of the products, (2) health effects posed by various ingredients of the products—not just the nano-ingredient(s), and (3) the effect of product application or dispensing method on the generated aerosol characteristics, and consequently, the potential health effects.

When Kleinmann GmbH first released "Magic-Nano" into the consumer market, it was in the form of a pump spray. The use of this product did not result in any reported health effects. Later, however, the company released "Magic Nano Glass & Ceramic" and "Magic Nano Bath" in the form of propellant sprays (Pauluhn et al. 2008). The use of the latter two products caused at least 153 people to report severe respiratory problems including strong cough, dyspnea, and severe pulmonary edema, resulting in some 13 people getting hospitalized (Pauluhn et al. 2008). The use of the original pump spray "Magic Nano" did not result in these health effects, but the propellant sprays "Magic Nano Glass & Ceramic" and "Magic Nano Bath" did cause health effects (Kanarek 2007). The original "Magic Nano," "Magic Nano Glass & Ceramic," and "Magic Nano Bath" were later examined in a study of acute lung injury following nasal inhalation in a rat model (Pauluhn et al. 2008): (1) The pump-dispensed spray ("Magic Nano") caused mortality at 81,222 mg/m^3 while one of the two propellant sprays ("Magic Nano Glass & Ceramic") appeared to be much more toxic causing mortality at only 2269 mg/m^3; and (2) the second propellant spray ("Magic Nano Bath") appeared not to cause lung irritation or injury and did not cause mortality up to the maximum tested concentration of 28,100 mg/m^3 achieved by regular spraying.

It was disclosed that, apparently, the manufacturer's (Kleinmann GmbH) use of the term "nano" in the product names referred to the thickness of the waterproofing coating produced by the sprays rather than the presence of any engineered nanomaterial in their formulations (Pauluhn et al. 2008). In the absence of regulations for labeling, one cannot be sure that a manufacturer's claim regarding the presence/absence of nanomaterial(s) in a product is representative of the actual product being sold. Thus, independent testing would be needed to verify the presence/absence of nanosized materials or objects in a particular product. At the same time, tests are usually able to detect only certain types of nano-objects and a negative result is not a guarantee that nano-objects are absent in a given product. Thus, accuracy of information about the content and concentration of nanomaterial ingredients provided by the manufacturers is crucial for exposure and safety assessment, and toxicological studies.

The current situation where manufacturers rarely present information about the content of nanomaterials in their products (Hansen et al. 2008) complicates our ability to determine the concentration of nanomaterial(s) in the product and, as a result, to determine the amount of nanomaterial, to which a user could be exposed. Composition of a final product and the form, in which it is delivered to the consumers, can lead to chemical modification of the initial nanosized ingredient(s), their coagulation, agglomeration, and other changes that can have a substantial effect on exposure and health effects of a given "nanoproduct." In other words, it is impossible to predict with certainty the nanomaterial exposure and health effects of a consumer "nanoproduct" solely based on the physicochemical characteristics of the nanosized and/or nanostructured components within it (Maynard 2007). Therefore, it is essential to characterize nanotechnology-based consumer products in the form they are delivered to the end users. Toxicological and other health effects studies; should not rely solely on data for pure nanomaterials or separate product components that will not adequately reflect actual exposures and health effects actual exposures and health effects.

9.3 POTENTIAL OF NANOMATERIAL EXPOSURE FROM NANOTECHNOLOGY-BASED CONSUMER PRODUCTS

Consumer spray and powder products containing nanomaterials present a potential for nanomaterial inhalation exposure due to their specific mode of application and use (Nazarenko et al. 2011; Quadros and Marr 2011). This exposure potential can be tested through a realistic simulation of product use accompanied by the measurement of the released aerosol particles that can be inhaled by a consumer, and analysis of the liquid or powder products themselves as well as released aerosol particles (Lioy 2010). The following text provides an overview of the current literature investigating exposure to nanotechnology-based consumer products as well as discusses issues noted regarding nanotechnology-based consumer sprays.

9.3.1 STUDIES INVESTIGATING NANOTECHNOLOGY-BASED CONSUMER PRODUCTS

A sample of consumer nanoproducts investigated by independent studies and relevant references are listed in Table 9.1, and are summarized as follows.

1. A study by Nazarenko et al. (2011) examined six nanotechnology-based and five regular spray products (Figure 9.1). All these products were pump sprays, and the nanosprays are included in Table 9.1. The products were realistically sprayed close to the breathing zone of a mannequin head, which had sampling inlets in its nostrils (Figure 9.2). These products were also aerosolized by standard nebulization techniques and their size distributions determined. One of the six investigated nanotechnology-based sprays was labeled as containing nanosilver (silver nanospray); another nanospray (facial nanospray) was labeled as containing nanosized particles of copper, calcium, magnesium, and zinc. The other three nanosprays (hair pump spray, disinfectant pump spray, skin hydrating pump spray) were simply labeled as

FIGURE 9.1 Nanotechnology-based and regular consumer spray products tested by Nazarenko et al. (2011).

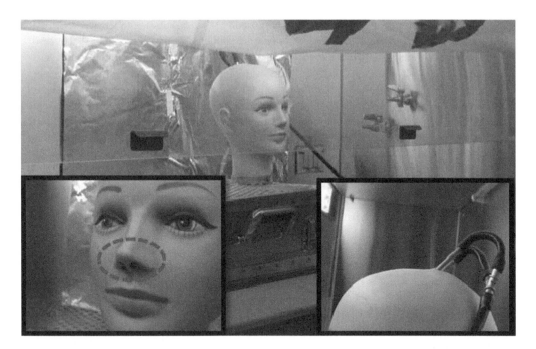

FIGURE 9.2   Mannequin head with aerosol sampling inlets installed in the nostrils used by Nazarenko et al. (2011, 2012b).

nanoproducts, but no specific nanomaterial ingredients were identified on their labels. The concentration and size distribution of the particles that were released and could potentially be inhaled from the investigated 11 spray products were measured using real-time aerosol instruments (a scanning mobility particle sizer [SMPS]) and an aerodynamic particle sizer (APS) (TSI, Inc., Shoreview, MN, USA) covering the aerosol size range between approximately 14 nm and 20 μm. TEM was used to determine the sizes and shapes of particles in desiccated product samples, while the photon correlation spectroscopy (PCS), also known as dynamic light scattering (DLS), allowed particle analysis in the original liquid products. Overall, the use of spray products resulted in the release of particles spanning the entire measurement range: from 14 nm to 20 μm. Nano-sized particles and their agglomerates were observed not only in the nanotechnology-based, but also in some regular products. Importantly, simulated application of both nanotechnology-based and regular spray products resulted in aerosol particles smaller than 100 nm in diameter (nanoparticles) as well as particles larger than 10 μm in diameter. It was concluded that the use of certain nanotechnology-based as well as regular consumer sprays would result in inhalation exposures to single nanosized particles, their agglomerates, and possibly complex nanomaterial-containing composites.

2. A study conducted by Hagendorfer et al. (2010) examined airborne particle size distributions and effect of aging for particles 10–500 nm (analyzed by an SMPS) produced by one commercial nanosilver spray product in both pump and propellant dispensers. The researchers also investigated aerosol particles using TEM after collecting them using an electrostatic sampler. In addition, energy-dispersive x-ray spectroscopy (EDX) was employed for elemental analysis of the collected particles. The results showed no measurable release of nanoparticles from a pump sprayer (no detection by both SMPS and TEM). However, a propellant sprayer generated a substantial amount of individual silver nanoparticles and nanosized clusters (agglomerates). The EDX confirmed the presence of silver in the collected nanoparticles and even showed traces of other chemical elements (Na and O) left from the specific chemistry of silver nanoparticle synthesis. The researchers concluded that silver

nanoparticles were dispersed in such large droplets when a pump sprayer was used that they settled out before reaching the employed instruments. It was also noted that aerosol dynamics played an important role. Specifically, the size distributions of the original aerosols changed over time, with a shift toward larger particle sizes. Particle morphology also changed during and after spraying: particles agglomerated as spray droplets desiccated.

3. Lorenz et al. (2011) looked at four commercial sprays, one of which was labeled as containing nanosilver (a pump spray for treatment of plants), the second and third ones—labeled as nanoproducts (shoe impregnation propellant sprays), and the fourth one—labeled to contain "antibacterial silver molecules" and silver citrate (antiperspirant propellant spray). The authors examined the original liquid products using inductively coupled plasma mass spectrometry (ICPMS) and scanning TEM (STEM) with EDX, as well as aerosols produced with the original spraying mechanisms using SMPS and STEM. They found a very high release of airborne nanoparticles from three propellant-spray products and a negligible release from one pump spray product. However, engineered nanoparticles were found in only one of the three propellant-spray products and, notably, in the original liquid pump spray product as well. Thus, the presence of nanosized particles in the produced aerosol did not always correlate with the presence of engineered nano-objects in the original liquid.

4. The study carried out by Chen et al. (2010) at the U.S. National Institute for Occupational Safety and Health (NIOSH) and the U.S. Consumer Product Safety Commission (CPSC) looked at one nano-TiO_2-containing bathroom cleaning product with a propellant sprayer that was investigated in a simulated home environment and in a testing chamber. An operator simulated a typical use of the product, and particles in the user's breathing zone were measured by a nephelometer (measurement range 0.08–10 μm), an optical aerosol spectrometer (0.3–20 μm), an SMPS (15–680 nm), and an APS (TSI, Inc., Shoreview, MN, USA) (0.5–20 μm). The mass of the produced aerosol, as well as the change in mass while drying as a function of time, and the mass of the final residue after evaporation were determined gravimetrically. In addition, a laser diffraction analyzer was used to measure the concentration of produced spray droplets in the range between 2 and 2000 μm. Scanning electron microscopy (SEM) with EDX was used to analyze the morphology and elemental composition of the aerosol spray particles collected on a polycarbonate filter. The EDX identified particles in the spray aerosol as TiO_2, which initially had a count median diameter (CMD) of 22 μm, but then reduced in size to mostly unagglomerated 75 nm TiO_2 nanoparticles due to drying. The researchers found that aerosols generated in the test chamber and by an operator in a simulated home environment were comparable with respect to the total mass and number aerosol particle concentration and particle size distribution. The total aerosol mass concentration in the breathing zone was measured at 3.4 mg/m^3, of which 170 μg/m^3 were in the nanosize range. By number, however, most particles were nanosized (1.2×10^5 out of the total concentration of 1.6×10^5 cm^{-3}).

5. Quadros and Marr (2011) focused on three silver-containing sprays, in which the manufacturers claimed inclusion of silver nanoparticles (antiodor pump spray for hunters), colloidal silver (throat treatment pump spray), or silver ions (a pump spray surface disinfectant). The pump spray surface disinfectant is not listed in Table 9.1, because it was not advertised or labeled as a nanoproduct, and the researchers did not find nanomaterials such as silver nanoparticles in it. It is necessary to note that the term "colloidal" refers to the particle size range between approximately 1 and 1000 nm (Levine 2002), so labeling a product as containing a colloid may or may not be indicative of its nano status. The products were manually sprayed inside a testing chamber, from where the aerosol was sampled using an SMPS (14 nm–0.7 μm) and an optical particle counter (OPC) measuring particles from 0.3 to 10 μm. Silver content in the original liquid products was analyzed by ICPMS and DLS, and by UV–vis spectrophotometry. Additionally, thermophoretically collected aerosol particles and

aliquots of the original liquid products were examined using TEM and an environmental SEM (ESEM) with EDX. In the only one investigated product labeled as containing nanosilver, the authors found most silver to be associated with larger particles (>450 nm). The product labeled to contain colloidal silver was found to actually contain most of the silver in the nanosized particulate form and was also associated with larger particles. The third product labeled to contain silver ions was confirmed to contain nearly all silver in the ionic form. Similar to Nazarenko et al. (2011), wide aerosol distributions, ranging from nanoparticles to 10 μm (and assumed even larger), were measured for all products. An interesting finding was a substantial difference in the appearance (color and turbidity) of the liquid and the size distributions of the two different batches of the same nanosilver product, which the authors noted as a challenge for regulators. The researchers concluded that both the characteristics of the liquid and spray mechanism affect the size distribution of produced aerosols. They also observed that there is little correlation between the particle size distribution of silver in the liquid products and the aerosol size distribution of the spray produced from them.

6. In the study by Nørgaard et al. (2009), four nanofilm spray products not labeled as containing engineered nanomaterials—except for one product labeled to contain nanoscale anatase TiO_2—were examined. Three of these products were dispensed in hand pump spray bottles, including the nano-TiO_2-containing spray. The fourth product was in a pressurized can. The products were intended to produce a coating layer of thickness on the order of nanometers for treatment of various household surfaces. They were evaluated using a test chamber to measure releases of volatile organic compounds (VOCs) and particles in the range from 6 nm to 20 μm. Silica particles were listed among the ingredients of one of the products (a pump spray). These particles may have been of nanosize as nanosized silica particles are used in preparation of hydrophobic surfaces as a component of chemical systems similar to those of the products investigated by Nørgaard and colleagues (Hikita et al. 2005). Size distributions of particles generated by spraying the four nanofilm spray products were measured using a fast mobility particle sizer (FMPS). These distributions were multimodal, with 1–4 modes in the 6–100 nm range alone. The aerosol size distributions by number for all four products were dominated by nanoparticles. The researchers concluded that the use of all four products would result in a very high airborne nanoparticle exposure. They also observed that the one product in a pressurized can generated much finer particles than those generated by the pump sprays. Specifically, for the three pump spray products, the initial number of particles measured after spraying 1 g of a product was on the order of 10^8, but for the pressurized can spray, 10^9.

Each of the earlier-mentioned studies investigated liquid spray products. To date, we are aware of only one study (Nazarenko et al. 2012b) where nanotechnology-based powder products were evaluated. These were cosmetic powders that were applied to the face of a mannequin head in order to realistically simulate their application.

7. Three nanotechnology-based powders (a moisturizing powder, a blusher, and a loose powder sunscreen), and three regular cosmetic powders tested in the study of Nazarenko et al. (2012b) are shown in Figure 9.3. For the cosmetic powders, the authors employed the same mannequin head setup shown in Figure 9.2, used in an earlier study with consumer sprays (Nazarenko et al. 2011). However, here the cosmetic powders were applied directly to the mannequin face while the produced aerosol was sampled through the nostril inlets and measured by SMPS and APS (TSI, Inc., Shoreview, MN, USA). The TEM and laser diffraction spectroscopy (LDS) were used to analyze the original bulk powders. In this study, the release and simulated inhalation of particles larger than 100 nm and as large as 20 μm during the cosmetic powder application was observed. TEM and aerosol measurements suggest that exposure to nanomaterial(s) due to the use of cosmetic powders will be predominantly in the form of nano-object-containing agglomerates or nanomaterials attached to larger particles.

FIGURE 9.3 Nanotechnology-based and regular cosmetic powder products tested by Nazarenko et al. (2012b).

9.3.2 SUMMARY OF TECHNIQUES USED TO INVESTIGATE NANOTECHNOLOGY-BASED CONSUMER PRODUCTS

Analysis of nanotechnology-based consumer products is commonly comprised of two parts: analysis of the original product (liquid, powder, etc.) and analysis of aerosol generated from the product. For the original products, researchers have employed TEM (Hagendorfer et al. 2010; Lorenz et al. 2011; Nazarenko et al. 2011, 2012a,b; Quadros and Marr 2011), SEM (Chen et al. 2010) including environmental SEM (Quadros and Marr 2011), and STEM (Lorenz et al. 2011) to determine the sizes and shapes of particles, and how they interact with the electron beam in powder products and desiccated spray product samples. These electron microscopy techniques are sometimes combined with EDX, allowing elemental analysis of particles (Chen et al. 2010; Hagendorfer et al. 2010; Lorenz et al. 2011). Additionally, ICPMS (Lorenz et al. 2011; Quadros and Marr 2011), DLS, also known as photon correlation spectrometry (PCS) (Nazarenko et al. 2011; Quadros and Marr 2011), and UV–vis spectrophotometry (Quadros and Marr 2011) were used for liquid products and laser diffraction spectrometry (LDS) (Nazarenko et al. 2012b) for powder products.

To analyze aerosol generated from the consumer products, which presents the inhalation exposure concern, researchers employed various aerosol spectrometers: SMPS (Chen et al. 2010; Hagendorfer et al. 2010; Lorenz et al. 2011; Nazarenko et al. 2011, 2012a,b; Quadros and Marr 2011), an APS (TSI, Inc., Shoreview, MN, USA) (Chen et al. 2010; Nazarenko et al. 2011, 2012a,b), an FMPS (Nørgaard et al. 2009), an OPC (Quadros and Marr 2011), and a nephelometer (Chen et al. 2010). In one study, in addition to aerosol particles alone, various VOCs were analyzed using mass spectrometry (MS) (Nørgaard et al. 2009), gas chromatography/mass spectrometry (GC/MS) (Nørgaard et al. 2009), and gas chromatography with a flame ionization detector (GC/FID) (Nørgaard et al. 2009).

9.3.3 OBSERVED CHALLENGES WHEN INVESTIGATING EXPOSURE POTENTIAL OF NANOTECHNOLOGY-BASED CONSUMER PRODUCTS

9.3.3.1 Effect of the Aerosol Dynamics

A study by Nørgaard et al. (2009) illustrates how substantially the aerosol size distribution can change between the time it is generated and inhaled. Nørgaard's is the only published study of consumer spray products where researchers used a high time resolution system and were able to investigate the dynamics of aerosol size distribution including particle formation, coagulation, and evaporation with 1 s time resolution. Changes in the particle size distributions were substantial and very different from product to product. The changes included an increase and decrease of concentrations in particle number modes in the nanosize range, accompanied by slow concentration decrease at other modes. These changes occurred on the timescale of seconds to several minutes. There was a notable difference in aerosol dynamics between the pump sprays and the pressurized spray. For the pump sprays, there was a general trend toward a decrease of particle concentrations by number at all modes of the multimodal aerosol size distributions following spraying. However, for two out of three pump products, this decrease was preceded by a short period of particle concentration increase (up to 30 s). However, for the pressurized spray, the mode diameters increased and reached the maximum concentration approximately 80 s after spraying, including the concentration of the smallest nanomode (11 nm). However, over the following ~90 s, the particle concentration in this mode decreased rapidly, followed by a slower decrease after that. These changes were accompanied by a substantial increase in the concentration of the 19.1 nm mode and a slow concentration increase in the 34 nm mode. The concentration of the latter mode peaked 280 s after spraying. This observation is a notable example of aerosol complexity and its dynamics presented by multicomponent consumer sprays. Because deposition of aerosol particles in the respiratory system is dependent on the size of particles, the aerosol dynamics is an important factor when evaluating inhalation exposure to consumer spray products, including the nanotechnology-based sprays. The aerosol dynamics can also have a substantial impact on the total number of particles inhaled and deposited, particularly within the nanosize range (1–100 nm). The authors of the study also made an important observation that aerosol particles released from consumer spray products may often be a "mixture of solid particles, condensed semivolatile compounds, and gas-phase reaction products" (Nørgaard et al. 2009).

Based on a study by Nørgaard et al. (2009), the dynamics of aerosol generated from consumer spray products even under normal usage conditions can be very complex. It can affect both the total particle concentration, especially in the nanosize range, and size distribution, which can cause major differences in the level of exposure and potential health effects due to deposition of inhaled particles in different regions of the respiratory system. These differences may depend on the temperature, humidity, and air exchange rate of the environment, in which a product is used. For example, if a product is sprayed in a well-ventilated room, inhalation exposure will occur to freshly formed aerosol, but in a poorly ventilated space, the aerosol will have an extended time to linger and age, thus possibly changing its characteristics sufficiently to cause different health effects.

9.3.3.2   Effect of the Aerosolization Method and the Sprayer Type

Experiments using sprayers, included with the products (Chen et al. 2010; Hagendorfer et al. 2010; Lorenz et al. 2011; Nazarenko et al. 2011; Nørgaard et al. 2009; Quadros and Marr 2011), as well as constant aerosal output atomizers, described by Nazarenko et al. (2011), showed that spraying technique substantially affects concentration and size distribution of the released particles. This is in line with other studies (Hagendorfer et al. 2010; Hansen et al. 2008; Lorenz et al. 2011; Nørgaard et al. 2009; Quadros and Marr 2011) that showed the effect of propellant vs. pump sprayer on the released particle characteristics, including the particle concentration and size distribution with smaller particles released when a propellant sprayer is used. Thus, the exposure to particles from either nanotechnology-based or regular products would be affected by the product composition and by the sprayer type.

Many existing propellant spray products, while not designed as nanotechnology based or nanoparticle producing, may in fact generate large amounts of nanosized aerosol particles. Nanosized aerosol particles can be released from certain consumer spray products even if they do not contain nanoparticles in the original liquid itself (Lorenz et al. 2011; Nørgaard et al. 2009; Quadros and Marr 2011). In other words, we are describing nanoparticles that do not exist in the liquid product and are only formed or produced in the process of aerosol generation. These nanoparticles can also be made up of anthropogenic chemicals, in which case their "engineered" status may be debated.

However, we could not identify any study that explored this subject. The nanosized aerosol particles generated from the consumer sprays may be (1) engineered nanoparticles, introduced into or formed in the original liquid products, (2) nanosized composite particles of nano-objects with other product ingredients, (3) nanoparticles formed from liquid product ingredients in the process of spraying, (4) nanoparticles resulting from evaporation of water or organic solvents from larger droplets, and (5) formation of new nanoparticles from gas-phase chemicals. Therefore, simple measurement of the nanosize aerosol fraction, as done in many studies of nanotechnology-based consumer spray products, may not be indicative of the presence and concentration of engineered nano-objects in the original liquid product itself. There is a possibility that health effects resulting from exposure to such particles would depend on their size as well as composition. Adverse health effects can be caused or aggravated when exposure occurs to nanosized particles compared with larger particles made of the same chemical constituents and of the same mass. Therefore, even if a product does not contain nanoscale materials and nanoscale particles form only during product use, they should be part of exposure assessment.

9.3.3.3   Labeling Issues Related to Nanotechnology-Based Consumer Products

As studies discussed earlier showed, labeling of products by manufacturers as "nano" or any related adjective, and even a mention of a nano-ingredient(s) among the contents, is not necessarily an indication that an engineered nanomaterial is in fact present in a product or will be present in aerosol that is generated from it during use. On the other hand, the opposite might be true when products not labeled as nanotechnology-based are in fact formulated with an engineered nanomaterial(s). Additionally, one should interpret manufacturer-reported contents of products with caution. For example, the facial nanospray investigated by Nazarenko et al. (2011) is labeled to contain nano-sized particles of copper, calcium, and magnesium, which cannot be chemically stable as highly dispersed metallic particles. It is possible that the manufacturer meant that nanoparticles in the product contained compounds of these metals.

9.3.3.4   Use of Various Metrics for Nanosized Particle Measurement

Studies discussed earlier demonstrated that the highest number of particles generated from certain consumer spray products during their use is in the nanosize range. At the same time, these nano-sized particles usually contribute minimally to the total aerosol mass (Chen et al. 2010; Nazarenko et al. 2011, 2012a). In this context, it must be mentioned that usage of different metrics for ultrafine (nanosized) aerosol measurements in order to best describe health effects is still debated (Dhawan et al. 2009). Particle number, surface area, and mass have all been shown to be adequate metrics for different types of nanoparticles in describing particular health effects. The surface area and number of nanoparticles deposited in the respiratory system have been shown to correlate well with toxic effects of airborne materials such as nanoparticulate quartz, metallic cobalt and nickel, and elemental carbon ^{13}C (Duffin et al. 2002; Oberdorster et al. 2004). However, particle mass had a good correlation with health effects due to nanoparticulate TiO_2, carbon black, polystyrene beads, and surface-modified quartz (Duffin et al. 2007; USEPA 2011b; Wittmaack 2007).

9.3.3.5   Effect of Particle Agglomeration

Agglomeration and deagglomeration of particles both in original products and aerosol generated from those products are important factors in the context of engineered nanomaterial exposure and potential health effects. Nano-objects can not only agglomerate with each other but also associate with larger particles, thus forming agglomerates that can contain a combination of engineered nanomaterial(s) and non-nanomaterials. Nanomaterial exposure can thus occur not only to nano-sized particles but also to larger agglomerates containing engineered nanomaterials.

Currently available instruments cannot measure accurately number and surface area concentrations of agglomerated nano-objects including nano-object agglomerates associated with larger particles. Such agglomerates are thought to represent an aerosol size fraction, through which much of the nanomaterial in consumer products is delivered to the respiratory system (Nazarenko et al. 2011, 2012b). In other words, a nanomaterial-containing aerosol generated from nanotechnology-based consumer products can hold nanomaterials distributed across a wide range of particle sizes, including agglomerates of supercoarse particle sizes. In this case, if particle number is measured, each agglomerate would be counted as one particle even if it contains a large number of nano-objects within itself. It is also possible that, similar to observations of other agglomerates (Park et al. 2003), some agglomerates from consumer products have spaces between the primary particles filled with organic and other components of multi-ingredient products. In this case, surface area of such an agglomerate may not be representative of the engineered nano-objects that may be embedded within it. Depending on the structure of agglomerates with nano-objects, they may dissociate after deposition in the respiratory system (Howard et al. 2006; Petitot et al. 2013; Wiebert et al. 2006a,b), which will result in engineered nano-object release and contact with live tissue at the sites of deposition. In this case, measurements of airborne particle number and surface area will not be representative of the actual dose and may result in biased interpretation of health effects.

The larger agglomerates containing nano-objects deposit predominantly in the upper airways of the human respiratory system rather than in the alveolar and tracheobronchial regions as would be expected if only individual primary nano-objects were considered (Nazarenko et al. 2012a). This kind of distribution of aerosol particle deposition across the respiratory system may cause different health effects than would be anticipated based on the behavior of single nano-objects and toxicology studies for the alveolar region. Therefore, the likelihood of predominant nanomaterial deposition in the upper airways must be considered in inhalation toxicology studies of nanotechnology-based consumer products.

9.3.3.6 Consistency of Consumer Product Composition

The issue of inconsistency of product composition was identified by Quadros and Marr (2011): The final contents and properties of the same product can differ as the product ages and also from batch to batch of the same product. This inconsistency would be of concern, for example, if one batch is found innocuous in a test conducted by a regulator or an independent investigator, and a product is approved for sale, while another batch, already in the hands of consumers, may actually have different properties rendering it hazardous. A similar problem arises due to aging of consumer products—the time between manufacture and use. Changes in physical properties and composition of the product due to the earlier-mentioned phenomena can affect the properties of the aerosol produced during product use, such as particle concentration and size distribution, leading to variability in potential health effects.

9.4 QUANTITATIVE INHALATION EXPOSURE ASSESSMENT FOR CONSUMER PRODUCTS

When application of the nanomaterial-containing consumer products is simulated and the inhalable aerosol is measured, inhalation exposure to aerosol particles can be quantified. It can be reported as the dose of mass, volume, number, or surface area of (1) inhaled particles in different size ranges, including the aerosol fraction below 100 nm, and (2) particles deposited in various regions of the respiratory system.

Chen et al. (2010) considered a human exposure scenario for the propellant bathroom cleaning spray product. They measured nanoparticle concentration as well as total aerosol concentration using both mass and number metrics. The scanning electron micrograph of the particles in the produced aerosol shows spherical TiO_2 particles of various sizes from ~100 nm and up to 2–3 μm that seem to be tight agglomerates containing a very high number of primary nanoparticles (<10 nm). The researchers described these particles as singlets with rough surface. These apparent agglomerates of primary nanoparticles exist predominantly as singlets (92%), but some of them are attached to each other in pairs or in higher numbers (8%). The researchers assessed exposure as lung burden after 1 min of the propellant spray application in terms of mass of TiO_2 particles of all sizes that deposit on 1 m² of alveolar epithelium and found the exposure to be ~0.075 µg/m²/min of TiO_2 for a person engaged in 33% sitting and 67% light exercise.

Lorenz et al. (2011) considered deposition of only nanosized aerosol (approximately 10–100 nm) in the alveolar, tracheobronchial, and nasal regions of the respiratory system. This is in contrast to the study by Nazarenko et al. (2012a) who considered deposition of nanomaterial-containing aerosols of a wide particle size range from less than 100 nm to approximately 20 µm. Lorenz and colleagues stated that mass is not a suitable metric for nanoparticles and based their model on particle number instead. The utilized model was adapted from an earlier version (ver. 1.1 published in July 2008) of the model suggested by the European Chemicals Agency (ECHA) for assessing inhalation exposure under "Registration, Evaluation, Authorisation and Restriction of Chemicals" (REACH). The current version is 2.1, published in October 2012 (European Chemicals Agency 2012). The ECHA model itself had not been validated for use with nanomaterials and is not presented here. Lorenz's model accounted for deposition of 32 different particle sizes in accordance with the SMPS measurement channels (midpoint channel diameters were used). The number-based concentration in each of these 32 size channels was used as input for the following formula (Lorenz et al. 2011):

$$E_{j\_region} = C_{inhj\_region}\, r_{depj\_region}\, F_{resp}\, IH_{air}\, T_{contact},\qquad(9.1)$$

where

E_{j_region} is the deposition of SMPS aerosol fraction j in either the alveolar, tracheobronchial, or the nasal region (#)

C_{inhj_region} is the aerosol concentration by number in SMPS fraction j (#/cm³)

r_{depj_region} is the deposition fraction for the particle size in SMPS fraction j according to ICRP model (International Commission on Radiological Protection 1994)

F_{resp} is the respirable fraction of nanosized aerosol in SMPS fraction j

IH_{air} is the inhalation rate of a female/male consumer (cm³/min)

$T_{contact}$ is the time of exposure during one product application event (min)

Therefore, these deposited dose calculations accounted for differences in breathing rate of men and women. Lorenz et al. (2011) used the worst case scenario length for duration of product application: 5 min for antiperspirant application obtained from Bremmer et al. (2006), and 8 min as the mean durations of shoe spray application (USEPA 1997). The researchers also extrapolated particle concentrations measured in the glove box during realistic spraying to match mass of spray liquid dispersed during the exposure times: 4 g for the antiperspirant spray, and 12.65 g (female) and 18.48 g (male) for the shoe impregnation sprays based on literature (Bremmer et al. 2006; Engelund and Sørensen 2005). The researchers reported deposition in the three regions of the respiratory system as well as the combined deposition in terms of number of nanoparticles in the investigated nanosize region (~10–100 nm) that deposited during one product application event. Deposition was found to be higher for men than for women, with the difference being greater for the shoe impregnation sprays due to higher dispersed aerosol mass (consistent with larger shoe size in men compared to women). The total deposition was on the order of 10^{10} particles for all sprays. The highest deposition of the nanosized aerosol was found in the alveolar region (2.4–4.8 × 10^{10} #/application) and the lowest—in the nasal region (2.6–4.9 × 10^9 #/application). Again, it must be emphasized that the work reported earlier refers to deposition of only the nanosized aerosol and does not account for engineered nanomaterial possibly delivered into the respiratory system as larger agglomerates. Thus, the results cannot be projected to describe the entire dynamics of particle deposition for many consumer product emissions.

Quadros and Marr (2011) approached estimation of inhaled dose in a different way. The first method to exposure estimation was for a silver product intended to be sprayed on clothes. They applied a mass balance to calculate how much silver aerosol mass can be inhaled based on the number of sprayer activations in a room of a given volume with a particular air exchange rate, and the duration of a person's presence in that room. For this inhaled silver aerosol, deposition in the nasopharyngeal, tracheobronchial, and the alveolar regions of the human respiratory system was calculated using the ICRP model as referenced by Hinds (1999). They estimated inhalation of 0.62 ng of silver particles and total silver deposition of 0.38 ng during 10 min if the spray was activated 20 times at a rate of 1 s^{-1} in a 10 m^3 room with an air exchange rate of 0.5 h^{-1}. Of the deposited 0.38 ng of silver, 77% would deposit in the nasopharyngeal region, 6% in the tracheobronchial region, and 17% in the alveolar region. The researchers recognized that actual exposure would differ due to incomplete aerosol mixing throughout the volume of the room and varying distance between the user's inhalation zone and the sprayer. They also estimated that less than 1% of the mass of deposited silver would be in the form of nanosized aerosol, but made no comment regarding deposition of agglomerated nanosilver. The authors used a different method to estimate inhalation exposure from a throat treatment spray. Here they assumed that all silver released during spraying would be inhaled. The dose was estimated to be 70 ng of silver per day if the spray was used according to the manufacturer's recommendation (1–2 times per day). The deposition was estimated to be 82% in the nasopharyngeal region, 2% in the tracheobronchial region, and 16% in the alveolar region.

Nørgaard et al. (2009) estimated exposure to both VOCs as well as aerosols including nanosized aerosols released during the realistic use of nanofilm spray products intended to coat various surfaces (see Table 9.1). The authors note the likelihood of synergistic toxic effects that the particles (including nanomaterials) and VOCs may induce. Among the four spray products tested, three were intended for indoor use where ventilation could be lower, resulting in high inhalation exposure potential compared with outdoor application. The researchers calculated exposure to some VOCs they had detected and measured in the sprays, specifically chloroacetones and perfluorinated silane. They assumed no ventilation in a 17.3 m^3 completely mixed room where 7 m^2 of the floor material was treated using 40 g/m^2 of one of the products intended for indoor use. This scenario would result in ~6700 μg/m^3 of chloroacetones and ~3300 μg/m^3 of perfluorinated silane in the air immediately after treatment. For the chloroacetones, this level exceeds the threshold limit value-ceiling (TLV-C) established by the American Conference of Governmental Industrial Hygienists (ACGIH) of 3800 μg/m^3 for skin absorption (Proctor et al. 2004), which Nørgaard et al. (2009) referenced in the absence of any established human inhalation limits (USEPA 2011a). Exposure to particles from 6 nm to 20 μm was calculated for the same scenario assuming no particle formation after spraying and no particle dynamics. It must be noted, however, that these phenomena had actually been observed for the products in this study. The calculations were based on the number of aerosol particles that the products generated, which was 3 × 10^8 per 1 g of product applied for the first two pump sprays: the ceramic tile and floor coating products. This would result in the aerosol concentration by number in the modeled indoor space exceeding 3 × 10^6 cm^{-3} and 100-fold higher for the third product intended for treatment of indoor surfaces. The authors compared these number concentrations with those in high-traffic streets where 1 × 10^6 cm^{-3} is rarely exceeded. The fourth spray product is intended for outdoor coating of window glass, so the researchers assumed rapid dilution of particles and VOCs released during its application. Exposure could not be estimated for this outdoor product, but the authors suggested it would not exceed the levels calculated for the products intended for indoor use in terms of the number of inhaled particles. At the same time, this last product was the only one, for which an engineered nanomaterial (nanosized anatase TiO$_2$) was listed among the contents. Hence, even despite the fact that this product is applied outdoors and a lower aerosol exposure is expected, it presents the greatest concern of exposure to engineered nano-objects.

Nazarenko et al. (2012a) investigated inhalation exposure to several aerosol particle size fractions ranging from 14 nm to 20 μm that were measured during simulated use of nanotechnology-based cosmetic powders. The study assumed that engineered nanomaterials were distributed across all aerosol size fractions, deposition of which in different regions of the respiratory system was calculated. Using measured aerosol concentrations, exposure was assessed by calculating inhaled and deposited mass doses of particles per 1 min of product use. The deposition was determined for the head airways, and the tracheobronchial and alveolar regions.

The following equation was used to calculate the inhaled dose:

$$ID = \frac{f_{nano}\, C_{inh}\, Q_{inh}\, T_{contact}}{Bw},\qquad(9.2)$$

where
   ID is the inhaled dose of particulate matter (ng/kg bw)
   C_{inh} is the mass concentration of particulate matter in inhaled air (ng/L)
   Q_{inh} is the inhalation flow rate for a given gender/activity scenario (L/min)
   $T_{contact}$ is the duration of contact per application (min)
   Bw is the body weight (kg)
   f_{nano} is the mass fraction of nanomaterial (s) in the inhaled aerosol

Mass concentration of particulate matter in the inhaled air (C_{inh}) used in Equation 9.2 is described as follows:

$$C_{inh} = IF \cdot C_{air},\qquad(9.3)$$

where C_{air} is mass concentration of aerosol particulate matter in the personal breathing cloud. Inhalability fraction (IF) used in Equation 9.3 represents the fraction of particulate matter in the personal breathing cloud that is actually inhaled into the respiratory system and is described by Hinds (1999) as follows:

$$IF = 1 - 0.5\left(1 - \frac{1}{1 + 0.00076 d_p^{2.8}}\right),\qquad(9.4)$$

where d_p is particle diameter. This equation is valid for particles up to 100 μm in diameter.

This study used a human mannequin head to realistically simulate application of powders, where released particles were sampled through the mannequin's nostrils at a realistic sampling flow rate. Thus, it could be assumed that the particle aspiration efficiency through the mannequin's nostrils approximately matches the IF. Under this assumption, C_{inh} was obtained directly from the real-time measurements of released particles using SMPS and APS (TSI, Inc., Shoreview, MN, USA). Because differently sized particles have different potential health impacts and have different deposition characteristics in the respiratory system, the inhaled particle dose was calculated for each of the size channels measured by those instruments. At the same time, the native size channels of the instruments were grouped into aerosol size fractions that are relevant in the context of a given study or health effect: 0.014–0.1, 0.1–1, 1–2.5, 2.5–10, and 10–20 μm.

Deposited dose, DD_i, was calculated as a product of the inhaled dose, ID, and the deposition fraction, DF_i, integrated over a particle size range, d_p (Nazarenko et al. 2012a):

$$DD_i = \int_{d_p} DF_i(d_p)\, ID(d_p),\qquad(9.5)$$

where
   $i = HA$ for the head airways
   $i = TB$ for the tracheobronchial region
   $i = AL$ for the alveolar region
   $i = T$ for the entire respiratory system

Deposition fraction DF_i is a fraction of inhaled airborne particulate matter removed from the air stream due to deposition in a particular region or the respiratory system. These deposition fractions can be calculated using different mathematical models. The most common model that can be used is based on the equations fitted to the ICRP (International Commission on Radiological Protection 1994) model for monodisperse spheres of standard density at standard conditions developed by Hinds (1999). If a mannequin head sampler is used, these equations need to be modified to exclude the *IF* because of the assumption that *IF* is already taken into account due to the sampling through the nostrils of the human mannequin head (see Equation 9.3). The modified equations for DF_i as a function of particle diameter are as follows:

$$DF_{HA}(d_p) = \left(\frac{1}{1+\exp\left(6.84+1.183\ln d_p\right)} + \frac{1}{1+\exp\left(0.924-1.885\ln d_p\right)} \right) \tag{9.6}$$

$$DF_{TB}(d_p) = \frac{\left(\dfrac{0.00352}{d_p}\right)\left[\exp\left(-0.234\left(\ln d_p+3.40\right)^2\right)+63.9\exp\left(-0.819\left(\ln d_p-1.61\right)^2\right)\right]}{1-0.5\left(1-\dfrac{1}{1+0.00076d_p^{2.8}}\right)} \tag{9.7}$$

$$DF_{AL}(d_p) = \frac{\left(\dfrac{0.0155}{d_p}\right)\left[\exp\left(-0.416\left(\ln d_p+2.84\right)^2\right)+19.11\exp\left(-0.482\left(\ln d_p-1.362\right)^2\right)\right]}{1-0.5\left(1-\dfrac{1}{1+0.00076d_p^{2.8}}\right)} \tag{9.8}$$

$$DF_T(d_p) = \left(0.0587 + \frac{0.911}{1+\exp\left(4.77+1.485\ln d_p\right)} + \frac{0.943}{1+\exp\left(0.508-2.58\ln d_p\right)} \right) \tag{9.9}$$

where
DF_{HA} is the deposition fraction for the head airways
DF_{TB} is the deposition fraction for the tracheobronchial region
DF_{AL} is the deposition fraction for the alveolar region
DF_T is the total deposition fraction, equal to the sum of DF_{HA}, DF_{TB}, and DF_{AL}

Because aerosol size distribution data are collected as concentrations of inhalable particles in particular size channels, the preceding equations could be used, with d_p being equated to a midpoint diameter of each native instrument size channel. The DF_{HA}, DF_{TB}, DF_{AL}, and DF_T are then calculated as a sum of deposited doses in each particle size channel:

$$DD_i = \sum_{d_p} DF_i\left(d_p\right) ID\left(d_p\right) \tag{9.10}$$

The results of these calculations for the three nanotechnology-based and three regular cosmetic powders showed that the highest inhaled dose in terms of aerosol mass was in the coarse aerosol fraction (2.5–10 μm) (Nazarenko et al. 2012a). At the same time, individual nanosized particles or nanosized agglomerates (agglomerates smaller than 100 nm) contributed minimally to the inhaled particle mass. For all six tested powders, 85%–93% of aerosol deposition occurred in the head airways, <10% in the alveolar region, and <5% in the tracheobronchial region. As mentioned earlier, in aerosols formed during the use of consumer spray and powder products, nanomaterials are likely distributed within agglomerates across the entire aerosol size range, including supercoarse particles. While the actual fraction of nanomaterials in those particles is usually unknown, based on the aerosol size distribution, one can assume that a substantial mass of nanomaterials is likely deposited in the head airways, while some nanomaterials will end up in the alveolar and the tracheobronchial regions. Moreover, it is mostly the agglomerated form of nano-objects and less individual nanosized particles that make up most mass of the deposited nanomaterials. Nanomaterial exposure through individually dispersed nano-objects or small nanosized agglomerates would likely be minimal. However, after deposition, larger agglomerates containing nanomaterials could dissociate or otherwise cause biological effects driven by their nanosize. Hence, for inhalation toxicology studies of nanotechnology-based consumer products, all inhalable aerosol size fractions must be considered as all particle sizes can be vehicles for engineered nanomaterial delivery to points of contact with living cells and tissues. Likewise, deposition across the entire respiratory system, the head airways (including nasopharyngeal region) and the tracheobronchial and alveolar regions, must be considered. Nanomaterial uptake and translocation to other systems and organs, and migration within the respiratory system after deposition, is another factor to take into account.

9.5 SUMMARY

Existing research demonstrates that nanomaterials in the form of nano-objects as well as agglomerates containing nanomaterials can be released as aerosol particles from nanotechnology-based products during their use. Two categories of nanotechnology-based consumer products—sprays and powders—present the highest potential for inhalation exposure to nanomaterial-containing aerosol. The results of inhalation exposure modeling conducted for various types of products by different research teams point to a large variety of consumer products that have a potential to release inhalable particles that contain nanomaterials. Such aerosol particles span an airborne particle size range from individual nanosized objects (at least one dimension smaller than 100 nm) to supercoarse particles as large as 20 μm (size limit of commonly used aerosol size spectrometers) and, possibly, larger. Nanomaterials in agglomerated state are expected to be present in all size fractions of such aerosol, leading to inhalation exposure and nanomaterial deposition in all regions of the human respiratory system with possible health consequences.

Moreover, in addition to engineered nanomaterials, nanomaterial-containing consumer products often contain other ingredients that can sometimes be toxic. Thus, particles and agglomerates released from the products are likely to contain ingredients of the product matrix, including components of different volatility. Those additional ingredients are likely to chemically modify surfaces of particles and agglomerates. In addition, aerosol dynamics during and after particle release from consumer products can lead to rapidly changing particle size distribution, which affects particle inhalability and deposition in various regions of the respiratory system. Likewise, aerosol dynamics post-inhalation due to evaporation of volatile organic content from particles and hygroscopic particle growth in the respiratory system will contribute to changing aerosol size distribution, which affects the deposition profile. Initial aerosol particle size distribution and its changes before and after inhalation, and consequently particle deposition in the respiratory system, depend on product composition and the way it is aerosolized, as well as temperature, humidity, and air circulation in the actual microenvironment where personal exposure occurs. Hence, realistic simulation of actual product use in a proper setting is the most appropriate method to estimate potential exposure.

Labeling of nanotechnology products is an issue, since products labeled as nanotechnology-based may in fact not contain engineered nanomaterials; likewise, engineered nanomaterials may be present in consumer products that are not labeled or reported as nanotechnology based. Most referenced studies selected products to investigate based on manufacturers' claims that a given product is nanotechnology based or includes nanomaterials in its composition. While for certain nanomaterials, for example, metallic nanosilver or nano-TiO_2, it is possible to determine their presence and even concentration in a consumer product, such analyses are not possible for all engineered nanomaterials. Additionally, certain consumer product formulations may render product analysis using certain analytical techniques impossible. For instance, solvent mixtures with unknown refractive indices or presence of very large particles as well as presence of unknown particles make PCS (dynamic light scattering) and LDS unreliable or impossible. Many highly volatile and organic components of consumer products can contaminate electron microscopes, and many materials decompose or change physicochemically under the electron beam. Additionally, nano-objects present in some consumer products may be formed during manufacturing processes rather than deliberately introduced into the products. Potential health effects of nanomaterials and the lack of information about the engineered status of nanomaterials in consumer products as well as analytical challenges of establishing this status suggest that there should be a debate regarding proper labeling of consumer products containing engineered nanomaterials.

In case of spray products, the method of liquid aerosolization itself greatly affects aerosol size distribution and concentration of nanosized aerosol particles, especially in the case of propellant-driven spraying, since they typically yield nanosized particles during their application. A distinction between nanosized aerosol exposure and engineered nanomaterial exposure is not clear in many studies published thus far, but must be established in the future exposure and toxicological studies. While the size of inhaled particles is a major factor determining their inhaled fraction and deposition location in the respiratory system, there is not necessarily a connection between the number of individual nanosized aerosol particles generated or inhaled during product use and engineered nanomaterial exposure. Future exposure studies focusing on nanotechnology-based consumer products must reliably identify engineered nanomaterials and measure them in the original products and in the particles that are released, inhaled, and deposited in the respiratory system during actual or realistically simulated product use. Human exposure will need to be assessed not only for the engineered nanomaterial(s) in a given product but also for the other non-nanomaterial ingredients. Given the presented issues on the release of nanoparticles and their agglomerates, including components of product matrix, during product use, it is critical that toxicology studies investigate effects of exposure to actual nanotechnology-based consumer products and not only to pure nanomaterials. In addition, toxicology studies should consider that during the actual product use, particles can agglomerate, and the majority of nanomaterial deposition would occur not in the alveolar region, as would be expected based on the primary nanoparticle size, but in the head airways. Thus, the health effects are likely to be different, and this should be taken into account during toxicology studies.

QUESTIONS

1. How does one define human exposure to nanomaterials?
   Answer: Contact between the outer boundary of a human body and nanomaterials.
2. What is the primary reason to be concerned about human contact with nanomaterials in consumer products and what types of consumer products are of major health concern to the general public?
   Answer: Danger of adverse health effects, a large number of people exposed, potential for unintended product use by consumers. Liquids intended for spraying and powders are the greatest concern.
3. How can the method of nanotechnology-based consumer product application and/or dispensing affect nanomaterial inhalation exposure?
   Answer: Different aerosolization methods may produce different size distribution of the generated aerosol, which affect particle transport into and through the breathing zone and particle deposition in the respiratory system. Product aerosolization closest to the breathing zone may lead to higher inhalation exposure.

4. Why is proper labeling of nanotechnology-based products important for exposure and toxicological assessments and eventually risk assessment?

Answer: It is an analytical challenge to characterize and quantify nanomaterials in consumer products. Knowing the chemical nature of nanomaterial ingredients and their concentration is required for exposure, and toxicological and risk assessments.

5. In the context of inhalation exposure, does the size distribution of the aerosol generated during use of a consumer product remain stable after an aerosolization event or does it change with time? What are the mechanisms of change, if any?

Answer: It changes with time. Among the mechanisms of change are changes in spatial distribution of aerosol particles of different size, particle coagulation, evaporation, and/or condensation.

6. Once aerosolized from a spray or another type of nanotechnology-based product, where in the respiratory system will nanomaterials likely deposit upon their inhalation by a consumer? What factors will influence the deposition profile in the respiratory system?

Answer: Entire respiratory system, distributed across a wide range of particle sizes. The deposition profile will be influenced by aerosol particle size distribution, and volatility and hygroscopicity of aerosol particles.

REFERENCES

Bakand S, Hayes A, Dechsakulthorn F. 2012. Nanoparticles: A review of particle toxicology following inhalation exposure. *Inhal Toxicol* 24(2): 125–135.

Balogh LP. 2010. Why do we have so many definitions for nanoscience and nanotechnology? *Nanomedicine* 6(3): 397–398.

Baroli B, Ennas MG, Loffredo F, Isola M, Pinna R, López-Quintela MA. 2007. Penetration of metallic nanoparticles in human full-thickness skin. *J Invest Dermatol* 127: 1701–1712.

Baxter J, Bian Z, Chen G, Danielson D, Dresselhaus MS, Fedorov AG et al. 2009. Nanoscale design to enable the revolution in renewable energy. *Energy Environ Sci* 2(6): 559–588.

Bekker C, Brouwer DH, Tielemans E, Pronk A. 2013. Industrial production and professional application of manufactured nanomaterials-enabled end products in Dutch industries: Potential for exposure. *Ann Occup Hyg* 57(3): 314–327.

Bermudez E, Mangum JB, Wong BA, Asgharian B, Hext PM, Warheit DB et al. 2004. Pulmonary responses of mice, rats, and hamsters to subchronic inhalation of ultrafine titanium dioxide particles. *Toxicol Sci* 77: 347–357.

Borm PJ, Robbins D, Haubold S, Kuhlbusch T, Fissan H, Donaldson K et al. 2006. The potential risks of nanomaterials: A review carried out for ECETOC. *Part Fibre Toxicol* 3(11): 11.

Bradford A, Handy RD, Readman JW, Atfield A, Mühling M. 2009. Impact of silver nanoparticle contamination on the genetic diversity of natural bacterial assemblages in estuarine sediments. *Environ Sci Technol* 43(12): 4530–4536.

Bremmer HJ, Prud'homme de Lodder LCH, Engelen van JGM. 2006. Cosmetics fact sheet. To assess the risks for the consumer. Updated version for ConsExpo 4. 320104001/2006.

Cedervall T, Lynch I, Lindman S, Berggård T, Thulin E, Nilsson H et al. 2007. Understanding the nanoparticle–protein corona using methods to quantify exchange rates and affinities of proteins for nanoparticles. *PNAS* 104(7): 2050–2055.

Chatterjee R. 2008. The challenge of regulating nanomaterials. *Environ Sci Technol* 42(2): 339–343.

Chen BT, Afshari A, Stone S, Jackson M, Schwegler-Berry D, Frazer DG et al. 2010. Nanoparticles-containing spray can aerosol: Characterization, exposure assessment, and generator design. *Inhal Toxicol* 22(13): 1072–1082.

Colvin VL. 2003. The potential environmental impact of engineered nanomaterials. *Nat Biotechnol* 21: 1166–1170.

Cross SE, Innes B, Roberts MS, Tsuzuki T, Robertson TA, McCormick P. 2007. Human skin penetration of sunscreen nanoparticles: In-vitro assessment of a novel micronized zinc oxide formulation. *Skin Pharma Physiol* 20(3): 148–154.

Dhawan A, Sharma V, Parmar D. 2009. Nanomaterials: A challenge for toxicologists. *Nanotoxicology* 3(1): 1–9.

Dionysios DD. 2004. Environmental applications and implications of nanotechnology and nanomaterials: ASCE. *J Environ Eng* 130(7): 723–724.

Drobne D. 2007. Nanotoxicology for safe and sustainable nanotechnology. *Arh Hig Rada Toksikol* 58(4): 471–478.

Duffin R, Tran CL, Brown D, Stone V, Donaldson K. 2007. Proinflammogenic effects of low-toxicity and metal nanoparticles in vivo and in vitro: Highlighting the role of particle surface area and surface reactivity. *Inhal Toxicol* 19(10): 849–856.

Duffin R, Tran CL, Clouter A, Brown DM, MacNee W, Stone V et al. 2002. The importance of surface area and specific reactivity in the acute pulmonary inflammatory response to particles. *Ann Occup Hyg* 46(Suppl. 1): 242–245.

El-Ansary A, Al-Daihan S. 2009. On the toxicity of therapeutically used nanoparticles: An overview. *J Toxicol* 2009: 754810.

Elder A, Gelein R, Silva V, Feikert T, Opanashuk L, Carter J et al. 2006. Translocation of inhaled ultrafine manganese oxide particles to the central nervous system. *Environ Health Perspect* 114(8): 1172–1178.

Engelund B, Sørensen H. 2005. Mapping and health assessment of chemical substances in shoe care products. (Survey of Chemical Substances in Consumer Products). No. 52 2005. Danish Toxicology Centre, Environmental Protection Agency, Danish Ministry of the Environment, Copenhagen, Denmark.

European Chemicals Agency. 2012. Chapter R.15: Consumer exposure estimation. In: *Guidance on Information Requirements and Chemical Safety Assessment*. Helsinki, Finland.

Fauss E. 2008. *The Silver Nanotechnology Commercial Inventory*. Washington, DC: Woodrow Wilson International Center for Scholars.

Fischer DB. 2008. Nanotechnology—Scientific and regulatory challenges. *Villanova Environ Law J* 19: 315–332.

Frater L, Stokes E, Lee R, Oriola T. 2006. *An Overview of the Framework of Current Regulation Affecting the Development and Marketing of Nanomaterials*. Cardiff, U.K.: Cardiff University, p. 192.

Geiser M, Rothen-Rutishauser B, Kapp N, Schürch S, Kreyling W, Schulz H et al. 2005. Ultrafine particles cross cellular membranes by nonphagocytic mechanisms in lungs and in cultured cells. *Environ Health Perspect* 113(11): 1555–1560.

Gomez-Mejiba S, Zhai ZA, Akram H, Pye Q, Hensley K, Kurien B, Scofield R et al. 2008. Inhalation of environmental stressors & chronic inflammation: Autoimmunity and neurodegeneration. *Mutat Res* 674(1–2): 62–72.

Grassian VH, Adamcakova-Dodd A, Pettibone JM, O'Shaughnessy PT, Thorne PS. 2007a. Inflammatory response of mice to manufactured titanium dioxide nanoparticles: Comparison of size effects through different exposure routes. *Nanotoxicology* 1(3): 211–226.

Grassian VH, O'Shaughnessy PT, Adamcakova-Dodd A, Pettibone JM, Thorne PS. 2007b. Inhalation exposure study of titanium dioxide nanoparticles with a primary particle size of 2 to 5 nm. *Environ Health Perspect* 115(3): 397–402.

Gruère GP. 2011. Labeling nano-enabled consumer products. *Nano Today* 6(2): 117–121.

Gruère GP. 2012. Implications of nanotechnology growth in food and agriculture in OECD countries. *Food Policy* 37(2): 191–198.

Gwinn MR, Vallyathan V. 2006. Nanoparticles: Health Effects—Pros and Cons. *Environ Health Perspect* 114(12): 1818–1825.

Hagendorfer H, Lorenz C, Kaegi R, Sinnet B, Gehrig R, Goetz NV et al. 2010. Size-fractionated characterization and quantification of nanoparticle release rates from a consumer spray product containing engineered nanoparticles. *J Nanopart Res* 12(7): 2481–2494.

Haglund Jr. RF. 1998. Ion implantation as a tool in the synthesis of practical third-order nonlinear optical materials. *Mater Sci Eng*: A 253(1–2): 275–283.

Hansen SF, Michelson ES, Kamper A, Borling P, Stuer-Lauridsen F, Baun A. 2008. Categorization framework to aid exposure assessment of nanomaterials in consumer products. *Ecotoxicology* 17: 438–447.

Hikita M, Tanaka K, Nakamura T, Kajiyama T, Takahara A. 2005. Super-liquid-repellent surfaces prepared by colloidal silica nanoparticles covered with fluoroalkyl groups. *Langmuir* 21(16): 7299–7302.

Hinds WC. 1999. *Aerosol Technology: Properties, Behavior, and Measurement of Airborne Particles*. 2, illustrated ed. New York: John Wiley & Sons, Inc.

Hoet PHM, Brüske-Hohlfeld I, Salata OV. 2004. Nanoparticles—Known and unknown health risks. *J Nanobiotechnol* 2: 12.

Holgate ST. 2010. Exposure, uptake, distribution and toxicity of nanomaterials in humans. *J Biomed Nanotechnol* 6(1): 1–19.

Howard KA, Rahbek UL, Liu X, Damgaard CK, Glud SZ, Andersen MØ et al. 2006. RNA interference in vitro and in vivo using a chitosan/siRNA nanoparticle system. *Mol Ther* 14(4): 476–484.

Hu C, Zhang Z, Liu H, Gao P, Wang ZL. 2006. Direct synthesis and structure characterization of ultrafine CeO2 nanoparticles. *Nanotechnology* 17(24): 5983–5987.

Iavicoli I, Calabrese EJ, Nascarella MA. 2010. Exposure to nanoparticles and hormesis. *Dose-Response* 8(4): 501–517.

International Commission on Radiological Protection. 1994. Human respiratory tract model for radiological protection. ICRP Publication 66. *Ann ICRP* 24(1–3).

Jackson P, Hougaard KS, Vogel U, Wu D, Casavant L, Williams A et al. 2012. Exposure of pregnant mice to carbon black by intratracheal instillation: Toxicogenomic effects in dams and offspring. *Mutat Res/Genet Toxicol Environ Mutagen* 745(1–2): 73–83.

Kanarek MS. 2007. Nanomaterial health effects Part 3: Conclusion—Hazardous issues and the precautionary principle. *Wis Med J* 106(1): 16–19.

Keelan JA. 2011. Nanotoxicology: Nanoparticles versus the placenta. *Nat Nano* 6(5): 263–264.

Keenan CR, Goth-Goldstein R, Lucas D, Sedlak DL. 2009. Oxidative stress induced by zero-valent iron nanoparticles and Fe(II) in human bronchial epithelial cells. *Environ Sci Technol* 43(12): 4555–4560.

Kim Y-G, Oh S-K, Crooks RM. 2004. Preparation and characterization of 1в€'2 nm dendrimer-encapsulated gold nanoparticles having very narrow size distributions. *Chem Mater* 16(1): 167–172.

Konan YN, Gurny R, Allemann E. 2002. Preparation and characterization of sterile and freeze-dried sub-200 nm nanoparticles. *Int J Pharm* 233(1–2): 239–252.

Kreyling WG, Möller W, Schmid O, Semmler-Behnke M, Oberdörster G. 2011. Translocation of inhaled nanoparticles. In: (Cassee FR, Mills NL, Newby D, eds) *Cardiovascular Effects of Inhaled Ultrafine and Nanosized Particles.* Hoboken, NJ: John Wiley & Sons, Inc., pp. 125–143.

Kreyling WG, Moller W, Semmler-Behnke M, Oberdorster G. 2007. Particle toxicology. In: *Particle Toxicology.* Part 1st (Donaldson K, Borm P, eds). Boca Raton, FL: CRC Press/Taylor & Francis Group, LLC, pp. 48–74.

Levine IN. 2002. *Physical Chemistry.* Boston, MA: McGraw-Hill Companies.

Lioy PJ. 2010. Exposure science: A view of the past and major milestones for the future. *Environ Health Perspect* 118: 1081–1090.

Lioy PJ, Nazarenko Y, Han TW, Lioy MJ, Mainelis G. 2010. Nanotechnology and exposure science— What is needed to fill the research and data gaps for consumer products. *Int J Occup Environ Health* 16(4): 376–385.

Lorenz C, Hagendorfer H, von Goetz N, Kaegi R, Gehrig R, Ulrich A et al. 2011. Nanosized aerosols from consumer sprays: Experimental analysis and exposure modeling for four commercial products. *J Nanopart Res* 13(8): 3377–3391.

Majestic B, Erdakos G, Lewandowski M, Oliver K, Willis R, Kleindienst T et al. 2010. A review of selected engineered nanoparticles in the atmosphere: Sources, transformations, and techniques for sampling and analysis. *Int J Occup Environ Health* 16(4): 488–507.

Maynard AD. 2007. Nanotechnology: The next big thing, or much ado about nothing. *Ann Occup Hyg* 51(1): 1–12.

Maynard AD, Aitken RJ. 2007. Assessing exposure to airborne nanomaterials: Current abilities and future requirements. *Nanotoxicology* 1(1): 26–41.

Maynard AD, Aitken RJ, Butz T, Colvin C, Donaldson K, Oberdörster G et al. 2006. Safe handling of nanotechnology. *Nature* 444: 267–269.

McCall MJ. 2011. Environmental, health and safety issues: Nanoparticles in the real world. *Nat Nano* 6(10): 613–614.

Michelson ES. 2008. Globalization at the nano frontier: The future of nanotechnology policy in the United States, China, and India. *Technol Soc* 30: 405–410.

National Science and Technology Council. 2007. The National Nanotechnology Initiative Strategic Plan (NNISP).

Nazarenko Y, Han TW, Lioy PJ, Mainelis G. 2011. Potential for exposure to engineered nanoparticles from nanotechnology-based consumer spray products. *J Exp Sci Env Epid* 21: 515–528.

Nazarenko Y, Zhen H, Han T, Lioy P, Mainelis G. 2012a. Nanomaterial inhalation exposure from nanotechnology-based cosmetic powders: A quantitative assessment. *J Nanopart Res* 14(11): 1–14.

Nazarenko Y, Zhen H, Han T, Lioy PJ, Mainelis G. 2012b. Potential for inhalation exposure to engineered nanoparticles from nanotechnology-based cosmetic powders. *Environ Health Perspect* 120: 885–892.

Nel A, Xia T, Mädler L, Li N. 2006. Toxic potential of materials at the nanolevel. *Science* 311: 622–627.

Nørgaard AW, Jensen KA, Janfelt C, Lauritsen FR, Clausen PA, Wolkoff P. 2009. Release of VOCs and particles during use of nanofilm spray products. *Environ Sci Technol* 43(20): 7824–7830.

Nowack B, Brouwer C, Geertsma RE, Heugens EH, Ross BL, Toufektsian M-C et al. 2012. Analysis of the occupational, consumer and environmental exposure to engineered nanomaterials used in 10 technology sectors. *Nanotoxicology* 7(6): 1152–1156.

Nowack B, Bucheli TD. 2007. Occurrence, behavior and effects of nanoparticles in the environment. *Environ Pollut* 150(1): 5–22.

Oberdörster G, Maynard A, Donaldson K, Castranova V, Fitzpatrick J, Ausman K et al. 2005. Principles for characterizing the potential human health effects from exposure to nanomaterials: Elements of a screening strategy. *Particle Fibre Toxicol* 2(8): 1–35.

Oberdorster G, Sharp Z, Atudorei V, Elder A, Gelein R, Kreyling W et al. 2004. Translocation of inhaled ultrafine particles to the brain. *Inhal Toxicol* 16(6–7): 437–445.

Park K, Cao F, Kittelson DB, McMurry PH. 2003. Relationship between particle mass and mobility for diesel exhaust particles. *Environ Sci Technol* 37(3): 577–583.

Pauluhn J, Hahn A, Spielmann H. 2008. Assessment of early acute lung injury in rats exposed to aerosols of consumer products: Attempt to disentangle the "Magic Nano" conundrum. *Inhal Toxicol* 20(14): 1245–1262.

Perrault SD, Chan WCW. 2009. Synthesis and surface modification of highly monodispersed, spherical gold nanoparticles of 50в€'200 nm. *J Am Chem Soc* 131(47): 17042–17043.

Petitot F, Lestaevel P, Tourlonias E, Mazzucco C, Jacquinot S, Dhieux B et al. 2013. Inhalation of uranium nanoparticles: Respiratory tract deposition and translocation to secondary target organs in rats. *Toxicol Lett* 217(3): 217–225.

Plummer LE, Pinkerton KE, Madl AK, Wilson DW. 2011. Effects of nanoparticles on the pulmonary vasculature. In: (Cassee, Mills NL, Newby D, eds.) *Cardiovascular Effects of Inhaled Ultrafine and Nanosized Particles*. Hoboken, NJ: John Wiley & Sons, Inc., pp. 317–350.

Proctor NH, Hughes JPMD, Hathaway GJ. 2004. *Proctor and Hughes' Chemical Hazards of the Workplace*. Hoboken, NJ: Wiley.

Quadros ME, Marr LC. 2011. Silver nanoparticles and total aerosols emitted by nanotechnology-related consumer spray products. *Environ Sci Technol* 45(24): 10713–10719.

Reijnders L. 2012. Human health hazards of persistent inorganic and carbon nanoparticles. *J Mater Sci* 47(13): 5061–5073.

Riediker M. 2009. Chances and risks of nanomaterials for health and environment nano-net. In: *Proceedings of the Fourth International ICST Conference, Nanonet 2009*, Lucerne, Switzerland, October 18–20, 2009, Vol. 20, (Schmid A, Goel S, Wang W, Beiu V, Carrara S, eds), Berlin, Germany: Springer, pp. 128–133.

Romig Jr. AD, Baker AB, Johannes J, Zipperian T, Eijkel K, Kirchhoff B et al. 2007. An introduction to nanotechnology policy: Opportunities and constraints for emerging and established economies. *Technol Forecast Soc Change* 74(9): 1634–1642.

Sarma B. 2008. Nanomaterials and PM industry. In: *Powder Metallurgy*. Part 1st (Ramakrishnan P, ed). New Delhi, India: New Age International Pvt Ltd Publishers, p. 396.

Schmid A, Goel S, Wang W, Beiu V, Carrara S, Riediker M. 2009. Chances and risks of nanomaterials for health and environment. In: *Nano-Net*. Vol. 20, (Akan O, Bellavista P, Cao J, Dressler F, Ferrari D, Gerla M et al., eds), Berlin, Germany: Springer, pp. 128–133.

Schummer J. 2007. Identifying ethical issues of nanotechnologies. In: *Nanotechnologies, Ethics and Politics*, (Have HAMJt, ed). Paris, France: UNESCO, pp. 79–98.

Segal SH. 2004. Environmental regulation of nanotechnology: Avoiding big mistakes for small machines. *Nanotechnol Law Bus* 1(3): 290–304.

Shaw GK. 2011. *"Principles" Issued. New Haven Independent.* New Haven, CT, July 1, 2011.

Shimada M, Wang W-N, Okuyama K, Myojo T, Oyabu T, Morimoto Y et al. 2009. Development and evaluation of an aerosol generation and supplying system for inhalation experiments of manufactured nanoparticles. *Environ Sci Technol* 43(14): 5529–5534.

Shrader-Frechette K. 2007. Nanotoxicology and ethical conditions for informed consent. *Nanoethics* 1: 47–56.

Singh S, Nalwa HS. 2007. Nanotechnology and health safety toxicity and risk assessments of nanostructured materials on human health. *J Nanosci Nanotechnol* 7(9): 3048–3070.

Tabuchi A, Kuebler WM. 2008. Endothelium-platelet interactions in inflammatory lung disease. *Vascul Pharmacol* 49: 141–150.

Tedgui A, Mallat Z. 2006. Cytokines in atherosclerosis: Pathogenic and regulatory pathways. *Physiol Rev* 86(2): 515–581.

Teow Y, Asharani PV, Hande MP, Valiyaveettil S. 2011. Health impact and safety of engineered nanomaterials. *Chem Commun* 47(25): 7025–7038.

USEPA 2006, Symposium on Nanotechnology and the Environment: Fate and Transport of Nanomaterials: Highlights, Question and Answer Session, available at:http://www.epa.gov/oswer/nanotechnology/events/OSWER2006/pdfs/mc-fate-and-transport-highlights-qa.htm. Accessed 3 May 2013.

USEPA. 1997. *Exposure Factors Handbook*. EPA/600/P-95/002Fa,b,c. Washington, DC.

USEPA. 2011a. Acute Exposure Guideline Levels (AEGLs) for Chloroacetone (CAS Reg. No.78-95-5). (NAC Interim: February 2011).

USEPA. 2011b. *Exposure Factors Handbook*. EPA/600/R-090/052F. Washington, DC.

Van Calster G. 2006. Regulating nanotechnology in the European union. *Nanotechnol Law Bus* 3(3): 359–374.

Vayssieres L. 2003. Growth of arrayed nanorods and nanowires of ZnO from aqueous solutions. *Adv Mater* 15(5): 464–466.

Wang J, Zhou G, Chen C, Yu H, Wang T, Ma Y et al. 2007. Acute toxicity and biodistribution of different sized titanium dioxide particles in mice after oral administration. *Toxicol Lett* 168(2): 176–185.

Wardak A, Gorman ME, Swami N, Deshpande S. 2008. Identification of risks in the life cycle of nanotechnology-based products. *J Industr Ecol* 12(3): 435–448.

Warheit DB, Webb TR, Reed KL, Frerichs S, Sayes CM. 2007. Pulmonary toxicity study in rats with three forms of ultrafine-TiO2 particles: Differential responses related to surface properties. *Toxicol Lett* 230(1): 90–104.

Wick P, Malek A, Manser P, Meili D, Maeder-Althaus X, Diener L et al. 2009. Barrier capacity of human placenta for nanosized materials. *Environ Health Perspect* 118(3): 432–436.

Wiebert P, Sanchez-Crespo A, Falk R, Philipson K, Lundin A, Larsson S et al. 2006a. No significant translocation of inhaled 35-nm carbon particles to the circulation in humans. *Inhal Toxicol* 18(10): 741–747.

Wiebert P, Sanchez-Crespo A, Seitz J, Falk R, Philipson K, Kreyling W et al. 2006b. Negligible clearance of ultrafine particles retained in healthy and affected human lungs. *Eur Respir J* 28(2): 286–290.

Wittmaack K. 2007. In search of the most relevant parameter for quantifying lung inflammatory response to nanoparticle exposure: Particle number, surface area, or what? *Environ Health Perspect* 115(2): 187–194.

Wolff RK, Niven RW. 1994. Generation of aerosolized drugs. *J Aerosol Med* 7(1): 89–106.

Woodrow Wilson International Center for Scholars. 2011. The Project on emerging nanotechnologies. Available at http://www.nanotechproject.org/ (accessed May 26, 2011).

Xiong J, Miller VM, Li Y, Jayachandran M. 2012. Microvesicles at the crossroads between infection and cardiovascular Diseases. *J Cardiovasc Pharmacol* 59(2): 124–132, 110.1097/FJC.1090b1013e31820c36254.

10 Influence of Physicochemical Properties on the Bioactivity of Carbon Nanotubes/ Nanofibers and Metal Oxide Nanoparticles*

Dale W. Porter and Vincent Castranova

CONTENTS

10.1 INTRODUCTION

Nanomaterials have already been integrated into many consumer products, and future use is projected to be a major source of economic development, affecting our lives in many aspects. The wide variety of possible applications for nanoparticles stem from their unique properties, that is, approaching nanometer size, extremely high surface areas, and unique physicochemical properties. However, these unique properties also cause concern for poor compatibility with biological systems. Unfortunately, the development of engineered nanomaterials has outpaced our ability to evaluate potential human

---

* The findings and conclusions in this report are those of the author and do not necessarily represent the views of the National Institute for Occupational Safety and Health.

health impacts. Thus, there is interest in understating the relationships between physicochemical properties of nanoparticles and their bioactivity so that their relative hazard can be estimated.

To date, several key physicochemical properties have been identified that play critical roles in the bioactivity of nanoparticles. In this chapter, we will discuss how these physicochemical properties influence bioactivity for two important types of nanomaterials: carbon nanotubes/nanofibers and metal oxides. In addition, we will consider the relative potency of several well-studied nanoparticles in relation to their physicochemical properties.

10.2   CARBON NANOTUBES AND CARBON NANOFIBERS

Methods have been perfected to arrange carbon atoms in a crystalline graphene lattice with a tubular morphology. A single-walled carbon nanotube (SWCNT) is composed of a single cylindrical sheet of graphene and has a diameter of 0.5–2 nm. Multiwalled carbon nanotube (MWCNT) consists of multiple tubes within a tube and have diameters of 10–150 nm, depending on the number of concentric tubes forming the structure. Carbon nanotubes (CNTs) can range in length from 0.5–30 μm (Shvedova et al. 2009). Carbon nanofibers (CNFs) are composed of graphene layers arranged at an angle to the fiber axis. Size of CNF ranges from 70 to 200 nm in diameter and 10 to over 100 μm in length (De Jong and Geus 2000). CNTs exhibit high tensile strength, possess unique electrical properties, are resistant to acid or high temperature, and can be easily functionalized. Therefore, application as structural materials, in electronics, as heating elements, in batteries, in production of conductive and stain resistant fabric, for bone grafting and dental implants, as well as in targeted drug delivery are being developed. CNFs are strong, flexible fibers, which are currently being used to produce strong but lightweight composite materials.

10.2.1   DISPERSION STATUS OF SWCNT

The initial reports on the pulmonary responses to SWCNT varied from study to study. Comparing the results of several studies suggests that the dispersion status of SWCNT may have a significant impact on the responses to these exposures. The first study (Lam et al. 2004) reported that after intratracheal (IT) instillation of a poorly dispersed suspension of SWCNT, granulomatous lesions were observed and that on an equal-weight basis, SWCNTs were more toxic than carbon black or quartz. A subsequent study (Warheit et al. 2004) conducted in rats exposed by IT instillation of SWCNT demonstrated non-dose-dependent granulomatous lesions. They also reported a number of deaths due to blockade of bronchioles by large SWCNT agglomerates. A later study in mice (Shvedova et al. 2005), which used pharyngeal aspiration exposure, showed that exposure to SWCNT (a suspension containing large agglomerates as well as smaller nanorope structures) produced two distinct responses, that is, a granulomatous response associated with agglomerates and a progressive alveolar interstitial fibrotic response not associated with deposition of SWCNT agglomerates visible by light microscopy. These data suggest that SWCNT structures not in the form of large agglomerates, that is, more dispersed SWCNT, might be responsible for the progressive interstitial fibrotic response.

To resolve whether responses to SWCNT differed depending on dispersion status of SWCNT, a study was conducted which compared pulmonary responses to more dispersed vs. more agglomerated SWCNT (Mercer et al. 2008). This study concluded that approximately 80% of the poorly dispersed SWCNT deposited as large agglomerates at the terminal bronchioles and proximal alveoli of the lung, forming granulomatous lesions. The other portion of these SWCNT, containing smaller nanoropes, distributed throughout the deep lung and migrated into the alveolar walls, where they stimulated excess collagen production. In contrast, exposure to a well-dispersed SWCNT preparation produced a more potent alveolar interstitial fibrotic response and a complete absence of granulomas. Thus, these data indicate that dispersed SWCNT are responsible for the progressive interstitial fibrotic response, whereas the larger agglomerates stimulate the formation of granulomatous lesions.

In vitro studies also suggest that well-dispersed SWCNTs exhibit greater direct effects on lung fibroblasts in culture than an equal mass of poorly dispersed SWCNT, causing greater fibroblast proliferation and collagen production as well as greater TGF-β production (a fibrogenic mediator) from lung epithelial cells in culture (Wang et al. 2010, 2011).

10.2.2 METAL CONTAMINANTS AND OXIDATIVE STRESS

One of the common methods to manufacture SWCNT is using the high-pressure CO disproportionation process (HiPCO). This method uses CO in a continuous-flow gas-phase as the carbon feedstock and $Fe(CO)_5$ as the iron-containing catalyst precursor. However, significant amounts of iron can remain after manufacture of the SWCNT by the HiPCO process. In vitro studies with keratinocytes (Shvedova et al. 2003) and the monocyte-macrophage cell line RAW 264.7 (Kagan et al. 2006) determined that a majority of the toxicity induced by exposure to SWCNT was due to the generation of reactive oxygen species by the residual metal catalyst. Thus, it was proposed that purification, that is, removal of residual metal catalysts from CNT, may reduce CNT toxicity.

In vivo comparisons of low- and high-metal-containing SWCNT have been made (Shvedova et al. 2005, 2008), thus allowing the hypothesis that removal of residual metal catalysts from SWCNT would reduce SWCNT toxicity to be tested. Analysis of unpurified SWCNT determined they contained iron (Fe, 17.7%), as well as other metals such as Cu (0.16%), Cr (0.049%), and Ni (0.046%). Acid washing was performed to remove the contaminating metals, and the resulting SWCNTs were 99.7% (wt) elemental carbon with only 0.23% (wt) of iron. After pharyngeal aspiration exposure to the unpurified or purified SWCNT, pulmonary inflammation showed dose- and time-dependent changes. Comparison of the lung inflammatory responses, gaged by polymorphonuclear leukocyte (PMN) levels obtained by bronchoalveolar lavage (BAL), showed that unpurified SWCNT did not produce a greater inflammatory response than purified SWCNT. Exposure to unpurified SWCNT did result in a transient depletion of total lung antioxidants and an elevation of lung protein carbonyl levels. Given that these SWCNTs contained significant amounts of iron, a redox-active metal, such a result was not surprising. However, exposure to purified SWCNT also resulted in transient accumulation of 4-HNE, a biomarker of oxidative stress in the BAL fluid, and depletion of GSH in the lung, despite a much lower level of iron in the SWCNT. Exposure to both unpurified and purified SWCNTs produced persistent granulomatous inflammation and interstitial fibrosis. Thus, metal contaminants in SWCNT do not appear to drive pulmonary inflammatory or fibrogenic responses in vivo. A similar conclusion was made by Lam et al. (2004) who reported granulomatous lesions after pulmonary exposure to either unpurified or purified SWCNT.

10.2.3 SWCNT VERSUS CNF AND ASBESTOS

Crocidolite is a specific form of asbestos, and pulmonary exposure has been linked to the development of lung disease. The pathological response to asbestos has been associated with the high aspect ratio (length: width) and durability of the asbestos fibers. CNT and CNF exhibit a high aspect ratio and are resistant to degradation by acid treatment or high temperatures. Therefore, it has been proposed that CNT and CNF may lead to asbestos-like lung disease (Donaldson et al. 2006).

A study comparing pulmonary toxicity of SWCNT, CNF, and crocidolite asbestos has been conducted (Murray et al. 2012). Mice were dosed with SWCNT (40 μg/mouse), CNF (120 μg/mouse), or crocidolite asbestos (120 μg/mouse) by pharyngeal aspiration and were sacrificed at 1, 7, and 28 days postexposure. The pulmonary inflammatory responses, on an equal lung burden basis, following SWCNT, CNF, and asbestos, measured as the number of PMN harvested by BAL, were 813-, 150-, and 50-fold above control, respectively, at 1 day postexposure (Figure 10.1). By 28 days postexposure, the inflammatory response of mice exposed to all three of the particles had declined and approached control levels. Pulmonary fibrosis, characterized as collagen accumulation, was determined at day 28 postexposure. On an equal lung burden basis, lung collagen levels of

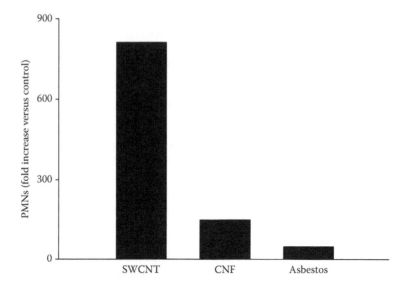

FIGURE 10.1 Increase in PMNs harvested by bronchoalveolar lavage 1 day after pharyngeal aspiration of SWCNT, CNF, or crocidolite asbestos (120 μg/mouse, equivalent lung burden).

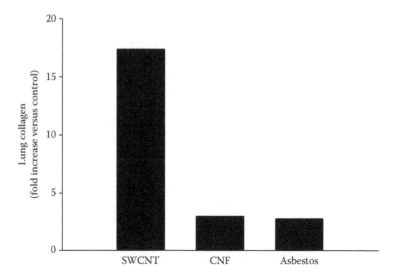

FIGURE 10.2 Increase in lung collagen 28 days after pharyngeal aspiration of SWCNT, CNF, or crocidolite asbestos (120 μg/mouse, equivalent lung burden).

SWCNT-exposed mice increased 17.4-fold above control, while CNF-exposed mice had a 3-fold increase and asbestos-exposed mice had a 2.8-fold increase (Figure 10.2). Pulmonary exposure to these fibers also increased levels of the fibrotic mediator, transforming growth factor-beta (TGF-β), in bronchoalveolar lavage fluid at 7 days postexposure. Fold increases above control, on an equal lung burden basis, were 13.5 (SWCNT), 2 (CNF), and 2 (asbestos). Therefore, in regard to the induction of alveolar interstitial fibrosis, the potency sequence was SWCNT > CNF = asbestos.

10.2.4 MWCNT Length

A pulmonary toxicity study of whole (5.9 μm long) vs. fragmented (0.7 μm long, produced by grinding) MWCNT reported both samples caused acute pulmonary inflammation at 3 days postexposure and induced pulmonary fibrosis at 60 days postexposure after IT instillation of 0.5–5 mg MWCNT/rat (Muller et al. 2005). At 3 days postexposure, 500 mg of ground MWCNT/rat was 2–3 times more potent in causing pulmonary inflammation (BAL neutrophils) and damage (BAL protein and LDH) than unground MWCNT. This may reflect the delivery of a greater number of structures per mass for ground vs. unground MWCNT. In contrast, 5 mg of unground MWCNT/rat was 2 times more fibrotic (lung hydroxyproline and lung collagen content) than ground MWCNT at 60 days postexposure. This may reflect the fact that ground MWCNTs were cleared from the lung 3 times faster than unground MWCNTs, that is, at 60 days postexposure, 81% of unground MWCNTs remained in the lung vs. 36% for ground MWCNT.

10.2.5 AGGLOMERATION STATUS OF MWCNT RELATED TO FIBER THICKNESS

Mice exposed by aspiration exposure to a well-dispersed preparation of relatively thick (49 µm) MWCNT exhibited dose-dependent pulmonary inflammation, damage, and granulomatous inflammation, as well as the rapid development of pulmonary fibrosis (Porter et al. 2010). In this study, MWCNT singlet structures were shown to penetrate epithelial cells and enter the alveolar walls to induce alveolar interstitial fibrosis. Likewise, recent data from our laboratory indicate that mice exposed by whole body inhalation exposure to a well-dispersed aerosol of these MWCNT exhibited dose-dependent pulmonary inflammation and damage as well as the rapid development of pulmonary fibrosis associated with the migration of singlet structures into the alveolar interstitium (Porter et al. 2013). In contrast, aerosolization of thin MWCNT (Baytubes, 10–15 nm) resulted in 3 µm consolidated agglomerates (Pauluhn 2010). A 13-week inhalation of rats to 0.4–6 mg/m³ MWCNT resulted in persistent pulmonary inflammation and granulomas with a low level of interstitial fibrosis associated with the absence of singlet structures in the lung tissue. These data suggest that relatively thick MWCNTs are more easily dispersed into smaller structures and are more fibrotic (Porter et al. 2010) than thin MWCNTs such as Baytubes, which form large condensed agglomerated structures (Pauluhn 2010).

10.2.6 EFFECT OF FUNCTIONALIZATION OF MWCNT

Modification of carbon nanotubes is key to extending their applications. These modifications can alter the charge, hydrophobicity, and reactivity of the surface of nanoparticles, which in turn may change their bioactivity. Functionalization of MWCNT with COOH groups enhances their water solubility and provides a terminal group for further bioconjugation or linkage to other molecules. However, as will be discussed, this modification also alters the bioactivity of MWCNT.

In comparison to bare (unmodified) MWCNT, COOH-MWCNT stimulated less excess collagen production in the lungs of mice exposed by pharyngeal aspiration (Wang et al. 2011). This result may be explained by the greater interstitialization of bare (hydrophobic) MWCNT. These initial data were confirmed and extended upon in a later study (Sager et al. 2014). In this study, mice exposed to bare MWCNT exhibited significantly more pulmonary inflammation, as well as fibrosis, in comparison to COOH-MWCNT-exposed mice (Figure 10.3). The data from these first two studies are supported from a recent interlaboratory evaluation of pulmonary responses to bare- vs. COOH-MWCNTs (Bonner et al. 2013). In this study, mice and rats were exposed to bare- or COOH-MWCNT by pharyngeal aspiration or IT instillation, respectively. Composite results indicated that bare MWCNTs were more inflammatory than the COOH-MWCNT.

In a recent study, Li et al. (2013) modified MWCNT with covalently bound functional groups, which resulted in five MWCNT types which varied from exhibiting a highly negative surface charge (carboxylated-MWCNT) to a highly positive surface charge (polyetherimide-MWCNT). Negative carboxylated (COOH)-MWCNT exhibited low bioactivity in vitro, that is, low production of the inflammasome cytokine interleukin-1 beta (IL-1β) and fibrotic mediators platelet-derived growth factor alpha (PDGF-α) and transforming growth factor-beta 1 (TGF-β1). In contrast, positive

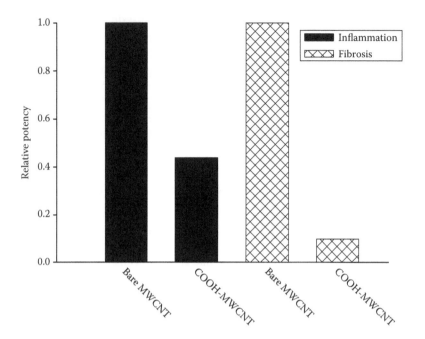

FIGURE 10.3 Decrease in inflammatory (1 day postexposure) and fibrotic (56 days postexposure) potency after pharyngeal aspiration of mice to 40 μg/mouse of COOH-MWCNT compared to unfunctionalized (bare) MWCNT.

polyetherimide (PEI)-MWCNT was strongly bioactive. After exposure of mice by pharyngeal aspiration, COOH-MWCNT exhibited a low potential to enhance BAL levels of PDGF-α and TGF-β1 and increase lung collagen. In contrast, PEI-MWCNT was more potent than bare MWCNT. These results indicate that functionalization of MWCNT with groups resulting in a strong positive surface charge results in a more fibrotic nanotube.

Another alteration of CNT is by "doping" them with another material or element. For example, MWCNT can be doped with nitrogen. MWCNTs are well known for their ability to induce pulmonary inflammation, damage, and fibrosis after aspiration and inhalation exposures (Porter et al. 2010, 2013). In comparison to MWCNT, N-doped MWCNT caused less acute pulmonary inflammation (Figure 10.4). Mice exposed to N-doped MWCNT still developed pulmonary fibrosis, but it was significantly less than that observed in mice exposed to unmodified MWCNT.

10.2.7 SWCNT VERSUS MWCNT

In general, pulmonary responses to SWCNT and MWCNT are qualitatively similar in that both induce transient pulmonary inflammation, persistent granulomatous lesions, and fibrosis (Porter et al. 2010; Shvedova et al. 2005). However, on a quantitative basis, differences emerge.

In a study which used well-dispersed MWCNT, the MWCNTs were found in every cell/cell layer of the mouse lung parenchyma (Porter et al. 2010). At 56 days postexposure to a dose of 80 μg of well-dispersed MWCNT, 68% of the total lung burden was found within and/or penetrating into alveolar macrophages; 8% was in the interstitium of the alveolar tissue; 1.6% of the total lung burden was in the subpleural tissue, which consists of mesothelial cells of the visceral pleura and the immediately adjacent interstitium; while granulomatous lesions in the alveolar airspace accounted for 20% of the lung burden (Mercer et al. 2010). Thus, alveolar macrophages from MWCNT-exposed lungs were highly loaded with MWCNT. This is in contrast with the distribution of well-dispersed SWCNT in mouse lungs exposed by aspiration where only 10% of the lung burden was in alveolar macrophages vs. 90% in the alveolar interstitium (Mercer et al. 2008). MWCNTs

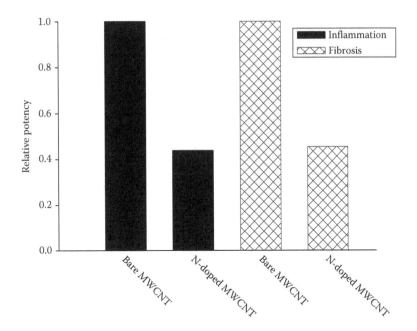

FIGURE 10.4 Decrease in inflammatory (1 day postexposure) and fibrotic (56 days postexposure) potency after pharyngeal aspiration of mice to 40 µg/mouse of nitrogen-doped MWCNT compared to undoped MWCNT.

were also observed in the subpleural interstitium and penetrating through the visceral pleura into the pleural space (Mercer et al. 2010). SWCNTs were observed in subpleural tissue but were not found in the intrapleural space (Mercer et al. 2008). The thin diameter of SWCNT (1–2 nm) compared to MWCNT (49 nm) may explain the lack of recognition of SWCNT by alveolar macrophages and the inability of SWCNT to pierce the outer surface of the lung.

As stated above, fibrotic responses to SWCNT and MWCNT were qualitatively similar in that both induce this response. However, quantitative morphometric analysis of the thickness of alveolar wall collagen in response to SWCNT and MWCNT suggests that SWCNTs are significantly more fibrogenic than MWCNTs (Mercer et al. 2011). The difference in pulmonary responses may be due to the delivery of a greater number of thin SWCNT than thick MWCNT on an equal mass lung burden basis. Additionally, different fibrogenic potency may result from the difference in the dose of SWCNT and MWCNT delivered to the alveolar interstitial tissue, that is, 90% of the SWCNT lung burden but only 8% of the MWCNT lung burden was distributed to the interstitium (Mercer et al. 2011). Of interest, in vitro studies have determined that SWCNT and MWCNT induce similar induction of collagen production by lung fibroblasts (Mishra et al. 2012; Wang et al. 2010).

10.3 METAL OXIDES

10.3.1 Dispersion Status

Studies have been conducted to assess the lung toxicity of IT instilled nanosized and fine-sized TiO_2 (Sager et al. 2008; Warheit et al. 2006, 2007). In the first two studies (Warheit et al. 2006, 2007), fine and ultrafine TiO_2 were suspended in PBS, which produced agglomerates of both the fine- and nanosized TiO_2. The mean structure diameters of the fine- and nano-TiO_2, measured by dynamic light scattering, both equaled 2.5 µm. As a result, both TiO_2 preparations caused similar acute levels of inflammation, with no significant long-term pulmonary toxicity. In contrast, when well-dispersed ultrafine TiO$_2$ was tested, UFTiO$_2$ caused substantially more inflammation and damage than well-dispersed fine TiO$_2$ over a 42-day postexposure study period (Sager et al. 2008). The discrepancies between these studies were attributed to differences in the dispersion of the ultrafine TiO$_2$ between the studies (Sager et al. 2008). Taken together, these studies indicate that the dispersion status of nanosized metal particles has a significant impact on bioactivity.

10.3.2 ROS GENERATION

In 2006, a three-tiered hierarchical oxidative stress model was proposed (Nel et al. 2006). The first tier was oxidative stress (Tier 1), followed by pro-inflammatory responses initiated by the oxidative stress in Tier 2, followed by apoptosis and necrosis via various pathways in Tier 3. This model has been supported by several studies since it was proposed.

Support for the ROS paradigm came from a study that examined eight spherical metal oxide particles as to their ability to generate ROS in vitro and cause pulmonary inflammation in vivo (Rushton et al. 2010). Initially, the ROS generation by these particles was assessed in an acellular system and in the presence of alveolar macrophages and the particles ranked as to potency. Next, rats were exposed to the same eight spherical particles by IT instillation, and the extent of particle-induced pulmonary inflammation was determined. This study concluded that when particle surface area was used as the dose-metric, both acellular and cellular ROS generation in vitro correlated with in vivo responses, with the correlation being stronger (R = 0.9) using cellular ROS generation.

In 2011, a theoretical model that described the relationship between metal oxide band gap and cellular redox potential was proposed (Burello and Worth 2011). Essentially, this model stated that when material and biological energetic states were similar, reactions involving electron transfer would result in oxidative stress, either by increased ROS production, decreased cellular antioxidants, or a combination of both. This model was tested in a study which used 24 metal oxide nanoparticles (Zhang et al. 2012). The authors demonstrated that those metal oxide nanoparticles that exhibited an overlap between conduction band gap energy levels and cellular redox potential shared a good correlation between in vitro particle-induced oxidative stress and in vivo inflammation. In addition, metal oxide nanoparticles with low in vitro ROS generation correlated with low acute neutrophilic inflammation and cytokine responses in the lungs of mice after aspiration.

10.3.3 SOLUBILITY

The role of nanoparticle dissolution can be seen in the outcome of a recent study that compared the effect of ZnO and CeO$_2$ nanoparticles on two cell lines, RAW 264.7 and BEAS-2B (Xia et al. 2008). This study determined that ZnO induced the production of ROS and concomitant oxidant injury, and also caused inflammation and cell death. CeO$_2$ nanoparticles were also taken up by BEAS-2B and RAW 264.7 cells, suppressed ROS production, and did not induce cytotoxicity or inflammation. ZnO dissolution in the cell culture medium, as well as after cellular uptake, disrupted cellular zinc homeostasis and ultimately caused cell death. In contrast, CeO$_2$ nanoparticles did not dissolve and, once phagocytized into cells, protected the cells from exogenous oxidative stress. The results of this initial study were confirmed in a subsequent study that used iron-doped ZnO nanoparticles (George et al. 2010). The dissolution of iron-doped ZnO nanoparticles was substantially lower than that of undoped ZnO. The rate of dissolution of ZnO correlated with in vitro toxicity of undoped vs. doped ZnO in BEAS-2B and RAW 264.7 cells. Lastly, the conclusion of these in vitro studies, that is, dissolution of ZnO is related to its toxicity, was confirmed in vivo. Specifically, mice and rats were exposed to ZnO and iron-doped ZnO nanoparticles by pharyngeal aspiration and IT instillation, respectively (Xia et al. 2011). In both species, exposure to undoped ZnO caused greater pulmonary damage and inflammation in comparison to iron-doped ZnO.

10.3.4 Role of Metal Oxide Nanoparticle Shape and Functionalization

Fibrosity of mineral particles has been related to their pathogenicity, as in the case of asbestos. It appears that aspect ratio (length to width) also influences the bioactivity of metal oxide nanoparticles. Using in vitro exposure of cultured cells, TiO_2 nanobelts were found to be more toxic than TiO_2 nanospheres of similar composition and diameter (Hamilton et al. 2009). These results were extended by a pulmonary exposure (aspiration; 30 µg/mouse) study of mice to TiO_2 nanospheres, short nanobelts (1–5 µm long), or long nanobelts (6–12 µg long) (Porter et al. 2013). The inflammatory potency of these TiO_2 nanoparticles at 3 and 7 days postexposure was long nanobelts > short nanobelts > nanospheres = control. At 112 days postexposure, the fibrotic potential of these TiO_2 nanoparticles was long nanobelts > short nanobelts > nanospheres = control. Therefore, fibrous shape influences the toxicity of TiO_2 nanoparticles.

The bioactivity of TiO_2 nanobelts is also affected by surface functionalization. When the surface of the TiO_2 nanobelts was altered by addition of COOH functionalities, the bare TiO_2 nanobelts were found to be more inflammatory in a mouse aspiration model than the COOH-modified TiO_2 nanobelts (Figure 10.5).

10.3.5 Surface Coating

Coating of metal oxide nanoparticles has been reported to alter bioactivity (Warheit et al. 2007). Rats were exposed by IT instillation to the following: (1) uf-1, having a titanium dioxide core with an alumina surface coating (~98% titanium dioxide and ~2% alumina), (2) uf-2, composed of a titanium dioxide core with a silica and alumina surface coating (~88 wt% titanium dioxide, ~7 wt% amorphous silica and ~5 wt% alumina), and (3), uf-3, consisting of uncoated 80% anatase/20% rutile TiO_2 particles. Chemical reactivity was measured using a vitamin C consumption assay that determined that the uf-3 sample had the greatest chemical reactivity in comparison to the uf-1 and uf-2 samples. Using the rat lung model, the uf-1 and uf-2 samples produced acute inflammatory responses that resolved within 7 days postexposure, and did not cause any significant pathological changes in the lung. In contrast, uf-3 exposure caused pulmonary inflammation and cytotoxicity up to 1 month postexposure, as well as pathological outcomes. This study demonstrates that coating TiO_2 nanoparticles decreases chemical reactivity and in vivo toxicity of these metal oxide nanoparticles.

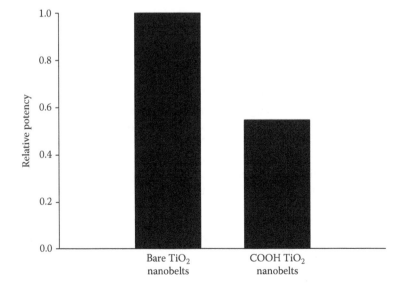

FIGURE 10.5 Functionalization of TiO_2 nanobelts with COOH groups decreases pulmonary inflammation in mouse lungs 24 h after pharyngeal aspiration.

Coating has also been reported to be effective in decreasing the inflammatory response of rats to inhaled nano CeO_2 (Demokritou et al. 2013). An aerosol of CeO_2 nanoparticles with a count mean diameter (CMD) of 89.5 nm was generated by a Versatile Engineered Nanomaterial Generation System (VENGES) using a flame spray pyrolysis reactor. Additionally, amorphous silica-coated CeO_2 nanoparticles (a CeO_2 core with a thin, uniform coating of amorphous silica, CMD = 100.4 nm) were generated to determine if toxicity could be dampened by coating metal oxide nanoparticles. At 1 day after inhalation exposure (2.7 mg/m³, 2 h/day, 4 days), CeO_2 increased neutrophils harvested by bronchoalveolar lavage by 5.5-fold and lavage fluid LDH activity by 2-fold. In contrast, silica-coated CeO_2 did not exhibit significant inflammation or lung injury.

10.4 SUMMARY

In regard to CNT, several physicochemical properties have now been established to impact their bioactivity and pathogenesis: (1) the dispersion status has a significant impact on the type of pathogenesis, that is, dispersed CNTs produce a potent interstitial fibrotic response whereas agglomerated CNTs cause primarily a granulomatous response; (2) although metal contaminants in CNT are linked to oxidative stress based on in vitro studies, when in vivo studies were conducted both SWCNT with and without metal contaminates exhibited similar levels of inflammation and fibrosis; and (3) long CNTs appear more bioactive than short CNTs. Fourth, functionalization of CNT can alter its bioactivity. This was demonstrated in vivo: COOH-MWCNT and N-doped MWCNT were less bioactive than their bare counterparts.

In regard to metal oxides, (1) ROS generation, which is related to band gap theory, is predictive of the toxicity resulting from exposure to metal oxide nanoparticles; (2) similar to CNT, dispersion status of metal oxide nanoparticles impacts on the inflammatory effect and pathological outcomes resulting from exposure; (3) the solubility of metal oxide nanoparticles can also influence their toxicity, increasing the magnitude while decreasing the duration of responses; (4) metal oxide nanobelts are more bioactive than spheres; and (5) functionalization or coating of the surface decreases the potency of metal oxide nanoparticles.

QUESTIONS

1. What pulmonary responses are characteristic of exposure of the lung to carbon nanotubes?
   Answer: Pulmonary responses to SWCNT and MWCNT are qualitatively similar in that both induce transient pulmonary inflammation, persistent granulomatous lesions, and fibrosis.
2. How do dispersion status, metal impurities, fiber length, and surface functionalization affect the pulmonary response to carbon nanotubes?
   Answer: Data indicate that dispersed SWCNT are responsible for the progressive interstitial fibrotic response, whereas the larger agglomerates stimulate the formation of granulomatous lesions. Metal contaminants in SWCNT do not appear to drive pulmonary inflammatory or fibrogenic responses in vivo. Shorter MWCNTs cause more pulmonary inflammation in comparison to longer MWCNT, possible due to greater number of structures per mass. Longer MWCNTs are more fibrogenic than shorter MWCNT, possibly due to slower clearance or reduced clearance of the longer MWCNT. Bare MWCNTs exhibit significantly more pulmonary inflammation and fibrosis in comparison to MWCNT whose surface has been functionalized with the –COOH moiety.
3. What is the relative potency of single-walled carbon nanotubes, carbon nanofibers, and asbestos in induction of fibrosis?
   Answer: The potency sequence is SWCNT > CNF, and CNFs are essentially equivalent to asbestos.

4. Do reactive oxygen species play a role in the pulmonary inflammatory response to spherical metal/metal oxide nanoparticles?

Answer: Yes. A three-tiered hierarchical oxidative stress model has been proposed, and this model has subsequently been supported by several studies.

5. How do dispersion status, shape, and surface functionalization or coating affect the pulmonary response to metal oxide nanoparticle exposure?

Answer: Several different studies indicate that the dispersion status of metal nanoparticles impacts their bioactivity. Similar to mineral fibers like asbestos, studies have also determined that aspect ratio (length to width) also influences the bioactivity of metal oxide nanoparticles. Coating metal nanoparticles surface can decrease chemical reactivity and/or in vivo bioactivity.

6. How does the pulmonary response to soluble versus insoluble metal oxide nanoparticles differ?

Answer: The dissolution of iron-doped ZnO nanoparticles was substantially lower than that of undoped ZnO. Pulmonary exposure of mice and rats to undoped ZnO caused greater pulmonary damage and inflammation in comparison to iron-doped ZnO, demonstrating dissolution rate affects pulmonary response.

REFERENCES

Bonner, J.C., R. Silva, A. Taylor et al. 2013. Inter laboratory evaluation of rodent pulmonary responses to engineered nanomaterials: The NIEHS Nano GO Consortium. *Environ Health Perspect* 121(6):676–682.

Burello, E. and A.P. Worth. 2011. A theoretical framework for predicting the oxidative stress potential of oxide nanoparticles. *Nanotoxicology* 5:228–235.

De Jong, K. and J. Geus. 2000. Carbon nanofibers: Catalytic synthesis and applications. *Catal Rev Sci Eng* 42:481–510.

Demokritou, P., S. Gass, G. Pyrgiotakis et al. 2013. An in vivo and in vitro toxicological characterisation of realistic nanoscale CeO$_2$ inhalation exposures. *Nanotoxicology* 7:1338–1350.

Donaldson, K., R. Aitken, L. Tran et al. 2006. Carbon nanotubes: A review of their properties in relation to pulmonary toxicology and workplace safety. *Toxicol Sci* 92:5–22.

George, S., S. Pokhrel, T. Xia et al. 2010. Use of a rapid cytotoxicity screening approach to engineer a safer zinc oxide nanoparticle through iron doping. *ACS Nano* 4:15–29.

Hamilton, R.F., N. Wu, D. Porter, M. Buford, M. Wolfarth, and A. Holian. 2009. Particle length-dependent titanium dioxide nanomaterials toxicity and bioactivity. *Part Fibre Toxicol* 6:35.

Kagan, V.E., Y.Y. Tyurina, V.A. Tyurin et al. 2006. Direct and indirect effects of single walled carbon nanotubes on RAW 264.7 macrophages: Role of iron. *Toxicol Lett* 165:88–100.

Lam, C.W., J.T. James, R. McCluskey, and R.L. Hunter. 2004. Pulmonary toxicity of single-wall carbon nanotubes in mice 7 and 90 days after intratracheal instillation. *Toxicol Sci* 77:126–134.

Li, R., X. Wang, Z. Ji et al. 2013. Surface charge and cellular processing of covalently functionalized multiwall carbon nanotubes determine pulmonary toxicity. *ACS Nano* 7(3):2352–2368.

Mercer, R.R., A.F. Hubbs, J.F. Scabilloni et al. 2010. Distribution and persistence of pleural penetrations by multi-walled carbon nanotubes. *Part Fibre Toxicol* 7:28.

Mercer, R.R., A.F. Hubbs, J.F. Scabilloni et al. 2011. Pulmonary fibrotic response to aspiration of multi-walled carbon nanotubes. *Part Fibre Toxicol* 8:21.

Mercer, R.R., J. Scabilloni, L. Wang et al. 2008. Alteration of deposition pattern and pulmonary response as a result of improved dispersion of aspirated single-walled carbon nanotubes in a mouse model. *Am J Physiol Lung Cell Mol Physiol* 294:L87–L97.

Mishra, A., Y. Rojanasakul, B.T. Chen et al. 2012. Assessment of pulmonary fibrogenic potential of multiwalled carbon nanotubes in human lung cells. *J Nanomaterials* Article ID 930931.

Muller, J., F. Huaay, N. Moreau et al. 2005. Respiratory toxicity of multi-walled carbon nanotubes. *Toxicol Appl Pharmacol* 207:221–231.

Murray, A.R., E.R. Kisin, A.V. Tkach et al. 2012. Factoring-in agglomeration of carbon nanotubes and nanofibers for better prediction of their toxicity versus asbestos. *Part Fibre Toxicol* 9:10.

Nel, A., T. Xia, L. Madler, and N. Li. 2006. Toxic potential of materials at the nanolevel. *Science* 311:622–627.

Pauluhn, S. 2010. Subchronic 13-week inhalation exposure of rats to multiwalled carbon nanotubes: Toxic effects are determined by density of agglomerated structures, not fibrillar structure. *Toxicol Sci* 113: 226–242.

Porter, D.W., A.F. Hubbs, B.T. Chen et al. 2013. Acute pulmonary dose-responses to inhaled multi-walled carbon nanotubes. *Nanotoxicology* 7:1179–1194.

Porter, D.W., A.F. Hubbs, R.R. Mercer et al. 2010. Mouse pulmonary dose- and time course-responses induced by exposure to multi-walled carbon nanotubes. *Toxicology* 269:136–147.

Porter, D.W., N. Wu, A.F. Hubbs et al. 2013. Differential mouse pulmonary dose and time course responses to titanium dioxide nanospheres and nanobelts. *Toxicol Sci* 131:179–193.

Rushton, E.K., J. Jiang, S. Leonard et al. 2010. Concept of assessing nanoparticle hazards considering nanoparticle dosemetric and chemical/biological response metrics. *J Toxicol Environ Health A* 73:445–461.

Sager, T.M., C. Kommineni, and V. Castranova. 2008. Pulmonary response to intratracheal instillation of ultrafine versus fine titanium dioxide: Role of particle surface area. *Part Fibre Toxicol* 5:17.

Sager, T.M., M.W. Wolfarth, M. Andrew et al. 2014. Effect of multi-walled carbon nanotube surface modification on bioactivity in the C57BL/6 mouse model. *Nanotoxicology* In press.

Shvedova, A.A., V. Castranova, E.R. Kisin et al. 2003. Exposure to carbon nanotube material: Assessment of nanotube cytotoxicity using human keratinocyte cells. *J Toxicol Environ Health A* 66:1909–1926.

Shvedova, A.A., E. Kisin, A.R. Murray et al. 2008. Inhalation vs. aspiration of single-walled carbon nanotubes in C57BL/6 mice: Inflammation, fibrosis, oxidative stress, and mutagenesis. *Am J Physiol Lung Cell Mol Physiol* 295:L552–L565.

Shvedova, A.A., E.R. Kisin, R. Mercer et al. 2005. Unusual inflammatory and fibrogenic pulmonary responses to single-walled carbon nanotubes in mice. *Am J Physiol Lung Cell Mol Physiol* 289:L698–L708.

Shvedova, A.A., E.R. Kisin, D. Porter et al. 2009. Mechanisms of toxicity and medical application of carbon nanotubes: Two faces of Janus? *J Pharmacol Ther* 121:192–204.

Wang, L.Y., R.R. Mercer, Y. Rojanasakul et al. 2010. Direct fibrogenic effects of dispersed single-walled carbon nanotubes on human lung fibroblasts. *J Toxicol Environ Health A* 73:410–422.

Wang, X., T. Xia, S. Ntim et al. 2011. Dispersal state of multiwalled carbon nanotubes elicits profibrogenic cellular responses that correlate with fibrogenesis biomarkers and fibrosis in the murine lung. *ACS Nano* 5:9772–9787.

Warheit, D.B., B.R. Laurence, K.L. Reed, D.H. Roach, G.A. Reynolds, and T.R. Webb. 2004. Comparative pulmonary toxicity assessment of single-wall carbon nanotubes in rats. *Toxicol Sci* 77:117–125.

Warheit, D.B., T.R. Webb, K.L. Reed, S. Frerichs, and C.M. Sayes. 2007. Pulmonary toxicity study in rats with three forms of ultrafine-TiO$_2$ particles: Differential responses related to surface properties. *Toxicology* 230:90–104.

Warheit, D.B., T.R. Webb, C.M. Sayes, V.L. Colvin, and K.L. Reed. 2006. Pulmonary instillation studies with nanoscale TiO$_2$ rods and dots in rats: Toxicity is not dependent upon particle size and surface area. *Toxicol Sci* 91:227–236.

Xia, T., M. Kovochich, M. Liong et al. 2008. Comparison of the mechanism of toxicity of zinc oxide and cerium oxide nanoparticles based on dissolution and oxidative stress properties. *ACS Nano* 2:2121–2134.

Xia, T.A., Y. Zhao, T. Sager et al. 2011. Decreased dissolution of ZnO by iron doping yields nanoparticles with reduced toxicity in the rodent lung and zebrafish embryos. *ACS Nano* 5:1223–1235.

Zhang, H., Z. Ji, T. Xia et al. 2012. Use of metal oxide nanoparticle band gap to develop a predictive paradigm for oxidative stress and acute pulmonary inflammation. *ACS Nano* 6:4349–4368.

11 Inhalation Toxicology of Riot Control Agents*

Harry Salem, Michael Feasel, and Bryan Ballantyne

CONTENTS

---

* The views and opinions expressed in this chapter are those of the author and should not be construed as an official Department of the Army position, policy, or decision, unless so designated by other official documentation. Citation of trade names in this chapter does not constitute an official Department of the Army endorsement or approval of the use of such commercial items.

11.1   INTRODUCTION AND BACKGROUND

The *Encyclopedia of Chemical Technology* lists riot control agents as one of the five categories of chemicals used in war. They are described as nonlethal tear agents most effective against unprotected personnel. The term chemical warfare has been used since 1917. Irritant chemicals such as lacrimators and sternutators that were used in World War I are traditional examples of harassing agents. The effects of these are briefly incapacitating and reversible (Harris, 1992). In 1975, the US president Gerald Ford signed Executive Order 11850 renouncing the first use of riot control agents in war except in defensive military modes to save lives. A more recent development is that the secretary of defense must ensure that they will not be used unless there is presidential approval in advance (SECDEF, 2003).

Riot control agents are those that cause disabling physiological effects when they come into contact with the eyes or skin, or when inhaled. They have the capacity to induce intense sensory irritation of the skin and mucous membranes of the eye and respiratory tract. Riot control agents are peripheral sensory irritants that pharmacologically interact with sensory nerve receptors in skin and mucosal surfaces at the site of contamination, resulting in local pain and discomfort sensations with associated reflexes (Salem et al., 2001). The reflex most associated with the inhalation exposure of irritants is the Kratschmer reflex, first reported in 1870 (Kratschmer, 1870). He described this reflex response following the exposure of rabbits to chemical irritants such as chloroform and carbon dioxide. On exposure, the immediate response was apnea or cessation of respiration. This reflex is one of the protective reflexes or defense mechanisms programmed to prevent or reduce the amount of the noxious chemical reaching the lower respiratory tract and maintain homeostasis. This effect is accompanied by bradycardia and a biphasic fall and rise in aortic blood pressure. The Kratschmer reflex is mediated by the olfactory (I), trigeminal (V), and glossopharyngeal (IX) cranial nerves and was also demonstrated to occur in humans (Allen, 1928–1929). The Kratschmer reflex also occurred in rodent and canine experiments following an exposure to volatile solvents (Aviado, 1971). The cardiopulmonary receptors involved in the defense mechanisms prevent the absorption and distribution of the inhaled irritant to the vital organs, as well as facilitate the expulsion of the irritant, while the extra-cardiopulmonary mechanisms promote metabolism and excretion of the absorbed chemicals. These have been described by Aviado and Salem (1968, 1987) and by Aviado and Aviado (2002). During apnea or cessation of respiration, blood levels of carbon dioxide increase and thus drive the respiratory center to restart breathing. Individuals with compromised immune systems, or nervous system depression as a result of alcohol or illicit drug consumption, or a combination of these, may prevent the restarting of respiration, resulting in death from asphyxia. This may be, in part, responsible for the over 100 in-custody deaths attributed by law enforcement agencies to positional asphyxia following exposure to *pepper* sprays.

11.1.1   Riot Control Agents

The Edgewood Chemical Biological Center and its predecessor organizations have over 95 years of experience in chemical and biological defense, including being the nation's primary nonlethal agent development laboratory (Salem, 2003). Although they are considered military chemicals, riot control agents are distinct from chemical warfare agents (CWAs). CWAs include nerve agents such as tabun (GA), sarin (GB), soman (GD), cyclosarin (GF), and VX, nontraditional agents; blister agents such as mustard (HD) and lewisite (L); choking agents/lung irritants such as phosgene (CG); blood agents such as hydrogen cyanide (AC) and cyanogen chloride (CK); incapacitating agents such as adamsite (DM), 3-quinuclidinyl benzylate (BZ); and centrally acting anesthetics such as the fentanyls. Riot control agents are also designated as nonlethal and less-than-lethal agents, as well as incapacitating, immobilizing, and calmative agents. Their pharmacological classes include irritants, lacrimators, sternutators, emetics, sedatives, hypnotics, serotonin antagonists, hypotensives, thermoregulatory disruptors, nauseants, vision disruptors, neuromuscular blockers, and malodorous substances. In addition to all of the details given earlier, military chemicals also include training agents, smoke materials, and herbicides (Salem et al., 2001). The United States does not recognize riot control agents as CWAs as defined in the 1925 Geneva Convention (Sidell, 1997), and this is still the official US-held policy (Office of the Press Secretary [The White House], 1994). Riot control agents or water under pressure can be used in civil disturbances to distract, deter, incapacitate, disorient, or disable disorderly people, to clear facilities, to deny areas, or for hostage rescue. They can also be used in peacekeeping operations (Ballantyne, 1977a,b).

11.1.2 HISTORICAL

Although lacrimatory and irritant chemicals are considered to have a history dating from World War I, it appears that the first recorded effort to overcome an enemy by the generation of suffocating and poisonous gases seems to have been in the wars of the Athenians and Spartans in 431–404 BCE. The Spartans, when besieging the cities of Plateau and Belgium, saturated wood with pitch and sulfur and burned it under the walls of these cities in the hope of choking and incapacitating the defenders and rendering the assault less difficult. Similar uses of poisonous gases are recorded during the Middle Ages. In effect, they were like our modern stink balls, but were projected by squirts or in bottles in a manner similar to the hand grenade (West, 1919; Fries and West, 1921). The use of capsaicin as an irritant also predates World War I. When Christopher Columbus and his crew landed on Hispaniola in 1492, their behavior brought them into conflict with the local tribe of Arawak Indians. For their protection, Columbus and his men built a stockade. They were confronted by the Indians, who filled bottle gourds, or calabashes, with a paste of wood ash and ground jalapeno peppers and lobbed them over the fence. The capsaicin-laden smoke, which was intended to encourage Columbus and his men to leave, was unsuccessful (Garrett, 2000).

Capsaicin, which is considered the active ingredient of *pepper sprays*, was synthesized at Edgewood Arsenal in the early 1920s and tested as an irritant. Following the synthesis of chlorobenzylidene malononitrile (CS) by British Chemists Carson and Stoughton in 1928, the United States abandoned their work on capsaicin to develop CS (Carson and Stoughton, 1928).

Chloropicrin (trichloronitromethane) also known as Green Cross and PS, first synthesized around 1850, was used both as a harassing agent and lethal chemical in World War I, along with the other lethal agents, chlorine, phosgene, and trichloroethyl-chloroformate. Adamsite (diphenylaminochloroarsine or DM), an arsenic-based compound, was developed for use in World War I. It is a vomiting and sneezing (sternutator) agent and was used as a riot control agent after the war. According to Swearengen (1966), ethyl bromoacetate (EBA) was considered the first riot control agent since it was used in grenades by the police in Paris, France, in 1912 to temporarily disable lawless gangs. This tear gas was again used in the 1970s (Royer and Gainet, 1973). The other tear gases used in World War I included acrolein (Papite), bromoacetone (BA, B-stoff), bromobenzyl cyanide (BBC, CA), chloroacetone (A-stoff), and xylylbromide (T-stoff). Bromoacetone was the most widely used potent lacrimatory agent used in World War I (Salem et al., 2001).

Chloroacetophenone (MACE) was invented by the German chemist, Graebe, in 1869. Other sources indicate it was originally synthesized in 1871 and 1881. The name MACE™ is derived from the chemical name, methyl chloroacetophenone (Methyl chloroACEtophenone). This was the original chemical MACE product made by Smith and Wesson and was widely regarded as the original *tear gas* (Graebe, 1881).

Although MACE was trademarked for CN, it is commonly used as the generic term for riot control agents. After the United States entered World War I, American chemists as well as the British investigated CN and found it to be one of the most effective lacrimators known. Its lacrimatory effects and persistency were equal to or slightly better than BBC, and it contained the less expensive chlorine instead of bromine. It is very stable under normal conditions and does not corrode steel. CN is a crystalline solid and can be dissolved in a solvent or delivered in thermal grenades. Although it was not produced in sufficient quantities to be used in World War I, it was used by Japanese forces as early as 1930 during the pacification campaign against Taiwan aboriginals.

CN was used as the tear gas of choice for the three decades following its introduction toward the later stage of World War I, but because of its relative toxic profile and the development of more effective and safer tear agent CS, its use markedly declined (CHPPM, 1996).

11.2 CHEMISTRY OF SELECTED RIOT CONTROL AGENTS

11.2.1 CN: (Chloroacetophenone, MACE)

Its CAS number is 532-27-4. It is also referred to as ω-chloroacetophenone, α-chloroacetophenone, phenacyl chloride, 2-chloro-1-phenylethanone, and phenyl chloromethyl ketone. It is a white solid with an apple blossom odor and a molar mass of 154.5 corresponding to a molecular formula of C_8H_7OCl. The molar solubility at 20°C is 4.4×10^{-3} mol/L (68 mg/100 mL). Melting and boiling points are 54°C and 247°C, respectively. Density of the solid is 1.318 g/cm^3 at 0°C, and density of the liquid is 1.187 g/m^3 at 58°C. The vapor is 5.3 times heavier than air. The vapor pressure of the solid is 2.6×10^{-3} Torr at 0°C, 4.1×10^{-3} Torr at 20°C, and 15.2×10^{-3} Torr at 50°C.

CN: Chloroacetophenone

11.2.2 CS: (Chlorobenzylidene Malononitrile)

Its CAS number is 2698-41-1, and it is also known as β,β-dicyano-ortho-chlorostyrene, 2-chloro-benzalmalononitrile. It is a white solid with a molar mass of 188.5 corresponding to a molecular formula of $C_{10}H_5N_2Cl$. The molar solubility in water at 20°C is 2.0×10^{-4} mol/L (=~4 mg/100 mL). Dissolved CS is rapidly hydrolyzed; however, it may persist in the environment because its solubility in water is limited. The melting and boiling points are 72°C and 335°C, respectively. The vapor is several times heavier than air, and the vapor pressure of the solid is 0.00034 mmHg at 20°C.

Chlorobenzylidene malononitrile

In addition to the nonpersistent form of CS, two hydrophobic variations were created, CS1 and CS2. CS1 is a micronized powder formulation containing 5% hydrophobic silica aerogel, which can persist for up to 2 weeks in normal weather conditions (Ballantyne, 1977a,b), and CS2 is a siliconized microencapsulated form of CS1 with a long shelf life, persistence, resistance to degradation, and ability to float on water, which could restrict or deny the use of water for military operations. CS is commonly used as a riot control agent and a stimulant for training (Thomas et al., 2002). Members of military organizations and law enforcement agencies are routinely exposed to heated CS during training. The heat vaporizes the CS for dispersion, which thus condenses to form an aerosol.

CS, CN, and CR are SN2-alkylating agents with activated halogen groups that react with nucleophilic sites and combines with intracellular sulfhydryl groups on enzymes such as lactic dehydrogenase to inactivate it. The effect is transient as the enzymes are rapidly reactivated. CS reacts rapidly with the disulfhydryl form of lipoic acid, a coenzyme in the pyruvate decarboxylase system. It has been suggested that tissue injury may be related to inactivation of certain of these enzyme systems.

Pain can occur without tissue injury and may be bradykinin mediated. CS has been shown to cause the release of bradykinin, both in vivo and in vitro, and the elimination of bradykininogen in vivo abolishes the systemic response to CS (USAMRICD, 2000). The release of chlorine atoms on contact with skin and mucous membranes that are reduced to hydrochloric acid can cause local irritation and burns (Anderson et al., 1996).

In addition to being SN2-alkylating agents, CS, CN, and CR are also now known to be some of the most potent transient receptor potential ankyrin 1 or TrpA1 receptor agonists (Brone et al., 2008). The TrpA1 receptor is a cation-selective channel that, upon activation by one of these riot control agents, rapidly depolarizes the sensory neuron and transduces pain signals to the brain. It is in the same family of receptors as the capsaicin receptor, TrpV1 (discussed later in this chapter), but is a different variant. The TrpA1 receptor is a thermoreceptor and chemoreceptor. It responds endogenously to cold temperatures and, traditionally, the isothiocyanates such as that found in mustard oil. CS, CN, and CR all activate this receptor much more potently, giving them their inherent incapacitating effects.

The sensory effects seen with tear gas exposures are due to the afferents formed by trigeminal neurons that express high concentrations of TrpA1 receptors. For example, stimulation of the tearing reflex transduces a signal through the trigeminal nerve and registers as pain. That being said, antagonists for the TrpA1 receptor could be potential therapeutic targets for tear gas over-exposure treatments. With regard to TrpA1 activation, the potencies of the tear agents are described in Table 11.1.

Repeat exposures of thermally dispersed CS were conducted in rats and dogs. They were exposed from 4 to 5 min per day, 5 days a week for 5 weeks. The 25-day cumulative dosage (Ct) to which the rats were exposed to was 91,000 mg min/m^3 (3,640 mg min/m^3 per day), while the dogs were exposed to a cumulative dosage of 17,000 mg min/m^3 (680 mg min/m^3 per day). No lethality occurred in the dogs, while the rats became hyperactive and aggressive, biting noses and tails of other rats and scratching their own noses. No changes were found in blood values for sodium, potassium, protein, albumin, or creatinine throughout the tests. Five of the thirty rats exposed died, two following the cumulative dosage of 25,000 mg min/m^3 and three died after 68,000 mg min/m^3. Gross pathological examination of the rats that died was negative, as were those of six other rats that were sacrificed after 5-week exposure. The exposed rats lost about 1% of body weight, while unexposed rats gained about 20% during the 5 weeks. There were no significant differences in organ to body weight ratios for heart, kidneys, lungs, liver, or spleen following the 5-week exposures. It was concluded that repeated exposures did not make the animals more sensitive to the lethal effects of CS. The animals that died after exposure to CS showed increased numbers of goblet cells in the respiratory and gastrointestinal tracts and conjunctiva, as well as necrosis in the respiratory and gastrointestinal tracts, pulmonary edema, and occasionally hemorrhage in the adrenals. Death appeared to result from poor transfer of oxygen from the lungs to the blood stream, probably because of edema,

TABLE 11.1

Tear Gas Potencies

Compound	TC$_{50}$ (mg/m^3)	IC$_{50}$ (mg min/m^3)
CR	0.004	0.7
CN	0.3	20–50
CS	0.004	3.6
OC	0.0003–0.003	NA

Source: Blain, P.G., *Toxicol. Rev.*, 22(2), 103, 2003.

TC$_{50}$ = threshold concentration.

IC$_{50}$ = incapacitating concentration.

OC included for comparison and is not a TrpA1 agonist.

and hemorrhage in the lungs, and obstruction of the airways. Marrs et al. (1983a,b) also studied the effects of repeated exposures to CS. Mice, rats, and guinea pigs were exposed to neat CS aerosols for 1 h per day, 5 days per week for 120 days. High concentrations of CS were fatal to the animals after only a few exposures, while mortality in the low and medium concentrations did not differ significantly from the controls. They concluded that CS concentrations below 30 mg/m³ were without deleterious effects.

11.2.3   CR: (DIBENZ (B,F) 1:4-OXAZEPINE)

Its CAS number is 257-07-8. It is a pale yellow solid with a pungent pepper-like odor and a molar mass of 195.3 corresponding to a molecular formula of $C_{13}H_9ON$. The molar solubility in water at 20°C is 3.5×10^{-4} mol/L (~7 mg/100 mL). The melting and boiling points are 72°C and 335°C, respectively. The vapor is 6.7 times heavier than air, and the vapor pressure of the solid is 0.00059 mmHg at 20°C. CR is a stable chemical and may persist for prolonged periods in the environment. It was first synthesized in 1962 and is considered a potent sensory irritant of low toxicity.

Dibenzoxazepine

11.2.4   OC

Oleoresin capsicum (OC) is a reddish-brown oily liquid obtained by extracting dried, ripe fruit of chili peppers, usually *Capsicum annuum* or *Capsicum frutescens*. The oleoresin is a mixture of many compounds. Its composition is variable and depends on factors such as maturity of the fruit and the environment in which the plants are grown, as well as the conditions of the extraction. More than 100 compounds have been identified in oleoresin capsicum. Among the branched- and straight-chain alkyl vanillylamides isolated from oleoresin capsicum, capsaicin is the major pungent component in many peppers, and it is particularly noted for its irritant properties. Depending on the variety of chili pepper, the oleoresin contains from 0.01% to 1.0% capsaicinoids on a dry mass basis. Some of the capsaicinoids found in the oleoresin are capsaicin (~70%) with CAS number of 404-86-4, dihydrocapsaicin (~20%) with CAS number of 19408-84-5, norhydrocapsaicin (~7%) with CAS number of 28789-35-7, homocapsaicin (~1%) with CAS number of 58493-48-4, and homodihydrocapsaicin (~1%) with CAS number of 279-0605. Other components of the oleoresin such as phenolic compounds, acids and esters, may also possess irritant properties.

Capsaicin

Dihydrocapsaicin

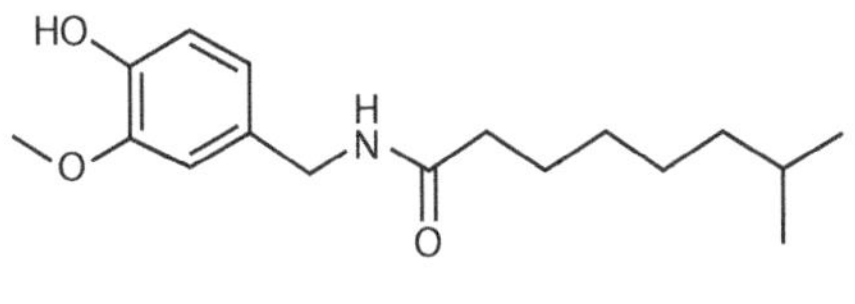

Nordihydrocapsaicin

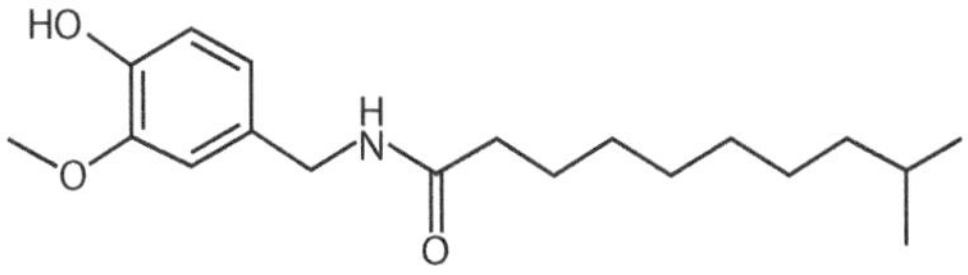

Homodihydrocapsaicin

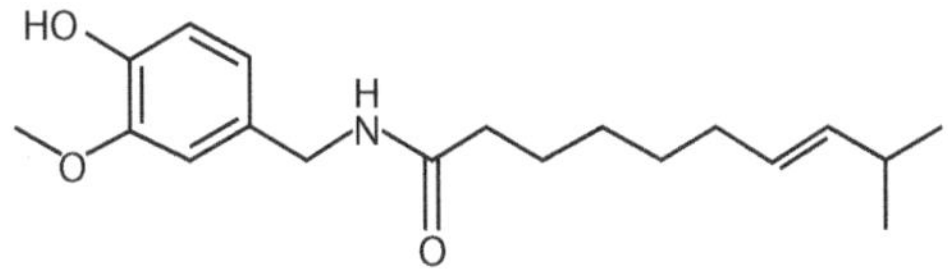

Homocapsaicin

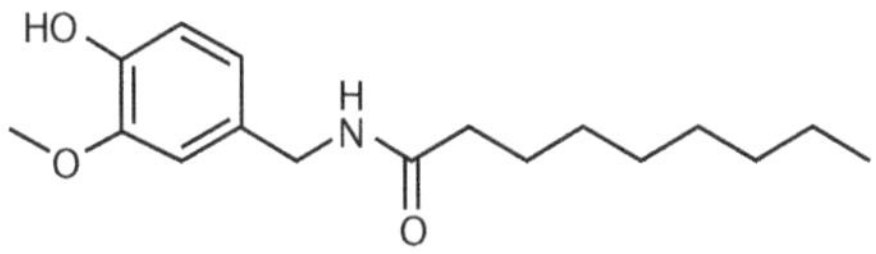

Nonivamide

OC is considered a highly effective irritant that has received much attention as a less-than-lethal agent in civilian, governmental, and military sectors. OC spray or pepper spray has gained popularity as a law enforcement weapon in recent years. Since OC is a natural product, it is considered safe—a viewpoint not necessarily accurate. OC has been incorporated into a variety of formulations and marketed as *pepper gas*; pepper mace and pepper spray for self-defense, criminal incapacitation, law enforcement, and riot control purposes. OC exposure induces involuntary closing of the eyes and lacrimation. It also causes respiratory-related effects such as severe coughing and sneezing, nasal irritation, bronchoconstriction, and shortness of breath. It causes burning sensations of the skin and loss of motor control. As a result, many exposed individuals can be easily subdued. Acute effects of capsaicin and casaicinoids cause edema, hypertensive crisis, and hypothermia. Since 1990, there have been over 100 deaths reported following the use of OC spray. Although a causal relationship has not been established, most of the reported deaths had occurred within 1 h following exposure. Although there is an extensive database on capsaicin, there is very little data on the oleoresin capsicum. Studies reported on capsaicin indicate that it is capable of producing mutagenic and carcinogenic effects, sensitization, cardiovascular and pulmonary toxicity, as well as neurotoxicity. It also has been reported to induce impairment in thermoregulation against overheating (Monsereenusorn et al., 1982; Obal et al., 1983; Suzuki and Iwai, 1984; Buck and Burks, 1986; Fuller, 1990; Govindarajan and Sathyanarayana, 1991; O'Neill, 1991; Holzer, 1992).

Capsaicin was prepared and evaluated in humans as early as the 1920s. However, interest in its development waned when CS was synthesized and research efforts were redirected to the development of CS. Unlike other riot control agents such as CS, CR, and CN, which have definite chemical compositions, OC is a mixture of compounds containing capsaicinoids, various acids and esters, alcohols, aldehydes, ketones, and carotenoid pigments (Teranishni et al., 1980; Games et al., 1984; Govindarajan, 1986; Cordell and Araujo, 1993). Capsaicin is the major component and is considered to be the active ingredient without consideration as to the activity of the other capsaicinoids. Although their activity is similar, they differ in potency (Cordell and Araujo, 1993).

Since OC is derived from chilies and thus is a food product, many studies have been done to analytically determine hotness, which helps categorize the hottest chilies in the world. In 2007, the *Guinness Book of World Records* named the *Bhut Jolokia* or *Ghost Chili* the hottest chili in the world with a Scoville Heat Unit (SHU) score of 1,041,427. As of 2012, the New Mexico State University's Chili Institute identified an even hotter chili. Hotness is measured by the SHU scale; originally based on how many dilutions it took a panel of judges to no longer taste the heat of a chili extract, the SHU scale is now based on more standardized high-performance liquid chromatography (HPLC) analysis of capsaicinoid content. Relative concentrations of various capsaicinoids are measured through HPLC, and a score is then derived from this analysis. Only one study, however, has analyzed the Bhut Jolokia in a scientific manner and determined which capsaicinoids are present in the pepper and their quantities. Previous studies have shown that depending on the chili species, capsaicinoids equate to 0.1%–1% of the chemical makeup of oleoresin. However, in the Bhut Jolokia, levels approaching 5× that amount are present. Per 100 g of fresh weight, 5364 mg of that weigh is capsaicinoid in the Bhut Jolokia. Of that 5364 mg of capsaicinoids, researchers identified and quantified what capsaicinoid species were present in this chili. Only two capsaicinoids were identified in this chili, the Scotch Bonnet and the Jalapeno: capsaicin (C) and dihydrocapsaicin (DHC). Per 100 g of material, 4090 mg of capsaicin and 1274 mg of dihydrocapsaicin was present. While Jalapenos and Scotch Bonnet chilies had capsaicin:dihydrocapsaicin ratios of 12 and 14, respectively, the Bhut Jolokia has a C:DHC ratio of 3.2, a significantly lower ratio, demonstrating much more DHC relative to C (Yunbao and Nair, 2010). This ratio, combined with the shear percentage by mass of capsaicinoid content, is responsible for the heat of the chili.

This pepper has been the subject of conversation recently, not only because of its heat in the food community, but due to its use in the military community as well. India, through its Ministry of Defense, has developed a riot control device around the Bhut Jolokia pepper. The pepper is indigenous to the states of Assam, Nagaland, and Manipur, India. The device that has been developed is a pyrotechnic disseminating device with pellets of the dry chili mash distributed throughout the pyrotechnic mixture. The device is housed in a plastic cylindrical shell with ported vents placed throughout the walls of the cylinder. It is presumed that upon detonation, the pyrotechnic mixture burns, and with it, the pellets of pepper mash are aerosolized, creating a cloud of respirable Bhut Jolokia particles. While no exploitation of the device by the United States has occurred to date, it can be assumed from the chemical makeup of the chili that aerosolized powders of the dry mash could be more potent than current OC grenade devices. Claims have been made by the Defense Research and Development Organization (DRDO), under the Indian Ministry of Defense, as to the potency of this device. It has been described as being able to cause incapacitation, vomiting, and choking, while having no long-term health effect (DRDO, 2010; Premier, 2012). These claims have gone unverified. Even more recently, a new chili has been said to be hotter than the Bhut Jolokia, according to the New Mexico State University's (NMSU) Chili Institute. The newly identified chili is the Trinidad Moruga Scorpion chili. NMSU claims that the chili has a Scoville heat rating of approximately 2,000,000 SHU (Banister, 2012). No capsaicinoid content literature has been published to date for this chili pepper.

11.2.5 DM (ADAMSITE)

DM (Adamsite) has a CAS number of 579-94-9 and is also known as diphenylaminochloroarsine and phenarsazine chloride. It is a yellowish and odorless solid that is very stable in pure form. The melting point is 195°C, and the vapor pressure is negligible (2×10^{-13} mmHg at 20°C). As a solid, the rate of hydrolysis is not significant, owing to the formation of any oxide coating. However, the rate of hydrolysis is rapid when it is an aerosol. DM has a molecular weight of 277.5 with the formula $C_6H_4(AsCl)(NH) C_6H_4$.

DM is among the group of compounds including diphenylchloroarsine (DA), diphenylcyano-arsine (DC), and chloropicrin, which are classified militarily as vomiting agents. DM has been characterized as both a vomiting agent and a sneezing agent (sternutator) and was used in World War I. Sidell (1997) has reported the estimated human LCt_{50} to be 11,000 mg min/m³. DM effects, unlike those of CN, CS, and CR, have a slightly delayed onset and a relatively long recovery period. DM effects occur in about 3 min after inhalation of the exposure and may last for several hours (British Ministry of Defence [BMOD], 1972; Ballantyne, 1977a). Also unlike the tear agents, DM is more likely to cause prolonged systemic effects. Signs and symptoms of DM exposure include eye irritation, upper respiratory tract irritation, uncontrolled sneezing and coughing, choking, headache, acute pain, tightness in the chest, nausea, and vomiting as well as unsteady gait, weakness in the limbs, and trembling. Ballantyne (1977a) also indicated that mental depression might follow DM exposure. Exposures to high concentrations can result in serious illness as a result of pulmonary damage, edema, and death (British Ministry of Defence, 1972).

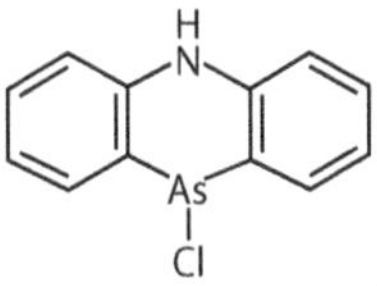

10-chloro-5,10-diphenylarsazine (DM)

11.2.6 FENTANYLS

Fentanyls are synthetic opioids recognized for their short acting and highly potent narcotic analgesic, anesthetic, and immobilizing properties in both animals and humans (Janssen, 1984; Hess and Knakel, 1985). Fentanyl or N-phenyl-N-[1-(2-phenylethyl)-4-piperidinyl] propanamide is a potent narcotic analgesic that has a molecular weight of 336.46. Its CAS Registry number is 437-38-7. The citrate of fentanyl is a crystalline powder with a bitter taste, and 1 g dissolves in about 40 mL of water. Fentanyl is also used as an adjunct to general anesthesia and as an anesthetic for induction and maintenance. It is primarily a mu-opioid agonist. Abuse of this drug leads to habituation or addiction. The CAS Registry numbers of some of the analogs of fentanyl are sufentanil, 56030-54-7; carfentanil, 59708-52-0; and remifentanil, 132875-61-7.

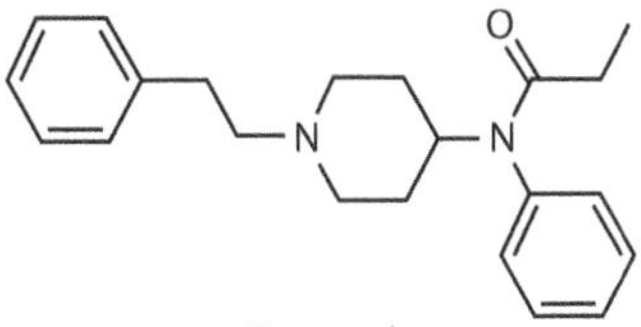

Fentanyl

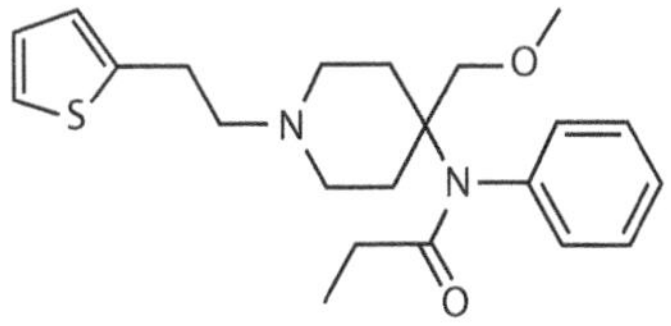

Sufentanil

Carfentanil

Remifentanil

During the Cold War (1945–1991), a great deal of research was directed to chemicals that were not necessarily lethal but would merely incapacitate enemy personnel. In particular, the United States and the former Soviet Union investigated a wide number of pharmacological agents for their potential as incapacitants, such as depressants, hallucinogens, belladonna drugs, and opiate derivatives (Chemical Biological Weapons Nonproliferation Treaty, 2002).

11.3   CS

11.3.1   TOXICOLOGICAL EFFECTS

Exposure to CS is highly irritating to the mucous membranes that line the tissues of the eyes, nose, throat, and respiratory and gastrointestinal tracts. Irritation of the eyes may cause pain, excessive tearing, conjunctivitis, and blepharospasm (uncontrolled blinking). The nose and mouth may perceive a stinging or burning sensation with excessive rhinorrhea or discharge of nasal mucus. Irritation of the respiratory tract may cause tightness of the chest, sneezing and coughing, as well as increased respiratory secretions. Severe lung injury and subsequent respiratory and circulatory failure characterize death in experimental animals following inhalation of very high dosages of CS. Irritation of the gastrointestinal tract may cause vomiting and/or diarrhea. Six minutes following exposure of the skin, a burning sensation may be experienced, with subsequent inflammation and redness. The irritation during exposure is so intense that the individual exposed seeks to escape. The lethal effect in animals following inhalation exposures is caused by lung damage leading to asphyxia and circulatory failure, or bronchopneumonia secondary to respiratory tract injury. Pathology involving the liver and kidneys following inhalation of high dosages of CS is also secondary to respiratory and circulatory failure. The animal and human toxicological effects have been described by Ballantyne (1977a,b), Ballantyne and Callaway (1972), Ballantyne and Swanston (1973, 1978), Ballantyne et al. (1976), Beswick et al. (1972), McNamara (1969), Owens and Punte (1963), and Punte et al. (1962a,b, 1963). Experimental studies included several animal species as well as human exposures to CS in actual use in Northern Ireland and in Waco, Texas, USA.

McNamara (1969) summarized some of the experimental animal studies in which aerosols of CS were generated by various methods, and several animal species were exposed in a single exposure from 5 to 90 min. The toxic signs observed in mice, rats, guinea pigs, rabbits, dogs, and monkeys were immediate and included hyperactivity, followed by copious lacrimation, and salivation within 30 s of exposure in all species except the rabbit. The initial level of heightened activity subsided and, within 5–15 min following initiation of the exposure, exhibited lethargy and pulmonary stress, which continued for about an hour following cessation of the exposure. All other signs had disappeared within 5 min following removal from the exposure. When toxic signs were observed, they occurred following exposure by all of the dispersion methods.

Lethality estimates were expressed by calculation of LCt_{50}s. From acute exposures to CS dispersed from a 10% CS in methylene dichloride, the LCt_{50}s were as follows: mice, 627,000 mg min/m^3; rats, 1,004,000 mg min/m^3; and guinea pigs, 46,000 mg min/m^3. No deaths occurred in rabbits exposed to up to 47,000 mg min/m^3. CS at dosages up to 30,000 mg min/m^3 did not cause any deaths in any of the monkeys with pulmonary tularemia. The combined LCt_{50} for mice, rats, guinea pigs, and rabbits was calculated to be 1,230,000 mg min/m^3 for CS dispersed from methylene dichloride. Goats, pigs, and sheep did not exhibit hyperactivity on exposure to CS, and they were also resistant to its lethal effect. Therefore, no LCt_{50} values could be calculated for goats, pigs, or sheep. However, a combined LCt_{50} was calculated for all of the species tested, mice, rats, guinea pigs, rabbits, dogs, monkey, goats, pigs, and sheep and was estimated to be 300,000 mg min/m^3. LCt_{50}s were also calculated for CS dispersed from M18 and M7A3 thermal grenades. These were 164,000 mg min/m^3 for rats and 36,000 mg min/m^3 for guinea pigs exposed to the M18 thermal grenade dissemination, and for the M7A3 thermal grenade, they were as follows: rats, 94,000 mg min/m^3; guinea pigs, 66,000 mg min/m^3; rabbit, 38,000 mg min/m^3; goat, 48,000 mg min/m^3; 0 pigs, 17,000 mg min/m^3; dog, 30,000 mg min/m^3; and monkey, 120,000 mg min/m^3.

All of the acute exposure results were combined, and LCt_{50}s were calculated for all rodents to be 79,000 mg min/m^3 and for all nonrodent species tested to be 36,000 mg min/m^3, and for all the species, it was 61,000 mg min/m^3. The LCt_{50}s for CS2 was also calculated. CS2 is 95% CS, 5% Cal-o-Sil®, and 1% hexamethyldisilazane, and the LCt_{50}s are as follows: rats, 68,000 mg min/m^3; guinea pigs, 49,000 mg min/m^3; dogs, 70,000 mg min/m^3; and monkeys, 74,000 mg min/m^3.

The effects of CS inhalation on embryonic development in rats and rabbits were studied at concentrations consistent with those expected in riot control situations (~10 mg/m^3). Although the concentrations were low and the duration of exposure (5 min) may not have been adequate to assess the fetotoxic and teratogenic potential of CS, no significant increase in the numbers of abnormal fetuses or resorptions was noted (Upshall, 1973). The mutagenic potentials of CS and CS2 were studied in microbial and mammalian bioassays. CS was positive in the Ames Assay (Von Daniken et al., 1981), while Zeiger et al. (1987) reported questionable genotoxicity for *Salmonella typhimurium*, and those of Reitveld et al. (1983) and Wild et al. (1983) reported CS negative when tested in *S. typhimurium* strains TA 98, TA 1535, and TA 1537 with and without metabolic activation (NTP, 1990a,b). The mutagenic potential for CS and CS2 was investigated in mammalian assays such as the Chinese hamster ovary test for the induction sister chromatid exchange (SCE) and chromosomal aberration (CA), and the mouse lymphoma L5178Y assay for the induction of trifluorothymidine (Tfi) resistance. The results of these assays indicated that CS2 induced SCE, CAs, and Tfi resistance (McGregor et al., 1988; Schmid et al., 1989; NIH [NTP], 1990a,b). The Committee on Toxicology of the National Research Council (1984) reported that taken in their totality, the test of CS for gene mutation and chromosomal damage provided no clear evidence of mutagenicity. Although most of the evidence is consistent with nonmutagenicity, in the committee's judgment, it is unlikely that CS poses a mutagenic hazard to humans. CS2 was evaluated for carcinogenicity in the NTP (1990a,b) 2-year rodent bioassay. Compound-related nonneoplastic lesions of the respiratory tract were observed. The pathologic changes observed in the exposed rats included squamous metaplasia of the olfactory epithelium, hyperplasia, and metaplasia of the respiratory epithelium. In mice, hyperplasia and squamous metaplasia of the respiratory epithelium were observed. Neoplastic effects were not observed in either rats or mice, and it was concluded that the findings suggest that CS2 is not carcinogenic to rats and mice. McNamara (1973) also tested CS in methylene chloride in mice and rats for carcinogenicity in a 2-year study. No tumorigenic effects were observed in the CS-exposed animals.

11.3.2 Metabolism

CS is absorbed very rapidly from the respiratory tract, and the half-lives of CS and its principal metabolic products are extremely short (Leadbeater, 1973). The disappearance of CS follows first-order kinetics over the dose range studied. CS spontaneously hydrolyses to malononitrile (Patai and Rappaport, 1962), which is transformed to cyanide in animal tissues (Nash et al., 1950; Stern et al., 1952). Metabolically, CS undergoes conversion to 2-chlorobenzyl malononitrile (CSH2), 2-chlorobenzaldehyde (oCB), 2-chlorohippuric acid, and thiocyanate (Cucinell et al., 1971; Feinsilver et al., 1971; Leadbeater, 1973; Leadbeater et al., 1973; Paradowski, 1979; Ballantyne, 1977a,b). CS and its metabolites can be detected in the blood after inhalation exposure, but only after large doses. Following inhalation exposure of CS in rodent and nonrodent species, CS and two of its metabolites, 2-chlorobenzaldehyde, 2-chlorobenzyl malononitrile, were detected in the blood (Leadbeater, 1973). In the Leadbeater study, human uptake by the respiratory tract, only 2-chlorobenzyl malononitrile was detected in trace amounts in the blood. CS and 2-chlorobenzaldehyde were not detected, even after high doses of CS up to 90 mg min/m^3. This finding is consistent with the CS uptake studies in animals, and with the maximum tolerable concentration in humans, which is below 10 mg/m^3, it is unlikely that significant amounts of CS would be absorbed by the inhalation route at or near the tolerable concentrations.

11.3.3 Human Volunteer Exposures

Human volunteers have been exposed to CS under varying conditions and concentrations in order to determine the dosage that will incapacitate 50% of the exposed population in 1 min (ICt50). Since the toxicity in all animal species studied was low, and the potency as a sternutator and lacrimator was high, it was necessary to study its effectiveness in humans, if CS could be a replacement for CN and DM as a riot control agent. Both normal military and civilian personnel volunteered for these studies and, following preexposure medical evaluation, were placed in one of four groups to establish base-lines for comparison to postexposure results. The first two groups were of untrained men, one group with masks available, one without; the other two groups of trained men (previously exposed), with and without masks available. Special categories of subjects included those with hypertension; those liable to hay fever, drug sensitivity, or bronchial asthma; those with jaundice or hepatitis; those with a history of peptic ulcer unaccompanied by gastrointestinal bleeding; and those who were between 50 and 60 years of age. Prior to exposure, all men received instructions in the use of the protective masks (M9A1 or M17). Four subjects at a time were exposed in a wind tunnel (8 ft^3) at a fixed speed of 5 mph at temperatures from 0°F to 95°F.

CS was disseminated from a 10% solution in acetone, from a 10% solution in methylene dichloride, or from a miniature M18 thermal grenade. CS solution dissemination was accomplished using spray nozzles (Spraying Co., Type J) placed in the air intake. Airborne samples of CS were taken isokinetically from before dissemination to 1 min following the exposure. Particle size determinations were made using a six-stage modified cascade impactor, and airborne concentrations were estimated by ultraviolet light absorption at 260 μm using a Beckman DU spectrophotometer.

The mass median diameter of CS produced was 3.0 μm for the CS in acetone, 1.0 μm for the CS in methylene dichloride, and 0.5 μm for the miniature M18 CS thermal grenade. These studies revealed the following: many untrained subjects were unable to don and retain their masks at low concentrations of CS, but at high concentrations were able to mask well enough to remain in the contaminated atmosphere. When properly fitted, these masks will fully protect against CS. In those who were unable to mask rapidly, panic was evident. Concentrations of 9–10 mg/m^3 forced 50% of the subjects to leave the chamber within 30 s, and that 99% left at approximately 17 mg/m^3, and 100% left were considered incapacitated at 40 mg/m^3 or greater. Persons who had been exposed previously to a high concentration developed a fear of the agent, and even though subsequently exposed to a lower concentration, the time to incapacitation for trained men was shorter than expected.

There were no significant differences noted in the time to incapacitation in subjects exposed to CS at 0°F–95°F, although it appeared that the subjects appeared unable to tolerate the agent as well as those exposed at ambient temperature. At 95°F and RH of 35% and 97%, the skin-burning effects were much more prominent, possibly because of the excessive diaphoresis. Hypertensive subjects reacted similarly to and tolerated CS as well as normotensive individuals. However, their blood pressure elevation was greater and lasted longer than in normotensives, possibly because of the stress of exposure. The hypertensive subjects recovered as rapidly as the normotensives. Subjects with a history of peptic ulcer, jaundice, or hepatitis, and those between the ages of 50 and 60 reacted similar to normal subjects. Persons with a history of drug allergy, hay fever, asthma, or drug sensitivity were able to tolerate CS exposure as well as the normal subjects; however, a higher percentage of this group had more severe chest symptoms than normal. Although many of these lay prostrate on the ground for several minutes, no wheezing or ronchi were heard on auscultation, and recovery time was as rapid as for any other group tested. Hyperventilating subjects were incapacitated at much lower concentrations than normally breathing subjects, and recovery time was slightly prolonged, but only by 1–2 min. Although not significantly different, subjects exposed to CS disseminated from methylene dichloride appeared to tolerate the agent for a slightly longer period that those subjected to CS in acetone solution, nor was there many difference from CS disseminated from the miniature M18 CS smoke grenade. There was also a group exposed to a combination of CS and DM. The effects of DM were negligible when CS was effective within 30 s (Gutentag et al., 1960).

11.3.4 CLINICAL SIGNS AND SYMPTOMS

In a group of seven volunteers given 10 exposures of from 1 to 13 mg/m³ of CS in a period of 15 days revealed no clinical abnormalities. The dominant effect of the first exposure remained the dominant effect on subsequent exposures. None of the volunteers developed a tolerance to CS during the 10 exposures.

The immediate effects upon exposure to aerosols of CS were on the eyes and were demonstrated by severe conjunctivitis accompanied by a burning sensation and pain that persisted from 2 to 5 min and usually disappeared abruptly rather than gradually. The conjunctivitis remained intense for up to 25 or 30 min. Erythema of the eyelids was generally present, persisted for an hour, and was occasionally accompanied by blepharospasm. Lacrimation was invariably present, tended to be profuse and lingering for up to 12–15 min. The occasional *tired feeling* in their eyes lasted for about 24 h. Photophobia, which was quite marked in 5%–10% of the volunteers, remained for up to an hour. On repeated exposures, the eye effects were reproduced. Rhinorrhea and salivation were profuse and persisted for up to 12 h.

The effects on the respiratory system appeared to be dependent on the duration of exposure and the depth of respiration. The first symptom was usually a burning sensation beginning in the nose and throat and then progressing down the respiratory tract, sometimes associated with coughing. As the exposure continued, the burning became painful and was rapidly followed by a *constricting sensation* throughout the chest, which caused incapacitation for several minutes. Panic usually accompanied and accentuated this symptom, and these volunteers appeared unable to inhale or exhale. Fresh air and encouragement abated these effects. Auscultation of the chest immediately after exposure did not reveal wheezing, rales, or rhonchi. Airway resistance measured by the Asthmometer showed no significant changes, and a portable breath recording apparatus measured breathing patterns of exposed individuals. The patterns verified that when the aerosol was inhaled, the subjects involuntarily gasped, and then held their breath or breathed slowly and shallowly. This was followed by short paroxysms of coughing that forced the individual to exit the exposure. An irregular respiratory rhythm was noted for several minutes after exposure was terminated. Many of the exposed individuals were aphonic for 1–2 min post exposure, and several were hoarse for 24 h. The authors concluded that the incapacitation cause by CS was due to the effects on the eyes, respiratory tract, or both, but regarded the effects on the respiratory system as potentially the most capable of causing incapacitation (Gutentag et al., 1960). This conclusion was followed up (Craig et al., 1960) who also exposed a group of volunteers in the wind tunnel, wearing a self-contained remotely controlled breath-recording system. They were exposed to 5–150 µg CS per liter for 110–120 s. Although the breathing patterns were disrupted by the CS exposure, adequate ventilation of the lungs was maintained, so they concluded that incapacitation is attributed to the unpleasant sensations rather than to any degree of respiratory failure. The apnea and cardiovascular changes observed following inhalation exposure to CS are not inconsistent with the Kratschmer reflex. These investigators also reported that sneezing was common among the observers exposed to small concentrations of CS at some distance from the exposure chamber (Craig et al., 1960).

Inhalation toxicity studies by aerosol dispersions of melted agents sprayed in the molten form, dry powder dispersion, sprayed from solutions of acetone or methylene dichloride, or dispersed from grenades by liberation of hot gases have been performed since World War I. Prior to the research on CS in 1958 and 1959, no toxicity studies were performed using munitions. In 1965, munition studies were conducted with CN and DM. All of these studies demonstrated that munition-dispersed agents were less toxic than dispersion by other methods. The human official LD_{50} value, based on the combined animal species toxicity data, is 52,000 mg min/m³ for CS by molten dispersion, and 61,000 mg min/m³ dispersed by the M7A3 grenade (McNamara, 1969).

11.3.5 HUMAN IN USE EXPOSURES

Park and Giammona (1972) reported the first documented case of pneumonia in the pediatric age group following an exposure to tear gas. A normal 4-month-old white male infant was exposed to CS gas for 2–3 h. He was in a house into which police fired CS tear gas canisters in order to subdue a disturbed adult. Immediately when taken from the house to the emergency room, the infant was observed to have copious nasal and oral secretions, sneezed and coughed frequently, and required suction to relieve upper airway obstruction. The pneumonitis was treated aggressively, and the patient was discharged from hospital on the 12th day since there was no growth noted in his blood and urine cultures, his respiratory problems had subsided, and his temperature had also returned to normal. However, within 24 h, the infant was returned to the emergency room and was rehospitalized. A repeat chest roentgenogram demonstrated a progression of the pulmonary infiltrates. Following treatment with antibiotics, the chest roentgenogram was clear on the 17th day, and improvement continued, and the patient was discharged after 28 days of hospitalization.

The next reported case of serious intoxication with CS teat gas was by Krapf and Thalmann (1981). Eleven days following a thorough internal medical examination that revealed no clinical or pathological findings, a 43-year-old male was in a room in a cloud of fumes from a CS canister that a friend had ignited as a joke. Immediately, he suffered from burning pains in the eyes and in his upper respiratory tract, lacrimation, and pains in his chest with dyspnea and coughing. This unusual exposure led to serious long-term complications such as toxic pulmonary edema, gastrointestinal difficulties, and indications of liver damage and passing right heart insufficiency. After 3 months of hospitalization, all tests were negative, and the patient was discharged to his home in a condition capable of work.

Hill et al. (2000) described what they considered to be the first reported case of major hepatitis attributable to CS inhalation exposure. They described a case where a 30-year-old incarcerated male was sprayed with CS and was hospitalized for 8 days later with erythroderma, wheezing, pneumonitis with hypoxemia, hepatitis with jaundice, and hypereosinophilia. For months, he continued to suffer from generalized dermatitis, recurrent cough, and wheezing consistent with reactive airway dysfunction syndrome, and eosinophilia. Systemic corticosteroids were successful, but abnormalities recurred off treatment. Although the dermatitis resolved gradually over 6–7 months, the asthma-like symptoms persisted a year after exposure. Patch testing confirmed sensitization to CS. The mechanism of the prolonged reaction is unknown, but may involve cell-mediated hypersensitivity, perhaps to adducts of CS, or a metabolite, and tissue proteins. The investigators report this as the first documented case in which CS apparently caused a severe multisystem illness by hypersensitivity rather than direct tissue toxicity.

Other human exposures were reported by Hu (1989). He reports on the alleged use of tear gas in almost every major city in South Korea in June 1987. It appeared that more than 350,000 canisters of tear gas were used by the government against civilians who exhibited cough and shortness of breath for several weeks. Hospitalized patients with asthma and chronic bronchitis, exposed to CS wafting through hospital wards through open windows, experienced deterioration in lung function. Persons close to the exploding tear gas canisters and grenades sustained penetrating trauma from plastic fragments that was exacerbated by the tear gas. Lack of information and objective as well as epidemiological studies was due to fear of serious government reprisals.

Hu (1989) also refers to the allegations that exposures to tear gas in Gaza and the West Bank of Israel have been associated with increases in miscarriages and stillbirths. Inquiries by groups such as Amnesty International and Physicians for Human Rights prompted a Government Accounting Office (GAO) investigation requested by Congressman Ronald Dellums. The GAO Report (April, 1989) concluded that the Physicians for Human Rights fact-finding trip could not confirm any deaths linked to tear gas inhalation, nor could they substantiate the rumors of increased miscarriages. There was also no verifiable evidence available to conclude linking tear gas exposure to fetal deaths. In addition, the US State Department reported that they did not have any medical evidence to support a direct causation between tear gas inhalation and the number of deaths and miscarriages alleged. The exaggerated number of almost 400 deaths attributed to the use of tear gas by the Israeli Defense Forces (IDF) has also been repudiated by the State Department. They have concluded that at least four deaths had resulted from tear gas use in enclosed areas and that the IDF was using primarily CN at the time (US GAO, 1989).

The use of CS by the US forces in Vietnam was to flush the enemy from bunkers and tunnels and reduce the ability of the enemy to deliver aimed fire while attacking and to deny fighting positions and infiltration routes for extended periods of time.

Interest and possible concern developed about the adverse effects of chemicals employed in peacekeeping operations in the United Kingdom following the use of CS by the Ulster Constabulary in Londonderry, Northern Ireland, on August 13, 1969, and August 14, 1969. As a result of the first use of CS for crowd control, a three-man Committee of Inquiry was established to determine the medical effects, if any, in persons exposed to CS. Their report published in 1969 recommended an expansion of the committee and the evidence in regard to CS assessed in the widest possible way. They reported that on exposure to various concentrations of CS, the effects vary from a slight prickly or peppery sensation in the eyes and nasal passages up to the maximum symptoms of streaming from the eyes and nose, spasm of the eyelids, profuse lacrimation, and salivation, retching and sometimes vomiting, burning of the mouth and throat, cough, and gripping pain in the chest. Even at low concentrations, the onset of symptoms is immediate, and they disappear when removed from the exposure. Of the many tens of thousands of military personnel in the United Kingdom who were exposed to CS in the course of their training, as well as those of the US military who undergo similar training, the signs and symptoms were similar to those described earlier, and there were no significant aftereffects. On exposure, the eyes are red, but this disappears on leaving the contaminated atmosphere. On the skin, CS causes a burning sensation on the exposed parts that can be followed by redness or the appearance of small blisters or vesicles at the point of friction. These effects are more prevalent in fair-skinned persons and if the skin is hot and moist. Infants asleep in rooms where CS entered via broken windows were sufficiently distressed to awaken them crying from sleep. On snatching them out of the contaminated atmosphere, they quieted rapidly and required no hospitalization. They also found no special susceptibility to CS associated with old age. Human volunteers and members of the Himsworth committee over 50 years of age were exposed to 35 mg/m^3, and the symptoms experienced and the time to recover from these were no different from those in young adults. Exposure to CS was determined not to have had any effect on pregnancy since comparison of the 9 months following exposure compared to the 9 months of the previous year demonstrated no difference in abortions, stillbirths, or congenital abnormalities. Middle-aged and elderly people who had chronic bronchitis, and had been significantly exposed to CS, did not show exacerbation different than that caused by natural causes. Following the riots of 1969, there was no increase in the death rate from chronic bronchitis and asthma. Asthmatics, especially children who were exposed to CS, did not show any difference in the number of attacks from their experience prior to the exposure. The committee reported that there is ample evidence that if CS causes unconsciousness in humans, it can do so only rarely and that many, if not all of the cases, reported are more probably the result of other conditions that occur in riot situations. In animals, unconsciousness does not occur after inhaling CS (Himsworth et al., 1971). The Himsworth reports (1969, 1971) considered to be the most extensive study of the use of CS agent on humans, by UK forces in Northern Ireland in the late 1960s, found that no deaths and no long-term injuries resulted from the widespread use of CS agent there (US Congress, 1996).

On April 19, 1993, CS was used unsuccessfully to induce the residents to leave the structure when it was injected through the walls of the Davidian residence in Waco, Texas. Had the building been airtight, which was not the case, the total amount delivered into the building would have been 411.92 mg/m^3. This concentration is far below the 61,000 mg/m^3 projected to be lethal to 50% of a given population of humans. In reality, the concentration of CS inside the Branch Davidian residence did not reach even these levels, since the residence was poorly constructed and allowed for air to move in and out of the residence continuously. The total amount of methylene chloride mixed with the CS inserted into the building was 1,924.87 mg/m^3. Since there does not appear to be a published human lethal estimate, it was compared to that for rats and found to be far below the estimated LCt_{50} of 2,640,000 mg min/m^3 for rats. Thus, the committee concluded that the methylene chloride could not have caused the death of any of the Branch/Davidians. During the insertion of the CS into the Davidian compound, their leader David Koresh did not advise his followers to leave; instead, they spread highly flammable liquids throughout the compound and set fire to the whole building. In the aftermath of the fire, the bodies of over 70 Branch Davidians were recovered. According to the autopsy reports from the Tarrant County, Texas, Coroner, 30 people died of asphyxiation due to smoke inhalation, 2 people died of injuries resulting from blunt force trauma, and 20 people including David Koresh and a 20-month-old infant, died of gunshot wounds inflicted at close range by themselves or others within the compound. Of the nine Branch Davidians who survived the fire, seven escaped through the walls and windows of the compound created by the combat engineering vehicles. The shoes and clothing of several of those who escaped contained concentrations of gasoline, kerosene, and other flammable liquids. CS is not a chemical accelerant or a flammable agent, but will support combustion if ignited. The methylene chloride and carbon dioxide used in the dissemination of the CS will not burn and will actually inhibit fire ignition (Union Calendar No. 395). On April 12, 1993, the FBI presented its tear gas plan to the attorney general Janet Reno. Between that date and April 17, she conducted at least eight meetings with military and civilian tear gas experts to debate the tear gas plan in a barricade situation, the properties of the tear gas chosen, and the scientific and medical information concerning the toxicity and flammability of the type of tear gas proposed and the effect of tear gas on vulnerable populations such as children, the elderly, and pregnant women. On April 17, the attorney general approved the tear gas insertion plan and informed the president of her decision.

As a result of the activities at Waco, Texas, the US Congress initiated an investigation by the Subcommittee on Crime of the House Committee on the judiciary and the Subcommittee on National Security, International Affairs, and Criminal Justice of the House Committee on Government Reform and Oversight on the actions of the Federal agencies involved in law enforcement activities in late 1992 and early 1993. As part of the investigation, 10 days of public hearings were held in which more than 100 witnesses appeared and provided testimony concerning all aspects of the government's actions. In addition, the subcommittees reviewed thousands of documents requested from and provided by the agencies involved in these actions. Included in these documents were many that are referenced in this chapter. In the report, it is stated that all of those consulted who had personal knowledge or professional expertise agreed that the use of tear gas was the only way to compel the Branch Davidians to leave the compound without the use of force or loss of life. Evidence and testimony during the hearing clearly indicated that the CS tear gas was not the direct or proximate cause of the ignition or acceleration of the fire. As previously stated, CS is a common riot control agent used in the United States and Europe, the purpose of which is to cause irritation of the eyes, skin, and respiratory system sufficient to encourage an individual to leave the premises or any open area. CS is considered the least toxic agent in the family of chemical tear gas irritants. In order to reach a level that would be lethal to 50% of the population, CS must reach a concentration of 25–150,000 mg min/m^3. The CS gas used was in a concentration that would reach only 16,000 mg min/m^3, if all the CS used was released at the same time in a single closed room, and that the residents were exposed continuously for 10 min. In reality, at Waco, then CS was released throughout different areas of the building while openings were created in the windows and walls. The CS was inserted for a total of 5 min over a 6 h period, and the wind velocity was 35 knots during the tear gas delivery. Therefore, given the amount of CS used, the presence of high winds, building ventilation, and the delivery of the gas at different areas of the compound, it is highly unlikely that anything close to the 50% lethality rate was reached. They further state that there are no documented cases in which the use of CS caused death in humans, and the reports that Amnesty International linked the use of CS to deaths of Palestinians in the West Bank and Gaza of Israel are an extremely biased reading of the report. The report discussed the use of both CS and CN. CN has been reported to be lethal in closed quarters, while the overwhelming majority of evidence of ill effects of CS was anecdotal. Therefore, there are no reliable scientific data that would lead to the conclusion that CS alone was implicated in any of the deaths. The Physicians for Human Rights also could not confirm the reports of deaths from CS inhalations. The Congressional Report also refers to the Himsworth Report issued by the British Government that found that there is no evidence of any special sensitivity of the elderly, children, or women. In addition, the Himsworth Commission chronicled the effects of CS exposure on an infant who, after being removed from the affected area, recovered rapidly. It is interesting that the Registrar General studied, under a pledge of medical secrecy, the individual health records of the employees who worked for at least 11 years, at Nancekuke in Cornwall where CS is manufactured, and compared them to a group of men of the same age from England and Wales. The results reported in Appendix 12 of the Himsworth et al. Report (1971) indicate the workers may undergo multiple exposures to small doses of CS without having any ill effect on them. There was no evidence of increased risk from any group of causes such as neoplasms, nervous system, circulatory system, respiratory system, digestive system, genitourinary system, accidents, poisoning, and violence as compared to the control group. From all causes, there were only 41 deaths compared to the expected deaths of 48.3 over a 20-year period. For respiratory diseases and all the other causes except digestive, the mortality of workers at Nancekuke was somewhat less than for those in England and Wales.

More recently, tear gas has received much publicity for its use by police against rioters in Taksim Square in Istanbul Turkey during political protests (Kurtz, 2013). To date, no RCA-associated deaths have occurred as a result of their use in this instance.

11.4 CR

11.4.1 TOXICOLOGICAL EFFECTS

Ballantyne (1977a,b) summarized the mammalian toxicology in various animal species. The acute toxicity of CR (LD_{50} and LCt_{50}) indicates that CR is less toxic than CS and CN by all routes of exposure. Animals exposed to CR exhibited ataxia or incoordination, spasms, convulsions, and tachypnea or rapid breathing. In the animals that survived, these effects gradually subsided over a period of 15–60 min. Death was preceded by increasing respiratory distress. The animals that died following intravenous and oral administration demonstrated congestion of liver sinusoids and alveolar capillaries. At necropsy, the surviving animals did not show any gross or histological abnormalities. The toxic signs following intraperitoneal administration included muscle weakness and heightened sensitivity to handling. These effects persisted throughout the first day of exposure. Some animals also exhibited central nervous system effects. Surviving animals did not exhibit any gross or histological abnormalities at necropsy. Several animal species were exposed to the acute inhalation of CR aerosols and smokes for various time periods. Rats exposed to aerosol concentrations from 13,050 to 428,400 mg min/m^3 manifested nasal secretions and blepharospasm or uncontrollable closure of the eyelids, which subsided within an hour after termination of the exposure. There were no deaths during or following these exposures. There were also no deaths in rabbits, guinea pigs, or mice exposed to CR aerosols of up to 68,000 mg min/m^3. Animals exposed to CR smoke generated pyrotechnically, had alveolar capillary congestion and intra-alveolar hemorrhage, as well as kidney and liver congestion.

The potential of CR aerosols to produce physiological and ultrastructural changes in the lungs was evaluated by Pattle et al. (1974). Electron microscopy of rats exposed to CR aerosol of 115,000 mg min/m^3 did not reveal any effects on organelles such as lamellated osmiophilic bodies. The studies by Colgrave et al. (1979) evaluated the lungs of animals exposed to aerosols of CR at dosages of 78,200, 140,900, and 161,300 mg min/m^3 and found them to appear normal on gross examination. On microscopic examination, however, the lungs revealed mild congestion, hemorrhage, and emphysema. Electron microscopy showed isolated swelling and thickening of the epithelium, as well as early capillary damage, as evidenced by ballooning of the endothelium. The authors concluded that these very high dosages of CR aerosols produced only minimal pulmonary damage.

The cardiovascular effects of CR were studied following intravenous administration, by Lundy and McKay (1975). They found a dose-dependent increase in blood pressure of short duration and an increased heart rate and arterial catecholamines. The investigators postulated that the cardiovascular effects of CR were related to sympathetic nervous system effects as evidenced by the abolition of CR-induced pressor effects by phentolamine and 6-hydroxydopamine.

Repeated inhalation exposures were conducted by Marrs et al. (1983a,b), who exposed mice and hamsters to concentrations of 204, 236, and 267 mg/m^3 CR for 5 days per week for 18 weeks. The high concentrations produced death in both species, but no single cause of death could be ascertained, although pneumonitis was present in many cases. Chronic inflammation of the larynx was observed in mice. Although alveologenic carcinoma was found in a single low-dose and a single high-dose group of mice, the findings and conclusions were questioned because the spontaneous occurrence of alveologenic carcinoma is high in many mouse strains (Grady and Stewart, 1940; Stewart et al., 1979). Further, this tumor type differs in many respects from human lung tumors. No lung tumors and no lesions were found in hamsters exposed to CR aerosols. Histopathology revealed hepatic lesions in mice, but these were of infectious origin and not CR related. The authors concluded that CR exposures at high concentrations reduced survivability and that CR produced minimal organ-specific toxicity at many times the intolerable human dose, which has been reported as 0.7 mg/m^3 (IC$_{50}$) within a minute (Ballantyne, 1977a,b) and 0.15 mg/m^3 (IC$_{50}$) within a minute (Marrs et al., 1983a,b; McNamara et al., 1972).

Repeated dermal administration of CR was conducted in rabbits and monkeys by Owens (1970) and in mice by Marrs et al. (1982), who applied it to the skin 5 days per week for 12 weeks. The investigators concluded that repeated dermal applications of CR had little effect on the skin. They further postulated that in view of the absence of any specific organ effects, absorption of even substantial amounts of CR would have little effect.

Higginbottom and Suschitzky (1962) were first to note the intense lacrimation and skin irritation caused by CR. Mild and transitory eye effects such as mild redness and mild chemosis were observed in rabbits and monkeys after a single dose of 1% CR solution. Multiple doses over a 5-day period of 1% CR solution to the eye produced only minimal effects. Biskup et al. (1975) reported no signs of eye irritation in animals following single- or multiple-dose applications of 1% CR solutions.

Moderate conjunctivitis following the application of 5% CR solution to the eyes of rabbits was reported by Rengstorff et al. (1975), although histological examination revealed normal corneal and eyelid tissues. Ballantyne and Swanston (1974) also studied the ocular irritancy of CR and arrived at a threshold concentration (TC_{50}) for blepharospasm in several species. Ballantyne et al. (1975) studied the effects of CR as a solid, as an aerosol, and as a solution in polyethylene glycol. Aerosol exposures of 10,800 and 17,130 mg min/m^3 resulted in mild lacrimation and conjunctival injection, which cleared in 1 h, while in solution produced reversible dose-related increases in corneal thickness. The authors concluded that CR produced considerably less damage to the eye than CN and that there was a much greater degree of safety for CR than for CN.

The effects on skin were reported by Ballantyne (1977a,b) and by Holland (1974) to produce transient erythema, but did not induce vesication or sensitization and did not delay the healing of skin injuries. The burning sensation on exposure to CR persists for 15–30 min, and the erythema may last for 1–2 h.

Upshall (1974) studied the reproductive and developmental effects of CR on rabbits and rats. They were exposed to inhalation of aerosolized CR at concentrations of 2, 20, and 200 mg/m^3 for 5 and 7 min. Groups of animals were also dosed intragastrically on days 6, 8, 10, 12, 14, 16, and 18 of pregnancy. No dose-related effects of CS were observed in any of the parameters measured and the number and types of malformations observed. No externally visible malformations were seen in any group, and no dose-related effects of CR were noted in any of the fetuses in any group. Based on the overall observations, the author concluded that CR was neither teratogenic nor embryotoxic to rabbits and rats.

There is only one study reported on the genotoxicity of CR. Colgrave et al. (1983) studied the mutagenic potential of technical-grade CR and its precursor (2-aminodiphenyl ether) in the various strains of *S. typhimurium*, as well as in mammalian assay systems. CR and its precursor were negative in all the assays, suggesting that CR is not mutagenic. Further testing is required to exclude the genetic threat to humans as well as to determine the carcinogenic potential and its ability to cause other chronic health effects. Husain et al. (1991) studied the effects of CR and CN aerosols in rats on plasma glutamic oxaloacetic transaminase, plasma glutamic pyruvic transaminase, acid phosphatase, and alkaline phosphatase. The rats exposed to CR exhibited no change in any of these parameters, while there were significant increases in all of these parameters in rats exposed to CN, suggesting that CN could cause tissue damage.

Upshall (1977) reported that CR aerosols are very quickly absorbed from the respiratory tract. Following inhalation, the plasma half-life (T1/2) is about 5 min, which is about the same following intravenous administration. French et al. (1983) studied CR metabolism in vitro and in vivo and supported the previous conclusions that the major metabolic fate of CR in the rat is the oxidation to the lactam, subsequent ring hydroxylation, sulfate conjugation and urinary excretion.

11.4.2 HUMAN TOXICOLOGY

Studies at Edgewood Arsenal and other research centers have been conducted to assess the effects of CR on humans following aerosol exposures, drenches, and local application (Weigand, 1970; Ballantyne et al., 1973; Ballantyne and Swanston, 1974; Holland, 1974; Ballantyne et al., 1976; Ashton et al., 1978). The human aerosol and cutaneous studies conducted at Edgewood Arsenal from 1963 to 1972 have been summarized in a National Academy of Sciences Report (NAS) 1984 on irritants and vesicants. The respiratory effects following aerosol exposures included respiratory irritation with choking and difficulty in breathing or dyspnea, while the ocular effects consisted of lacrimation, irritation, and conjunctivitis. Ballantyne et al. (1973) reported the effects of dilute CR solution on humans following splash contamination of the face or facial drench. These exposures resulted in an immediate increase in blood pressure concomitant with decreased heart rate. In subsequent studies by Ballantyne et al. (1976), humans were exposed to whole-body drenches that resulted in the same effects of immediate increase in blood pressure and bradycardia. The authors concluded that the cardiovascular effects in both studies were not due to the CR, since they theorized that there was insufficient CR uptake to produce the systemic effects on the heart. However, they did not provide an explanation for the cardiovascular changes. Lundy and McKay (1977) suggested that these cardiovascular changes resulted from the CR effects on the heart via the sympathetic nervous system. Ashton et al. (1978) exposed human subjects to a mean CR aerosol concentration of 0.25 mg/m^3 (particle size 1–2 μm) for 1 h. Expiratory flow rate was decreased approximately 20 min after the onset of exposure. The investigators theorized that CR stimulated the pulmonary irritant receptors to produce bronchoconstriction and increasing pulmonary blood volume by augmenting sympathetic tone.

11.5 OC

11.5.1 Toxicological Effects

Not much is known about the toxicology of OC, but because it is a natural product and much utilized food component, it is considered to be relatively safe, with a low order of toxicity (Clede, 1993). The pharmacology and toxicology of capsaicin on the other hand have been well characterized in both animals (Glinsukon et al. 1980) studied and human studies. The pharmacological actions of capsaicin and capsaicinoids were characterized in the 1950s (Issekutz et al., 1950; Toh et al., 1955). Glinsukon et al. (1980) studied the acute toxicity of capsaicin in several species and found capsaicin to be highly toxic by all routes of administration except gastric, rectal, and dermal. The intravenous doses of capsaicin caused convulsions within 5 s, and times to death were from 2 to 5 min. The toxic signs observed included excitement, convulsions with limbs extended, dyspnea, and death due to respiratory failure. We compared the acute intravenous LD$_{50}$s in mice with other well-known chemicals. This was done since there was no known comparative inhalation LD$_{50}$s in mice, for all of these chemicals, and the intravenous route has been considered very close to the inhalation route of exposure. This demonstrates that capsaicin's acute toxicity in mice is between that of nicotine and strychnine, two well-known potent poisons. Additionally, Glinsukon et al. (1980) also compared the intraperitoneal acute toxicity of capsaicin to the oleoresin capsicum in female mice and found the extract to be four times more toxic than the capsaicin with LD$_{50}$s of 1.51 and 6.50 mg/kg, respectively. Guinea pigs appeared to be more susceptible than mice and rats, while hamsters and rabbits were less vulnerable to the toxic actions of capsaicin.

The pulmonary pharmacology and toxicology of capsaicin has been studied in some detail. Inhalation of capsaicin is consistent with the induction of the Kratschmer reflex, which is apnea, bradycardia, and a biphasic fall and rise in aortic blood pressure. Exposure to capsaicin causes bronchoconstriction in animals and humans and the release of substance P, a neuropeptide, from sensory nerve terminals, as well as mucosal edema (Jansco et al., 1977; Russel and Lai-Fook, 1979; Davis et al., 1982; Fuller et al., 1985; Hathaway et al., 1993). The pulmonary effects of capsaicin appear to be species related. In guinea pigs, intravenous and intra-arterial administration causes bronchoconstriction (Biggs and Goel, 1985). The bronchoconstriction in the dog and cat following intravenous capsaicin is dependent on a vagal cholinergic reflex, as is the bronchoconstriction in the cat following aerosol exposure (Adcock and Smith, 1989). In guinea pigs, the bronchoconstriction following aerosol exposure suggests both a vagal-cholinergic and noncholinergic local axon reflex (Buchan and Adcock, 1992). The cardiorespiratory effects following intravenous administration resulted in a triphasic effect on blood pressure and altered cardiac parameters. The complex effects on the cardiovascular system consist of tachypnea, hypotension (Bezold–Jarisch reflex), bradycardia and apnea (Chahl and Lynch, 1987; Porszasz and Szolesanyi, 1991/1992).

There is speculation as to the acute respiratory toxicity of OC due to many deaths associated with OC use by police. Possible reasons for these unfortunate incidents have been linked to drug/alcohol use, positional asphyxiation, or loss of the Kratschmer reflex. However, recent insight into the transient receptor potential cation channel subfamily V member 1 (TrpV1) receptor shows a

TABLE 11.2

Mouse Intravenous LD_{50} mg/kg (RTECS)[a]

Botulinum toxin	0.00001	Parathion	13
VX	0.012	DM	35
GB	0.10	CR	37
Nicotine	0.30	CS	48
Capsaicin	0.40	Caffeine	62
Strychnine	0.41	CN	81
Potassium cyanide	2.60	Cocaine	161
Mustard gas	3.30	Isopropyl alcohol	1509
Methamphetamine	10	Ethyl alcohol	1973

[a] Registry of toxic effects of chemical substances.

potentiating effect that ethanol has on the tussive response, or cough. TrpV1 receptors in the respiratory tract largely mediate cough with regard to chemical stimulation by various exogenous compounds. Capsaicin is a potent agonist of this receptor, causing burning, itching, and coughing upon exposure. However, in riot control devices, specifically sprayers, OC is not the only material that is aerosolized from the container. Along with OC, there are various propellants and stabilizers. One used very frequently as a medium in OC sprayers is an aqueous ethanol solution. A study conducted by the Italian Universities of Florence and Ferrara concludes that ethanol potentiates the cough response in patients with airway hyperreactivity (Gatti et al., 2009). Airway hyperreactivity is one of the initial symptoms of OC exposure and manifests itself much like an asthma attack. Swelling of the bronchial tubes and increased mucous secretion occur, making breathing more difficult. When ethanol is added to this initial insult, cough response increases nearly fourfold in guinea pigs. This potentiation occurs through TrpV1 channel phosphorylation via PKC activation. Upon ethanol exposure, PKC phosphorylates the TrpV1 receptor, lowering its activation threshold to capsaicin, heat, or other endogenous compounds, allowing for a more sensitive tussive response.

Table 11.2 shows the toxic effects of various chemical substances on mice.

11.6 DM

11.6.1 TOXICOLOGICAL EFFECTS

The NAS (1984) reported on the toxicology of DM, based on the studies by Owens et al. (1963) and McNamara et al. (1972). Various animal species including monkeys have been exposed to DM. Following acute exposures, the animals exhibited ocular and nasal irritation, hyperactivity, salivation, labored breathing, ataxia, and convulsions. Punte et al. (1962a) found these same effects at high-concentration exposures for up to 90 min, as well as lethargy.

Histopathology did not reveal any abnormalities at exposure dosages of below 500 mg min/m³. At higher dosages, animals that died or were killed demonstrated hyperemia of the trachea, pulmonary congestion and edema, and pneumonia. These effects were consistent to exposure to pulmonary irritants. DM toxicity values are presented in Table 11.3.

Striker et al. (1967) exposed monkeys to varying concentrations and durations. At a Ct dosage of 2565 mg min/m³, only one animal responded and that was with oral and nasal discharge and diminished response to stimuli. A Ct of 8540 mg min/m³ resulted in ocular and nasal conjunctival congestion, facial erythema, and decreased responses, all of which were resolved within 24 h. Exposure to the high dosage of 28,765 mg min/m³ resulted in hyperactivity, copious nasal discharge, conjunctival congestion, marked respiratory distress, as well as gasping and gagging in all the exposed monkeys. Eight of these exposed monkeys died within 24 h of exposure. Necropsy of these animals

TABLE 11.3
DM Toxicity

Species	LCT50 (mg min/m³)	LD (mg/kg)[a]
Mice	22,400	17.9
Rats	3,700	14.1
Guinea pigs	7,900	2.4

[a] Theoretical dose calculated from respiratory volume, LCt_{50}, and estimated percent retention.

revealed congestion and extremely edematous lungs. Microscopic examination revealed ulceration of the tracheobronchial tree and pulmonary edema. Studies were also conducted in which monkeys were exposed to low target concentrations of 100 and 300 mg/m³ DM for 2–60 and for 2–40 min, respectively. As the exposure duration increased, toxic signs increased, which is characteristic of exposure to irritants.

At the maximum dosage of 13,200 mg min/m³, the animals exhibited nausea and vomiting, oral and nasal discharge, and conjunctival congestion. Below 1296 mg min/m³, the only signs were blinking.

The effects of DM on the gastrointestinal tract were suggested as a possible cause of death. Dogs were given lethal doses of DM both intravenously and orally, while the following parameters were monitored: central venous pressure, right ventricular pressure, cortical electric activity, alveolar CO_2, respiratory rate, heart rate, electrocardiogram, and gastric activity. DM caused a marked elevation of both amplitude and rate of gastric activity for 15–20 min and then returning to normal. Pretreatment with trimethobenzamide, an effective antiemetic for peripheral and centrally acting emetics, did not prevent DM gastric activity, but chlorpromazine was effective. The authors concluded that DM affects the stomach directly and that the primary cause of death following exposure to DM is its effects on the lungs as demonstrated in the studies of Striker et al. (1967).

The effects of DM on the eyes and skin of rabbits were studied with DM suspended in corn oil instilled into the eyes of rabbits in doses of 0.1, 0.2, 0.5, 1.0, and 5.0 mg. No effect was observed at 0.1 mg, but at 0.2 mg, mild conjunctivitis was observed. At 0.5 mg, mild blepharitis was also seen. Corneal opacity persisted over the 14 days of observation period in rabbit eyes that were given 1.0 and 5.0 mg DM. Corn oil suspensions of DM (100 mg/mL) were placed on the clipped backs of rabbits at doses of 1, 10, 50, 75, and 100 mg. At 10 mg and higher, necrosis of the skin was observed. Rothberg (1969) studied the skin sensitization potential of riot control agents in guinea pigs. For DM, his results were negative, indicating that DM does not have the potential for skin sensitization.

11.6.2 Human Exposures

The human toxicology of DM was reviewed by Ballantyne (1977a,b). He described the effects of inhalation exposures to DM as beginning with acute pain in the nose and sinuses. This was followed by pain in the throat and chest, with sneezing and violent coughing, and also eye pain, lacrimation, blepharospasm, rhinorrhea, salivation nausea, and vomiting. Recovery is usually complete in 1–2 h after exposure. The onset of signs and symptoms is delayed for several minutes, unlike the onset for CS and CN, which is almost immediate. The slow onset of DM allows for the absorption of much more DM before a warning is perceived. Threshold doses were estimated for irritation of the throat, lower respiratory tract, and initiation of the cough reflex to be 0.38, 0.5, and 0.75 mg/m³, respectively. Punte et al. (1962a) and Gongwer et al. (1958) studied the effects of varying concentrations of DM on human subjects and agreed that humans

TABLE 11.4
Human Ocular Irritancy

Compound	Onset/Action	Threshold Concentration (mg/m³)	Intolerable Concentration (mg/m³)	10 min Exposure Lethal Concentration (mg/m³)
CN	Immediate	0.3	5–30	850
CR	Immediate	0.002	1	10,000
CS	Immediate	0.004	3	2,500
DM	Rapid	1	5	650
BBC	Rapid	0.15	0.8	350
Acrolein	Rapid	2–7	50	350
OC	Rapid	—	—	—

Sources: Ballantyne, B. and Swanston, D.W., *Acta Pharmacol. Toxicol.*, 32, 266, 1973; Ballantyne, B. and Swanston, D.W., *Acta Pharmacol. Toxicol.*, 35, 412, 1974; Ballantyne, B., Riot-control agents (biomedical and health aspects of the use of chemicals in civil disturbances), in *Medical Annual*, Scott, R.B. and Frazer, J., eds., Wright and Sons, Bristol, U.K., 1977a.

could tolerate concentrations of 22–92 mg/m³ for 1 min or more, and concentration range of 22–220 mg/m³ would appear to be intolerable to 50% of a population for 1 min.

Although a dosage of 49 mg min/m³ was sufficient to cause nausea and vomiting, based on studies in humans exposed to DM at Ct values between 7 and 236 mg min/m³, high confidence is lacking because the estimate is founded on highly variable data. Ballantyne (1977a) estimated that a dosage of 370 mg min/m³ was required to cause nausea and vomiting. Inhalation of high concentrations of DM has resulted in severe pulmonary damage and death (British Ministry of Defence, 1972). DM is less effective as a riot control or incapacitating agent than CS and CN, and it was conjectured that there might be greater differences among peoples' susceptibility to DM than to the other agents.

Castro (1968) found DM and CS to be cholinesterase inhibitors and suggested that this might explain their lacrimatory effect. Although DM has a direct effect on gastric activity, Striker et al. (1967) found evidence that the lethal effect of DM is respiratory (Table 11.4).

11.7 FENTANYLS

11.7.1 Toxicological Effects

A major breakthrough in opioid drugs for use in medicine was the synthesis of fentanyl in Belgium in the late 1950s; it was first patented by Janssen in France in 1963. Its primary use in medicine was alone or in combination for anesthesia. However, its major complication is respiratory depression, which can be monitored and reversed in an operating room but can be a problem if used operationally in the field. Since 1996, several different analogs of fentanyl have been introduced for use in anesthesia, such as carfentanil, sufentanil, alfentanil, and remifentanil. Their pharmacological activity is the characteristic of opiates, and they produce all the effects of heroin, including analgesia, euphoria, miosis, and respiratory depression. Because of their high lipid solubility, regardless of the route of administration, the fentanyls reach the brain very quickly, thus providing a very fast onset of action. Some of the analogs have been synthesized specifically for sale as Persian white, China white, Mexican brown, and synthetic heroin, on the illicit drug market and to circumvent regulations on controlled substances. These illicit drugs are also called designer fentanyls and are used by abusers via intravenous injection, or they can be smoked or snorted.

The feasibility of dissociating the respiratory depressant effect from the opioid-induced sedative activity of alfentanil and fentanyl with naloxone was studied in the rabbit by Brown and Pleury (1981). They found that naloxone was more effective as an antagonist to alfentanil than to fentanyl. Later studies by Mioduszewski and Reutter (1991) also suggested that in the rat and ferret, dissociation of the opioid-induced sedation and respiratory depression was feasible. This was accomplished by coadministration of the opiate agonist with antagonists. The opioid-induced effects were akinesia, catalepsy, loss of righting reflex, light anesthesia, and apnea. The pharmacodynamic mechanism of the coadministration may involve competitive displacement of the opioid agonist by the antagonist at their common receptor sites within the central nervous system. A pharmacokinetic mechanism may also be involved such that the opiate uptake, distribution, and clearance are affected, either directly or indirectly, by the antagonist. Changes in respiratory frequency, oxygen consumption, and apnea were monitored in ferrets following the intravenous coadministration of the opioid agonist sufentanil and the antagonist nalmefene. These studies demonstrated a dissociation of the sufentanil-induced sedation/anesthesia and severe respiratory depression. Nalmefene coadministration shortened the duration but did not significantly delay the onset by the opiate-induced sedative/anesthetic effect. Narcotic antagonists such as nalmefene, naltrexone, and naloxone have clinical application in the diagnosis of addiction, prophylactic treatment of narcotic abuse, and emergency treatment of narcotic overdosage. These antagonists displace either previously assimilated opiates from their receptor sites or, if administered prior to the narcotic, will preclude the narcotic agonist from acting at these sites (Langguth et al., 1990).

More recently, Manzke et al. (2003) reported that the serious adverse effects of opiate analgesia such as depression of breathing are caused by direct inhibition of rhythm-generating respiratory neurons in the pre-Bötzinger complex (PBC) of the brain stem. Serotonin 4(a) or 5-HT4(a) receptors are strongly expressed in these neurons, and their selective activation protects spontaneous respiratory activity. Rats treated with a 5-HT4 receptor–specific agonist overcame the fentanyl-induced respiratory depression and reestablished stable respiratory rhythm without loss of fentanyl's analgesic effect. However, this discovery did not translate into similar functionality in humans. Lotsch et al. attempted to recreate the dissociate effect of respiration and analgesia seen in rats, in humans. His study, in 2005, could not use the same 5-HT4 agonist that Manzke had used, BIMU8, as it is not approved for human use, so instead mosapride was used as this study's agonist. Mosapride is known to be less centrally acting overall and has more peripheral effects such as promoting gastrointestinal motility. With mosapride included in this study instead of BIMU8, the study failed to show dissociation of respiratory function from analgesia in the human model. Perhaps future studies with other 5-HT4 agonists might show more promising results.

A more novel class of drugs known as ampakines have recently received much acclaim for their ability to attenuate respiratory depression seen with opioid use in a clinical setting and, with further study, could be a possible treatment for such types of toxicity. Ampakines act to modulate the amino-3-hydroxyl-5-methyl-4-isoxazolepropionate (AMPA) receptor, augmenting conductance through glutamatergic excitation (Ren et al., 2009). This excitation at the AMPA receptor in the PBC has shown to be enough stimulation to overcome the suppressive opioid effect in the same region, allowing for spontaneous respiration. Of course, Ren's studies were conducted in the rat model; so once again, testing was carried out to determine if this phenomenon held true in the human physiology. Oertel et al. (2010) showed that in fact, ampakine CX717 (proprietary designator) did in fact reverse opioid-induced respiratory depression in human volunteers. It should be noted that these effects were seen at clinical levels of fentanyl and their anologs, and not at overdoses of the drug, but the principle is that these compounds can act to overcome the most dangerous of side effects of this class of drugs, while keeping their analgesic properties completely intact.

Based on his studies of the dissociation of the sedative/anesthetic effects and respiratory depression of sufentanil coadministration with nalmefene in ferrets, Mioduszewski (1994) was awarded a US patent for an opiate analgesic formulation with improved safety. The invention described is a homogeneous mixture of an opiate drug agonist and an opiate drug antagonist suitable for in vivo administration to a patient by intravenous injection or by inhalation as an aerosol. The opioid agonist of the fentanyl series (fentanyl, carfentanil, alfentanil, and sufentanil) and the antagonist nalmefene, naloxone, or naltrexone are in the ideal ratios so that the duration of agonist and antagonist effects are the same and will not result in renarcotization and the associated respiratory depression if the agonist effects outlast those of the antagonist.

Opiate effects are mediated via multiple opioid receptors such as the mu, kappa, delta, and sigma (Martin, 1983). The mu receptors mediate analgesia, euphoria, physical dependence, and depression of ventilation, whereas kappa receptors mediate sedation and dieresis. Drugs may act at more than one opiate receptor with varying effect. Traditionally, narcotic antagonists such as naloxone and naltrexone have been used to reverse opioid agonist effects, whereas in these studies, naltrexone did not appear to have any effect on the potency and onset of carfentanil-induced sedative/hypnotic effects in the ferret. When used clinically, longer-acting opioids such as fentanyl may produce renarcotization because of differences in the pharmacokinetics of agonists and antagonist (McGilliard and Takemori, 1978).

11.7.2 HUMAN IN USE EXPOSURES

Because fentanyl is not listed in any of the schedules of the Chemical Warfare Convention (CWC) and is traditionally characterized by the rapid onset and short duration of 15–30 min of analgesia, some can argue that it should be legally considered a riot control agent according to the definition set forth in the CWC (Chemical Biological Weapons Non-proliferation Treaty [CBWNP], 2002).

On October 23, 2002, at least 129 of the almost 800 hostages died in the Moscow Dubrovka Theater Center when authorities subdued the hostage takers (the Chechnian terrorist group) there by pumping what many believe was fentanyl into the building (Couzin, 2003), while some believe that a mixture of fentanyl and halothane was used (CBWNP, 2002; Couzin, 2003; Mercadente, 2003; Miller and Lichtblau, 2003). It was also considered that the Russians might have used remifentanil because it is rather unique and extremely potent with relatively fast action and short duration. The chemical structure of remifentanil also allows the body to quickly metabolize the substance into nontoxic and water-soluble forms, thus minimizing risks for both hostages and hostage takers. This is due to the presence of an ester linkage in the structure that is easily metabolized independently of the liver by circulating esterases. Although Russian authorities insisted that emergency personnel were prepared with 1000 antidotes in anticipation of the raid, controversy still exists about whether local hospitals and physicians were adequately informed about the gas used during the operation (CBWNP, 2002; Glasser and Baker, 2002). It has also been suggested that the Russian government revealed that a mixture of fentanyl and halothane was used to incapacitate the Chechnian terrorists in the attempt to liberate the hostages in Moscow. It was further suggested that it was likely that massive doses of carfentanil were used to saturate the theater so that the maximal effect by inhalation could be achieved. Carfentanil is a potent opioid used to rapidly immobilize large wild animals, horses, and goats. It produces rapid catatonic immobilization, characterized by limb and neck hyperextension. Adverse effects include muscle rigidity, bradypnea, and oxygen desaturation. Recycling and renarcotization have been reported as possible causes of death when low doses of antagonist are used (Miller et al., 1996).

It appears that the mixture used at the Dubrovka Theater over a decade ago has been determined by researchers in the United Kingdom. Riches et al. published a paper in 2012 that analyzed clothing and urine from exposed hostages at the 2002 incident, and the analysis revealed that both remifentanil and carfentanil were present in the theater at the time of the siege. Further information regarding its dissemination or whether or not there were additional agents present as carriers or for supplemental activity is unknown still, but evidence does point to remi- and car-analogs as the agents of choice by the Russian Spetsnaz forces.

Although there were naloxone syringes found in the theater, it is possible that the doses were insufficient to reverse the respiratory depression (Mercadente, 2003). Opioid overdose and toxicity was not the only thing that was responsible for the deaths of the 129 who expired that night in 2002. Those that were pulled to safety from the theater were not placed in a good recovery positions on their side. Instead they were placed on their backs for prolonged periods of time. Vomiting occurred that led to aspiration due to the loss of reflex from opioid administration, which led subsequently to asphyxiation. Some taken to hospitals were treated for nerve agent exposure due to poor communication. They were given atropine, which worked only to slow their heart and breathing further leading to expiration. These are not the only things that led to the deaths of these people, but they are the unfortunate result of the logistical errors of planning covert use of agents such as this class of compounds. For more effective use, responders need to be alerted as to what precisely was given in order to properly triage and care for those exposed.

11.8 CONCLUSIONS

The toxicological effects, which are actually the pharmacological effects, of exposure to riot control agents, but are perceived as adverse or toxicological, can be local or topical, as well as systemic following absorption. In addition, the effects can be acute or long term. In addition, the exposure can be either acute or repeated. The disposition of the agent in the exposed individual also needs to be considered. That is the absorption, distribution, metabolism (biotransformation), and excretion (ADME).

Riot control agents have been described as nonlethal or less-than-lethal agents. Exposures to these compounds involve ocular, dermal, and inhalation effects and indirectly oral or gastrointestinal. Their primary action is their local or topical effect on the eye that appears to be the most sensitive target organ. They also act on the skin and respiratory tract. The immediate effects on exposure to irritants include intense irritation of the eyes, marked irritation of the nose, throat and lungs, as well as irritation of the skin. The margin of safety between the dose eliciting the intolerable effect and that which causes serious adverse effects is large. For example, the estimated lethal human dose is 2600 times as great as the dose required to cause temporary disabling and that of BBC is 3000 times as great. Riot control agents do not usually cause long-term or permanent toxic effects, although the risk for serious toxic effects, long-term sequelae, or even death increases with higher exposure concentration, and greater exposure duration, or in susceptible individuals. Overall, however, the toxicity of acute and short-term repeated exposures to riot control agents is well characterized.

QUESTIONS

1. Modern research into mechanisms of action of CS, CN, and CR has led to their characterization as what kinds of compounds?
   a. SN2-alkylating agents
   b. TrpA1 receptor agonists
   c. β-Blockers
   d. SSRIs
   e. Both a and b

   Answer: e, both a and b

2. OC elicits its response through which sensory receptor located both throughout the pulmonary tract and in ocular organs?
   a. μ-Opioid receptor
   b. TrpV1 thermo/chemoreceptor
   c. G-protein coupled receptors
   d. BRCA1 receptors
   e. FSH receptors

   Answer: b

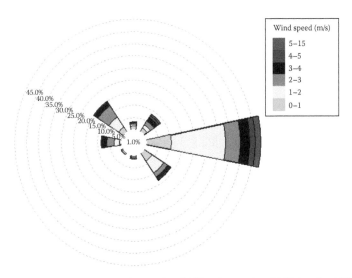

FIGURE 1.12 Wind rose diagrams for the December 2007 to November 2008 (station EET–ARSO).

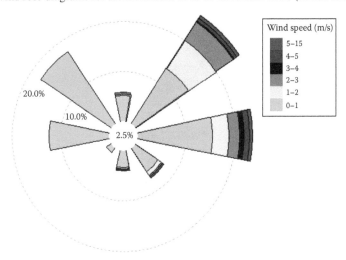

FIGURE 1.13 Wind rose diagrams for the December 2007 to November 2008 (inside of the Port of Koper–measured by PINT).

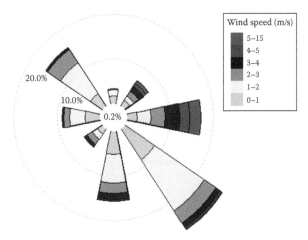

FIGURE 1.14 Wind rose diagrams for the December 2007 to November 2008 (meteorological station Markovec, Koper—ARSO).

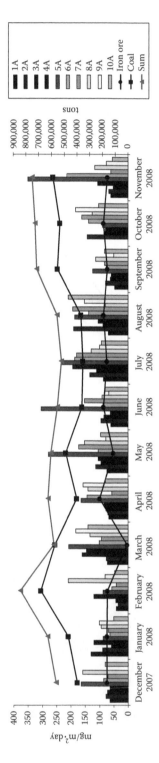

FIGURE 1.25 Correlation among particulate matter deposition and coal and iron ore manipulation at the EET—sampling site A.

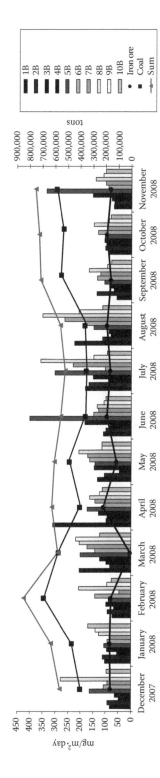

FIGURE 1.26 Correlation among particulate matter deposition and coal and iron ore manipulation at the EET—sampling site B.

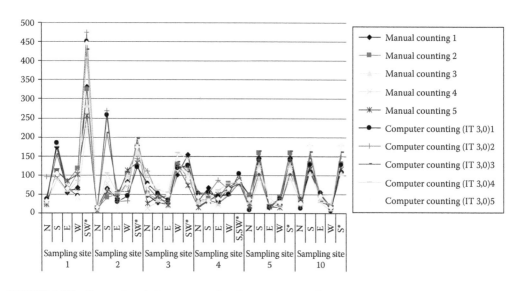

FIGURE 1.29 Comparison between manual and computer counting.

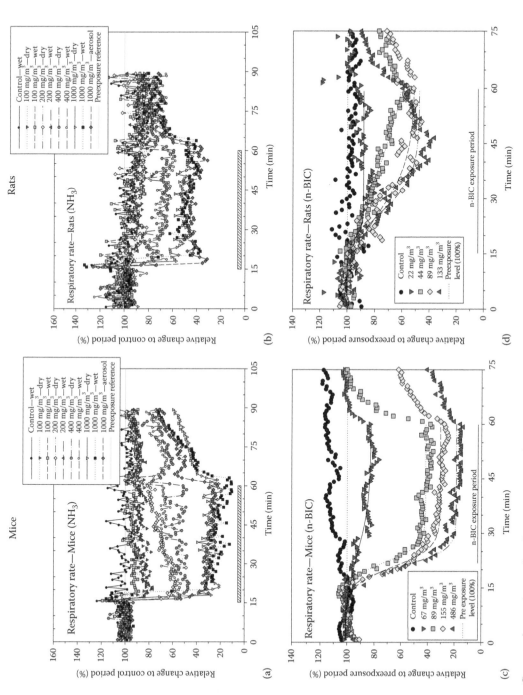

FIGURE 6.2 Comparative respiratory function measurements in rats and mice before (15 min), during (45 min), and after exposure to (a and b) the upper respiratory tract water-soluble irritant ammonia (NH_3) gas and (c and d) the respiratory tract moderately water-soluble irritant n-BIC (vapor). Data represent means of four animals/group.

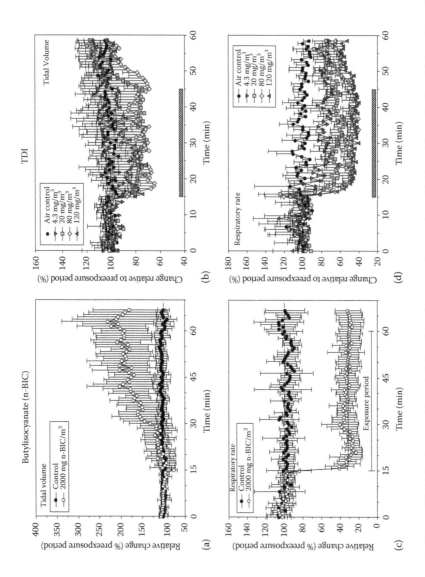

FIGURE 6.4 Measurements of (a and b) tidal volume and (c and d) breathing frequency in conscious tidal breathing rats placed into volume displacement plethysmographs before, during, and after exposure to the vapor phase of aliphatic n-butyl(mono)isocyanate (n-BIC) and aromatic toluene diisocyanate (TDI). The n-BIC- and TDI-exposed animals were exposed to the sensory irritant for 45 and 30 min, respectively. Data represent averages ±SD of at least four rats/group.

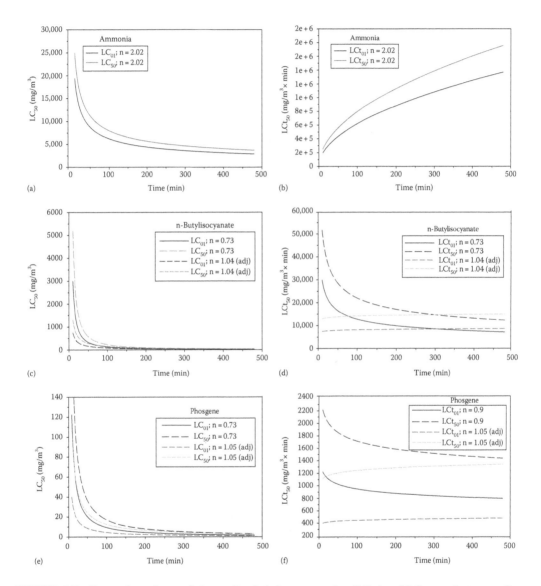

FIGURE 6.7 $C^n \times t$–dependence of the median lethal concentration (LC_{50}) and LCt_{50} products as well as the respective nonlethal threshold concentrations (LC_{01} and LCt_{01}) of rats exposed to the upper respiratory tract irritant ammonia (gas), the respiratory tract irritant n-BIC (vapor), and lower respiratory tract irritant phosgene (gas). The actual exposure concentrations were analyzed as measured or were adjusted for the time-dependent depression in ventilation (adj.) as shown in Figure 6.4.

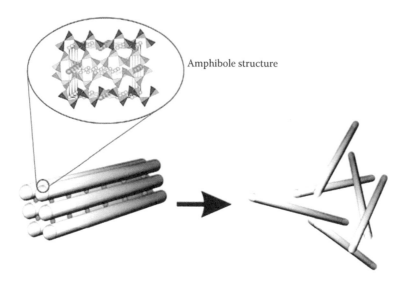

Amphibole structure

FIGURE 14.6 With amphiboles, the soluble cations shown as small circles are located between the fibers which are formed with double chain silicate. When the soluble cations dissolve as can happen in the lung, the amphibole fibers in these bundles are released as individual fibers. The double chain silicate amphibole fibers themselves are highly insoluble in both the lung fluids and in the macrophages. (From Bernstein, D.M. et al., *Crit. Rev. Toxicol.*, 43(2), 154, 2013.)

FIGURE 14.7 Illustration of the tightly bound silica-based structure on exterior surfaces of amphibole fibers. (Adapted from Department of Geology and Geophysics, University of Wisconsin, Crystal Structure Movies, http://www.geology.wisc.edu.)

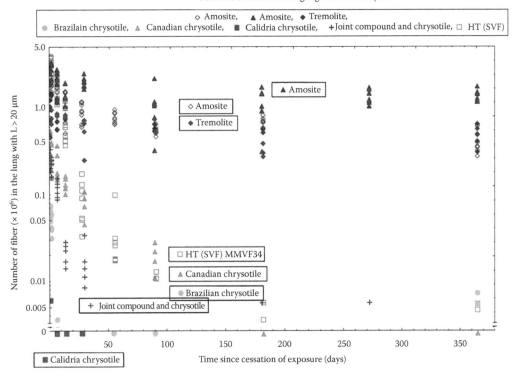

FIGURE 14.10 Summary of studies showing the number of fibers remaining in the rat's lungs is shown as a function of the time in days following cessation of the 5-day exposure. Included are the three amphibole asbestos studies including tremolite asbestos and 2 amosite asbestos, a soluble synthetic vitreous (SVF) fiber, HT, and 4 chrysotile fibers from Brazil, the United States (Calidria), and Canada and one study on a commercial joint compound mixed with chrysotile fibers. (Data presented are from Bernstein et al., 2003a, 2005c; Bernstein, D.M. et al., *Inhal. Toxicol.*, 23(7), 372, June, 2011.)

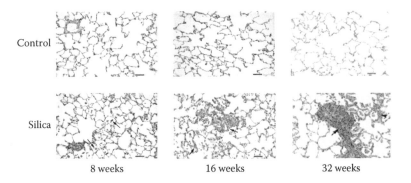

FIGURE 18.3 Progression of lung damage in crystalline silica–exposed rats. Rats were exposed to Min-U-Sil 5 crystalline silica (15 mg/m³, 6 h/day, 5 days) or air (control) as described in detail in our original publications (Sellamuthu et al., 2011b, 2012). Lung sections prepared at the post–silica exposure time intervals of 0, 1, 2, 4, 8, 16, and 32 weeks were stained with hematoxylin and eosin. Only photomicrographs of lung sections of 8, 16, and 32 weeks post–silica exposure time intervals are presented, and those of earlier time intervals can be found in our original publication (Sellamuthu et al., 2011b). Arrows indicate AMs in alveolar space (8 weeks) and type II pneumocyte hyperplasia (16 and 32 weeks). Magnification = 20×. (Modified from Sellamuthu, R. et al., *Toxicol. Sci.*, 122(2), 253, 2011b; Sellamuthu, R. et al., *Inhal. Toxicol.*, 24(9), 570, 2012. With permission.)

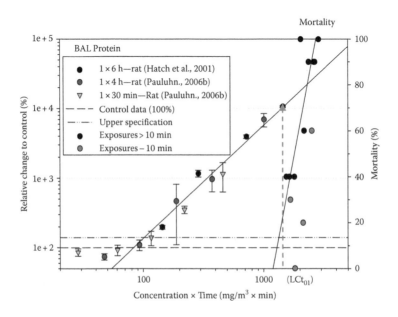

FIGURE 19.1 Comparison of BALF protein and mortality in phosgene-exposed rats at different C × t relationships. BAL-protein data were from whole-body-exposed Fischer 344 rats (Hatch et al., 2001) or nose-only-exposed Wistar rats (Pauluhn, 2006b) to multiple concentrations of phosgene bracketing exposure durations from 30 min to 6 h. Rats were sacrificed approximately 24 h following inhalation exposure. Data represent group means ± SDs. (Data from Pauluhn, J., *Inhal. Toxicol.*, 18, 609, 2006c.) Mortality data were from rat inhalation studies utilizing exposure durations from 30 to 360 min. (From Pauluhn, J., *Inhal. Toxicol.*, 18, 423, 2006a.)

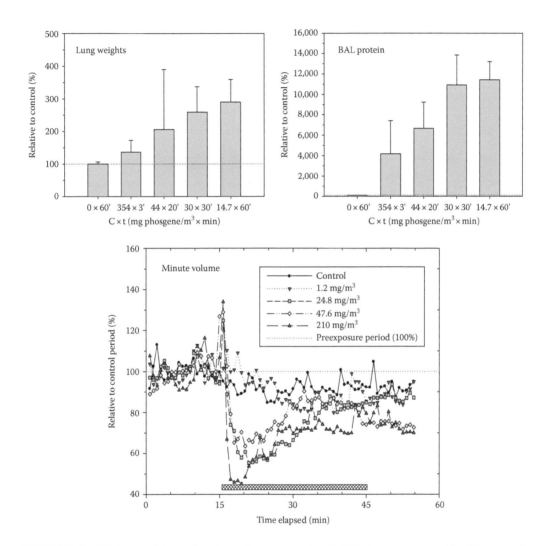

FIGURE 19.2 Wet lung weights and total protein concentrations in BAL from rats exposed for 1 h to air only (control), 354, 44, 30, and 14.7 mg phosgene/m³ at nose-only exposure durations of 3, 20, 30, and 60 min. Rats were sacrificed on postexposure day 1. Data represent group means ± SD (n = 6–8). Data were normalized to the conditioned air nose-only-exposed control group (= 100%). The average respiratory minute volume of 8 rats/group exposed for 30 or 10 min (210 mg/m³) to phosgene is shown in the lower panel. Each exposure was preceded by a 15 min exposure period to air. Data were averaged for time periods of 45 s. Measurements utilized volume-displacement nose-only plethysmographs. (Data modified from Li, W.-L. et al., *Inhal. Toxicol.*, 23, 842, 2011; Pauluhn, J., *Inhal. Toxicol.*, 18, 423, 2006a.)

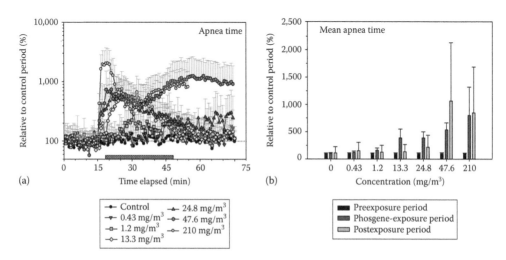

FIGURE 19.6 Average duration of apnea periods (changes in apnea time relative to exposure duration (a) and the mean apnea times during the respective preexposure, exposure, and postexposure (b) segment) between individual breathing cycles of rats exposed nose-only to phosgene in volume-displacement plethysmographs. All data were normalized to the pre–phosgene exposure data collection period of 15 min (baseline data = 100%). With the exception of the 210 mg/m³ group (exposure duration 10 min), all rats were exposed to phosgene for 30 min (hashed bar), followed by an air-only exposure period of 30 min. The averaged values for each time period are given in the bar chart. Data represent means ± SD (n = 4). (Data modified from Pauluhn, J., *Inhal. Toxicol.*, 18, 423, 2006a.)

3. What Department of Defense organization first synthesized and tested capsaicin as a riot control agent in the 1920s?

   a. US Army Ballistic Research Laboratory

   b. US Naval Research Laboratory

   c. Idaho National Laboratories

   d. Edgewood Arsenal

   e. US Army Medical Research Institute for Chemical Defense

   Answer: d

4. A potential therapy for RCA exposure could be blocking the receptors at which RCA's elicit their response, that is, the TrpA1 receptor for CS, CN, and CR and the TrpV1 receptor for OC or capsaicin. Current research has shown promise for antagonists of the TrpA1 receptor, but not the TrpV1. In fact, TrpV1 antagonism has led to organ failure and death. Formulate a hypothesis for why blocking the TrpV1 receptor is difficult to do safely.

   Answer: Blocking the TrpV1 receptor has been done. The easy part is blocking the receptor. However, the TrpV1 receptor plays a very large role in thermoregulation. By antagonizing this receptor, severe and unrecoverable hyperthermia was induced and ultimately led to systemic organ failure and death in animals. TrpA1 antagonism does not appear to cause any thermo-regulatory distress and is thus being pursued as a possible treatment route for CS, CN, and CR exposure and/or prophylaxis.

5. Explain how capsaicin can be used as a therapy for pain, that is, the mechanism of action of the compound and how this alleviates localized pain.

   Answer: Capsaicin exposure (acute) leads to the release of Substance-P, a compound involved in the transduction of pain. Persistent exposure to capsaicin, as in an over-the-counter (OTC) topical cream, causes localized depletion of Substance-P from sensory neurons in the skin, which ultimately attenuates or eliminates the sensation of pain at the site.

6. Casualties of the Moscow Theatre Siege should have been given what drug as a method of treatment for opioid exposure/overdose? What new drugs are being considered for possible treatments of opioid-induced respiratory depression and what is their mechanism of action?

   Answer: Naloxone should be given and patient monitored for renarcotization. Ampakines and 5-HT4 agonists are being sought after for their therapeutic use in opioid administration to alleviate respiratory depression while keeping desirable opioid activity intact.

REFERENCES

Adcock, J.J. and Smith, T.W., Inhibition of reflex bronchoconstriction by the opioid peptide BW 443C81 in the anaesthetized cat, *Br. J. Phamacol.*, 96, 596, 1989.

Allen, W.F., Effect on respiration, blood pressure, and carotid pulse of various inhaled and insufflated vapors when stimulating one cranial nerve and various combinations of cranial nerves, *Am. J. Physiol.*, 87, 319–325, 1928–1929.

Anderson, P.J., Lau, G.S.N., Taylor, W.R.J., and Critchley, J.A., Acute effects of the potent lacrimator o-chlorobenzylidene malononitrile (CS) tear gas, *Human Exp. Toxicol.*, 15, 461–465, 1996.

Ashton, L., Cotes, J.E., Holland, P., Johnson, G.R., Legg, S.J., Saunders, M.J., and White, R.G., Acute effect of dibenz (b,f)-1:4-oxazepine upon the lung function of healthy young men, *J. Physiol.*, 215, 85, 1978.

Aviado, D.M. and Aviado, D.G., The Bezold-Jarisch reflex a historical perspective of cardiopulmonary reflexes, *Ann. N. Y. Acad. Sci.*, 940, 48–58, 2002.

Aviado, D.M. and Salem, H., Acute effects of air pollutants on the lungs, *Arch. Environ. Health*, 16, 903–907, 1968.

Aviado, D.M. and Salem, H., Respiratory and bronchomotor reflexes in toxicity studies, in *Inhalation Toxicology*, Salem, H., ed., Marcel Dekker, New York, 1987.

Aviado, D.M., Kratschmer reflex induced by inhalation of aerosol propellants, in *Conference on Toxic Hazards of Halocarbon Propellants*, U.S. Food and Drug Administration, Washington, DC, 1971, pp. 63–77.

Ballantyne, B. and Callaway, S., Inhalation toxicology and pathology of animals exposed to o-chlorobenzylidene malononitrile (CS), *Med. Sci. Law*, 12, 43, 1972.

Ballantyne, B. and Swanston, D.W., The comparative acute mammalian toxicity of 1-chloroacetophenone (CN) and 2-chlorobenzylidene malononitrile (CS), *Arch. Toxicol.*, 40, 75, 1978.

Ballantyne, B. and Swanston, D.W., The irritant effects of dilute solutions of dibeozox-azepine (CR) on the eye and tongue, *Acta Pharmacol. Toxicol.*, 35, 412, 1974.

Ballantyne, B. and Swanston, D.W., The irritant potential of dilute solutions of ortho-chlorobenzylidene malononitrile (CS) on the eye and tongue, *Acta Pharmacol. Toxicol.*, 32, 266, 1973.

Ballantyne, B., Beswick, F.W., and Thomas, D., The presentation and management of individuals contaminated with solutions of dibenzoxazepine (CR), *Med. Sci. Law*, 13, 265, 1973.

Ballantyne, B., Gall, D., and Robson, D.C., Effects on man of drenching with dilute solutions of o-chlorobenzylidene malononitrile (CS) and dibenz (b,f)-1:4-oxazepine (CR), *Med. Sci. Law*, 16, 159, 1976.

Ballantyne, B., Gazzard, M.F., Swanston, D.W., and Williams, P., The comparative ophthalmic toxicology of 1-chloroacetophenone (CN) and dibenz (b,f)-1:4-oxazepine (CR), *Arch. Toxicol.*, 34, 183, 1975.

Ballantyne, B., Riot-control agents (biomedical and health aspects of the use of chemicals in civil disturbances), in *Medical Annual*, Scott, R.B. and Frazer, J., eds., Wright and Sons, Bristol, U.K., 1977a, pp. 7–41.

Ballantyne, B., The acute mammalian toxicology of dibenz (b,f)-1:4-oxazepine, *Toxicology*, 8, 347, 1977b.

Bannister, Justin. NMSU's Chile Pepper Institute names the Trinidad Moruga Scorpion hottest pepper on Earth. February 13, 2012. NMSU News Center. Available at http://newscenter.nmsu.edu/Articles/view/8341. Accessed April 2013.

Barclay, M., India's hot chili pepper grenade: The bhut jolokia is so potent it may be used for crowd control. Available at http://www.macleans.ca/news/world/indias-hot-chili-pepper-grenade. Maclean's. Published: Thursday, April 8, 2010. Accessed April 23, 2013. Publisher: Penny Hicks. Editor: Mark Stevenson. Maclean's is part of Rogers Media, Inc.

Beswick, F.W., Holland, P., and Kemp, K.H., Acute effects of exposure to ortho-chlorobenzylidene malononitrile (CS) and the development of tolerance, *Br. J. Ind. Med.*, 29, 298, 1972.

Biggs, D.F. and Goel, V., Does capsaicin cause bronchospasm in guinea-pigs? *Eur. J. Pharmacol.*, 115, 71, 1985.

Biskup, R.C., Swentzel, K.C., Lochner, Ma.A., and Fairchild, D.G., Toxicity of 1% CR in propylene glycol/water (80/20), Technical report EB-TR-75009, Edgewood Arsenal, Aberdeen Proving Ground, Aberdeen, MD, May 1975.

Blain, P.G., Tear gases and irritant incapacitants, *Toxicol. Rev.*, 22(2), 103–110, 2003.

Boyd, K., Military Authorized to Use Riot Control Agents in Iraq. May 2003. Arms Control Association. Available at http://www.armscontrol.org/act/2003_05/nonlethal_May03. Accessed April 23, 2014.

Boyd, K., Military Authorized to Use Riot Control Agents in Iraq, Arms Control Today, May 2003, p. 36.

British Ministry of Defence, Medical manual of defence against chemical agents, JSP 312, Her Majesty's Stationary Office, London, U.K., 1972.

Brone, B., Peeters, P., Marrannes, R., Mercken, M., Nuydens, R., Meert, T., and Gijsen, H., Tear gasses CN, CR and CS are potent activators of the human TrpA1 receptor, *Toxicol. Appl. Pharmacol.*, 231, 150–156, 2008.

Brown, J.H. and Pleury, B.J., Antagonism of the respiratory effects of alfentanil and fentanyl by natoxone in the conscious rabbit, *Br. J. Anaesth.*, 53, 1033–1037, 1981.

Buchan, P. and Adcock, J.J., Capsaicin-induced bronchoconstriction in the guinea pig: Contribution of vagal cholinergic reflexes, local axon reflexes and their modulation by BW 443C81, *Br. J. Pharmacol.*, 105, 448, 1992.

Buck, S.H. and Burks, T.F., The neuropharmacology of capsaicin: Review of some recent observations, *Pharmacol. Rev.*, 38, 179, 1986.

Carson, B.B. and Stoughton, R.W., Reactions of alpha, beta unsaturated dinitriles, *J. Am. Chem. Soc.*, 50, 2825, 1928.

Castro, J.A., Effects of alkylating agents on human plasma cholinesterase, *Biochem. Pharmacol.*, 17, 295, 1968.

Chahl, L.A. and Lynch, A.M., The acute effects of capsaicin on the cardiovascular system, *Acta Physiol. Hung.*, 69, 413, 1987.

Chemical Biological Weapons Non-Proliferation Treaty, 2002. Available at http://www.Cws.gov/cwc_treaty.html. Accessed on April 24, 2014.

CHPPM (Center for Health Promotion and Preventive Medicine), Detailed and general facts about chemical agents, TG 218, U.S. Army Center for Health Promotion and Preventive Medicine, Aberdeen Proving Ground, Aberdeen, MD, 1996.

Clede, B., Oleoresin capsicum, *Law Order*, March, 63, 1993.

Colgrave, H.F., Brown, R., and Cox, R.A., Ultrastructure of rat lungs following exposure to aerosols of dibenzoxazepine (CR), *Br. J. Exp. Pathol.*, 60, 130, 1979.

Colgrave, H.F., Lee, C.G., Marrs, T.C., and Morris, B., Repeated-dose inhalation toxicity and mutagenicity status of CR (dibenz (b,f)-1:4-oxazepine), *Br. J. Pharmacol.*, 78, 169, 1983.

Cordell, G.A. and Araujo, D.E., Capsaicin: Identification, nomenclature and pharmacology, *Ann. Pharmacother.*, 27, 330, 1993.

Couzin, J., A sigh of relief for painkillers, *Science*, 301, 150, 2003.

Craig, F.N., Blevins, W.V., and Cummings, E.G., Breathing patterns during human exposure to CS, Report AD318487, Aberdeen Proving Ground, Aberdeen, MD, 1960.

Cucinell, S.A., Swentzel, K.C., Biskup, R., Snodgrass, H., Lovre, S., Stark, W., Feinsilver, L., and Vocci, R., Biochemical interactions and metabolic fate of riot control agents, *Fed. Proc.*, 30, 86, 1971.

Davis, B., Roberts, A.M., Coleridge, H.M., and Coleridge, J.C.B., Reflex tracheal gland secretion evolved by stimulation of bronchal c-fiber in dogs, *J. Appl. Physiol.*, 53, 985, 1982.

Feinsilver, L., Chambers, H.A., Vocci, F.J., Daasch, L., and Berkowitz, L.M., Some metabolites of CS from rats, Edgewood Arsenal Technical Report, Series 4521, 1971.

French, M.C., Harrison, J.M., Inch, T.D., Leadbeater, L., Newman, J., and Upshall, D.G., The fate of dibenz (b,f)-1,4-oxazepine (CR) in the rat, rhesus monkey, and guinea pig, Part I, metabolism in vivo, *Xenobiotica*, 13, 345, 1983.

Fries, A.A. and West, C.J., The history of poison gases, in *Chemical Warfare*, McGraw-Hill Book Company, New York, 1921, pp. 1–9.

Fuller, R.W., Dixon, C.M.S., and Barnes, P.J., Bronchoconstrictor response to inhaled capsaicin in humans, *J. Appl. Physiol.*, 58, 1080, 1985.

Fuller, R.W., The human pharmacology of capsaicin, *Arch. Int. Pharmacodyn.*, 303, 147, 1990.

Games, D.E., Alcock, N.J., van der Greef, J., Nyssen, L.M., Maarse, H., and Ten, M.C., Analysis of pepper and capsicum oleoresins by high performance liquid chromatography–mass spectrometry and field desorption mass spectrometry, *J. Chromatogr.*, 294, 269, 1984.

GAO Report, Report to the Honorable Ronald V. Dellums, House of representatives, NSIAD-89-128, Israel's Use of Tear Gas, 1989.

Garrett, B., Toxin of the Month: Capsaicin, 2000. Available at www.wgsiwg:/main.l88/http://chembio/subscribe/features/capsaicin.html.

Gatti, R., Andre, E., Barbara, C., Dinh, T., Fontana, G., Fischer, A., Geppetti, P., and Trevisani, M., Ethanol potentiates the TRPV1-mediated cough in the guinea pig, *Pulm. Pharmacol. Ther.*, 22, 33–36, 2009.

Glasser, S.B. and Baker, P., Russia confirms suspicions about gas used in raid, Washington Post Foreign Service, p. A15, October 31, 2002.

Glinsukon, T., Stitmunnaithum, V., Toskulkao, C., Buranawuti, T., and Tangkrisanavinot, V., Acute toxicity of capsaicin in several animal species, *Toxicon*, 18, 215, 1980.

Gongwer, L.E., Ballard, T.A., Gutentag, P.J., Punte, C.L., Owens, E.J., Wilding, J.L., and Hart, J.W., The comparative effectiveness of four riot-control agents, Technical Memorandum 24–18, U.S. Army Chemical Warfare Laboratories, Army Chemical Center, Edgewood Arsenal, MD, 1958.

Govindarajan, V.S. and Sathyanarayana, M.N., Capsicum-production, technology, chemistry, and quality. Part V. Impact on physiology, pharmacology, nutrition and metabolism; structure pungency, pain and desensitization sequences, *Crit. Rev. Food Sci. Nutr.*, 29, 435, 1991.

Govindarajan, V.S., Capsicum-production, technology, chemistry, and quality. Part II. Processed products standards, world production and trade, *Crit. Rev. Food Sci. Nutr.*, 29, 207, 1986.

Grady, H.G. and Stewart, H.L., Histogenesis of induced pulmonary tumors in strain A mice, *Am. J. Pathol.*, 16, 417, 1940.

Graebe, C., *Berichte des Deutschem Chemisch Gesellscheft*, Commissionsverlog von R. Friedlander and Sohn, Berlin, Germany, 55, 35, 1881.

Gutentag, P.J., Hart, J., Owens, E.J., and Punte, C.L., The evaluation of CS aerosol as a riot control agent in man, Technical Report CWLR 2365, U.S. Army Chemical Warfare Laboratories, Army Chemical Center, April 1960.

Harris, B.L., Chemicals in war, in *Encyclopedia of Chemical Technology*, 4th edn., Vol. 5, Jacqueline I.K., and Mary Howe-Grant., eds., John Wiley & Sons, New York, 1992.

Hathaway, T.J., Higenbottam, T.W., Morrison, J.F., Clelland, C.A., and Wallwork, J., Effects of inhaled capsaicin in heart-lung transplant patients and asthmatic subjects, *Am. Rev. Respir. Dis.*, 148, 123, 1993.

Hess, L. and Knakel, J., First exposure with immobilization with carfentanil at Zoo Prague, *Gazella*, 3, 87–90, 1985.

Higginbottom, R. and Suschitzky, H., Synthesis of heterocyclic compounds. II. Cyclization of O-nitrophenyl oxygen, *J. Chem. Soc.*, 456, 2367, 1962.

Hill, A.R., Silverberg, N.B., Mayorga, D., and Baldwin, H.E., Medical hazards of the tear gas CS, *Medicine*, 79, 234–240, 2000.

Himsworth, H., Black, D.A.K., Crawford, T., Dornhorst, A.C., Gilson, J.C., Neuberger, A., Paton, W.M.D., and Thompson, R.H.S., Report of the enquiry into the medical and toxicological aspects of CS (orthochlorobenzylidene malononitrile). Part II. Enquiry into the toxicological aspects of CS and its use for civil purposes, Cmnd 4775, H.M.S.O., London, U.K., 1971.

Himsworth, H., Thompson, R.H.S., and Dornhorst, A.C., Report of the enquiry into the medical aspects of CS (orthochlorobenzylidene malononitrile). Part I. Enquiry into the medical situation following the use of CS in Londonderry on 13th and 14th August, 1969, Cmnd 4173, H.M.S.O., London, U.K., 1969.

Holland, P., The cutaneous reactions produced by dibenzoxazepine (CR), *Br. J. Dermatol.*, 90, 657, 1974.

Holzer, P., Capsaicin: Selective toxicity for thin primary sensory neurons, *Handb. Exp. Phamacol.*, 102, 420, 1992.

Hu, H., Tear gas—Harassing agent or toxic chemical weapon? *JAMA*, 262, 660, 1989.

Husain, K., Kumar, P., and Malhotra, R.C., A comparative study of biochemical changes induced by inhalation of aerosols of o-chloroacetophenone and dibenz (b,f)-1,4-oxazepine in rats, *Indian J. Med. Res.*, 94, 76, 1991.

Issekutz, B., Lichtneckert, I., and Nagy, H., Effect of capsaicin and histamine on heat regulation, *Arch. Int. Pharmacodyn.*, 81, 35, 1950.

Jansco, G., Kiraly, E., and Jansco-Gabor, A., Pharmacologically-induced selective degeneration of chemosensitive primary sensory neurons, *Nature*, 270, 741, 1977.

Janssen, P., The development of new synthetic narcotics. In: *Opioids in Anesthesia*, Esafenous, E.G., ed. Boston, MA: Butterworth Publishers, 1984.

Krapf, R. and Thalmann, H., Akute exposition durch CS-rauchgas und linische beobachtungen, *Schweiz. Med. Wochenschr*, 11, 2056, 1981.

Kratschmer, F., Uber Reflexe von der Nasenschleimhaut auf athmung und Kreislaub, *Sirzungsber. Akad. Wiss.*, 62, 147–170, 1870.

Kurtz, J. Arwa Damon reports on clash between police and protesters in Turkey: "Incredibly intense fire with the tear gas, water cannons…when that happens there's complete chaos", 11 June 2013. http://piersmorgan.blogs.cnn.com/2013/06/11/arwa-damon-reports-on-clash-between-police-and-protesters-in-turkey-incredibly-intense-fire-with-the-tear-gas-water-cannons-when-that-happens-theres-complete-chaos/?iref=allsearch. Accessed on April 24, 2014.

Langguth, P., Khan, P.J., and Garrett, E.R., Pharmacokinetics of morphine and its surrogates. XI. Effects of simultaneously administered naltrexone and morphine on the pharmacokinetics and pharmacodynamics of each in the dog, *Biopharm. Drug Dispos.*, 11, 419–444, 1990.

Leadbeater, L., Sainsbury, G.L., and Utley, D., o-Chlorobenzyl malononitrile, a metabolite formed from o-chlorobenzylidene malononitrile (CS), *Toxicol. Appl. Pharmacol.*, 25, 111, 1973.

Leadbeater, L., The absorption of ortho-chlorobenzylidenemalononitrile (CS) by the respiratory tract, *Toxicol. Appl. Phamacol.*, 25, 101, 1973.

Lopez, S. L., Bhut Jolokia Hottest Chili in the World. 2007. NMSU News Release. Available at http://web.archive.org/web/20070219124128/http://www.nmsu.edu/~ucomm/Releases/2007/february/hottest_chile.htm. Accessed April 1, 2013.

Lotsch, J., Skarke, C., Schneider, A., Hummel, T., and Geisslinger, G., The 5-hydroxytryptamine 4 receptor agonist mosapride does not antagonize morphine-induced respiratory depression, *Clin. Pharm. Ther.*, 78(3), 278–287, 2005.

Lundy, P.M. and McKay, D.H., Mechanism of the cardiovascular activity of dibenz [b,f] [1,4] oxazepine (CR) in cats, Suffield Technical Paper 438, Defence Research Establishment, Ralston, Alberta, Canada, 1975.

Manzke, T., Guenther, U., Ponimaskin, E.G., Haller, M., Dutschmann, M., Schwarzacher, S., and Richter, D.W., 5-HT4(a) receptors avert opioid-induced breathing depression without loss of analgesia, *Science*, 301, 226–229, 2003.

Marrs, T.C., Colgrave, H.V., and Cross, N.L., A repeated dose study of the toxicity of technical grade dibenz-(b,f)-1.4 oxazepine in mice and hamsters, *Toxicol. Lett.*, 17, 13, 1983a.

Marrs, T.C., Colgrave, H.V., Cross, N.L., Gazzard, M.F., and Brown, R.F.R., A repeated dose study of the toxicity of inhaled 2-chlorobenzylidene malononitrile (CS) aerosol in three species of laboratory animals, *Arch. Toxicol.*, 52, 183, 1983b.

Marrs, T.C., Gray, M.L., Colgrave, H.F., and Gall, D., A repeated dose study of the toxicity of CR applied to the skin of mice, *Toxicol. Lett.*, 13, 259, 1982.

Martin, W.R., Pharmacology of opioids, *Pharm. Rev.*, 35, 285–323, 1983.

McGilliard, K. and Takemori, A.E., Antagonism by naloxone of narcotic-induced respiratory depression and analgesia, *J. Pharmacol. Exp. Ther.*, 207, 454–503, 1978.

McGregor, D.B., Brown, A., Cattanach, P., Edwards, I., Mcbride, D., and Caspary, W.J., Responses of the L5178Y tk.+/tk– mouse lymphoma cell forward mutation assay II: 18 coded chemicals, *Environ. Mol. Mutagen.*, 11, 91, 1988.

McNamara, B.P., Owens, E.J., Weimer, J.T., Ballard, T.A., and Vocci, F.J., *Toxicology of Riot Control Chemicals—CS, CN, and DM*. Edgewood Arsenal, MD: US Army Biomedical Laboratory, 1969. Technical Report EATR 4309.

McNamara, B.P., Renne, R.A., Rozmiarek, H., Ford, D.F., and Owens, E.J., *CS: A Study of Carcinogenicity*. Aberdeen Proving Ground, MD: US Army Biomedical Laboratory, 1973. Edgewood Arsenal Technical Report, EATR 4760.

McNamara, B.P., Vocci, F.J., Owens, E.J., Ward, D.M., and Anson, N.M., The search for an effective riot control agent solvent system for large volume dispersers, Edgewood arsenal technical report 4675, October 1972.

Miller, J. and Lichtblau, E., U.S. Investigates Moscow Theatre Seige Seeking Qaeda Link, 2003. Available at wystwyg://17/http://www.nytirnes.com/200...DI.html?8br = &pagewanted = print&position =. Accessed on April 24, 2014.

Miller, M.W., Wild, M.A., and Lance, W.R., Efficacy & Safety of naltrexone hydrochloride for antagonizing carfentanil citrate immobilization in captive Rocky Mountain elk (Cervus elaphus nelsoni), *J. Wild Dis.*, 32, 234–239, 1996.

Mioduszewski, R. and Reutter, S., Dissociation of opiate-induced sedation and respiratory depression in ferrets by opiate antagonistic co administration: Potential pharmacological mechanisms, in *Proceedings of the 1991 U.S. Army Chemical Research, Development and Engineering Center Scientific Conference on Chemical Defense Research*, Natick, MA, November 19–22, 1991, AD-A269 546, ADE479569, 1991, pp. 691–698.

Mioduszewski, R., US Patent 5,834,477, 1994.

Monsereenusorn, Y., Kongasamut, S., and Pezalla, P.D., Capsaicin—A literature survey, *CRC Crit. Rev. Toxicol.*, 22, 321, 1982.

Nash, J.B., Doucet, B.M., Ewing, P.L., and Emerson, G.A., Effects of cyanide antidotes and inanition on acute lethal toxicity of malononitrile in mice, *Arch. Int. Pharmacodyn.*, 84, 395, 1950.

National Academy of Sciences, *Possible Long-Term Health Effects of Short-Term Exposure to Chemical Agents*, Vol. 2. Cholinesterase Reactivators, Psychochemicals, and Irritants & Vesicants, National Academy Press, Washington, DC, 1984.

National Institutes of Health (National Toxicology Program), Toxicology and Carcinogenesis Studies of 2-Chloroacetophenone (CAS 532-27-4) in F344/N Rats and B6C3F I Mice (Inhalation Studies), NTP TR 379, March 1990, National Toxicology Program, Research Triangle Park, NC, 1990a.

National Institutes of Health (National Toxicology Program), Toxicology and Carcinogenesis Studies of CS2 (94% o-chlorobenzalmalononitrile) (CAS 2698-41-1) in F344/N Rats and B6C3F1 Mice (Inhalation Studies), NTP TR 377, March 1990, National Toxicology Program, Research Triangle Park, NC, 1990b.

National Research Council, Committee on Toxicology, Cholinesterase reactivators, psychochemicals, and irritants and vesicants, in, *Possible Long-Term Health Effects of Short-Term Exposure to Chemical Agents*, Vol. 2. Washington, DC: National Academy Press, 1984, pp. 221–119.

Obal Jr., F., Obal, F., Benedek, G., and Jansco-Gabor, A., Central and peripheral impairment of thermoregulation after capsaicin treatment, *J. Therm. Biol.*, 8, 203–206, 1983.

Oertel, B.G., Tran, P.V., Bradshaw, M.H., Angst, M.S., Schmidt, H., Johnson, S., Greer, J.J., Geisslinger, G., Varney, M.A., and Lotsch, J., Selective antagonism of opioid-induced ventilatory depression by an ampakine molecule in humans without loss of opioid analgesia, *Nature*, 87(2), 202–211, February 2010.

Owens, E.J. and Punte, C.L., Human respiratory and ocular irritation studies utilizing o-chlorobenzylidene malononitrile aerosols, *Am. Ind. Hyg. Assoc. J.*, 24, 262–264, 1963.

Owens, E.J., Weimer, J.T., Ballard, T.A. et al., Ocular, Cutaneous, Respiratory, *Intratracheal Toxicity of Solutions of CS and EA 3547 in Glycol and Glycol Ether in Animals*. Edgewood Arsenal, MD: Edgewood Arsenal Medical Research Laboratory, 1970. Technical Report EATR 4446.

O'Neill, T.P., Mechanism of capsaicin action: Recent learnings, *Respir. Med.*, 85, 35, 1991.

Paradowski, M., Metabolism of toxic doses of o-chlorobenzylidene malononitrile (CS) in rabbits, *Polish J. Pharmacol. Pharm.*, 31, 563, 1979.

Park, S. and Giammona, S.T., Toxic effects of tear gas on an infant following prolonged exposure, *Am. J. Dis. Child.*, 123, 245, 1972.

Patai, S. and Rappaport, Z., Nucleophilic attacks on carbon-carbon double bonds. Part II. Cleavage of arylmethylenemalonitriles by water in 95% ethanol, *J. Chem. Soc. (Lond.)*, Part 1 (71), 383, 1962.

Pattle, R.E., Schock, C., Dirnhuber, P., and Creasy, J.M., Lung surfactant and organelles after an exposure to dibenzoxazepine (CR), *Br. J. Exp. Pathol.*, 55, 213, 1974.

Porszasz, R. and Szolesanyi, J., Circulatory and respiratory effects of capsaicin and resiniferatoxin on guinea pigs, *Acta Biochem. Biophys. Hung.*, 26, 131, 1991/1992.

Premier Explosives Limited, Mob control device. http://www.pelgel.com/mobcontrol.htm Trinidad Moruga Scorpion Hottest Pepper 2012. Accessed April 24, 2014.

Punte, C.L., Gutentag, P.J., Owens, E.J., and Gongwer, L.E., Inhalation studies with chloroacetophenone, diphenylaminochloroarsine, and pelargonic morpholide. II. Human exposures, *Am. Ind. Hyg. Assoc. J.*, 23, 199, 1962a.

Punte, C.L., Owens, E.J., and Gutentag, P.J., Exposures to ortho-chlorobenzylidene malononitrile, *Arch. Environ. Health*, 6, 72, 1963.

Punte, C.L., Weimer, J.T., Ballard, T.A., and Wilding, J.L., Toxicologic studies on o-chlorobenzylidene malononitrile, *Toxicol. Appl. Pharmacol.*, 4, 656, 1962b.

Punte, C.L., Owens, E.J., and Gutentag, P.J., Exposures to orthochlorobenzylidene malononitrile. Controlled human exposures. *Arch. Environ. Health*, 6, 366–374, 1963.

Reitveld, E.C., Delbressine, L.P., Waegemaekers, T.H., and Seutter-Berlage, F., 2-Chlorobenzylmercapturic acid, a metabolite of the riot control agent 2-chlorobenzylidene malononitrile (CS) in the rat, *Arch. Toxicol.*, 54, 139, 1983.

Ren, J., Ding, X., Funk, G.D., and Greed, J.J., Ampakine CX717 protects against fentanyl-induced respiratory depression and lethal apnea in rats, *Anesthesiology*, 110, 1364–1370, 2009.

Rengstorff, R.H., Petrali, J.P., Merson, M., and Sim, V.M., The effect of the riot control agent dibenz (b,f)-1:4-oxazepine (CR) in the rabbit eye, *Toxicol. Appl. Pharmacol.*, 34, 45, 1975.

Riches, J.R., Read, R.M., Black, R.M., Cooper, N.J., and Timperley, C.M., Analysis of clothing and urine from Moscow theatre siege casualties reveals carfentanil and remifentanil use, *J. Anal. Toxicol.*, 36(9), 647–656, November–December 2012.

Rothberg, S., Skin sensitization potential of the riot control agents CA, DM, CN, and CS in guinea pigs, Technical report EATR 4219, U.S. Army Medical Research Laboratory, Edgewood Arsenal, MD, 1969, 19pp.

Royer, J. and Gainet, F., Ocular effects of ethyl bromoacetate tear gas, *Bull. Soc. Ophthalmol. (France)*, 73, 1165, 1973.

Russel, J.A. and Lai-Fook, S.J., Reflex bronchoconstriction induced by capsaicin in the dog, *J. Appl. Physiol.*, 47, 961, 1979.

Salem, H., Issues in chemical and biological terrorism, *Int. J. Toxicol.*, 22, 465–471, 2003.

Salem, H., Olajos, E.J., and Katz, S.A., Riot control agents, in *Chemical Warfare Agents: Toxicity at Low Levels*, Somani, S.M. and Romano, Jr. J.A., eds., CRC Press, Boca Raton, FL, 2001.

Schmid, E., Bauchinger, M., Ziegler-Skylakaksis, K., and Andrae, U., 2-Chloro-benzylidene malononitrile (CS) causes spindle disturbances in V79 Chinese hamster cells, *Mutat. Res.*, 226, 133, 1989.

Sidell, F.R., Riot control agents, in *Textbook of Military Aspects of Chemical and Biological Agents*, Washington, DC: Office of the Surgeon General, US Army, TMM Publications, Bordon Institute, pp. 307–324, 1997.

Stern, J., Weil-Malherbe, H., and Green, R.H., The effects and the fate of malononitrile and related compounds in animal tissues, *Biochem. J.*, 52, 114, 1952.

Stewart, H.L., Dunn, T.B., Snell, K.C., and Deringer, M.K., Tumors of the respiratory tract, in *WHO Monographs on the Pathology of Laboratory Animals*, Vol. 1. The Mouse, Turusov, S.U., ed., International Association for Research on Cancer (IARC), Lyons, France, 1979, p. 251.

Striker, G.E., Streett, C.S., Ford, D.F., Herman, L.H., and Welland, D.R., A clinicopathological study of the effects of riot-control agents in monkeys. V. Low concentrations of diphenylaminochloroarsine (DM) or o-chlorobenzylidene malononitrile (CS) for extended periods, Technical Report 4072, U.S. Army Medical Research Laboratory, Edgewood Arsenal, MD, 1967.

Suzuki, T. and Iwai, K., Constituents of red pepper species: Chemistry, biochemistry, pharmacology and food science of the pungent principal of capsicum species, in *The Alkaloids*, Chap. 4, 23, Brossi, A., ed., Academic Press, Orlando, FL, 1984.

Swearengen, T.F., *Tear Gas Munitions*, Charles C. Thomas, Springfield, IL, 1966.

Teranishni, R., Keller, U., Flath, R.A., and Mon, T.R., Comparison of batchwise and continuous steam distillation-solvent extraction recovery of volatiles from oleoresin capsicum, African type (*Capsicum frutescens*), *J. Agric. Food Chem.*, 28, 156, 1980.

The White House, Office of the Press Secretary, Statement by President Clinton to the Senate of the United States, Subject: Riot Control Agents, Press Release Dated June 23, 1994, The White House, Washington, DC.

Thomas, R.J., Smith, P.A., Rascona, D.A., Louthan, J.D., and Gumpert, B., Acute pulmonary effects from o-chlorobenzylidenarnalonitrile "tear gas": A unique exposure outcome unmasked by strenuous exercise after a military training event, *Mil. Med.*, 167, 136–139, 2002.

Toh, C.C., Lee, T.S., and Kiang, A.K., The pharmacological actions of capsaicin and analogues, *Br. J. Pharmacol.*, 10, 175, 1955.

U.S. Congress, Investigation into the activities of federal law enforcement agencies toward the Branch Davidians, Union Calendar 395, House of Representatives Report 104–749, U.S. Government Printing Office, Washington, DC, 1996.

U.S. General Accounting Office (GAO), Use of U.S. Manufactured Tear Gas in the Occupied Territories, GAO/NSIAD-89–128, Washington, DC, 1989.

Upshall, D.G., Effects of o-chlorobenzylidene malononitrile (CS) and the stress of aerosol inhalation upon rat and rabbit embryonic development, *Toxicol. Appl. Pharmacol.*, 24, 45–59, 1973.

Upshall, D.G., Riot control smokes: Lung adsorption and metabolism of peripheral sensory irritants, in *Clinical Toxicology*, Duncan, W.A. and Leonard, B.J., eds., Excerpta Medica, Amsterdam, the Netherlands, 1977, p. 121.

Upshall, D.G., The effects of dibenz (b,f)-1,4-oxazepine (CR) upon rat and rabbit embryonic development, *Toxicol. Appl. Pharmacol.*, 29, 301, 1974.

USAMRICD, *Medical Management of Chemical Casualties Handbook*, 3rd edn., MCMR-UV-ZM, Aberdeen Proving Ground, Aberdeen, MD, 2000.

Von Daniken, A., Friederich, U., Lutz, W.K., and Schlatter, C., Tests for mutagenicity in salmonella and covalent binding to DNA and protein in the rat of the riot control agent o-chlorobenzylidene malononitrile (CS), *Arch. Toxicol.*, 49, 15, 1981.

West, C.J., The history of poison gases, *Science*, 49, 412–417, 1919.

Wild, D., Eckhardt, K., Harnasch, D., and King, M.T., Genotoxicity study of CS (ortho-chlorobenzylidene malononitrile) in salmonella, drosophila, and mice, *Arch. Toxicol.*, 54, 167, 1983.

Yunbao, L. and Nair, M.G., Capsaicinoids in the Hottest Pepper Bhut Jolokia and its antioxidant and antiinflammatory activities, *Nat. Prod. Commun.*, 5(1), 91–94, 2010.

Zeiger, E., Anderson, B., Haworth, S., Lawlor, T., Mortelmans, K., and Speck, W., Salmonella mutagenicity test. III. Results from the testing of 255 chemicals, *Environ. Mutagen.*, 9(Suppl. 9), 1, 1987.

12 Incapacitating Agents

Joseph L. Corriveau and Michael Feasel

CONTENTS

12.1 INTRODUCTION

Incapacitating agents (ICAs) are pharmaceutical-based agents developed as nonlethal weapons for law enforcement or warfare purposes. Notable examples of ICAs that were developed for use as nonlethal weapons include depressants such as cannabinols, psychedelics such as D-lysergic acid diethylamide (LSD), deliriants such as BZ, and opioids such as fentanyls (Field Manual 3-11.9, 2005). The intent to use these ICAs in weapons is, for the most part, a relic of the Cold War. For example, it has been decades since depressants and psychedelics were researched, and promptly dismissed, as potential ICAs. BZ and fentanyls, on the other hand, more advanced in their development as ICAs, and fentanyls were actually used in 2002 for law enforcement purposes. Further, the rise in fentanyl-related deaths resulting from smoking transdermal fentanyl patches (e.g., Duragesic) warrants the inclusion of these opioids in this discussion of inhalation toxicology.

The inhalation toxicology of BZ and fentanyls is significant because aerosols generated by ICA dissemination devises are likely to be respirable. Indeed, the US M43 cluster bomb and M44 generator cluster devise were pyrotechnic munitions designed to generate aerosolized BZ (Smart, 1997). These BZ-filled weapons were demilitarized during the Cold War because they were assessed to be ineffective. Yet the reality of ICA-filled weapons still exists. Indeed, a forensic analysis has revealed that a chemical aerosol containing carfentanil and remifentanil was used by the Russian Federal Security Service (FSB) to bring an end to the Chechen terrorists' siege of some 800 hostages at the Dubrovka Theater in Moscow on October 26, 2002 (Riches et al., 2012). The FSB rescued hundreds of hostages, but at a tragic price. There were 125 hostages that died from a combination of exposure to lethal doses of the carfentanil/remifentanil cocktail and/or inadequate medical treatment following the rescue. Interest in pharmaceutical-based agents as potential ICAs appears to have proliferated to other nations.

The Defence Research and Development Establishment (DRDE) in Gwalior, India, has reported on studies related to the thermal generation of fentanyl aerosols, to include the thermal stability of fentanyls during flash pyrolysis (Manral et al., 2009a), as well as the vapor pressure and related thermodynamic properties of fentanyl (Gupta et al., 2008). In a related DRDE study on the effect of exposure of fentanyl aerosols (generated by a nebulizer using solutions of fentanyl in DMSO) in mice on respiratory depression, Manral et al. (2009b) concluded that fentanyls should not be used as ICAs because of their low margin of safety.

At the *Third European Symposium on Non-Lethal Weapons*, Hess et al. (2005) reported on a series of studies using fentanyl analogs and other pharmaceutical-based agents with the potential to safely *immobilize*. For example, one *immobilization* study was described in which the triple combination of the sedatives midazolam (dose of 0.08 μg/kg) and dexmedetomidine (dose 1.5 μg/kg) in combination with the anesthetic ketamine (dose 3 μg/kg) were introduced by syringe to 10 volunteer nurses. Among the parameters measured were onset of effect, immobilization time, respiration depression, and recovery. In another study, an immobilizing dose of remifentanil (50 μg/kg) in rabbit was demonstrated to produce apnea as a side effect within 2–3 min of administration. Subsequent studies were done using the coadministration of an immobilizing dose of either remifentanil or alfentanil with the antagonist naloxone to mitigate respiratory depression. From these and other studies, the authors concluded that many agents used in anesthesiology can be used in *pharmacological nonlethal weapons*.

12.1.1 Definition of ICAs

There is a lack of consensus on the definition of ICAs. In particular, there has been confusion in the literature regarding the difference between the ICAs and riot control agents (RCAs) such as the tearing agents O-chlorobenzylidene malononitrile (CS) or capsaicin (OC). The NATO Glossary of Terms and Definitions (AAP-6, 2008) provides definitions that allow clear discrimination between ICAs and RCAs. ICAs are defined as "a chemical agent that produces temporary disabling conditions which (unlike those caused by RCA) can be physical or mental and persist for hours or days after exposure to the agent has ceased. Medical treatment, while not required, facilitates a more rapid recovery." RCAs, on the other hand, are defined as "a chemical not listed in a schedule of the 1993 Chemical Weapons Convention (CWC), which can rapidly produce irritant or disabling physical effects that disappear shortly after termination of exposure." Further discrimination between ICAs and RCAs that must be considered is the toxicity of the agents. Indeed, RCAs are generally accepted as *nonlethal* agents, whereas ICAs can be lethal, as evidenced by the large number of accidental deaths resulting from the carfentanil/remifentanil cocktail used to end the hostage crisis at the Russian opera house in 2002. In short, ICAs are not *nonlethal* agents and should not be confused with RCAs that are used for law enforcement purposes.

12.2 BZ AS AN ICA

BZ (3-quinuclidinyl benzilate) is an anticholinergic compound discovered in the 1950s as a possible antispasmodic for the treatment of gastrointestinal issues (Kirby, 2006). The structural formula and key chemical and physical properties of BZ are provided in Figure 12.1 and Table 12.1, respectively. During the earlier Cold War years of the twentieth century, BZ was adapted for use as an ICA for warfare purposes because of its effect of producing delirium at a very low dosage with a high safety margin. The human estimated incapacitation ICt_{50} is reported to be 112 mg min/m^3 and the LCt_{50} to be 200,000 mg min/m^3 (Salem et al., 2001). BZ was weaponized by the United States during the Cold War, but demilitarized years before the United States ratified the CWC. The US M43 and M44 cluster munitions contained BZ as a white crystalline powder, which was intended to be pyrotechnically disseminated as an aerosol (Smart, 1997). Today, BZ is classified as a CWC Schedule 2 toxic chemical that is prohibited from use in a chemical weapon, and its only legitimate use is as a pharmacological tool, the muscarinic antagonist QNB.

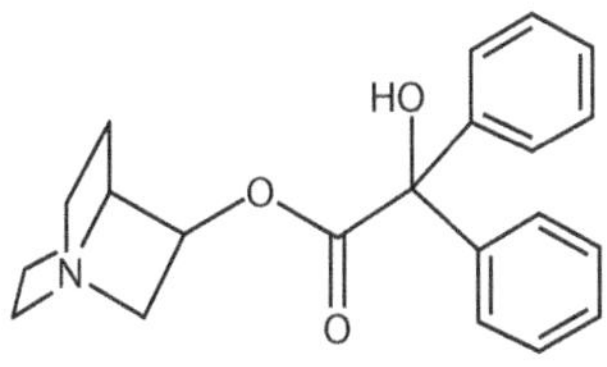

FIGURE 12.1 Chemical structure of BZ.

TABLE 12.1
Select Chemical and Physical Properties of BZ

Chemical formula	$C_{21}H_{23}NO_3$
Molecular weight	337.4
Physical state	Solid
Melting point	167.5°C
Boiling point	320°C
Relative vapor density	11.6
Solubility	Slightly soluble in water and soluble in most organic solvents
Vapor pressure	0.03 mm Hg at 70°C

Source: Hoenig, S.L., Incapacitating agents: Agent BZ, in *Compendium of Chemical Warfare Agents*, Springer, New York, 2007, pp. 73–76.

12.2.1 TOXICODYNAMICS AND CLINICAL EFFECTS

BZ as an ICA would be expected to be disseminated as an aerosol that would enter the body by inhalation or percutaneous absorption (see Ketchum et al., 1997; USAMRICD, 2000; Ketchum and Salem, 2008). BZ acts by blocking the action of acetylcholine on the central and peripheral nervous systems. It is a tertiary amine and crosses the blood–brain barrier. BZ on acute exposure increases both heart and respiratory rates, dilates the pupils, and causes paralysis of the eye muscles necessary for near focusing. It also causes dry mouth and skin, elevates body temperature, impairs coordination, and causes flushing of the skin, hallucinations, stupor, forgetfulness, and confusion. Within 15 min to 4 h following exposure, the principal effects are dizziness, involuntary muscle movements, near-vision difficulty, and total incapacitation. From 6 to 10 h after exposure, the effects are psychotropic, and full recovery is expected after 4 days.

The peripheral nervous system effects of BZ are considered as understimulation of the end organs. This decreased stimulation of eccrine and apocrine sweat glands in the skin results in dry skin and a dry mouth. The reduction in the ability to dispel heat by evaporative cooling decreases sweating, and the compensatory cutaneous vasodilation causes the skin to become warm. This is similar to the atropine flush. The decreased heat loss also results in an increased core temperature.

The peripheral effects usually precede the central nervous system effects of BZ, which includes dose-dependent *mad as a hatter* mental changes. These effects fluctuate between a conscious state and delirium that ranges from drowsiness to coma. Disorientation, decreased social restraints, inappropriate behavior, and decreased short-term memory are common. Speech becomes slurred and indistinct.

12.2.2 INHALATION TOXICOLOGY

Recently, the National Advisory Council for Acute Exposure Guideline Levels for Hazardous Substances and the Committee on Acute Exposure Guideline Levels (AEGLs) utilized available human and animal inhalation toxicity data to establish AEGLs for BZ (National Research

TABLE 12.2
AEGL Values for BZ

Classification	10 min	30 min	1 h	4 h	8 h	End Point (Reference)
AEGL-1	NR[a]	NR	NR	NR	NR	Insufficient data
AEGL-2	0.067[b]	0.022	0.011	NR	NR	Estimated threshold (20 mg min/m³) for incapacitation in human volunteers (Ketchum et al., 1967)
AEGL-3	1.2	0.41	0.21	NR	NR	Estimated lethality threshold (3700 mg min/m³) in monkeys (US Dept Army, 1974)

[a] Not recommended.
[b] mg/m³.

Council, 2013). AEGLs represent threshold exposure levels for the general public and are applicable to emergency exposures. The levels are determined for 10 min, 30 min, 1 h, 4 h, and 8 h exposures. There are three AEGL classifications. AEGL-1 (nondisabling) is an airborne concentration expressed as parts per million (ppm) or mg/m³ of a substance above which it is predicted that the general population, including susceptible individuals, could experience discomfort, irritation, or certain asymptomatic nonsensory effects. Effects are not disabling and are transient and reversible upon cessation of exposure. AEGL-2 (disabling) is an airborne concentration above which individuals could experience irreversible or other long-lasting adverse health effects. AEGL-3 (lethal) is an airborne concentration above which individuals could experience life-threatening adverse health effects or even death. The AEGL values for BZ are shown in Table 12.2. There were insufficient data to develop AEGL-1 values for BZ.

Inhalation exposure experiments conducted by Ketchum et al. (1967) with human volunteers were used to determine an ICt_{50} value (concentration–time product causing incapacitation in 50% of the test subjects) of 60.1 mg min/m³. For AEGL-2 calculations, a one-third reduction of the ICt_{50} (20 mg min/m³) was used as the threshold for incapacitating effects (see Table 12.2).

No data on the lethality of BZ in humans after inhalation exposure are available. However, LCt_{50} (concentration–time product causing lethality in 50% of the test subjects) values have been reported for the monkey, dog, mouse, rat, rabbit, and guinea pig (US Department of the Army, 1974). Details of the experiments are not available. AEGL-3 values (see Table 12.2) were derived using 3700 mg min/m³ as the point of departure. This value was determined by reducing the LCt_{50} for monkey (37,000 mg min/m³) 10-fold. The LCt_{50} for monkey was neither the highest nor the lowest of the six species tested, but the monkey was considered a better model for aerosol inhalation exposure in humans than the other species.

12.2.3 CLINICAL MANAGEMENT

The BZ clinical profile closely resembles that of atropine and scopolamine, differing in the duration of action and potency (Ketchum et al., 1997; USAMRICD, 2000; Ketchum and Salem, 2008). BZ toxicity can occur by inhalation, ingestion, or percutaneous exposure routes. Signs and symptoms of exposure include hallucinations; agitation; mydriasis (dilated pupils); blurred vision; dry, flushed skin; urinary retention; ileus; tachycardia; hypertension; and elevated temperature (>101°F). The onset of incapacitation is dose dependent; therefore, it might occur as early as 1 h after exposure and continue up to 48 h.

Physostigmine salicylate can reverse the delirium induced by BZ and enable individuals to achieve normal patterns of behavior in about 8 h (Ketchum and Salem, 2008).

BZ is excreted via the urine following metabolic processing in the liver. A gas chromatography/ mass spectroscopy-based analytical method has been developed to detect BZ and its major metabolites in urine (Byrdt et al., 1992).

12.3   FENTANYLS AS ICAS

Fentanyl represents a major breakthrough in synthetic opioid drugs for use in medicine as an analgesic and anesthetic (Salem et al., 2006). It was first patented by Janssen in 1963. There is now a variety of fentanyl analogs used for the management of chronic pain, as an adjunctive anesthetic agent and as an immobilizing agent for large animals. Examples include fentanyl that is delivered by transdermal patches (e.g., Duragesic), oral lozenges (e.g., Actiq), and intravenous/intramuscular injections (e.g., Sublimaze); sufentanil and short-acting remifentanil that are usually delivered by the intravenous route; and carfentanil, the immobilizing drug of large animals that is delivered by intramuscular injection, to include small darts (e.g., Wildnil). See chemical structures in Figure 12.2.

Worthy of mention, fentanyl and a number of its analogs are abused as recreational drugs. Illegally manufactured analogs such as alpha-methylfentanyl (street name *China White*) are sold on the black market, and fentanyl patches are a common source for illicit use. Indeed, fentanyl patches have been abused through the intravenous injection of patch contents (Lelleng et al., 2004), oral ingestion (Arvanitis and Satonik, 2002), rectal insertion (Coon et al., 2005), and volatilization followed by inhalation (Marquardt and Tharratt, 1994).

In addition to their medical and veterinarian applications, fentanyl analogs have been considered for use as an aerosolized ICA that would act as a fast nonlethal immobilizing agent (see Riches et al., 2012). Indeed, fentanyl analogs are powerful synthetic opioids, some with relatively wide

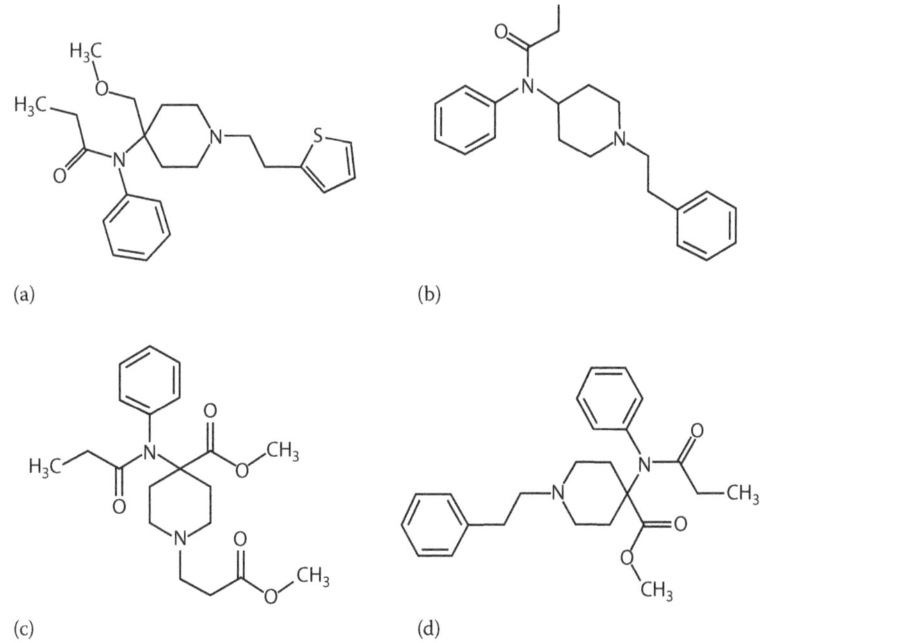

(a)   (b)

(c)   (d)

FIGURE 12.2   Chemical structure of select fentanyl derivatives: (a) sufentanil (SAN:BAN:INN) (RN: 56030-54-7), (b) fentanyl (RN: 437-38-7), (c) remifentanil (RN: 132875-61-7), and (d) carfentanil (RN: 59708-52-0).

TABLE 12.3
Effectiveness and Safety of Select Fentanyls[a]

Compound	Lowest ED_{50}	LD_{50}	Safety Margin	Potency Ratio
Pethidine	6.2	29.0	4.8	1
Fentanyl	0.01	3.1	277.0	560
Alfentanil	0.04	48.0	1,080.0	140
Lofentanil	0.0006	0.07	112.0	10,400
Sufentanil	0.0007	18.0	26,716.0	9,200
Carfentanil	0.0004	3.1	8,460.0	16,600

Source: Janssen, P.A.J., The development of new synthetic narcotics, in *Opioids in Anesthesia*, Estafanous, F.G., ed., Butterworth Publishers, Stoneham, MA, 1984, pp. 37–44.

[a] Lowest ED_{50} values (mg/kg) from tail-withdrawal test in rats, and LD_{50} values (mg/kg), safety margins, and potency ratio after intravenous administration.

margins of safety[*] as shown in Table 12.3 (Janssen, 1984). For example, based on test results from rats, sufentanil and carfentanil have potency ratios of 9,200 and 16,600, respectively, as compared to pethidine (i.e., Demerol). Their respective safety margins are 26,716 and 8,460, respectively, and Wax et al. (2003) reported a safety margin of 33,000 for remifentanil. Nonetheless, the effects of fentanyl exposure are dosage dependent; hence, there is a risk of overdosage that can lead to inhibition of respiratory control centers within the brain stem, resulting in decreased respiration frequency and tidal volume, dyspnea, apnea, and finally the cessation of breathing. This side effect of opioid overdosage was dramatically highlighted by the death of the 125 hostages that were exposed to an aerosol containing carfentanil and remifentanil in the Dubrovka Theater in Moscow (Riches et al., 2012).

Studies have been conducted to determine if coadministration of an immobilizing dose of a fentanyl analog with an antagonist can mitigate respiratory depression. Hess et al. (2005) reported on the coadministration of an immobilizing dose (via intramuscular injection) of either remifentanil (50 µg/kg) or alfentanil (150 µg/kg) with the antagonist naloxone (0.1 mg) in rabbit. In another study, Mioduszewski et al. (1990) reported on the pharmacological effects of coadministered (via intraperitoneal injection) carfentanil and the antagonist naltrexone in ferrets. In both studies, coadministration of the opioid (agonist) and antagonist mitigated adverse side effects.

12.3.1 TOXICODYNAMICS AND CLINICAL EFFECTS

Fentanyl and its analogs can be absorbed into the body via inhalation, oral exposure or ingestion, or skin contact. As an ICA, these compounds can be disseminated as an aerosol that enters the body by inhalation. A number of studies, to include those of Mather et al. (1998) and McLeod et al. (2012), demonstrated that the pharmacokinetic profile of an inhaled fentanyl aerosol for an effective treatment of pain is comparable to intravenous administration. Further, Rabinowitz et al. (2004) demonstrated that thermally generated aerosols of fentanyl can be utilized for treating a variety of acute and episodic conditions, including breakthrough pain. Although it is unclear how the carfentanil/remifentanil cocktail was generated by the Russian Special Forces to immobilize the Chechen terrorists in 2003, it is apparent that a respirable aerosol was utilized.

---

[*] Margin of safety is calculated as the ratio of the dose that is just within the lethal range (LD_{01}) to the dose that is 99% effective (ED_{50}). The margin of safety = LD_{01}/ED_{99}.

Fentanyl and its analogs are strong synthetic piperidine opioid agonists that interact primarily with μ-opioid receptors located in the brainstem and medial thalamus (Trescot et al., 2008). Their pharmacological activity is the characteristic of opioids, and they produce all of the effects of heroin, including analgesia, euphoria, miosis, and respiratory depression. Due to their high lipid solubility, regardless of the route of administration, fentanyls reach the brain very quickly, thus providing a very fast onset of action. Mather et al. (1998) reported that subjects experienced the effects of fentanyl immediately following both intravenous and aerosolized pulmonary drug delivery. Noteworthy, however, is that transdermal formulations used in fentanyl patches have a lag time of 6–12 h to the onset of action following application (Trescot et al., 2008).

12.3.2 INHALATION TOXICOLOGY

Unlike BZ, AEGLs have not been established for fentanyl and its analogs. Nonetheless, since the pharmacokinetics of fentanyl are similar whether administered by the intravenous or aerosolized pulmonary route, it seems reasonable to use data from intravenous-based delivery studies to make some general observations about the toxicity of fentanyl and its analogs in humans. The reported LD_{50} for fentanyl via the intravenous route in rat is 3.1 mg/kg (Janssen, 1984) and in the monkey is 0.03 mg/kg (Sutlovic et al., 2007). These results reflect known interspecies differences regarding sensitivity to fentanyl exposure, with the nonhuman primates being more sensitive. Further, since the monkey is considered a better model for aerosol inhalation exposure in humans than other species, one would estimate the human LD_{50} to be 0.03 mg/kg or 2.1 mg for a 70 kg person. By comparison, therapeutic doses of fentanyl aerosols administered for the treatment of pain is generally 100–300 μg (Mather et al., 1998; MacLeod et al., 2012). Recent studies have also shown that in healthy human volunteers, the pharmacokinetic profile of inhaled fentanyl almost perfectly mimics PK of IV administration (MacLeod et al., 2012). Thus, with near-identical pharmacokinetic profiles, inhalation toxicity is thought to be highly comparable to IV toxicity based on identical doses.

12.3.3 CLINICAL MANAGEMENT

The most common serious adverse reactions resulting from overdosage of fentanyl or one of its analogs are respiratory depression, apnea, rigidity, and bradycardia. If untreated, then respiratory arrest, circulatory depression, or cardiac arrest can occur. Other adverse reactions that have been reported are hypertension, hypotension, dizziness, blurred vision, nausea, emesis, laryngospasm, and diaphoresis.

Narcotic antagonists such as nalmefene, naltrexone, and naloxone have clinical application in the emergency treatment of fentanyl overdosage (Trescot et al., 2008). These antagonists displace either previously assimilated opioids from their receptor sites or, if administered prior to the narcotic, will preclude the narcotic agonist from acting at these sites.

Fentanyl and most fentanyl analogs are excreted via the urine following metabolic processing in the liver, with the exception of remifentanil. Remifentanil is more easily broken down by circulating blood esterases due to the presence of the ester bond in this analog's structure (Davis et al., 2002). This makes the remi-analog more tolerable to hepatocompromised patients and also contributes to its short half-life. A liquid chromatography/mass spectrometry-based analytical method has been developed to detect fentanyl and a number of its analogs and major metabolites in urine (Wang and Bernert, 2006).

12.4 SUMMARY AND DISCUSSION

In summary, a number of compounds were examined during the Cold War as potential *nonlethal* ICAs, to include depressants, psychedelics, deliriants, and opioids. We examined BZ in this chapter because the deliriant was actually weaponized and stockpiled by the United States. BZ-filled pyrotechnic-type weapons were intended to disseminate the incapacitating chemical via respirable aerosols. The United States abandoned this ICA weapon concept and destroyed its stockpile before the Cold War ended (Smart, 1997).

Opioids were also examined in this chapter in large part because of the use of the remifentanil/carfentanil cocktail by the FSB to end the Chechen terrorists' siege of the opera house. This historic event in counterterrorism may have been a tipping point regarding the decision by some to investigate opioid-based ICAs, as exemplified by the research of Hess et al. (2005) in the Czech Republic and Manral et al. (2009) in India. Furthermore, in December 2011, the European Court of Human Rights ruled that the *gas* used by the Russian authorities was not intended to kill either the terrorists or the hostages (Vandova, 2013). In other words, it appears that the Russian use of an ICA for law enforcement purposes was found to be appropriate because of its intended use.

There is an alternate view on the feasibility of using opioids as ICAs. For example, a recent British Medical Association report (BMA Board of Science, 2007) refers to ICAs as "drugs as weapons." The authors of the report argue that drugs (e.g., fentanyl and its analogs) as weapons that are intended to incapacitate in a tactical scenario without the risk of generating a significant mortality do not exist. In short, the authors argue that ICAs are in fact *lethal* as opposed to *nonlethal*. Indeed, fentanyl and its analogs are powerful opioids used therapeutically under very controlled circumstances that do not exist in a tactical scenario. For example, if a person were to receive an adequate incapacitating dose via aerosol, the incapacitated person would remain in the presence of the drug where he/she would continue to inhale more agent and start to display adverse side effects, perhaps even death, from overdosing.

Finally, some would argue that eventually science will bring forth compounds that have the desired effects of an ICA but lacking adverse side effects, for example, respiratory depression. Recently, Majumdar et al. (2011) used μ-opioid receptor knockout mice to study the opioid action of iodobenzoylnaltrexamide (IBNtxA), a naltrexone analog. IBNtxA is a potent analgesic that is 10 times more potent than morphine. IBNtxA was found to be a potent analgesic yet lacking adverse side effects such as respiratory depression. Evidence indicates that IBNtxA acted through a truncated, six-transmembrane variant of the G protein-coupled μ-opioid receptor. In addition, a new class of compounds could potentially help alleviate the undesirable respiratory depression associated with fentanyl use. In human volunteers, ampakines have shown to reverse respiratory depression without losing the other desirable effects of the drug (Oertel et al., 2010). In rats, these compounds have shown to have an excitatory effect on the AMPAergic neurons in the pre-Botzinger Complex of the brain, a locale that is highly involved in spontaneous respiration (Ren et al., 2009). By stimulating these non-opioid-affected neurons, respiratory depression is overcome by AMPAergic overstimulation, and breathing continues. Innovative approaches such as these will contribute to advances in the design and development of new opioids, no doubt. But whether or not there is ever a *holy grail* drug found to be palatable as a truly nonlethal ICA may be a bridge too far.

QUESTIONS

Fill in the blank

1. _____ is an airborne concentration above which individuals could experience life-threatening adverse health effects or even death.
   a. Acute Exposure Guideline Level-1
   b. Acute Exposure Guideline Level-2
   c. Acute Exposure Guideline Level-3
   d. Acute Exposure Guideline Level-4
   Answer: c

2. The pharmacokinetics of fentanyl are similar whether administered by the intravenous or _____ route.
   a. Percutaneous
   b. Oral
   c. Aerosolized pulmonary
   d. All of the above
   Answer: c
3. The aerosolized chemical cocktail used to end the Chechen terrorists' siege of some 800 hostages at the Dubrovka Theater in Moscow was _____.
   a. Fentanyl and naloxone
   b. Remifentanil and naloxone
   c. Midazolam and dexmedetomidine
   d. Remifentanil and carfentanil
   Answer: d
4. Fentanyl and its analogs should not be considered as nonlethal incapacitating agents because _____.
   a. The effects are dose dependent
   b. An adverse side effect is respiratory depression
   c. Fentanyls are very potent opioids
   d. All of the above
   Answer: d

DISCLAIMER

The views and opinions expressed in this chapter are those of the author and should not be construed as an official Department of the Army position, policy, or decision, unless so designated by other official documentation. Citation of trade names in this chapter does not constitute an official Department of the Army endorsement or approval of the use of such commercial items.

REFERENCES

Arvanitis, M.L. and Satonik, R.C., Transdermal fentanyl abuse and misuse, *Am. J. Emerg. Med.*, 20(1): 58–59, 2002.

British Medical Association Board of Scientists, The use of drugs as weapons: The concerns and responsibilities of healthcare professionals, British Medical Association, London, U.K., 2007.

Byrdt, G.D., Paule, R.C., Sander, L.C., Sniegoski, L.T., White, E., and Bausum, H.T., Determination of 3-Quinuclidinyl Benzilate (QNB) and its major metabolites in urine by isotope dilution gas chromatography/mass spectrometry, *J. Anal. Toxicol.*, 16(3): 182–187, 1992.

Coon, T.P., Miller, M., Kaylor, D., and Jones-Spangle, K., Rectal insertion of fentanyl patches: A new route of toxicity, *Ann. Emerg. Med.*, 46(5): 473, 2005.

Davis, P.J., Stiller, R.L., Wilson, A.S., McGowan, F.X., Egan, T.D., and Muir, K.T., In vitro remifentanil metabolism: The effects of whole blood constituents and plasma butyrylcholinesterase, *Anesth. Analg.*, 95: 1305–1307, 2002.

Field Manual 3-11.9, Potential military chemical/biological agents and compounds, 2005.

Gupta, P.K., Ganesan, K., Gutch, P.K., Manral, L., and Dubey, D.K., Vapor pressure and enthalpy of vaporization of fentanyl, *J. Chem. Eng. Data*, 53: 841–845, 2008.

Hess, L., Schreiberova, J., and Fusek, J., Pharmacological non-lethal weapons, in *Third European Symposium on Non-Lethal Weapons*, Ettlingen, Germany, May 2005.

Hoenig, S.L., Incapacitating agents: Agent BZ, in *Compendium of Chemical Warfare Agents*, Springer, New York, pp. 73–76, 2007.

Janssen, P.A.J., The development of new synthetic narcotics, in *Opioids in Anesthesia*, Estafanous, F.G., ed., Butterworth Publishers, Stoneham, MA, pp. 37–44, 1984.

Ketchum, J.S. and Salem, H., Incapacitating agents, in *Textbook of Military Medicine: Medical Aspects of Chemical Warfare*, Tuorinsky, S.D. ed., Office of the Surgeon General at TMM Publications, Borden Institute, Walter Reed Army Medical Center, Washington, DC, 2008.

Ketchum, J.S. and Sidell, F.R., Incapacitating agents, in *Textbook of Military Medicine; Military Aspects of Chemical and Biological Warfare*, Sidell, F.R., Takafuji, E.T., and Franz, D.R., eds., Office of the Surgeon General at TMM Publications, Borden Institute, Walter Reed Army Medical Center, Washington, DC, pp. 287–305, 1997.

Ketchum, J.S., Tharp, B.R., Crowell, E.B., Sawhill, D.L., and Vancil, M.E., The human assessment of BZ disseminated under field conditions, Edgewood Arsenal Report EATR 4140, US Department of the Army, Medical Research Laboratory, Edgewood Arsenal, MD, 1967.

Kirby, R., Paradise lost; the psycho agents, *The CBW Conventions Bulletin*, May 2006, pp. 2–3.

Lelleng, P.K., Mehlum, L.I., Bachs, L., and Morild, I., Deaths after intravenous misuse of transdermal fentanyl, *J. Forensic Sci.*, 49(6): 1364–1366, 2004.

MacLeod, D.B., Habib, A.S., Ikeda, K., Spyker, D.A., Cassella, J.V., Ho, K.Y., and Gan, T.J., Inhaled fentanyl aerosol in healthy human volunteers: Pharmacokinetics and pharmacodynamics, *Anesth. Analg.*, 115(5): 1071–1077, November 2012.

Majumdar, S., Grinnell, S., Le Rouzic, V., Burgman, M., Polikar, L., Ansonoff, M., Pintar, J., Pan, Y., and Pasternak, G.W., Truncated G protein-coupled mu opioid receptor MOR-1 splice variants are targets for highly potent opioid analgesics lacking side effects, *PNAS*, 108: 19778–19783, 2011.

Manral, L., Gupta, P.K., Suryanarayana, M.V.S., Ganesan, K., and Malhotra, R.C., Thermal behavior of fentanyl and its analogues during flash pyrolysis, *J. Therm. Anal. Calorim.*, 96(2): 531–534, 2009a.

Manral, L., Muniappan, N., Gupta, P.K., Ganesan, K., Malhotra, R.C., and Vijayaraghavan, R., Effect of exposure to fentanyl aerosol in mice on breathing pattern and respiratory variables, *Drug Chem. Toxicol.*, 32(2): 108–113, 2009b.

Marquardt, K.A. and Tharratt, R.S., Inhalation abuse of fentanyl patch, *J. Toxicol. Clin. Toxicol.*, 32(1): 75–78, 1994.

Mather, L.E., Woodhouse, A., Ward, M.E., Farr, S.J., and Rubsamen, R.A., Pulmonary administration of aerosolized fentanyl: Pharmacokinetic analysis of systemic delivery, *Br. J. Clin. Pharmacol.*, 46: 37–43, 1998.

Mioduszewski, R., Reutter, S., and Berg, D., Pharmacological effects of opioid agonist and antagonist coadministered in ferrets, in *Proceedings of the 1989 U.S. Army Chemical Research, Development and Engineering Center Scientific Conference on Chemical Defense Research*, Aberdeen, MD, November 14–17, 1989, pp. 903–909, 1990.

National Research Council, *Acute Exposure Guideline Levels for Selected Airborne Chemicals*, Vol. 14, National Academy Press, Washington, DC, 2013.

NATO Glossary of Terms and Definitions, AAP-6, 2008. Available at http://www.fas.org/irp/doddir/other/nato2008.pdf. Accessed on April, 2014.

Oertel, B.G., Felden, L., Tran, P.V., Bradshaw, M.H., Angst, M.S., Schmidt, H., Johnson, S., Greer, J.J., Geisslinger, G., Varney, M.A., and Lotsch, J., Selective antagonism of opioid-induced ventilatory depression by an ampakin molecule in humans without loss of opioid analgesia, *Nature*, 87(2): 204–211, February 2010.

Rabinowitz, J.D., Wensley, M., Lloyd, P., Myers, D., Shen, W., Lu, A., Hodges, C., Hale, R., Mufson, D., and Zaffaroni, A., Fast onset medications through thermally generated aerosols, *JPET*, 309: 769–775, 2004.

Ren, J., Ding, X., Funk, G.D., and Greer, J.J., Ampakine CX717 protects against fentanyl-induced respiratory depression and lethal apnea in rats, *Anesthesiology*, 110(6): 1364–1370, June 2009.

Riches, J.R., Read, R.W., Black, R.M., Cooper, N.J., and Timperley, C.M., Analysis of clothing and urine from Moscow theatre siege casualties reveals carfentanil and remifentanil use, *J. Anal. Toxicol.*, 36(9): 647–656, 2012.

Salem, H., Ballantyne, B., and Katz, S.A., Inhalation toxicology of riot control agents, in *Inhalation Toxicology*, 2nd edn., Salem, H. and Katz, S.A., eds., CRC Press, Boca Raton, FL, pp. 485–520, 2006.

Salem, H., Olajos, E.J., and Katz, S.A., Riot control agents, in *Chemical Warfare Agents: Toxicity at Low Levels*, Somani, S.M. and Romano, Jr. J.A., eds., CRC Press, Boca Raton, FL, 2001.

Smart, J.K., History of chemical and biological warfare: An American perspective, in *Textbook of Military Medicine; Military Aspects of Chemical and Biological Warfare*, Sidell, F.R., Takafuji, E.T., and Franz, D.R., eds., Office of the Surgeon General at TMM Publications, Borden Institute, Walter Reed Army Medical Center, Washington, DC, pp. 9–86, 1997.

Sutlovic, D. and Definis-Gojanovic, M., Suicide by fentanyl, *Arh. Hig. Rada. Toksikol.*, 58: 317–321, 2007.

Trescot, A.M., Datta, S., Lee, M., and Hansen, H., Opioid pharmacology, *Pain Phys.*, Opioid Special Issue, 11: S133–S153, 2008.

USAMRICD, *Medical Management of Chemical Casualties Handbook*, 3rd edn., MCMR-UV-ZM, Aberdeen Proving Ground, Aberdeen, MD, 2000.

US Department of the Army, Chemical agent data sheets, Vol. 1, Edgewood Arsenal Special Report AD0030, US Department of the Army Edgewood Arsenal, Aberdeen Proving Ground, Aberdeen, MD, pp. 109–113, 1974.

Vandova, V., The European court of human rights' judgement in the case Finogenov and others v. Russia, in *Report of an Expert Meeting, "Incapacitating Chemical Agents": Law Enforcement, Human Rights Law and Policy Perspectives*, Montreux, Switzerland, April 24–26, 2012, ICRC, Geneva, Switzerland, pp. 46–49, 2013.

Wang, L. and Bernert, J.T., Analysis of 13 fentanils, including sufentanil and carfentanil, in Human urine by liquid chromatography-atmospheric-pressure ionization-tandem mass spectrometry, *J. Anal. Toxicol.*, 30: 335–341, 2006.

Wax, P.M., Becker, C.E., and Curry, S.C., Unexpected "Gas" casualties in Moscow: A medical toxicology perspective, *Ann. Emerg. Med.*, 41: 700–705, 2003.

13 Exposure to Ammonia

Finis Cavender and Glenn Milner

CONTENTS

13.1 INTRODUCTION

Ammonia is a colorless, corrosive alkaline gas that has a very pungent odor. Exposures to ammonia occur as a result of accidents during highway and railway transportation, by releases at manufacturing facilities, and from farming accidents. Ammonia is very soluble in water. Because of its exothermic properties, ammonia forms ammonium hydroxide and produces heat when it contacts moist surfaces, such as mucous membranes. The corrosive and exothermic properties of ammonia can result in immediate damage (severe irritation and burns) to eyes, skin, and mucous membranes of the oral cavity and respiratory tract. In addition, ammonia is effectively scrubbed in the nasopharyngeal region of the respiratory tract because of its high solubility in water. Reversible effects are manifested by irritation to the eyes, throat, and nasopharyngeal region of the respiratory tract.

More than 500 million pounds of ammonia is manufactured each year, with 80% being used in fertilizers, 6% in explosives, 3% in animal feeds, and 2% in household cleaners. Anhydrous ammonia is a gas that can be compressed to condense as a liquid and is usually stored and shipped in this form. The odor of ammonia is familiar to most people because ammonia is used in household cleaners, smelling salts, and window cleaning products. Ammonia easily dissolves in water to form ammonium hydroxide, which is slightly ionized to form ammonium and hydroxide ions. Ammonium ions are not gaseous and have no odor. Ammonia is produced by mammals and other living organisms. In humans, ammonia is metabolically produced and utilized via the glutamine cycle and the urea cycle. It is found in water, soil, and air and is a source of much-needed nitrogen for plants and animals. Most of the ammonia in the environment comes from the natural breakdown of manure, decaying plants, and animal carcasses.

13.2 ANIMAL TOXICITY DATA

13.2.1 Odor Threshold

Odor has long been recognized as a valuable, but not perfect, aid to chemical safety. Odor is not reliable as a warning system in cases where health effects occur at concentrations below the odor threshold or in cases where adaptation or olfactory fatigue occurs. The literature for the odor threshold of ammonia varies greatly. One of the first reports on the odor of ammonia was by Fieldner et al. (1931) in an investigation of the odor of numerous compounds, including ammonia, to determine their suitability as warning agents for fuel gases. Odor thresholds for various chemicals have frequently been used by industrial hygienists as a rough estimate of their airborne concentration (Amoore and Hautala, 1983). For some chemicals, odor is a valuable warning property to alert those at home, at work, or in the outdoor environment. Unfortunately, for some individuals in the general population, an odor may trigger an inappropriate fear of poisoning (chemphobia), resulting in unwarranted emotional distress and/or an irrational emergency response.

The odor threshold is reported in the range of 0.037–103 ppm (Henderson and Haggard, 1943). One of the earliest published values for the ammonia odor threshold was 0.037 mg/L (53 ppm) (Fieldner et al., 1931). This level was also reported by Henderson and Haggard (1927, 1943), and this level dominated the literature for more than 40 years. The level of 53 ppm (0.037 mg/L) was first published in 1921 (Fieldner et al., 1921) and was cited as undated and unpublished data from the Chemical Warfare Service, American University Experiment Station, Washington, DC. Unfortunately, the actual report from the Chemical Warfare Service is not readily available, and the experimental basis, if any, for the level of 53 ppm remains a mystery. However, this concentration has continued to be cited despite several challenges to the validity of the data.

The first challenge of the 53 ppm was provided (Carpenter et al., 1948) in which the authors stated that the odor threshold for ammonia was 1 ppm. Later that year, in *Industrial Hygiene and Toxicology*, Patty (1948) stated that "ammonia is detectable by odor in concentrations of less than 5 p.p.m, and concentrations of 20 p.p.m. are easily noticeable." This statement in Patty (1948) appears to be based on the Carpenter report. The details of the Carpenter et al. (1948) methodology for determining the odor threshold were not provided until Smyth (1956) reported that a panel of 10 subjects detected and identified ammonia odor at 1 ppm. Since the details of the Carpenter et al. (1948) work were never published in the peer-reviewed literature, the level was never taken seriously, apparently because it was vastly different from the 53 ppm level, which had gone unchallenged since 1921. Interestingly, a study by Leonardos et al. (1969) using a trained panel reported a level of 47 ppm, which seemed to support the level of 53 ppm. In 2001, the American Conference of Governmental Industrial Hygienists (ACGIH, 2013) reported the odor threshold for ammonia as ≪ 5 ppm, but ACGIH did neither cite any experimental data nor provide a reference for the level. In 2002, the Dutch reported an odor detection threshold of 0.63 ppm using a trained panel (van Doorn et al., 2002). Similarly in 2003, results of an odor panel study conducted by the Japan Ministry of the Environment stated that the odor threshold for ammonia was 1.5 ppm (Nagata, 2003). Finally, in a series of field studies involving 7065 samples, the odor detection threshold was given as 1.08 ppm (Cawthon et al., 2009) as shown in Table 13.1.

The field study data and the data derived using trained panelists all agree that the odor threshold for ammonia is 1.0 ± 0.5 ppm. When these data are viewed in light of several controlled human

TABLE 13.1
Ammonia Sampling and Odor Results

Concentration Range (ppm)	Number of Samples (n)	Odor Detects	Odor Detectsᵃ (%)	Technicians Detecting Odor (%)	Number of Technicians Detecting Odor
0–0.5	6222	95	1.5	65.7	23
0.6–1	317	113	35.6	77.4	24
1.1–1.5	65	51	78.5	80	16
1.6–2	85	74	87.1	95.7	22
2.1–2.5	20	18	90.0	90.9	10
2.6–3	53	45	84.9	88.2	15
>3	303	300	99.0	100	33
Total	7065				

[a] The percent technician indicates the detected odor within that concentration range. Some technicians recorded both odor and no odor within each individual concentration range.

exposure studies (MacEwen et al., 1970; Industrial Bio-Test, 1973; Ferguson et al., 1977; Verbeck, 1977) that clearly demonstrate that eye, nose, and throat irritation are not observed until the concentration reaches 30–50 ppm, the evidence is overwhelming that the 53 ppm reported by Fieldner et al. (1931) and cited by Henderson and Haggard (1927) is not valid for the general population. At the time of Fieldner's (1931) study, ammonia workers were routinely exposed to concentrations in excess of 200 ppm. In such inured workers, the odor detection could have easily been 53 ppm as many of these workers were at least partially anosmic. Ammonia is one of the industrial gases that helped justify the need for worker protection and industrial hygiene.

Although there is no experimental data available to support the 53 ppm, still one might ask, "What about the Leonardos et al. (1969) study that reported an odor threshold for ammonia of 47 ppm?" Interestingly, in the Leonardos et al. study (1969), the authors reported the odor *recognition* threshold, which they defined as the concentration at which 100% of the panel could positively identify the odor as being that of ammonia. This is vastly different than the usual definition of odor threshold or the level of detection of an odor and, more importantly, different from the definition that others were using, especially the definitions used in recent odor panel studies.

Current references cite the odor *detection* threshold as the level at which 50% of the panel members detect the presence of an odor (Carpenter et al., 1948; van Doorn et al., 2002; Nagata, 2003). This detection level for 50% of the panel is a much lower concentration than the recognition (conformation that the odor is that of ammonia) level for which 100% of the panel identifies the odor as ammonia. A random selection of a panel will include some who may be sensitive to the odorant as well as some who are tolerant to the odor as seen in any set of values that fit a normal distribution. To illustrate this further, in the field study (Cawthon et al., 2009), there were two readings of 10 ppm for which the technician did not detect an odor. The requirement of 100% detection greatly elevates the concentration required to set the level. The requirement for 100% recognition would be even higher, perhaps by an order of magnitude. In the field study, the requirement for 100% detection for the sampling personnel would raise the odor threshold from 1.08 to 10 ppm. Given the heterogeneity in the general population in detecting odors, use of a 100% recognition threshold is clearly inappropriate in the development of the odor threshold. Many individuals in the population would be calling the authorities to complain of a strong odor and nasal irritation long before the 100% detection or recognition levels were attained in the environment.

In perusing the historical literature on ammonia odor threshold, one finds that the literature is rife with transposition errors and incorrect conversions between units of measurement. Sullivan (1969), who cites Sayers (1927), lists an odor threshold of 0.037 ppm; however, this is simply an error in recording the concentration units because the original data are recorded as 0.037 mg/L (53 ppm) in Fieldner et al. (1931). In Verschueren (1977), the erroneous 0.037 ppm figure is converted to 0.026 mg/m^3, which is later cited by Ruth (1986). The Ruth (1986) report also transposes 36.9 mg/m^3 (53 ppm) to 39.6 mg/m^3 (56.8 ppm), and the 56.8 ppm value is later cited in the Hazardous Substances Data Bank (HSDB, 2013) and Michaels (1999). Amoore and Hautala (1983) calculated an odor threshold for ammonia of 5.2 ppm; however, when the Agency for Toxic Substances and Disease Registry (ATSDR) (2004) cites Amoore and Hautala (1983), the authors mistakenly record the TLV value of 25 ppm instead. There are also many references to earlier sources purported to contain ammonia odor data, which, if you look in those sources, there are no ammonia odor data present.

In addition, there is considerable variability in ammonia odor threshold data in the literature. For instance, Ruth (1986) indicated that the odor threshold data for ammonia range from 0.0266 to 39.6 mg/m^3 (0.037–56.8 ppm); CalEPA (1999) cites odor threshold data ranging from 0.4 to 103 ppm; the ATSDR (2004) reported odor thresholds for ammonia of 25, 48, and 53 ppm; and the American Conference of Governmental Industrial Hygienists (2005) reported the odor threshold was much less than 5 ppm.

Fortunately, new testing methods have improved the manner in which the odorant is presented to the test subject, and the use of panelists, both trained and naive, has helped standardize the responses to an odorant. Instrumentation has evolved from simple glass syringes to highly technical dynamic dilution olfactometers. The use of panels has allowed a test to distinguish between an odor detection threshold (defined as the concentration at which 50% of the panel can detect an odor) and an odor recognition threshold (defined as the concentration at which 50% of the panel can identify the odor as the test chemical). To determine the odor recognition threshold, the panelists must be familiar with the odorant, thus eliminating the use of naive panelists.

It is interesting to note that Carpenter et al. (1948) correctly identified the odor threshold of 1.0 ppm for ammonia using a trained panel in the 1948, but no one supported this level at the time. These newer data exonerate these pioneer workers whose data in many areas have served as a major portion of the foundation of good industrial hygiene practice. Their work again reminds us that there is no substitute for hard work and good science.

Incredibly, the historical literature on the ammonia odor threshold contains levels varying over several orders of magnitude. In recent years, odor testing has evolved considerably, and changes in terminology, methods, and instrumentation help to reduce the variability in odor threshold data. Ambient air sampling data following the rapid response to hazardous chemical spills also provide valuable data that can be used to more accurately identify the ammonia odor threshold. In recent years, the Dutch have developed rigorous methodology for quantifying odor data (van Doorn et al., 2002).

13.2.2 Pulmonary/Sensory Irritation

The RD_{50} is the concentration expected to elicit a 50% reduction in respiratory rate in Swiss Webster mice (Barrow et al., 1978). The authors stated that the RD_{50} concentration would elicit intense sensory irritation and is expected to be rapidly incapacitating to humans. Groups of four outbred male Swiss Webster mice were exposed to ammonia by inhalation for 30 min. The authors did not report the concentration of ammonia inhaled by the mice, but judging by the graphic representations, the concentrations were 100, 200, 400, and 800 ppm. The maximum depression in respiratory rate was achieved within the initial 2 min of exposure, after which the response diminished. The RD_{50} was 303 ppm (95% confidence limits = 188–490 ppm) for a 30 min inhalation exposure to ammonia. There was no microscopic examination of the respiratory tract.

In a follow-up study (Buckley et al., 1984), the histopathologic effects of repeated exposures to ammonia at 303 ppm were evaluated. Groups of 16–24 male Swiss Webster mice were exposed to 0 or 303 ppm of ammonia 6 h/day for 5 days. The respiratory tract was examined in one half of the animals that were killed immediately after terminating exposure. The other one half were killed

3 days later. Histopathological findings, which were confined to the respiratory epithelium of the nasal cavity, included minimal exfoliation, erosion, ulceration, and necrosis; moderate inflammatory changes; and slight squamous metaplasia.

In a similar study, groups of 10 male Swiss mice were exposed to ammonia to target concentrations of 90.9, 303, and 909 ppm ($0.3 \times RD_{50}$, the RD_{50}, or $3 \times RD_{50}$), 6 h/day for 4, 9, or 14 days (Zissu, 1995). The analytically measured concentrations were 78.0, 271, and 711 ppm. The three control mice were exposed to filtered air. The entire respiratory tract was examined microscopically. No clinical signs of toxicity were noted for mice exposed to ammonia. Rhinitis and pathologic lesions with metaplasia and necrosis were seen only in the respiratory epithelium of the nasal cavity of mice inhaling 711 ppm, the severity of the lesions increased with duration of exposure, ranging from moderate on day 4, severe on day 9, to very severe on day 14. No lesions were seen in the controls or in mice inhaling the 78 ppm. No effects were seen at 271 ppm, even after 9 days of exposure.

Generally, an RD_{50} of 300 ppm would suggest that significant irritation would be seen in humans at 3 ppm. However, the respiratory epithelium of the nasal tract in rodents is much more sensitive to chemical exposure than human respiratory epithelium. Thus, instead of taking a 100-fold safety factor in extrapolating to humans, only a factor of 3 is sufficient, giving an irritation level of 100 ppm. This supports the data seen in humans.

In a 2010 study, young adult male Wistar rats, a species commonly used in inhalation toxicity studies, and OF1 mice, an alternate mouse species used in some sensory irritation studies, were simultaneously nose-only exposed for 45 min to ammonia in concentrations from 92 to 1243 mg/m^3 (Li and Pauluhn, 2010). This study examined airway reflexes by the changes in respiratory patterns elicited by ammonia in either dry, steam-humidified (approximately 95% relative humidity), or aqueous aerosol–containing atmospheres. This served the objective to explore whether high concentrations of anhydrous ammonia and/or high humidity and aqueous aerosols change the predominant nasal deposition site to more distal locations in the lung. Animals from all groups tolerated the exposure without any evidence of respiratory tract irritation and changes in body and lung weights. The evoked changes on breathing patterns resembled those known to occur following exposure to upper respiratory tract sensory irritants, rapid in onset and reversibility. Reflex stimulation from the lower airways was not observed in any group. While mice showed some adaption during the 45 min exposure period, rats displayed more stable changes in respiratory patterns. In this species, humidity-related or aqueous aerosol–related changes in sensory irritation potency did not occur to any appreciable extent. The RD_{50} was 972 and 905 mg/m^3 (corresponding to 680 and 634 ppm, respectively) in rats in dry and wet air, respectively. In contrast, mice appeared to be more susceptible to ammonia in the presence of dry air (the RD_{50} was 582 [407 ppm] and 732 mg/m^3 [547 ppm] in dry and wet air, respectively). It was shown that the sensory irritation potency of ammonia does not increase when inhaling wet atmospheres, and ammonia does not penetrate into the lower airways up to 1243 mg/m$^3 \times 45$ min. The authors concluded that even in a dry atmosphere, sensory irritation–related effects do not appear to be offensive enough to impair escape as a result of respiratory epithelial effects.

In rodent RD_{50} studies, the depression of the respiratory rate is a defense mechanism. Ammonia can reduce the respiratory rate to 30% of preexposure levels. The rats are effectively inhaling one-third as much ammonia compared with breathing at their preexposure rate. Humans react to ammonia quite differently: they increase the breathing rate to 250% preexposure levels (Silverman et al., 1949). Thus, humans increase their exposure levels while rats lower their exposure levels.

It should be noted that the RD_{50} was developed using Swiss Webster mice. The OF1 mouse has been used in Europe, but has not been calibrated against Swiss Webster mice. The RD_{50} data that have been examined using both strains do not calibrate very well. The OF1 data can be higher or lower than Swiss Webster data depending on the toxicant. Unfortunately, a comparison of the pathology of the nasal epithelium in the two species following exposure to identical ammonia concentrations has not been conducted. In the case of ammonia, the OF1 mouse ($RD_{50} = 400$ ppm; Li and Pauluhn, 2010) is not as sensitive as the Swiss Webster mouse ($RD_{50} = 300$ ppm; Barrow et al., 1978). Based on either data set, the risk number for humans based on the RD_{50} concentrations would not impair escape in humans as the equivalent level in humans be approximately 100 ppm (a factor of 3 below the RD_{50} value). Based on data presented in the next section, lacrimation will start within 5 min in naive human subjects exposed to 140 ppm (Industrial Bio-Test, 1973). Thus, the data for both strains of mice support the level of 140 ppm identified in humans as the threshold for lacrimation.

13.2.3 ACUTE LETHALITY STUDIES

Several LC_{50} studies in rats and mice have been conducted using ammonia.

Species	Exposure Time (min)	LC_{50} (ppm)	Reference
Rat			
M&F	10	40,300	Appelman et al. (1982)
M&F	20	28,600	Appelman et al. (1982)
M&F	40	20,300	Appelman et al. (1982)
M&F	60	16,600	Appelman et al. (1982)
F only	60	19,830	Appelman et al. (1982)
M only	60	14,180	Appelman et al. (1982)
M only	60	7,340	MacEwen and Vernot (1972)
M only	60	17,700	Pauluhn (2013)
M only	240	7,100	Pauluhn (2013)
Mouse			
M&F	10	10,100	Silver and McGrath (1948)
M only	30	21,400	Hilaldo et al. (1977)
M only	60	4,840	MacEwen and Vernot (1972)
M only	60	4,230	Kapegian et al. (1982)

13.2.3.1 Mouse Data

These data show that mice are much more sensitive to ammonia than are rats. The Hilaldo et al. (1977) data seem out of sync with the other mouse data. It turns out that this study was conducted in a static chamber (as opposed to the preferred dynamic chamber) and the exposure level is the nominal concentration of ammonia based on the amount injected into the chamber. The actual concentration in the breathing zone of the animals was not verified using analytical methods. Thus, the average ammonia exposure concentration is not known. Since ammonia is very reactive, the concentration will decrease over time as it reacts with animal fur, chamber walls, and caging as well as being inhaled by the animals.

In the mouse bioassays, the mice exhibited signs and symptoms, which included eye irritation (blinking and scratching), dyspnea, frothing, convulsions, excitation/escape behavior, coma, and death. Signs of irritation started immediately and lasted for up to 10 min and were followed by other signs of toxicity. All animals that died on study did not survive the exposure. The lungs of animals from every treatment group sacrificed on day 14 displayed a mild-to-moderate degree of chronic focal pneumonitis, which was increased with increasing concentrations. The lungs of animals that died during the exposure were all extremely congested with evidence of hemorrhage on both gross and microscopic examination. Focal atelectatic changes were also evident in the lungs of the survivors of the groups exposed to the highest concentrations. Histopathology of the lungs of mice that died showed alveolar disruption and loss of septal continuity (Silver and McGrath, 1948; MacEwen and Vernot, 1972; Hilaldo et al., 1977; Kapegian et al., 1982).

13.2.3.2 Rat Data

In the 1 h rat studies, while there are differences in the 1 h LC_{50} studies (MacEwen and Vernot, 1972; Appelman, 1982; Pauluhn, 2013), the three studies are considered equal; that is, they are within the range of experimental variability, even though two were whole-body exposures and one was a nose-only exposure. There was some concern about the Appelman study because of high animal loading (~10% of the chamber volume) and the fact that the analytical concentration was measured only in a mixing chamber prior to entry into the horizontal exposure chamber. However, these potential sources of experimental problems were overcome in this instance because of the very high flow rate through the chamber, $t_{99} = 1.86$.

In an exposure to evaluate the sex differences in chambers with analytical concentrations measured in the breathing zone of the rats (Pauluhn, 2013), rats were exposed to ammonia in nose-only chambers for 1 or 4 h. In the more susceptible male rats, the 1 and 4 h LC_{50}s were 12,303 mg/m^3 (17,700 ppm) and 4,923 mg/m^3 (7,100 ppm), respectively, and the calculated LC_{01}s were 10,067 mg/m^3 (14,500 ppm) and 4,028 mg/m^3 (5,800 ppm), respectively. At sublethal exposure levels, the ventilation of rats was about one-third of normal minute volume. This change in ventilation and inhalation dosimetry was adjusted for C × t–dependent lethal end points, whereas sensory irritation–related phenomena were C dependently adjusted. In summary, the outcome of this study shows that C- and C × t–dependent causes of toxicity require serious consideration when extrapolating across species with species-specific inhalation dosimetry. It also appears to be indispensable that each exposure metric must be disentangled when translating C × t–dependent lethality and reflexively induced, sensation-based C-dependent point of departures (PODs). For 1 h exposure periods, these PODs are 1500 ppm for the threshold of lethality in man and 500 ppm for the threshold of serious or irreversible toxicity in man, respectively.

For both rats and mice, the lethality data exhibit an unusually steep concentration–response as seen later:

Selected rat and mouse lethality data with lowest lethal and highest nonlethal levels are presented in the following.

Species	Gender	1 h LC_{50} (ppm)	Lowest Lethal Level (ppm)	Highest Nonlethal Level (ppm)	HNL/LC_{50}[a]	Referances
Rat	M	7,340	6,820	6,210	0.85	MacEwen and Vernot (1972)
Rat	M	14,180	14,210	ND	—	Appelman et al. (1982)
Rat	F	19,830	12,500	16,270	0.82	Appleman et al. (1982)
Rat	M	17,700	15,860	13,280	0.75	Pauluhn (2013)
Mouse	M	4,840	4,500	3,600	0.74	MacEwen and Vernot (1972)
Mouse	M	4,230	3,950	3,440	0.81	Kapeghian et al. (1982)

[a] HNL/LC_{50} = highest nonlethal level divided by the LC_{50}.

The highest nonlethal levels were 13,280 ppm (R) for male rats and 3,600 ppm (M) for mice. For most irritants, the threshold for lethality in humans is a factor of 10 lower than the highest nonlethal level in rodents (Rusch et al., 2009). Ammonia causes a respiratory depression of up to 70% in rats and causes a rise in the respiratory rate of 250% in humans. For this reason, the threshold for human lethality is only one-third of the highest nonlethal level in rats or one-third of the calculated LC_{01}. For these data, the ratio of highest nonlethal level to the LC_{50} ranges from 0.85 to 0.74 indicating a very steep dose–response curve.

13.2.4 Nonlethal Effects

Rats were exposed to ammonia at vapor concentrations of 15, 32, 310, or 1157 ppm for 24 h. No clinical signs or evidence of irritation to the eyes or mucous membranes were observed. Blood gases (pO$_2$ and pCO$_2$) and pH were within the normal range for rats (Schaerdel et al., 1983).

Rabbits were exposed to a mean ammonia vapor concentration of 9,800 ppm (range 4,900–12,180 ppm) for 1 h before or after intratracheal cannulation. The mean survival time was 33 h for rabbits exposed before cannulation and 18 h for rabbits exposed after cannulation. Toxic signs included marked excitation during the early stages of exposure followed by a curare-like paralysis. Microscopically, the trachea and bronchi appeared normal in rabbits exposed before cannulation, but were severely damaged in animals exposed after cannulation. Bronchiolar (damage to epithelial lining) and alveolar effects (which the authors described as congestion, edema, atelectasis, hemorrhage, and emphysema) were similar in both groups (Boyd et al., 1944).

Cats were exposed to 1000 ppm of ammonia for 10 min using endotracheal tubes. The effect of the exposure on pulmonary function and lung pathology was assessed 1, 7, 21, and 35 days following exposure. Signs of toxicity included poor general condition, severe dyspnea, anorexia, dehydration, bronchial breath sounds, sonorous and sibilant bronchi, and coarse rales. Pulmonary function tests showed evidence of airway damage throughout the experiment and central lung damage on day 21. Gross examination showed the lungs to be congested, hemorrhagic, and edematous; evidence of interstitial emphysema and collapse was also seen. Bronchopneumonia, which caused the death of one animal, was commonly seen after day 7. Microscopic examination showed necrosis and sloughing of the bronchial epithelium on day 1 with healing of the bronchi on day 7 (Dodd and Gross, 1980).

13.3 TOXICITY OF AMMONIA IN HUMANS

The data for deriving a toxicological dossier in humans are obtained primarily from the case studies of accident victims, experimental studies in human volunteers, and experimental studies on lethality and toxic end points in animals. While human exposure data are desirable in determining acute and chronic toxicity for any chemical, case studies are generally of limited use for quantitative evaluation because reliable analytical ammonia levels are rarely available. What has proven useful in documenting ammonia toxicity is experimental studies in humans and animals because the accompanying quantitative data allow establishing various points of departure for ammonia. This chapter is not a comprehensive study of ammonia toxicity; rather it details odor threshold, sensory irritation, and inhalation toxicology, especially in relating specific exposure levels to specific clinical signs. A more complete rendering of general ammonia toxicity can be found in several references (ATSDR, 2004; ACGIH, 2013; NRC, 2007; HSDB, 2012; US EPA, 2012; AIHA, 2014).

No reliable quantitative lethality data were available for humans dying as a result of exposure to ammonia. One case study reported the death of an individual exposed to a high but unknown concentration of ammonia. Other case studies also contained no exposure estimates, but showed that high concentrations of ammonia cause severe damage to the respiratory tract, particularly in the tracheobronchial and pulmonary regions. Death, however, is most likely to occur when damage includes pulmonary edema. Nonlethal, irreversible, or long-term effects occur when damage progresses to the tracheobronchial region, manifested by reduced performance on pulmonary function tests, bronchitis, bronchiolitis, emphysema, and bronchiectasis.

The purpose of this analysis is to develop a method of estimating the airborne concentration in accidental exposures for which no analytical samples were taken by using the symptoms and frank effects reported by victims exposed during accidental releases of ammonia. This is to be accomplished by developing a scale of known effects that have been reported in volunteer studies and accidental exposures to known concentrations. As volunteer studies cannot include lethal or near-lethal levels, an extrapolation beyond the known data is required. Once the scale is complete based on human exposure data, the scale will be compared with data from animal studies. If necessary, the animal data will be used to tweak the effects scale, particularly for high exposure concentrations.

Considerable information on human and animal exposures to ammonia has been generated. In the reports of human exposure, the data are complicated by the fact that (1) some exposures were to liquid ammonia rather than to gaseous ammonia alone and (2) some exposures were on acclimated workers and others were on the general public which had no routine, significant exposure to ammonia. Liquid ammonia causes severe burns in addition to the effects of inhaling the gaseous ammonia, especially in cases where the liquid ammonia was sprayed on the individual or group of individuals. Such burns can have a considerable effect on the long-term consequences of exposure. For the analysis of the effects of gaseous ammonia alone, exposures involving direct effects of liquid ammonia will not be discussed. Similarly, since workers routinely exposed to ammonia are up to 10-fold less sensitive to the effects of ammonia, the reports dealing with workers or highly motivated individuals such as young healthy military personnel will be separated from those dealing with the general public or uninured volunteers.

13.3.1 Human Exposures: Results from Accidental Exposures (Chronological Order)

1. Forty-seven persons were accidentally exposed to ammonia in an enclosed area (air raid shelter) (Caplin, 1941). The subjects were divided into three groups depending on the degree to which they were affected: *mildly*, *moderately*, or *severely*. No deaths occurred among the nine *mildly* affected individuals. Three of the twenty-seven *moderately* affected patients showed signs and symptoms similar to pulmonary edema and died within 36 h. Nine *moderately* affected individuals developed bronchopneumonia within 2–3 days, and three died 2 days after the onset. The mortality rate for the *moderately* affected individuals was 22% (6/27). The 11 *severely* affected individuals developed pulmonary edema; 7 died within 48 h. The mortality rate for the *severely* affected individuals was 63% (7/11).

2. Five or six laboratory workers inhaled the exhaust fumes generated in an exposure chamber for an inhalation study and stated that the disagreeable odor and respiratory distress would prevent a person from voluntarily remaining in an atmosphere containing 170 ppm of ammonia (concentration range: 140–200 ppm) for an appreciable length of time (Weatherby, 1952).

3. In a farm accident, liquid ammonia splashed three workers and one refrigeration technician in the face and upper body. They suffered damage to the tracheobronchial region causing upper airway obstruction and injury to the respiratory tract persisting for 2 years after the accident (Levy et al., 1964).

4. A worker was exposed to an estimated ammonia concentration of 10,000 ppm. Duration of exposure was not reported, but it could have been a few minutes; nevertheless, the worker continued to perform his duties for an additional 3 h after his initial exposure to ammonia. He experienced coughing, dyspnea, and vomiting soon after exposure. Three hours after initial exposure, his face was *red and swollen*, his mouth and throat were *red and raw*, his tongue was swollen, his speech was difficult, and he had conjunctivitis. He died of cardiac arrest 6 h after exposure. An autopsy revealed marked respiratory irritation, denudation of the tracheal epithelium, and pulmonary edema (Mulder and Van der Zalm, 1967).

5. In a railroad tanker accident, 33,000 gal of anhydrous ammonia were released resulting in 8 deaths and an additional 70 being injured (Kass et al., 1972). A heavy fog kept the ammonia vapors close to the ground for a long period of time after the accident. Two young women who were exposed for 30 or 90 min continued to show effects more than 2 years after exposure. One woman was found unconscious 90 min after the accident, and the other woman was exposed when she went outdoors for 30 min after the accident. Damage to the eyes caused marked visual deterioration. Bronchiectasis was detected

2 years after exposure, and pulmonary function tests showed abnormalities indicative of small airways obstruction. Various tests and examinations showed areas of atelectasis and emphysema in the lungs, thickened alveolar walls with histiocytic infiltration into the alveolar spaces, as well as mucous and desquamated cells in the bronchiolar lumen. Some of these effects may be secondary to the damage caused by ammonia. The woman exposed for 90 min was carrying her 1-year-old child, who was exposed at the same time. The child became *quite ill* but recovered completely except for a chemical scar on his abdomen.

6. Seven workers were exposed to ammonia in an industrial accident that resulted in one death (Walton, 1973). The autopsy report noted marked laryngeal edema, acute congestion, pulmonary edema, and denudation of the bronchial epithelium.

7. Three workers died immediately after exposure to a high concentration of ammonia in a farming accident (Sobonya, 1977). A 25-year-old man died 60 days after the accident. The autopsy report noted damage to the bronchial epithelium, bronchiectasis, mucous and mural thickening of the smallest bronchi and bronchioles, fibrous obliteration of small airways, and a purulent pneumonia characterized by large numbers of *Nocardia asteroides* (nocardial pneumonia).

8. In a truck accident in Houston, a tanker released 17.2 tons of pressurized anhydrous ammonia. The ammonia cloud extended 1500 m downwind and 550 m wide. Five people were killed and 178 were injured, some with permanent disabling injuries (not otherwise described). The fatalities and disabling injuries occurred within about 70 m of the accident (NTSB, 1979). A 30-year-old woman died 3 years after exposure to ammonia during an accident involving a tanker truck carrying anhydrous ammonia in Houston (Hoeffler et al., 1982). Her injuries resulted in severe immediate respiratory effects including pulmonary edema. She required mechanically assisted respiration throughout her remaining life. Bronchiectasis was detected 2 years after exposure and confirmed upon autopsy. The autopsy examination also showed bronchopneumonia and cor pulmonale (heart disease secondary to pulmonary disease). According to the authors, the bronchiectasis may have been due to bacterial bronchitis or the chemical injury. In this accident, three children and a 17-year-old female suffered second- or third-degree burns to the body, damage to the eyes, burns to the oral mucosa, upper airway obstruction (probably due to damage to the laryngeal and tracheobronchial regions), and some pulmonary damage (Hatton et al., 1979). All the patients recovered between 7 and 32 days. Nine of fourteen patients exposed to ammonia for only a few seconds or few minutes showed moderate symptoms of chest abnormalities or airway obstruction and recovered within 6.3 days (average) (Montague and MacNeil, 1980).

9. In another accident, several workers were exposed to ammonia for 5 min (O'Kane, 1973). One patient developed necrotizing pneumonia described as *chronic infective lung disease*, one had a persistent hoarseness and a productive cough for several months, and a third developed impaired pulmonary diffusion (75% of normal).

10. In a refrigeration accident, a man splashed with liquid ammonia showed evidence of peripheral (possibly bronchiolitis) and central airway obstruction 5 years after the accident (Flury et al., 1983).

11. In Potchefstroom, South Africa, an accident involved a pressurized ammonia storage tank that failed and instantaneously released 38 tons of anhydrous ammonia into the atmosphere. Eighteen people died, and an unknown number were injured (Lonsdale, 1975). A visible cloud extended about 300 m wide and about 450 m downwind; all deaths occurred within 200 m of the release point (Pedersen and Selig, 1989).

12. Tubular bronchiectasis was detected 8 years after exposure of a 28-year-old man to a high concentration of anhydrous ammonia in an industrial accident (Leduc et al., 1992).

Twelve years after exposure, the man continued to have a productive cough, frequent bronchial infections, dyspnea upon exertion, and severe airflow obstruction (62% reduction in forced expiratory volume at 1 s [FEV$_1$]).

13. From these accidental exposures, the development of pulmonary edema after inhaling ammonia is a consistent finding in cases resulting in lethality. The acute effects in survivors exposed to high ammonia concentrations include burns to the eyes and oral cavity, damage to the nasopharyngeal region, and damage to the tracheobronchial region of the respiratory tract. The toxicological responses to ammonia in these cases include conjunctivitis, corneal burns, visual impairment, pain in the pharynx and chest, cough, dyspnea, hoarseness, aphonia, rales, wheezing, rhonchi, hyperemia, and edema of the pharynx and larynx, tracheitis, bronchiolitis, and purulent bronchial secretions (Levy et al., 1964; Walton, 1973; Hatton et al., 1979; Montague and Macneil, 1980; Flury et al., 1983; O'Kane, 1973). Cyanosis, tachycardia, convulsions, and abnormal electroencephalograms have also been described for some victims (Kass et al., 1972; Walton, 1973; Hatton et al., 1979; Montague and Macneil, 1980). Pulmonary edema occurred in some patients who survived (Caplin, 1941), but most often resulted in lethality.

14. In an unpublished report, 31 cars of a rail freight train derailed on January 18, 2002, at about 1:25 a.m. releasing 380 tons of anhydrous ammonia in 1 min and an additional 115 tons in the next 5 min (Fthenakis, 2004). An additional 55 tons were slowly released over the next 5 days. The resulting dense cloud engulfed the city and remained in place for about 6 h. There was one death and numerous injuries. As the derailment happened in the middle of the night on a very cold night, most of the residents were boarded in their homes, thereby escaping the much higher concentrations out of doors. No other details of injuries were provided.

While these reports provide insight into the signs and symptoms following ammonia exposure, not very useful in documenting human toxicity because both the concentration and the duration of exposure are required to do so. Fortunately, considerable human toxicity studies have been conducted on ammonia.

13.3.2 Toxicity Studies in Volunteers (Chronological Order)

1. The responses of humans exposed to ammonia based on industrial exposures are summarized in the following text (Henderson and Haggard, 1943):

Conc. (ppm)	Response
53	Least detectable odor
408	Least conc. causing irritation of the throat
698	Least conc. causing irritation of the eyes
1720	Least conc. causing coughing

Based on these results and reported accidental exposures, the authors postulated the following:

300–500	Max. conc. allowable for short exposure (½–1 h)
2,500–6,500	Dangerous even for short exposure (½ h)
5,000–10,000	Rapidly fatal for short exposure

This summary indicates that concentrations greater than 2500 ppm are dangerous for an exposure of 30 min or more. These levels are based on data available at the time they were written, which included data based on signs and symptoms seen in inured workers, routinely exposed to 250 ppm.

2. Seven male subjects were exposed to 500 ppm of anhydrous ammonia by means of a nose and mouth mask; six subjects were exposed for 30 min and one for 15 min (Silverman et al., 1949). The inspired ammonia concentration was calculated, and the expired ammonia concentration was analyzed in grab samples taken every 3 min. Respiratory rates and minute volumes were measured for each subject. Throat irritation was reported by two subjects. Nasal irritation with stuffiness similar to that of a cold or nasal dryness and irritation was reported by six subjects. The stuffiness lasted for about 24 h. Only two subjects were able to continue nasal breathing for the full 30 min, the others changing to mouth breathing on account of nasal dryness and irritation. Hypoesthesia (decreased sensitivity) of the skin around the nose and mouth was experienced by all subjects, and excessive lacrimation was reported by two. Hyperventilation (increases in the respiratory rates and minute volumes) occurred in all subjects. Hyperventilation occurred immediately in three subjects, was delayed for 10–30 min in the remaining four, and fluctuated with a 25% decrease at 4–7 min intervals. The increase in the minute volume was 141%–289%. No coughing was reported; the authors noted that 1000 ppm causes immediate coughing. This study showed that irritation of upper respiratory tract and throat occurred in subjects inhaling 500 ppm of anhydrous ammonia for 15–30 min. There was no difference in the effects noted in the subject inhaling ammonia for 15 min and those inhaling ammonia for 30 min.

3. Six human volunteers were exposed head-only to ammonia at concentrations of 30 or 50 ppm for 10 min (MacEwen et al., 1970). Each subject rated his irritation responses on a scale of 0–4 (not detectable, just perceptible, moderate irritation, discomforting or painful, and exceedingly painful) and odor perception on a scale of 0–5 (not detectable, positively perceptible, readily perceptible, moderate intensity, highly penetrating, and intense or very strong). At 30 ppm, two subjects reported irritation as *just perceptible* and three as not detectable, and one gave no response. The odor was highly penetrating for three subjects and moderately intense in two subjects exposed to 30 ppm; no response was given by the sixth subject. At 50 ppm, four subjects reported the irritation as moderate, one just perceptible, and one not detectable. The odor was highly penetrating or strong for all six subjects inhaling 50 ppm of ammonia.

4. Industrial Bio-Test Laboratories, Inc. (1973) determined the irritation threshold in 10 human volunteers exposed to ammonia at concentrations of 32, 50, 72, or 143 ppm for 5 min. Irritation was defined as any annoyance to the nose, throat, eyes, mouth, or chest. The results are given in Table 13.2. The subjects showed dose-related responses for dryness of the nose and eye, nose, throat, and chest irritation. The severity of the effects was not noted. This study was cited from NIOSH (1974).

5. In one company in 1972, workers did not voluntarily use gas masks until ammonia concentrations reached 400 or 500 ppm (Ferguson et al., 1977). They also reported that prior to 1951, workers were continuously exposed to concentrations ranging from 150 to 200 ppm. To establish the bounds for controlled exposure studies, they conducted two reconnaissance experiments. In the first experiment, they reported that four male subjects were able to tolerate *continuous exposure* to 130–150 ppm (duration not reported) after exposure to lower concentrations for <2 h. In the second experiment, they noted that in a bicarbonate plant, after 30 min of acclimation to 100 ppm of ammonia, a 30 s exposure at 300 ppm was just barely tolerable. In the controlled exposure study, the authors assessed the effect of ammonia on six (three groups of two) human volunteers (industrial workers) exposed to concentrations of 25, 50, or 100 ppm after exposure to the same concentrations during a 1-week practice period in a plant in areas where concentrations of ammonia

TABLE 13.2

Effect of Ammonia Inhalation on Human Volunteers Exposed for 5 min

Effects	32 ppm	50 ppm	72 ppm	143 ppm
Dryness of the nose	+ (1)[a]	+ (2)		
Nasal irritation			+ (2)	+ (7)
Eye irritation			+ (3)	+ (5)
Lacrimation				+ (5)
Throat irritation			+ (3)	+ (8)
Chest irritation				+ (1)

Source: Industrial Bio-Test Laboratories, Inc., Irritation threshold evaluation study with ammonia, Report to International Institute of Ammonia Refrigeration by Industrial Bio-Test Laboratories, Inc., IBT 663-03161, 1973, March 23, 1973.

[a] Number of volunteers showing a response out of a total of 10 participating.

varied between 25 and 50 ppm. The workers were then exposed to 100 ppm in an exposure chamber. Ammonia concentrations were monitored each half hour using NIOSH-certified detector tubes that had an overall accuracy of ±10%. Exposure periods ranged from 2 to 6 h/day for 5 weeks. There was no adverse effect on respiratory function and no increase in the frequency of eye, nose, or throat irritation at 100 ppm. The only complaints were lacrimation and nasal dryness during brief excursions above 150 ppm. There was no interference with performance of work duties and no effect on pulse rate or respiratory function during exercise (i.e., no effect on physical or mental ability to perform work duties) that was consistent with concentration or duration. Definite redness of the nasal mucosa occurred in one subject exposed to 100 ppm with an excursion to 200 ppm, but the effect had cleared by the next morning (i.e., no lasting effects occurred). Four of the six subjects were exposed to different concentrations, making it difficult to establish trends related to exposure concentration or duration.

6. In studying the effects of exercise, 18 servicemen were exposed to ammonia at concentrations of 71, 106, 144, or 235 mg/m³ (102, 152, 206, or 336 ppm) (Cole et al., 1977). The duration of exposure was not reported. The same men served as the exposed groups and controls. Measurements of respiratory parameters (respiratory rate, minute volume, tidal volume, and oxygen uptake) and cardiac frequency were taken during the morning for control conditions (subjects inhaled air only) and during the afternoon for experimental conditions (subjects inhaled ammonia). During exposure to ammonia, the subjects noted only a prickling sensation in the nose and a slight dryness of the mouth. Minute volume showed a dose-related decrease between 71 and 144 mg/m³ (102–206 ppm) compared with control measurements; statistical significance was achieved for 106 and 144 mg/m³ (206–336 ppm). The tidal volume was significantly decreased (9% and 8%, respectively), and respiratory frequency was increased (11% and 8%, respectively) at 144 and 235 mg/m³ compared with the control values, but there was no clear dose–response relationship. It is doubtful that the small changes in tidal volume and respiratory frequency are clinically significant. Therefore, the effects on respiratory function are considered equivocal.

7. The effects of ammonia on respiratory function and subjective responses were studied in two groups of subjects (Verberk, 1977). One group consisted of eight individuals familiar with the effects of ammonia and had no previous exposure (expert group, 29–53 years old),

whereas the other group consisted of eight individuals unfamiliar with the effects of ammonia and also had no previous exposure (nonexpert group, 18–30 years old). Four members of each group were smokers. Each group was exposed to ammonia at concentrations of 50, 80, 110, or 140 ppm for 2 h. Subjective responses (smell, eye irritation, throat irritation, cough, etc.) were recorded every 15 min, and parameters of respiratory function (vital capacity, forced expiratory volume [FEV_{1s}] and forced inspiratory volume [FIV_{1s}]) were measured before exposure and after the 2 h exposure. Subjective responses were rated on a scale of 0–5 (0 = no sensation, 1 = just perceptible, 2 = distinctly perceptible, 3 = nuisance, 4 = offensive, and 5 = unbearable). Chamber concentrations were monitored instantaneously using an infra-red spectrometer. There was no effect on respiratory functions in either group inhaling any concentration of ammonia. Table 13.3 summarizes the combined responses for both groups. Generally, the expert group scored their responses lower than those of the nonexpert group; therefore, the scores at the lower end of the range are the expert scores. Four nonexpert subjects exposed to 140 ppm left the exposure chamber between 30 min and 1 h, and none remained in the chamber for the full 2 h. The greatest difference in responses between the expert and nonexpert groups was in general discomfort. The expert group perceived no general discomfort even after exposure to the highest concentration for 2 h, whereas the four nonexpert subjects perceived their general discomfort to range from *distinctly perceptible* to *unbearable* after 1 h. This study showed dose– and duration–response relationships for the effects of ammonia particularly for the

TABLE 13.3
Scores for Subjective Responses of Expert and Nonexpert Subjects Exposed to Ammonia[a]

Response	50 ppm	80 ppm	110 ppm	140 ppm[c]
Smell	1–3[b], ½ h	1–4, ½ h	2–4, ½ h	1–4, ½ h
	1–4, 1 h	1–4, 1 h	2–4, 1 h	1–4, 1 h
	1–4, 2 h	1–4, 2 h	2–4, 2 h	1–4, 2 h
Eye irritation	0–3, ½ h	0–4, ½ h	0–4, ½ h	1–5, ½ h
	0–3, 1 h	0–3, 1 h	0–4, 1 h	1–5, 1 h
	0–3, 2 h	0–4, 2 h	0–4, 2 h	1–5, 2 h
Nose irritation	Similar to eyes	Similar to eyes	Similar to eyes	Similar to eyes
Throat irritation	0–2, ½ h	0–3, ½ h	0–4, ½ h	0–5, ½ h
	0–3, 1 h	0–3, 1 h	0–4, 1 h	0–5, 1 h
	0–3, 2 h	0–4, 2 h	0–4, 2 h	0–5, 2 h
Urge to cough	0–1, ½ h	0–2, ½ h	0–2, ½ h	0–5, ½ h
	0–2, 1 h	0–2, 1 h	0–3, 1 h	0–3, 1 h
	0–2, 2 h	0–4, 2 h	0–4, 2 h	0–4, 2 h
Irritation of chest	Similar to cough	Similar to cough	Similar to cough	Similar to cough
General discomfort	0–1, ½ h	0–3, ½ h	0–3, ½ h	0–4, ½ h
	0–1, 1 h	0–3, 1 h	0–3, 1 h	0–5, 1 h
	0–2, 2 h	0–3, 2 h	0–4, 2 h	0–5, 2 h

Source: Verberk, M.M., *Int. Arch. Occup. Environ. Health*, 39, 73, 1977.

[a] Expert subjects were familiar with the effects of ammonia; the nonexpert subjects were not.

[b] Based on a scale of 1–5; 0 = sensation; 1 = just perceptible; 2 = distinctly perceptible; 3 = nuisance; 4 = offensive; 5 = unbearable.

[c] Only four of the nonexpert subjects tolerated the ammonia for 1 h; none of the nonexpert subjects tolerated the ammonia for 2 h, the upper range of the score in the table is the same as recorded at 1 h or 110 ppm after 2 h (urge to cough).

nonexpert subjects. This study also showed that general knowledge about the chemical may help to alleviate the concern about exposure and the intensity of the symptoms experienced during exposure.

8. The effect of ammonia on nasal airway resistance (NAR) was studied in atopic and nonatopic human subjects (McClean et al., 1979). Ammonia (100 ppm at a pressure of 9 N/cm^2) was introduced into each nostril for 5, 10, 15, 20, and 30 s. NAR was measured using a pneumotachograph attached to a face mask every minute for 5 min then every 2 min for 10 min (total of 10 measurements over a 15 min period). The same subjects were used for each successive ammonia exposure, which immediately followed the NAR measurements. The nonatopic subjects were screened based on strict criteria that included a questionnaire, physical examination, spirometry, nasal smear for eosinophils, and a battery of 19 prick and 6 intracutaneous tests. Nonatopic subjects could have no personal or immediate family history of atopic disease (allergic rhinitis, asthma, or atopic dermatitis), have no more than 5% eosinophils in their nasal smears, and have a negative prick test reaction. Atopic subjects were screened based on a characteristic (not otherwise specified) history of allergic rhinitis and at least one 3+ or 4+ prick test reactions. Some of the atopic subjects had a history of asthma. All subjects had been symptom free for several weeks before the study, and none were taking medications that would influence skin or mucosal tests. Baseline NAR measurements were made for a 15 min period before introducing the ammonia. Additional test included introducing 0.1 mL aerosolized phosphate-buffered saline, 0.1 mL atropine, or 0.1 mL chlorpheniramine maleate into the nostrils, each followed by ammonia for 20 s. The NAR after ammonia exposure to nonatopic and atopic subjects increased significantly with the time of exposure from 5 to 20 s. Only a small further increase was noted for subjects exposed for 30 s compared with 20 s. The percent increased for atopic compared with nonatopic subjects was similar, and there was no difference between the allergic rhinitis subjects with or without a history of asthma. Atropine inhibited the response to ammonia in atopic and nonatopic by up to 89%, whereas chlorpheniramine had no effect on the NAR induced by ammonia. The study authors noted that the results of the inhibitor study suggest that the effects of ammonia are mediated primarily by parasympathetic reflex on the nasal vasculature and not via histamine release.

9. The respiratory effects in a group of 58 workers (51 production and 6 maintenance) chronically exposed to ammonia vapor (9.2 ± 1.4 ppm, mean ± standard deviation) were compared with the effects in a group of 31 plant workers with essentially no exposure to ammonia (0.3 ± 0.1 ppm, mean ± standard deviation) (Holness et al., 1989). During a 1-week period, the workers were assessed based on a questionnaire, sense of smell, and pulmonary function. There were no differences between the two groups.

10. The threshold concentration of ammonia required to elicit reflex glottis closure, which is a protective response stimulated by inhaling irritant or noxious vapors at concentrations too small to produce cough, was studied in volunteers of varying age (Erskine et al., 1993). It is accompanied by a brief pause in inspiration. The investigators measured glottis closure in 102 healthy nonsmoking subjects, ranging from 17 to 96 years old, after single intermittent breaths of ammonia vapor using an inspiratory pneumotachograph. The results showed a strong positive correlation coefficient of 0.85 between age and the threshold concentration. The younger subjects were more sensitive, with the reflex response occurring at 571 ± 41.5 ppm (±standard error) in subjects 21–30 years old and 1791 ± 52 ppm (±standard error) in subjects 86–95 years old. The threshold was about 1000 ppm for 60-year-old subjects. The data showed that younger people are about three times more sensitive to the induction of this protective mechanism (glottis closure) by ammonia than the elderly.

11. Twelve healthy volunteers were randomly exposed to 0, 5, or 25 ppm ammonia, 3 h/exposure on three occasions (Sundblad et al., 2004). During the exposure periods, one half of the time was spent resting, and the remainder spent exercising. Symptoms were monitored throughout exposure, and bronchial responsiveness to methylcholine, pulmonary functions tests, and exhaled nitric oxide were conducted before and 7 h after exposure. In addition, nasal lavage and peripheral blood samples were collected 7 h after the exposure. The incidence of symptoms increased in all categories in the 25 ppm subjects when compared with controls. No differences were noted for methylcholine treatment, pulmonary function tests, or exhaled nitric oxide levels. At 25 ppm, there was no indication of adaption to ammonia during exposure. Summary of volunteer studies on ammonia is given in Table 13.4.

13.3.3 RESULTS OF MODELING STUDIES

In addition to human toxicity studies, modeling exercises have been conducted for three major accidental releases. These provide some assistance in determining lethality levels for humans.

13.3.3.1 For the Houston and Potchefstroom Accidents

The WHAZAN gas dispersion model incorporated meteorological data and physicochemical data for ammonia to predict the concentration isopleths for ammonia released during the Houston and Potchefstroom accidents (Pedersen and Selig, 1989).

In the Houston accident, the 10,000 ppm isopleth extended 600 m long and 350 m wide, the 5000 ppm isopleth to 835 m long and 430 m wide, the 2500 ppm isopleth to 875 m long and 420 m wide, and the 1200 ppm isopleth to 1130 m long and 400 m wide. The investigators reported that their model overestimated distance to zero death (200 m) by 2.9 times for the Houston accident and 2.5 times for the Potchefstroom accident. The authors concluded that, based on the distance of zero fatalities, the risk for a *few minutes of exposure to ammonia* for the four isopleths would be as follows:

10,000 ppm	Very high risk to the general population (g.p.)
5,000 ppm	High risk to g.p.; very high risk to vulnerable population (v.p.)
2,500 ppm	Some risk to g.p.; high risk to v.p.
1,200 ppm	Predicted limit of cloud for emergency planning purposes

The authors also derived a probit equation that estimated the LC_{50} for exposure of the general population to ammonia as follows: Probit = 1.85 ln D − 35.9, where the dose is equal to the concentration squared × time of exposure as described elsewhere (Withers et al., 1986). The LC_{50} for a 30 min exposure to the general population was calculated to be 11,500 ppm. The probit equation for vulnerable members of the population (elderly people, children, and people with respiratory or heart disorders) was as follows: Probit = 3.04 ln D − 59.1. The probit equations are valid for exposures times of 5–60 min as there are no data supporting its use for exposure times greater than 60 min (Pedersen and Selig, 1989). The details for deriving the probit equation were not provided in this abbreviated form of a much longer report that assessed the consequences of accidental releases of ammonia and submitted to the Danish National Agency for Environmental Protection. The WHAZAN model as well as other dose reconstruction models contains many uncertainties in its predictions of atmospheric concentrations. The WHAZAN model cannot be adequately evaluated because of the lack of input data used to derive the probit equation and LC_{50} value.

The HGSYSTEM gas dispersion model was developed by Shell Research Ltd. with support from industry and government (Mudan and Mitchell, 1996). It has been subjected to testing and validation and benchmarked against field tests with irritant and flammable gases. Concentration estimates derived using the HGSYSTEM and other gas dispersion models are associated with a substantial level of uncertainty, because only a few variables are known for the reconstruction of

TABLE 13.4
Summary of Studies on Volunteers Based on Concentration × Time

Conc. (ppm)	C × t (ppm·min)	Effect	Reference
32	160	Nasal dryness	Industrial Bio-Test Lab (1973)
50	250	Nasal dryness	Industrial Bio-Test Lab (1973)
30	300	Moderately intense to penetrating odor; barely detectable irritation	MacEwen et al. (1970)
72	360	Nasal, eye, and throat irritation	Industrial Bio-Test Lab (1973)
50	500	Highly penetrating odor; moderate irritation	MacEwen et al. (1970)
143	715	Nose, eye, throat, and chest irritation, lacrimation	Industrial Bio-Test Lab (1973)
50	1,500	Moderately intense odor, moderate irritation to eyes and nose, mild irritation to throat and chest, slight urge to cough, slight general discomfort	Verberk (1977)
80	2,400	Highly intense odor, highly intense eye and nose irritation, moderate throat and chest irritation, mild urge to cough, moderate general discomfort	Verberk (1977)
50	3,000	Highly intense odor; moderate irritation to eyes, nose, throat, and chest; mild urge to cough; slight general discomfort	Verberk (1977)
110	3,300	Highly intense odor; highly intense eye, nose, throat, and chest irritation; mild urge to cough; and moderate general discomfort	Verberk (1977)
140	4,200	Highly intense odor; unbearable eye, nose, throat, and chest irritation; mild urge to cough; and moderate general discomfort	Verberk (1977)
80	4,800	Highly intense odor; moderate eye, nose, throat, and chest irritation; mild urge to cough; moderate general discomfort	Verberk (1977)
1000	5,000	Immediate urge to cough	Silverman et al. (1949)
50	6,000	Highly intense odor; moderate irritation to eyes, nose, throat, and chest; mild urge to cough; mild general discomfort	Verberk (1977)
110	6,600	Highly intense odor; highly intense eye, nose, throat, and chest irritation; moderate urge to cough; and moderate general discomfort	Verberk (1977)
500	7,500–15,000	Nose and throat irritation, nasal dryness and stuffiness, excessive lacrimation, hyperventilation, unbearable	Silverman et al. (1949)
140	8,400	Highly intense odor; highly intense eye, nose, throat, and chest irritation; intense urge to cough; moderate general discomfort	Verberk (1977)
80	9,600	Highly intense odor; highly intense eye, nose, throat, and chest irritation; highly intense urge to cough; and moderate general discomfort	Verberk (1977)
100	12,000–36,000 per day	No adverse effects on respiratory function and no increase in frequency of eye, nose, throat irritation	Ferguson et al. (1977)
110	13,200	Highly intense odor; highly intense eye, nose, throat, and chest irritation; urge to cough; and moderate general discomfort	Verberk (1977)
140	16,800	Highly intense odor; unbearable eye, nose, highly intense throat, and chest irritation; highly intense urge to cough; unbearable general discomfort	Verberk (1977)

historical accidents. Uncertainty is attributable to a "lack of complete understanding of the complex mass and heat transport phenomena that take place during the atmospheric dispersion" and an inability to "accurately characterize all necessary input parameters required for a physical model."

The HGSYSTEM gas dispersion model was used to estimate atmospheric ammonia concentrations generated at the time of the ammonia accident in Potchefstroom, South Africa (Mudan and Mitchell, 1996). They provided upper-bound (wind speed = 1 m/s) and lower-bound (wind speed = 2 m/s) estimates of ammonia concentration based on distance from the release point and the time after release. Instantaneous concentrations were estimated to be in excess of 500,000 ppm (upper bound) within 50 m of the release point. The model predicted rapidly decreasing concentrations, such that, by 1 min after the release, concentrations would fall below 100,000 ppm. It was estimated that personnel were exposed to ammonia concentrations exceeding 50,000 ppm for the first 2 min, decreasing to 10,000 ppm during the next 3–4 min. The charts provided by the authors of the South Africa accident showed that 10 workers were within 50 m of the release point at the time of release (Zone 1); 7 died (100% mortality for workers exposed outside). All survivors in Zone 1 remained sheltered inside buildings and therefore would not have experienced the outside atmospheric ammonia concentrations predicted by the model. Five deaths occurred in Zone 2 (50–100 m). Workers in Zone 2 who were upwind and outside at the time of the release survived, and those who escaped in an upwind direction also survived. Workers in Zone 2 who were downwind and outside at the time of release or attempted to escape downwind did not survive (except for one worker who escaped downwind) (83% mortality of workers exposed). All Zone 2 victims who died were outside, whereas individuals who were inside buildings survived. Five deaths occurred in Zone 3 (100–~200 m). Four victims were found downwind and >150 m from the release point, and another victim was found <150 m from the release point and in a crosswind location. The charts did not show the location or number of any survivors downwind and inside or outside buildings in Zone 3; that is, no data were available from the charts to determine if there were individuals who remained outside buildings in Zone 3 and survived. Therefore, the mortality rate cannot be calculated for Zone 3. It appears that within 150 m of the release point, individuals downwind of the ammonia cloud and outside a building were not likely to survive, but individuals downwind and sheltered indoors or those upwind whether or not they were sheltered indoors were likely to survive. Thus, the lack of data on survivors in the path of the plume precludes estimating ammonia concentrations associated with zero mortality.

Using the results of the HGSYSTEM gas dispersion model, the 5 min ammonia concentrations of 87,479 ppm for a 60% mortality rate, 73,347 ppm for a 26% mortality rate, and 33,737 ppm for zero mortality were calculated (RAM TRAC, 1996). In predicting these rates, RAM TRAC assumed that exposures to individuals sheltered indoors were equal to exposures to individuals outdoors (i.e., modeled atmospheric concentrations); RAM TRAC also assumed that individuals located upwind experienced the same exposures as those downwind of the release. By including individuals sheltered inside buildings and those located outside and upwind of the release, RAM TRAC underpredicted the mortality responses. Additionally, the HGSYSTEM gas dispersion model does not predict indoor concentrations from accidental releases (Mazzola, 1997). RAM TRAC's interpolation or extrapolation of modeled values also predicted a 5 min LC_{50} of 83,322 ppm. Predicted concentrations were derived from the arithmetic average of the upper- and lower-bound atmospheric estimates provided by Mudan and Mitchell (1996) in their gas dispersion model of the ammonia accident in Potchefstroom, South Africa. RAM TRAC (1996) used the arithmetic mean to calculate the time-weighted-average (TWA) concentration for the total dispersion time after the release, which ranged from 7.458 to 9.333 min. The 5 min concentrations were based on modeled concentrations averaged over theoretical durations of 7–9 min. It appears, however, that RAM TRAC (1996) applied the Haber's rule, $C \times t$, rather than ten Berge's equation (ten Berge et al., 1986), when extrapolating the TWA ammonia concentrations associated with various mortality rates to 5 min exposure scenarios. In case of accidental releases, concentrations downwind of the release are a function of time; however, if the exposure–time relationship is converted to a mortality response over a period of time different from that of the actual exposure time, then ten Berge's equation is used to determine the exposure–time–mortality relationship.

Based on their assessment of the data on the human response to ammonia, the following summary was prepared (Markham, 1986; Pedersen and Selig, 1989):

Conc. (ppm)	Exp. Time (min)	Response
72	5	Some irritation
330	30	Concentration tolerated
600	1–3	Eyes streaming within 30 s
1000	1–3	Eyes streaming immediately, vision impaired but not lost, breathing intolerable to most participants
1500	1–3	Instant reaction is to get out of the environment

These data suggest that humans cannot or will not tolerate exposure to ammonia at irritating concentrations for extended periods of time. However, no actual human exposure response data were provided to substantiate these observations.

13.4 CONCLUSIONS BASED ON HUMAN DATA

A comparison of data from humans is given in the following text using the data of industrial experience (Henderson and Haggard, 1943); studies on volunteers, both experienced healthy subjects and those with no previous exposure to ammonia as tabulated; and data derived from modeling studies (Markham, 1986; Pedersen and Selig, 1989; Fthenakis, 2004).

	Industrial Experience (ppm)	Experienced Volunteers (ppm)	Naive Volunteers (ppm)	Modeling Data (ppm)
No odor	—	5	—	—
Odor perception	53	25	—	—
Nasal irritation	—	140	30	72
Throat irritation	408	143	50	—
Eye irritation	698	140	50	72
Cough reflex	1720	140	110	1000

Based on observed toxicity and accident reconstruction or extrapolation, the ability to escape and lethality levels are as follows:

Could escape	1000 ppm	200–300 ppm	140 ppm	600 ppm
Onset of lethality	2500–5000 ppm	—	—	1200 ppm

The levels for the naive volunteers are low because they could leave the chamber whenever they felt uncomfortable. The ability to escape levels for experienced volunteers may be low because no higher levels were tested. The levels for industrial experience are high because of adaptation to ammonia and motivation to get the job done. The Emergency Response Planning committee (AIHA, 2014) and the Acute Emergency Guideline Level (NRC, 2007) committee (a FACA committee) published the following levels for a 1 h exposure:

(Onset of irritation)	ERPG-1	25 ppm	AEGL-1	30 ppm
(Ability to escape)	ERPG-2	150 ppm	AEGL-2	160 ppm
(Onset of lethality)	ERPG-3	1500 ppm	AEGL-3	1100 ppm

TABLE 13.5
Summary Dose-Response Data For Human Exposure To Ammonia

Toxic End Point	4 h (ppm)	60 min (ppm)	30 min (ppm)	15 min (ppm)
Odor perception	5	5	5	5
Onset of irritation	25	25	25	25
Severe irritation	80	100	110	200
Intolerable (escape)	110	140	150	400
Lethality	750	1200	1600	2400

These levels are designed for the general public including children, the elderly, and infirmed. The ERPG-1/AEGL-1 levels agree with the summary data given earlier—25 ppm is the threshold for irritation.

The ERPG-2 level agrees well with the summary data, while the AEGL-2 level is low because the discussion relied heavily on *urge-to-cough* data from Verberk (1977). Based on incapacitating eye irritation, the level of 110 ppm is low.

The ERPG-3 of 1,500 ppm level may be low based on the modelers' calculation of a 30 min LC_{50} for healthy humans as 11,500 ppm, but not for sensitive subpopulations. The rule of thumb is that the onset of lethality is generally a factor of 3 below the LC_{50} (Rusch et al., 2009). For healthy workers and volunteers, that would be 3833 ppm for 30 min, while for sensitive populations, the onset of lethality would be lower, perhaps an additional factor of three. To extrapolate to a 1 h exposure, the level would be approximately 1900 ppm for acclimated workers. The AEGL-3 level is 1100 ppm.

Using all of these data and analyses, the dose–response for the general population for exposure to ammonia is summarized in Table 13.5.

These lethality levels do not match what was calculated by the modelers. The modelers acknowledged that some of their calculations were considerably different from what was observed, for example, the distance to zero deaths (Pederson and Selig, 1989). A recent report by the Nordic Expert Group agreed that the NOAEL for ammonia is 5 ppm and the LOAEL is 25 ppm (Johanson et al., 2006).

13.5 SUPPORTING ACUTE LETHALITY AND IRRITATION DATA FROM ANIMAL STUDIES

13.5.1 PULMONARY IRRITATION DATA

The RD_{50} is the concentration expected to elicit a 50% reduction in respiratory rate (Barrow et al., 1978). The authors stated that the RD_{50} concentration would elicit intense sensory irritation and is expected to be rapidly incapacitating to humans. Groups of four outbred male Swiss Webster mice were exposed to ammonia by inhalation for 30 min. The authors did not report the concentration of ammonia inhaled by the mice, but judging by the graphic representations, the concentrations were 100, 200, 400, and 800 ppm. The maximum depression in respiratory rate was achieved within the initial 2 min of exposure, after which the response diminished. The RD_{50} was 303 ppm (95% confidence limits = 188–490 ppm) for a 30 min inhalation exposure to ammonia. There was no microscopic examination of the respiratory tract.

In a follow-up study (Buckley et al., 1984), the histopathologic effects of repeated exposures to ammonia at 303 ppm were evaluated. Groups of 16–24 male Swiss Webster mice were exposed to 0 or 303 ppm of ammonia 6 h/day for 5 days. The respiratory tract was examined in one half of the animals that were killed immediately after terminating the exposure. The other half were killed 3 days later. Histopathological findings, which were confined to the respiratory epithelium of the nasal cavity, included minimal exfoliation, erosion, ulceration, and necrosis; moderate inflammatory changes; and slight squamous metaplasia.

In a similar study, groups of 10 male OF1 mice were exposed to ammonia at analytically measured concentrations of 0.3 of the RD_{50} (78.0 ppm), the RD_{50} (257 ppm), or three times the RD_{50} (711 ppm) 6 h/day for 4, 9, or 14 days (Zissu, 1995). The three target concentrations were 90.9, 303, and 909 ppm. Control mice were exposed to filtered air. The entire respiratory tract was examined microscopically. No clinical signs of toxicity were noted for mice exposed to ammonia. Pathologic lesions including rhinitis with metaplasia and necrosis were seen only in the respiratory epithelium of the nasal cavity of mice inhaling 711 ppm (three times the RD_{50}); the severity of the lesions increased with the duration of exposure, ranging from moderate on day 4, severe on day 9, to very severe on day 14. No lesions were seen in the controls or in mice inhaling the lower concentrations of ammonia. In contrast to the study conducted using Swiss Webster mice (Buckley et al., 1984), this study showed no lesions in the nasal cavity of OF1 mice exposed to 257 ppm.

It should be noted that in a comparison of RD_{50} levels in Swiss Webster mice and OF1 mice, data for the OF1 mouse are often different from data for Swiss Webster mice. The difference seen in the OF1 mouse can be higher or lower than the data for the Swiss Webster mouse. For ammonia, it should be noted that the OF1 mouse is significantly different (up to 10-fold different) from Swiss Webster mice in response to many sensory irritants. For ammonia, the RD_{50} in OF1 mice is 257 ppm and in Swiss Webster mice is 303 ppm, essentially the same number; however, the differences in histopathology point out that differences between the strains of mice is significant even when the RD_{50} values are similar.

An RD_{50} of 300 ppm would suggest that significant irritation would be seen in humans at 3 ppm. However, the respiratory epithelium of rodents is much more sensitive to chemical exposure than human respiratory epithelium. Thus, instead of taking a 100-fold safety factor in extrapolating to humans, only a factor of 3 is sufficient, giving an irritation level of 100 ppm. This supports the data seen in humans.

13.5.2 Acute Lethality Data

Groups of 10 male CFE rats were exposed to 0, 6210, 7820, or 9840 ppm (4343, 5468, and 6881 mg/m^3) of ammonia for 1 h; surviving animals were observed for 14 days (MacEwen and Vernot, 1972). Signs of eye and nasal irritation were seen immediately, followed by labored breathing and gasping. Surviving animals exposed to the low concentration weighed less than controls on day 14, and gross examination showed mottling of the liver and fatty changes at the two highest concentrations. All rats exposed to 6210 ppm survived, and eight exposed to 7820 ppm and nine exposed to 9840 ppm died. The LC_{50} was 7338 ppm (95% CI = 6822–7893 ppm).

The LC_{50} was determined for mice exposed to ammonia (8,723–12,870 ppm) for 10 min and observed for 10 days (Silver and McGrath, 1948). Each group consisted of 20 mice (sex and strain not specified). During exposure, the mice closed their eyes, exhibited great excitement initially but soon became quiet, gasped, pawed, scratched their noses, and convulsed before dying. At the lowest concentration of 8,723 ppm, 25% of the animals died, and 80% died at the highest concentration of 12,870 ppm. Overall, 90/180 mice died during the second 5 min of exposure, and another 8 died during the observation period. The other animals surviving exposure recovered rapidly. The LC_{50} for the 10 min exposure was 10,096 ppm.

Groups of 10 male CF1 mice were exposed to ammonia at analytically measured concentrations of 0, 3600, 4550, or 5720 ppm (0, 2520, 3185, and 4004 mg/m^3) for 1 h (MacEwen and Vernot, 1972). Immediately upon exposure, the animals showed signs of nasal and eye irritation followed by labored breathing and gasping. Animals surviving the low and intermediate concentrations lost weight during the 14-day observation period. Gross examination of surviving mice showed mild congestion of the liver at the intermediate and high concentrations. Three mice exposed to 4500 ppm died, nine exposed to 5720 ppm died, but none exposed to 3600 ppm died. The LC_{50} was 4837 ppm (95% CI = 4409–5305 ppm).

The 1 h LC_{50} value for male albino ICR mice (12/group) was determined using concentrations of 1190–4860 ppm of ammonia (Karpeghian et al., 1982). The animals were observed for 14 days following exposure. A control group exposed to air only was included for comparison. Clinical signs, which were noted immediately and lasted 5–10 min, included excitation/escape behavior, rapid vigorous tail revolution, blinking and scratching (eye and nose irritation), and dyspnea. As signs of irritation decreased, the animals became less active, and other signs of toxicity were noted, including tremors, ataxia, clonic convulsions, frothing, coma, final tonic extensor seizure, and death. At the higher concentrations, almost all deaths (90%) occurred during the first 15–20 min of exposure and as late as 45 min at the lower concentrations. Additional deaths occurred during the first 3 days following exposure. All deaths occurred at concentrations ≥3950 ppm (25%–100% mortality). The mortality response was 22/24 at 4860 ppm, 8/12 at 4490 ppm, 5/12 at 4220 ppm, 3/12 at 3950 ppm, and 0/12 at 3440, 2130, 1340, and 1190 ppm. The 1 h LC_{50} was 4230 ppm for ammonia. Other effects observed during the 14-day observation period included lethargy, dyspnea, weight loss, and a *humped-back* appearance. Pathologic lesions occurring in mice dying during exposure included acute vascular congestion, intra-alveolar hemorrhage, disruption of alveolar septal continuity, and acute congestion of hepatic sinusoids and blood vessels. In animals surviving the 14-day observation period, pathologic lesions included mild-to-moderate pneumonitis (dose-related severity), focal atelectasis in the lungs (4860 ppm), and degenerative hepatic lesions (dose-related severity, 3440–4860 ppm). The author did not discuss specific effects in animals exposed to concentrations less than 3440 ppm.

These data are summarized in the following array.

Species	1 h LC_{50} (ppm)	Lowest Lethal Concentration (ppm)	Highest Nonlethal Concentration (ppm)
Rat	7340	7820	6210
Mouse	4840	4500	3600
Mouse	4230	3950	3440

In each study, the highest nonlethal level is not very much below the LC_{50} value, certainly not a factor of 3 below. For sensory irritants that lead to pulmonary edema, the mice and rats are not much different from humans in sensitivity to the chemical. Thus, if we use a factor of 3 below the lowest of the nonlethal levels (Rusch et al., 2009) given earlier, the threshold for human lethality might be 1150 ppm, which is very close to the 1 h AEGL-3, although the AEGL number was derived using much different logic.

It is interesting to note that the modelers (Pedersen and Selig, 1989) came to the conclusion that the concentration of interest, 1200 ppm for ammonia (which appears to be the threshold of lethality), agrees with the human data for a 1 h exposure and animal data for a 1 h exposure. Thus, the summary dose–response data given in Table 13.6 appear valid for human exposures to ammonia based on human toxicity studies, extrapolation from animal studies, and modeling studies.

13.6 CONCLUSIONS BASED ON ANIMAL AND HUMAN EXPOSURES

For human exposures, all symptoms and toxicity end points are listed by exposure concentration in Table 13.6.

A comparison of an abbreviated list of symptoms for the general public and for workers is given in Table 13.7. The worker data are taken from Henderson and Haggard (1943), which included data originating with Lehmann (1886). Some of data appear to be anecdotal as well. None the less, the data appear to be valid for long-term acclimated workers. At lower levels, the concentrations for specific effects are 10-fold higher for acclimated workers compared to effects reported for uninured volunteers. At higher concentrations, the end points, for example, lethality, are not well documented, and the effects reported in workers are only twofold higher than for the general public.

TABLE 13.6

Scale of Concentration and Associated Symptoms and Toxic Effects in Uninured Humans for Exposures Lasting At Least 5 min

Concentration (ppm)	Effect	Reference
Odor		
5	Odor threshold	Mac
Irritation—eye and nose		
25	Very mild eye and nose irritation—30 min	Sund
	Mild chest irritation, urge to cough, and nausea—180 min	Sund
30	Irritation threshold—10 min	Mac
32	No irritation—5 min	IBT
40	Headache, nausea, reduced appetite—chronic	Naka[a]
Irritation—throat and chest		
50	Mild eye and nose irritation—10 min	Mac
	Mild throat and chest irritation—30 min	Ver
	Moderate throat and chest irritation—60 min	Ver
	Mild urge to cough—120 min	Ver
72	Mild eye, nose, and throat irritation—5 min	IBT
80	Intense eye and nose irritation—30 min	Ver
	Mild urge to cough—30 min	Ver
	Moderate throat and chest irritation—30 min	Ver
	Intense throat and chest irritation—120 min	Ver
	Intense urge to cough—120 min	Ver
	Moderate discomfort—120 min	Ver
110	Intense eye, nose, throat, and chest irritation—30 min	Ver
	Mild urge to cough—30 min	Ver
	Moderate urge to cough—60 min	Ver
	Intense urge to cough—120 min	Ver
	Moderate discomfort—30 min	Ver
134	Moderate eye, nose, and throat irritation—5 min	IBT
	Mild chest irritation—5 min	IBT
	Moderate lacrimation—5 min	IBT
Intolerable		
140	Unbearable eye, nose, throat, and chest irritation—30 min	Ver
	Mild urge to cough—30 min	Ver
	Intense urge to cough—120 min	Ver
	Unbearable discomfort—60 min	Ver
170	Intolerable in a few minutes	Weat
200	Vision affected, but not seriously impaired—3 min	Wall
400	Vision not seriously impaired—1 min	Wall
Hyperventilation/incapacitation		
500	Hyperventilation, lacrimation (face mask)—30 min	Silv
	Hypesthesia around mouth and nose, unbearable	Silv
600	Eyes streaming in 30 s	Wall
700	Immediate lacrimation; incapacitating in a few minutes	Wall
Skin irritation		
1000	Immediate lacrimation, cough, intolerable; panic level	Wall
	Skin irritation in few minutes	Wall
	Coughing	Silv
Approaching lethality		
1500	Immediately incapacitating	Wall
1700	Convulsive coughing	CIA[a]

[a] Do not have the original reference.

TABLE 13.7

Comparison of the Concentration at Which Selected Ammonia Effects Are Noted in Ammonia Workers versus the General Public in 30 min

Effect	Concentration	
	Workers (ppm)	General Public (ppm)
Odor threshold	50	5
Throat irritation	400	50
Eye irritation	700	25
Cough	1,700	500
Dangerous in 30 min (pulmonary edema)	4,500	2,000
Rapidly fatal (convulsions)	10,000	5,000

For patients suffering from the more serious toxic effects, five reports divide the exposures into categories of mild, moderate, and severe (Caplin, 1941; O'Kane, 1973; Montague and McNeill 1980; Arwood et al., 1985; Leung and Loo, 1992). The five sources give similar descriptions for the three categories (the patients reported by Montague were assigned to the mild and moderate categories only). A combined summary of effects for these three levels of exposure is given in Table 13.8.

TABLE 13.8

Toxic Effects Noted in the General Public Following Mild, Moderate, or Severe Exposure to Ammonia

Exposure Intensity	Toxic Effects Reported—One or More of the Following
Mild	Inflammation of conjunctiva and upper respiratory tract; swelling of the eyelids; burns on the skin; pain on swallowing; hoarseness; X-ray clear; no respiratory distress, coughing, or shock; discharged in hours.
Moderate	Any of the signs or symptoms from the mild category; sore eyes; continued lacrimation and swelling of the eyelids; blood-stained sputum; no vomiting; difficulty swallowing; denuded areas of the oral cavity; wheezing; productive cough; respiratory distress; may require intubation; bronchospasm; nonspecific X-ray changes; moderate dyspnea followed by bronchopneumonia; inflammation of the bronchi and bronchioles; fatal to approximately 25% within 3–4 days; survivors discharged in 6–8 days.
Severe	Incapacitated; severe respiratory distress; slight cyanosis; nausea; pulmonary edema; massive nasal discharge; retching/convulsing; wheezing; persistent hoarseness and productive cough; severe dyspnea with frothy sputum; inflamed eyes, lips, tongue; edema of pharynx; inflamed lung parenchyma; shock; fatal to approximately 67% within 2 days; survivors discharged in 9–10 days.

In comparing Table 13.5 and Table 13.7, the exposure concentration that matches the three categories are the following:

Mild exposure	Approximately 500+ ppm based on excessive lacrimation, skin irritation and burns, hyperventilation, but no cough.
Moderate exposure	Approximately 1000+ ppm based on pulmonary edema, cough, hyperventilation, and bronchospasm.
Severe exposure	Approximately 1500+ ppm based on frothy sputum, incapacitation, and lung parenchyma inflammation.

Armed with these concentration effect levels, animal data will be compared with these results to see if concordance exists for the two sets of data. In reviewing animal studies, a number of studies were rejected based on the methodology. The Hilado et al. (1977) and the Boyd et al. (1944) studies were conducted as static exposures instead of dynamic exposures. In static exposures, the starting concentration decreases over time so that the average exposure concentration is not known. As the average concentration during exposure is considerably less than the starting concentration, this methodology often reports higher values for the study. Thus, an LC_{50} based on static exposures may be considerably higher than the LC_{50} based on dynamic, analytically determined concentrations. In addition, the Appelman et al. (1982) study was also rejected because the chamber loading (total animal weight) was too high for the chamber size. According to good laboratory practices and standard practice in inhalation toxicology, the animal loading should not exceed 5% of the chamber volume. This minimizes the effects of the test material depositing on the animal fur, caging, and the chamber walls. This shows up in the data by great differences in lethality of males and females, as the females were always loaded away from the entry port of the chamber. Specific effects are reported at higher concentrations in studies with high chamber loading when compared to studies in which the loading is below the 5% level. Based on the remaining animal inhalation studies, an effects table for the concentrations studied is given in Table 13.9.

It is quite clear from the data of Coon et al. (1970) that rats and mice are much more sensitive to ammonia than are guinea pigs, rabbits, dogs, and monkeys. Based on lethality studies, it is also clear that mice are about twice as sensitive as rats, but the levels are not significantly different. In Table 13.10, the lethality data for rats and mice are compared.

The LC_{50} values are not significantly different for 1 and 2 h exposures for both rats and mice. Most deaths occur during exposure, which suggests that there may be a threshold of lethality based on the product of concentration and time ($C \times t$). Using the highest nonlethal levels, the product of $C \times t$ for rats is 372,600 ppm·min and for mice is 216,000 ppm·min based on the 1 h studies of MacEwen, which provided the highest nonlethal levels. These numbers would be just below the threshold for lethality. As the mouse is the most sensitive species, the threshold for uninured humans may be a factor of 3 below this level. For a 30 min exposure, the lethal concentration for mice would be 7200 ppm, and after dividing by 3, the threshold of lethality for humans would be 2400 ppm for a 30 min exposure. A similar level, 2484 ppm, is derived using the level for rats and dividing by a factor of 5 for the next most sensitive species.

A threshold for lethality in humans of 2400–2500 for a 30 min exposure based on extrapolation from animal studies compares favorably with the dangerous level of 2000 ppm for the general public given in Table 13.3. Exposure to higher levels would lead to desquamation of the trachea and bronchi leading to rapid death. Thus, the proposed concentration ranges for mild, moderate, and severe exposure seem appropriate for exposures lasting at least 30 min. These ranges are as follows:

Mild exposure	Threshold = 500+ ppm based on excessive lacrimation, skin irritation, and burns but no cough.
Moderate exposure	Threshold = 1000+ ppm based pulmonary edema, cough, hyperventilation, and bronchospasm.
Severe exposure	Threshold = 1500+ ppm based on frothy sputum, incapacitation, and lung parenchyma inflammation.

There would be a range of concentrations above the threshold that would allow for biological variation within the general public. Lung-compromised individuals, for example, asthmatics, might exhibit these symptoms at lower concentrations.

TABLE 13.9

Toxic Effects Noted in Animal Exposures to Ammonia

Species Conc. (ppm)	Duration (min)	Irritation	Mild	Moderate	Severe	NOAEL (ppm)	Nonlethal (ppm)	Lethal (ppm)	LC$_{50}$ (ppm)	Reference
Rat										
2000	240							2000		Carpenter et al. (1949)
1000	960							1000		Weedon et al. (1940)
Varied	1440+		377	654			377	654		Coon et al. (1970)
Varied	480+		1107		223		1107			Coon et al. (1970)
660	1440+				660			660		Coon et al. (1970)
60	1440+					60				Coon et al. (1970)
Male	Varied	240		6210	7820		6210	7820	7338	MacEwen and Vernot (1972)
Male	Varied	60			9870			9870	9830	Appelman et al. (1982)
	Varied	480–570	800–1070	820–1430	4000		2000–2 h	4000–2 h	10,930–4 h	Kimmerle (1974)
	500	1440+	500				500			Richard et al. (1978a)
	Varied	1440	1157			1157	1157			Schaerdel et al. (1983)
	Varied	1440+	714			165	714			Schaerdel et al. (1983)
	Varied	60	300			100	300			Tepper et al. (1985)
Guinea pigs	Varied	480+	200			140	220	680		Coon et al. (1970)

Species									Reference
Mice	1000	960						1000	Weedon et al. (1940)
	Varied	10		6100		6100	7060		Silver and McGrath (1948)
	Varied	60				3600	4550	4837–1 h	MacKwen et al. (1970)
	Varied	120		3420			3420	4760–2 h	Kimmerle (1974)
	Varied	60				1190	3950	4230–1 h	Karpeghian et al. (1982)
	Varied	60				2130	4860		
	Varied	60				3440	4220		
	Varied	360	711	5000	78	711			Zissu (1995)
Rabbits	Varied	480+	257		220	680			Coon et al. (1970)
	50/100	180	680		100	100			Mayan and Merilan (1972)
	Varied	180	1000	3000	6000+	3000	5000		Richard et al. (1978b)
Dogs	Varied	480+	680		220	680			Coon et al. (1970)
Monkeys	Varied	480+	680		220	680			Coon et al. (1970)

TABLE 13.10
Lethality Data for Rats and Mice

Species	LC$_{50}$	Concentration Lowest Lethal (ppm)	Concentration Highest Nonlethal (ppm)	Reference
Rat	7,340 ppm–1 h	7,820	6,210	MacEwen and Vernot (1972)
Rat	10,930 ppm–2 h			Kimmerle (1974)
Mouse	4,840 ppm–1 h	4,500	3,600	MacEwen and Vernot (1972)
Mouse	4,230 ppm–1 h	3,950	3,440	Kapeghian et al. (1982)
Mouse	4,760 ppm–2 h		3,420	Kimmerle (1974)

Using these levels, it is possible to estimate the level of exposure for subjects who reported to a physician or hospital because of their symptoms and discomfort. For example, using a representative summary for a 2002 train derailment, the exposures for 12 individuals may be estimated using one or more critical symptoms as follows:

Excessive lacrimation and skin lesions—mild	500+ ppm
Bronchospasm—moderate	1000+ ppm
Cough—moderate	1000+ ppm
Cough—moderate	1000+ ppm
Excessive lacrimation—mild	500+ ppm
Lacrimation and difficulty breathing—mild	500+ ppm
Cough—moderate	1000+ ppm
Lacrimation and difficulty breathing—mild	500+ ppm
Cough—moderate	1000+ ppm
Cough—mild (asthmatic)	500+ ppm
Cough—mild (asthmatic)	500+ ppm
Cough—mild (asthmatic)	500+ ppm

This procedure has produced a scale of exposure concentrations and associated clinical signs that can be used to estimate victim exposure concentrations in accidental releases of ammonia. This is a powerful tool useful to first responders, hospital and clinic emergency room personnel, and victims themselves. The scale does not consider those who may be hypersensitive to ammonia or those who are resistant to the effects of ammonia. The data are particularly useful in determining in-house concentrations following catastrophic releases of ammonia.

13.7 ENVIRONMENTAL CONSIDERATIONS

Environmental consideration from ammonia releases described in this chapter was obtained from actual incidents the author(s) were at or worked on. A description of these events, the environmental issues, and damage associated with these releases serves as real-life examples of what can happen when ammonia enters the environment.

13.7.1 MINOT, ND, AMMONIA RELEASE

At approximately 1:37 a.m. on January 18, 2002, a Canadian Pacific Railway freight train derailed 31 of its 112 cars about half a mile west of the city limits of Minot, North Dakota (Figure 13.1). Five tank cars carrying about 30,000 gal each of anhydrous ammonia catastrophically ruptured, and a vapor plume covered the derailment site and surrounding area.

FIGURE 13.1   Scene of the derailment in Minot, North Dakota.

When the tanks ruptured, sections of the fractured tanks were propelled as far as 1200 ft from the tracks. About 146,700 gal of anhydrous ammonia were released from the five cars, and a cloud of hydrolyzed ammonia formed almost immediately. Ammonia is liquefied under pressure, and since the pressure and temperature were sufficiently high, the sudden release of ammonia became and remained airborne as a mixture of vapor and very fine liquid droplets that do not fall to the ground. The droplets evaporate quickly cooling the air so that a cold mixture of air and ammonia vapor is formed. The mixture is initially denser than air. Accordingly, the plume rose an estimated 300 ft and gradually expanded 5 miles downwind of the accident site and over a population of about 11,600 people (NTSB, 2004). Over the next 5 days, another 74,000 gallons of anhydrous ammonia were released from six other anhydrous ammonia tank cars. About 11,600 people occupied the area affected by the vapor plume. One resident was fatally injured, and 60–65 residents of the neighborhood nearest the derailment site were rescued. As a result of the accident, 11 people sustained serious injuries, and 322 people, including the 2 train crewmembers, sustained minor injuries. Damages exceeded $2 million, and more than $8 million has been spent for environmental restoration.

Very early on in the incident, 911 operators were telling residents to stay in their homes and close their windows. Because of the ammonia vapor cloud and the dangers it posed to the residents of both the Tierracita Vallejo neighborhood close to the derailment and to the city of Minot, emergency responders had decided not to evacuate residents. This response, called "sheltering-in-place," differs from an evacuation in that people who shelter-in-place take precautions such as breathing through a wet cloth. But remain within the *hot zone*. The emergency responders then issued additional guidelines and implemented the public notification procedures by contacting the local media and sounding the outdoor warning system.

The conductor and engineer sustained minor injuries as a result of this accident. They were both taken to a local hospital after the derailment. The conductor was admitted for 1 day approximately 3 h after the accident and treated for chest tightness, shortness of breath, eye irritation, and anxiety. The engineer was treated for difficulty breathing and released the same day. Of the 122 firefighters who responded to the accident, 7 sustained minor injuries consisting of headaches, sore throats, eye irritation, and/or chest pain. An additional 11 police officers sustained minor injuries while blocking and directing traffic around the perimeter of the accident scene. Their injuries were eye irritation, chest discomfort, respiratory distress, and/or headaches.

A resident attempted to flee their home in a truck, but the driver (who apparently could not see very well because of the thick ammonia cloud and/or had severe eye irritation) crashed into a house in the Tierracita Vallejo neighborhood. This 38-year-old male sustained fatal injuries.

The Ward County coroner determined that the cause of death was prolonged exposure to anhydrous ammonia. Three residents of the Tierracita Vallejo neighborhood sustained serious injuries as a result of the accident and were admitted to Trinity Hospital. Their injuries included chemical burns to the face and the feet, respiratory failure, and erythema of the eyes and the nose.

Eight other residents sustained serious injuries as a result of the movement of the ammonia cloud over parts of the city of Minot. The injuries, which included shortness of breath, difficulty breathing, and/or burning of the eyes, were determined to have been complicated by preexisting health problems such as asthma and heart conditions.

A total of 301 other persons sustained minor injuries as a result of the accident. Of these, 11 were admitted to Trinity Hospital for less than 48 h. The remaining 290 individuals were treated and released at either a local hospital or temporary triage center. For the first few hours after the accident, the residents of Tierracita Vallejo were effectively trapped in their homes. Because of the nature of anhydrous ammonia, however, being trapped in their homes was actually the best possible situation for those so close to the derailment.

As a result of the derailment, two houses in the Tierracita Vallejo neighborhood were damaged.

One section of tank car GATX 47982 (car 22) was propelled approximately a quarter mile east of the derailment, crashing into a room in which two people were sleeping. The second house damaged was the one struck by the truck in which two residents were attempting to leave the area.

The release of liquid from the cars and ensuing vapor cloud resulted in soil, water/ice, and concerns over groundwater contamination. There were also reports of wildlife fatalities including five deer. A large quantity of ammonia was released into soil, which served as an ongoing source of odors and off-gassing to the nearby community. Consequently, the environmental cleanup consisted of soil borings to guide soil excavation. Almost 100,000 tons of soil with >500 mg/kg ammonia was removed and disposed of. In addition, 25,000 ft^2 of ice from the Souris River was removed and disposed of. Groundwater collection sumps were installed in topographically low areas located south and north of the track as well as a groundwater extraction system. A site-wide groundwater monitoring program was also implemented.

As a follow-on to this incident, our group published two studies worth briefly describing herein.

The first is a study by Tarkington et al. (2009) regarding the effectiveness of sheltering in place, and the second is a study of the odor threshold of ammonia under ambient conditions (Cawthon et al., 2009). Shelter-in-place recommendations have been made to and used by community members exposed to anhydrous ammonia after catastrophic release of ammonia gas due to a derailment or other accidents. Such incidents have resulted in fatalities and serious injury to exposed individuals; however, other individuals within the same area have escaped injury and, in many cases, sustained no injuries as a result of sheltering-in-place. The Tarkington et al. (2009) study was designed to simulate sheltering-in-place inside a typical bathroom with the shower running. The effectiveness of breathing through a damp cloth was also evaluated using a CPR mannequin placed inside a chamber built to represent a typical household bathroom. Ammonia gas at 300 or 1000 ppm was added to the chamber until the concentration peaked and stabilized, then the shower was turned on and the ammonia gas concentration was continuously monitored.

In the mannequin studies, using a damp cloth reduced exposure to ammonia gas by 2- to 18-fold.

Turning on the shower was even more effective at reducing ammonia levels. After 27 min, the ammonia concentration in the chamber was reduced to 2% of the initial concentration. The results of this study indicated that shelter-in-place substantially reduces ammonia exposure and comports with observations found for the Minot ammonia incident.

Cawthon et al. (2009) collected 6539 readings of ammonia between 0 and 1 ppm and reported that odor was detected only in 208 samples (3.2%). Of 65 readings between 1.1 and 1.5 ppm, odor was detected in 51 samples (78.5%). These data are consistent with an ammonia odor threshold within a concentration range of 1.1–1.5 ppm. Furthermore, a review of the ammonia literature demonstrates that the ammonia odor threshold is significantly lower than levels that produce eye, nose, or throat irritation.

13.7.2 PIPELINE AMMONIA RELEASE

On the evening of January 12, 2010, an anhydrous ammonia (NH_3) pipeline ruptured, creating an NH_3 vapor release near Skedee, Oklahoma (Figure 13.2). The 6 in. pipeline was located approximately 8 ft beneath the soil and was pressured at approximately 400 pounds per square inch (psi) prior to the release. Due to the relative proximity to the rupture site, nine nearby residences were evacuated by local emergency service personnel. At the time of the incident, the recorded wind conditions were 4–9 miles per hour (mph) and blowing from the east southeast. Air monitoring was performed at community and worker locations to determine the extent of NH_3 distribution in the ambient air. Fixed-location air monitors hand-held air monitoring, and mobile real-time monitoring for NH_3 was conducted throughout the community and at the incident site among workers.

Real-time instantaneous ammonia measurements in the work area averaged between 6 and 20 ppm with peak levels of 289 ppm, whereas ammonia levels in the community were mainly nondetectable.

However, the average of the detected levels was 4–6 ppm with a peak value of 40 ppm. Fixed-station analytical sampling results in the community were also mainly nondetectable (<0.38 ppm).

Only two samples resulted in measurable concentrations of ammonia of 0.57 ppm.

The release of NH_3 vapor and liquid resulted in impacted soil in areas adjacent to the leak source. The soil near the leak source was sampled and confirmed the presence of NH_3 in the upper two feet of topsoil, which required excavation. The soil also represented an ongoing source of vapors.

Engineering controls consisting of multiple high-velocity pneumatic fans were effectively used to direct the NH_3 vapors from the leak source or impacted soil away from nearby workers. While the workers were equipped with appropriate respiratory protection, the fans were used to reduce NH_3 levels below the action level of 300 ppm.

13.7.3 JACK RABBIT PROJECT AMMONIA RELEASE

We participated in the Jack Rabbit Project, sponsored by the US Department of Homeland Security Transportation Security Administration. This was a study designed to improve the understanding of

FIGURE 13.2 Scene of ammonia pipeline release.

(a)

(b)

FIGURE 13.3 Scene of ammonia release after (a) 2 s and after (b) 1 min. (Reprinted from Fox, S. and Storwald, D., Project jack rabbit: Field tests, CASC 11-006, Department of Homeland Security, Washington, DC, July 2011. With permission.)

rapid, large-scale releases of pressurized, liquefied toxic inhalation hazard gases including ammonia. The project involved outdoor release trials of 1 and 2 ton quantities of chlorine and anhydrous ammonia in 10 successful trials occurring in April and May 2010. One objective of the study was to characterize the ammonia vapor/aerosol cloud movement, behavior, and physiochemical characteristics and compare those characteristics with known observations and testing of large-scale releases of the testing materials. Detection instrumentation was deployed out to a range of 500 m, with some point detection MiniRAE instruments recording data at multiple heights, including 1, 3, and 6 m (Fox and Storwald, 2011).

The detailed conclusions and observations of this study are found in Fox and Storwald (2011) and briefly summarized in the following text. Ammonia exhibited dense gas behavior upon release from the pressurized containment due to the cooling of the gas as it is forced through the opening in the containment. This process is known as the Joule–Thomson effect, which makes the ammonia much denser than the surrounding ambient air. Consequently, ammonia tends to accumulate near the ground and collect in low-lying areas. Ammonia created a billowing white cloud that was propelled outward in all directions from the tank by the force of the escaping gas (Figure 13.3a).

The white cloud consisted of condensed water vapor generated by humidity present in the ambient air coming into contact with very cold anhydrous ammonia gas. While ammonia gas is invisible by itself, the condensed water vapor served in this case and in general based on field studies as a good indicator as to the location and behavior of the cloud.

At lower wind speeds, the cloud rapidly surrounded the tank and expanded beyond the borders of the 50 m diameter depression basin. The ammonia cloud reached a height of about 5 m, and radial spreading extended to a diameter of over 100 m, which flattened out as it vertically slumped after the termination of the release, consistent with expectations of a dense gas (Figure 13.2).

Under low wind speeds observed (<~1.5 m/s), initially ammonia exhibits dense gas behavior, whereas at higher wind speeds (>~3.5 m/s), the wind clearly has an immediate and more prominent effect on how ammonia dissipates and behaves after the release. After a certain time, ammonia becomes neutrally buoyant and is easily carried downwind. The observations found in the Jack Rabbit study are also consistent with those found by Bouet et al. (2005), who reported that the ammonia cloud behaves like a heavy gas, and no elevation of the cloud is observed. Temperatures near the release can fall to −70°C, which creates a dermal burn hazard. The authors also report that solid obstacles (wall or ground) located at a few meters from the discharge point have a considerable effect on the concentration values measured downwind, which agree with our observations that modeling results do not agree with real-world releases of dense gases like ammonia.

QUESTIONS

1. In an effort to utilize fewer animals in inhalation toxicity studies, one suggestion has been to utilize only one sex of the species—usually females since they are considered more sensitive than males. Ammonia is one chemical for which females are more resistant than males in acute inhalation studies. For the chemical you want to study, which has no other acute inhalation data, discuss briefly the advantages of the following options:
   a. Consider ammonia to be an anomaly and utilize females only.
   b. Perform preliminary exposures to groups of three males and three females to rule out or identify differences in sensitivity.
   c. Complete the study using both sexes in all exposures.
   d. Perform the test via pulmonary instillation to eliminate sex differences.
      Option (b) is best if the number of animals is critical. Option (a) could be dangerous, and option (d) does not eliminate sex differences.
2. Some chemicals, for example, sulfuric acid and chlorosilanes, take up water immediately upon contact. This makes it difficult to measure airborne levels at most levels of relative humidity since water changes the analyte. For ammonia, which of the following is the best procedure?
   a. Slow the flow rate through the chamber to allow the reaction with water to be complete.
   b. Proceed as though ammonium hydroxide is equally as toxic as ammonia.
   c. Reduce the humidity in the exposure chamber to eliminate the effects of water.
   d. Trap all of the ammonia and ammonium hydroxide, convert to a single analyte, and report measured concentration as total ammonia.
   e. Bubble the air sample through a solvent to trap the ammonium hydroxide.
      Option (d) is the best procedure when one does not know how ammonia will react in a chamber full of animals.
3. Ammonia is heavier than air. Discuss how this helps and/or hurts first responders answering a call to an environmental release.
4. In the Minot train derailment in which seven cars of ammonia were involved, the only person that died panicked and ran to his car to escape. Why did the persons that boarded in place survive this exposure in which the concentration reached 4000–6000 ppm? How would you relate this to your fellow employees or neighbors if a similar spill happened at your location?
5. In most cases, a salt form of a chemical is safer to handle than the vapor form of the chemical. However, in the summer of 2013, the town of West, Texas, was nearly wiped off the map when a fertilizer company storing large quantities of ammonium nitrate exploded when volunteer firemen were attempting to extinguish a fire at the facility. As a toxicologist, how do you relate the dangers of chemicals so as to prevent or minimize the effect of a fire or other dangers?
6. Ammonia has excellent warning properties, and certain forms are used to evoke a response in subjects that are unconscious. Which of the following is the best procedure for ammonia?
   a. Respect the toxicity of ammonia and provide sufficient protective equipment to all employees.
   b. Utilize ammonia in a closed system to eliminate or minimize the potential exposure to ammonia.
   c. Respect the toxicity of ammonia and provide all employees with a copy of the MSDS for ammonia.
   d. Because of the toxicity of ammonia, seek a less toxic chemical to replace ammonia in your plant or laboratories.
   e. Respect the toxicity of ammonia and fully disclose the nature and consequences of ammonia exposure and how to effectively utilize the available protective equipment.
   All answers are good answers, but (e) is the best procedure to follow.

REFERENCES

ACGIH. 2013. Ammonia. Documentation of the TLVs and BEIs with other worldwide occupational exposure values. Cincinnati, OH: American Conference of Governmental Industrial Hygienists.

AIHA (American Industrial Hygiene Association). 2014. Emergency response planning guidelines for ammonia. Fairfax, VA: AIHA Emergency Response Planning Committee.

Amoore, J.E. and Hautala, E. 1983. Odor as an aid to chemical safety: Odor thresholds compared with threshold limit values and volatilities for 214 industrial chemicals in air and water dilution. *J. Appl. Toxicol.* 3(6):272–290.

Appelman, L.M., ten Berg, W.F., and Reuzel, P.G.J. 1982. Acute inhalation toxicity of ammonia in rats with variable exposure periods. *Am. Ind. Hyg. Assoc. J.* 43(9):662–665.

Arwood, R., Hammond, J., and Ward, G.G. 1985. Ammonia inhalation. *J. Trauma* 25(5):444–447.

ATSDR (Agency for Toxic Substances and Disease Registry). 2004. *Toxicological Profile for Ammonia (Update)*. Atlanta, GA: Agency for Toxic Substances and Disease Registry.

Barrow, C.S., Alarie, Y., and Stock, M.F. 1978. Sensory irritation and incapacitation evoked by thermal decomposition products of polymers and comparisons with known sensory irritants. *Arch. Environ. Health* 33:79–88.

Bouet, R., Duplantier, S., and Salvi, O. 2005, July. Ammonia large scale atmospheric dispersion experiments in industrial configurations. *J. Loss Prev. Process Ind.* 18(4–6):512–519.

Boyd, E.M., MacLachlan, M.L., and Perry, W.F. 1944. Experimental ammonia gas poisoning in rabbits and cats. *J. Ind. Hyg. Toxicol.* 26:29–34.

Buckley, L.A., Jiang, X.Z., James, R.A., Morgan, K.T., and Barrow, C.S. 1984. Respiratory tract lesions induced by sensory irritants at the median respiratory rate decrease concentration. *Toxicol. Appl. Pharmacol.* 74:417–429.

CalEPA. 1999. Acute toxicity summary: Ammonia. In *Air Toxics Hot Spots Program. Risk Assessment Guidelines. Part 1. Determination of Acute Reference Exposure Levels for Airborne Toxicants*. CALEPA, pp. C13–C22. Sacramento, CA: Office of Environmental Health Hazard Assessment, California Environmental Protection Agency.

Caplin, M. 1941. Ammonia-gas poisoning—Forty-seven cases in a London shelter. *Lancet* 2:95–96 (cited in NIOSH, 1974).

Carpenter, C.P., Smyth, H.F., and Pozzani, U.C. 1949. The assay of acute vapor toxicity and the grading and interpretation of results on 96 chemical compounds. *J. Ind. Hyg. Toxicol.* 31:343–346.

Carpenter, C.P., Smyth, H.F., and Shaffer, C.B. 1948. The acute toxicity of ethylene imine to small animals. *J. Ind. Hyg. Toxicol.* 30(1):2–6.

Cawthon, D., Hamlin, D., Steward, A., Davis, C., Cavender, F., and Goad, P. 2009, June. Field studies on the ammonia odor threshold based on ambient air-sampling following accidental releases. *Toxicol. Environ. Chem.* 91(4):597–604.

Chemical Industry Association. 1974. Code of Practice for Large Scale Storage of Fully Refrigerated Ammonia in the United Kingdom.

Cole, T.J., Cotes, J.E., Johnson, G.R. et al. 1977. Ventilation, cardiac frequency and pattern of breathing during exercise in men exposed to *o*-chlorobenzylidene malononitrile (CS) and ammonia gas in low concentrations. *Q. J. Exp. Physiol. Cogn. Med. Sci.* 62:341–351.

Coon, R.A., Jones, R.A., Jenkins, L.J. Jr. et al. 1970. Animal inhalation studies on ammonia, ethylene, glycol, formaldehyde, dimethylamine and ethanol. *Toxicol. Appl. Pharmacol.* 16:646–655.

Dodd, K.T. and Gross, D.R. 1980. Ammonia inhalation toxicity in cats: A study of acute and chronic respiratory dysfunction. *Arch. Environ. Health* 35:6–14.

Erskine, R.J., Murphy, P.J., Langston, J.A., and Smith, G. 1993. Effect of age on the sensitivity of upper airways reflexes. *Br. J. Anaesth.* 70:574–575.

Ferguson, W.S., Koch, W.C., Webster, L.B., and Gould, J.R. 1977. Human physiological response and adaption to ammonia. *J. Occup. Med.* 19:319–326.

Fieldner, A.C., Katz, S.H., and Kinney, S.P. 1921. Ammonia and Table 9—Poison doses of industrial gases and vapors in air. In *Gas Masks for Gases Met in Fighting Fire* (Technical Paper; 248), A.C. Fieldner, S.H. Katz, and S.P. Kinney (eds.), pp. 13–14 and Table 9. Washington, DC: Bureau of Mines, U.S. Department of the Interior.

Fieldner, A.C., Sayers, R.R., Yant, W.P., Katz, S.H., Shohan, J.B., and Leith, R.D. 1931. *Bureau of Mines Monograph 4. Warning Agents for Fuel Gases*. New York: US Bureau of Mines and the American Gas Association.

Flury, K.E., Dines, D.E., Rodarte, J.R., and Rodgers, R. 1983. Airway obstruction due to inhalation of ammonia. *Mayo Clin. Proc.* 58:389–393.

Fox, S. and Storwald, D. 2011, July. Project jack rabbit: Field tests, CASC 11-006. Washington, DC: Department of Homeland Security.

Fthenakis, V.M. 2004. *Derailment and Ammonia Release at Minot, SD: Source Characterization and Atmospheric Dispersion.* Upton, NY: EnviroConsultants Inc.

Hatton, D.V., Leach, C.S., Beaudet, A.L. et al. 1979. Collagen breakdown and ammonia inhalation. *Arch. Environ. Health* 34:83–87.

Henderson, Y. and Haggard, H.W. 1927. Ammonia gas. In *Noxious Gases and the Principles of Respiration Influencing their Action*, Y. Henderson and H.W. Haggard (eds.), pp. 125–126. New York: Chemical Catalog Company, Inc.

Henderson, Y. and Haggard, H.W. 1943a. Special characteristics of various irritant gases. In *Noxious Gases and the Principles of Respiration Influencing Their Action*, 2nd revised edn., Y. Henderson, and H.W. Haggard (eds.), pp. 124–141. New York: Reinhold, 1943.

Hilado, C.J., Casey, C.J., and Furst, A. 1977. Effect of ammonia on swiss albino mice. *J. Combust. Toxicol.* 4:385–388.

Hoeffler, H.B., Schweppe, H.I., and Greenberg, S.D. 1982. Bronchiectasis following pulmonary ammonia burn. *Arch. Pathol. Lab. Med.* 106:686–687.

Holness, D.L., Purdham, J.T., and Nethercott, J.R. 1989. Acute and chronic respiratory effects of occupational exposure to ammonia. *Am. Ind. Hyg. Assoc. J.* 50:646–650.

HSDB (Hazardous Substance Data Bank). Ammonia [online]. Available at: http://toxnet.nlm.nih.gov/ (accessed April 17, 2012).

Industrial Bio-Test Laboratories, Inc. 1973, March 23. Irritation threshold evaluation study with ammonia, IBT 663-03161. Report to International Institute of Ammonia Refrigeration by Industrial Bio-Test Laboratories, Inc., Northbrook, IL.

Johanson, G., Alexandrie, A., Jarnberg, J., and Lieivuori, J. 2006. Nordic expert group: Evaluation of health risks of ammonia. Karolinska Institute, Stockholm, Sweden. *Toxicol. Sci.* 90:332.

Kapeghian, J.C., Mincer, H.H., Hones, A.B. et al. 1982. Acute inhalation toxicity of ammonia in mice. *Bull Environ. Contam. Toxicol.* 29:371–378.

Kass, I., Zame, N., Bobry, C.A., and Holzer, M. 1972. Bronchiectasis following ammonia burns of the respiratory tract. *Chest* 62:282–285.

Kimmerle, G. 1974. Aspects and methodology for the evaluation of toxicological parameters during fire exposures. *JFF/Combust. Toxicol.* 1:4–50.

Leduc, D., Gris, P., Lheureux, P. et al. 1992. Acute and long term respiratory damage following inhalation of ammonia. *Thorax* 47:755–757.

Lehman, K.B. 1886. Experimentelle Studien Uber Den Einfluss Technish Und Hygienisch Wichtiger Gase Und Dampfe Auf Den Organismus (Teil I and II – Amoniak Und Saltraurega). *Arch. Hyg.* 5:1–126.

Leonardos, G., Kendall, D.A., and Barnard, N.J. 1969. Odor threshold determinations of 53 odorant chemicals. *APCA J.* 19(2):91–95.

Leung, C.M. and Loo, C.L. 1992. Mass ammonia inhalational burns—Experience in the management of 12 patients. *Ann. Acad. Med.* 21:624–629.

Levy, D.M., Bivertie, M.B., and Henderson, J.W. 1964. Ammonia burns of the face and respiratory tract. *J. Am. Med. Assoc.* 190:95–98.

Li, W.-L. and Pauluhn, J. 2010. Comparative assessment of sensory irritation in rats and mice nose-only exposed to dry and humidified atmospheres. *Toxicology* 276:135–142.

Lonsdale, H. 1975. Ammonia tank failure—South Africa. American Institute of Chemical Engineers. *Ammonia Plant Saf.* 17:126–131 (cited by RAM TRAC, 1996).

MacEwen, J.D. and Vernot, E.H. 1972. Toxic hazards research unit annual technical report: 1972. SysteMed Report No. W-72003, AMRL-TR-72-62. Sponsor: Aerospace Medical Research Laboratory, Wright-Patterson AFB, OH. AD-755 358.

MacEwen, J.D., Theodore, J., and Vernot, E.H. 1970. Human exposure to EEL concentrations of mono-methylhydrazine, AMRL-TR-70-102, Paper no. 23. In: *Proceedings of the First Annual Conference on Environmental Toxicology*, Wright-Patterson AFB, OH, September 9–11, 1970, pp. 355–363.

Markham, R.S. 1986, August. A review of damage from ammonia spills. Paper presented at *the 1986 Ammonia Symposium, Safety in Ammonia Plants and Related Facilities*. Boston, MA: A.I.Ch.E.

Mayan, M.H. and Merilan, C.P. 1972. Effects of Ammonia inhalation on respiration rate of rabbits. *J. Anim. Sci.* 34:448–452.

Mazzola, C. 1997. Potchefstroom dose reconstruction: Inherent uncertainties that significantly limit effective application to human health standards process. Prepared for the National Advisory Committee for Acute Exposure Guideline Levels for Hazardous Substances. Stone & Webster Engineering Corporation, May 1997.

McClean, J.A., Mathews, K.P., Solomon, W.R. et al. 1979. Effect of ammonia on nasal resistance in atopic and nonatopic subjects. *Ann. Otol. Rhinol. Laryngol.* 88:228–234.

Michaels, R.A. 1999. Emergency planning and the acute toxic potency of inhaled ammonia. *Environ. Health. Perspect.* 107(8):617–627.

Montague, T.J. and Macneil, A.R. 1980. Mass ammonia inhalation. *Chest* 77:496–498.

Mudan, K. and Mitchell, K. 1996. Report on the Potchefstroom, South Africa ammonia incident. Columbus, OH: Four Elements, Inc., 14pp.

Mulder, J.S. and Van der Zalm, H.O. 1967. A fatal case of ammonia poisoning. *Tijdsch Soc. Geneeskd* 45:458–460 (in Dutch) (cited in NIOSH, 1974).

Nagata, Y. 2003. Measurement of odor threshold by triangle odor bag method. In *Odor Measurement Review*, pp. 118–127. Tokyo, Japan: Office of Odor, Noise and Vibration, Government of Japan, Ministry of the Environment.

NIOSH (National Institute for Occupational Safety and Health). 1974. Criteria for recommended standard..... Occupational exposure to ammonia. Rockville, MD: NIOSH. PB246699.

NRC. 2007. Ammonia. In: *Committee on Acute Exposure Guideline Levels, Committee on Toxicology National Research Council. Acute Exposure Guideline Levels for Selected Airborne Chemicals,* Vol. 6. Washington, DC: National Academy Press, pp. 44–85. www.epa.gov/oppt/aegl/.

NTSB (National Transportation Safety Board). 1979. Survival in hazardous materials transportation accidents. Report NTSC-HZM-79-4 (cited in Pedersen and Selig, 1989).

NTSB. 2004. Derailment of Canadian Pacific Railway Freight Train 292-16 and subsequent release of anhydrous ammonia near Minot, North Dakota January 18, 2002. Washington, DC: National Transportation Safety Board, March 9, 2004; NTSB/RAR-04/01.

O'Kane, G.J. 1973. Inhalation of ammonia vapour: A report on the management of eight patients during the acute stages. *Anesthesia* 38:1208–1213.

Patty, F.A. 1948. Alkaline materials: Ammonia. In *Industrial Hygiene and Toxicology,* F.A. Patty (ed.), p. 560. New York: Interscience.

Pauluhn, J. 2013. Acute inhalation toxicity of ammonia: Revisiting the importance of RD_{50} and $LCt_{01/50}$ relationships for setting emergency response guideline values. *Regul. Toxicol. Toxicol.* 66:315–325.

Pedersen, F. and Selig, R.S. 1989. Predicting the consequences of short-term exposure to high concentrations of gaseous ammonia. *J. Hazard. Mater.* 21:143–159.

RAM TRAC. 1996. Acute inhalation risk potentially posed by anhydrous ammonia. Robert A. Michaels, Project Director, RAM TRAC Corporation, Schenectady, NY, 99pp.

Richard, D., Bouley, G., and Boudene, C. 1978a. Effects of continuous inhalation of ammonia in the rat and mouse. *Bull. Eur. Physiopathol. Respir.* 14:573–582 (in French).

Richard, D., Jouany, J.M., and Boudene, C. 1978b. Acute toxicity of ammonia gas in the rabbit by inhalation. *C. R. Acad. Hebd. Seances Acad. Sci. Ser. D* 287:375–378 (in French).

Rusch, G.M., Bast, C.B., and Cavender, F.L. 2009. Establishing a point of departure for risk assessment using acute inhalation toxicology data. *Reg. Toxicol. Pharmacol.* 54:247–255.

Ruth, J.H. 1986. Odor thresholds and irritation levels of several chemical substances: A review. *Am. Ind. Hyg. Assoc. J.* 47(3):A142–A151.

Sayers, R.R. 1927. Toxicology of gases and vapors. In *International Critical Tables of Numerical Data, Physics, Chemistry and Technology,* E.W. Washburn (ed.), pp. 318–321. New York: McGraw-Hill.

Schaerdel, A.D., White, W.J., Lang, C.M. et al. 1983. Localized and systemic effects of environmental ammonia in rats. *Lab. Anim. Sci.* 33(1):40–45.

Silver, S.D. and McGrath, F.P. 1948. A comparison of acute toxicities of ethylene imine and ammonia to mice. *J. Ind. Hyg. Toxicol.* 30:7–9.

Silverman, L., Whittenberger, J.L., and Muller, J. 1949. Physiological response of man to ammonia in low concentrations. *J. Ind. Hyg. Toxicol.* 31:74–78.

Smyth, H.F. Jr. 1956. Hygienic standards for daily inhalation. *Am. Ind. Hyg. Assoc. Q.* 17(2):129–185.

Sobonya, R. 1977. Fatal anhydrous ammonia inhalation. *Hum. Pathol.* 8:293–299.

Sullivan, R.J. 1969. Appendix B. Table 2. Recognition odor threshold of odorants. In *Air Pollution Aspects of Odorous Compounds,* R.J. Sullivan (ed.), pp. 157–169. Bethesda, MD: National Air Pollution Control Administration, U.S. Department of Health, Education and Welfare.

Sundblad, B.M., Larsson, B.M., Acevedo, F., Ernstgard, L., Johanson, G., Larsson, K., and Palmberg, L. 2004. Acute respiratory effects of exposure to ammonia on healthy persons. *Scand. J. Work Environ. Health* 30:313–321.

Tarkington, B., Harris, A.J., Barton, P.S., Chandler, B., and Goad, P.T. 2009, April. Effectiveness of common shelter-in-place techniques in reducing ammonia exposure following accidental release. *J. Occup. Environ. Hyg.* 6(4):248–255.

ten Berge, W.F., Zwart, A., and Appelman, L.M. 1986. Concentration–time mortality response relationship of irritant and systemically acting vapours and gases. *J. Hazard. Mater.* 13:301–309.

Tepper, J.S., Weiss, B., and Wood, R.W. 1985. Alterations in behavior produced by inhaled ozone or ammonia. *Fundam. Appl. Toxicol.* 5:1110–1118.

USEPA. 2012. Integrated Risk Information System, US Environmental Protection Agency, Washington, DC. www.epa.gov/iris/.

van Doorn, R., Ruijten, M., and van Harreveld, T. 2002. Guidance for the application of odor in chemical emergency response. The Hague, the Netherlands: RIVM.

Verberk, M.M. 1977. Effects of ammonia on volunteers. *Int. Arch. Occup. Environ. Health* 39:73–81.

Verschueren, K. 1977. *Handbook of Environmental Data on Organic Chemicals.* New York: Van Nostrand Reinhold Company.

Wallace, D.P. 1978. Atmospheric emissions and control safety in ammonia plants and related facilities. In *AICHE Symposium*, Miami, FL, November 12–17, 1978.

Walton, M. 1973. Industrial ammonia gassing. *Br. J. Ind. Med.* 30:78–86.

Weatherby, J.H. 1952. Chronic toxicity of ammonia fumes by inhalation. *Proc. Soc. Exp. Biol. Med.* 81:300–301.

Weedon, F.R., Hartzell, A., and Setterstrom, C. 1940. Toxicity of ammonia, chlorine, hydrogencyanide, hydrogen sulphide, and sulphur dioxide gases. *V. Anim. Contrib. Boyce Thompson Inst.* 11:365–385.

Withers, J., ten Berge, W., Gordon, J. et al. 1986. The lethal toxicity of ammonia. A report to the Major Hazards Advisory Panel. Institution of Chemical Engineers, Rugby, U.K. North Western Branch Papers 1986, No. 1 (cited by Pedersen and Selig, 1989).

Zissu, D. 1995. Histopathological changes in the respiratory tract of mice exposed to ten families of airborne chemicals. *J. Appl. Toxicol.* 15:207–213.

14 Serpentine and Amphibole Asbestos

David M. Bernstein

CONTENTS

14.1 INTRODUCTION

The principles of fiber toxicology are based upon three important criteria: dose, dimensions, and durability. This chapter addresses the importance of these criteria for asbestos and provides detailed understanding and support of how these influence the toxicity of what is commonly referred to as asbestos.

To assess these criteria, it is important to realize that asbestos is a term that refers to two minerals, serpentine and amphibole, occurring in fibrous form with very different mineralogical properties, which in the past have often been used and referred to interchangeably. The serpentine form is chrysotile, while the commercial amphibole forms are crocidolite, amosite, and tremolite. In addition, there are, as discussed, several other amphibole forms.

While asbestos has been known for centuries, dating back to Greek and Roman times (Browne and Murray, 1990; Ross and Nolan, 2003), the references did not differentiate the mineral type of asbestos. The first reference to chrysotile, the serpentine form of asbestos, was in 1834 by von Kobell (1834) in which he described that chrysotile is distinguished by its behavior of being decomposed by acid.

The name amphibole (Greek αμφιβολος—amphibolos meaning *ambiguous*) was used by René Just Haüy in 1801 to include tremolite, actinolite, tourmaline, and hornblende. The group was so named by Haüy in allusion to the protean variety, in composition and appearance, assumed by its minerals (Leake, 1978).

Both serpentine and amphibole asbestos are naturally occurring minerals, which are extracted from the earth in surface, open pit, or underground mines. The fibers are intertwined with adjacent rock and are separated through crushing and milling with subsequent filtration/separation steps. Although the production of chrysotile asbestos has probably always exceeded that of amphibole asbestos by more than 10:1, the amphibole minerals are more abundant. They are frequently constituents of igneous rocks, often major components of metamorphic rocks and thought to account for approximately 20% of the shield area of the earth. However, fortunately for us, most of these rock-forming amphiboles are not at all asbestos like (Whittaker, 1979).

The mineralogical distinction between the two fibrous minerals was blurred, with the commercial mining and use of asbestos starting in the Western Alps of Italy and in England in the 1800s where serpentine and amphibole asbestos was very often found in close proximity. In Canada, the first chrysotile mine was opened in 1879 at Thetford, in the Quebec province. This was followed shortly thereafter with commercial chrysotile mining in Russia.

While both chrysotile and amphibole asbestos are usually of dimensions which can be quite respirable in humans, the mineralogical differences between these two minerals result in a very different formation of the fibers themselves, which has a large impact on the potential toxicity.

In understanding the toxicology of serpentine and amphibole asbestos, the question of dose has an important impact on the evaluation of both toxicology studies and epidemiology studies. As discussed in the following, the early toxicology studies often used very high doses which resulted in a lung overload effect rather than a true evaluation of the fiber. Similarly, many of the epidemiological studies which have been used for evaluation of the relative toxicity of chrysotile versus amphibole asbestos were performed based upon exposures in the first half of the 1900s when there was little understanding and virtually no control of fiber levels in the work environments. It is interesting to note, as discussed in the following, that in a large majority of the mines and plants that have been studied, the dust levels were reduced dramatically due to the implementation of control technology by the time the publications were written.

Based upon lung measurements in humans in various work environments, and more recently on biopersistence and toxicology studies in rats, chrysotile fibers are considerably less biopersistent than amphibole asbestos. This is a result of the very different structural makeup and mineralogical composition of chrysotile versus amphibole asbestos. As with other fiber types, this impacts considerably on the potential of the fibers to cause disease.

14.1.1 CHRYSOTILE CHARACTERISTICS

The words for the golden and fibrous nature of the fibers were used by the Greeks to derive the name chrysotile.

Von Kobell (1934) first described chrysotile stating that it was distinguished by its behavior of being decomposed by acid. As discussed in the following, this is one of the characteristics which differentiate chrysotile from the amphibole asbestos. The other characteristic which differentiate chrysotile is that it is formed of a curved structure of the Mg—analog of kaolinite. This was first suggested by Pauling (1930) due to the misfit between the octahedral and tetrahedral sheets. The crystal structure of chrysotile has been investigated extensively over the years, starting with Warren and Bragg (1930) and subsequently Noll and Kircher (1951) and Bates et al. (1950) who published electron micrographs which showed the cylindrical and apparently hollow chrysotile fibers.

The chrysotile fiber is a sheet silicate, monoclinic in crystalline structure, and has a unique rolled form. The chemistry of chrysotile is composed of a silicate sheet of composition $(Si_2O_5)_n^{-2n}$, in which three of the O atoms in each tetrahedron are shared with adjacent tetrahedra and a nonsilicate sheet of composition $[Mg_3O_2(OH)_4]_n^{2n-}$. In chrysotile, the distances between apical oxygens in a regular (idealized) silicate layer are shorter (0.305 nm) than the O–O distances in the ideal Mg-containing layer (0.342 nm), which may account for the curling of the layers, which results in the rolling up, like a carpet to form concentric hollow cylinders (Skinner et al., 1988). The walls of the chrysotile fiber are made up of approximately 12–20 of these layers in which there is some mechanical interlocking. It is important to note however that there is no chemical bonding between the layers. Each layer is about 7.3 Å thick, with the magnesium surface facing the outside of the curl and the silica and oxygen tetrahedron inside the curl (Whittaker, 1963, 1957; Tanji et al., 1984). The mineralogical structure is illustrated in Figure 14.1 (adapted from Skinner et al., 1988). The polyhedral model of the chrysotile structure shown in Figure 14.2 illustrates one cylindrical curved layer of a chrysotile fiber (Rakovan, 2011). The Mg atom is on the outside of the curl and is thus exposed to the surrounding environment. This layered construction of chrysotile is illustrated in Figure 14.3 (Bernstein et al., 2013). High-resolution transmission electron photomicrographs of chrysotile are shown in Figure 14.4 (Kiyohara, 1991). Figure 14.5 shows the two forms that occur with chrysotile, one with concentric cylindrically curve layers and the other with apparently rolled layers (Rakovan, 2011).

Table 14.1 summarizes the chemical composition of typical serpentine and the amphiboles tremolite and amosite asbestos. The chemical composition and the structure of chrysotile are notably different from that of amphibole asbestos (Hodgson, 1979).

Commercial chrysotile is usually subdivided into groups using the Canadian Quebec Screening Scale (QSS). These groups are determined using an apparatus with a nest of four rotating trays superimposed one above the other. A known quantity of fiber is placed on the top tray, and the trays are rotated for a fixed time to produce a sifting action. The longest/thickest fibers stay on the top screen (tray) which has the largest openings and the shorter/thinner fibers fall through to lower screens. The grade is determined based upon the weight fractions deposited upon each screen and ranges from 3 to 9, with 3 being the longest (Cossette and Delvaux, 1979).

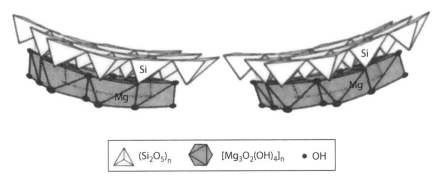

$(Si_2O_5)_n$ $[Mg_3O_2(OH)_4]_n$ • OH

FIGURE 14.1 Chrysotile structure: thin rolled sheet (7.3 A) with Mg on the outside and Si on the inside. (Reproduced from Bernstein, D.M. et al., *Crit. Rev. Toxicol.*, 43(2), 154, 2013.)

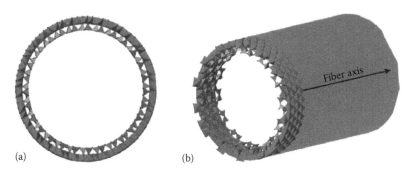

(a)   (b)

FIGURE 14.2 (a) Polyhedral model of the chrysotile structure with one cylindrically curved 1:1 layer; viewed down the fiber axis (a-crystal axis). (b) A perspective view of the chrysotile structure. (From Rakovan, J., *J. Rocks Miner.*, 86(1), 63, 2011.)

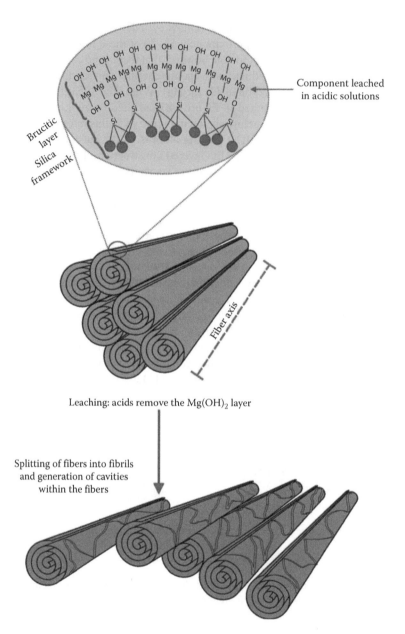

FIGURE 14.3 Structural formation of the sheet silica chrysotile asbestos. (From Bernstein, D.M. et al., *Crit. Rev. Toxicol.*, 43(2), 154, 2013.)

Nagy and Bates (1952) reporting on the stability of chrysotile, showed that it has a high solubility in hydrochloric acid. They also observed that chrysotile has a relatively low thermal stability compared to other hydrous silicate minerals. The heat of the electron beam of an electron microscope caused a very rapid change in the morphology of the fibers, and prolonged exposure to electron bombardment resulted in complete disintegration of the material (Noll and Kircher, 1952). Hargreaves and Taylor (1946) reported that if fibrous chrysotile is treated with dilute acid, the magnesia can be completely removed, and the hydrated silica remaining, though fibrous in form, completely loses the elasticity characteristic of the original chrysotile and gives an x-ray pattern of one or perhaps two diffuse broadbands, indicating that the structure is "amorphous" or "glassy" in type.

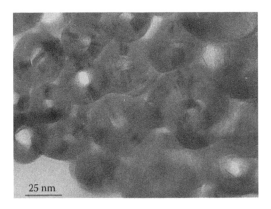
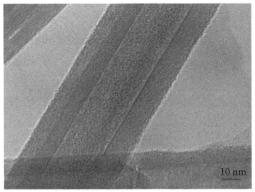

FIGURE 14.4 Transmission electron micrographs of chrysotile. (From Kiyohara, P.K., Estudo da interface crisotila-cimento Portland em compósitos de fibro-cimento por métodos óptico-eletrônicos, Tese de Doutorado, apres. EPUSP, São Paulo, Brazil, 1991.)

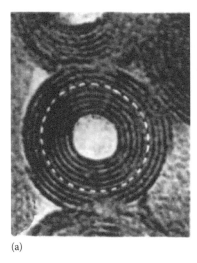
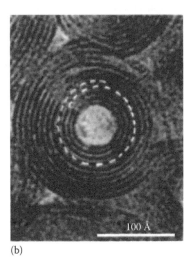

(a)   (b)

FIGURE 14.5 High-resolution transmission electron photomicrographs of chrysotile fibers composed of (a) multiple concentric cylindrically curved 1:1 layers looking directly down the fiber axis, and (b) a 1:1 layer rolled up in a spiral. The dark bands are individual 1:1 layers. Dashed white lines follow a 1:1 layer. (Modified from Yada (1971) by Rakovan, J., *J. Rocks Miner.*, 86(1), 63, 2011.)

14.1.2 AMPHIBOLE CHARACTERISTICS

In contrast to chrysotile, with amphibole asbestos, the basic structure is in the form of a double silica chain which appears as an I-beam with corner-linked $(SiO_4)^{-4}$ tetrahedra linked together in a double-tetrahedral chain that sandwiches a layer with the Ca_2Mg_5. These chains are paired, "back-to-back," with a layer of hydrated cations in between to satisfy the negative charges of the silica chains. The final structure is formed by the stacking of these sandwich ribbons in an ordered array (Speil and Leineweber, 1969). This amphibole structure is illustrated in Figure 14.6 (Bernstein et al., 2013). As shown below the structure, the fibers can be connected by soluble cations shown as small circles which are located between the double chain silicate fibers. When the soluble cations dissolve, as can happen in the lung, the amphibole fibers in these bundles are released as individual fibers; however, the fibers themselves are not affected. The double chain silicate amphibole fibers themselves are highly insoluble in both the lung fluids and in the macrophages.

TABLE 14.1
Typical Chemical Composition (Percent)

Compound	Chrysotile[a]	Tremolite[b]	Amosite[b]
SiO_2	40.6	55.10	49.70
Al_2O_3	0.7	1.14	0.40
Fe_2O_3	2.3	0.32	0.03
FeO	1.3	2.00	39.70
MnO	—	0.10	0.22
MgO	39.8	25.65	6.44
CaO	0.6	11.45	1.04
K_2O	0.2	0.29	0.63
Na_2O	—	0.14	0.09
H_2O	—	3.52	1.83
H_2	—	0.16	0.09
CO_2	0.5	0.06	0.09
Ignition loss	14.0	—	—
Total	100	99.93	100.26

Source: Hodgson, A.A., Chemistry and physics of asbestos, in *Asbestos: Properties, Applications and Hazards*, L.M.a.S.S. Chissick (ed.), John Wiley & Sons, New York, 1979, pp. 80–81.
[a] Typical chemical analysis of Canadian chrysotile from the Quebec Eastern.
[b] Townships (LAB Chrysotile, Inc., Quebec, Canada).

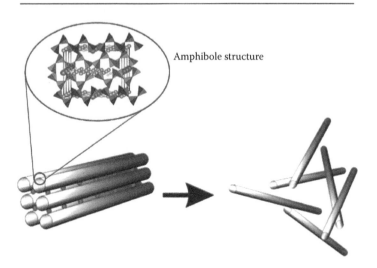

Amphibole structure

FIGURE 14.6 **(See color insert.)** With amphiboles, the soluble cations shown as small circles are located between the fibers which are formed with double chain silicate. When the soluble cations dissolve as can happen in the lung, the amphibole fibers in these bundles are released as individual fibers. The double chain silicate amphibole fibers themselves are highly insoluble in both the lung fluids and in the macrophages. (From Bernstein, D.M. et al., *Crit. Rev. Toxicol.*, 43(2), 154, 2013.)

There are five asbestiform varieties of amphiboles: anthophyllite asbestos, grunerite asbestos (amosite), riebeckite asbestos (crocidolite), tremolite asbestos, and actinolite asbestos. Of these, crocidolite and amosite were the only amphiboles with significant industrial uses (Virta, 2002). Tremolite, while not used commercially, has been found as a contaminant in other fibers or in other industrial minerals (e.g., chrysotile and talc).

FIGURE 14.7 **(See color insert.)** Illustration of the tightly bound silica-based structure on exterior surfaces of amphibole fibers. (Adapted from Department of Geology and Geophysics, University of Wisconsin, Crystal Structure Movies, http://www.geology.wisc.edu.)

Depending upon the type of amphibole, the principal cations are magnesium, iron, calcium, and sodium. The principal types are as follows:

Crocidolite	$(Na_2Fe_3^{2+}Fe_2^{3+})Si_8O_{22}(OH)_2$
Amosite	$(Fe^{2+}, Mg)_7Si_8O_{22}(OH)_2$
Tremolite	$Ca_2Mg_5Si_8O_{22}(OH)_2$
Actinolite	$Ca_2(Mg, Fe^{2+})_5Si_8O_{22}(OH)_2$
Anthophyllite	$(Mg, Fe^{2+})_7Si_8O_{22}(OH)_2$

The exterior surfaces of the amphiboles are tightly bound silica-based structures. This is illustrated with tremolite in Figure 14.7.

As a result of their structure, amphibole fibers have negligible solubility at any pH that would be encountered in an organism (Speil and Leineweber, 1969). Some service metals associated with the fibers such as iron can become ionized and released under certain conditions (Aust et al., 2011).

14.2   FACTORS INFLUENCING FIBER TOXICOLOGY

As mentioned earlier, mineral fiber toxicology has been associated with three key factors: dose, dimension, and durability. The dose is determined by the fiber's physical characteristics/dimensions, how the fibrous material is used, and the control procedures that are implemented. The thinner and shorter fibers will weigh less and thus can remain suspended in the air longer than thicker and longer fibers. Most asbestos fibers are thinner than commercial insulation fibers; however, they are thicker than the new nanofibers, which are currently being developed. Control procedures in mining and manufacturing have changed dramatically over the years, resulting in markedly reduced exposure concentrations.

The fiber dimensions govern two factors: whether the fiber is respirable, and secondly, if it is respirable, whether the dimensions are also a factor in determining their response in the lung milieu once inhaled. Shorter fibers of the size which can be fully engulfed by the macrophage will be cleared by mechanisms similar to those for nonfibrous particles. These include clearance through the lymphatics and macrophage phagocytosis and clearance. It is only the longer fibers which the macrophage cannot fully engulf; if they are persistent, this will lead to disease.

This leads to the third factor: durability. Those fibers whose chemical structure renders them wholly or partially soluble once deposited in the lung are likely to either dissolve completely, or dissolve until they are sufficiently weakened focally to undergo breakage into shorter fibers. The remaining short fibers can then be removed through successful phagocytosis and clearance.

In addition, because of the differences in the mineralogical composition and structure of serpentine asbestos (chrysotile) in comparison to amphibole asbestos (e.g., crocidolite and amosite), both the physical structure of the fiber and its ability to dissolve in acid are important criteria in determining the potential to have toxicological effects.

These factors have been shown to be important determinants for both synthetic mineral fibers (Hesterberg et al., 1998a,b; Miller et al., 1999; Oberdöster, 2000; Bernstein et al., 2001a,b) and asbestos (Bernstein et al., 2013).

14.3   IN VITRO TOXICOLOGY

In vitro toxicology studies are often very helpful in elucidating possible mechanisms involved in pathogenesis. However, as used in the assessment of fiber toxicology, they are difficult to interpret. This stems from several factors. The in vitro test system is a static system and thus is not sensitive to differences in fiber solubility. High doses of fibers are used to obtain a positive response, and it is difficult to extrapolate from these large short-term cellular exposures to the considerably lower-dose chronic exposures that occur in vivo. In addition, the number of fibers and size distribution are often not quantified. Most important, however, is that these end points have not been validated as screening assays that are predictive of long-term pathological effects in vivo. While in vitro tests may be useful tools to identify and evaluate possible mechanisms, with fibers, these in vitro test systems are of limited use in differentiating different fiber types (Bernstein et al. 2005a).

14.4   IN VITRO BIODURABILITY

Within the biological system of the lung, upon deposition, fibers are exposed to two types of environmental conditions. These are the lung surfactant which occurs through the tracheobronchial tree and the alveolar region, and the pulmonary macrophage which is the first line of defense once the fiber is deposited within the lung. The lung surfactant has a pH of 7.4 (neutral), and within the macrophage phagolysosomes, the pH is as low as 4 (acid).

With chrysotile being a thin rolled sheet with the magnesium hydroxide layer on the outside of the fiber, the chrysotile fiber has poor acid resistance compared to the amphibole fibers which is encapsulated by silica. With the amphibole fibers, the silicate oxygen atoms are on the outside of the layers and the hydroxides are masked within the fiber resulting in a fiber, very resistant to solubility at either neutral or acid pH. Von Kobell (1834) was the first to describe the acid solubility of chrysotile as being an important characteristic. Hargreaves and Taylor (1946) described how with treatment of chrysotile fibers with dilute acid, the magnesium can be completely removed, leaving a hydrated silica which has lost the elasticity characteristic of the original fiber. The resulting structure was characterized as amorphous or grassy in type. Similar findings were reported by Wypych et al. (2005) who described how the leached products consisted of layered hydrated disordered silica with a distorted structure similar in form to the original fiber. They also described the removal of the brucite-like (magnesium hydroxide) sheets leaving silica with an eminently amorphous structure. Suquet (1989) also reported that "Acid leaching transformed chrysotile into porous, noncrystalline hydrated silica, which easily fractured into short fragments. If the acid attack was too severe, these fragments converted into shapeless material."

The ability of an acid environment to break apart long chrysotile fibers into shorter fibers in vitro has been reported by Osmon-McLeod et al. (2011). The authors assessed the durability of a number of fibers including long fiber amosite (LFA) and long fiber chrysotile (LFC) in a Gambles solution that was adjusted to a pH of 4.5 to mimic that inside macrophage phagolysosomes, which the authors described as "potentially the most degradative environment that a particle should encounter following lung deposition and macrophage uptake." Figures 14.8 and 14.9 (modified from Figures 3 and

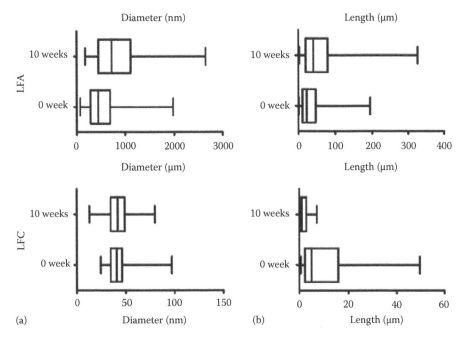

FIGURE 14.8 Effect of incubation in Gambles solution on (a) fiber widths and (b) lengths. Boxplots showing the distribution of fiber widths and lengths (nm) in samples that had been incubated in Gambles solution for 0 week or 10 weeks. The line in the box represents the median value of measurements from TEM images, and the edges of the box represent the lower and upper quartiles. The ends of the whiskers represent minimum and maximum values. Note that the scale of the horizontal axis is different in the LFA and LFC figures. (Modified from Figure 3 in Osmon-McLeod, M.J. et al., *Part Fibre Toxicol.*, 8, 15, May 13, 2011 to show only the results for long fiber chrysotile [LFC] and long fiber amphibole [LFA].)

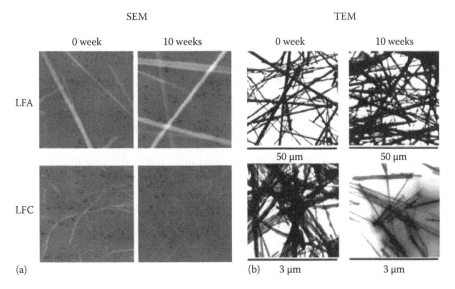

FIGURE 14.9 Appearance of fibers before and after incubation in Gambles solution. (a) Representative SEM images of samples after 0-week or 10-weeks incubation in Gambles solution are shown at 5.0 K magnification in the two panels on the left. (b) TEM images of equivalent samples, at indicated magnifications, are on the right. Note that the scale of the horizontal axis is different in the LFA and LFC figures. (Modified from Figure 4 in Osmon-McLeod, M.J. et al., *Part Fibre Toxicol.*, 8, 15, May 13, 2011 to show only he results for long fiber chrysotile [LFC] and long fiber amphibole [LFA].)

4 in Osmon-McLeod et al. [2011] to show only the results for long fiber chrysotile [LFC] and long fiber amphibole [LFA]) show the marked reduction in length for the long fiber chrysotile following 10 weeks of treatment. The chrysotile fibers following 10 weeks of treatment in vitro break apart so that there are no fibers longer than 10 µm in length. In the lung, such shorter fibers can be readily cleared by the macrophage.

14.5 BIOPERSISTENCE

A fiber is unique among inhaled particles in that the fibers aerodynamic diameter is largely related to three times the fiber diameter. Because of this, long thin fibers can penetrate into the deep lung effectively bypassing the filtration which occurs for nonfibrous particles. Within the lung, fibers which can be fully engulfed by the macrophage can be removed as with any other particle. However, those fibers which are too long to be fully engulfed by the macrophage cannot be cleared by this route.

Fibers less than 5 µm in length are effectively not different than nonfibrous particles and are cleared with similar kinetics and mechanism as particles. While longer fibers may also be cleared effectively by the macrophage and as a result not be different kinetically than particles, the 5 µm cut-off was chosen to mirror the use by the WHO of a 5 µm cut-off in their counting schemes for fibers. As is discussed later, recent reviews of these size fibers have concluded that they present very little or no risk to human health (ATSDR, 2003).

Fibers between 5 and 20 µm in length represent the transition range between those fibers which are cleared as particles and the longer fibers that the macrophage cannot fully phagocytize. The actual limit as to what length fiber can be fully phagocytized has been proposed for the rat, as ranging from 15 µm (Miller, 2000) to 20 µm (Morimoto et al., 1994; Luoto et al., 1995).

In the lung, extensive work on modeling the dissolution of synthetic vitreous fibers (SVFs) using in vitro dissolution techniques and inhalation biopersistence has shown that the lung has a very large fluid buffer capacity (Mattson, 1994). These studies have shown that an equivalent in vitro flow rate of up to 1 mL/min is required to provide the same dissolution rate of SVF as that which occurs in the lung. This large fluid flow within the lung results in the dissolution of fibers which are soluble at pH 7.4. Recent publications have shown that the biopersistence of the fibers longer than 20 µm is an excellent predictor of the pathological response to fibers following chronic inhalation studies and chronic intraperitoneal studies (Hesterberg et al., 1998a,b; Bernstein et al., 2001a,b). The value of 20 µm is used as an index for fibers that cannot be fully phagocytized and cleared by the macrophage. The protocol used in these biopersistence studies was developed by a working group for the European Commission and involves a 5-day inhalation exposure, followed by analysis of the lungs at periodic intervals of up to 1 year postexposure (Bernstein and Riego-Sintes, 1999; Bernstein et al., 2005).

The clearance half-time of SVFs longer than 20 µm ranges from a few days to less than 100 days. This is illustrated in Table 14.2. Highlighted in this table are those studies performed on chrysotile using the same protocol which is within the lower range of the SVFs. In contrast, amphiboles have biopersistence half-times considerably longer than the SVFs. For synthetic vitreous fibers, the European Commission has established a directive which states that if the inhalation biopersistence clearance half-time of a fiber is less than 10 days, then it is not classified as a carcinogen.

Clearly, there is a large difference in biopersistence between serpentine and the amphibole asbestos. In addition, as the serpentine asbestos, chrysotile, is a naturally occurring mined fiber, there appears to be some differences in biopersistence, depending upon from where it is mined. However, chrysotile lies on the soluble end of this scale and ranges from the least biopersistent fiber to a fiber with biopersistence in the range of glass and stonewools. It remains less biopersistent than ceramic and special purpose glasses and more than an order of magnitude less biopersistent than amphiboles.

TABLE 14.2

Clearance Half-Time of Synthetic Vitreous Fibers, Chrysotile, and Amphibole Asbestos Longer than 20 μm

Fiber	Type	Weighted $T_{1/2}$ Fibers L > 20 μm (Days)	Reference
Calidria chrysotile	Serpentine asbestos	0.3	Bernstein et al. (2005b)
Brazilian chrysotile	Serpentine asbestos	2.3	Bernstein et al. (2004)
Fiber B	B01.9	2.4	Bernstein et al. (1996)
Fiber A	Glasswool	3.5	Bernstein et al. (1996)
Fiber C	Glasswool	4.1	Bernstein et al. (1996)
Fiber G	Stonewool	5.4	Bernstein et al. (1996)
Chrysotile combined with sanded joint compound	Serpentine asbestos		Bernstein et al. (2011)
MMVF34	HT stonewool	6	Hesterberg et al. (1998a)
MMVF22	Slagwool	8	Bernstein et al. (1996)
Fiber F	Stonewool	8.5	Bernstein et al. (1996)
MMVF11	Glasswool	9	Bernstein et al. (1996)
Fiber J	X607	9.8	Bernstein et al. (1996)
Canadian chrysotile	Serpentine asbestos	11.4	Bernstein et al. (2005c)
MMVF 11	Glasswool	13	Bernstein et al. (1996)
Fiber H	Stonewool	13	Bernstein et al. (1996)
MMVF10	Glasswool	39	Bernstein et al. (1996)
Fiber L	Stonewool	45	Bernstein et al. (1996)
MMVF33	Special purpose glass	49	Hesterberg et al. (1998a)
RCF1a	Refractory ceramic	55	Hesterberg et al. (1998a)
MMVF21	Stonewool	67	Hesterberg et al. (1998a)
MMVF32	Special purpose glass	79	Hesterberg et al. (1998a)
MMVF21	Stonewool	85	Bernstein et al. (1996)
Amosite	Amphibole asbestos	418	Hesterberg et al. (1998a)
Crocidolite	Amphibole asbestos	536	Bernstein et al. (1996)
Tremolite	Amphibole asbestos	∞	Bernstein et al. (2005b)
Amosite	Amphibole asbestos	>1000 days	Bernstein et al. (2011)

The rapid clearance of chrysotile is thought to be characterized not by congruent dissolution as with many SVF but rather with the loss of structural integrity of the serpentine sheet silicate and the subsequent disintegration into smaller pieces as a result of the action of the lung surfactant and the acid environment of the macrophage.

This difference between chrysotile and amphiboles is better illustrated, with the actual lung burden data for the fibers longer than 20 μm from the inhalation biopersistence studies. In Figure 14.10, the number of fibers remaining in the rat's lungs is shown as a function of the time in days following cessation of the 5-day exposure (Bernstein et al., 2003a, 2005b, 2011c). Included are the three amphibole asbestos studies including tremolite asbestos and 2 amosite asbestos, a SVF fiber, HT, which has a clearance half-time of 6 days and which showed no tumors or fibrosis in a chronic inhalation toxicology study and 4 chrysotile fibers from Brazil, the United States (Calidria), and Canada and one study on a commercial joint compound mixed with chrysotile fibers. The inhalation exposure aerosol in terms of the number of fibers longer than 20 μm was in the range of 150–200 fibers (L > 20 μm)/cm³ for all fibers except the Brazilian chrysotile which was 400 fibers (L > 20 μm)/cm³.

The amphiboles are very durable with only a small amount of clearance in the days following the cessation of exposure with virtually no further clearance thereafter. In the tremolite biopersistence study,

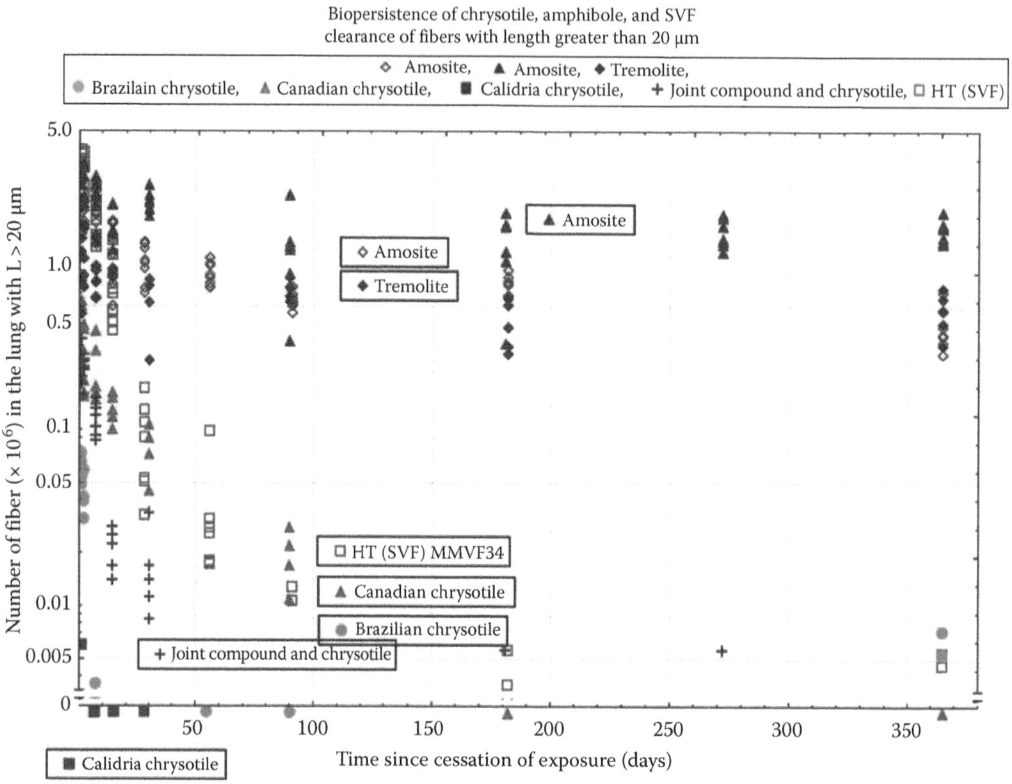

Biopersistence of chrysotile, amphibole, and SVF
clearance of fibers with length greater than 20 μm

FIGURE 14.10 (See color insert.) Summary of studies showing the number of fibers remaining in the rat's lungs is shown as a function of the time in days following cessation of the 5-day exposure. Included are the three amphibole asbestos studies including tremolite asbestos and 2 amosite asbestos, a soluble synthetic vitreous (SVF) fiber, HT, and 4 chrysotile fibers from Brazil, the United States (Calidria), and Canada and one study on a commercial joint compound mixed with chrysotile fibers. (Data presented are from Bernstein et al., 2003a, 2005c; Bernstein, D.M. et al., *Inhal. Toxicol.*, 23(7), 372, June, 2011.)

the histopathological response of the lungs was examined following the 5-day exposure. A pronounced inflammatory response with the rapid development of granulomas was seen at day 1 postexposure, followed by the development of fibrosis characterized by collagen deposition within these granulomas and by 90 days, even mild interstitial fibrosis. In the same study, chrysotile showed no inflammatory or pathological response following the 5-day exposure (Bernstein et al., 2003b).

While all the chrysotiles cleared relatively quickly, differences were observed between the three types studied. The Calidria chrysotile which is known to be a short fiber chrysotile cleared the fastest, with a clearance half-time of 0.3 days for the fibers longer than 20 μm.

The clearance half time of the Brazilian chrysotile was 2.3 days. At the end of 12 months, 2–3 long fibers were measured following the lung digestion procedure. However, the exposure concentration for the Brazilian chrysotile was 400 fibers L > 20 μm/cm³, rather than the 150–200 fibers L > 20 μm/cm³ for the other fibers evaluated, thus resulting in a very high aerosol concentration of 7 million WHO fibers/cm³ and more than 32 million total fibers/cm³. It certainly is possible that this extremely high total exposure resulted in a response very different from what might be expected at lower exposure concentrations. Even so, the number of fibers observed at 12 months was not statistically different than that which was observed for the HT fiber which had a 6-day clearance half-time for the long fibers.

For the Canadian chrysotile study, textile grade chrysotile was evaluated. This grade was chosen as it was specifically produced to have thin long fibers, which facilitated the production of textiles.

The clearance of the Canadian chrysotile long fibers was 11.4 days. By 365 days, there were no long Canadian chrysotile fibers remaining in the lung.

The pathological response and translocation in the lung and pleura of a commercial chrysotile product similar to that which was used through the mid-1970s in a joint compound intended for sealing the interface between adjacent wall boards was evaluated in comparison to amosite asbestos (Bernstein et al., 2010, 2011). The study design was enhanced to include procedures for the quantitative evaluation of fibers and pathological response not only in the lung but also in the pleural space while limiting procedural artifacts.

In this study, rats were exposed by inhalation for 5 days (6 h/day) to either sanded joint compound consisting of a mixture of added chrysotile fibers and sanded joint compound particles or amosite asbestos. The mean exposure concentration for fibers longer than 20 μm was 295 fibers/cm^3 for chrysotile and 201 fibers/cm^3 for amosite. The mean number of WHO fibers in the chrysotile fibers and sanded joint compound particle atmosphere was 1496 fibers/cm^3. The amosite-exposure atmosphere had fewer shorter fibers, with a mean of 584 WHO fibers/cm^3. While the exposure concentrations were controlled to avoid the effect of lung overload, the chrysotile concentration was still more than 10,000 times the OSHA occupational exposure limit of 0.1 fibers/cm^3.

The study included examination of the diaphragm as a parietal pleural tissue and the in situ examination of the lungs and pleural space obtained from freeze-substituted tissue in deeply frozen rats. The diaphragm was chosen as a representative parietal pleural tissue because at necropsy, it could be removed within minutes of sacrifice with minimal alteration of the visceral lung surface with the area of the diaphragm chosen for examination that included an important lymphatic drainage site (stomata) on the diaphragmatic surface. Both confocal microscopy and SEM were used to identify fibers as well as examine the pleural space, in situ, for possible inflammatory response. The examination of the pleural space in situ including the lung, visceral pleura, and parietal pleura in rats deeply frozen immediately after termination provided a unique noninvasive method for determining fiber location and inflammatory response.

The results of this study showed that there was no pathological response observed at any time point in the chrysotile fibers and sanded joint compound particles exposure group. As with the other studies reported earlier, the chrysotile long fibers in the lung (L > 20 μm) cleared rapidly ($T_{1/2}$ of 4.5 days) and were not observed in the pleural cavity at any time point. In contrast, following the 5-day exposure to amosite asbestos, a rapid inflammatory response occurred in the lung which resulted in a Wagner grade 4 interstitial fibrosis within 28 days and which persisted through 90 days. (Histopathology was evaluated through 90 days postexposure as the animals were allocated to the confocal microscopic analyses from 181 through 365 days postexposure.) The amosite fibers longer than 20 μm had a biopersistence of $T_{1/2}$ > 1000 days in the lung and were observed in the pleural cavity within 7 days postexposure. In the pleural cavity, a marked inflammatory response was observed on the parietal pleural surface by 90 days postexposure. In contrast to the amosite asbestos exposure, this study provides support that exposure to the chrysotile fibers and joint compound particles following short-term inhalation would not initiate any inflammatory response in the lung and that the chrysotile fibers in the lung do not migrate or cause an inflammatory response in the pleural cavity, the site of mesothelioma formation.

14.5.1 Clearance Mechanism of HT and Chrysotile

Kamstrup et al. (2001) described possible mechanisms that could account for the rapid clearance half-time of the long HT fibers. He stated that

> The HT fiber is characterized by relatively low silica and high alumina content, with a high dissolution rate at pH 4.5 and relatively low rate at pH 7.4 (Guldberg et al. 2002; Knudsen et al., 1996). Apart from possible exposure to the acidic environment of the phagolysosomes within the macrophages (Oberdörster, 1991), measurements have shown that the microenvironment at the surface of activated

macrophages is acidic with pH < 5 between attached macrophages and a nonporous glass surface (Etherington et al., 1981). It is therefore probable that long HT fibers, highly soluble at pH 4.5, are subject to extracellular dissolution and consequent breakage when exposed to the acidic environment of attached macrophages without being engulfed completely.

As reviewed earlier, the ability of chrysotile to be broken apart in the acid environment has been known since the publication by von Kobell in 1834. As discussed earlier, a similar process to that described for HT fibers has been demonstrated for chrysotile fibers by Osmond-McLeod et al. (2011) who showed that the long chrysotile fibers break apart into shorter fibers less than 10 μm in length following treatment in Gambles solution that was adjusted to a pH of 4.5.

14.5.2 SHORT FIBER CLEARANCE

For all fiber exposures, there are many more shorter fibers less than 20 μm in length and even more less than 5 μm in length. The clearance of the shorter fibers in these studies has been shown to be either similar to or faster than the clearance of insoluble nuisance dusts (Stoeber et al., 1970; Muhle et al., 1987). In a recent report issued by the Agency for Toxic Substances and Disease Registry entitled "Expert Panel on Health Effects of Asbestos and Synthetic Vitreous Fibers: The Influence of Fiber Length," the experts stated that "Given findings from epidemiologic studies, laboratory animal studies, and in vitro genotoxicity studies, combined with the lung's ability to clear short fibers, the panelists agreed that there is a strong weight of evidence that asbestos and SVFs (synthetic vitreous fibers) shorter than 5 μm are unlikely to cause cancer in humans" (ATSDR, 2003; EPA, 2003). In addition, Berman and Crump (2003) in their technical support document to the EPA on asbestos-related risk also found that shorter fibers do not appear to contribute to disease.

14.6 CHRONIC INHALATION TOXICOLOGY STUDIES

The studies presented earlier indicate that there is a large difference in the biopersistence between the serpentine chrysotile and the amphiboles, tremolite, and amosite. These differences appear to be related to the differences in chemical structure between the serpentines and amphiboles and possibly the influence of the acidic pH associated with the macrophage on the chrysotile fiber.

Yet when the chronic inhalation studies that have been performed on chrysotile and amphiboles are examined, these differences are not always apparent.

In an analysis by Berman et al. (1995) of 13 inhalation studies that have been performed on nine different types of asbestos, they concluded that

- Short fibers (less than somewhere between 5 and 10 μm in length) do not appear to contribute to cancer risk.
- Beyond a fixed, minimum length, potency increases with increasing length, at least up to a length of 20 μm (and possibly up to a length of as much as 40 μm).
- The majority of fibers that contribute to cancer risk are thin, with diameters less than 0.5 μm and the most potent fibers may be even thinner. In fact, it appears that the fibers that are most potent are substantially thinner than the upper limit defined by respirability.
- Identifiable components (fibers and bundles) of complex structures (clusters and matrices) that exhibit the requisite size range may contribute to overall cancer risk, because such structures likely disaggregate in the lung. Therefore, such structures should be individually enumerated when analyzing to determine the concentration of asbestos.
- For asbestos analyses to adequately represent biological activity, samples need to be prepared by a direct-transfer procedure.
- Based on animal dose–response studies alone, fiber type (i.e., fiber mineralogy) appears to impart only a modest effect on cancer risk (at least among the various asbestos types).

Concerning the lack of differentiation seen in the dose–response studies, the authors stated that this may be due at least in part to the limited lifetime of the rat relative to the biodurability of the asbestos fiber types evaluated in these studies.

More important in understanding these results are the specifics in the study design of these studies in light of the more recent understanding of the effect of high concentrations of insoluble particles on the rat lung. The majority of the inhalation toxicology studies evaluated by Berman et al. (1995; Table 14.1) were performed at very high exposure concentrations (10 mg/m^3).

The chronic inhalation studies that have been performed on asbestos are summarized in Tables 14.3 and 14.4. The exposure regime was similar in most studies and ranged from 5 to 7 h/day, 5 days per week for either 12 months or 24 months. While it is difficult to determine how this was derived, the exposure concentration was set for most studies based upon mass concentration at 10 mg/m^3. Davis et al. (1978) referencing Wagner et al. (1974) states that 10 mg/m^3 was considered to be high enough to cause significant pathological change; however, there is no rational presented in the Wagner et al. paper as to why 10 mg/m^3 was chosen. While the differential toxicity of chrysotile and amphibole asbestos was not well understood at the time, with a common name *asbestos*, a common toxicity may have been considered. Amphibole asbestos produced toxic effects at lower concentrations; however, with the very different mineral chrysotile, it may have been necessary to increase the dose to what is now considered *lung overload* concentrations to produce a similar effect as seen for amphibole asbestos.

Bernstein et al. (2013) have reported on the methods used to prepare the two most commonly used chrysotile samples for toxicology studies, the UICC chrysotile and the NIEHS chrysotile samples. With both of these chrysotile samples, the fibers were extensively ground using large-scale commercial milling devices.

The UICC chrysotile sample was milled using a "Classic Mill designed by R. F. Bourne, at The Asbestos Grading Equipment Company, Johannesburg" (Timbrell et al., 1968). Timbrell and Rendall (1972) describes "The Classic mill is an airswept attrition mill fitted with a disc rotor (16 in. diam.) which carries four beaters and is mounted on a horizontal shaft driven by an electric motor at speeds up to 5000 r.p.m." The patent (Patent number GB 3,490,704) on the mill provides greater detail.

The NIEHS chrysotile was prepared from a grade 4 chrysotile used in the plastics industry which was prepared by passing the material through a hurricane pulverizer (Campbell et al., 1980; Pinkerton et al., 1983). The hurricane pulverizer is a high-speed impact hammer industrial mill with a size classifier which recycled larger fibers/particles back into the device for continued milling (Work, 1962; Perry and Chilton, 1973).

Suquet (1989) has reported that severe grinding of chrysotile fibers "converted them into fragments cemented by a shapeless, noncrystalline material." The authors explained that the comminution treatment apparently broke atomic bonds and produced strong potential reaction sites, which were able to adsorb CO_2 and H_2O molecules from the atmosphere."

The issue of using equivalent fiber number for exposure was approached in a study reported by Davis et al. (1978) where chrysotile, crocidolite, and amosite were compared on an equal mass and equal number basis; however, the fiber number was determined by phase contrast optical microscopy (PCOM) and thus, the actual number of the chrysotile fibers was probably greatly underestimated. The 10 mg/m^3 exposure to chrysotile was reported by Davis et al. (1978) using PCOM as approximately 2000 fibers/cm^3 (length greater than 5 µm) while when a similar mass concentration of another chrysotile was measured by SEM 10,000 fibers/cm^3 length greater than 5 µm were reported with a total fiber count of 100,000 fibers/cm^3 (Mast et al., 1995). There is little quantitative data presented in these publications on the nonfibrous particle concentration of the test substances to which the animals were exposed. Pinkerton et al. (1983) presents summary tables of length measurements of Calidria chrysotile by SEM in which the number of nonfibrous particles counted is stated; however, from the data presented, the aerosol exposure concentration of nonfibrous particles cannot be extracted. In all studies, the asbestos was ground prior to aerosolization,

TABLE 14.3
Inhalation Toxicology Studies Using Rats

Fiber Type	Exposure Time (Hours/Day, Day/Week, Total Months)	Exposure Type, Exposure Concentration (mg/m³)	Fiber Concentration (Fibers/cm³) (Determined by Electron Microscopy Unless Otherwise Noted)	Equivalent Fiber (Concentration/cm³) Based on TEM Evaluation	Type and Total No. of Rats	Number of Pulmonary Tumors	% Pulmonary Tumors	Number of Mesotheliomas	References
Chrysotile Canadian (nickel, cobalt, chromium, and lead contamination)	6, 5, 14	w.b. 86	Nd	8,600,000	NS, 41	10	24	1	Gross et al. (1967)
Chrysotile UICC Canadian	7, 5, 24	w.b. 10	Nd	970,000	W, 21	10	48	1	Wagner et al. (1974)
Chrysotile UICC Rhodesian	7, 5, 24	w.b. 10	Nd	1,470,000	W, 17	11	65	0	Wagner et al. (1974)
Chrysotile Canadian 714–7D (friction linings)	5, 5, 24	w.b. 15	1.7×10^5 SEM 9978 > 5 μm	1,500,000	W, 45	9	20	0	Le Bouffant et al. (1984–1987)
SFA chrysotile	7, 5, 24	w.b. 10	430 > 5 μm pcom 669 particles pcom	1,080,000	W, 22	8	36	0	Wagner et al. (1980)
Grade 7 chrysotile	7, 5, 24	w.b. 10	1020 > 5 μm pcom 745 particles pcom	1,080,000	W, 24	3	13	0	Wagner et al. (1980)
UICC chrysotile	7, 5, 24	w.b. 10	3750 > 5 μm pcom 338 particles pcom	1,080,000	W, 23	5	22	0	Wagner et al. (1980)
Chrysotile Calidria	5, 5, 12	w.b. 6	241 131 > 5 μm reported as *thick bundles*	600,000	W, 50	0	0	0	Muhle et al. (1987)

Chrysotile long	7, 5, 12	w.b. 10	1170 > 5 µm pcom 33 > 20 µm pcom	1,000,000	W, 40	20	50	2	Davis and Jones (1988)
Chrysotile court	7, 5, 12	w.b. 10	5510 > 5 µm pcom 670 > 20 µm pcom	1,000,000	W, 40	7	17	1	Davis and Jones (1988)
Chrysotile UICC A	7, 5, 12	w.b. 10	2560 > 5 µm pcom	1,000,000	Included for comparative fiber numbers without animal exposure				Davis and Jones (1988)
Chrysotile NIEHS	6, 5, 24	n-o 10	1.02×10^5 SEM 1.06×10^4 > 5 µm	1,000,000	F, 69	13	18	1	Mast et al. (1995)
Chrysotile	7, 5, 12	w.b. 10	1950 > 5 µm pcom 360 > 20 µm pcom	1,000,000	W, 40	15	38	0	Davis et al. (1978)
Chrysotile	7, 5, 12	w.b. 2	390 > 5 µm pcom 72 > 20 µm pcom	200,000	W, 42	8	19	1	Davis et al. (1978)
Chrysotile Calidria	7, 5, 12	w.b. 10	Nd	778,000	F, 51	2	4	0	Ilgren and Chatfield (1997, 1998) and Pinkerton et al. (1983)
Chrysotile Jeffrey	7, 5, 12	w.b. 10	Nd	1,136,000	F, 49	11	22	0	Ilgren AND Chatfield (1997, 1998) and Pinkerton et al. (1983)
Chrysotile UICC/B	7, 5, 12	w.b. 10	Nd	1,099,000	F, 54	13	24	0	Ilgren and Chatfield (1997, 1998) and Pinkerton et al. (1983)
Amosite	7, 5, 12	w.b. 10	550 > 5 µm pcom 6 > 20 µm pcom	—	W, 43	2	5	0	Davis et al. (1978)
Tremolite Korean	7, 5, 12	w.b. 10	1600 pcom	—	39	18	46	2	Davis et al. (1985)
Amosite UICC	7, 5, 24	w.b. 10	Nd	—	W, 21	13	62	0	Wagner et al. (1974)

(continued)

TABLE 14.3 (continued)
Inhalation Toxicology Studies Using Rats

Fiber Type	Exposure Time (Hours/Day, Day/Week, Total Months)	Exposure Type, Exposure Concentration (mg/m³)	Fiber Concentration (Fibers/cm³) (Determined by Electron Microscopy Unless Otherwise Noted)	Equivalent Fiber (Concentration/cm³) Based on TEM Evaluation	Type and Total No. of Rats	Number of Pulmonary Tumors	% Pulmonary Tumors	Number of Mesotheliomas	References
Amosite long	7, 5, 12	w.b. 10	2060 > 5 μm pcom / 70 > 10 μm pcom	—	W, 40	11	28	3	Davis et al. (1986)
Amosite short	7, 5, 12	w.b. 10	70 > 5 μm pcom / 12 > 10 μm pcom	—	W, 42	0	0	1	Davis et al. (1986)
Amosite		w.b. 300	Nd	—	SD, 16	3	19	0	Lee et al. (1981)
Crocidolite UICC	7, 5, 24	w.b. 10	Nd	—	W, 18	13	72	0	Wagner et al. (1974)
Crocidolite	7, 5, 12	w.b. 10	860 > 5 μm pcom estimated figure / 34 > 20 μm pcom	—	W, 40	1	3	0	Davis et al. (1978)
Crocidolite	7, 5, 12	w.b. 5	430 > 5 μm pcom / 17 > 20 μm pcom	—	W, 43	2	5	1	Davis et al. (1978)
Crocidolite	5, 5, 12	w.b. 2.2	2011 / 162 > 5 μm	—	W, 50	1	2	0	Muhle et al. (1987)
Crocidolite UICC	6, 5, 24	w.b. 7	3000 / 90 > 10 μm	—	OM, 60	3	5	1	Smith et al. (1987)
Crocidolite exposure truncated	6, 5, 10	n-o 10	1.6 × 10⁴ > 5 μm SEM	—	F, 106	15	14	1	McConnell et al. (1994)

Exposure types: w.b., whole body; n-o, nose only.
Type of rat: F, Fisher 344; OM, Osborne Mendel; SD, Sprague Dawley; W, Wistar.
ND, not determined.
PCOM, phase contrast optical microscopy.
SEM, scanning electron microscopy.

TABLE 14.4
Inhalation Toxicology Studies Using Hamsters

Fiber Type	Exp. Time (Hours/Day, Day/Week, Max. Months)	Exposure Concentration (mg/m³)	Fiber Concentration (Fibers/cm³)	Total Number of Hamsters	Number of Pulmonary Tumors	% Pulmonary Tumors Adenoma, Carcinomas	Number of Mesotheliomas	References
Amosite		w.b. 300	ND	7	0	0		Lee et al. (1981)
Amosite—low	6, 5, 18	n-o 0.8	36 > 5 µm 10 > 20 µm	83	0	0	3	McConnell et al. (1999)
Amosite—mid	6, 5, 18	n-o 3.7	165 > 5 µm 38 > 20 µm	85	0	0	22	McConnell et al. (1999)
Amosite—high	6, 5, 18	n-o 7.1	263 > 5 µm 69 > 20 µm	87	0	0	17	McConnell et al. (1999)
Crocidolite UICC	6, 5, 18	w.b. 7	3000 90 > 10 µm	58	0	0	0	Smith et al. (1987)
Chrysotile NIEHS	6, 5, 18	n-o 10	1.02×10^5 $1.06 \times 10^4 > 5$ µm	Not reported	0	0	0	Mast et al. (1994)

Exposure types: w.b., whole body; n-o, nose only.

a procedure which when applied to chrysotile would produce a large number of short fibers and particles. Bernstein et al. (2013) have estimated number of chrysotile fibers that would have been present in the aerosol if measured by transmission electron microscope (TEM) in the chrysotile studies listed in Table 14.3. These values have been derived from the gravimetric concentration which was reported for the studies and a conversion based upon the SEM measurements reported by Mast et al. (1995) and Hesterberg et al. (1993), with an extrapolation to TEM (Breysse et al., 1989). The gravimetric exposure concentrations shown in Table 14.3 ranged from 2 to 86 mg/m^3, which based upon this extrapolation corresponds to between 200,000 and 8,600,000 fibers/cm^3.

In the study reported by Mast et al. (1995) and Hesterberg et al. (1993), the total chrysotile lung burden following 24 months of exposure was 5.5×10^{10} fibers/lung, as measured by SEM (Bernstein, 2007). With extrapolation to that which would have been observed by TEM, the lung burden would have been approximately 9.4×10^{11} fibers/lung. This would correspond to an average of 2300 fibers per alveoli (assuming 10% deposition) (Bernstein et al., 2013).

In addition, most of the studies prior to Mast et al. (1995) used for aerosolization of the fibers an aerosol generation apparatus based on the design of Timbrell et al. (1968) in which a rotating steel blade pushed/chopped the fibers off a compressed plug into the airstream. As some authors have stated, the steel used in the grinding apparatus and as well the aerosolization apparatus often wore, resulting in sometimes considerable exposure to the metal fragments as well. These factors contribute significantly to the difficulty in interpreting the results of the serpentine chrysotile inhalation exposure studies.

In these studies, a tumorigenic response to amphibole's response is observed as would be expected from the biopersistence results; however, as mentioned, there is also a tumorigenic response to some of the chrysotile exposures even though the biopersistence results would suggest otherwise. Eastes and Hadley (1996) developed a model which related the dose of fibers in the lung to potential pathogenicity.

However, as many studies have now shown, in the rat, another factor can also influence the inflammatory and pathological responses. High concentrations of insoluble nuisance dusts have been shown to compromise the clearance mechanisms of the lung, cause inflammation, and a tumorigenic response in the rat, a phenomenon often referred to as lung overload (Bolton et al., 1983; Morrow, 1988; Muhle et al., 1988; Oberdörster, 1995).

The biopersistence studies conducted at exposure concentrations that did not exceed lung overload conditions elucidate two kinetic patterns with chrysotile. They show that the long fibers are not biopersistent. As a result of the fiber chemistry and structure, the longer fibers are attacked, disintegrating into the smaller pieces. The biopersistence studies also show that these smaller pieces clear at a rate which is similar to the rate of clearance of insoluble nuisance dusts. Chrysotile has also been shown to split longitudinally. In most of the chronic inhalation studies, the total aerosol concentration was probably in the order of 10^6 particles and fibers/cm^3 and if the fibers upon contact with the lung begin to split and break apart, the effective dose in terms of the total number of particles will be increased even further.

With such a breakdown of chrysotile into shorter particles, the question remains as to whether the resulting concentration of particles can result in a nonspecific inflammatory reaction and an overload effect in the rat lung. In a recent study, Bellmann et al. (2003) reported on a calibration study to evaluate the end points in a 90-day subchronic inhalation toxicity study of man-made vitreous fibers with a range of biopersistence and amosite. One of the fibers was a calcium–magnesium–silicate (CMS) fiber for which the stock preparation due to method of preparation had a large concentration of particulate material in addition to the fibers. The aerosol exposure concentration for the CMS fiber was 286 fibers/cm^3 length < 5 μm, 990 fibers/cm^3 length > 5 μm, and 1793 particles/cm^3, a distribution which is not observed in manufacturing. The total CMS exposure concentration was 3069 particles and fibers/cm^3. The authors point out that "The particle fraction of CMS that had the same chemical composition as the fibrous fraction seemed to cause significant effects." The number of polymorphonuclear leukocytes (PMN) in the bronchoalveolar lavage fluid

(BALF) was higher, and interstitial fibrosis was more pronounced than had been expected on the basis of biopersistence data. In addition, interstitial fibrosis persisted through the 14-week recovery period following the 90-day exposure. In a separate study on X607, a fiber chemically similar to CMS with, however, considerably fewer particles present in the aerosol was evaluated in a chronic inhalation toxicity study and produced no lung tumors or fibrosis at any time point (Hesterberg et al., 1998b).

This effect attributed to particles in the rat CMS study was observed with an exposure concentration of 3069 particles and fibers/cm^3, 50% of which were particles or short fibers. It would follow directly from this and the many publications on overload to expect that a dramatically more pronounced effect would occur if the exposure concentration was 1,000,000 particles and fibers/cm^3, 90% of which were particles or short fibers as was the case with chrysotile.

These discrepancies in study design put in question the value of especially the chrysotile studies listed in Tables 14.3 and 14.4. McConnell et al. (1999) reported on a well-designed multiple-dose study of amosite asbestos in the hamster where particle and fiber number were well controlled (Table 14.4). In this study, the aerosol concentration ranged from 10 to 69 fibers/cm^3 and were chosen based upon a previous, multi-dose 90-day subchronic longer than 20 µm (Hesterberg et al., 1999).

14.6.1 Fiber Length

In an analysis that provided the basis for the European Commission's Directive on synthetic mineral fibers, Bernstein et al. (2001a,b) reported that there exists for SVF an excellent correlation between the biopersistence of fibers longer than 20 µm and the pathological effects following either chronic inhalation or chronic intraperitoneal injection studies. This analysis showed that it was possible using the clearance half-time of the fibers longer than 20 µm as obtained from the inhalation biopersistence studies to predict the number of fibers longer than 20 µm remaining following 24 month chronic inhalation exposure. These studies, however, only included synthetic mineral fibers.

As mentioned earlier, Berman et al. (1995) analyzed statistically 9 different asbestos types in 13 separate studies. Due to limitations in the characterization of asbestos structures in the original studies, new exposure measures were developed from samples of the original dusts that were regenerated and analyzed by transmission electron microscopy. The authors reported that while no univariate model was found to provide an adequate description of the lung tumor responses in the inhalation studies, the measure most highly correlated with tumor incidence was the concentration of structures (fibers) $\geq$20 µm in length. However, using multivariate techniques, measures of exposure were identified that do adequately describe the lung tumor responses.

The potency appears to increase with increasing length, with structures (fibers) longer than 40 µm being about 500 times more potent than structures between 5 and 40 µm in length. Structures <5 µm in length do not appear to make any contribution to lung tumor risk. As discussed earlier, while this analysis also did not find a difference in the potency of chrysotile and amphibole toward the induction of lung tumors, most of the studies included were performed at very high exposure concentrations.

14.6.2 Purity of the Samples

In most inhalation studies on both amphiboles and serpentines, there was no analytical confirmation that the fibers that were aerosolized were uniquely of the type stated.

In addition, an issue which has been discussed at length is whether the presence of tremolite in the chrysotile samples can account for some of its carcinogenic potential as well. This is especially pertinent to the mesotheliomas that have been observed in some of the rat inhalation studies (Churg, 1994; McDonald et al., 1999; Roggli et al., 2002). Using microscopic analysis, Frank et al. (1998) have reported the absence of tremolite in the UICC chrysotile sample which has often been used in the chronic studies. However, when present with chrysotile, tremolite is usually found in very low concentrations which could be missed during microscopic analysis.

To resolve this issue of method sensitivity, Addison and Davies (1990) developed a method of chemical digestion of chrysotile in which the chrysotile is dissolved using acid, leaving behind the amphiboles such as tremolite. This method was applied to a sample UICC chrysotile obtained from Dr. Fred Pooley who has a repository of the original UICC preparation. In conjunction with Gesellschaft für Schadstoffmessung und Auftragsanalytik GmbH (GSA, Neuss, Germany), 2.13 g of UICC chrysotile was digested in acid, following a procedure similar to that of Addison and Davies (1990). Following digestion, the bivariate size distribution was determined for all residual fibers by transmission electron microscopy and the chemical composition of each fiber was determined by EDAX in order to clearly identify if it is amphibole, chrysotile, or something else.

In the 2 mg sample analyzed, the results indicated that there were 3400 tremolite fibers per mg UICC chrysotile. These fibers ranged in length from 1.7 to 14.4 μm and had a mean diameter of 0.65 μm. Forty one percent of the fibers were longer than 5 μm, with 1394 WHO tremolite fibers per mg of UICC chrysotile. These results indicate that tremolite is present in the UICC sample at low concentrations. As no dose–response studies have been performed at low amphibole concentrations, quantification of the effect of these fibers is not possible in the rat. However, as discussed earlier, amphibole asbestos fibers are very biopersistent in the lung and will persist once inhaled. Davis et al. (1985) performed a chronic inhalation toxicity study on tremolite in order to determine the effect of commercial tremolite in comparison to other asbestos types. The authors reported that tremolite was the most dangerous mineral that they have studied, producing 16 carcinomas and 2 mesotheliomas in a group of 39 animals. As described earlier, even short exposure to tremolite produces a notable response in the lung. Bernstein et al. (2003b) reported that following a 5-day exposure to tremolite, a pronounced inflammatory response was observed, with the rapid development of granulomas, collagen deposition within these granulomas, and by 90 days, even mild interstitial fibrosis.

14.7 EPIDEMIOLOGY

Both chrysotile and amphibole asbestos were used extensively, often in uncontrolled situations through a large part of the twentieth century. With the understanding of the danger in the use of amphibole asbestos, governments gradually prohibited its use starting in the 1960s, with France being one of the last countries in 1996 to implement such a prohibition.

While some countries have prohibited chrysotile as well, other countries are still mining and using chrysotile largely for high density cement products such as cement roofing and pipes. The understanding of the importance of industrial hygiene controls often termed *controlled use* has resulted in markedly cleaner work environments in the mines and manufacturing facilities.

Many studies have shown that chrysotile is not of the same potency as the amphiboles and is cleared from the lung more rapidly than amphibole (Howard, 1984; Churg and DePaoli, 1988; Mossman et al., 1990; Churg, 1994; Morgan, 1994; McDonald and McDonald, 1995, 1997; McDonald, 1998; McDonald et al., 1999, 2002, 2004; Rodelsperger et al., 1999; Hodgson and Darnton, 2000; Berman and Crump, 2003). Still other studies have stated the opposite.

Two reviews (Hodgson and Darnton, 2000; Berman and Crump, 2003) have reported on quantification of the potency of chrysotile and amphiboles based upon the statistical analysis of epidemiology studies that were available at the time. However, as discussed in the following, the studies characterized as chrysotile exposure were in actuality studies with *predominately chrysotile* exposure. The authors stated that very small quantities of amphibole fiber were ignored as being important to the findings in some cohorts.

Hodgson and Darnton (2000) provided a review of potency of asbestos for causing lung cancer and mesothelioma in relation to fiber type. They concluded that amosite and crocidolite were, respectively, on the order of 100 and 500 times more potent for causing mesothelioma than chryso-tile. They regarded the evidence for lung cancer to be less clear cut, but concluded nevertheless that amphiboles (amosite and crocidolite) were between 10 and 50 times more potent in causing lung cancer than chrysotile.

Berman and Crump (2003) reviewed and analyzed as part of a technical support document for the USEPA an epidemiology database consisting of approximately 150 studies, of which approximately 35 contained exposure data sufficient to derive quantitative exposure–response relationships.

However, due to the state of occupational hygiene measurements at the time, none of the stud-ies were able to use exposure measurements which included fiber number or fiber type. The asso-ciations to disease were attributed to the fiber most used without consideration of the criteria that have been understood more recently to determine fiber potency: fiber mineralogy, biopersistence, and fiber length. In addition, the lack of complete occupational histories is a significant limitation in the early epidemiology studies, resulting at times in improper characterization of fiber-specific exposure. The limitations in these earlier studies have been reviewed by Berman and Crump (2003). However, this review could not take into account the more recent toxicology studies that have been published since, which provide a basis for assessing the importance of even small amounts of amphibole fibers compared to chrysotile. In addition, there has been no systematic analysis of fiber dimensions in these epidemiological studies due to the state of the art at the time. This further compounds understanding the importance of the exposure estimates as the more recent toxicology studies have shown that fiber length is an important determinant in toxicity.

Bernstein et al. (2013) have reviewed the studies characterized as predominately chrysotile and found that amphibole asbestos was often present as well and that other sources of amphibole asbes-tos were sometimes not considered in the evaluations.

This presence of amphibole asbestos is best supported by fiber lung burden analyses in cohorts such as the Charleston, South Carolina, and Quebec and Italy. Sebastien et al. (1989) reported on the analysis of 161 lung tissue samples taken at necropsy from asbestos textile workers in Charleston, South Carolina, and Quebec miners and millers, both exposed to chrysotile. The authors reported that while chrysotile, tremolite, amosite, crocidolite, talc-anthrophyllite, and other fiber types (including rutile, micas, iron, silica, and unidentified silicates) were found, in both cohorts, tremolite predominated. Churg et al. (1984) analyzed the fiber lung content from 6 cases, with mesothelioma derived from a series of approximately 90 autopsies of long-term workers in the Quebec chrysotile industry. The authors reported that the patients with mesothelioma having only chrysotile ore components had a much higher ratio of tremolite group amphiboles (9.3) than chrysotile fibers (2.8), compared to the control group. Pooley and Mitha (1986) in a report on the determination and interpretation of the levels of chrysotile in lung tissue included result from the South Carolina textile workers. Chrysotile, crocidolite, and amosite fibers were found. In addi-tion, in the lungs from the control population, chrysotile and amosite were also found. Case et al. (2000) evaluated asbestos fiber type and length in lungs of fibers longer than 18 μm in length in chrysotile textile plant from the South Carolina cohort and chrysotile miners/milers from the Thetford Mines portion of the Quebec cohort. The Case et al. (2000) results indicated that the *chrysotile only* textile workers had a high proportion of individuals with lung tissue containing amosite and/or crocidolite. Fornero et al. (2009) assessed fiber lung burden in cattle lungs from two areas in Italy's Western Alps, the Susa Valley and Lanzo Valley. This is the same region in which the chrysotile mine at Balangero is located which was the subject of the epidemiological evaluation by Piolatto et al. (1990), where effect was attributed to chrysotile. Fornero et al. (2009) reported that fibers of tremolite/actinolite, chrysotile, grunerite, and crocidolite were found in the cattle lungs.

The studies included in the reviews by Hodgson and Darnton (2000) and Berman and Crump (2003) were not of chrysotile as used today without amphibole asbestos present and were from periods when the exposure concentration was very high. Today the situation is remarkably different in that only chrysotile is used commercially. In those chrysotile mines where tremolite veins may be present, the veins can be readily avoided during the mining process as they can be easily differentiated by color (Williams-Jones et al., 2001). As reviewed by Bernstein et al. (2013), the Cana Brava chrysotile mine in Brazil routinely monitors for the presence of amphiboles and has found no detectable amphibole asbestos. Studies on the Calidria (New Idria, California) chrysotile mine have also found only very rarely cleavage fragments away from the ore zone. Reports on the Uralasbest mine in Asbest, Russia, which is the largest mine currently in production, have found no tremolite in air samples.

In chrysotile mines today, the exposure levels have been greatly reduced through the use of water control spraying technology and closed-circuit systems (Williams et al. 2008; 2011).

It is interesting to note that many of the facilities characterized as predominately chrysotile that were studied in the Hodgson and Darnton (2000) and Berman and Crump (2003) evaluations had achieved marked reduction in exposure concentrations prior to their closures. As an example, at the Balangero chrysotile mine in Italy, Silvestri et al. (2001) reported exposure concentrations were reduced from over 100 fibers/mL in the 1930s to 0.19 fibers/mL in the mine; 0.54 fibers/mL in the crushing area; 0.93 fibers/mL in the fiber selection area and 0.78 fibers/mL in the bagging area in the 1980s. In a study of the chrysotile miners and millers in Quebec, Liddell et al. (1998) reported that "On the other hand, modern dust conditions are well below the average even of dust category 1 and so there can be considerable confidence that the risk of lung cancer as a result of such exposure has become vanishingly small."

As mentioned earlier, the toxicology studies indicate that even short-term exposure to amphiboles can lead to important pathological response, with transfer of fibers to the pleural cavity. The importance of amphibole point sources to the induction of mesothelioma has been reported in several studies. Musti et al. (2009) and Barbieri et al. (2012) reported on the relationship of increased mesothelioma risk of individuals who lived near an amphibole asbestos plant for over 50 years. Kurumatani and Kumagai (2008) reported that residents who lived within a 300 m radius of a cement pipe plant that used crocidolite and chrysotile had an standard mortality ratio (SMR) for mesothelioma of 13.9 (5.6–28.7) for men and 41.1 (15.2–90.1) for women. Case and Abraham (2009) reported that an industrial legacy use exposure area in which crocidolite and amosite were used was high in mesothelioma incidence and mortality. Pan et al. (2005) reported that people living in proximity to ultramafic rock deposits had an independent and dose–response association with mesothelioma risk.

Epidemiological studies of workers exposed to chrysotile in high density cement plants have been reviewed (Bernstein et al., 2013). Weill et al. (1979) reported that no excess mortality was observed in asbestos cement manufacturing workers following exposure for 20 years to chrysotile at levels equal to or less than 100 MPPCF·years (corresponding to approximately 15 fibers/cm^3 years). Thomas et al. (1982) reported on a cohort within an asbestos-cement factory that used chrysotile. The authors stated that: "Thus the general results of this mortality survey suggest that the population of the chrysotile asbestos-cement factory studied are not at any excess risk in terms of total mortality, all cancer mortality, cancers of the lung and bronchus, or gastrointestinal cancers." Gardner et al. (1986) reported on a cohort study carried out at an asbestos cement factory in England. The authors reported that at mean fiber concentrations below 1 fiber/cm^3 (although higher levels had probably occurred in certain areas of the asbestos cement factory), there were no excess of lung cancers or other asbestos-related excess death. Ohlson and Hogstedt (1985) reported on a cohort study of asbestos cement workers in a Swedish plant using chrysotile asbestos in which no excess work-related mortality was observed at cumulative exposures estimated at about 10–20 fibers/cm^3 years.

The evaluation of chrysotile should be based upon exposure scenarios which occur in production and use currently. Based upon the science reviewed earlier, in the absence of amphibole asbestos, the use of chrysotile at the current permissible exposure limits in the workplace has not been associated with a statistically detectable increase in risk as observed epidemiologically.

14.8 SUMMARY

The mineralogy of the serpentine chrysotile fibers and amphiboles fibers shows distinct differences in the structure and chemistry of these two minerals. The curled layered construction of the sheet silicate chrysotile combined with the susceptibility to acid attack results in the ability for this fiber to be degraded and broken apart in the lung and cleared by the macrophage. In contrast, the amphibole fibers are rigid impermeable structures which are resistant to degradation at any pH encountered in the lung. These differences are reflected in the inhalation biopersistence studies which clearly differentiate chrysotile from the amphiboles and show that longer chrysotile fibers rapidly disintegrate in the lung, while the longer amphiboles once deposited remain.

There is no question that amphibole asbestos is highly carcinogenic. Both animal studies and epidemiology studies indicate the potency of amphibole asbestos. Inhalation toxicology studies on tremolite and amosite asbestos show that even short exposure can produce a pathogenic response in the lung. The recent work with amosite asbestos has shown that after a 5-day exposure, fibers are translocated to the pleural cavity within seven days where they initiate a pathological response. This is in contrast to chrysotile which does not initiate a pathological response in the lung and is not translocated to the pleural cavity.

There is an excellent correlation between the biopersistence of the longer synthetic vitreous fibers and chronic toxicity data. Due to the difficulties in study design and the large particle/fiber exposure concentrations used at the time, the chronic inhalation studies with chrysotile asbestos are difficult to interpret in part due to the nonspecific lung-overload effects of the very large particle concentrations in the exposure aerosols. A 90-day chronic inhalation toxicology study of chrysotile performed at lower doses to minimize lung-overload effects has shown that the chrysotile does not produce pathological response at an exposure concentration 5000 times greater than the US total lung volume (TLV) of 0.1 fibers (WHO)/cm^3.

Recent quantitative reviews which analyzed the data of available epidemiological studies to determine potency of asbestos for causing lung cancer and mesothelioma in relation to fiber type also differentiated between chrysotile and amphibole asbestos. The most recent analysis also concluded that it is the longer thinner fibers which have the greatest potency. The quantitative experimental results provide additional support for this differentiation. However, even studies characterized as predominately chrysotile in these evaluations also had amphibole asbestos exposure and were based upon situations in the past where there were high uncontrolled exposures. Many of the same studies showed that exposures can be effectively controlled.

Today chrysotile is used predominantly in manufacturing of high density cement products in situations where the potential exposure is greatly reduced through the implementation of industrial hygiene controls.

With heavy and prolonged occupational exposure to chrysotile there is evidence, as with other respirable particles, such exposure can produce lung cancer. The recent inhalation toxicology studies of chrysotile and the epidemiology studies reporting on use of chrysotile alone in high density cement products and the implementation of controls in mining and manufacturing provide a framework for establishing safe use.

It would be most helpful if future studies on chrysotile and amphiboles asbestos, whether in vitro or in vivo, would be performed using size distributions and at doses approaching those to which humans have been exposed.

QUESTIONS

1. Explain the mineralogical characteristics of the two minerals families which are called asbestos.
   Answer: Asbestos is a generic name which refers to two different mineral families: serpentine and amphibole asbestos.
   - Chrysotile is the most common serpentine asbestos. The chrysotile fiber is a sheet silicate, monoclinic in crystalline structure, and has a unique rolled form resulting from the molecular spacing of the silica and magnesium atoms. The walls of the chrysotile fiber are made up of approximately 12–20 layers in which there is some mechanical interlocking. Each layer is about 7.3 Å thick, with the magnesium surface facing the outside of the curl and the silica and oxygen tetrahedron inside the curl. Chrysotile is distinguished by its behavior of being decomposed by acid. The macrophage which clears foreign material from the lung produces an acidic environment which attacks the chrysotile fiber.
   - Amphibole asbestos has a basic structure consisting of a double silica chain which appears as an I-beam with corner-linked $(SiO_4)_{-4}$ tetrahedra linked together in a double-tetrahedral chain that sandwiches a layer with the Ca_2Mg_5. The double chain silicate amphibole fibers themselves are highly insoluble in both the lung fluids and in the acidic environment of the macrophages.

2. What are the principal criteria which determine fiber toxicity?
   Answer: Mineral fiber toxicology has been associated with three key factors: dose, dimension, and durability.
   - The dose is determined by the fiber's physical characteristics/dimensions, how the fibrous material is used, and the control procedures that are implemented. The thinner and shorter fibers will weigh less and thus can remain suspended in the air longer than thicker and longer fibers.
   - The fiber dimensions govern two factors: whether the fiber is respirable and, secondly, if it is respirable, the dimensions are also a factor in determining their response in the lung milieu once inhaled. Shorter fibers of the size which can be fully engulfed by the macrophage will be cleared by mechanisms similar to those for nonfibrous particles. These include clearance through the lymphatics and macrophage phagocytosis and clearance. It is only the longer fibers which the macrophage cannot fully engulf that have the potential for causing disease if they are persistent in the lung.
   - The durability is important especially for fibers longer than the macrophage (~20 μm). Fibers longer than the macrophage inhibit the macrophage's mobility, preventing clearance. Fibers which either dissolve completely, or dissolve until they are sufficiently weakened to undergo breakage into shorter fibers can be subsequently cleared. Longer fibers which do not can lead to inflammation, fibrosis, and eventually cancer.

3. How long after a 5-day exposure in rats does it take for serpentine and amphibole asbestos fibers to reach the pleural cavity?
   - The serpentine asbestos fiber chrysotile has not been found to translocate to the pleural cavity in such studies and does initiate any pathological response in the pleural cavity.
   - The amphibole asbestos fiber amosite has been found to translocate to the pleural space of the rat within 7 days following a 5-day exposure. A marked inflammatory response was observed on the parietal pleural surface by 90 days postexposure.

4. What are the primary difficulties in interpreting the older epidemiological studies that attempt to differentiate chrysotile from amphibole asbestos.
   - The occupational hygiene measurements used at the time of exposure in these studies did not use exposure measurements which included fiber number, fiber dimensions, or fiber type (chrysotile, amosite, crocidolite, etc.). The associations to disease were attributed to the fiber most used without consideration of the criteria that have been understood more recently to determine fiber potency: fiber mineralogy, biopersistence, and fiber length.

- The lack of complete occupational histories was also a significant limitation in many of the early epidemiology studies, resulting at times in improper characterization of fiber-specific exposure.

  The toxicology studies indicate that even short-term exposure to long fiber (>~20 μm) amphiboles can lead to important pathological responses, with transfer of fibers to the pleural cavity. The importance of amphibole point sources to the induction of mesothelioma has been reported more recently in several studies.

REFERENCES

Addison, J. and L.S. Davies, 1990. Analysis of amphibole asbestos in chrysotile and other minerals. *Ann Occup Hyg* 34(2):159–175.

ATSDR, 2003. Report on the Expert Panel on Health effects of asbestos and synthetic vitreous fibers: The influence of fiber length. Atlanta, GA: Prepared for Agency for Toxic Substances and Disease Registry Division of Health Assessment and Consultation.

Aust, A.E., P.M. Cook, and R.F. Dodson, 2011. Morphological and chemical mechanisms of elongated mineral particle toxicities. *J Toxicol Environ Health B Crit Rev* January–June; 14(1–4):40–75.

Barbieri, P.G., D. Mirabelli, A. Somigliana, D. Cavone, and E. Merler, 2012. Asbestos fibre burden in the lungs of patients with mesothelioma who lived near asbestos-cement factories. *Ann Occup Hyg* 56(6):660–670.

Bates, T.F., L.B. Sand, and J.F. Mink, 1950. Tubular crystals of chrysotile asbestos. *Science* 3:512.

Bellmann, B., H. Muhle, O. Creutzenberg, H. Ernst, M. Müller, D.M. Bernstein, and J.M. Riego-Sintes, 2003. Calibration study on subchronic inhalation toxicity of man-made vitreous fibers in rats. *Inhal Toxicol* October;15(12):1147–1177.

Berman, D.W. and K.S. Crump, 2003. Draft technical support document for a protocol to assess asbestos-related risk. Washington, DC: Office of Solid Waste and Emergency Response U.S. Environmental Protection Agency.

Berman, D.W., K.S. Crump, E.J. Chatfield, J.M. Davis, and A.D. Jones, 1995. The sizes, shapes, and mineralogy of asbestos structures that induce lung tumors or mesothelioma in AF/HAN rats following inhalation. *Risk Anal* 15(2):181–195.

Bernstein, D., V. Castranova, K. Donaldson, B. Fubini, J. Hadley, T. Hesterberg, A. Kane et al., 2005a. ILSI Risk Science Institute Working Group. Testing of fibrous particles: Short-term assays and strategies. *Inhal Toxicol* September;17(10):497–537.

Bernstein, D.M., J. Chevalier, and P. Smith, 2005b. Comparison of calidria chrysotile asbestos to pure tremolite: Final results of the inhalation biopersistence and histopathology examination following short-term exposure. *Inhal Toxicol* 17(9):427–424.

Bernstein, D.M., R. Rogers, and P. Smith, 2005c. The biopersistence of Canadian chrysotile asbestos following inhalation: Final results through 1 year after cessation of exposure. *Inhal Toxicol* 17(1):1–14.

Bernstein, D.M. and J.M.R. Riego-Sintes, 1999. Methods for the determination of the hazardous properties for human health of man made mineral fibers (MMMF). Vol. EUR 18748 EN, April 93, tsar.jrc.ec.europa.eu/documents/Testing-Methods/mmmfweb.pdf. European Commission Joint Research Centre, Institute for Health and Consumer Protection, Unit: Toxicology and Chemical Substances, European Chemicals Bureau, Ispra, Italy. Accessed May 19, 2014.

Bernstein, D.M., R. Rick, and S. Paul, 2003a. The biopersistence of Canadian chrysotile asbestos following inhalation. *Inhal Toxicol* 15(13):101–128.

Bernstein, D.M., 2003b. Fiber biopersistence, toxicity and asbestos. *J Occup Environ Health (UOEH)* 25(1):237–243.

Bernstein, D.M., 2007. Synthetic vitreous fibers: A review toxicology, epidemiology and regulations. *Crit Rev Toxicol* 37(10):839–886.

Bernstein, D.M., C. Morscheidt, H.G. Grimm, and U. Teichert, 1996. The evaluation of soluble fibers using the inhalation biopersistence model, a nine fiber comparison. *Inhal Toxicol* 8:345–385.

Bernstein, D.M., J. Dunnigan, T. Hesterberg, R. Brown, J.A. Legaspi-Velasco, R. Barrera, J. Hoskins, and A. Gibbs, 2013. Health risk of chrysotile revisited. *Crit Rev Toxicol* 43(2):154–183.

Bernstein, D.M., J.M. Riego-Sintes, B.K. Ersboell, and J. Kunert, 2001a. Biopersistence of synthetic mineral fibers as a predictor of chronic inhalation toxicity in rats. *Inhal Toxicol* 13(10):823–849.

Bernstein, D.M., J.M. Riego-Sintes, B.K. Ersboell, and J. Kunert, 2001b. Biopersistence of synthetic mineral fibers as a predictor of chronic intraperitoneal injection tumor response in rats. *Inhal Toxicol* 13(10):851–875.

Bernstein, D.M., R. Rick, and S. Paul, 2004. The biopersistence of Brazilian chrysotile asbestos following inhalation. *Inhal Toxicol* 16(9):745–761.

Bernstein, D.M., R.A. Rogers, R. Sepulveda, K. Donaldson, D. Schuler, S. Gaering, P. Kunzendorf, J. Chevalier, and S.E. Holm, 2010. The pathological response and fate in the lung and pleura of chrysotile in combination with fine particles compared to amosite asbestos following short term inhalation exposure—Interim results. *Inhal Toxicol* 22(11):937–962.

Bernstein, D.M., R.A. Rogers, R. Sepulveda, K. Donaldson, D. Schuler, S. Gaering, P. Kunzendorf, J. Chevalier, and S.E. Holm, 2011. Quantification of the pathological response and fate in the lung and pleura of chrysotile in combination with fine particles compared to amosite-asbestos following short-term inhalation exposure. *Inhal Toxicol* June;23(7):372–391.

Bolton, R.E., J.H. Vincent, A.D. Jones, J. Addison, and S.T. Beckett, 1983. An overload hypothesis for pulmonary clearance of UICC amosite fibres inhaled by rats. *Br J Ind Med* 40:264–272.

Bragg, G.M., 2001. Fiber release during the handling of products containing chrysotile asbestos using modern control technology. *Can Mineral* 5(Spec. Publ.):111–114.

Breysse, P.N., J.W. Cherrie, J. Addison, and J. Dodgson, 1989. Evaluation of airborne asbestos concentrations using TEM and SEM during residential water tank removal. *Ann Occup Hyg* 33(2):243–256.

Browne, K. and R. Murray, 1990. Asbestos and the Romans. *Lancet* 336(8712):445.

Campbell, W.J., C.W. Huggins, and A.G. Wylie, 1980. Chemical and physical characterization of amosite, chrysotile, crocidolite, and nonfibrous tremolite for oral ingestion studies by the National Institute of Environmental Health Sciences. Report of Investigations 8452. United States Department of the Interior, U.S. Bureau of Mines, Avondale, MD.

Case, B.W. and J.L. Abraham, 2009. Heterogeneity of exposure and attribution of mesothelioma: Trends and strategies in two American counties. *J Phys Conf Series* 151:012008.

Case, B.W., A. Dufresne, A.D. McDonald, J.C. McDonald, and P. Sebastien, 2000. Asbestos fibre type and length in lungs of chrysotile textile and production workers fibers longer than 18 μm. *Inhal Toxicol* 12(Suppl. 3):411–418.

Churg, A. and L. DePaoli, 1988. Clearance of chrysotile asbestos from human lung. *Exp Lung Res* 14(5):567–574.

Churg, A., 1994. Deposition and clearance of chrysotile asbestos. *Ann Occup Hyg* 38(4):625–633, 424–425.

Churg, A., B. Wiggs, L. Depaoli, B. Kampe, and B. Stevens, 1984. Lung asbestos content in chrysotile workers with mesothelioma. *Am Rev Respir Dis.* December;130(6):1042–1045.

Cossette, M. and P. Delvaux, 1979. Technical evaluation of chrysotile asbestos ore bodies. In *Short Course in Mineralogical Techniques of Asbestos Determination*, R.C. Ledoux (ed.). Toronto, Ontario, Canada: Mineralogical Association of Canada. pp. 79–110.

Davis, J.M. and A.D. Jones, 1988. Comparisons of the pathogenicity of long and short fibres of chrysotile asbestos in rats. *Br J Exp Pathol* 69(5):717–737.

Davis, J.M., J. Addison, R.E. Bolton, K. Donaldson, A.D. Jones, and B.G. Miller, 1985. Inhalation studies on the effects of tremolite and brucite dust in rats. *Carcinogenesis* 6(5):667–674.

Davis, J.M., J. Addison, R.E. Bolton, K. Donaldson, A.D. Jones, and T. Smith, 1986. The pathogenicity of long versus short fibre samples of amosite asbestos administered to rats by inhalation and intraperitoneal injection. *Br J Exp Pathol* 67(3):415–430.

Davis, J.M., S.T. Beckett, R.E. Bolton, P. Collings, and A.P. Middleton, 1978. Mass and number of fibres in the pathogenesis of asbestos-related lung disease in rats. *Br J Cancer* 37(5):673–688.

Eastes, W. and J.G. Hadley, 1996. A mathematical model of fiber carcinogenicity and fibrosis in inhalation and intraperitoneal experiments in rats. *Inhal Toxicol* 8:323–342.

EPA, 2003. Report on the Peer Consultation Workshop to Discuss a Proposed Protocol to Assess Asbestos-Related Risk. Prepared for: US Environmental Protection Agency, Office of Solid Waste and Emergency Response, Washington, DC 20460. EPA Contract No. 68-C-98-148, Work Assignment 2003–2005. Prepared by: Eastern Research Group, Inc. Lexington, MA 02421. Final Report: May 30, 2003.

Etherington, D.J., D. Pugh, and I.A. Silver, 1981. Collagen degradation in an experimental inflammatory lesion: Studies on the role of the macrophage. *Acta Biol Med Ger* 40(10–11):1625–1636.

Fornero, E., E. Belluso, S. Capella, and D. Bellis, 2009. Environmental exposure to asbestos and other inorganic fibres using animal lung model. *Sci Total Environ* January 15;407(3):1010–1018.

Frank, A.L., R.F. Dodson, and M.G. Williams, 1998. Carcinogenic implications of the lack of tremolite in UICC reference chrysotile. *Am J Ind Med* 34(4):314–317.

Gardner, M.J., P.D. Winter, B. Pannett, and C.A. Powell, 1986. Follow up study of workers manufacturing chrysotile asbestos cement products. *Br J Ind Med* 43:726–732.

Gross, P., R.T. DeTreville, E.B. Tolker, M. Kaschak, and M.A. Babyak, 1967. Experimental asbestosis. The development of lung cancer in rats with pulmonary deposits of chrysotile asbestos dust. *Arch Environ Health* 15(3):343–355.

Guldberg, M., S.L. Jensen, T. Knudsen, T. Steenberg, and O. Kamstrup, April 2002. High-alumina low-silica HT stone wool fibers: A chemical compositional range with high biosolubility. *Regul Toxicol Pharmacol* 35(2 Pt 1):217–226.

Hargreaves, A. and W.H. Taylor, 1946. An X-ray examination of decomposition products of chrysotile (asbestos) and serpentine. *Mineral Mag* 27:204–216.

Hesterberg, T.W., C. Axten, E.E. McConnell, G.A. Hart, W. Müller, J. Chevalier, J. Everitt, P. Thevenaz, and G. Oberdörster, 1999. Studies on the inhalation toxicology of two fiberglasses and amosite asbestos in the syrian golden hamster. Part I. Results of a subchronic study and dose selection for a chronic study. *Inhal Toxicol* 11(9):747–784.

Hesterberg, T.W., G. Chase, C. Axten, W.C. Miller, R.P. Musselman, O. Kamstrup, J. Hadley, C. Morscheidt, D.M. Bernstein, and P. Thevenaz, 1998a. Biopersistence of synthetic vitreous fibers and amosite asbestos in the rat lung following inhalation. *Toxicol Appl Pharmacol* 151(2):262–275.

Hesterberg, T.W., G.A. Hart, J. Chevalier, W.C. Müller, R.D. Hamilton, J. Bauer, and P. Thevenaz, 1998b. The importance of fiber biopersistence and lung dose in determining the chronic inhalation effects of X607, RCF1, and chrysotile asbestos in rats. *Toxicol Appl Pharmacol* 153(1):68–82.

Hesterberg, T.W., W.C. Miiller, E.E. McConnell, J. Chevalier, J.G. Hadley, D.M. Bernstein, P. Thevenaz, and R. Anderson, 1993. Chronic inhalation toxicity of size-separated glass fibers in Fischer 344 rats. *Fundam Appl Toxicol* 20(4):464–476.

Hodgson, A.A., 1979. Chemistry and physics of asbestos. In *Asbestos: Properties, Applications and Hazards*, L.M.a.S.S. Chissick (ed.), pp. 80–81. New York: John Wiley & Sons.

Hodgson, J.T. and A. Darnton, 2000. The quantitative risks of mesothelioma and lung cancer in relation to asbestos exposure. *Ann Occup Hyg* 44(8):565–601.

Howard, J.K., 1984. Relative cancer risks from exposure to different asbestos fibre types. *N Z Med J* 97(764):646–649.

Ilgren, E. and E. Chatfield, 1997. Coalinga fibre—A short, amphibole-free chrysotile. Evidence for lack of fibrogenic activity. *Indoor Built Environ* 6:264–276.

Ilgren, E. and E. Chatfield, 1998. Coalinga fibre—A short, amphibole-free chrysotile. Part 2: Evidence for lack of tumourigenic activity. *Indoor Built Environ* 7:18–31.

Kamstrup, O., A. Ellehauge, J. Chevalier, J.M. Davis, E.E. McConnell, and P. Thévenaz, July 2001. Chronic inhalation studies of two types of stone wool fibers in rats. *Inhal Toxicol* 13(7):603–621.

Kamstrup, O., A. Ellehauge, C.G. Collier, and J.M. Davis, 2002. Carcinogenicity studies after intraperitoneal injection of two types of stone wool fibres in rats. *Ann Occup Hyg* 46(2):135–142.

Kiyohara, P.K., 1991. Estudo da interface crisotila-cimento Portland em compósitos de fibro-cimento por métodos óptico-eletrônicos, Tese de Doutorado, apres. EPUSP, São Paulo, Brazil.

Kurumatani, N. and S. Kumagai, 2008. Mapping the risk of mesothelioma due to neighborhood asbestos exposure. *Am J Respir Crit Care Med* September 15;178(6):624–629.

Leake, B.E., 1978. Nomenclature of amphiboles. *Can Mineral* 16:501–520.

Le Bouffant, L., H. Daniel, J.P. Henin, J.C. Martin, C. Normand, G. Tichoux, and F. Trolard, 1987. Experimental study on long-term effects of inhaled MMMF on the lungs of rats. *Ann Occup Hyg* 31(4B):765–790.

Lee, K.P., C.E. Barras, F.D. Griffith, R.S. Waritz, and C.A. Lapin, 1981. Comparative pulmonary responses to inhaled inorganic fibers with asbestos and fiberglass. *Environ Res* 24(1):167–191.

Leineweber, J.P., 1982. Solubility of fibres in vitro and in vivo. In *Proceedings of WHO//ARC Conference in Biological Effects of Man-Made Mineral Fibers*. Copenhagen, Denmark: World Health Organization. pp. 87–101.

Liddell, F.D.K., A.D. McDonald, and J.C. McDonald, 1998. Dust exposure and lung cancer in quebed chrysotile miners and millers. *Ann Occup Hyg* 42(1):7–20.

Luoto, K., M. Holopainen, J. Kangas, P. Kalliokoski, and K. Savolainen, 1995. The effect of fiber length on the dissolution by macrophages of rockwool and glasswool fibers. *Environ Res* 70(1):51–61.

Mast, R.W., T.W. Hesterberg, L.R. Glass, E.E. McConnell, R. Anderson, and D.M. Bernstein, 1994. Chronic inhalation and biopersistence of refractory ceramic fiber in rats and hamsters. *Environ Health Perspect* 102(Suppl. 5):207–209.

Mast, R.W., E.E. McConnell, R. Anderson, J. Chevalier, P. Kotin, D.M. Bernstein, P. Thevenaz, L.R. Glass, W.C. Miiller, and T.W. Hesterberg, 1995. Studies on the chronic toxicity (inhalation) of four types of refractory ceramic fiber in male Fischer 344 rats. *Inhal Toxicol* 7(4):425–467.

Mattson, S.M., 1994. Glass fibres in simulated lung fluid: Dissolution behavior and analytical requirements. *Ann Occup Hyg* 38:857–877.

McConnell, E.E., 1995. Fibrogenic effect of wollastonite compared with asbestos dust and dusts containing quartz. *Occup Environ Med* 52:621–624.

McConnell, E.E., C. Axten, T.W. Hesterberg, J. Chevalier, W.C. Müller, J. Everitt, G. Oberdörster, G.R. Chase, P. Thevenaz, and P. Kotin, 1999. Studies on the inhalation toxicology of two fiberglasses and amosite asbestos in the Syrian golden hamster. Part II. Results of chronic exposure. *Inhal Toxicol* 11(9):785–835.

McConnell, E.E., O. Kamstrup, R. Musselman, T.W. Hesterberg, J. Chevalier, W.C. Müller, and P. Thievenaz, 1994. Chronic inhalation study of size-separated rock and slag wool insulation fibers in Fischer 344/N rats. *Inhal Toxicol* 6:571–614.

McDonald, J.C., 1998. Mineral fibre persistence and carcinogenicity. *Ind Health* 36(4):372–375.

McDonald, J.C. and A.D. McDonald, 1995. Chrysotile, tremolite, and mesothelioma. *Science* 267(5199):776–777.

McDonald, J.C. and A.D. McDonald, 1997. Chrysotile, tremolite and carcinogenicity. *Ann Occup Hyg* 41(6):699–705.

McDonald, J.C., A.D. McDonald, and J.M. Hughes, 1999. Chrysotile, tremolite and fibrogenicity. *Ann Occup Hyg* 43(7):439–442.

McDonald, J.C., J. Harris, and B. Armstrong, 2002. Cohort mortality study of vermiculite miners exposed to fibrous tremolite: An update. *Ann Occup Hyg* 46(suppl 1):93–94.

McDonald, J.C., J. Harris, and B. Armstrong, April 2004. Mortality in a cohort of vermiculite miners exposed to fibrous amphibole in Libby, Montana. *Occup Environ Med* 61(4):363–366.

Miller, B.G., A.D. Jones, A. Searl, D. Buchanan, R.T. Cullen, C.A. Soutar, J.M. Davis, and K. Donaldson, 1999. Influence of characteristics of inhaled fibres on development of tumours in the rat lung. *Ann Occup Hyg* 43:167–179.

Miller, F.J., 2000. Dosimetry of particles: Critical factors having risk assessment implications. *Inhal Toxicol* 12(Suppl. 3):389–395.

Mohr, U., F. Pott, and F.J. Vonnahme, 1984. Morphological aspects of mesotheliomas after intratracheal instillations of fibrous dusts in Syrian golden hamsters. *Exp Pathol* 26(3):179–183.

Morgan, A., 1994. The removal of fibres of chrysotile asbestos from lung. *Ann Occup Hyg* 38(4):643–646.

Morimoto, Y., H. Yamato, M. Kido, I. Tanaka, T. Higashi, A. Fujino, and Y. Yokosaki, 1994. Effects of inhaled ceramic fibers on macrophage function of rat lungs. *Occup Environ Med* 51(1):62–67.

Morrow, P.E., 1988. Possible mechanisms to explain dust overloading of the lungs. *Fundam Appl Toxicol* 10(3):369–384.

Mossman, B.T., J. Bignon, M. Corn, A. Seaton, and J.B. Gee, 1990. Asbestos: Scientific developments and implications for public policy. *Science* 247(4940):294–301.

Muhle, H. and F. Pott, 2000. Asbestos as reference material for fibre-induced cancer. *Int Arch Occup Environ Health* 73 Suppl.:S53–S59.

Muhle, H. and I. Mangelsdorf, 2003. Inhalation toxicity of mineral particles: Critical appraisal of endpoints and study design. *Toxicol Lett* 140–141:223–228.

Muhle, H., B. Bellman, and U. Heinrich, 1988. Overloading of lung clearance during chronic exposure of experimental animals to particles. *Ann Occup Hyg* 32(Suppl. 1):141–147.

Muhle, H., B. Bellmann, O. Creutzenberg, C. Dasenbrock, H. Ernst, R. Kilpper, J.C. MacKenzie et al., 1991. Pulmonary response to toner upon chronic inhalation exposure in rats. *Fundam Appl Toxicol* 17(2):280–299.

Muhle, H., F. Pott, B. Bellmann, S. Takenaka, and U. Ziem, 1987. Inhalation and injection experiments in rats to test the carcinogenicity of MMMF. *Ann Occup Hyg* 31(4B):755–764.

Musti, M., A. Pollice, D. Cavone, S. Dragonieri, and M. Bilancia, 2009. The relationship between malignant mesothelioma and an asbestos cement plant environmental risk: A spatial case-control study in the city of Bari (Italy). *Int Arch Occup Environ Health* March;82(4):489–497 [Epub 2008 Sep 23].

Nagy, B. and T.F. Bates, 1952. Stability of chrysotile asbestos. *Am Mineral* 37:1055–1058.

Noll, W. and H. Kircher, 1951. Über die Morphologie von Asbesten und ihren Zusammenhang mit der Kristallstruktur. *Neues Jb Mineral Mh* 1951:219–240.

Oberdörster, G., 1991. Deposition, elimination and effects of fibers in the respiratory tract of humans and animals, pp. 17–37. VDI Ber.: Düsseldorf, Germany.

Oberdörster, G., 1995. Lung particle overload: Implications for occupational exposures to particles. *Regul Toxicol Pharmacol* 21(1):123–135.

Oberdörster, G., 2000. Determinants of the pathogenicity of man-made vitreous fibers (MMVF). *Int Arch Occup Environ Health* 73 Suppl.:S60–S68.

Ohlson, C.G. and C. Hogstedt, 1985. Lung cancer among asbestos cement workers. A Swedish cohort study and a review. *Br J Ind Med* 42(6):397–402.

Osmond-McLeod, M.J., C.A. Poland, F. Murphy, L. Waddington, H. Morris, S.C. Hawkins, S. Clark, R. Aitken, M.J. McCall, and K. Donaldson, 2011. Durability and inflammogenic impact of carbon nanotubes compared with asbestos fibres. *Part Fibre Toxicol* May 13;8:15.

Pan, X.L., H.W. Day, W. Wang, L.A. Beckett, and M.B. Schenker, 2005. Residential proximity to naturally occurring asbestos and mesothelioma risk in California. *Am J Respir Crit Care Med* October 15;172(8):1019–1025.

Pauling, L., 1930. The structure of chlorites. *Proc Natl Acad Sci* 16:578–582.

Perry, R.H. and C.H. Chilton (eds.), 1973. *Chemical Engineers' Handbook*, 5th edn. New York: McGraw-Hill.

Pinkerton, K.E., A.R. Brody, D.A. McLaurin, B. Adkins, Jr., R.W. O'Connor, P.C. Pratt, and J.D. Crapo, 1983. Characterization of three types of chrysotile asbestos after aerosolization. *Environ Res* 31(1):32–53.

Piolatto, G., E. Negri, C. La Vecchia, E. Pira, A. Decarli, and J. Peto, 1990. An update of cancer mortality among chrysotile asbestos miners in Balangero, northern Italy. *Br J Ind Med* 47(12):810–814.

Pooley, F.D. and R. Mitha, 1986. Determination and interpretation of the levels of chrysotile asbestos in lung tissue. In *Biological Effects of Chrysotile* (Accomplishments in Oncology, Vol. 1, No. 2), Wagner, J.C. (ed.), pp. 12–18. Philadelphia, PA: Lippincott.

Rakovan, J., 2011. Serpentine, California's state rock. *J Rocks Mineral* 86(1):63–68.

Rakovan, J., January/February 2011. Serpentine California's state rock. *Rocks Minerals* 86:63–68.

Rodelsperger, K., H.J. Woitowitz, B. Bruckel, R. Arhelger, H. Pohlabeln, and K.H. Jockel, 1999. Dose–response relationship between amphibole fiber lung burden and mesothelioma. *Cancer Detect Prev* 23(3):183–193.

Roggli, V.L., R.T. Vollmer, K.J. Butnor, and T.A. Sporn, 2002. Tremolite and mesothelioma. *Ann Occup Hyg* 46(5):447–453.

Ross, M. and R.P. Nolan, 2003. History of asbestos discovery and use and asbestos-related disease in context with the occurrence of asbestos within ophiolite complexes. In *Ophiolite Concept and the Evolution of Geological Thought*, Y. Dilek and S. Newcomb (eds.). Boulder, CO: Geological Society of America. pp. 447–470.

Sebastien, P., J.C. McDonald, A.D. McDonald, B. Case, and R. Harley, 1989. Respiratory cancer in chrysotile textile and mining industries: Exposure inferences from lung analysis. *Br J Ind Med* March;46(3):180–187.

Silvestri, S., C. Magnani, R. Calisti, and C. Bruno, 2001. The experience of the Balangero chrysotile asbestos mine in Italy: Health effects among workers mining and milling asbestos and the health experience of persons living nearby, Canadian Mineralogist, pp. 177–186. (The Health Effects of Chrysotile Asbestos: Contribution of Science to Risk-Management Decisions Can. Mineral., Spec. Pub. 5, pp. 177–186 (2001).)

Skinner, H.C.W., M. Ross, and C. Frondel, 1988. *Asbestos and Other Fibrous Materials—Mineralogy, Crystal Chemistry, and Health Effects*, 204pp. New York, Oxford University Press.

Smith, D.M., L.W. Ortiz, R.F. Archuleta, and N.F. Johnson, 1987. Long-term health effects in hamsters and rats exposed chronically to man-made vitreous fibres. *Ann Occup Hyg* 31(4B):731–754.

Speil, S. and J.P. Leineweber, 1969. Asbestos minerals in modern technology. *Environ Res* 2:166–208.

Stoeber, W., H. Flachsbart, and D. Hochrainer, 1970. Der Aerodynamische Durchmesser von Latexaggregaten und Asbestfassern. *Staub-Reinh Luft* 30:277–285.

Suquet, H., 1989. Effects of dry grinding and leaching on the crystal structure of chrysotile. *Clays Clay Mineral* 37: 439–445.

Tanji, T., K. Yada, and Y. Akatsuka, 1984. Note: Alternation of clino- and orthochrysotile in a single fiber as revealed by high-resolution electron microscopy. *Clays Clay Minerals* 32:429–432.

Thomas, H.F., I.T. Benjamin, P.C. Elwood, and P.M. Sweetnam, 1982. Further follow-up study of workers from an asbestos cement factory. *Br J Ind Med* 39(3):273–276.

Timbrell, V. and R.E.G. Rendall, 1972. Preparation of the UICC standard reference samples of asbestos. *Powder Technol* 5:279.

Timbrell, V., A.W. Hyett, and J.W. Skidmore, 1968. A simple dispenser for generating dust clouds from standard reference samples of asbestos. *Ann Occup Hyg* October;11(4):273–281.

Virta, R.L., 2002 USGS Open file 02-149. *Asbestos: Geology, Mineralogy, Mining, and Uses.* Prepared in cooperation with *Kirk-Othmer Encyclopedia of Chemical Technology*, online edition. New York: Wiley-Interscience, a division of John Wiley & Sons, Inc.

von Kobell, F., 1834. Ueber den schillernden Asbest von Reichenstein in Schlesian: Jour. Prakt. *Chemie* 2:297–298.

Wagner, J.C., G. Berry, J.W. Skidmore, and F.D. Pooley, 1980. The comparative effects of three chrysotiles by injection and inhalation in rats. *IARC Sci Publ* 1980(30):363–372.

Wagner, J.C., G. Berry, J.W. Skidmore, and V. Timbrell, 1974. The effects of the inhalation of asbestos in rats. *Br J Cancer* 29(3):252–269.

Warren, B.E. and W.L. Bragg, 1930. The structure of chrysotile, $H_4Mg_3Si_2O_9$. *Z Krystallographie* 76:201–210.

Weill H., J. Hughes, and C. Waggenspack, 1979. Influence of dose and fiber type on respiratory malignancy risk in asbestos cement manufacturing. *Am Rev Respir Dis* 120(2):345–354.

Whittaker, E.J.W., 1957. The structure of chrysotile. V. Diffuse reflexions and fibre texture. *Acta Crystallogr* 10:149.

Whittaker, E.J.W., 1963. Research report: Chrysotile fibers—Filled or hollow tubes? Mathematical interpretation may resolve conflicting evidence, Chem. Eng. News-41(39):34–35, September 30.

Whittaker, E.J.W., 1979. Mineralogy, chemistry and crystallography of amphibole asbestos. In *Short Course in Mineralogical Techniques of Asbestos Determination*, R.C. Ledoux (ed.). Toronto, Ontario, Canada: Mineralogical Association of Canada. pp. 1–34.

Williams, M., P. Larorche, and R. Jauron, 2008. *The Basics of Chrysotile Asbestos Dust Control*, 4th edn., Chrysotile Institute, Montreal, Quebec, Canada.

Williams, M., P. Larorche, and R. Jauron, 2011. *Safe Use of Chrysotile Asbestos: A Manual on Preventive and Control Measures.* Chrysotile Institute, Montreal, Quebec, Canada.

Williams-Jones, A.E., C. Normand, J.R. Clark, H. Vali, R.F. Martin, A. Dufresne, and A. Nayebzadeh, 2001. Controls of amphibole formation in chrysotile deposits evidence from the Jeffrey mine. Canadian Mineralogist, Asbestos, Quebec, Canada, Special Publication, 5, 89–104.

Work, L.T., 1962. Size reduction gets a new stature. *Ind Eng Chem* 54(3):52–54.

Wypych, F., L.B. Adad, N. Mattoso, A.A. Marangon, and W.H. Schreiner, 2005. Synthesis and characterization of disordered layered silica obtained by selective leaching of octahedral sheets from chrysotile. *J Colloid Interface Sci* 283(1):107–112.

Yada, K., 1971. Study of microstructure of chrysotile asbestos by high resolution electron microscopy. *Acta Cryst* A27:659–664.

15 Pulmonary Primary Cells Cocultured in a Novel Cell Culture System, the Integrated Discrete Multiple Cell–Type Coculture System (IdMOC)*

Pulmonary Cytotoxicity of Eight Cigarette Smoke Condensates and Nicotine

Albert P. Li, Patricia Richter, Stephanie Cole, Janna Madren-Whalley, Jonathan Oyler, Russell Dorsey, and Harry Salem

CONTENTS

---

* The views and opinions expressed in this chapter are those of the author and should not be construed as an official Department of the Army position, policy, or decision, unless so designated by other official documentation. Citation of trade names in this chapter does not constitute an official Department of the Army endorsement or approval of the use of such commercial items.

15.1 INTRODUCTION

Cigarette smoke condensate (CSC) is believed to contain ingredients that may be responsible for the in vivo health effects of smoking. CSC, akin to other combustion-related environmental pollutants such as diesel exhaust extracts and fly ash extracts, is a complex mixture with a myriad of ingredients (McCoy and Rosenkranz 1982, Adams et al. 1984, Ramdahl et al. 1986, Chepiga et al. 2000, Lu et al. 2004a,b, Sepetdjian et al. 2008). For these complex mixtures, in vitro experimental systems are frequently applied to evaluate their overall biological effects. In vitro experimental systems, when applied in combination with analytical chemistry, may allow the identification of the biologically active components in these complex mixtures (Mizusaki et al. 1977, Wakabayashi et al. 1995, Smith and Hansch 2000).

We report here an evaluation of the cytotoxicity of CSCs produced under eight different conditions as part of a research program to develop data for the range of biological effects for a diverse set of CSCs. The genotoxicity of the same CSCs studied here has been previously published (DeMarini 1981, DeMarini et al. 2008). The cytotoxicity of these CSCs was assessed in primary lung cells, using a novel in vitro model for toxicity evaluation: the Integrated Discrete Multiple Organ Coculture (IdMOC) experimental system (Li et al. 2004, 2012, Li 2007a,b, 2008, 2009). The IdMOC utilizes a wells-in-a-well concept, allowing the coculturing of multiple cell types as physically discrete cultures that can interact through a common overlaying medium.

In this study, the IdMOC system was used as an in vitro model of the human lung (IdMOC-lung) via coculturing of three major pulmonary cell types: normal bronchial epithelial cells (NHBE), small airway epithelial cells (SAEC), and human lung microvascular endothelial cells (HMVEC-L). We report here the first application of the IdMOC-lung to evaluate the pulmonary cytotoxicity of CSCs as well as a known cytotoxic and pharmacologically active ingredient: nicotine.

15.2 MATERIALS AND METHODS

15.2.1 IdMOC Experimental System

Collagen-coated Integrated Discrete Multiple Organ Coculture (IdMOC™) plates (APSciences Inc., Columbia, MD) were used in the study. The IdMOC plates used were IdMOC-96, which had the footprint of a 96-well plate, with 16 chambers (containing wells), with 6 inner wells per chamber. A schematic representation and a photograph of the IdMOC plate are shown in Figures 15.1 and 15.2, respectively.

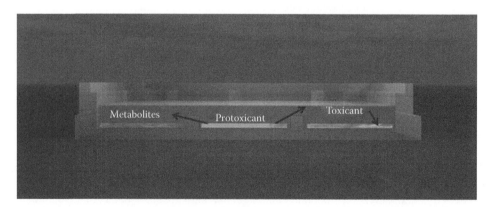

FIGURE 15.1 Schematic representation of the IdMOC experimental system. The IdMOC consists of small, shallow wells, within a large containing well (chamber). The system allows cell–cell interaction via the overlying medium. Metabolic interaction is shown here where the metabolically competent cell type (e.g., hepatocytes; center well) generates toxic metabolites from a relative nontoxic parent chemical (protoxicant), which can cause cytotoxicity in the cocultured metabolically incompetent cell type (e.g., fibroblasts). The IdMOC is a convenient and relevant model to include hepatic metabolism in the evaluation of toxic potential of a chemical toward nonhepatic cell types.

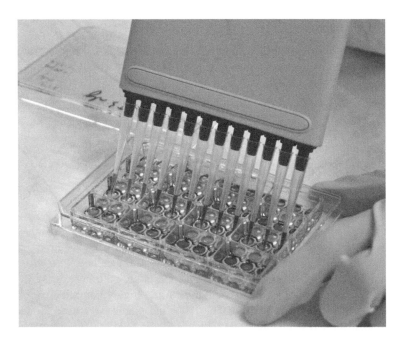

FIGURE 15.2 A photograph of an IdMOC-96 plate. The plate has an identical footprint of a 96-well plate and thereby is compatible with laboratory equipment such as the multichannel pipette as shown here.

15.2.2 Cigarette Smoke Condensates

The CSCs were produced generally under standard smoking-machine conditions established in the 1960s by the Federal Trade Commission (FTC; Pillsbury et al. 1996) with the exception of one (CSC-3) which was prepared under an *intense* condition. The preparation of the CSCs used in this study, including the source of the cigarettes, smoking conditions and condensate preparations, and the FTC and *intense* smoking conditions, were as previously published (DeMarini 2008). The CSCs were supplied to our laboratory as 20 mg/mL solutions in DMSO. They were stored at −20°C until experimentation. The eight CSC studied are as follows:

- CSC-1. Low tar reference cigarette (2R4F) smoked under FTC conditions
- CSC-2. A commercial light nonmenthol cigarette smoked under FTC conditions
- CSC-3. A commercial light nonmenthol cigarette smoked under intense conditions
- CSC-4. A commercial full flavor nonmenthol cigarette smoked under FTC conditions
- CSC-5. 100% reconstituted cigarette smoked under FTC conditions
- CSC-6. 100% flue cured cigarette smoked under FTC conditions
- CSC-7. 100% burley cigarette smoked under FTC conditions
- CSC-8. Charcoal filter cigarettes smoked under FTC conditions

On the day of the experiment, the 20 mg/mL CSC solutions were diluted serially (1:1, v/v) in DMSO to yield 10, 5, and 2.5 mg/mL. These solutions (at concentrations of 20, 10, 5, and 2.5 mg/mL) were diluted 1:100 to yield the final concentrations of 200, 100, 50, and 25 µg/mL, respectively, in medium for cytotoxicity evaluation (see below).

15.2.3 Nicotine

Nicotine was obtained commercially from Sigma-Aldrich (St. Louis, MO) and was dissolved in dimethylsulfoxide (DMSO) to yield 100× concentrated stock solutions on the day of experimentation. The final concentrations of nicotine evaluated were 7.8, 15.6, 31.3, 62.5, 125, 250, 500, and 1000 µM.

15.2.4   PRIMARY HUMAN CELLS

Human microvascular endothelial cells from the lungs (HMVEC-L), normal human bronchial epithelial cells (NHBE), and human small airway epithelial cells (SAEC) and their respective media were commercially obtained (Lonza, Inc., Walkersville, MD).

15.2.5   OTHER CHEMICALS

DMEM-F12 medium, fetal calf serum, DMSO, MTT reagent, and nicotine were commercially obtained from Sigma-Aldrich, Inc. (St. Louis, MO).

15.2.6   CELL CULTURE AND TREATMENT

The three human pulmonary cell types were obtained from the supplier as monolayer cultures in 75 cm^2 tissue culture flasks. The cells were cultured for approximately 3 days before they were used for experimentation. On the day before the experiment, the cells were trypsinized from the flasks and plated directly onto IdMOC plates (APSciences, Inc., Baltimore, MD). Cell culture was performed in a 37°C cell culture incubator maintained in a highly humidified atmosphere of 5% carbon dioxide and 95% air (Napco 6300). The following media were used for the initial seed-ing (all from Lonza Inc., Walkersville, MD): BEBM Basal Medium supplemented with BEGM SingleQuot Kit for NHBE; SABM Basal Medium supplemented with SAEC SingleQuot Kit for SAEC; and EBM Basal Medium supplemented with EGM-MV SingleQuot Kit for HMVEC-L. Each cell type (5000 cells per well) was plated in its respective medium in a volume of 20 µL per well (two wells per cell type per chamber). After overnight culturing (approximately 24 h), a vol-ume of 600 µL of serum-free DMEM-F12 medium (Sigma–Aldrich) was added into each chamber. Treatment was started with the addition of 7.2 µL of the 100× stock solutions or DMSO as solvent control. The treatment period was 24 h. All treatments were performed in triplicate (three chambers per treatment).

15.2.7   VIABILITY MEASUREMENT

At the end of the treatment period, a volume of 50 µL of 5 mg/mL MTT was added per chamber for the evaluation of MTT metabolism. After incubation with the MTT reagent, the IdMOC cultures were inverted on absorbent paper to remove all medium. A volume of 20 µL of DMSO was added per well to dissolve the blue crystals formed, yielding a blue solution. The DMSO solution was then removed and placed into a 96-well plate containing 100 µL DMSO per well for reading the inten-sity of the blue solution at an absorbance of 490 nm. A plate reader (Wallac Victor 1420 Multilabel Counter) was used to quantify the blue color intensity.

15.2.8   DATA ANALYSIS

The results with the test articles are expressed as relative viability that is calculated as the following:

$$\text{Relative viability } (\%) = \frac{\text{Absorbance (treatment)}}{\text{Absorbance (solvent control)}} \times 100$$

The concentration for 50% cytotoxicity (EC-50) was calculated from the dose–response equation derived by linear regression analysis of a plot of the logarithm of the relative survival versus concentration. Linear regression analysis was performed using the KaleidaGraph 3.6 software (Synergy Software).

15.3 RESULTS

15.3.1 CYTOTOXICITY OF CSCs

Dose-dependent cytotoxicity was observed for all eight CSCs in the three cell types: NHBE (Table 15.1), SAEC (Table 15.2), and HMVEC-L (Table 15.3). The calculated EC50 values are shown in Table 15.5.

TABLE 15.1
Relative Viability of NHBE after Treatment with CSCs

µg/mL	CSC-1 Mean	sem	CSC-2 Mean	sem	CSC-3 Mean	sem	CSC-4 Mean	sem
1% DMSO	100.0	11.5	100.0	6.6	100.0	0.9	100.0	7.8
25	93.1	4.8	88.9	6.3	104.6	6.3	91.9	6.4
50	91.7	4.2	93.3	3.4	71.9*	10.4	91.5	6.3
100	41.3*	4.4	40.7*	3.3	40.1*	6.2	34.3*	0.4
200	16.7*	2.2	21.9*	1.3	35.5*	10.8	24.6*	7.1

µg/mL	CSC-5 Mean	sem	CSC-6 Mean	sem	CSC-7 Mean	sem	CSC-8 Mean	sem
1% DMSO	100.0	10.7	100.0	13.3	100.0	5.2	100.0	3.2
25	77.3	5.0	31.5*	3.3	100.2	14.3	89.3	1.7
50	91.5	5.0	27.7*	2.6	78.3*	6.6	76.3*	2.5
100	65.4*	4.9	31.9*	3.7	69.7*	2.9	69.0*	2.8
200	32.2*	2.1	49.4*	6.4	66.3*	0.7	58.6*	0.8

Note: The results represent mean and standard errors of triplicate observations. Asterisk (*) represents results that are statistically significant to be different (p < 0.05) from that of the 1% DMSO (solvent control) treatment.

TABLE 15.2
Relative Viability of SAEC after Treatment with CSCs

µg/mL	CSC-1 Mean	sem	CSC-2 Mean	sem	CSC-3 Mean	sem	CSC-4 Mean	sem
1% DMSO	100.0	2.4	100.0	5.0	100.0	1.7	100.0	9.5
25	192.8*	0.9	180.0*	5.9	148.5	11.3	132.4*	2.9
50	207.8*	2.6	200.6*	2.9	125.4	6.8	133.6*	4.8
100	141.3	7.5	139.0	10.5	80.0	8.0	86.5	1.6
200	11.3	5.8	80.0	4.3	24.3	2.9	24.0	1.3

µg/mL	CSC-5 Mean	sem	CSC-6 Mean	sem	CSC-7 Mean	sem	CSC-8 Mean	sem
1% DMSO	100.0	5.6	100.0	12.2	100.0	5.2	100.0	3.3
25	77.8	11.0	81.1	4.6	95.5	6.0	106.3	6.8
50	88.3	2.9	35.2*	2.8	79.0	2.8	101.6	5.5
100	93.0	19.3	31.6*	16.3	80.1	1.8	95.9	9.6
200	60.5*	1.8	20.6*	3.6	48.7*	0.3	51.1*	2.7

Note: The results represent mean and standard errors of triplicate observations. Asterisk (*) represents results that are statistically significant to be different (p < 0.05) from that of the 1% DMSO (solvent control) treatment.

TABLE 15.3

Relative Viability of HMVEC-L after Treatment with CSCs

µg/mL	CSC-1 Mean	CSC-1 sem	CSC-2 Mean	CSC-2 sem	CSC-3 Mean	CSC-3 sem	CSC-4 Mean	CSC-4 sem
1% DMSO	100.0	17.8	100.0	18.7	100.0	2.1	100.0	3.5
25	172.8*	5.1	200.7*	7.3	131.3	4.8	131.8	10.5
50	172.2*	3.2	176.1*	3.7	106.4	8.8	118.9	4.0
100	121.0	8.3	133.1	21.3	81.5	6.2	88.6	3.5
200	29.0*	3.4	11.3*	5.8	35.7*	1.1	26.5*	3.3

µg/mL	CSC-5 Mean	CSC-5 sem	CSC-6 Mean	CSC-6 sem	CSC-7 Mean	CSC-7 sem	CSC-8 Mean	CSC-8 sem
1% DMSO	100.0	2.6	100.0	9.3	100.0	0.3	100.0	5.2
25	106.8	3.0	67.7*	8.8	98.7	4.8	115.0	6.5
50	141.9*	6.2	24.3*	3.4	93.4	5.3	120.6	10.5
100	116.2	0.9	23.4*	2.9	74.2*	0.8	108.3	5.5
200	67.1*	1.5	34.4*	3.3	65.0*	0.5	73.3*	1.7

Note: The results represent mean and standard errors of triplicate observations. Asterisk (*) represents results that are statistically significant to be different ($p < 0.05$) from that of the 1% DMSO (solvent control) treatment.

TABLE 15.4

Relative Viability of the Three Pulmonary Cell Types after Treatment with Nicotine

µM	NHBE Mean	NHBE sem	SAEC Mean	SAEC sem	HMVEC-L Mean	HMVEC-L sem
1% DMSO	100.00	30.05	100.00	19.00	100.00	25.13
7.8	115.25	25.85	97.24	2.41	156.79	11.88
15.6	49.46*	17.73	84.75	9.02	87.11	7.76
31.3	31.93*	14.48	72.27*	1.91	64.11*	4.52
62.5	21.37*	4.59	67.23*	2.07	60.63*	4.07
125	30.01*	2.06	65.07*	2.65	47.39*	1.72
250	32.41*	3.18	67.71*	3.27	58.19*	3.85
500	30.97*	14.20	56.66*	1.80	48.78*	1.07
1000	24.73*	1.79	57.62*	2.30	60.28*	2.43

Note: The results represent mean and standard errors of triplicate observations. Asterisk (*) represents results that are statistically significant to be different ($p < 0.05$) from that of the 1% DMSO (solvent control) treatment.

15.3.2 NICOTINE

Nicotine treatment at 7.8–1000 µM yielded dose-dependent cytotoxicity in all three cell types (Table 15.4).

15.4 DISCUSSION

In vitro experimental systems represent a well-accepted approach in the evaluation of the toxicity of xenobiotics. The research effort in our laboratory is to develop in vitro approaches with improved accuracy in the prediction of in vivo effects (Li 2007a,b, 2008). One drawback of in vitro cytotoxicity studies is that often a single cell type is used; while in vivo, when a toxicant is introduced into the systemic circulation, multiple cell types in multiple organs are exposed. As the multiple cell types and organs in vivo are connected via the systemic circulation, the ultimate toxicity observed in a single cell type could be the results of metabolites or cellular factors generated from different cell types. The IdMOC system developed in our laboratory is intended to overcome this lack of multiple cell type interactions in our current unicell type in vitro experimental models. We have recently reported the use of IdMOC cocultures of metabolically active hepatocytes and mouse 3T3 cells to demonstrate this major application of IdMOC, namely, the incorporation of hepatic metabolism in the evaluation of a toxicant that is metabolically activated (Li et al. 2012).

In the IdMOC, multiple cell types can be cultured as physically separated (discrete) cultures that are interconnected (integrated) by an overlying medium, akin to the in vivo situation whereas multiple cell types are interconnected by the systemic circulation. The advantage of the IdMOC system is that one can evaluate differential cytotoxicity of a toxicant toward multiple cell types under near-identical experimental conditions, with interaction among multiple cell types allowed via the overlying medium. As the multiple cell types are physically separated, the effect of a toxicant on each particular cell type can be evaluated individually after experimentation. An IdMOC culture has advantages over coculturing of multiple cell types as mixed cultures in the same cell culture vessel as the cell types evaluated may not have physical contacts in vivo, and that it would be very challenging, if not physically impossible, to separate the different cell types after experimentation to evaluate the effects of the toxicants being studied (Li et al. 2004, Li 2008, 2009).

This chapter focuses on the application IdMOC as an in vitro model of the lung for the evaluation of a relevant inhaled toxicant, cigarette smoke condensate, and one of its major chemical ingredients, nicotine. (A comparison of IdMOC results to other experimental systems has been previously published [Richter et al. 2010].) That CSC has cytotoxicity toward primary cells in culture has been reported for gingival fibroblasts (Zhang et al. 2009), vascular endothelial cells (Nordskog et al. 2003), human bronchial/tracheal epithelial cells, coronary artery endothelial cells, coronary artery smooth muscle cells, foreskin keratinocytes (McKarns et al. 2000), bronchial epithelial cell line NCI-H292, subcultures of primary bronchial epithelial cells (Luppi et al. 2005), and bronchial epithelial cells (Hellermann et al. 2002, Luppi et al. 2005). The cytotoxicity is also observed in various transformed cell lines (Curvall et al. 1984, 1985, Matsukura et al. 1991, Bombick et al. 1997, 1998, Foy et al. 2004, Kato et al. 2007, Guo et al. 2011, Chen et al. 2012). All the studies were performed in single cell type systems. The novelty of our finding is the evaluation of cigarette smoke condensate cytotoxicity in an in vitro model of the human lung via the coculturing of three major cell types in the IdMOC system: bronchial epithelial cells (NHBE), small airway epithelial cells (SAEC), and human lung microvascular endothelial cells (HMVEC-L). The cells were cultured in a condition allowing communication via the overlying medium akin to the lung in vivo.

A range of biological activities were observed for the eight CSCs used in this study. Cytotoxicity ranking based on the EC50 values (Table 15.5) estimated from the dose response curves is as follows:

1. NHBE: CSC-6 (most cytotoxic; EC50 = 15 µg/mL) > CSC-1, CSC-2, CSC-4, CSC-5 (EC50 ranged from 92 to 132 µg/mL) > CSC-7 = CSC-8 (EC50 ≥ 200 µg/mL)
2. SAEC: CSC-6 (most cytotoxic; EC50 = 46 µg/mL) > CSC-1 = CSC-3 = CSC-4 (EC50 ranged from 128 to 170 µg/mL) > CSC-2 = CSC-5 = CSC-7 = CSC-8 (EC50 > 200 µg/mL)
3. HMVEC-L: CSC-6 (most cytotoxic; EC50 = 32 µg/mL) > CSC-1 = CSC-2 = CSC-3 = CSC-4 (EC50 ranged from 125 to 160 µg/mL > CSC-5 = CSC-7 = CSC-8 (EC50 > 200 µg/mL)

The results therefore show that the eight CSCs studied can be classified into three groups based on cytotoxicity. CSC-6 represents the most cytotoxic group, with EC50 ranging from 15 to 46 µg/mL for the three cell types. CSC-1, CSC-2, CSC-3, CSC-4, and CSC-5 belong to the group with intermediate cytotoxicity, with EC50s between 92 and 170 µg/mL. CSC-7 and CSC-8 belong to the least toxic

TABLE 15.5
EC50 Values (in µg/mL) of the Eight CSCs
in the Three Pulmonary Cell Types

	NHBE	SAEC	HMVEC-L
CSC-1	92.26	127.75	159.76
CSC-2	98.59	>200 (ca. 295.86)	125.46
CSC-3	116.96	135.49	157.51
CSC-4	97.65	136.91	141.08
CSC-5	132.23	>200 (ca. 356.95)	>200 (ca. 309.21)
CSC-6	15.00	46.25	32.37
CSC-7	>200 (316.94)	>200 (ca. 202.09)	>200 (ca. 295.28)
CSC-8	>200 (ca. 261.99)	>200 (ca. 219.44)	>200 (ca. 349.54)

group, with EC50 values consistently greater than the highest concentration evaluated of 200 µg/mL for all three cell types.

The results show that there are some cell type differences in response to CSCs. With the exception of CSC-7, NHBE in general appears to be more sensitive to the CSCs, as illustrated by the relatively lower EC50 values than the other two cell types. It is interesting to note that more than 150% relative viability was observed for treatments with the lower concentrations of CSC-1 and CSC-2 for both SAEC and HMVEC-L. This finding suggests that treatment led to increases in MTT activity, presumably due to increases in cell number after the 24 h treatment period. A stimulatory effect was also observed for CSC-3 and CSC-4 in SAEC and HMVEC-L. No apparent stimulatory effects by any of the CSCs were observed for NHBE.

Our results therefore show that the CSC samples tested had two apparently dose-dependent major cellular effects based on MTT metabolism: cytotoxicity and stimulatory effects. Cytotoxicity of the CSC may be responsible for toxic pulmonary injuries which may ultimately lead to diseases such as emphysema. The stimulatory effects are consistent with the known increased airway epithelial proliferation in smokers. That CSC at low concentrations can promote cell proliferation has been previously reported for the bronchial epithelial cell line NCI-H292 (Ray et al. 2002) and in transformed lung epithelial type II cells (Kaushik et al. 2008). Our data confirmed the proliferative effects of CSC in normal airway epithelial cells and microvascular endothelial cells, which may play critical roles in the promotion of carcinogenic events in the lungs upon initiation due to genotoxic components of the cigarette smoke (Curvall et al. 1984, 1985, Dertinger et al. 2001, Andreoli et al. 2003, Aufderheide and Gressmann 2007, DeMarini et al. 2008, Jianlin et al. 2009a,b, Lou et al. 2010).

Nicotine is a component of cigarette smoke, which is known to have cytotoxicity toward cells in culture including periodontal ligament fibroblasts (Chang et al. 2002), lung fibroblasts (Jin et al. 2003), and gingival fibroblasts (Park et al. 2013). This cytotoxic effect of nicotine was also observed for all three pulmonary cell types cocultured in IdMOC. As for the SCSs, the bronchial-derived cells, NHBE, was apparently more sensitive to the cytotoxicity of nicotine than SAEC and HMVEC-L. Unlike some of the CSCs, nicotine did not have stimulatory effects on MTT metabolism. The results therefore suggest that the stimulatory effects of CSC on MTT metabolism may not be a property of their nicotine content. It is interesting to note that the most cytotoxic CSC was CSC-6, an experimental 100% flue-cure tobacco cigarette, which has been found to have the most tar per cigarette, as well as the highest nicotine concentration (DeMarini et al. 2008), implying a role for both tar and nicotine in its cytotoxicity.

The observations made here with the IdMOC-lung system are consistent with the known effects of cigarette smoking in experimental animals and in the human population (Clapp et al. 1977, Tuyns and Esteve 1983, Melikian et al. 1989, Smith et al. 2006, Carpagnano et al. 2010). The cytotoxic properties are likely to be related to smoking-induced lung obstructive diseases, while the proliferation stimulatory effects may be related to the tumor-promotional events. We previously reported that IdMOC can be applied toward the evaluation of the toxic potential of an agent toward multiple cell types (Li et al. 2004, 2012, Li 2008, 2009). The IdMOC experimental system used here with the three major pulmonary cell types, namely, bronchial epithelium, small airway epithelium, and microvascular endothelium, may be applied to the routine evaluation of the toxicity of inhaled toxicants.

QUESTIONS

1. Recently, there is an emphasis in academic institutions, industrial laboratories, and regulatory agencies to decrease the use of animals in research. One of the approaches is to use in vitro systems instead of whole animals. What are in vitro systems?
   a. Nonhuman animals
   b. Fish
   c. Birds
   d. Cell culture
   e. Bacteria
   Answer: d and e

2. Cell cultures are routinely used for in vitro toxicity testing. The cell culture systems used are either cell lines or primary cells. What are primary cells?
   a. Cells from primates
   b. An immortalized cell line
   c. A transformed cell line
   d. A commonly used cell line
   e. Cells developed from organs that retain organ-specific properties
   Answer: e

3. Primary cells are useful, because they retain properties of the organ from which they are derived. However, in vitro toxicity evaluation with primary cells may not adequately reflect toxic response in the organ in vivo. This is because of which of the following?
   a. The organ has more than one cell types, and the primary cells used may not contain all the cell types.
   b. The use of primary cell of a single cell type lacks interactions with other cell types, which may be important for the ultimate toxicity in the organ.
   c. Primary cells may dedifferentiate in culture and, thereby, no longer possess organ-specific properties.
   d. Exposure of cells in vitro to a toxicant may not be equivalent to that in vivo.
   e. All of the above.
   Answer: e

4. IdMOC is an experimental system designed to provide an in vitro test system more representative of a whole organ or a whole organism. What does IdMOC stand for?
   a. Integrated discrete multiple organ coculture
   b. Independent differentiated model of cells
   c. Information derived mammalian organ culture
   d. Independently developed multiple organ cultures
   e. Identifiable differentiated multiple of cells
   Answer: a

5. How does IdMOC mimic a whole organism?
    a. It contains primary cells from different organs, which, like the organism in vivo, are situated as physically separated entities but interconnected by the systemic circulation.
    b. It can move around due to robotic components.
    c. It eats, drinks, and excretes.
    d. It has only one cell type.
    e. It is kept at 37°C.
    Answer: a
6. Which of the following studies can be performed with IdMOC
    a. Evaluation of the toxicity of metabolites generated by a metabolically competent cell type (e.g., hepatocytes) toward a metabolically incompetent cell type (e.g., 3T3 fibroblasts)
    b. Evaluation of the paracrine interactions of multiple cell types from the same organ
    c. Evaluation of endocrine interactions of cell types from multiple organs
    d. Evaluation of differential sensitivity of multiple cell types toward a toxicant
    e. All of the above
    Answer: e

REFERENCES

Adams, J. D., S. J. Lee, and D. Hoffmann (1984). Carcinogenic agents in cigarette smoke and the influence of nitrate on their formation. *Carcinogenesis* **5**(2): 221–223.

Andreoli, C., D. Gigante, and A. Nunziata (2003). A review of in vitro methods to assess the biological activity of tobacco smoke with the aim of reducing the toxicity of smoke. *Toxicol In Vitro* **17**(5–6): 587–594.

Aufderheide, M. and H. Gressmann (2007). A modified Ames assay reveals the mutagenicity of native cigarette mainstream smoke and its gas vapour phase. *Exp Toxicol Pathol* **58**(6): 383–392.

Bombick, D. W., B. R. Bombick, P. H. Ayres, K. Putnam, J. Avalos, M. F. Borgerding, and D. J. Doolittle (1997). Evaluation of the genotoxic and cytotoxic potential of mainstream whole smoke and smoke condensate from a cigarette containing a novel carbon filter. *Fundam Appl Toxicol* **39**(1): 11–17.

Bombick, D. W., K. Putnam, and D. J. Doolittle (1998). Comparative cytotoxicity studies of smoke condensates from different types of cigarettes and tobaccos. *Toxicol In Vitro* **12**(3): 241–249.

Carpagnano, G. E., A. Spanevello, G. P. Palladino, C. Gramiccioni, C. Ruggieri, F. Carpagnano, and M. P. Foschino Barbaro (2010). Cigarette smoke and increased COX-2 and survivin levels in exhaled breath condensate of lung cancer patients: How hot is the link? *Lung Cancer* **67**(1): 108–113.

Chang, Y. C., F. M. Huang, K. W. Tai, L. C. Yang, and M. Y. Chou (2002). Mechanisms of cytotoxicity of nicotine in human periodontal ligament fibroblast cultures in vitro. *J Periodontal Res* **37**(4): 279–285.

Chen, H., L. Cui, X. Y. Jiang, Y. Q. Pang, G. L. Tang, H. W. Hou, J. H. Jiang, and Q. Y. Hu (2012). Evaluation of the cytotoxicity of cigarette smoke condensate by a cellular impedance biosensor. *Food Chem Toxicol* **50**(3–4): 612–618.

Chepiga, T. A., M. J. Morton, P. A. Murphy, J. T. Avalos, B. R. Bombick, D. J. Doolittle, M. F. Borgerding, and J. E. Swauger (2000). A comparison of the mainstream smoke chemistry and mutagenicity of a representative sample of the US cigarette market with two Kentucky reference cigarettes (K1R4F and K1R5F). *Food Chem Toxicol* **38**(10): 949–962.

Clapp, M. J., D. M. Conning, and J. Wilson (1977). Studies on the local and systemic carcinogenicity of topically applied smoke condensate from a substitute smoking material. *Br J Cancer* **35**(3): 329–341.

Curvall, M., C. R. Enzell, T. Jansson, B. Pettersson, and M. Thelestam (1984). Evaluation of the biological activity of cigarette-smoke condensate fractions using six in vitro short-term tests. *J Toxicol Environ Health* **14**(2–3): 163–180.

Curvall, M., T. Jansson, B. Pettersson, A. Hedin, and C. R. Enzell (1985). In vitro studies of biological effects of cigarette smoke condensate. I. Genotoxic and cytotoxic effects of neutral, semivolatile constituents. *Mutat Res* **157**(2–3): 169–180.

DeMarini, D. M. (1981). Mutagenicity of fractions of cigarette smoke condensate in *Neurospora crassa* and *Salmonella typhimurium*. *Mutat Res* **88**(4): 363–374.

DeMarini, D. M., R. Gudi, A. Szkudlinska, M. Rao, L. Recio, M. Kehl, P. E. Kirby, G. Polzin, and P. A. Richter (2008). Genotoxicity of 10 cigarette smoke condensates in four test systems: Comparisons between assays and condensates. *Mutat Res* **650**(1): 15–29.

Dertinger, S. D., D. A. Nazarenko, A. E. Silverstone, and T. A. Gasiewicz (2001). Aryl hydrocarbon receptor signaling plays a significant role in mediating benzo[a]pyrene- and cigarette smoke condensate-induced cytogenetic damage in vivo. *Carcinogenesis* **22**(1): 171–177.

Foy, J. W., B. R. Bombick, D. W. Bombick, D. J. Doolittle, A. T. Mosberg, and J. E. Swauger (2004). A comparison of in vitro toxicities of cigarette smoke condensate from Eclipse cigarettes and four commercially available ultra low-"tar" cigarettes. *Food Chem Toxicol* **42**(2): 237–243.

Guo, X., T. L. Verkler, Y. Chen, P. A. Richter, G. M. Polzin, M. M. Moore, and N. Mei (2011). Mutagenicity of 11 cigarette smoke condensates in two versions of the mouse lymphoma assay. *Mutagenesis* **26**(2): 273–281.

Hellermann, G. R., S. Nagy, X. Kong, R. F. Lockey, and S. S. Mohaptra (2002). Mechanism of cigarette smoke condensate induced acute inflammatory response in human bronchial epithelial cells. *Respir Res* **3**: 22–30.

Jianlin, L., C. Guohai, Z. Guojun, J. Jian, H. Fangfang, X. Juanjuan, Z. Shu et al. (2009a). Assessing cytogenotoxicity of cigarette smoke condensates using three in vitro assays. *Mutat Res* **677**(1–2): 21–26.

Jianlin, L., C. Guohai, Z. Guojun, J. Jian, H. Fangfang, X. Juanjuan, Z. Shu et al. (2009b). Studying the impact of S9 on cyto-genotoxicity of cigarette smoke in human peripheral blood lymphocytes in vitro. *Environ Toxicol Pharmacol* **28**(2): 275–279.

Jin, J. S., M. S. Kim, J. M. Yi, J. H. Lee, S. J. Moon, K. P. Jung, J. K. Lee, N. H. An, and H. M. Kim (2003). Inhibitory effect of Sejin-Eum I/II on nicotine- and cigarette extract-induced cytotoxicity in human lung fibroblast. *J Ethnopharmacol* **86**(1): 15–20.

Kato, T., H. Nagasawa, C. Warner, R. Okayasu, and J. S. Bedford (2007). Cytotoxicity of cigarette smoke condensate is not due to DNA double strand breaks: Comparative studies using radiosensitive mutant and wild-type CHO cells. *Int J Radiat Biol* **83**(9): 583–591.

Kaushik, G., T. Kaushik, S. Khanduja, C. M. Pathak, and K. L. Khanduja (2008). Cigarette smoke condensate promotes cell proliferation through disturbance in cellular redox homeostasis of transformed lung epithelial type-II cells. *Cancer Lett* **270**(1): 120–131.

Li, A. P. (2007a). Human hepatocytes: Isolation, cryopreservation and applications in drug development. *Chem Biol Interact* **168**(1): 16–29.

Li, A. P. (2007b). Human-based in vitro experimental systems for the evaluation of human drug safety. *Curr Drug Saf* **2**(3): 193–199.

Li, A. P. (2008). In vitro evaluation of human xenobiotic toxicity: Scientific concepts and the novel integrated discrete multiple cell co-culture (IdMOC) technology. *ALTEX* **25**(1): 43–49.

Li, A. P. (2009). The use of the Integrated Discrete Multiple Organ Co-culture (IdMOC) system for the evaluation of multiple organ toxicity. *Altern Lab Anim* **37**(4): 377–385.

Li, A. P., C. Bode, and Y. Sakai (2004). A novel in vitro system, the integrated discrete multiple organ cell culture (IdMOC) system, for the evaluation of human drug toxicity: Comparative cytotoxicity of tamoxifen towards normal human cells from five major organs and MCF-7 adenocarcinoma breast cancer cells. *Chem Biol Interact* **150**(1): 129–136.

Li, A. P., A. Uzgare, and Y. S. LaForge (2012). Definition of metabolism-dependent xenobiotic toxicity with cocultures of human hepatocytes and mouse 3T3 fibroblasts in the novel integrated discrete multiple organ co-culture (IdMOC) experimental system: Results with model toxicants aflatoxin B1, cyclophosphamide and tamoxifen. *Chem Biol Interact* **199**(1): 1–8.

Lou, J., G. Zhou, G. Chu, J. Jiang, F. Huang, S. Zheng, Y. Lu, X. Li, and J. He (2010). Studying the cyto-genotoxic effects of 12 cigarette smoke condensates on human lymphoblastoid cell line in vitro. *Mutat Res* **696**(1): 48–54.

Lu, X., M. Zhao, H. Kong, J. Cai, J. Wu, M. Wu, R. Hua, J. Liu, and G. Xu (2004a). Characterization of cigarette smoke condensates by comprehensive two-dimensional gas chromatography/time-of-flight mass spectrometry (GC × GC/TOFMS). Part 2: Basic fraction. *J Sep Sci* **27**(1–2): 101–109.

Lu, X., M. Zhao, H. Kong, J. Cai, J. Wu, M. Wu, R. Hua, J. Liu, and G. Xu (2004b). Characterization of complex hydrocarbons in cigarette smoke condensate by gas chromatography–mass spectrometry and comprehensive two-dimensional gas chromatography-time-of-flight mass spectrometry. *J Chromatogr A* **1043**(2): 265–273.

Luppi, F., J. Aarbiou, S. van Wetering, I. Rahman, W. I. de Boer, K. F. Rabe, and P. S. Hiemstra (2005). Effects of cigarette smoke condensate on proliferation and wound closure of bronchial epithelial cells in vitro: Role of glutathione. *Respir Res* **6**: 140.

Matsukura, N., J. Willey, M. Miyashita, B. Taffe, D. Hoffmann, C. Waldren, T. T. Puck, and C. C. Harris (1991). Detection of direct mutagenicity of cigarette smoke condensate in mammalian cells. *Carcinogenesis* **12**(4): 685–689.

McCoy, E. C. and H. S. Rosenkranz (1982). Cigarette smoking may yield nitroarenes. *Cancer Lett* **15**(1): 9–13.

McKarns, S. C., D. W. Bombick, M. J. Morton, and D. J. Doolittle (2000). Gap junction intercellular communication and cytotoxicity in normal human cells after exposure to smoke condensates from cigarettes that burn or primarily heat tobacco. *Toxicol In Vitro* **14**(1): 41–51.

Melikian, A. A., S. S. Hecht, and D. Hoffmann (1989). Mechanistic studies of tobacco carcinogenesis in mouse epidermis and lung tissues. *Prog Clin Biol Res* **298**: 331–345.

Mizusaki, S., H. Okamoto, A. Akiyama, and Y. Fukuhara (1977). Relation between chemical constituents of tobacco and mutagenic activity of cigarette smoke condensate. *Mutat Res* **48**(3–4): 319–325.

Nordskog, B. K., A. D. Blixt, W. T. Morgan, W. R. Fields, and G. M. Hellmann (2003). Matrix-degrading and pro-inflammatory changes in human vascular endothelial cells exposed to cigarette smoke condensate. *Cardiovasc Toxicol* **3**(2): 101–117.

Park, G. J., Y. S. Kim, K. L. Kang, S. J. Bae, H. S. Baek, Q. S. Auh, Y. H. Chun, B. H. Park, and E. C. Kim (2013). Effects of sirtuin 1 activation on nicotine and lipopolysaccharide-induced cytotoxicity and inflammatory cytokine production in human gingival fibroblasts. *J Periodontal Res* **48**(4): 483–492.

Pillsbury, H. C., C. C. Bright, K. J. O'Connor, and F. W. Irish (1969). Tar and nicotine in cigarette smoke. *J Assoc Off Anal Chem* **52**: 458–462.

Ramdahl, T., B. Zielinska, J. Arey, R. Atkinson, A. M. Winer, and J. N. Pitts (1986). Ubiquitous occurrence of 2-nitrofluoranthene and 2-nitropyrene in air. *Nature* **321**(6068): 425–427.

Ray, S., D. N. Watkins, N. L. Misso, and P. J. Thompson (2002). Oxidant stress induces gamma-glutamylcysteine synthetase and glutathione synthesis in human bronchial epithelial NCI-H292 cells. *Clin Exp Allergy* **32**(4): 571–577.

Richter, P. A., A. P. Li, G. Polzin, and S. K. Roy (2010). Cytotoxicity of eight cigarette smoke condensates in three test systems: Comparisons between assays and condensates. *Regul Toxicol Pharmacol* **58**(3): 428–436.

Sepetdjian, E., A. Shihadeh, and N. A. Saliba (2008). Measurement of 16 polycyclic aromatic hydrocarbons in narghile waterpipe tobacco smoke. *Food Chem Toxicol* **46**(5): 1582–1590.

Smith, C. J. and C. Hansch (2000). The relative toxicity of compounds in mainstream cigarette smoke condensate. *Food Chem Toxicol* **38**(7): 637–646.

Smith, C. J., T. A. Perfetti, R. Garg, and C. Hansch (2006). Utility of the mouse dermal promotion assay in comparing the tumorigenic potential of cigarette mainstream smoke. *Food Chem Toxicol* **44**(10): 1699–1706.

Tuyns, A. J. and J. Esteve (1983). Pipe, commercial and hand-rolled cigarette smoking in oesophageal cancer. *Int J Epidemiol* **12**(1): 110–113.

Wakabayashi, K., I. S. Kim, R. Kurosaka, Z. Yamaizumi, H. Ushiyama, M. Takahashi, S. Koyota, A. Tada, H. Nukaya, and S. Goto (1995). Identification of new mutagenic heterocyclic amines and quantification of known heterocyclic amines. *Princess Takamatsu Symp* **23**: 39–49.

Zhang, W., F. Song, and L. J. Windsor (2009). Cigarette smoke condensate affects the collagen-degrading ability of human gingival fibroblasts. *J Periodontal Res* **44**(6): 704–713.

16 Animal Models of Chronic Obstructive Pulmonary Disease (COPD)

Steven G. Kelsen

CONTENTS

16.1 INTRODUCTION

Chronic obstructive pulmonary disease (COPD) is an increasingly important major public health concern in the United States and worldwide. The disease affects an estimated 16 million people in the United States or approximately 5% of the population and anywhere from 15% to 35% of heavy cigarette smokers are affected (Rennard and Vestbo, 2006; Rabe et al., 2007). In fact, COPD is the third leading cause of death in the United States and fourth leading cause of death worldwide. It is estimated that 170,000 Americans die of COPD annually in the United States. In addition to its effects on the mortality and morbidity of cigarette smokers, COPD also represents a serious financial burden upon the health system of the United States, costing in excess of $193 billion dollars a year. It is estimated that only 50% of cases have been formally diagnosed. Two important factors appear to explain the significant underdiagnosis of the disease: (1) Symptoms develop only relatively late in the disease and (2) direct testing of lung function by the technique of spirometry is not performed routinely in all smokers.

Pathophysiologically, COPD is characterized by an abnormal inflammatory response in the lung and conducting airways that causes structural changes in lung terminal lung gas exchanging units, that is, the alveolar region, the small and large airways, and the pulmonary vasculature (Henson et al., 2006; Yoshida and Tuder, 2007; Taraseviciene-Stewart and Voelkel, 2008). Moreover, structural cells in the airways and parenchyma undergo apoptosis and programmed cell death at a greater than normal rate (Henson et al., 2006; Yoshida and Tuder, 2007). The remodeling process affects lung function in complex fashion. Within the gas exchanging regions of the lung, destruction of alveolar septa causes loss of alveolar walls. The resultant coalescence of alveoli into larger than normal saclike structures, that is, emphysema, decreases the alveolar surface area available for gas exchange in the lung. Emphysema also affects the pressure–volume (P–V) relationship of the lung. The elastic recoil of the lung, which is the driving pressure for expiratory airflow increases and the force that tethers the intrapulmonary airways and helps to maintain their patency, is decreased.

Changes in the small, membranous intrapulmonary airways are characterized by inflammation and increases in thickness of the airway wall including the mucosal (epithelial), submucosal, and adventitial layers, that is, chronic bronchitis/bronchiolitis. The epithelial lining of the airway undergoes both hyperplasia and metaplasia with increases in epithelial cell size and a change in the phenotype of some cells toward a mucous-secreting or goblet cell form.

Quite interestingly, the remodeling process that takes place in the small intrapulmonary airways appears to be qualitatively different from that which takes place in the lung parenchyma. In contrast to emphysema, remodeling in the small intrapulmonary airways appears to be a proliferative response that results in thickening of the airway wall. Airway remodeling is characterized by hyperplasia of the airway epithelial cells with increase in size and metaplasia toward a mucus secreting phenotype, deposition of connective tissue in the airway wall, and increases in the mass smooth muscle. Increased deposition of matrix and connective tissue and increases in the mass of smooth muscle result in increased thickness of the submucosal layer. The inflammatory infiltrate present in the submucosa and to a lesser extent in the epithelial layer is characterized by increased numbers of CD8+/Tc1 and CD4+/Th1 lymphocytes, B cells, macrophages, and mast cells and hence is qualitatively different from the inflammatory infiltrate present in asthma (DiStefano et al., 1996, 1998; Saetta et al., 1997, 1998, 1999, 2002; Panina-Bordignon et al., 2001; Turato et al., 2002; Willemse et al., 2004; Taraseviciene-Stewart et al., 2006). Mucous plugs may form within the lumen and compromise luminal caliber. In severe cases, airways become stenotic or even obliterated as a result of fibroblast proliferation (McDonough et al., 2011). Increases in airway wall connective tissue, mucous secretion, and epithelial cell hyperplasia cause airway luminal narrowing, increased resistance to airflow during inspiration and expiration, and maldistribution of inspired gas in the lung.

In the large extrapulmonary airways, the multicellular, submucosal mucous glands undergo an expansion in size and number in COPD. Hyperplasia and hypertrophy of submucosal glands contribute importantly to the chronic cough and mucous production, which are important and early symptoms of the human disease.

Changes in the pulmonary vasculature are characterized by thickening of the intima and media of the vessel wall with increased connective tissue and smooth muscle content. Changes in pulmonary vascular structure are accompanied by an increase in vascular tone and diminished responsiveness to vasodilator stimuli. Reductions in pulmonary vascular responsiveness appear to be explained in part at least by decreases in nitric oxide production with the vessel wall (Wright et al., 2006). These changes lead to increases in pulmonary vascular resistance and ultimately pulmonary hypertension. In turn, right ventricular strain and right ventricular failure secondary to pulmonary hypertension contributes to COPD morbidity and mortality. Of interest, changes in the pulmonary vasculature appear to be of independent of structural changes in the lung parenchyma and airways.

Of considerable importance, the inflammatory infiltrate in COPD is present in the small airways, lung parenchyma, submucosal glands, and pulmonary arteries (Finkelstein et al., 1995; DiStefano et al., 1996, 1998; Saetta et al., 1997, 1998, 1999). In addition, T cells, B cells, macrophages, and dendritic cells aggregate into organized lymphoid follicles in close proximity to the airways and within the lung parenchyma (Hogg et al., 2004, 2007; van der Strate et al., 2006). It has been reported that the number of airways containing lymphoid follicles increases as the severity of the disease progresses (McDonough et al., 2011). Of considerable importance, the inflammatory process persists long after cigarette smoking (CS) has ceased (Willemse et al., 2004; Gamble et al., 2007). Continuation of the process of inflammation is poorly understood but is believed to contribute to the accelerated rate of decline of lung function that occurs in subjects with COPD compared to those who have never smoked.

The precise mix of inflammation, lung parenchymal destruction (i.e., emphysema), and airway remodeling (i.e., chronic obliterative bronchitis) varies across individuals leading to distinctive phenotypes in human smokers (Han et al., 2010; Nussbaumer-Ochsner and Rabe, 2011). Also for reasons that are not clear, within a given individual, emphysema and airway remodeling are not homogeneously distributed but rather are patchy in their distribution. Human COPD also has systemic manifestations that are reflected in abnormalities in extrapulmonary systems like the systemic vasculature, that is, accelerated atherosclerosis and coronary artery disease, and skeletal musculature, that is, muscle wasting and inflammation (Barnes et al., 2003; Elias et al., 2006). Differences in the relative magnitude of emphysema, airway remodeling, and systemic complications have led to an awareness of the presence of distinct COPD phenotypes (Han et al., 2010). Distinctive phenotypes have led to the idea produced by diverse molecular mechanisms (Barnes et al., 2003; Elias et al., 2006).

Finally, the propensity to develop COPD varies widely across cigarette smokers and correlates only weakly with the smoking history as reflected in the number of cigarette pack years (Burrows et al., 1977; Rennard and Vestbo, 2006). In fact, it is estimated that only a minority (i.e., 15%–35%) of chronic, continuous cigarette smokers develop COPD (Fletcher, 1976; Rennard and Vestbo, 2006). That the majority of long-term smokers do not develop COPD suggests that failure of compensatory mechanisms that protect the lung from reactive oxygen species (ROS) or xenobiotic materials contributes to development of the disease. In support of this concept, the expression of antioxidant genes believed to be important in protection of the lung from cigarette smoke-induced injury (e.g., thioredoxin, peroxiredoxin and glutathione S-transferase [GST], glutathione peroxidase [GP]) varies widely in airway epithelial cells harvested from chronic cigarette smokers (Hackett et al., 2003).

16.2 ROLE OF CIGARETTE SMOKING IN COPD

CS is the main preventable cause of COPD worldwide and especially in the United States. In the United States alone, more than 45 million people or 20.9% of the population are chronic smokers and an estimated 95% of COPD cases are attributed to smoking. The gaseous and particulate phases of cigarette smoke expose the lung to heightened oxidant stress and a large variety of toxic substances including toxic gases (nitrogen dioxide, nitric oxide, carbon monoxide), alpha and beta aldehydes (acrolein), heavy metals (cadmium, zinc, iron), aromatic and nonaromatic hydrocarbons (dioxin), and known carcinogens (dioxin, benzopyrene, etc.) (Church and Pryor, 1985; Pryor and Stone, 1993; Zang et al., 1995; MacNee, 2005). In fact, smoke from a burning cigarette contains 4500 separate compounds and approximately 10^{15} ROS per puff (Cosgrove et al., 1985). Moreover, cigarette smoke contains a variety of bacterial-derived substances (lipopolysaccharides [LPSs]), which also induce important biological effects on the innate and adaptive immune systems.

CS also increases oxidant stress in the lung by promoting the generation of ROS and reactive nitrogen species (RNS) in resident lung structural and inflammatory cells such as epithelial cells, macrophages, and pulmonary endothelial cells (Cheng et al., 2001; Lambeth, 2004; Ranjan et al., 2006; Fink et al., 2008; Nagai et al., 2008; Zhang et al., 2011). For example, CS activates endogenous NADPH oxidase (NOX) isoforms that transport electrons from cytoplasmic high-energy electron donor NADPH to generate O_2^- and hydrogen peroxide (H_2O_2) (Lambeth, 2004). In addition, NOX-derived ROS can induce mitochondrial ROS production indicating the possibility of a positive feedback loop (Hawkins et al., 2007; Zinkevich and Gutterman, 2011). Of interest, nicotine- and tar-free cigarette smoke is capable of inducing NOX activation indicating that gaseous phase ROS as

well as cigarette tar contribute to oxidant stress (Asano et al., 2012). Many of the compounds in cigarette smoke also induce an inflammatory response by activating redox-sensitive, proinflammatory
pathways including nuclear factor-kappa B (NF-κB), AP-1, and MAPKs.

16.3   ANIMAL MODELS OF COPD: AN OVERVIEW

A variety of approaches have been used to model COPD in animals. These include chronic exposure
to cigarette smoke, installation of proteolytic enzymes into the airway to digest lung connective tissue, and exposure to redox-active heavy metals like cadmium to overwhelm the oxidant defenses
of the lung (Snider et al., 1986; Hantos et al., 2008; Takahashi et al., 2008; Fievez et al., 2009). In
general, however, most studies have utilized chronic cigarette smoke exposure as the intervention
because of its obvious relevance to the human condition (Churg et al., 2008). Recent studies have
utilized a *double-hit* approach in which chronic exposure to cigarette smoke is combined with
administration to the airway of a microbial product such as like an LPS or double-stranded RNA to
activate the innate immune system and mimic respiratory tract infection (Kang et al., 2008).

In general, a variety of rodent species including a mouse, rat, and guinea pig have been used as
models of COPD. A limited number of studies have been performed utilizing larger animals including dogs, pigs, and primate species. Most studies involving chronic cigarette smoke-induced COPD
have utilized rodent species, in particular mouse. This is the case for cost reasons, the ability to
study large numbers of animals in a given experiment, the availability of a large variety of reagents
for use and in this species, and the ease with which discrete genes and their signaling pathways can
be assessed in transgene animals or strains deficient in a specific protein. Studies in larger animals
are far more costly to perform and are limited by the relative lack of reagents.

Animal models of emphysema have generally utilized the mouse, whereas models of chronic
bronchitis and excessive mucus secretion have largely utilized the rat. Of interest, the guinea pig
appears to be useful for studies of emphysema, airway remodeling, and pulmonary vascular changes.
Unfortunately, at the present time, no well-accepted model of acute exacerbations of COPD is available.

Unfortunately, no perfect animal model of COPD exists and no model incorporates all of the main
features of the human disease, that is, emphysema, chronic bronchitis with small airway remodeling,
pulmonary hypertension, and acute exacerbations (Shapiro, 2000, 2007; Churg et al., 2008; Wright
and Churg, 2008). Moreover, the disease produced in rodents, at least, tends to be considerably
milder than the human variety. In addition, the disease in mouse and rats also tends to be reversible
once exposure to smoke ceases unlike the human condition that progresses even after cigarette exposure has ceased. These limitations in the severity of the disease and differences in involvement of
the several lung structures affected by the human condition likely stem from differences in the gross
and microscopic anatomy and cellular composition of the lung in rodents and man (see Section 16.4).

Despite these limitations, animal models of COPD especially those performed in the mouse or
the rat have been extremely useful in elucidating the importance of a variety of pathogenetic mechanisms of human COPD. In particular, rodent studies have demonstrated the importance of imbalance
in oxidant/antioxidant systems (Rangasamy et al., 2004; Singh et al., 2006, 2009; Yao et al., 2010),
proinflammatory/anti-inflammatory systems (Rajendrasozhan et al., 2010; Kang et al., 2012;
Sundar et al., 2012), and protease/antiprotease systems (Hautamaki et al., 1997; Churg et al., 2007;
Atkinson et al., 2011). They have also demonstrated the importance of reductions in the activity of
histone deacetylases (HDACs) and the role of accelerated aging in the development of emphysema
and airway remodeling (Malhotra et al., 2011; Yao et al., 2012).

16.4   COMPARATIVE ANATOMY OF THE HUMAN AND ANIMAL LUNG

The gross and microscopic anatomy of the lungs varies considerably between humans and other
animals especially rodents. Moreover, lung growth and alveolar septation also differs considerably between man and rodents. In particular, rats and mice have no true alveoli at birth. In fact,

the majority of the alveoli form between 4 and 14 days after birth (McGowan, 2004). In contrast, the guinea pig lung is well alveolated at birth and demonstrates only a small increase in alveolar number with aging. Alveolar size also increases progressively throughout life in guinea pigs and rats. In mice, alveolar size changes with development in a strain-dependent manner. For example, BALB/C mice demonstrate progressive increases in alveolar size up to 19 months of age while strain 129 demonstrates constant alveolar size 4 weeks after birth (Kawakami et al., 1984). Of considerable importance, mice and rats demonstrate continued alveolar septation throughout life. These factors need to be taken into account when choosing both an animal species and the timing of smoke exposure since they may have an important confounding effect on the quantitation of the severity of emphysema.

The human lung demonstrates several generations of membranous and respiratory bronchioles. In man, emphysema is believed to begin in the respiratory bronchioles and small airway remodeling is believed to begin in the membranous intrapulmonary bronchioles (Yoshida and Tuder, 2007). In contrast, mice and rats demonstrate less airway branching then do humans and do not have respiratory bronchioles.

In human subjects, the airway epithelium demonstrates a pseudostratified appearance from the trachea and bronchi up to the subsegmental bronchus at which point a simple epithelial lining is present (Jeffery, 1983). The majority of epithelial cells are ciliated throughout the bronchial tree in man. However, mucous-secreting goblet cells are present in the large airways and continue with decreasing frequency as the airways branch but are present even into the terminal bronchiole. On the other hand, rats and mice have a single columnar layer and Goblet cells are rare in the tracheobronchial tree of mice and very infrequent in rats. In contrast, mice and rats have much larger numbers of nonciliated Clara cells. The greater numbers of Clara cells in these rodents are likely of importance since these cells participate in metabolism of xenobiotic materials and in production of surfactant. Of interest, guinea pigs resemble the human epithelial pattern.

In humans, submucosal mucous glands are present in considerable numbers in the large airways and increases in their size and number account for most of the excessive mucus secretion in human COPD (Yoshida and Tuder, 2007). In contrast, submucosal mucous glands are absent in the airways of mouse and rabbit, almost absent in hamsters, and concentrated only in the upper portion of the trachea in the rat. These structural differences in mucous glands make small animals poor models for the study of the pathophysiology of mucus overproduction in human chronic bronchitis.

The microscopic anatomy of the pulmonary vasculature also differs between man and rodents (Wright and Churg, 1991; Wright et al., 2006). In humans, arterial and venous vessels have separate pathways in the lung. The arteries follow the path of the airways while the veins travel through the interlobular septa, which are lacking in rodents. In man, the main pulmonary artery and vessels greater than 1000 μm in diameter are elastic in nature and have medial walls that are comprised of short, branching elastic fibrils. In contrast, vessels ranging in size from 500 to 100 μm are muscularized with a medial layer present between two elastic laminae. In humans, muscularization becomes progressively less apparent as vessels decrease in size and muscle cells disappear at the level of the distal arterioles. Pulmonary veins have a single elastic lamina and gradually acquire a muscular layer.

In rats, guinea pigs, and mice, the main and large pulmonary arteries are muscular rather than elastic. Also in rodents, the generation of vessel at which the muscle content disappears varies. Most vessels adjacent to alveolar ducts are nonmuscular. Unlike humans, the pulmonary veins of rodents demonstrate a large amount of cardiac muscle at the hilum of the lung.

16.5　CIGARETTE SMOKE EXPOSURE

More than 150 published studies have examined the effect of cigarette smoke in animal models. These studies have been summarized in recent reviews (Wright, 2010; Goldklang, 2013; Leberl, 2013). In general, short-term exposures over the course of hours to several days have been used to study the acute inflammatory response to cigarette smoke in the lung. In contrast, chronic exposures occurring over 2–6 months have been used to produce the structural and functional changes in the lung that mimic human COPD. Animals exposed to room air in the same exposure system typically have served as controls. A variety of cigarettes and exposure chambers have been employed. Moreover, the intensity and duration of the exposure have varied considerably. In many instances, insufficient detail has been provided in this regard making it difficult to compare and contrast the results obtained in different studies.

The composition of cigarette smoke varies considerably across cigarette brands, the types of tobacco used, and the presence or absence of a filter. Beginning in 1969, the University of Kentucky (U of Kentucky) has provided cigarettes for research in which the composition of total particulate matter (TPM), nicotine, and carbon monoxide is defined (Roemer, 2012). Accordingly, the U of Kentucky research cigarettes are now preferred for cigarette smoke exposure experiments in animal models. Recently, an attempt has been made to define the composition of cigarette smoke for research purposes for at least two important constituents of cigarette smoke, that is, the total amount of particulates and the amount of nicotine.

Of note, the composition and biological effects of cigarette smoke also depend on whether smoke exposure is direct, that is, mainstream, or indirect, that is, side stream, or a combination of both mainstream and sidestream smoke. In man, mainstream smoke is directly inhaled into the mouth while taking a puff on a cigarette. In the animal, this corresponds to nose- or head-only exposure of smoke generated by an inhalation system. Sidestream smoke, on the other hand, enters the air from the tip of a burning cigarette. In the animal, this corresponds to smoke delivered into the chamber of a freely ambulating animal. Finally, environmental smoke is a combination of mainstream and sidestream, that is, a combination of smoke from the burning end of the cigarette and that which is drawn through the cigarette. In contrast to mainstream cigarette smoke, which is hotter and has a higher concentration of particulate matter, sidestream cigarette smoke is cooler, has a higher concentration of volatile ROS, and a 10-fold higher concentration of the toxic, highly redox-active, heavy metal, cadmium (Huang et al., 2005).

Smoke exposure has been performed using a variety of devices that deliver either mainstream (nose- or head-only) or sidestream (whole-body) exposures. Mainstream exposure generally has the advantage of providing a more intense exposure directly to the respiratory tract but has the disadvantage that the animal must be restrained and thereby subjected to stress. Moreover, this form of exposure is quite labor intensive since each animal must be placed into and removed from the device individually. In contrast, devices that deliver sidestream smoke allow the animal to move about the exposure chamber freely. Sidestream exposure systems allow a larger number of animals to be studied since the procedures are generally less labor intensive. The disadvantage of sidestream exposure is that smoke is delivered to the body surface where it may be ingested as the animal grooms itself.

Earlier CS exposure systems were largely homemade and utilized head- or nose-only exposure to mimic mainstream smoke exposure. These systems did not maintain consistent levels of total particulate material or have the ability to vary the concentration of total particulate material. Early studies of sidestream smoke used self-constructed chambers that are allowed environmental smoke to diffuse within a confined space; had to be operated manually; and lacked the ability to measure the parameters needed to assess cigarette smoke dose. Consequently, many of these earlier studies do not allow the nature and intensity of exposure to be determined thereby making the comparison of results across studies difficult.

More recently, a variety of commercially produced (e.g., the Teague and SCIREQ) cigarette smoke exposure systems have become available, which allow either sidestream or mainstream smoke exposure to relatively large numbers of animals for prolonged periods. In addition, these machines are automated and programmable. For example, these machines allow cigarettes to be loaded and lit robotically and the number of cigarettes and smoking duration to be preset. These devices also deliver consistent levels of particulates and have the capability of varying the concentration of these particulates. These newer systems also allow measurement of lung function in the animals.

Consequently, it is now recommended that studies should be performed using standardized research-grade cigarettes having a specified content of total suspended particulates (TSPs), nicotine, and carbon monoxide and with a known level of TSPs in the smoke in the chamber. The intensity of smoke exposure has also been assessed from measurements of blood carboxyhemoglobin in the animal immediately following exposure. Levels of blood carboxyhemoglobin of ~10% immediately postexposure are desirable since they represent values present in human subjects who are moderate to heavy smokers, that is, 1–1.5 packs per day.

Most studies utilizing sidestream smoke have set exposure levels of TSPs to between 70 and 150 mg/m^3. In contrast, most mainstreamed studies have been performed with TSPs of 140 mg/m^3. However, mainstream smoke exposures have utilized TSPs that have ranged from as low as 75 mg/m^3 to as high as 600 mg/m^3.

16.6   ASSESSING THE RESPONSE TO SMOKE EXPOSURE

16.6.1   EMPHYSEMA

Emphysema is perhaps the most studied aspect of COPD in animal model and develops in mice, rats, and guinea pigs typically after 4–6 months of chronic cigarette smoke exposure (Hautamaki et al., 1997; Rangasamy et al., 2004; Churg et al., 2008; Wright et al., 2008; Atkinson et al., 2011). This prolonged period of exposure is clearly a disadvantage. Encouraging in this regard, however, is a recent study (Beckett et al., 2013) using a nose-only exposure system that allows robust changes to be observed in the lung of the CD57BL/6J and BALB/c mice within 2 months.

Emphysema observed in laboratory animals generally resembles a rather mild form of human centrilobular emphysema. The more severe forms of bullous emphysema observed in human subjects or the heterogeneity in the distribution and types of emphysema, that is, panlobular or paraseptal forms, are not present. Moreover, in human emphysema, the disease begins in the respiratory bronchioles and there is an early loss of membranous bronchioles. In contrast, in rodents, much of the air space enlargement occurs in alveolar ducts. Moreover, there is a little evidence of alveolar destruction in the animal models compared to human subjects in which there is considerable evidence of disruption of alveolar septa.

Emphysema is generally quantitated morphometrically in tissue sections obtained from inflated lung from measurements of mean alveolar diameter expressed as the mean linear intercept determined (Lm). Lm is determined in lung fixed at total lung capacity. The magnitude of the effect of chronic CS depends among other things on the intensity and duration of smoke exposure and the animal strain. In general, the magnitude of the increase in Lm generally ranges between 15% and 40% compared to air-exposed controls.

Emphysema has also been assessed from the P–V characteristic of the lung in situ, that is, the transpulmonary pressure generated at a given lung volume. Measurements of lung P–V curve are performed typically with the lungs in situ and the chest either open or closed. Transpulmonary pressure is measured using catheters in the trachea and, if the chest is closed, the esophagus. The lung is inflated with a volumetric syringe. Chronic smoke exposure alters the P–V characteristic of the lung such that the recoil pressure at any given lung volume is reduced and the position of the P–V curve is shifted up and to the left. The slope of the relationship is also increased indicating an increase in lung compliance. In general changes in the P–V curve are observed only with severe changes in Lm. Milder forms of emphysema do not appear to affect this aspect of lung mechanics.

Other aspects of lung mechanics that have been used to define the severity of emphysema include measurements of lung volume at full inflation, that is, total lung capacity; at end expiration, that is, functional residual capacity; and at the end of a maximal exhalation, that is, residual volume. These approaches are typically performed with the lungs in situ and the airway connected to a source of positive or negative pressure. Measurements of lung volume have also been measured by water displacement. Expiratory airflow rates can also be measured by rapidly deflating the lungs from total lung capacity in response to application of 50 cm of water pressure to the airway. The volume of gas removed is measured with a volumetric syringe or flow transducer.

Because of animal size, measurement of lung mechanics is technically more challenging in mice than in guinea pigs or rats and generally requires the use of a small animal ventilator equipped with necessary hardware and software. These are commercially available. However, lung mechanics have been measured repeatedly in live mice and have the ability to detect smaller changes than measurements made across animals because within animal variability is generally less than across animal variability. Of note, measurements performed in dead animals are not as accurate or reproducible as measurements performed in life.

Of interest, considerable variation in response to chronic smoke exposure has been observed across mouse strains such that some strains appeared to be more susceptible to the effects of smoke and develop both more severe and earlier onset of emphysema.

16.6.2 Small Airway Remodeling

Small airway remodeling is an important aspect of the human disease. Recently, highly sophisticated studies in man utilizing micro-CT imaging of inflated fixed lung tissue cores have provided new insight into the extent and nature of the pathology in the small airway intrapulmonary (McDonough et al., 2011). The use of micro-CT has provided greater precision in identifying the location of affected airways, their branching pattern, and changes in appearance as they branch. These studies have demonstrated that as the disease progresses, the number of small membranous airways decreases by 1–2 orders of magnitude with a resultant 100-fold decrease in total luminal area.

Small airway remodeling and bronchiolitis have not been as well studied as emphysema in animal models of COPD. However, the limited number of studies performed in both mice and guinea pigs indicate that wall thickening does occur and correlates with functional assessment of airflow obstruction. Assessment of small airway remodeling requires careful analysis of sections obtained in inflated lung and morphometric quantitation of the thickness of the epithelial layer, the submucosal layer, the muscular layer, and the luminal area. The use of micro-CT of the lung in animal models is likely to prove to be a powerful way of assessing changes in the small airways of the lung and their relationship to changes in the emphysema.

Of considerable interest, the technique of laser capture microscopy of small airways and adjacent lung parenchyma has been used in the mouse model to discover that CS exposure rapidly upregulates a number of genes involved in connective matrix formation and repair (Churg et al., 2006). However, while the expression of these genes increased progressively in the small airways over a 6-month exposure period, they rapidly returned to baseline in the lung parenchyma. These observations suggest that the remodeling processes that take place in the airway are qualitatively different from those that occur in the parenchyma. Moreover, they appear to explain why the response to injury in the small airways is proliferative in nature with excess connective tissue laid down, while the response in the parenchyma is characterized by destruction and tissue loss. They also appear to provide an explanation for the finding that an intervention that protects against emphysema may not affect small airway remodeling or vice versa.

16.6.3 Pulmonary Hypertension

Relatively few models have addressed the tissue of pulmonary hypertension. The effects of chronic exposure to cigarette smoke have been best studied in the guinea pig model (Wright et al., 2006). Cigarette smoke increases pulmonary arterial pressure relatively early in the course of chronic CS exposure. Changes observed involved muscularization of small peripheral vessels adjacent to alveolar ducts that are ordinarily not muscularized. In addition, increased expression of vasoactive mediators at the mRNA and protein level is present within the vessel wall. Accordingly, in the animal model, vascular remodeling represents a proliferative response to smoke exposure. Of interest, strain differences in susceptibility to the vascular injury induced by chronic cigarette smoke exposure have been demonstrated in mice (Nadziejko et al., 2007).

16.6.4 INFLAMMATION

Exposure of rodents to either sidestream or mainstream CS induced for seven short periods of hours to days induces an inflammatory response in the lung. The inflammatory response in rodents has been assessed from measurements of cell number and type in bronchoalveolar lavage fluid (BAL) and lung tissue. Inflammation has also been assessed from measurements of pro- and anti-inflammatory cytokines and chemokines in lung and BAL. In general, the pattern of the inflammatory cell infiltrate, that is, macrophages, T cells, polymorphonuclear leukocytes, and Tc1 and Th1 cytokines/chemokines upregulated during acute and chronic CS exposure in the rodent, in general, and the mouse, in particular, resembles the pattern observed in human COPD. However, in mice, the inflammatory response tends to resolve after CS exposure ceases unlike the human subjects with COPD in which heightened levels of inflammation and oxidant stress continue long after CS exposure has ceased. Strain differences in susceptibility of the inflammatory response to acute and chronic cigarette smoke exposure have been demonstrated in mice. However, strain differences in inflammation do not appear to correlate with strain differences in susceptibility to emphysema.

16.7 STRAIN DIFFERENCES IN RESPONSE TO CIGARETTE SMOKE

The remodeling responses to chronic cigarette smoke exposure and the inflammatory response to acute exposure is strain specific in the mouse (Cavarra et al., 2001; Takubo et al., 2002; Guerassimov et al., 2004; Ito et al., 2006). For example, Guerrasimov et al. (2004) examined the inflammatory, structural, and functional responses to chronic cigarette smoke exposure in five separate mouse strains. Male mice aged 3 months were exposed to smoke from two standard, unfiltered U of Kentucky research cigarettes daily (i.e., 2R1), 5 days a week for 6 months using a nose-only exposure system. Mouse strains studied included NZWlac/j, C57BL6/j, A/J, SJL/J, and AKR/J. Carbon monoxide concentrations measured immediately postexposure ranged between 10% and 12%. The supersusceptible, AKR/J strain showed the greatest increase in Lm and alterations in the P–V with an increase in compliance. AKR mice also demonstrated the greatest inflammatory response with increases in polymorphonuclear leukocytes, alveolar macrophages, and CD8 and CD4 T cells and greatest increases in cytokine and chemokine mRNA. In contrast, the highly resistant NZW mice showed no change in Lm and lung P–V curve. NZW mice did show a significant increase in alveolar macrophages but no increase in other inflammatory cells. Moreover, chemokine and cytokine mRNA in the lung actually decreased. The intermediately sensitive C57BL6/j strain showed a small increase in Lm but no change in P–V curve, an actual decrease in CD4 and CD8 T cells. In C57BL/6j, there was no change in cytokine or chemokine mRNA except for an isolated increase in IP-10, ligand for the CXCR3 chemokine receptor expressed on CD4 and CD8 T cells. Strain differences appear to depend on the levels of key antiproteases like α-1-antitrypsin and genetic variations in the expression of antioxidant gene and proinflammatory gene programs.

Of interest, strains susceptible to the effects of CS such as C57BL/6j and DBA/2 manifest a very general, genome-wide change in expression involving hundreds of genes as assessed by Affymetrix Murine Genome Gene Chip array (Cavarra et al., 2009). In contrast, a resistant strain, ICR, demonstrates a different and more limited pattern of response. In general, alterations in the expression of a variety of proemphysematous (e.g., serine proteases and TIMPs) and cell-adhesion genes (e.g., cadherin genes) occurred in C57BL/6j and DBA/2 but not in the ICVR strain.

More recently, the acute inflammatory response to cigarette smoke has been found to vary across mouse strains (Yao et al., 2008). Five mouse strains, C57Bl/6j, A/J, AKR/J, CD-1, and 129SvJ, were exposed for 3 consecutive days to the smoke generated from U of Kentucky research cigarettes

(2R4F) using a Baumgartner–Jaeger smoking machine. Mainstream smoke was diluted with room air to achieve a TSP concentration of 80 or 300 mg/m^3. Mice in the 300 mg/m^3 group were exposed for two, 1 h periods daily and had a carboxyhemoglobin concentration of 17%. Mice in the 80 mg/m^3 group were exposed for 6 h daily and had a carboxyhemoglobin concentration of 11%.

At 24 h after the last exposure, neutrophils in BAL were increased in C57Bl/6j, A/J, AKR/J, and CD-1 but not 129SvJ mice, and levels of the cytokines, KC, MCP-1, TNF-α, MIP-2, IL-6, and IL-13 in the lung showed a similar pattern. The total number of inflammatory cells and the expression of most cytokines were greatest in C57Bl/6j and least in 129SvJ mice. Changes in cells and cytokines were greater with the 300 mg/m^3 than the 80 mg/m^3 indicating that the inflammatory response was concentration dependent. Changes in inflammatory cells and mediators across strains correlated with increases in the expression of total and acetylated NF-κB and its DNA binding activity. That is, levels of expression of NF-κB tended to be highest in C57BL/6j and least in 129SvJ mice. (Of note, increases in NF-κB acetylation augment its DNA binding activity.) Decreases in HDAC2 expression demonstrated a similar pattern. These results suggest that differences in the magnitude of CS-induced upregulation of NF-κB activity and downregulation of HDAC2 expression contribute to mouse strain-dependent lung inflammatory responses. Of interest, for reasons that are not clear, the susceptibility to emphysema with chronic smoke exposure appears to differ from the acute inflammatory response to CS, that is, increases in Lm in A/J and AKR are greater than in C57BL/6j.

Also of interest, mouse alveolar macrophages treated acutely with cigarette smoke extract (CSE) in vitro demonstrate strain-type differences in oxidative and inflammatory responses (Vecchio et al., 2010). Specifically, ROS and H$_2$O$_2$ levels were greater in C57BL/6j than ICR mice. Moreover, protein levels of Nrf2 and the antioxidant enzymes, HO-1 and GP, were lower both at baseline and after treatment with CSE in C57BL/6 than in ICR mice. In contrast, HDAC2 expression was lower at baseline and decreased more with CSE treatment in CSE in C57BL/6 than in ICR mice. Finally, NF-κB activity and expression of the proinflammatory cytokines, KC, TNF-α, IL-6, MIP-2, and matrix metalloproteinases, MMP-9 and MMP-2, increased to a greater extent with CSE in C57BL/6j than ICR mice. These data indicate that in isolated alveolar macrophages, strain differences in both the oxidant and inflammatory responses to cigarette smoke are likely mediated by differences in the expression or activity of the transcription factors, Nrf2 and NF-κB.

16.8 ROLE OF Nrf2 AND NF-κB IN CIGARETTE SMOKE-INDUCED LUNG

16.8.1 MASTER ANTIOXIDANT TRANSCRIPTION FACTOR, Nrf2

The regulation of antioxidant defense in the lung is largely under the control of the redox-sensitive transcription factor, Nrf2 (Wakabayashi et al., 2010). Nrf2 regulates the expression of a large number of genes involved in adaptive responses to intrinsic and extrinsic cellular stresses. Specifically, Nrf2 regulates expression of enzymes that inactivate oxidants, increase NADPH synthesis, and enhance toxin degradation and export (Wakabayashi et al., 2010). Nrf2 also enhances the repair and removal of damaged proteins and inhibits cytokine-mediated inflammation (Wakabayashi et al., 2010). Of particular interest in the setting of cigarette smoke exposure, Nrf2 binds to antioxidant response elements in the promoter region of a variety of genes coding for important antioxidant enzymes (e.g., heme oxygenase-1 [HO-1], GST, GP, and superoxide dismutase [SOD]) (Wakabayashi et al., 2010). In fact, Nrf2 regulates two major redox systems, the glutathione and thioredoxin systems, by promoting expression of enzymes involved in glutathione synthesis, transfer, and reduction and thiodoxin synthesis and reduction. In addition, Nrf2 regulates several glutathione-dependent (e.g., UDP-glucuronosyltransferase) and glutathione-independent enzymes (e.g., NAD(P)H–quinone oxidoreductase 1 [NQO1]), which are important in the detoxification of tobacco smoke products.

When present in the cytoplasm attached to its actin-tethered, redox-sensitive inhibitor, Kelch like-ECH-associated protein 1 (Keap1), Nrf2 has a short half-life as a result of its susceptibility to ubiquitination and proteasomal degradation. Oxidation of Keap1 allows Nrf2 to dissociate and migrate to the nucleus where it binds to a specific DNA consensus sequence (Goven et al., 2008; Nguyen et al., 2000) found in the antioxidant response element [5′-NTGAG/CNNNGC-3′]. Nrf2 activity is also regulated by the cytosolic protein, DJ-1, and the nuclear protein, Bach1 (Goven et al., 2008). DJ-1 enhances Nrf2 expression by preventing its degradation by the proteasome (Clements et al., 2006). The transcriptional inhibitor, Bach-1, on the other hand, inhibits Nrf2 transcriptional activity by competing with Nrf2 for available transcriptional cofactors such as Maf K in the nucleus (Niture et al., 2014).

Of importance, posttranslational modifications of Nrf2 (i.e., phosphorylation and acetylation) affect its functional activity in terms of binding to its inhibitors, half-life, nuclear import and export, DNA binding affinity, and transcriptional activation (Apopa et al., 2008; Mercado et al., 2011; Niture et al., 2014). Nrf2 is also acetylated by histone acetyltransferase (HAT) and deacetylated by HDAC2 (Malhotra et al., 2011). Acetylation of Nrf2 diminishes his transcriptional activity and enhances its export from the nucleus. Accordingly, increases in the level of acetylated Nrf2 are associated with decreases in Nrf2 activity. Cigarette smoke-induced reductions in HDAC2 expression or activity (Kode et al., 2008) thereby reduce Nrf2-regulated HO-1 expression and increase sensitivity to oxidative stress in human airway epithelial cells as well as those of mice.

Oxidant stress and ER stress are heightened in subjects with COPD and persist for prolonged periods even after subjects have stopped smoking (Kinnula et al., 2007; Malhotra et al., 2009). In part, oxidant stress and ER stress is heightened because Nrf2 expression is reduced in subjects with COPD (Malhotra et al., 2008, 2009). Reductions in Nrf2 in lung tissue and in alveolar macrophages appear to explain reductions in both glutathione-dependent and glutathione-independent antioxidant defense, in particular HO-1, which is transcriptionally regulated by Nrf2 and ATF4 (He et al., 2001). ER stress appears to be enhanced in subjects with COPD because of increased oxidant burden and reductions in the rate of degradation of aberrant or misfolded proteins. Nrf2 upregulates the expression of the components of the 26 S proteasome (Malhotra et al., 2009). Accordingly, decreased Nrf2 expression decreases proteasomal activity and slows protein degradation.

Direct evidence of the importance of Nrf2 in the pathogenesis of CS-induced emphysema and lung inflammation has been confirmed in animal models and in cultured lung cells and human subjects. For example, Nrf2 knockout mice are more susceptible to CS-induced emphysema and inflammation while transcriptional induction of Nrf2 by CDDO reduces oxidative stress and alveolar destruction in wild-type mouse but not in Nrf2 knockout mice (Rangasamy et al., 2004). Mice deficient in Nrf2 demonstrate increased numbers of macrophages in BAL and lung tissue following cigarette smoke exposure. Moreover, type II pneumocytes from Nrf2 knockout mice demonstrate impaired growth and increased sensitivity to oxidant-induced cell death (Reddy et al., 2007). In addition, deletion of KEAP1 in Clara cells in the airways of mice attenuates CS-induced inflammation and oxidative stress (Blake et al., 2010). On the other hand, knockdown of DJ-1 in mouse lungs, mouse embryonic fibroblasts, and human airway epithelial cells impairs antioxidant induction in response to CS (Malhotra, 2008). Of considerable interest, recent studies (Malhotra et al., 2008) indicate that expressions of Nrf2 and several Nrf2-regulated antioxidant enzymes, for example, NQO1, HO-1, and glutamate cysteine ligase modifier subunit, are reduced in subjects with advanced COPD. CS-induced lung inflammation involves the reduction of HDAC2 abundance, which is associated with steroid resistance in patients with COPD and in individuals with severe asthma who smoke cigarettes (Adenuga et al., 2009). Nrf2 expression correlates with HDAC2 expression in monocyte-derived macrophages from healthy volunteers (nonsmokers and smokers) and subjects with COPD (Mercado et al., 2011). Reduced HDAC2 activity in COPD may explain increases in Nrf2 acetylation, reduced Nrf2 stability, and impaired antioxidant defense (Mercado et al., 2011).

The potential importance of Nrf2 in cigarette smoke-induced lung inflammation and tissue injury has recently prompted trials of substances that increase antioxidant gene expression in subjects with COPD (Kosmider et al., 2011). For example, sulforaphane, a substance contained in broccoli sprouts, and resveratrol, a polyphenolic phytoalexin in grapes, enhance Nrf2 expression (Kode et al., 2008).

16.8.2 Master Proinflammatory Transcription Factor, NF-κB

NF-κB is a redox-sensitive transcriptional factor and an important regulator of the inflammatory and cell stress responses in the lung (Rahman and Fazal, 2011; Wullaert et al., 2011). Specifically, NF-κB regulates expression of a variety of cytokines, chemokines, immunoreceptors, cell-adhesion molecules, stress-response genes, regulators of apoptosis, growth factors, and transcription factors. NF-κB is a family of homo- or heterodimers, which contain a conserved Rel homology domain responsible for dimerization and binding to the consensus sequence [5'-GGGRNNYYCC-3'] (Rahman and Fazal, 2011; Wullaert et al., 2011). The NF-κB family of proteins can be divided into two distinct families based on the presence of a transactivation domain. RelA (p65), RelB, and c-Rel all contain transactivation domains, while p50 and p52 do not and require heterodimerization with the Rel proteins for this function. In the absence of stimulation, NF-κB is inhibited in the cytosol by association with Ikb (Thompson et al., 1995; Whiteside et al., 1997; Zabel and Baeuerle, 1990). In response to appropriate stimuli, IKb is phosphorylated by IKB kinases (IKKs) at two separate serine residues, which leads to its ubiquitination and subsequent proteasomal degradation. Release of NF-κB from IKb allows its translocation to the nucleus and subsequent binding to the promoter region of over 100 target genes (Pahl, 1999). In particular, NF-κB regulates the expression of over 30 cytokines and chemokines, immune recognition receptors, and cell-adhesion molecules required for neutrophil migration including TNF-α, inducible NOS (iNOS), interleukin-1 (IL-1), intracellular adhesion molecule-1 (ICAM-1), and cyclooxygenase (COX-2) (Pan et al., 2000). A wide range of agents involved in oxidant stress, immune system activation, and bacterial infection stimulate IKK to activate NF-κB including H_2O_2, TNF-α, IL-1, phorbol esters, or microbial infection or Pathogen Associated Molecular Patterns (PAMPs) (Pan et al., 2000).

The importance of NF-κB in the inflammatory response of the lung is demonstrated by the fact that NF-κB knockout mice manifest less lung inflammation and cytokine levels in the BAL compared to wild-type animals in response to inhaled toxic substances. For example, NF-κB knockout mice demonstrate less neutrophil infiltration and cytokine expression in the lung in response to LPS (Poynter et al., 2003). Moreover, ROS in cigarette smoke such as hydrogen peroxide activate NF-κB in several cell lines in vitro (Schreck et al., 1991; Vollgraf et al., 1999; True et al., 2000). In fact, H_2O_2 treatment leads to phosphorylation and activation of IKK. Oxidants may also directly phosphorylate the p65 subunit of NF-κB.

Of interest, NF-κB is also regulated by Nrf2. For example, NF-κB activity is increased in Nrf2 knockout mice after treatment with TNF-α, LPS, and respiratory syncytial virus (Thimmulappa et al., 2006; Cho et al., 2009). In fact, Nrf2 attenuates IkBβ phosphorylation and IKK activity in response to TNF-α or LPS (Thimmulappa et al., 2006). In addition, Nrf2 appears to regulate expression of at least subsets of the NF-κB family directly. For example, p50 and p65 are reduced in Nrf2$^{-/-}$ fibroblasts while c-Rel is increased in Nrf2$^{-/-}$ fibroblasts (Yang et al., 2005). Greater expression of NF-κB and its targets in Nrf2-deficient animals may be a result of diminished ability to scavenge ROS and, hence, to greater oxidant stress or to more direct interactions between the two transcription factors. Of note, since a variety of stimuli such as ROS and LPS induce both Nrf2 and NF-κB activity, an entirely antagonistic relationship between the two transcription factors under all circumstances is unlikely (Ahn and Aggarwal, 2005; Anwar et al., 2005; Rushworth et al., 2005; Carayol et al., 2006).

16.9 CONCLUSIONS/FUTURE DIRECTIONS

A great deal of information vital to our understanding of the pathogenesis of COPD has been generated using animal models, chiefly rodents, which have been exposed acutely (days) and chronically (months) to cigarette smoke. In particular, the use of the chronic cigarette smoke-exposed mouse, in which gene expression has been altered experimentally or naturally (e.g., the pallid mouse deficient in α-1-antitrypsin), has greatly expanded our understanding of the mechanisms involved in emphysema and the airway wall remodeling. In particular, these studies have substantiated the important role played by imbalance in the functional activity of the protease/antiprotease and the oxidant/antioxidant defense systems. In particular, they have demonstrated the critical importance of the Nrf2 transcription factor, which regulates expression of key antioxidant enzymes and protein metabolism and the NF-κB transcription factor, which regulates the intensity of the innate and adaptive immune response in the lung.

Nonetheless, cigarette exposure models used presently are not ideal since they require 4–6 months of exposure and hence represent a considerable investment in time and money. Novel approaches that generate the structural changes of COPD in less than the 4–6 months presently required are needed. In addition, description of the key variables involved in smoke exposure is needed such that studies can be compared more easily. In particular, definition of the concentration of total particulates in the smoke, the concentration of nicotine in the cigarette used, and the carbon monoxide levels achieved in the animals is required.

Also desirable is the development of a model of acute exacerbations of COPD. At present, no well-accepted model of COPD exacerbations exists. Most COPD exacerbations occur as a result of tracheobronchitis in the setting of acute viral or bacterial infections. Accordingly, recent studies have utilized *a double-hit* approach in which an infectious agent or PAMP is administered in the setting of chronic cigarette smoke exposure. Whether any of these models will achieve a wide degree of acceptance remains to be seen. In this regard, identification of species that have both a susceptibility to the adverse effects of cigarette smoke on the lung and susceptibility to human respiratory tract viruses would be quite useful.

QUESTIONS

1. The ideal animal model of COPD will demonstrate the processes of emphysema, airway remodeling, and pulmonary hypertension acute exacerbations. True or False?
   Answer: True
2. Chronic exposure of rodents to cigarette smoke is the most common method of inducing COPD. True or False?
   Answer: True
3. Studies in the cigarette smoke-exposed mouse model have helped to elucidate basic mechanisms involved in COPD applicable to the human disease such as the role played by the transcription factors, NRF2 and NF-κB. True or False?
   Answer: True
4. Acute exacerbations of COPD can be modeled in the mouse. True or False?
   Answer: False

REFERENCES

Adenuga, D., Yao, H., March, T. H., Seagrave, J., and Rahman, I. (2009) Histone deacetylase 2 is phosphorylated, ubiquitinated, and degraded by cigarette smoke. *Am J Respir Cell Mol Biol*, 40, 464–473.

Ahn, K. S. and Aggarwal, B. B. (2005) Transcription factor NF-kappaB: A sensor for smoke and stress signals. *Ann N Y Acad Sci*, 1056, 218–233.

Anwar, A. A., Li, F. Y., Leake, D. S., Ishii, T., Mann, G. E., and Siow, R. C. (2005) Induction of heme oxygenase 1 by moderately oxidized low-density lipoproteins in human vascular smooth muscle cells: Role of mitogen-activated protein kinases and Nrf2. *Free Radic Biol Med*, 39, 227–236.

Apopa, P. L., He, X., and Ma, Q. (2008) Phosphorylation of Nrf2 in the transcription activation domain by casein kinase 2 (CK2) is critical for the nuclear translocation and transcription activation function of Nrf2 in IMR-32 neuroblastoma cells. *J Biochem Mol Toxicol*, 22, 63–76.

Asano, H., Horinouchi, T., Mai, Y., Sawada, O., Fujii, S., Nishiya, T., Minami, M. et al. (2012) Nicotine- and tar-free cigarette smoke induces cell damage through reactive oxygen species newly generated by PKC-dependent activation of NADPH oxidase. *J Pharmacol Sci*, 118, 275–287.

Atkinson, J. J., Lutey, B. A., Suzuki, Y., Toennies, H. M., Kelley, D. G., Kobayashi, D. K., Ijem, W. G. et al. (2011) The role of matrix metalloproteinase-9 in cigarette smoke-induced emphysema. *Am J Respir Crit Care Med*, 183, 876–884.

Barnes, P. J., Shapiro, S. D., and Pauwels, R. A. (2003) Chronic obstructive pulmonary disease: Molecular and cellular mechanisms. *Eur Respir J*, 22, 672–688.

Beckett, E. L., Stevens, R. L., Jarnicki, A. G., Kim, R. Y., Hanish, I., Hansbro, N. G., Deane, A. et al. (2013) A new short-term mouse model of chronic obstructive pulmonary disease identifies a role for mast cell tryptase in pathogenesis. *J Allergy Clin Immunol*, 131, 752–762.

Blake, D. J., Singh, A., Kombairaju, P., Malhotra, D., Mariani, T. J., Tuder, R. M., Gabrielson, E., and Biswal, S. (2010) Deletion of Keap1 in the lung attenuates acute cigarette smoke-induced oxidative stress and inflammation. *Am J Respir Cell Mol Biol*, 42, 524–536.

Burrows, B., Knudson, R. J., Cline, M. G., and Lebowitz, M. D. (1977) Quantitative relationships between cigarette smoking and ventilatory function. *Am Rev Respir Dis*, 115, 195–205.

Carayol, N., Chen, J., Yang, F., Jin, T., Jin, L., States, D., and Wang, C. Y. (2006) A dominant function of IKK/NF-kappaB signaling in global lipopolysaccharide-induced gene expression. *J Biol Chem*, 281, 31142–31151.

Cavarra, E., Bartalesi, B., Lucattelli, M., Fineschi, S., Lunghi, B., Gambelli, F., Ortiz, L. A., Martorana, P. A., and Lungarella, G. (2001) Effects of cigarette smoke in mice with different levels of alpha(1)-proteinase inhibitor and sensitivity to oxidants. *Am J Respir Crit Care Med*, 164, 886–890.

Cavarra, E., Fardin, P., Fineschi, S., Ricciardi, A., De Cunto, G., Sallustio, F., Zorzetto, M. et al. (2009) Early response of gene clusters is associated with mouse lung resistance or sensitivity to cigarette smoke. *Am J Physiol Lung Cell Mol Physiol*, 296, L418–L429.

Cheng, G., Cao, Z., Xu, X., van Meir, E. G., and Lambeth, J. D. (2001) Homologs of gp91phox: Cloning and tissue expression of Nox3, Nox4, and Nox5. *Gene*, 269, 131–140.

Cho, H. Y., Imani, F., Miller-Degraff, L., Walters, D., Melendi, G. A., Yamamoto, M., Polack, F. P., and Kleeberger, S. R. (2009) Antiviral activity of Nrf2 in a murine model of respiratory syncytial virus disease. *Am J Respir Crit Care Med*, 179, 138–150.

Church, D. F. and Pryor, W. A. (1985) Free-radical chemistry of cigarette smoke and its toxicological implications. *Environ Health Perspect*, 64, 111–126.

Churg, A., Cosio, M., and Wright, J. L. (2008) Mechanisms of cigarette smoke-induced COPD: Insights from animal models. *Am J Physiol Lung Cell Mol Physiol*, 294, L612–L631.

Churg, A., Tai, H., Coulthard, T., Wang, R., and Wright, J. L. (2006) Cigarette smoke drives small airway remodeling by induction of growth factors in the airway wall. *Am J Respir Crit Care Med*, 174, 1327–1334.

Churg, A., Wang, R., Wang, X., Onnervik, P. O., Thim, K., and Wright, J. L. (2007) Effect of an MMP-9/MMP-12 inhibitor on smoke-induced emphysema and airway remodelling in guinea pigs. *Thorax*, 62, 706–713.

Clements, C. M., Mcnally, R. S., Conti, B. J., Mak, T. W., and Ting, J. P. (2006) DJ-1, a cancer- and Parkinson's disease-associated protein, stabilizes the antioxidant transcriptional master regulator Nrf2. *Proc Natl Acad Sci U S A*, 103, 15091–15096.

Cosgrove, J. P., Borish, E. T., Church, D. F., and Pryor, W. A. (1985) The metal-mediated formation of hydroxyl radical by aqueous extracts of cigarette tar. *Biochem Biophys Res Commun*, 132, 390–396.

Di Stefano, A., Capelli, A., Lusuardi, M., Balbo, P., Vecchio, C., Maestrelli, P., Mapp, C. E., Fabbri, L. M., Donner, C. F., and Saetta, M. (1998) Severity of airflow limitation is associated with severity of airway inflammation in smokers. *Am J Respir Crit Care Med*, 158, 1277–1285.

Di Stefano, A., Turato, G., Maestrelli, P., Mapp, C. E., Ruggieri, M. P., Roggeri, A., Boschetto, P., Fabbri, L. M., and Saetta, M. (1996) Airflow limitation in chronic bronchitis is associated with T-lymphocyte and macrophage infiltration of the bronchial mucosa. *Am J Respir Crit Care Med*, 153, 629–632.

Elias, J. A., Kang, M. J., Crothers, K., Homer, R., and Lee, C. G. (2006) State of the art. Mechanistic heterogeneity in chronic obstructive pulmonary disease: Insights from transgenic mice. *Proc Am Thorac Soc*, 3, 494–498.

Fievez, L., Kirschvink, N., Zhang, W. H., Lagente, V., Lekeux, P., Bureau, F., and Gustin, P. (2009) Effects of betamethasone on inflammation and emphysema induced by cadmium nebulisation in rats. *Eur J Pharmacol*, 606, 210–214.

Fink, K., Duval, A., Martel, A., Soucy-Faulkner, A., and Grandvaux, N. (2008) Dual role of NOX2 in respiratory syncytial virus- and sendai virus-induced activation of NF-kappaB in airway epithelial cells. *J Immunol*, 180, 6911–6922.

Finkelstein, R., Fraser, R. S., Ghezzo, H., and Cosio, M. G. (1995) Alveolar inflammation and its relation to emphysema in smokers. *Am J Respir Crit Care Med*, 152, 1666–1672.

Fletcher, C. M. (1976) *The Natural History of Chronic Bronchitis and Emphysema: An Eight-Year Study of Early Chronic Obstructive Lung Disease in Working Men in London*. New York: Oxford University Press.

Gamble, E., Grootendorst, D. C., Hattotuwa, K., O'Shaughnessy, T., Ram, F. S., Qiu, Y., Zhu, J. et al. (2007) Airway mucosal inflammation in COPD is similar in smokers and ex-smokers: A pooled analysis. *Eur Respir J*, 30, 467–471.

Goven, D., Boutten, A., Lecon-Malas, V., Marchal-Somme, J., Amara, N., Crestani, B., Fournier, M. et al. (2008) Altered Nrf2/Keap1-Bach1 equilibrium in pulmonary emphysema. *Thorax*, 63, 916–924.

Guerassimov, A., Hoshino, Y., Takubo, Y., Turcotte, A., Yamamoto, M., Ghezzo, H., Triantafillopoulos, A., Whittaker, K., Hoidal, J. R., and Cosio, M. G. (2004) The development of emphysema in cigarette smoke-exposed mice is strain dependent. *Am J Respir Crit Care Med*, 170, 974–980.

Hackett, N. R., Heguy, A., Harvey, B. G., O'Connor, T. P., Luettich, K., Flieder, D. B., Kaplan, R., and Crystal, R. G. (2003) Variability of antioxidant-related gene expression in the airway epithelium of cigarette smokers. *Am J Respir Cell Mol Biol*, 29, 331–343.

Han, M. K., Agusti, A., Calverley, P. M., Celli, B. R., Criner, G., Curtis, J. L., Fabbri, L. M. et al. (2010) Chronic obstructive pulmonary disease phenotypes: The future of COPD. *Am J Respir Crit Care Med*, 182, 598–604.

Hantos, Z., Adamicza, A., Janosi, T. Z., Szabari, M. V., Tolnai, J., and Suki, B. (2008) Lung volumes and respiratory mechanics in elastase-induced emphysema in mice. *J Appl Physiol*, 105, 1864–1872.

Hautamaki, R. D., Kobayashi, D. K., Senior, R. M., and Shapiro, S. D. (1997) Requirement for macrophage elastase for cigarette smoke-induced emphysema in mice. *Science*, 277, 2002–2004.

Hawkins, B. J., Madesh, M., Kirkpatrick, C. J., and Fisher, A. B. (2007) Superoxide flux in endothelial cells via the chloride channel-3 mediates intracellular signaling. *Mol Biol Cell*, 18, 2002–2012.

He, C. H., Gong, P., Hu, B., Stewart, D., Choi, M. E., Choi, A. M., and Alam, J. (2001) Identification of activating transcription factor 4 (ATF4) as an Nrf2-interacting protein. Implication for heme oxygenase-1 gene regulation. *J Biol Chem*, 276, 20858–20865.

Henson, P. M., Vandivier, R. W., and Douglas, I. S. (2006) Cell death, remodeling, and repair in chronic obstructive pulmonary disease? *Proc Am Thorac Soc*, 3, 713–717.

Hogg, J. C., Chu, F., Utokaparch, S., Woods, R., Elliott, W. M., Buzatu, L., Cherniack, R. M. et al. (2004) The nature of small-airway obstruction in chronic obstructive pulmonary disease. *N Engl J Med*, 350, 2645–2653.

Hogg, J. C., Chu, F. S., Tan, W. C., Sin, D. D., Patel, S. A., Pare, P. D., Martinez, F. J. et al. (2007) Survival after lung volume reduction in chronic obstructive pulmonary disease: Insights from small airway pathology. *Am J Respir Crit Care Med*, 176, 454–459.

Huang, M. F., Lin, W. L., and Ma, Y. C. (2005) A study of reactive oxygen species in mainstream of cigarette. *Indoor Air*, 15, 135–140.

Ito, S., Bartolak-Suki, E., Shipley, J. M., Parameswaran, H., Majumdar, A., and Suki, B. (2006) Early emphysema in the tight skin and pallid mice: Roles of microfibril-associated glycoproteins, collagen, and mechanical forces. *Am J Respir Cell Mol Biol*, 34, 688–694.

Jeffery, P. K. (1983) Morphologic features of airway surface epithelial cells and glands. *Am Rev Respir Dis*, 128, S14–S20.

Kang, M. J., Choi, J. M., Kim, B. H., Lee, C. M., Cho, W. K., Choe, G., Kim, D. H., Lee, C. G., and Elias, J. A. (2012) IL-18 induces emphysema and airway and vascular remodeling via IFN-gamma, IL-17A, and IL-13. *Am J Respir Crit Care Med*, 185, 1205–1217.

Kang, M. J., Lee, C. G., Lee, J. Y., Dela Cruz, C. S., Chen, Z. J., Enelow, R., and Elias, J. A. (2008) Cigarette smoke selectively enhances viral PAMP- and virus-induced pulmonary innate immune and remodeling responses in mice. *J Clin Invest*, 118, 2771–2784.

Kawakami, M., Paul, J. L., and Thurlbeck, W. M. (1984) The effect of age on lung structure in male BALB/cNNia inbred mice. *Am J Anat*, 170, 1–21.

Kinnula, V. L., Ilumets, H., Myllarniemi, M., Sovijarvi, A., and Rytila, P. (2007) 8-Isoprostane as a marker of oxidative stress in nonsymptomatic cigarette smokers and COPD. *Eur Respir J*, 29, 51–55.

Kode, A., Rajendrasozhan, S., Caito, S., Yang, S. R., Megson, I. L., and Rahman, I. (2008) Resveratrol induces glutathione synthesis by activation of Nrf2 and protects against cigarette smoke-mediated oxidative stress in human lung epithelial cells. *Am J Physiol Lung Cell Mol Physiol*, 294, L478–L488.

Kosmider, B., Messier, E. M., Chu, H. W., and Mason, R. J. (2011) Human alveolar epithelial cell injury induced by cigarette smoke. *PLoS One*, 6, e26059.

Lambeth, J. D. (2004) NOX enzymes and the biology of reactive oxygen. *Nat Rev Immunol*, 4, 181–189.

Macnee, W. (2005) Pathogenesis of chronic obstructive pulmonary disease. *Proc Am Thorac Soc*, 2, 258–266; discussion 290–291.

Malhotra, D., Thimmulappa, R., Navas-Acien, A., Sandford, A., Elliott, M., Singh, A., Chen, L. et al. (2008) Decline in NRF2-regulated antioxidants in chronic obstructive pulmonary disease lungs due to loss of its positive regulator, DJ-1. *Am J Respir Crit Care Med*, 178, 592–604.

Malhotra, D., Thimmulappa, R., Vij, N., Navas-Acien, A., Sussan, T., Merali, S., Zhang, L. et al. (2009) Heightened endoplasmic reticulum stress in the lungs of patients with chronic obstructive pulmonary disease: The role of Nrf2-regulated proteasomal activity. *Am J Respir Crit Care Med*, 180, 1196–1207.

Malhotra, D., Thimmulappa, R. K., Mercado, N., Ito, K., Kombairaju, P., Kumar, S., Ma, J. et al. (2011) Denitrosylation of HDAC2 by targeting Nrf2 restores glucocorticosteroid sensitivity in macrophages from COPD patients. *J Clin Invest*, 121, 4289–4302.

McDonough, J. E., Yuan, R., Suzuki, M., Seyednejad, N., Elliott, W. M., Sanchez, P. G., Wright, A. C. et al. (2011) Small-airway obstruction and emphysema in chronic obstructive pulmonary disease. *N Engl J Med*, 365, 1567–1575.

Mercado, N., Thimmulappa, R., Thomas, C. M., Fenwick, P. S., Chana, K. K., Donnelly, L. E., Biswal, S., Ito, K., and Barnes, P. J. (2011) Decreased histone deacetylase 2 impairs Nrf2 activation by oxidative stress. *Biochem Biophys Res Commun*, 406, 292–298.

Nadziejko, C., Fang, K., Bravo, A., and Gordon, T. (2007) Susceptibility to pulmonary hypertension in inbred strains of mice exposed to cigarette smoke. *J Appl Physiol*, 102, 1780–1785.

Nagai, K., Betsuyaku, T., Suzuki, M., Nasuhara, Y., Kaga, K., Kondo, S., and Nishimura, M. (2008) Dual oxidase 1 and 2 expression in airway epithelium of smokers and patients with mild/moderate chronic obstructive pulmonary disease. *Antioxid Redox Signal*, 10, 705–714.

Nguyen, T., Huang, H. C., and Pickett, C. B. (2000) Transcriptional regulation of the antioxidant response element. Activation by Nrf2 and repression by MafK. *J Biol Chem*, 275, 15466–15473.

Niture, S. K., Khatri, R., and Jaiswal, A. K. (2014) Regulation of Nrf2-an update. *Free Radic Biol Med*, 66, 36–44.

Nussbaumer-Ochsner, Y. and Rabe, K. F. (2011) Systemic manifestations of COPD. *Chest*, 139, 165–173.

Pahl, H. L. (1999) Activators and target genes of Rel/NF-kappaB transcription factors. *Oncogene*, 18, 6853–6866.

Pan, M. H., Lin-Shiau, S. Y., and Lin, J. K. (2000) Comparative studies on the suppression of nitric oxide synthase by curcumin and its hydrogenated metabolites through down-regulation of IkappaB kinase and NFkappaB activation in macrophages. *Biochem Pharmacol*, 60, 1665–1676.

Panina-Bordignon, P., Papi, A., Mariani, M., Di Lucia, P., Casoni, G., Bellettato, C., Buonsanti, C. et al. (2001) The C-C chemokine receptors CCR4 and CCR8 identify airway T cells of allergen-challenged atopic asthmatics. *J Clin Invest*, 107, 1357–1364.

Poynter, M. E., Irvin, C. G., and Janssen-Heininger, Y. M. (2003) A prominent role for airway epithelial NF-kappa B activation in lipopolysaccharide-induced airway inflammation. *J Immunol*, 170, 6257–6265.

Pryor, W. A. and Stone, K. (1993) Oxidants in cigarette smoke. Radicals, hydrogen peroxide, peroxynitrate, and peroxynitrite. *Ann N Y Acad Sci*, 686, 12–27; discussion 27–28.

Rahman, A. and Fazal, F. (2011) Blocking NF-kappaB: An inflammatory issue. *Proc Am Thorac Soc*, 8, 497–503.

Rajendrasozhan, S., Chung, S., Sundar, I. K., Yao, H., and Rahman, I. (2010) Targeted disruption of NF-{kappa}B1 (p50) augments cigarette smoke-induced lung inflammation and emphysema in mice: A critical role of p50 in chromatin remodeling. *Am J Physiol Lung Cell Mol Physiol*, 298, L197–L209.

Rangasamy, T., Cho, C. Y., Thimmulappa, R. K., Zhen, L., Srisuma, S. S., Kensler, T. W., Yamamoto, M., Petrache, I., Tuder, R. M., and Biswal, S. (2004) Genetic ablation of Nrf2 enhances susceptibility to cigarette smoke-induced emphysema in mice. *J Clin Invest*, 114, 1248–1259.

Ranjan, P., Anathy, V., Burch, P. M., Weirather, K., Lambeth, J. D., and Heintz, N. H. (2006) Redox-dependent expression of cyclin D1 and cell proliferation by Nox1 in mouse lung epithelial cells. *Antioxid Redox Signal*, 8, 1447–1459.

Reddy, N. M., Kleeberger, S. R., Cho, H. Y., Yamamoto, M., Kensler, T. W., Biswal, S., and Reddy, S. P. (2007) Deficiency in Nrf2-GSH signaling impairs type II cell growth and enhances sensitivity to oxidants. *Am J Respir Cell Mol Biol*, 37, 3–8.

Rennard, S. I. and Vestbo, J. (2006) COPD: The dangerous underestimate of 15%. *Lancet*, 367, 1216–1219.

Rushworth, S. A., Chen, X. L., Mackman, N., Ogborne, R. M., and O'Connell, M. A. (2005) Lipopolysaccharide-induced heme oxygenase-1 expression in human monocytic cells is mediated via Nrf2 and protein kinase C. *J Immunol*, 175, 4408–4415.

Saetta, M., Baraldo, S., Corbino, L., Turato, G., Braccioni, F., Rea, F., Cavallesco, G. et al. (1999) CD8+ve cells in the lungs of smokers with chronic obstructive pulmonary disease. *Am J Respir Crit Care Med*, 160, 711–717.

Saetta, M., Di Stefano, A., Turato, G., Facchini, F. M., Corbino, L., Mapp, C. E., Maestrelli, P., Ciaccia, A., and Fabbri, L. M. (1998) CD8+ T-lymphocytes in peripheral airways of smokers with chronic obstructive pulmonary disease. *Am J Respir Crit Care Med*, 157, 822–826.

Saetta, M., Mariani, M., Panina-Bordignon, P., Turato, G., Buonsanti, C., Baraldo, S., Bellettato, C. M. et al. (2002) Increased expression of the chemokine receptor CXCR3 and its ligand CXCL10 in peripheral airways of smokers with chronic obstructive pulmonary disease. *Am J Respir Crit Care Med*, 165, 1404–1409.

Saetta, M., Turato, G., Facchini, F. M., Corbino, L., Lucchini, R. E., Casoni, G., Maestrelli, P., Mapp, C. E., Ciaccia, A., and Fabbri, L. M. (1997) Inflammatory cells in the bronchial glands of smokers with chronic bronchitis. *Am J Respir Crit Care Med*, 156, 1633–1639.

Schreck, R., Rieber, P., and Baeuerle, P. A. (1991) Reactive oxygen intermediates as apparently widely used messengers in the activation of the NF-kappa B transcription factor and HIV-1. *EMBO J*, 10, 2247–2258.

Shapiro, S. D. (2000) Animal models for COPD. *Chest*, 117, 223S–227S.

Shapiro, S. D. (2007) Transgenic and gene-targeted mice as models for chronic obstructive pulmonary disease. *Eur Respir J*, 29, 375–378.

Singh, A., Ling, G., Suhasini, A. N., Zhang, P., Yamamoto, M., Navas-Acien, A., Cosgrove, G. et al. (2009) Nrf2-dependent sulfiredoxin-1 expression protects against cigarette smoke-induced oxidative stress in lungs. *Free Radic Biol Med*, 46, 376–386.

Singh, A., Rangasamy, T., Thimmulappa, R. K., Lee, H., Osburn, W. O., Brigelius-Flohe, R., Kensler, T. W., Yamamoto, M., and Biswal, S. (2006) Glutathione peroxidase 2, the major cigarette smoke-inducible isoform of GPX in lungs, is regulated by Nrf2. *Am J Respir Cell Mol Biol*, 35, 639–650.

Snider, G. L., Lucey, E. C., and Stone, P. J. (1986) Animal models of emphysema. *Am Rev Respir Dis*, 133, 149–169.

Sundar, I. K., Chung, S., Hwang, J. W., Lapek, J. D., Jr., Bulger, M., Friedman, A. E., Yao, H., Davie, J. R., and Rahman, I. (2012) Mitogen- and stress-activated kinase 1 (MSK1) regulates cigarette smoke-induced histone modifications on NF-kappaB-dependent genes. *PLoS One*, 7, e31378.

Takahashi, S., Nakamura, H., Seki, M., Shiraishi, Y., Yamamoto, M., Furuuchi, M., Nakajima, T. et al. (2008) Reversal of elastase-induced pulmonary emphysema and promotion of alveolar epithelial cell proliferation by simvastatin in mice. *Am J Physiol Lung Cell Mol Physiol*, 294, L882–L890.

Takubo, Y., Guerassimov, A., Ghezzo, H., Triantafillopoulos, A., Bates, J. H., Hoidal, J. R., and Cosio, M. G. (2002) Alpha1-antitrypsin determines the pattern of emphysema and function in tobacco smoke-exposed mice: Parallels with human disease. *Am J Respir Crit Care Med*, 166, 1596–1603.

Taraseviciene-Stewart, L., Burns, N., Kraskauskas, D., Nicolls, M. R., Tuder, R. M., and Voelkel, N. F. (2006) Mechanisms of autoimmune emphysema. *Proc Am Thorac Soc*, 3, 486–487.

Taraseviciene-Stewart, L. and Voelkel, N. F. (2008) Molecular pathogenesis of emphysema. *J Clin Invest*, 118, 394–402.

Thimmulappa, R. K., Lee, H., Rangasamy, T., Reddy, S. P., Yamamoto, M., Kensler, T. W., and Biswal, S. (2006) Nrf2 is a critical regulator of the innate immune response and survival during experimental sepsis. *J Clin Invest*, 116, 984–995.

Thompson, J. E., Phillips, R. J., Erdjument-Bromage, H., Tempst, P., and Ghosh, S. (1995) I kappa B-beta regulates the persistent response in a biphasic activation of NF-kappa B. *Cell*, 80, 573–582.

True, A. L., Rahman, A., and Malik, A. B. (2000) Activation of NF-kappaB induced by H(2)O(2) and TNF-alpha and its effects on ICAM-1 expression in endothelial cells. *Am J Physiol Lung Cell Mol Physiol*, 279, L302–L311.

Turato, G., Zuin, R., Miniati, M., Baraldo, S., Rea, F., Beghe, B., Monti, S. et al. (2002) Airway inflammation in severe chronic obstructive pulmonary disease: Relationship with lung function and radiologic emphysema. *Am J Respir Crit Care Med*, 166, 105–110.

van der Strate, B. W., Postma, D. S., Brandsma, C. A., Melgert, B. N., Luinge, M. A., Geerlings, M., Hylkema, M. N., van den Berg, A., Timens, W., and Kerstjens, H. A. (2006) Cigarette smoke-induced emphysema: A role for the B cell? *Am J Respir Crit Care Med*, 173, 751–758.

Vecchio, D., Arezzini, B., Pecorelli, A., Valacchi, G., Martorana, P. A., and Gardi, C. (2010) Reactivity of mouse alveolar macrophages to cigarette smoke is strain dependent. *Am J Physiol Lung Cell Mol Physiol*, 298, L704–L713.

Vollgraf, U., Wegner, M., and Richter-Landsberg, C. (1999) Activation of AP-1 and nuclear factor-kappaB transcription factors is involved in hydrogen peroxide-induced apoptotic cell death of oligodendrocytes. *J Neurochem*, 73, 2501–2509.

Wakabayashi, N., Slocum, S. L., Skoko, J. J., Shin, S., and Kensler, T. W. (2010) When NRF2 talks, who's listening? *Antioxid Redox Signal*, 13, 1649–1663.

Whiteside, S. T., Epinat, J. C., Rice, N. R., and Israel, A. (1997) I kappa B epsilon, a novel member of the I kappa B family, controls RelA and cRel NF-kappa B activity. *EMBO J*, 16, 1413–1426.

Willemse, B. W., ten Hacken, N. H., Rutgers, B., Lesman-Leegte, I. G., Timens, W., and Postma, D. S. (2004) Smoking cessation improves both direct and indirect airway hyperresponsiveness in COPD. *Eur Respir J*, 24, 391–396.

Wright, J. L. and Churg, A. (1991) Effect of long-term cigarette smoke exposure on pulmonary vascular structure and function in the guinea pig. *Exp Lung Res*, 17, 997–1009.

Wright, J. L. and Churg, A. (2008) Animal models of COPD: Barriers, successes, and challenges. *Pulm Pharmacol Ther*, 21, 696–698.

Wright, J. L., Cosio, M., and Churg, A. (2008) Animal models of chronic obstructive pulmonary disease. *Am J Physiol Lung Cell Mol Physiol*, 295, L1–L15.

Wright, J. L., Tai, H., and Churg, A. (2006) Vasoactive mediators and pulmonary hypertension after cigarette smoke exposure in the guinea pig. *J Appl Physiol*, 100, 672–678.

Wullaert, A., Bonnet, M. C., and Pasparakis, M. (2011) NF-kappaB in the regulation of epithelial homeostasis and inflammation. *Cell Res*, 21, 146–158.

Yang, H., Magilnick, N., Lee, C., Kalmaz, D., Ou, X., Chan, J. Y., and Lu, S. C. (2005) Nrf1 and Nrf2 regulate rat glutamate-cysteine ligase catalytic subunit transcription indirectly via NF-kappaB and AP-1. *Mol Cell Biol*, 25, 5933–5946.

Yao, H., Arunachalam, G., Hwang, J. W., Chung, S., Sundar, I. K., Kinnula, V. L., Crapo, J. D., and Rahman, I. (2010) Extracellular superoxide dismutase protects against pulmonary emphysema by attenuating oxidative fragmentation of ECM. *Proc Natl Acad Sci U S A*, 107, 15571–15576.

Yao, H., Chung, S., Hwang, J. W., Rajendrasozhan, S., Sundar, I. K., Dean, D. A., McBurney, M. W. et al. (2012) SIRT1 protects against emphysema via FOXO3-mediated reduction of premature senescence in mice. *J Clin Invest*, 122, 2032–2045.

Yao, H., Edirisinghe, I., Rajendrasozhan, S., Yang, S. R., Caito, S., Adenuga, D., and Rahman, I. (2008) Cigarette smoke-mediated inflammatory and oxidative responses are strain-dependent in mice. *Am J Physiol Lung Cell Mol Physiol*, 294, L1174–L1186.

Yoshida, T. and Tuder, R. M. (2007) Pathobiology of cigarette smoke-induced chronic obstructive pulmonary disease. *Physiol Rev*, 87, 1047–1082.

Zabel, U. and Baeuerle, P. A. (1990) Purified human I kappa B can rapidly dissociate the complex of the NF-kappa B transcription factor with its cognate DNA. *Cell*, 61, 255–265.

Zang, L. Y., Stone, K., and Pryor, W. A. (1995) Detection of free radicals in aqueous extracts of cigarette tar by electron spin resonance. *Free Radic Biol Med*, 19, 161–167.

Zhang, W. J., Wei, H., Tien, Y. T., and Frei, B. (2011) Genetic ablation of phagocytic NADPH oxidase in mice limits TNFalpha-induced inflammation in the lungs but not other tissues. *Free Radic Biol Med*, 50, 1517–1525.

Zinkevich, N. S. and Gutterman, D. D. (2011) ROS-induced ROS release in vascular biology: Redox–redox signaling. *Am J Physiol Heart Circ Physiol*, 301, H647–H653.

17 Toxic Inhalation Injury
Management and Medical Treatment

Arik Eisenkraft and Avshalom Falk

CONTENTS

17.1 INTRODUCTION

The syndrome of toxic inhalation injury refers to the respiratory consequences of inhalation exposure to chemicals causing damage to lung tissue. The clinical spectrum is broad, ranging from mild irritation to temporary respiratory functional dysfunction and, in extreme cases, to acute respiratory failure, requiring intensive respiratory support and may be fatal. The underlying pathology is composed of neurogenic sensory response, cell and tissue damage, inflammation, pulmonary edema, and oxidative stress that stem from modification of cellular constituents by these highly reactive materials. In some cases, chronic respiratory disturbances may develop and persist long after the initial exposure event.

The causative chemical materials are mainly industrial materials used as precursors in the chemical and plastic industries, disinfectants, and bleaching agents. At ambient conditions, they are gases or volatile liquids and cause inhalation hazards when dispersed into the air during production, storage, or transportation accidents. Some of these materials like chlorine gas or ammonia are produced, stored, and transported in huge quantities, and their release may cause mass poisoning of workers, bystanders, and population residing in the adjoining areas affected by the toxic cloud. The most devastating toxic material dispersion was the Bhopal disaster in 1984. In that event, an explosion at the Union Carbide pesticide plant in Bhopal, India, caused the release of 30–40 tons of the toxic gas methyl isocyanate (MIC), killing 2,500–6,000, debilitating more than 200,000, and leaving a total number of surviving victims suffering mainly from ocular and respiratory symptoms (Dhara and Dhara, 2002; Mishra et al., 2009). Several massive dispersions of toxic industrial materials occurred during transportation. A notable event occurred in Graniteville, North Carolina, in 2005, in which 60 tons of chlorine were released from a ruptured derailed car. This resulted in nine deaths. Seventy-two people were hospitalized (42 for more than 3 nights, including 15 requiring ventilation or ICU support), and 525 people were examined as outpatients in hospital emergency departments (EDs) or private clinics (Wenk et al., 2007). Other events, like malfunctions in industrial facilities or swimming pool disinfection systems, end with lower numbers of victims, but the accumulated yearly numbers are high, typically between 1000 and 2000 injured and a few tens of deaths in the United States alone (ATSDR, 2006, 2009).

The availability of these materials in huge quantities, their toxic or irritant properties, and ease of dispersion made some of these materials agents of chemical warfare in World War I, most notably chlorine, phosgene, chloropicrin, and sulfur mustard among many others (Russell et al., 2006; Tourinsky and Sciuto, 2008; Jones et al., 2010). These materials may be released in chemical terror scenarios, as evidenced in the chlorine gas attacks against US and British troops positioned in Iraq (Jones et al., 2010). Inhalation toxicants—ammonia, chlorine, phosgene, MIC, phosphine, as well as other materials having respiratory toxicity—the vesicants sulfur mustard and nitrogen mustards, hydrocyanic acid, and hydrogen sulfide are listed as high-risk and top-priority chemical terror threat agents (CDC, 2012).

Another common form of toxic inhalation injury is caused by smoke inhalation in fires. This may be exceptionally severe and complex, with respiratory tract thermal and chemical injury from combustion products like hydrogen sulfide, aldehydes, and phosgene and systemic intoxication by toxic gases, mainly carbon monoxide and hydrocyanic acid, and difficult to treat (Cancio, 2005; Mlcak et al., 2007; Rehberg et al., 2009).

These topics demonstrate that release of inhalation toxicants, either by mistake, equipment malfunction, or malicious intent, may be a significant public health and security challenge. This in turn may emanate not only from death and injury but also from the social and economic disruption resulting from preventive measures like evacuation or sheltering of a large population at risk or from the load on hospitals in caring for unusual numbers of critically ill patients and severalfold more mild and *worried well* care seekers. Familiarity with the properties, toxicity mechanisms, and medical management of inhalation toxicants are crucial in the successful handling of mass or individual poisonings in the community. Lack of knowledge on the toxicology and treatment of MIC poisoning was a major factor in the devastating consequences of the Bhopal disaster (Mishra et al., 2009). These events invoked a surge in research, regulatory, and preparedness activities in the domain of major toxic events in general and inhalation toxicants in particular (Kales and Christiani, 2004). The topic of this chapter is the medical management of toxic inhalation injury. We shall review here the general pathophysiological and current medical treatments as well as more thorough discussion of some individual agents representing the different clinical variations of this injury, with emphasis on current as well as future directions in therapy.

17.2 TOXIC INHALATION INJURY: GENERAL CONSIDERATIONS

17.2.1 PROBLEM CHARACTERIZATION

The number of materials that cause toxic inhalation injury is large, and they belong to different chemical and toxicological categories. The occupational health database of the NIH, Haz-Map® (US National Library of Medicine, 2010), lists nearly 450 chemicals that cause lung damage, of these, about 220 are toxic gases and vapors (Bessac and Jordt, 2010). They are classified by their chemical nature and mechanism of toxicity into five categories: (1) anticholinergic agents; (2) simple asphyxiants; (3) chemicals that target metal ions in vital proteins like cytochrome c oxidase, hemoglobin, and other metalloproteins (cyanides, carbon monoxide, hydrogen sulfide); (4) blistering agents (arsine, sulfur, and nitrogen mustards); and (5) chemicals that react indiscriminately with biological molecules—acids, corrosive gases, oxidizers, reducers, nucleophiles, and electrophiles. A chemical may belong to more than one category; that is, cyanides and hydrogen sulfide are systemic toxicants whose major toxicity results from inhibition of cytochrome c oxidase, but they were also reported to cause lung edema in humans and animal models (Graham et al., 1997; Okolie and Osagie, 2000; Guidotti, 2010). Many of the toxic industrial chemicals—ammonia, MIC, chlorine, bromine, phosgene, and aldehydes—belong to category 5. Even though they may have different chemistries, their effects on the respiratory tract are similar with differences that may stem from their physical properties. The apparent lack of specificity in mode of action and cellular or anatomical targets prompts generic medical strategies that are defined by the type of injury or physiological effect and not the type of chemical agent.

The numbers of casualties in hazardous material incidents, and their distribution by type and severity, are important factors in planning and establishing preparedness. According to the data in the Hazardous Substances Emergency Events Surveillance (HSEES) database of the Agency for Toxic Substances and Disease Registry (ATSDR), respiratory and eye irritations are the most frequent injuries in hazardous material emergencies. The majority of patients were treated on scene or in hospitals without admission (Hick et al., 2003; ATSDR, 2006, 2009). High-impact events like the Bhopal disaster are a rarity, as are chemical terror attacks in the urban setting like the 1995 Tokyo subway sarin attack. However, if it happens, the impact on public health and social order may be so devastating when caught unprepared that investment in preparedness is prompt even if the likelihood of their occurrence may be low. In such events, the occurrence of serious health effects, fatalities, and long-term sequelae among survivors is assumed to be high, and so does the burden on society. Long-term respiratory sequelae are still evaluated and reported among survivors of the Bhopal tragedy (Mishra et al., 2009) and the Iranian victims of Iraqi chemical attacks with sulfur mustard during the Iran–Iraq War (Weinberger et al., 2011), exposure events that occurred nearly three decades ago.

A local event in the industrial setting is characterized by a defined distribution of ages, gender, and health status of the persons involved. According to a sample of HSEES reports (ATSDR, 2006, 2009), the majority of the victims in recorded hazardous material (HAZMAT) incidents were in the working age groups. The same is true for a chemical warfare attack against combat troops. In the other extreme such as a terror event or a major incident like the Bhopal tragedy, a broader and more heterogeneous population may be involved. These may include elderly people and children who are regarded more susceptible to toxic insults, with a higher frequency of severe injuries and special medical needs. The picture, however, is not so sharp. Pediatric victims were involved in chlorine gas leaks in swimming pools (Babu et al., 2008) or in chemical warfare attacks against civilian population during the Iran–Iraq War (Dompeling et al., 2004). The victims in the Tokyo underground sarin attacks in 1995 were mostly commuters, as determined by the timing and location of the attacks (Ohbu et al., 1997; Okudera, 2002). Thus, it is impossible to identify a sector of the population that is more vulnerable or immune in the broad scope of toxic inhalation injury scenarios.

Predicting the outcome of mass toxic events in terms of victim numbers and severity for planning purposes is not handy: Modeling may sometimes predict massive numbers of casualties, especially in terms of dead and severely injured, but in actual events, the figures may be much lower (Barrett and Adams, 2011). As Barrett and Adams suggest, it is more important to look at trends rather than the predicted numbers. A typical event may result in a relatively small number of deaths and severely injured victims, a larger number of victims requiring primary treatment with a short or no hospital stay, and many more may be affected by stress (Kales and Christiani, 2004; see examples in ATSDR, 2006, 2009; Wenk et al., 2007).

17.2.2 Pathogenesis of Toxic Inhalation Injury

The different materials listed as agents causing toxic inhalation injury have different biochemical activities, but the toxicological effects, clinical manifestations, and required therapies have common features. The acute manifestations include sensory irritation, airway and alveolar epithelial injury, inflammation, and edema. These pathological processes lead to pulmonary dysfunction of different degree and type, of which the most severe are acute lung injury (ALI) and acute respiratory distress syndrome (ARDS) (Ware and Matthay, 2000; Wheeler and Bernard, 2007; Tourinsky and Sciuto, 2008; Matthay et al., 2012). In many cases, long-term sequelae due to tissue remodeling may develop.

The respiratory tract can be affected by inhalation toxicants all along from the nasal cavity to the alveoli. The major location of tissue damage, and resulting clinical picture, is determined by the solubility of the toxicant in the fluid lining the walls of the airways and alveoli. Gases that are readily soluble in water, for example, ammonia, are trapped in the upper respiratory tract and cause the main damage there, while the water-insoluble gases, like phosgene, ozone, or nitrogen dioxide, are carried deeper and affect the smallest bronchioles and alveoli, causing damage to the alveolar epithelium and resulting in pulmonary edema and a high risk of ARDS. Toxicants with intermediate solubility like chlorine or sulfur mustard vapors affect different parts of the respiratory tract depending on the dose, damaging the deeper parts only at high exposure doses. The water-soluble toxicants are referred to as central pulmonary toxicants, while the less water-soluble toxicants are referred to as peripheral pulmonary toxicants (Tourinsky and Sciuto, 2008). In the case of centrally acting toxicants, the symptoms may start immediately, while in peripherally acting toxicants, the symptoms will appear with a latency of 2–24 h, depending on the intensity of exposure (Tourinsky and Sciuto, 2008).

The underlying common factors in the pathogenesis of all forms of toxic inhalation injury are inflammation and oxidative stress, two interconnected processes (Ware and Matthay, 2000; Wheeler and Bernard, 2007; Tourinsky and Sciuto, 2008). Inhalation exposure at low levels (e.g., less than 50 ppm of chlorine) causes nasal, orolaryngeal, and deeper airway irritation (Bessac and Jordt, 2008, 2010; Yadav et al., 2010). The common symptoms are painful or burning sensation in the throat and nose, cough, bronchospasm, and chest tightness with breathing difficulties. These early symptoms are invoked by activation of chemical sensor receptors in sensory nerves, termed transient response potential ankyrin 1 (TRPA1) (Bessac and Jordt, 2010). Activation of the TRPA1 receptor releases mediators that induce protective reflex actions like cough, increased mucous secretion, bronchial constriction, and inflammation. Oxidizing toxicants like chlorine react directly with water, extracellular components of the epithelial lining fluid, and cellular components, resulting in oxygen free radical formation, modification of proteins and enzymes, membrane lipid peroxidation, and depletion of pulmonary surfactants and low-molecular-weight antioxidants like reduced glutathione and ascorbic acid (Lang et al., 2002; Squadrito et al., 2010; White and Martin, 2010; Yadav et al., 2010). These primary interactions affect pulmonary functions directly and trigger a cascade of cytotoxic, pathophysiological, and mainly inflammatory processes that characterize the progressive phase of the injury.

Airway-centered injury is characterized by epithelial cell death and sloughing, peribronchial edema, and airway obstruction with mucous, fibrin, and cell debris. This is characteristic in the case of chlorine (Yadav et al., 2010) and smoke inhalation (Rehberg et al., 2009). Alveolar damage, which occurs in phosgene poisoning (Pauluhn et al., 2007), severe smoke inhalation (Rehberg et al., 2009), and high-level chlorine exposure (Batchinsky et al., 2006; Yadav et al., 2010), is characterized by alveolar cell injury, alveolar edema, and inflammation, which are all hallmarks of ALI/ARDS (Ware and Matthay, 2000; Wheeler and Bernard, 2007). Inflammatory mediators are released to the injured site in response to stimulation by TRPA1, as well as from injured endothelial cells, parenchymal cells, and activated alveolar macrophages. These factors recruit neutrophils, which infiltrate into the injured tissue from the circulation, release proteases, reactive oxygen and nitrogen species (ROS and RNS), that modify proteins and lipids, with additional exacerbation of tissue damage (Lang et al., 2002; Chow et al., 2003; Tourinsky and Sciuto, 2008; Yadav et al., 2010). In the airways, the main impairment in pulmonary function results from obstruction and bronchoconstriction (Yadav et al., 2010). In the alveoli, the inflammatory mediators and oxidative products impair endothelial and epithelial barrier functions, increase permeability to water and proteins, and impair fluid clearance, leading to edema, impaired gas exchange, and severe hypoxemia (Ware and Matthay, 2000; Wheeler and Bernard, 2007). Some mechanistic aspects of respiratory dysfunction in lung injury will be discussed in later sections, in the context of therapeutic measures. Another factor in the pathophysiology of both airway injury and alveolar injury is activation of the coagulation cascade, leading to fibrin deposition in the alveoli or in the airways, forming casts (Demling, 2008; Rehberg et al., 2009; Rancourt et al., 2012; Glas et al., 2013). The tissue injury and inflammation initiate responses of repair and regeneration, which, in a highly inflammatory environment, are diverted to pathologic remodeling, resulting in hyperreactivity and fibrosis (Beers and Morrisey, 2011).

Some of the individuals exposed to lung toxicants develop reactive airway dysfunction syndrome or RADS (Alberts and do Pico, 1996; Shakeri et al., 2008). It is also classified at present as irritant-induced asthma, a type of occupational asthma in which there is no latency and no evidence of immunological sensitization, and it occurs after a single high-level exposure to respiratory irritants (Francis et al., 2007; White and Martin, 2010). This response manifests in asthma-like symptoms (cough, wheezing, chest tightness, and dyspnea), which appears within 24 h after exposure and persists for at least 3 months. Pulmonary function tests may show airway obstruction, including increased airflow resistance or hyperreactivity to methacholine challenge.

17.2.3 Toxic Inhalation Injury and ALI/ARDS

The clinical entities known as ALI and ARDS are conditions of lung disease with a variety of underlying etiologies but a common pathological–clinical denominator of alveolar and interstitial edema, pulmonary inflammation, and severe hypoxemia (Ware and Matthay, 2000; Wheeler and Bernard, 2007). The two conditions have been defined by the American-European Consensus Conference (AECC), by the following criteria: (1) acute onset, (2) bilateral diffuse pulmonary infiltrates in frontal chest radiography, and (3) severe hypoxemia, discriminated by the depth of blood oxygenation impairment as ALI, with a ratio of arterial oxygen tension and fraction of inspired oxygen (PaO_2/FiO_2) between 200 mmHg and 300 mmHg (40 kPa), and the more severe ARDS, with a PaO_2/FiO_2 <200 mmHg (26.7 kPa) (Bernard et al., 1994). The definitions were revised recently by a multinational expert task force, initiated by the European Society of Intensive Care Medicine (ESICM) (Ferguson et al., 2012). The revised or Berlin definition used the same criteria of the AECC definition but sharpened some of the clinical definitions (time of onset of underlying etiology as 7 days at most and clarification of confounding chest radiography and clinical findings) and redefined the severity criteria of oxygenation as mild ($200 < PaO_2/FiO_2 \leq 300$ mmHg, positive end-expiratory pressure [PEEP] or CPAP ≥ 5 cm H_2O), moderate ($100 < PaO_2/FiO_2 \leq 200$ mmHg, PEEP or CPAP ≥ 5 cm H_2O), and severe ARDS ($PaO_2/FiO_2 \leq 100$ mmHg, PEEP or CPAP ≥ 5 cm H_2O). The risk or predisposing factors to ALI/ARDS are numerous. The most prevalent direct causes are pneumonia and aspiration of gastric content, while the most prevalent indirect causes are sepsis and severe trauma (Ware and Matthay, 2000). Toxic inhalation injury is listed among the less common direct causes of ALI/ARDS (Ware and Matthay, 2000), but the risk of ALI from smoke inhalation was found to be exceptionally high compared to more common predisposing factors (Gajic et al., 2011). Therefore, treatment strategies tested for ALI/ARDS, regardless of underlying etiology, but tested and found relevant in the context of toxic inhalation injury, will be discussed later on together with the more direct toxic inhalation treatment modalities.

17.2.4 Principles of Management and Medical Treatment

The principles and procedures of medical treatment will be discussed in the context of mass casualty scenarios, but most of the clinical and technical accounts are applicable to occupational, community, and domestic incidents, which involve smaller numbers of casualties or individuals, from which most of the clinical knowledge and experience were obtained. The principles of patient management in case of a toxic mass casualty event (TMCE) are described in Table 17.1, adapted from the ATSDR guidelines for medical management of HAZMAT incidents (ATSDR, 2001b).

17.2.4.1 Control Zones

The chemical prehospital scene is divided to three control zones, on the basis of agent concentrations in air or hazard levels, in order to define the personal protection levels and operational and medical activities (Kenar and Karayilanoglou, 2004; OSHA, 2005; Byers et al., 2008). The *hot zone* (*red zone* in the OSHA/NIOSH terminology) is the zone of highest contamination and hazard level or if the actual hazard is unknown. Due to personal protection demands, missions are restricted to the most urgent and essential. The medical emphasis is on the rescue of casualties for decontamination and treatment outside, limiting patient care to the most urgent life-preserving actions (Byers et al., 2008). The *warm zone* (or *yellow zone* by OSHA/NIOSH) is defined as "areas where contamination with CBRN agents is possible but active release has ended and initial monitoring exists." The definition applies to areas in the vicinity of the release area where airborne risk is below a certain level or secondary contamination may result from certain operational activities. From the medical point of view, this is the decontamination or contamination reduction zone, where the main activities are decontamination of patients and life-support care of the critically ill casualties. The area where contamination is unlikely is designated as the *cold zone* or *green zone*. This area is the site of support activities (*support zone*), allowing for life-preserving and stabilizing treatments that were precluded in the *warm zone* and prepare the severely injured for evacuation to health-care facilities. Traffic of rescue and medical personnel and evacuated victims is done between the operational zones through control points, where exiting personnel are monitored for residual contamination and undergo decontamination if necessary. Patient decontamination facilities should be deployed and operated at the health-care facility entrance, in order to decontaminate patients arriving independently or patients that have not been decontaminated at the event scene, especially at the early stage, before the establishment of on-site decontamination facilities (Macintyre et al., 2000). The delimitation of operational zones is possible if complete information on hazard levels, based on agent monitoring, meteorological data, and hazard evaluation with dispersion models, is available. The limits of zones are defined by predetermined exposure thresholds or criteria like EPA's AEGL (Watson et al., 2006), carried out and announced by HAZMAT experts. At the initial stage, before this information becomes available, an initial isolated zone should be established by the first responders on a default basis.

17.2.4.2 Exposure Control and Contamination Avoidance

The first mandatory action in the management of a toxic gas casualty is exposure termination. This is attained by removal of the victim to a clean environment or, as some authors suggest, placing a protective mask on the patient's face (Russell et al., 2006; Tourinsky and Sciuto, 2008). Another measure is decontamination. Decontamination starts with removal of the patient's clothes, which removes 85%–90% of the chemical from the patient (Kales and Christiani, 2004). Later, rinsing with ample amount of water followed by washing with water and soap may be generally adequate (ATSDR, 2001c; Kales and Christiani, 2004). For people at risk, the viable options are sheltering or evacuation from the hazard zone, and modeling studies have demonstrated the importance of

TABLE 17.1

Elements of Toxic Inhalation Injury Management

Operation Zone	Activity/Intervention	Prehospital Management	
		Action Taken	Comments
I. Hot zone	Rescuer protection	• *Respiratory protection*: Pressure-demand, self-contained breathing apparatus (SCBA) should be used in all response situations. • *Skin protection*: Chemical-protective clothing should be worn when local and systemic effects are unknown.	• Rescuers should be trained and appropriately attired before entering the hot zone. • See text and references for more information on protection levels and their impact on medical care capacity.
	ABC reminder	• Quickly ensure a patent airway. If trauma is suspected, maintain cervical immobilization manually and apply a cervical collar and a backboard when feasible.	• Only minimal patient care can be given in the hot zone while wearing level A or B PPE. • The ultimate goal is quick removal of the victim to avoid continued exposure and risk.
	Victim removal	• If victims can walk, lead them out of the hot zone to the decontamination zone. Victims who are unable to walk may be removed on backboards or gurneys; if these are not available, carefully carry or drag victims to safety.	
II. Decontamination zone	Rescuer protection	• If the chemical or concentration is unidentified, personnel in the decontamination zone should wear the same protective equipment used in the hot zone. • If exposure risk is determined to be lower, decontamination can be conducted while wearing lower-level PPE than in the hot zone.	• Wearing respirators and heavy gloves may cause difficulty to provide advanced medical care, e.g., inserting an intravenous line or performing endotracheal intubation; therefore, this care is not administered until the victim is transferred to the support zone.
	ABC reminder	• Quickly ensure a patent airway. Establishment of artificial airway may be required. • Stabilize the cervical spine with a collar and a backboard if trauma is suspected. • Administer supplemental oxygen or nebulized bronchodilators as required. • Assist ventilation with a bag-valve-mask device if necessary.	• Electronic equipment, such as cardiac monitors, generally is not taken into the decontamination zone because the equipment may not be safe to operate and may be difficult to decontaminate.

(continued)

TABLE 17.1 (continued)

Elements of Toxic Inhalation Injury Management

Operation Zone	Activity/Intervention	Prehospital Management	
		Action Taken	**Comments**
	Basic decontamination	• Remove and double-bag contaminated clothing and personal belongings. • Flush exposed or irritated skin and hair with water for 3–5 min. For oily or otherwise adherent chemicals, use mild soap on the skin and hair. • Flush exposed or irritated eyes with plain water or saline for at least 5 min. Remove contact lenses if present and easily removable without additional trauma to the eye. If a corrosive material is suspected or if pain or injury is evident, continue irrigation while transferring the victim to the support zone.	• Victims exposed only to gas or vapor and have no skin or eye irritation may be transferred immediately to the support zone. • Victims who are able and cooperative may assist with their own decontamination.
	Transfer to support zone	• As soon as basic decontamination is complete, move the victim to the support zone.	
III. Support zone	Avoidance of secondary contamination	• Be certain that victims have been decontaminated properly. • Victims who have undergone decontamination or have been exposed only to gas or vapor and who have no evidence of skin or eye irritation generally pose no serious risks of secondary contamination. *In such cases, support zone personnel require no specialized protective gear.*	
	ABC reminder	• Quickly ensure a patent airway. • If trauma is suspected, maintain cervical immobilization manually and apply a cervical collar and a backboard when feasible. • Ensure adequate respiration and administer supplemental oxygen as required. • Ensure a palpable pulse. • Establish intravenous access if necessary. • Attach a cardiac monitor. • Watch for signs of airway swelling and obstruction such as progressive hoarseness, stridor, or cyanosis.	

	Additional decontamination	• Continue irrigating exposed skin and eyes, as appropriate.	
	Advanced treatment	• *Intubate the trachea in cases of respiratory compromise.* When the patient's condition precludes endotracheal intubation, perform cricothyroidotomy if equipped and trained to do so. • Treat patients who have bronchospasm with aerosolized bronchodilators. • Patients who are comatose, hypotensive, or have seizures or cardiac dysrhythmias should be treated according to ALS protocols.	
	Transfer to medical facility	• Report to the base station and the receiving medical facility the condition of the patient, treatment given, and estimated time of arrival at the medical facility.	

ED Management

Operation Area	Activity/Intervention	Action Taken	Comments
I. Decontamination area	Avoidance of secondary contamination	• Previously decontaminated patients and patients exposed only to gas or vapor who have no evidence of skin or eye irritation may be transferred immediately to the critical-care area. Other patients will require decontamination. • Basic decontamination is safely and practically performed outside, in a naturally ventilated area adjacent to the ambulance entrance.	• As a rule, patients will be decontaminated before reaching the hospital, but in some cases, e.g., patients brought directly to the ED by coworkers, relatives, or occasional volunteers may arrive without decontamination.
	ABC reminder	• Evaluate and support airway, breathing, and circulation. Intubate the trachea in cases of respiratory compromise. If the patient's condition precludes intubation, surgically create an airway. • Treat patients who have bronchospasm with aerosolized bronchodilators. • Patients who are comatose, hypotensive, or have seizures or ventricular dysrhythmias should be treated in the conventional manner.	
	Basic decontamination	• Same procedure as described in part a, decontamination zone.	

(continued)

TABLE 17.1 (continued)
Elements of Toxic Inhalation Injury Management

Operation Area	Activity/Intervention	ED Management Action Taken	Comments
II. Critical-care area	Avoidance of secondary contamination	• Be certain that appropriate decontamination has been carried out.	• Standard infection or communicable disease control items (water-resistant gowns or aprons, latex gloves, and eye-splash protection) may provide adequate protection if some risk, from residual skin contamination, exists.
	ABC reminder	• Evaluate and support airway, breathing, and circulation. Establish intravenous access in seriously ill patients. Continuously monitor cardiac rhythm.	
		• Patients who are comatose, hypotensive, or have seizures or ventricular dysrhythmias should be treated in the conventional manner.	
		• Watch for signs of laryngeal edema and respiratory system compromise, such as progressive hoarseness, strider, hypoventilation, or cyanosis.	
		• Consider the possibility of exposure to multiple chemicals as well as multiple-system injuries, e.g., smoke inhalation.	
	Management of inhalation exposure	• Administer supplemental oxygen by mask to patients who have respiratory complaints. Treat patients who have bronchospasm with aerosolized bronchodilators.	• Patients who have chemically induced pulmonary edema do not benefit from digoxin, morphine, afterload reduction, or diuretics. Supplementary oxygen, delivered by mechanical ventilation and PEEP, if needed, are standard treatments for chemically induced (noncardiogenic) pulmonary edema.
		• Watch for signs of respiratory distress and intubate if necessary.	• Corticosteroids and antibiotics have been commonly recommended for treatment of chemical pneumonitis, but their effectiveness has not been substantiated.
		• Poorly soluble irritants (e.g., phosgene and some nitrogen oxides) produce slow onset of airway irritation and respiratory distress and a delayed (12–72 h) onset pulmonary edema.	

Laboratory tests	• Routine laboratory studies for all exposed patients include CBC, glucose, and electrolyte determinations. • Additional studies for patients exposed to an unidentified chemical include ECG monitoring, renal-function tests, and liver-function tests. • Chest radiography and pulse oximetry (or arterial blood gas measurements) are recommended for severe inhalation exposure.	• Laboratory test results are often within normal range immediately after an exposure but may become abnormal after a delay of several hours or even days, depending on the specific chemical exposure. • Chest radiography may not show signs of pulmonary edema for 12–24 h following exposure. • Pulse oximetry and routine arterial blood gas determination of PO tests may provide falsely normal, unreliable, or misleading results in patients with abnormal hemoglobin states (e.g., methemoglobinemia or carboxyhemoglobinemia).
Disposition and follow-up	• Consider hospitalizing patients who have a suspected serious exposure and persistent or progressive symptoms. • When the chemical has not been identified, the patient should be observed for an extended period or admitted to the hospital. • Asymptomatic patients who have minimal exposure, normal initial examinations, and no signs of toxicity after 6–8 h of observation may be discharged with instructions to seek medical care promptly if symptoms develop.	

Source: Adapted from Agency for Toxic Substances and Disease Registry (ATSDR), *Managing Hazardous Material Incidents, Vol. III: Medical Management Guidelines for Acute Toxic Exposures*, Atlanta, GA, http://www.atsdr.cdc.gov/MHMI/mhmi-v3.pdf (last accessed on March, 2013), pp.19–44, 2001b.

time as a factor in reducing casualties (Barrett and Adams, 2011). If at home, protection can be obtained by closing all doors and windows and turning off all heating, ventilation, and air-conditioning (HVAC) systems (Kales and Christiani, 2004). As for evacuation, it is important to follow instructions and directions by the incident commanders. These should be based on the advice of HAZMAT professionals depending on the nature of the agent and monitoring data of agent levels and meteorological conditions. This is an important lesson from the Bhopal incident, where agitated and panicked people, seeking safety or help, left their homes and swarmed into the toxic cloud in the absence of proper guidance (Dhara and Dhara, 2002).

A common occurrence in workplace or mass intoxication events was that unprotected or inadequately protected responders became affected themselves during rescue operations in the contaminated zones. Medical personnel, when treating contaminated casualties without appropriate protection, may be at risk of secondary contamination, as happened in the Tokyo sarin incident of 1995 (Ohbu et al., 1997; Okudera, 2002). In an incident of chlorine gas release in Tacoma, partially protected responders were exposed because of an unexpected change in wind direction (Jones et al., 2010). Therefore, rescue and medical personnel should use personal protective equipment (PPE) compatible with the mission and hazard. Patients have to be decontaminated as soon as possible before entry to the ED. It is important to note that patients exposed to gases or vapors that do not present chemical signs of injury do not pose secondary contamination risk and can be transferred without decontamination. Decontamination is mandatory in the case of exposure to liquids like bleach or sulfur mustard.

The levels of protection in toxic environments have been defined in the United States by the EPA according to the hazard levels. The US definitions, which other countries employ similarly, are the following (ATSDR, 2001a):

1. *Level A* protection should be worn where the highest levels of respiratory, skin, and eye protection are needed. It consists of a fully encapsulated, vapor-tight, chemical-resistant suit, chemical-resistant boots and gloves, and a pressure-demand self-containing breathing apparatus (SCBA).
2. *Level B* protection should be selected when the airborne agent concentration poses a high respiratory hazard but the skin exposure hazard is low, or other conditions precluding the use of air-purifying respirators exist, for example, oxygen concentration below 19.5% or agent type or airborne concentrations are unknown. Respiratory protection is provided by SCBA, and skin protection is provided by nonencapsulating chemical-resistant clothing with chemical-resistant gloves and footwear as mentioned earlier.
3. *Level C* protection should be selected when the same level of skin protection but a lower level of respiratory protection than level B is required. Respiratory protection is provided by any NIOSH-approved full-facepiece powered air-purifying respirator (PAPR) equipped with appropriate air-purifying canister. Skin protection is the same as in level B. Level C personal protection is generally recommended in the *warm zone* (ATSDR, 2001a).
4. *Level D* is appropriate for the *cold zone*, when no respiratory protection but only minimal skin protection is required. It includes standard work outfit, and for medical personnel, it includes standard infection control precautions as appropriate.

It should be stressed that chemical protection imposes physical and mental stress on the user and limits the operational performance and mission duration. To overcome these challenges, responders have to be physically fit and trained in wearing PPE and performing their duties in protective posture. Level C protection is appropriate for decontamination or treatment of patients evacuated from the *hot zone* (Macintyre et al., 2000; Hick et al., 2003; Baker, 2005). Full face protective masks have to be properly fitted to the user's face to minimize leakage. The military chemical, biological, radiological, and nuclear (CBRN) full protective gear is approximately equivalent to level C PPE (Byers et al., 2008; Jones et al., 2010). The standard CBRN canister is effective against organic vapors (e.g., sulfur mustard) and nerve agents and protects fairly against acidic gases (e.g., cyanides) and particulates. It is not efficient against alkaline gases, for example, ammonia, for which specified canisters are required. Therefore, the military canisters may be adequate against most of the materials at low levels. In any case, the choice of the suitable canister and other protection level decisions should be done in consultation with HAZMAT professionals.

The aforementioned information outlines general and schematic guidelines for chemical emergency preparedness and management. Site- and scenario-specific plans and standing operation procedures should be developed, and operational decisions should be based on assessment of multiple factors that include agent properties, agent and meteorological monitoring, characteristics of the terrain, demography of the affected population, and location and availability of medical and logistic assets.

17.2.4.3 Current Therapy: Respiratory Support and Pharmacological Interventions

The basic treatment of inhalation toxic injury is supportive care, as no specific antidotes for most of these agents exist. The core intervention, after exposure termination, is aimed at respiratory and circulatory support (Russell et al., 2006). Other aims are to ease the burden of symptoms, control inflammation, and minimize the risk of sequelae. Some pharmacological interventions, like administration of antioxidants, steroids, and other anti-inflammatory drugs, are employed on an empirical basis without a rigorous proof of clinical efficacy, although animal studies with promising results and case reports with successful therapy with these drugs are reported (Russell et al., 2006; Tourinsky and Sciuto, 2008; Yadav et al., 2010).

It should be kept in mind that life-support procedures may have to be performed on contaminated patients while wearing PPE, which may compromise the performance and narrow the scope of possible medical procedures. However, several tests have determined that performing basic and advanced lifesaving procedures can be accomplished with low to moderate levels of difficulty (Coats et al., 2000), which can be overcome by proper training, improved protective equipment like protective gloves, and switching to alternative procedures that are less vulnerable to the impact of PPE on manual dexterity and visual acuity (Ben-Abraham et al., 2003; Eisenkraft et al., 2007).

The clinical respiratory manifestations that require intervention are airway obstruction (laryngeal spasm and secretions), pulmonary dysfunction (bronchospasm, pulmonary edema, impaired lung compliance), and the resulting hypoxemia, hypercapnia, and acidosis. Patient assessment and triage are essential elements in the management of toxic inhalation events, in order to insure proper, timely, and correct use of medical resources, which may be limited in multiple-casualty events. The clinical scene may be heterogeneous and dynamic, in which asymptomatic or mild casualties may become critical within a few hours. This requires that triage will be continuous or repeated at every transition point of the patient's evacuation course from the site of the event to the health-care facility. Recent recommendations for mass critical-care triage define three stages of triage: *primary triage*, which is performed in the prehospital setting by emergency medical personnel and is limited to basic decisions; *secondary triage* at the ED, which is done by ED medical staff but is still limited in its diagnostic resources; and *tertiary triage*, performed by intensive care physicians to prioritize patients for definitive care at the ICU (Devereaux et al., 2008). Primary triage at the decontamination zone should mainly discriminate the walking, who are asymptomatic or with mild injuries, from the lying, who are severely injured. The latter should also be triaged for the need for urgent, life-support care like intubation and ventilation, to be done concomitantly with decontamination (Macintyre et al., 2000; Markel et al., 2008; Talmor, 2008). The walking should be directed to a decontamination facility in which they may get help from the station personnel if needed. This discrimination is required since passive decontamination of a disabled casualty is labor intensive and requires at least two attendants per patient. Patients should be triaged again at the support zone, in order to prioritize transportation to health-care facilities, determine the need of additional care or observation, or discharge. Upon arrival to the medical facility, the patients should be triaged at the ED to determine admission to hospital wards, ICU, or discharge (Macintyre et al., 2000;

Talmor, 2008). The triage system used by the military medical units includes immediate, delayed, minimal, and expectant (Kenar and Karayilanoglou, 2004). The immediate casualties (T1) are those who require medical care and advanced life support within a short time in the incident scene or the hospital. The delayed casualties (T2) are those who require medical care and hospitalization, but a delay of these will not affect their condition. Minimal casualties (T3) suffer from minor injuries, will not be evacuated, and may return to duty within a short time. Expectant (T4) casualties are injured so severely that they may not survive with available medical care until reaching definitive care. The US medical corps criteria for management of toxic inhalation casualties according to this categorization system are summarized in Box 17.1 (Tourinsky and Sciuto, 2008). In the case of critically injured victims in a mass casualty event, the aim of tertiary triage is to identify the patients whose prospects for survival are best and prioritize them to intensive care (Devereaux et al., 2008). Several algorithms for outcome prediction of critically ill patients were developed, like the sequential organ failure assessment (SOFA) scoring algorithm (Vincent et al., 1996; Ferreira et al., 2001), which is relevant to emergency situations. The algorithm is simple and requires only five input parameters that are routinely measured at every institution, representing respiratory (PaO_2/FiO_2 ratio), coagulation (platelet count), hepatic (blood bilirubin), cardiovascular (hypotension), CNS (Glasgow Coma Score), and renal (blood creatinine or urine output) functional status (Vincent et al., 1996; Ferreira et al., 2001). Emergency plans should be flexible enough to accommodate for the type of agent, number of casualties, and operational conditions. In an event involving a small number of casualties, a volatile material like chlorine, and a well-developed health-care facility network, casualties may be evacuated directly to hospitals and primary triage should be done there (Baxter et al., 1989). This has been experienced in Israel in the summer of 2005, where 40 casualties, mostly children, were exposed to chlorine in a swimming pool located in the center of the country and evacuated efficiently and successfully treated in four hospitals (Lehavi et al., 2008).

Securing a clear airway is one of the first steps to be taken (Baker, 1999). Hoarseness or stridor may indicate laryngeal edema and spasm, which requires urgent intubation or tracheostomy (Russell et al., 2006; Tourinsky and Sciuto, 2008; Rehberg et al., 2009). When secretions are watery and copious, suction and drainage should be performed. In severe cases with extensive tissue damage, respiratory airways may become occluded with debris from exfoliated epithelial cells, neutrophils, exudates, and mucous, with subsequent formation of fibrin casts (Rehberg et al., 2009). The use of nebulized anticoagulants is being studied in animal models and clinically (Mlacek et al., 2007; Rehberg et al., 2009). As a first-priority intervention, airway management is one of the few interventions that must be performed in parallel and even prior to patient decontamination—generally in the *warm zone* but also in the *hot zone* (ATSDR, 2001b; Byers et al., 2008; Talmor, 2008). Studies on the performance of airway management by medical personnel wearing chemical PPE have shown some degradation, especially lengthening of the time required to complete endotracheal intubation (Berkenstadt et al., 2003; Flaishon et al., 2004a; Garner et al., 2004; Talmor, 2008). Most of the studies were done in level C or military CBRN protective gear (e.g., Flaishon et al., 2004a,b), but the study of Garner et al. compared the performance of medical caregivers wearing levels A–D PPE. They have found no significant differences in the time required to ventilate a manikin lung using various devices, apart from slowing down of the time required for endotracheal intubation while wearing level A PPE, recommending to use laryngeal mask apparatus (LMA) while in level A PPE and switching to endotracheal tube after the patient has been decontaminated, if the conditions are favorable (Garner et al., 2004). Aside from LMA, other supraglottic airway devices are being evaluated for performance by caregivers wearing chemical PPE (Castle et al., 2011). Another important difficulty in chemical-protective posture is the use of auscultation to diagnose respiratory patterns, which can be partially overcome by external observation for color change, respiratory rate, and form (Baker, 1999).

Bronchial constriction, or bronchospasm, is managed with inhalation of β2-adrenergic agonists (β2-agonists). These agents act pharmacologically by elevating cellular cyclic AMP levels, which cause smooth muscle relaxation and bronchodilation. Other beneficial effects of β2-agonists, relevant

BOX 17.1 TRIAGE FOR VICTIMS OF TOXIC INHALATION INJURY: U.S. ARMY MEDICAL CORPS GUIDELINES

Patients seen within 12 h of exposure

- *Immediate*
  - Patients with pulmonary edema only and if intensive pulmonary care is immediately available. In general, a shorter latent period portends a more serious illness.
- *Delayed*
  - Patients who are dyspneic without objective signs should be observed closely and retriaged hourly if not sooner.
- *Minimal*
  - Asymptotic patients with known exposure—these patients must be observed and triaged every 2 h.
  - If a patient remains asymptotic 24 h after exposure, the patient may be discharged.
  - If exposure is doubtful, and the patient remains asymptotic 12 h following putative exposure, consider discharge.
- *Expectant*
  - Patients who present with pulmonary edema, cyanosis, and hypotension are classified as expectant.
  - A casualty with onset of these signs within 6 h after exposure generally will not survive.
  - A casualty with onset of these signs 6 h or longer after exposure may possibly survive with immediate, intensive medical care.
  - If ventilator support is not available, but adequate evacuation assets are, these patients should have priority for urgent evacuation to a facility where adequate ventilator support is available.

Patients seen more than 12 h after exposure

- *Immediate*
  - Patients who present with pulmonary edema and if they can receive intensive care treatment within 7 h.
- *Delayed*
  - Patients who are dyspneic should be observed closely and triaged at least every 2 h.
  - Patients who are recovering from exposure could be discharged 24 h after exposure.
- *Minimal*
  - Patients who are asymptotic or have resolving dyspnea are classified as minimal.
  - Patients who are asymptotic 24 h should be discharged.
- *Expectant*
  - Patients who present with persistent hypotension despite intensive medical care interventions.
  - If cyanosis and hypotension are present with pulmonary edema, triage the patient as expectant.

Sources: The material is based on Exhibit 10-3, p. 365 in Tourinsky, S.D. and Sciuto, A.M., Toxic inhalation injury and toxic industrial chemicals, Chapter 10, in: *Medical Aspects of Chemical Warfare*, Tourinsky, S.D. (senior editor), *Textbook of Military Medicine*, Office of the Surgeon General, US Army, Borden Institute, Walter Reed Army Medical Center, Washington, DC, pp. 339–370, 2008; US Army Medical Research Institute of Chemical Defense, *Medical Management of Chemical Casualties Handbook*, 4th edn., Aberdeen Proving Ground, Aberdeen, MD, 2007.

to pulmonary chemical injury and ARDS, are enhancement of alveolar fluid clearance, anti-inflammatory activity, and stimulation of alveolar endothelial repair (Hoyle, 2010; Bassford et al., 2012). The beneficial effects of β2-agonists in resolution of pulmonary edema were demonstrated in animal models of chemical or inflammatory lung injury (Sciuto and Hurt, 2004). However, in clinical trials, results were less encouraging with high risk of adverse cardiovascular effects, higher mortality rate, and smaller number of ventilator-free days (Matthay et al., 2011; Bassford et al., 2012; Gao Smith et al., 2012). On the basis of these trials, the use of β2-agonists was not recommended in ARDS (Bassford et al., 2012). In line with these recent results, nebulized salbutamol had no effect on mortality and worsened blood oxygenation, ventilation–perfusion mismatch, and heart rate in a pig model of phosgene-induced ALI (Grainge et al., 2009). In this study, nebulized salbutamol reduced the number of neutrophils in bronchoalveolar lavage (BAL) fluid, unlike the lack of effect on neutrophil numbers or activity in human patients (Perkins et al., 2007). This different effect on inflammation was attributed to the different injury courses in the model and clinical studies: The human studies enrolled ventilated patients with established ARDS, while in the animal study, treatment was started early enough to address the inflammation while in progress (Grainge et al., 2009). In ARDS patients, damage to alveolar epithelium may render β2-agonists useless in enhancement of fluid clearance, which is dependent on intact alveolar type II epithelium cells (Levitt and Matthay et al., 2012). The reduction of lung extracellular water level was started to be seen, contrary to expectation, only after 72 h of treatment, and the delay was attributed to increased tissue repair in the drug-treated patients (Perkins et al., 2006, 2008). Therefore, these results may not preclude the use of β-agonists in the management of lung injury in unventilated patients, which may have a prophylactic value that should be studied in prophylactically designed trials (Levitt and Matthay et al, 2012).

Supplemental oxygen therapy is definitely indicated in the management of respiratory compromised victims (Russell et al., 2006; Tourinsky and Sciuto, 2008). Supplementary oxygen may require positive airway pressure in order to achieve inspired oxygen fraction (FiO$_2$) of 0.3–1.9 (Russell et al., 2006). Supplementary oxygen, at the maximum achievable concentration, should be given by mask (Baker, 1999). Hyperbaric oxygen (HBO) was found useful especially in smoke inhalation (Hart et al., 1985), in order to alleviate carbon monoxide and cyanide poisoning. The mechanism of the beneficial effects of HBO is not completely known, but current evidence makes clear that apart from enhanced oxygenation, it has many relevant beneficial effects to include anti-inflammatory activity (Gill and Bell, 2004), whose role in HBO therapy for these conditions still has to be clarified. A study in rat model of smoke inhalation injury has shown that HBO treatment inhibited neutrophil infiltration into the lungs, as well as blood protein permeability into the alveoli, indicating inhibition of lung injury by anti-inflammatory activity (Thom et al., 2001). Like normobaric oxygen, HBO has to be administered at a moderate pressure (generally between 2 and 3 atm absolute or ATA) and for short periods (90–120 min per session) in order to prevent adverse effects of oxygen toxicity (Gill and Bell, 2004). The other drawbacks of HBO are limited availability, the risks of transferring critically ill patients to HBO facilities, and difficulties in management of such patients in HBO chambers. The occurrence of lung injury due to hyperoxia is mainly recognized in animal models, where prolonged exposure to high levels of FiO$_2$ induces excessive generation of ROS and an inflammatory process similar to ARDS, when the oxidative load overwhelms the existing protective capacity (Jackson, 1985; Fisher and Beers, 2008). The risks for humans are less clear, though a recent review (Kallett and Matthay, 2013) suggested that it may be real for patients undergoing prolonged ventilation at FiO$_2$ >0.7. Because of the poorly characterized risk of hyperoxia in humans, current guidelines advise to administer supplementary oxygen with caution (ATSDR, 2001c; Noltkamper and Burgher, 2012; Segal and Lang, 2012).

Patients with acute respiratory failure, in which adequate blood oxygenation or pH blood correction cannot be achieved with noninvasive oxygen therapy, require positive-pressure mechanical ventilation. These cases include ALI/ARDS, proximal airway injury, and bronchospasm or bronchorrhea (Rubinson et al., 2006). The ideal place for this intervention is the hospital, preferentially the ICU with the availability of expert staff, suitable equipment, and opportunities for additional supportive care. A major toxic event may impose the challenges of the need to begin treatment of respiratory failure in the prehospital setting and the scarcity of qualified personnel and equipment to care for the unusual numbers of victims requiring assisted ventilation (Markel et al., 2008). Ventilation in the field can be done with a manual bag-valve-mask device or a field-expedient ventilator (Baker, 1999; Rubinson et al., 2006). In case of ventilation in a toxic environment, which may have to begin in the *warm zone* in parallel with patient decontamination, a filter-bag-mask device or a ventilator with a filtering capacity may be needed (Baker, 1999). An evaluation of a bag-valve-mask device fitted with a multiple-agent canister (A2B2E2K1-P3D designation) has shown that, in spite of the reduction in tidal volume because of the filter resistance, 15 out of 20 experimenters could deliver a tidal volume of >5 L/min to a simulator manikin. This stressed the need for proper training in order to insure adequate performance (Brinker et al., 2008). The challenge of lack of ventilators may be met by collaborative or national programs to stockpile suitable ventilators and ancillary equipment that will be deployed where and when needed. Such ventilators should be deployable outside the conventional ICU or in the field; be operated by nonexpert medical staff; endure power and medical gas shortage; have reliable control of tidal volume, respiration rate, PEEP, and FiO_2; and have built-in alarms in order to minimize the staff required for patient monitoring (Rubinson et al., 2006; Branson et al., 2008). The suitability of available portable ventilators to the requirement of mass casualty events was analyzed in view of their performance characteristics and cost. Portable ventilators, either compressed gas or pneumatically powered, are preferable to EMS transport ventilators mainly because of alarm and control capabilities although they may be more expensive (Rubinson et al., 2006). Gas-powered machines have the limitation of dependence on compressed oxygen supply, which may be compromised in disaster situations but may be advantageous in a toxic environment as they do not use ambient air. An alternative solution is a machine capable of using filtered air as driving gas (Baker, 1999).

The impaired pulmonary function in ALI/ARDS patients dictated the use of positive-pressure ventilation at higher than normal tidal volumes, usually 10–15 mL/kg of predicted body weight compared with the normal 7–8 mL/kg, and high peak pressures in order to achieve proper blood oxygenation and pH (Ware and Matthay, 2000; Wheeler and Bernard, 2007). The mechanical stress from distention of alveolar walls and cyclic intertidal opening and closing of collapsed alveoli result in tissue injury (volutrauma and atelectrauma, respectively), which invokes an inflammatory response (biotrauma) resulting in ventilation-induced lung injury (VILI), which exacerbates the condition of the patients. The conditions leads to ARDS if it has not existed initially and to systemic multiple-organ failure (Gattinoni et al., 2010; de Prost et al., 2011). The risk of VILI was addressed by the American College of Chest Physicians (ACCP) consensus conference in 1993 (Slutsky, 1993) that published guidelines for ventilation in ARDS, which balance the need to provide adequate oxygenation and to protect the lungs from ventilation injury (Box 17.2). The most important strategies studied and employed are low tidal volume or ARDSNet strategy, high PEEP, and high-frequency oscillatory ventilation (HFOV).

The ARDS Clinical Trials Network (ARDSNet) of the National Heart, Lung, and Blood Institute conducted a randomized, controlled, multicenter trial, which compared a low tidal volume ventilation protocol (6 mL/kg predicted body weight) to the traditional protocol (12 mL/kg) for patients with ALI/ARDS. This led to the establishment of a standard protocol for lung-protective ventilation in ALI/ARDS (Brower et al., 2000). The trial was terminated after enrollment of 861 patients as a clear advantage of the low tidal volume protocol was evident: a significantly lower mortality (31.0% in the low tidal volume arm vs. 39.8% in the traditionally ventilated arm, $p = 0.007$) and a higher number of ventilator-free days during the first 28 days after randomization (12 ± 11 vs. 10 ± 11, respectively, $p = 0.007$). Significant differences between the two arms were seen in physiological parameters and inflammation measured between days 1 and 3 of ventilation: the mean tidal volumes were 6.2 ± 0.8 and 11.8 ± 0.8 mL/kg ($p < 0.001$) in the low tidal volume and traditionally ventilated arms; the mean plateau pressures were 25 ± 6 and 33 ± 8 cm H_2O, as well as significantly lower plasma IL-6 in the low tidal volume arm (Brower et al., 2000). The resultant protocol includes a

**BOX 17.2   ACCP CONSENSUS CONFERENCE
GUIDELINES FOR VENTILATION IN ARDS**

1. The clinician should choose a ventilation mode that has been shown to support oxygenation and ventilation in ARDS patients and that he has experience in its use.
2. Acceptable blood oxygen saturation (SaO_2, $\geq 90\%$) should be targeted.
3. Based primarily on animal data, a plateau pressure of ≥ 35 mmHg is a cause for concern. If this level of plateau pressure is reached or exceeded, tidal volume should be reduced to levels as low as 5 mL/kg or less. In clinical conditions that are associated with decreased chest compliance, plateau pressures ≥ 35 mmHg may be acceptable.
4. To accomplish the goal of limiting plateau pressure, $PaCO_2$ should be permitted to rise (permissive hypercapnia), unless other contraindications exist that demand more normal $PaCO_2$ or pH.
5. PEEP is useful in supporting oxygenation. An appropriate level of PEEP may be helpful in preventing lung damage. The level of PEEP, however, should be minimized as PEEP may also be associated with deleterious effects. The level of PEEP required should be established by empirical trial and reevaluated on a regular basis.
6. The current opinion is that FiO_2 should be minimized. The trade-off, however, may be a higher plateau pressure and relative risks of these two factors are not known. In some clinical conditions where the significant concern over both elevated plateau pressure and FiO_2 exists, consideration for accepting a SaO_2 slightly less than 90% is reasonable.
7. When oxygenation is inadequate, sedation, paralysis, and position change are possible therapeutic measures. Other factors in oxygen delivery (i.e., Q_T and hemoglobin) should also be considered.

Source: Based on Slutsky, A.S., *Chest*, 104, 1833, 1993.

tidal volume of 4–8 mL/kg predicted body weight, plateau pressure ≤ 30 cm H_2O, and a moderate PEEP level, which is set individually (Ware and Matthay, 2000; Mlcak et al., 2007; Wheeler and Bernard, 2007; Esan et al., 2010; Dushiantan et al., 2011).

High PEEP is used in order to increase the recruitment of collapsed alveoli and thus prevent atelectrauma. The recent randomized, controlled, multicenter studies failed to show improvement in the mortality in ALI/ARDS patients ventilated with high PEEP (14–15 cm H_2O) compared to lower PEEP levels, using low tidal volume ventilation, but detected improvement in lung function, reduced duration of ventilation, reduced duration of organ failure, and reduced need for rescue therapies (Brower et al., 2004; Meade et al., 2008; Mercat et al., 2008). A meta-analysis of the data of these three studies showed no significant advantage in mortality between the high PEEP group (1136 patients, 32.9% mortality) and the low PEEP group (1163 patients, 35.2% mortality) for all patients, but an improved survival by high PEEP was seen in patients with ARDS ($n = 1892$), where mortalities were 34.1% in the high PEEP group and 39.1% in the low PEEP group. In the non-ARDS patients ($n = 404$), the mortalities were 27.2% and 19.4% in the higher PEEP and lower PEEP groups, respectively, leading to the conclusion that high PEEP may be beneficial for patients with ARDS but harmful for patients without ARDS (Briel et al., 2010), probably due to increased alveolar strain (Caironi et al., 2010). The computerized tomography (CT) studies of Gattinoni et al. have shown that the important parameter in the response to high PEEP may be the percentage of recruitable lung tissue, which is higher in the more severely ill patients, and in these patients, the benefit of reduced cyclic alveolar opening and closing exceeds that of the alveolar strain caused by high PEEP (Gattinoni et al., 2006; Caironi et al., 2010). The lung-protective ventilation protocols employing the low tidal volume and high PEEP strategy were shown to reduce ventilation-induced pulmonary and systemic inflammation and increase survival in other clinical trials (Amato et al., 1998; Ranieri et al., 1999; Villar et al., 2006). The low tidal volume/high PEEP protocol was studied in an anesthetized pig model of phosgene inhalation, demonstrating increased 24 h survival in the animals ventilated in the protective protocol compared to conventionally ventilated controls (Grainge and Rice, 2010). The strategy is recommended for pulmonary edema induced by inhalation of lung-damaging materials in general (Russell et al., 2006; Grainge and Rice, 2010; Jugg et al., 2011) and smoke inhalation (Mlcak et al., 2007; Rehberg et al., 2009).

17.2.4.3.1 High-Frequency Oscillatory Ventilation

The technique of HFOV was developed in order to reduce lung overdistention by delivering very small tidal volumes through a very high respiration frequency (between 180 and 600 breaths/min for adult human) and preventing cyclical opening and closing of alveoli or in other words *keeping an open lung* by maintaining a high mean airway pressure without the injurious peak pressures of conventional ventilation (Chan et al., 2007). Mechanically, the oscillatory tidal waves are generated by a piston, oscillating at an adjustable frequency in the range of 3 to >15 Hz against an orthogonal bias gas flow, whose composition and supply rate can be set, thus enabling separate control of oxygenation and ventilation–oxygenation is determined by the FiO_2 and mean airway pressure and ventilation by the tidal volume, which is inversely proportional to the oscillatory frequency (Chan et al., 2007). Animal model studies of ALI have shown that HFOV was superior to the conventional lung-protective strategy of low tidal volume/high PEEP in improving oxygenation and reducing inflammation, lung pathology, and oxidative stress (Imai et al., 2001; Ronchi et al., 2011). Notably, HFOV did not cause hypercapnia and acidosis like the conventional protective strategy (Imai et al., 2001). Several randomized, controlled trials were done in order to compare the efficacy of HFOV to conventional or lung-protective ventilation in patients with ARDS ($PaO_2/FiO_2 < 200$ or 150 mmHg). A meta-analysis of 8 clinical trials (9419 patients totally) demonstrated a significantly lower 30-day hospital mortality (risk ratio 0.77, 95% CI 0.61–0.98, $p = 0.03$) and an advantage in blood oxygenation on days 1–3 of each trial in the HFOV-treated patients compared to controls (Sud et al., 2010a). Despite those encouraging results, the major limitations of the past studies were the harmful ventilation strategies used in the controls (Derdak et al., 2002) and the relatively small numbers of patients in the later trials that did use the ARDSNet ventilation strategy in their controls. Recently reported randomized, controlled, multicenter trials, comparing HFOV to the ARDSNet lung-protective ventilation strategy in early moderate to severe ARDS, did not show a survival benefit to HFOV (Ferguson et al., 2013; Young et al., 2013), and in one of these (Ferguson et al., 2013), the in-hospital mortality in the HFOV group was significantly higher (47% in the HFOV group vs. 35% in the control, risk ratio 1.33, 95% CI 1.09–1.64, $p = 0.005$). In this trial, which was stopped early due to the higher mortality in the HFOV group, HFOV was also associated with a higher use of sedatives, neuromuscular blockers, and vasoactive drugs (Ferguson et al., 2013). The authors of these studies recommended not to use HFOV in early ARDS in adults, and one of them, the Canadian OSCILLATE study (Ferguson et al., 2013), even questioned the use of HFOV as a rescue strategy for severe refractory hypoxemia.

Some cases of ARDS deteriorate to refractory hypoxemia, a condition characterized by a severe decline in blood oxygenation ($PaO_2/fiO_2 < 100$ mmHg) with no improvement in response to conventional therapy within 24 h of lung-protective ventilation (Esan et al., 2010; Pipeling and Fan, 2010; Raoof et al., 2010). These conditions require *rescue strategies* of ventilation, designed to improve alveolar recruitment and minimize tissue damage (Esan et al., 2010), and nonventilator adjunct measures (Raoof et al., 2010). The rescue strategies include high PEEP and recruitment maneuvers, airway pressure release ventilation (APRV) (Esan et al., 2010), and HFOV (Chan et al., 2007). The nonventilatory adjuncts in current use are local vasodilators—inhaled nitric oxide (iNO) and nebulized prostacyclins (Siobal and Hess, 2010), prone positioning (Pelosi et al., 2002), and extracorporeal membrane oxygenation (ECMO) in extreme cases. Randomized, controlled studies on the clinical benefit of these strategies, mainly in the direction of adoption as routine strategies for ALI/ARDS, showed that, in general, they were beneficial only for the more severe patients, and they were still recommended as rescue therapies for refractory hypoxemia (Esan et al., 2010; Liu et al., 2010; Pipeling and Fan, 2010; Raoof et al., 2010; Collins and Blank, 2011). Some of these strategies, which were also tested in the context of toxic inhalation injury, will be discussed in the following.

17.2.4.3.2 Inhaled Vasodilators: Nitric Oxide and Prostacyclins

The rationale for treatment with inhaled vasodilators—nitric oxide (NO) and prostacyclins—is their selective penetration into well-ventilated alveoli, causing localized vasodilation and diversion of perfusion into these regions, thus improving ventilation–perfusion matching and blood oxygenation in ALI/ARDS patients (Moloney and Evans, 2003; Griffiths and Evans, 2005; Siobal and Hess, 2010). Hypoxic pulmonary vasoconstriction is a normal protective response, aimed at decreasing perfusion of poorly ventilated lung regions and maintenance of proper ventilation–perfusion matching, but it plays an important detrimental role in the pathophysiology of ARDS (Moloney and Evans, 2003). In ARDS, hypoxic pulmonary vasoconstriction reduces blood oxygenation and increases the left to right shunt by decreasing the perfusion of the better-ventilated alveoli. Pulmonary hypertension results from the vasoconstriction and increases the load on the right heart, which may lead to a fatal right heart failure (Moloney and Evans, 2003). Inhalation of 5–80 ppm of NO was shown to reverse hypoxic pulmonary vasoconstriction and pulmonary hypertension (Frostell et al., 1991, 1993) without systemic vasodilation, prompting studying the effects of inhaled NO in ARDS. Short-term treatment of severe ARDS patients with inhaled NO resulted in temporary reduction in pulmonary artery pressure, pulmonary shunt, and increased PaO_2/FiO_2 ratio, while systemic (IV) administration of prostacyclin (PGI_2 analog) caused an opposite effect. Prolonged treatment of patients with 5–20 ppm NO for 3–53 days resulted in survival and discharge of all (Rossaint et al., 1993). Randomized, controlled trials and meta-analyses of inhaled NO in ALI/ARDS showed no improvement in mortality, in shortening ventilation duration or in more ventilator-free days. Other effects were a transitory improvement in blood oxygenation but also increased risk of renal dysfunction and higher mortality in NO-treated patients (Adhikari et al., 2007; Siobal and Hess, 2010; Afshari et al., 2011). These studies ended in a recommendation not to use inhaled NO for ALI/ARDS unless it is considered as a rescue therapy (Siobal and Hess, 2010). The few studies of inhaled NO in animal models of toxic lung injury were similarly unimpressive. Inhaled NO in phosgene-exposed rats was found to be detrimental (Li et al., 2011). Earlier studies of inhaled NO in an ovine model of smoke inhalation demonstrated reversal of pulmonary hypertension, improvement in pulmonary shunt, and a modest improvement in blood oxygenation and ventilation–perfusion matching but with no improvement in lung morphology, inflammation, or edema (Ogura et al., 1994a,b).

17.2.4.3.3 Prone Positioning

Ventilation in prone position is employed as a *rescue therapy* for severe and poorly responsive ARDS patients (Pelosi et al., 2002; Raoof et al., 2010). The beneficial effects of this strategy are improved oxygenation, improved ventilation–perfusion matching, facilitated drainage of secretions, and reduction of ventilation-associated lung injury. Physiological and CT studies in animals and human subjects have shown that the prone positioning relieves the gravity-caused collapse of the dependent, dorsal regions of the lungs. This leads to a more even distribution of transpulmonary pressure gradients, improved ventilation and perfusion, and decreased overdistension and atelectasis in these regions (Pelosi et al., 2002). The first randomized, controlled, multicenter study of ventilation in prone or supine positioning failed to show a survival advantage to prone position but did show a sustained effect of improved oxygenation in the prone group (Gattinoni et al., 2001). However, post hoc analysis demonstrated a significant 10-day mortality reduction compared to supine positioning in a subgroup of more severe patients, with $PaO_2/FiO_2 \leq 88$ mmHg (23.1% vs. 47.2%, risk ratio 0.49, 95% CI 0.25–0.95). A later study by this group, with stratification of the patients by severity and extension of the prone positioning duration from 7 h per day up to 17–20 h per day, did not find a significant survival advantage to prone positioning (Taccone et al., 2009). A meta-analysis encompassing these and 8 additional trials, with a total of 1867 patients, showed that prone positioning reduced mortality in patients with PaO_2/FiO_2 <100 mmHg (risk ratio 0.84, 95% CI 0.74–0.96, p = 0.01) but not in patients with PaO_2/FiO_2 ≥100 mmHg (risk ratio 1.07, 95% CI 0.93–1.22, p = 0.36) (Sud et al., 2010b). As for other outcomes, prone positioning increased the PaO_2/FiO_2 ratio taken on days 1–3 from randomization and reduced the occurrence of ventilation-associated pneumonia, but no effect was seen on the duration of mechanical ventilation or ventilation-free days in 28 days (Sud et al., 2010b). The analysis detected also an increased risk of pressure ulcers, endotracheal tube obstruction, and chest tube dislodgement in prone compared to supine patients, adverse effects known as a major limitation of this strategy, together with being a labor-intensive strategy (Pelosi et al., 2002). An updated meta-analysis reached the same conclusions and detected a significant effect of the duration of prone positioning on the decrease in mortality (Abroug et al., 2011). On the basis of these data, some unresolved issues, and the risk of adverse effects, prone positioning was not recommended for routine use but as a therapy for the most severe patients, whose hypoxemia is life threatening (Fessler and Talmor, 2010; Marini, 2010; Sud et al., 2010b). When looking at toxic inhalations, ventilation of chlorine-exposed pigs in prone position resulted in partial improvement of respiratory and cardiovascular functions compared to supine-positioned animals (Wang et al., 2002b).

17.2.4.3.4 Corticosteroids in Toxic Inhalation Injury

Corticosteroids have therapeutic potential in toxic inhalation injury and its complications—RADS, ALI, or ARDS—due to their anti-inflammatory activity and suppression of fibroproliferative tissue remodeling (Meduri, 1996; Meduri et al., 2009). Human and animal studies demonstrated beneficial effects in ALI and ARDS of different etiologies, including toxic inhalation injury, but the weight of the favorable evidence did not counterbalance the risk of adverse effects, which include prolonged neuromuscular weakness, deregulation of glucose metabolism, susceptibility to infections, and sepsis. Administration of corticosteroids, either systemic or by inhalation, has been reported in humans exposed to chlorine, especially in those hospitalized for severe respiratory injury (Babu et al., 2008). Even though the outcome in these cases was generally good, there is no conclusive evidence for their therapeutic value in chemical inhalation toxicity, as controlled studies in humans are impossible (Russell et al., 2006). A recent literature review of animal model studies concluded that corticosteroid treatment, either intravenous or inhaled, was beneficial in lung injury caused by water-soluble toxicants like ammonia or chlorine, but not in lung injury caused by water-insoluble toxicants like phosgene, ozone, or nitrogen dioxide (De Lange and Meulenbelt, 2011). Moreover, the beneficial effect could be demonstrated if treatment was started immediately or shortly after exposure (Wang et al., 2002a). In view of the current human and animal studies, corticosteroids are not recommended as first-line medications in toxic inhalation injury but should be considered as second-line medications in moderate and severe cases of exposure to chlorine (ATSDR, 2001c; Segal and Lang, 2012) or phosgene (Grainge and Rice, 2010; Noltkamper and Burgher, 2012).

17.2.4.3.5 Corticosteroids in the Treatment of ARDS

The role of corticosteroids in the treatment of ARDS is also not conclusively established (Sessler and Gay, 2010), but in view of promising results in clinical trials, further studies are planned (Reade and Milbrandt, 2007; Litell et al., 2011). Initial trials with high dosage (30 mg/kg/day) and short duration failed, but two small-scale, randomized, controlled trials that tested low dosages (0.5–2 mg/kg/day) of IV methylprednisolone for long durations (32 days to 6 weeks) were more successful. One included patients with late-stage ARDS (Meduri et al., 1998), and a later trial included patients with early-stage ARDS (Meduri et al., 2007). They have shown reduced mortality and improvements in clinical course and physiological parameters in the treatment groups of severe and unresolving ARDS patients. The ARDSNet conducted a randomized, controlled, multicenter trial to define the efficacy of low-dose, continuous IV methylprednisolone in late-stage ARDS (Steinberg et al., 2006). The trial showed an improvement in physiological parameters and clinical course in the treatment arm but failed to show improvement in survival and raised a safety issue due to higher mortality in a subgroup of patients enrolled after 14 days of ARDS. Meta-analysis studies (Meduri et al., 2008; Tang et al., 2009) corroborated the beneficial efficacy of prolonged, low-dose IV corticosteroid treatment in both early- and late-stage ARDS. In summary, beneficial effects are obtained in treatment starting with infusion of 1 mg/kg body weight of methylprednisolone and gradual tapering of the dose in 4 weeks for early, severe ARDS (PaO_2/FiO_2 <200 mmHg on PEEP of 10 cm H_2O). For unresolving ARDS, it is advisable to start with 2 mg/kg body weight of methylprednisolone and gradual tapering of the dose for 2 weeks after extubation (Meduri et al., 2009; Marik et al., 2011). In addition, secondary measures for prevention of treatment complications should be taken. These are intensive infection surveillance, avoidance of neuromuscular blocking agents to prevent neuromuscular weakness, and avoidance of rebound inflammation due to premature termination of treatment (Meduri et al., 2009; Marik et al., 2011). Early termination of treatment led to worsening of the physiological parameters and reintubation in the ARDSNet trial (Steinberg et al., 2006; Meduri et al., 2008). As for ALI (PaO_2/FiO_2 >200 mmHg), the role of glucocorticoids is less clear (Marik et al., 2008).

17.2.4.4 Experimental and Future Therapeutic Strategies

The current supportive treatment strategies in ALI/ARDS are resource intensive and costly and mortality is still high (Ware and Matthay, 2000; Wheeler and Bernard, 2007). As outlined previously, research and development efforts were made in order to improve the clinical efficacy, resolve safety concerns, prevent and minimize sequelae, and reduce costs, time of intensive care, and hospital stay in general. In spite of these efforts and progress in the quality of critical care, the decrease in mortality of ALI/ARDS patients in recent years is small (Zambon and Vincent, 2008; Phua et al., 2009), and the results of large, randomized, controlled trials show an overall trend in decreasing mortality from ARDS but without significant differences between the treatment and control arms (Spragg et al., 2010; Brower and Fessler, 2011). The approaches toward these goals where in the improvement of traditional modalities like mechanical ventilation, physical nonventilator techniques, pharmacological strategies, and, in recent years, novel modalities including growth factors (Lindsay, 2011), stem-cell and gene therapy, are being pursued.

17.2.4.4.1 Ventilatory Support

The principal challenge in ventilator support is to ensure blood oxygenation and minimize or prevent ventilation-associated lung injury. The low tidal volume, limited plateau pressure, and high PEEP protective strategies are not totally safe as some ARDS patients may not improve and even worsen with the ARDSNet protocol. The ARDS lung is not homogenous with respect to ventilation—some regions are normally ventilated, while others are poorly ventilated or not at all. In the patients whose lungs are not protected by the ARDSNet protocol, the normally ventilated regions are smaller than in the protected patients and thus undergo overdistension even in low tidal volumes, resulting in a higher tissue strain and inflammation (Terragni et al., 2007). For these patients, further reduction of tidal volume with increased PEEP and other maneuvers may protect the lungs and maintain adequate oxygenation but also result in unwanted hypercapnia and acidosis. Extracorporeal carbon dioxide removal ($ECCO_2R$) to afford *lung rest* by gentle ventilation enabled lung protection in animal models (Gattinoni et al., 1978) and by very low tidal volumes in patients who are not efficiently protected by the ARDSNet strategy (Terragni et al., 2009). The study included ARDS patients who were divided, after 72 h of standard ARDSNet protocol, to two groups on the basis of their plateau pressure, which was determined in a previous study (Terragni et al., 2007) as a surrogate for lung protectiveness. One group of patients, with plateau pressure between 25 and 28 cm H_2O, was treated with the ARDSNet protocol for 72 h until weaning. For the unprotected patients, with plateau pressure between 28 and 30 cm H_2O, tidal volume was reduced from 6.3 ± 0.2 to 4.2 ± 0.3 mL/kg and plateau pressure from 29.1 ± 1.2 to 25.2 ± 1.2 cm H_2O. These patients developed hypercapnia and acidosis, which were corrected by $ECCO_2R$. Efficient lung protection in these patients was evident by significant decrease in proinflammatory cytokines ($p = 0.001$) and improved lung morphology in CT scans. No patient-related complications were observed. The $ECCO_2R$ device used was a pump-driven veno-venous bypass consisting of a membrane oxygenator connected in series to a commercial hemofiltration unit. This system was demonstrated to be efficacious and safe in a prior study with sheep (Livigni et al., 2006) and simple to operate and is more similar in operation, logistic, and supportive care requirements to renal displacement therapy than common ECMO systems. A different system in current clinical testing is the pumpless interventional lung assist (iLA). This is a miniaturized single-use extracorporeal gas-exchange system in which blood flow through the membrane gas-exchange unit (iLA Membrane Ventilator®, Novalung, Talheim, Germany) is driven by the cardiac output in an arteriovenous (AV) bypass circuit (Bein et al., 2006). The system does not require any energy source or substrates, apart from oxygen. The only adverse effects that may occur are lower extremity ischemia because of the arterial cannulation, and it cannot be used in patients with hemodynamic instability. Increased blood oxygenation, reversal of hypercapnia, reduction in inflammatory cytokines, and enabling of gentler ventilation were noted when used as *rescue therapy* for patients with severe and refractory ARDS (Bein et al., 2006, 2009). A prospective randomized study compared the clinical efficacy of an *ultraprotective strategy*, based on a low tidal volume (~3 mL/kg predicted body weight) combined with $AV-ECCO_2R$ using the iLA system, to the standard ARDSNet strategy (tidal volume of 6 mL/kg) without $AV-ECCO_2R$ (Bein et al., 2013). The results showed that the AV-ECCO2R is easy to implement and safe, and although no significant change was seen in the primary clinical outcomes of 28-day and 60-day ventilator-free days, secondary outcomes of blood oxygenation and serum levels of the proinflammatory cytokine IL-6 were improved in the lower tidal volume and AV-ECCO2R group. In patients with more severe hypoxemia (PaO_2/FiO_2 <150 mmHg), the number of 60-day ventilator-free days was significantly higher in the AV-ECCO2R group compared to the controls (40.9 ± 1.8 vs. 28.2 ± 16.4, $p = 0.033$).

Interest in ECMO as a treatment option in ARDS rose recently after the CESAR trial in the United Kingdom (Peek et al., 2009) and the outcome of the use of this technology in the treatment of severe ARDS caused by the pandemic influenza A H1N1 2009 virus (Davies et al., 2009; Noah et al., 2011). In the CESAR trial, 180 patients with severe ARDS were randomized between 2001 and 2006 to conventional ventilation or referral to one specialist ECMO center. The conventionally treated controls were treated in their referring centers with no imposing of ventilation protocol by the trial directors. Of the 90 patients randomized to ECMO, 68 received the treatment and 57 (63% of the 90 ECMO intended patients) survived for 6 months without disabilities, compared to 41 of 87 (47%) among the conventionally managed patients (relative risk 0.69, 95% CI 0.05–0.97, $p = 0.03$). The study of Noah et al. (2011) compared the hospital mortality from influenza A (H1N1) 2009 virus in ARDS patients referred to four adult ECMO centers and similar patients receiving conventional therapy, and found a significantly lower mortality in the ECMO-referred patients. Although the evidence in favor of ECMO for severe ARDS is not complete, there is a general vision of ECMO as a future adjunct and even substitute for mechanical ventilation (Gattinoni et al., 2012). This is also because the technology is improving in the direction of miniaturized, more efficient, safe, and simple to use equipment like magnetic-driven centrifugal pumps, hollow-fiber membrane oxygenators, and advanced cannulation techniques from both $ECCO_2R$ and ECMO (Cove et al., 2012; MacLaren et al., 2012).

17.2.4.4.2 Pharmacological Strategies

Pharmacological strategies are regarded as attractive alternatives or supplements to the current supportive therapies. This need is obvious for the normal hospital or ICU environments but more so for TMCEs. Here, resources become not only inadequate but also of limited utility under field or prehospital conditions, and great efforts and investments are needed in stockpiling and mobilizing medical means like oxygen, oxygen delivery systems, and mechanical ventilators to the event arena, as well as health-care personnel with the required expertise (Baker, 1999; Rubinson et al., 2006;

Branson et al., 2008; Markel et al., 2008). In this regard, pharmacological interventions are less expensive and more field expedient from both operational and logistic standpoints (Grainge et al., 2009, 2010b). Pharmacological agents that target the multitude pathophysiological mechanisms of ALI/ARDS have been tested in vitro, small animal models, and clinical studies (Cepkova and Matthay, 2006). These include agents that target endothelial permeability, enhance alveolar fluid clearance, and target oxidative stress, inflammation, and lung tissue remodeling. Unfortunately, most of the large, randomized phase III clinical trials showed no significant improvement in primary clinical outcomes, even though phase II and some phase II/III trials showed improvement in gas exchange, edema, or inflammation (Cepkova and Matthay, 2006; Frank and Thompson, 2010; Brower and Fessler, 2011; Matthay et al., 2012). In the case of toxic inhalation injury, randomized, controlled trials are impractical and human studies are limited to case reports and observational cohort studies. Experimental testing of therapies can be done in in vitro, ex vivo, and small animal in vivo studies for discovery, proof of concept, and mechanism, and large animal models are employed as surrogates to human clinical studies (Jugg et al., 2011). In fact, some therapeutic options found efficacious in small animal models were not so and sometimes deleterious in large animal studies (Grainge et al., 2009; De Lange and Muelenbelt, 2010; Grainge and Rice, 2010), in analogy to the picture in the ALI/ARDS phase III clinical trials (Cepkova and Matthay, 2006; Frank and Thompson, 2010; Brower and Fessler, 2011). However, the *negative* results in human phase III and large animal studies did not result in categorical exclusion of interventions, especially those with a positive clinical experience in the past, but rather invoked second thoughts about their use like a shift in emphasis toward preventive treatment of patients with high risk of ALI/ARDS or at early stage, prior to acute respiratory failure (Litell et al., 2011; Levitt and Matthay et al., 2012). Even though most cases of ALI/ARDS enrolled in the clinical studies had etiologies of little or of no relevance to toxic inhalation injury, some of the animal experiments and phase II clinical studies, especially those that are mechanistically based or target pathological processes downstream from the primary insult, may be highly valuable in considering and evaluating potential interventions for toxic inhalation injury. This section will discuss examples of novel pharmacological strategies explored in the field of ALI/ARDS that may be relevant to toxic inhalation injury.

17.2.4.4.3   Statins in Prevention and Therapy of ALI/ARDS

The statins, as a group, are inhibitors of the enzyme hydroxymethylglutaryl-coenzyme A (HMG-CoA) reductase, a key enzyme in the synthesis of steroids and lipid moieties that control the function of enzymes and regulatory proteins (Singala and Jackson, 2012). These mechanisms underlie their pleiotropic pharmacological effects, of which the most relevant to ALI/ARDS are maintaining the integrity of the endothelial barrier by the cytoskeleton, modulation of ROS production by arrest of NOX and eNOS membrane binding, and modulation of ALI-relevant gene expression in endothelial cells (Singala and Jackson, 2012). Simvastatin treatment in mice attenuated lung injury induced by bacterial endotoxin (lipopolysaccharide, LPS) (Jacobson et al., 2005; Grommes et al., 2012) or mechanical ventilation (Müller et al., 2010). The prophylactic potential of simvastatin for lung injury was demonstrated in two human studies. In one study, prophylaxis with oral simvastatin (40 or 80 mg daily) attenuated the inflammatory response to inhalation of low dose of LPS (Shyamsundar et al., 2009). A cross-section study enrolling patients admitted to the ICU showed that 26% of patients who were on statin therapy before admission had a lower risk for ALI/ARDS (OR 0.60, 95% CI 0.36–0.39) compared to nonusers (O'Neal et al., 2011). A phase II randomized trial of simvastatin use in ALI showed that treatment was safe and showed a trend of improvement in blood oxygenation, lung mechanics, and inflammation (Craig et al., 2011). These animal and human studies justify continued testing in prevention of ALI and toxic lung injury.

17.2.4.4.4   α7 Nicotinic Acetylcholine Receptor Agonists

Immunological homeostasis is regulated by neural circuits, of which the most important is the anti-inflammatory reflex (Tracey, 2002; Andersson and Tracey, 2012). In this reflex pathway, afferent input invoked by inflammatory molecular signals, transferred to the brain stem through the vagus nerve, is relayed as efferent output via the vagal motor nerve fibers to the splenic macrophages. In these cells, signaling via the α7 nicotinic acetylcholine receptor (α7nAChR) downregulates the expression of the proinflammatory mediators TNFα, IL-1, IL-18, and HMGB1. α7nAChR agonists like nicotine and others have been tested as medications for nonresolving inflammatory disorders (de Jong and Ulloa, 2007). Systemic administration of α7nAChR agonists like nicotine attenuated ALI induced in several rat models (Su et al., 2007, 2010; Kox et al., 2011). Further studies are needed in order to establish their role in toxic inhalation injuries.

17.2.4.4.5 Neuromuscular Blocking Agents

Neuromuscular blocking agents have been used as adjuncts to mechanical ventilation or *rescue therapy* for severe and refractory hypoxemia, in order to promote patient–ventilator synchrony and improve oxygenation (Raoof et al., 2010). Although widely used, their use has been controversial as their therapeutic utility was not rigorously ascertained and there was a safety concern of muscular weakness, especially if given together with corticosteroids (Raoof et al., 2010). The efficacy of a short-term (48 h) treatment with the neuromuscular blocking agent cisatracurium besylate as an adjunct to conventional therapy in early severe ARDS was demonstrated in two multicenter, randomized, controlled studies in France (Gainnier et al., 2004; Forel et al., 2006). A subsequent large, multicenter, randomized, controlled study showed a clinical benefit of cisatracurium besylate in early severe ARDS (Papazian et al., 2010). The 90-day mortalities were 31.6% and 40.7% in the cisatracurium and placebo arms ($p = 0.08$), and the respective 28-day mortalities were 23.75% and 33.3% ($p = 0.05$). Although this was one of the few large, multicenter, randomized, controlled studies in ARDS clinical research that ended with a favorable result (Brower and Fessler, 2011), further studies are needed to replicate and extend these results before this intervention can be introduced as a standard clinical practice.

17.2.4.4.6 Liposome Delivery of Antioxidants and Radical Scavengers

Redox balance in lung tissue is maintained by enzymatic antioxidants like catalase, superoxide dismutase, glutathione peroxidase, glutathione S-transferase, as well as the low-molecular-weight antioxidant compounds reduced glutathione (GSH), ascorbic acid, vitamin E (α- and γ-tocopherol), and uric acid (Chabot et al., 1998; Lang et al., 2002; Chow et al., 2003). Although the oxidant–antioxidant balance is perturbed in ALI/ARDS in favor of the oxidative species (Lang et al., 2002; Chow et al., 2003), antioxidant therapy has received little attention in clinical research as the few small-scale clinical trials, testing the commonly used antioxidants N-acetylcysteine and vitamin E, had controversial results (Crimi et al., 2006; Mlcak et al., 2007; Wheeler and Bernard, 2007; Dushiantan et al., 2011). The reasons for this limited efficacy are attributable to limited bioavailability or biological stability of these medications (Suntres, 2011) and the systemic delivery (Crimi et al., 2006), which may be suboptimal for these materials. Local or direct delivery into the lungs has been tested with nebulized formulations or microencapsulation in liposomes. Liposomes are phospholipid bilayer membrane microparticles (around 100 nm in diameter), which can be used as drug vehicles. Antioxidant-containing liposomes have been tested in preclinical models of lung and other oxidative stress–driven diseases. Lipophilic materials like vitamin E can be incorporated into the membranous lipid outer layer, while water-soluble materials like N-acetylcysteine or antioxidant enzymes are entrapped in the aqueous inner side of the liposomes (Suntres, 2011). Liposome encapsulation stabilizes the drugs and allows the formation of localized high concentration of the drugs or allows the slow release of the drugs in the target organs. Liposomes are readily phagocytosed by macrophages and other phagocytic cells and therefore afford internalization of the materials inside inflammatory cells whose cellular oxidative balance is an important factor in the pathogenesis of lung injury (Hoesel et al., 2008; Suntres, 2011). Liposome-encapsulated antioxidants were tested in animal models for their potency in ameliorating both acute and long-term effects of lung exposure to alkylating agents, used as sulfur mustard surrogates (McClintock et al., 2006; Hoesel et al., 2008;

Wigenstam et al., 2009). In a rat model of lung injury induced by intratracheal (IT) instillation of the half-mustard analog 2-chloroethyl ethyl sulfide (CEES), N-acetylcysteine or N-acetylcysteine gluta-thione liposomes attenuated the short-term effect of increased lung permeability at 4 h postexposure, whereas α-/β-tocopherol or α-/β-tocopherol–N-acetylcysteine liposomes attenuated the long-term lung fibrosis at 3 weeks postexposure (Hoesel et al., 2008). In the same study, liposomes containing N-acetylcysteine attenuated the production of proinflammatory cytokines by alveolar macrophages exposed to CEES or LPS in vitro, thus indicating a role for macrophages and other phagocytes in the response to liposomal antioxidants. Systemic administration of liposomes containing α-tocopherol attenuated lung inflammatory responses including BAL fluid (BALF) polymorphonuclear leuko-cytes proinflammatory cytokines, long-term lymphocyte recruitment and pulmonary fibrosis in lung injury induced in mice by the nitrogen mustard derivative melphalan (Wigenstam et al., 2009). In this model, α-tocopherol liposomes were less efficacious than dexamethasone, which could also attenuate both early and long-term inflammatory responses and fibrosis. Still, this does not preclude their potential therapeutic value for chemical lung injury (Wigenstam et al., 2009).

17.2.4.4.7 Growth Factors, Gene and Stem-Cell Therapy

The limited efficacy of many therapeutic strategies in ALI/ARDS clinical studies has prompted the exploration of novel therapeutic modalities, modulating pathogenic and resolution processes by delivery of genes or gene products known to control these processes (Devaney et al., 2011; Lindsay, 2011; Zhu et al., 2011). The injury to the tissue initiates responses of repair and regeneration, which, in a highly inflammatory environment, are diverted to pathologic remodeling, resulting in hyper-reactivity and fibrosis (Beers and Morrisey, 2011). Candidate genes or gene products have been identified through molecular studies on the pathophysiology of lung injury and repair or remodeling of lung tissue (Lindsay, 2011).

17.2.4.4.8 Mesenchymal Stem-Cell Therapy

Mesenchymal stem cells (MSCs) have been studied as therapeutic agents for a multitude of tissue-injury diseases on the basis of their ability to promote tissue protection and repair processes and modulate immune responses without being immunogenic (Ren et al., 2012). Murine bone marrow and human umbilical cord–derived MSCs were shown to ameliorate lung injury in several in vivo and ex vivo lung disease models, including bacterial endotoxin (LPS)-induced lung injury and bleomycin-induced lung fibrosis (Lee et al., 2011; Sinclair et al., 2013). MSCs were shown to affect lung injury by secretion of soluble factors with the following activities: (1) modulation of inflammation by the release of the anti-inflammatory cytokine IL-10 in LPS-induced lung injury in mice (Gupta et al., 2007), reduction of proinflammatory cytokines, and expansion of the CD4+CD25+Focp3+ regulatory T cell population (Sun et al., 2011), which were shown to have a role in the control of inflammation in ALI (D'Allesio et al., 2009); (2) enhancement of lung fluid clearance via secreted keratinocyte growth factor in LPS-injured human lungs ex vivo (Lee et al., 2009); and (3) restoration of the per-meability barrier in inflammatory cytokine-injured cultured human alveolar epithelial type II cells by secretion of the endothelial growth and maintenance factor angiopoietin-1 (Ang-1) (Fang et al., 2010). In these studies, the MSCs were effective unmodified, but their protective efficacy against ALI was augmented further if they were engineered for enhanced expression of the lung-protective factors angiopoietin-1 (Mei et al., 2007) or keratinocyte growth factor (Aguilar et al., 2009). It may be postulated that MSCs in the injured lung provide the necessary microenvironment for tissue repair rather than remodeling (Sinclair et al., 2013). Clinical studies in MSC-based therapy for many nonpulmonary diseases are under way or have been completed (Ren et al., 2012), while clinical studies in lung diseases at present are very few, and the only completed COPD study did not show impressive results (Sinclair et al., 2013). We are not aware of ALI/ARDS clinical stem-cell studies, as the major impediments are knowledge gaps in the cell biology of MSCs, optimal dosage and mode of delivery, and incomplete knowledge on the interactions of pulmonary progenitor cells with MSCs in the injured lung (Hayes et al., 2012; Sinclair et al., 2013).

17.3 PROPERTIES AND THERAPY OF SELECTED TOXICANTS

17.3.1 CHLORINE

17.3.1.1 General Properties and Exposure Assessment

Chlorine gas (Cl_2) is a strongly oxidizing and highly corrosive material. Due to its extreme chemical reactivity, it has found extensive use in the chemical industry, especially in the production of chlorinated organic compounds used as precursors in the plastics and other industries. Chlorine and its derivatives, some of which release gaseous chlorine following decomposition, are used as disinfectants and bleaching agents at home, in industry, and for water treatment due to their oxidative and biocide capacities. In industrialized countries, chlorine is manufactured in large quantities and transported, mainly as compressed liquid, by rail and road. Human exposure incidents have occurred as a result of transport or industrial accidents, chlorination system malfunction, or unsafe use of chlorine-containing materials. Mass casualty incidents may occur in transportation accidents that result in release of huge quantities of chlorine, as exemplified by the tanker car collision in Graniteville, GA, in 2005 (Wenk et al., 2007; Van Sickle et al., 2009). Multiple-casualty events, although in a much smaller scale, resulted from chlorine emissions in malfunction of disinfection systems in swimming pools (Babu et al., 2008; Lehavi et al., 2008). As chlorine is 2.5-fold heavier than air, it tends to concentrate at low places, where the hazard may be greater, and this may aggravate the exposure consequences. In the Graniteville rail accident, the release of a huge quantity—approximately 46 tons—of chlorine gas in conditions of a stable atmosphere and low wind (≤ 2 m/s) led to high concentrations along a creek within a radius of 0.25 mile from the accident site, resulting in more severe cases, including nine deaths of which eight were attributed to asphyxia (Wenk et al., 2007; Van Sickle et al., 2009). Other cases with severe health consequences, up to ALI or ARDS, resulted from emission of chlorine in confined spaces like indoor swimming pools, home, or workplace (Mapp et al., 2000; Ho et al., 2010; Weiselberg and Nelson, 2011). Other cases of exposure occurred while mixing hypochlorite bleach with acid or alkaline cleaning materials (Mapp et al., 2000; Cevik et al., 2009). The outcome of acute chlorine exposure is affected by host factors like age—children are more vulnerable than adults due to their small size and different respiratory physiology (Hilmas and Hilmas, 2009). Another well-documented host factor is preexisting pulmonary disease such as asthma or increased lung responsiveness, which may aggravate the outcome in high-level exposure (Mapp et al., 2000) or amplify the response to low-level exposure (D'Alessandro et al., 1996). Due to its strong irritant activity, toxicity, and availability in huge quantities, chlorine gas has been the first chemical warfare agent used in combat in World War I (Tourinsky and Sciuto, 2008; Jones et al., 2010). Since then, it has been replaced by deadlier and more operational chemical weapons, but it is relevant as a terrorist threat—it has been used in attacks against British and US troops in Iraq (Jones et al., 2010) and is conceived as a credible threat agent in nonconventional terror scenarios (Kales and Christiani, 2004; Barrett and Adams, 2011).

Chlorine is slightly soluble in water (0.7% at 20°C) but more soluble than other inhalation toxicants like phosgene or ozone. Upon contact with water, it hydrolyzes to hypochlorous (HClO) and hydrochloric (HCl) acids; the former is a potent oxidizing agent by itself (Squadrito et al., 2010; White and Martin, 2010; Yadav et al., 2010). HClO reacts with other reactive oxygen and nitrogen species to form hydroxyl radicals (HO·), peroxynitrite (ONOO·), and other products. These react with functional groups in biomolecules, for example, sulfhydryl or hydroxyl and amino groups in proteins, and peroxidation of lipids, reactions that are detrimental to cellular functions (Squadrito et al., 2010; White and Martin, 2010; Yadav et al., 2010). HClO is regarded as the leading toxic species, as the acidity of HCl is buffered by bicarbonate in the airway epithelial lining fluid (Yadav et al., 2010). Chlorine (Cl_2) can react by itself with cellular constituents, and a recent analysis has concluded that it initially reacts with the low-molecular-weight antioxidants in the epithelial lining fluid—glutathione, ascorbate, and uric acid—and undergoes hydrolysis after their depletion (Squadrito et al., 2010).

17.3.1.2 Clinical Manifestations and Pathogenesis

The respiratory tract is the main organ system affected by exposure to chlorine. Depending on the dose, the acute effects range from a noxious odor and local irritation to airway obstruction and inflammation and in extreme cases to alveolar edema and ARDS (White and Martin, 2010; Yadav et al., 2010). The effects of low-level exposure are odor detection from 0.1 to 0.3 ppm; mild mucous membrane irritation tolerable for up to 1 h at 1–3 ppm; moderate mucous membrane irritation at 5–15 ppm; and immediate chest pain, dyspnea, and cough at 30 ppm, while at 40–60 ppm, toxic pneumonitis or pulmonary edema may develop. Higher exposure doses lead to a more severe lung injury that may be life threatening or lethal: concentrations of 400 ppm or above may be lethal within 30 min, and 1000 ppm or above may kill within minutes (Baxter et al., 1989; White and Martin, 2010). Death at high-level exposure is mainly from respiratory failure or cardiac arrest due to severe acute pulmonary hypertension (Baxter et al., 1989; Gunnrasson et al., 1998).

The irritating effects of low-level chlorine are dose dependent and include respiratory depression and increased airway resistance (Morris et al., 2005). At low exposure concentrations, inhaled chlorine is scavenged efficiently by the nasal and upper respiratory epithelial lining fluid, confining the direct damage to these sites (Nodelman and Ullman, 1999). At higher chlorine intake, the more distal airways, especially the bronchioles but also the alveoli, may be involved (White and Martin, 2010; Yadav et al., 2010). High-level, nonlethal exposure to chlorine was studied in rats and mice to model irritant-induced asthma or RADS (Demnati et al., 1998a; Martin et al., 2003). Exposure of rats to 1500 ppm chlorine for 5 min caused increases in lung resistance and reactivity to methacholine challenge that persisted for months. Histological changes in airway epithelium included initial necrosis and epithelial flattening, followed by increase in smooth muscle mass and epithelial regeneration with mucous cell hyperplasia, and increased number of neutrophils in BAL (Demnati et al., 1998a). In a study by the same group in mice (Martin et al., 2003), exposure to 400 or 800 ppm for 5 min, but not to 100 or 200 ppm, led to an increase in lung reactivity to methacholine, seen at 24 h and 7 days postexposure. Histologically, animals exposed to 100 ppm had mild airway damage, while in animals exposed to 800 ppm, the damage was more extensive—epithelial cell loss in all airways and alveolar edema and inflammation, including an increase in macrophages and neutrophils (but not eosinophils) in BAL. An important finding in this work was the biochemical demonstration of oxidative stress in chlorine-exposed mice, with a possible role for NO production by inducible NO synthase (iNOS) in the development of lung pathology (Martin et al., 2003). Lung tissue remodeling, evidenced by enhanced airway epithelium proliferation, smooth muscle hyperplasia, and fibrosis, together with a second increase in methacholine responsiveness was observed in mice 5–10 days after exposure to 800 ppm of chlorine (Tuck et al., 2008). Similar histological and functional findings indicating airway epithelial damage, inflammation, decreased pulmonary function, and increased methacholine reactivity were observed in human irritant–induced asthma due to acute high-level chlorine exposure (Deschamps et al., 1994; Lemiére et al., 1997). Other studies used anesthetized and mechanically ventilated large animals. The effects of inhalation of 140 ppm chlorine for 20 min in anesthetized and mechanically ventilated pigs resulted in an immediate respiratory distress and cardiovascular failure that caused the death of five out of six animals within 6 h of observation (Gunnarsson et al., 1998). The physiological pulmonary and hemodynamic effects were the main causes of early death from cardiac arrest in this model. Gross morphology and histological examinations showed interstitial edema, bronchial epithelial sloughing, and neutrophil infiltration, but the alveoli were intact. Exposure of anesthetized and mechanically ventilated sheep to chlorine, in which the determined LD_{50} was 280 ppm for 30 min, was similar to human ARDS, including the involvement of bronchial and alveolar damage and edema (Batchinsky et al., 2006). The overall picture of the pathogenesis in high-level exposure to chlorine features damage to airway epithelial cells, inflammation, and oxidative stress resulting from direct action of chlorine and its oxidant products, which may be exacerbated by oxidants produced by inflammatory cells (White and Martin, 2010; Yadav et al., 2010). The animal models described exemplified the broad spectrum of pathological effects of chlorine exposure in humans—irritation and asthma-like involvement of airway epithelia, ARDS in case of alveolar involvement, and cardiac failure due to pulmonary hypertension and the resulting hemodynamic disturbances. However, the detailed mechanisms of chlorine-invoked lung injury are not yet known and are an active field of research.

17.3.1.3 Medical Treatment of Chlorine Inhalation: Current Therapy

As in other cases of toxic inhalation injury, the current medical treatment is supportive and symptomatic (Baxter et al., 1989; Russell et al., 2006; Tourinsky and Sciuto, 2008; White and Martin, 2010). As the most severe complications of respiratory irritations are bronchospasm and pulmonary edema, patients with breathlessness at rest or mild effort and crackles on auscultation should be sought and prioritized for treatment. Others, with milder symptoms, can be released after an observation period of several hours if further worsening has not occurred (Baxter et al., 1989). The basic initial therapy for chlorine-induced lung damage is supplemental oxygen to obtain proper blood oxygenation and inhaled β2-agonists (generally albuterol/salmeterol) for bronchospasm (Baxter et al., 1989; ATSDR 2001c; Russell et al., 2006; Tourinsky and Sciuto, 2008; White and Martin, 2010; Segal and Lang, 2012). Inhaled or systemic corticosteroids are given when more severe symptoms occur or when further deterioration in the next few hours is imminent (Baxter et al., 1989; Babu et al., 2008; Cevik et al., 2009; Ho et al., 2010; Segal and Lang, 2012). In some cases, inhaled sodium bicarbonate was also used (Cevik et al., 2009; Van Sickle et al., 2009). In most cases, either individual or mass casualty events, most of the patients were discharged after their symptoms resolved with no evidence of further worsening. Varying proportions of patients, with more severe respiratory distress, hypoxemia, and radiographic or clinical evidence of pulmonary edema, were hospitalized for several days (Agabiti et al., 2001; Bonetto et al., 2006; Lehavi et al., 2008; Cevik et al., 2009; Mohan et al., 2010). Rare cases that present with ALI/ARDS require intubation and mechanical ventilation (Mapp et al., 2000; Bonetto et al., 2006; Babu et al., 2008; Van Sickle et al., 2009; Wieselberg and Nelson, 2011). Most of the hospitalized patients recover within a few days and are discharged without apparent symptoms, but longitudinal studies employing lung function tests, biochemical markers in exhaled air condensates, and methacholine or physical exertion challenges show that complete recovery may take several weeks to months (Agabiti et al., 2001; Bonetto et al., 2006; Mohan et al., 2010), and in a significant proportion of the patients, especially older or more severely injured, symptoms of decreased lung function or methacholine hyperresponsiveness may persist for years after exposure to chlorine (Schwartz et al., 1990).

Prehospital care is basic and should follow the guidelines outlined previously and in the references (Baxter et al., 1989; ATSDR, 2001c; Segal and Lang, 2012). When the number of casualties is small (usually less than a hundred), and the health-care facility networks are well developed, casualties may be evacuated directly to the hospitals (Baxter et al., 1989). This has been experienced in Israel, Turkey, Italy, and India (Agabiti et al., 2001; Lehavi et al., 2008; Cevik et al., 2009; Mohan et al., 2010). All patients were treated at the hospital with supplementary oxygen and inhaled β2-agonists. Most of them were discharged after several hours, and the more severe cases required hospitalization, to receive additional treatment with IV hydrocortisone and aminophylline. Patients with severe chest pain and severe bronchospasm were admitted to ICU. They received also IV furosemide for pulmonary edema. None of the patients died. In the 2005 Graniteville accident, most of the patients, including the severely ill and those who sought care immediately after the collision, reached the hospitals in privately owned vehicles (Wenk et al., 2007). Other methods of transportation included EMS ambulances and police cars. The Aiken county emergency services established three decontamination stations at the Graniteville area, which were operative within 1.5 h from the event, and decontaminated people fleeing from the scene. People were also decontaminated at four hospitals, where they were given a private place to change their clothes, hosed with cold water, and given disposable clothing or blankets (Wenk et al., 2007). Hospital management of chlorine inhalation victims is described in Box 17.3.

BOX 17.3   HOSPITAL MANAGEMENT OF CHLORINE EXPOSURE

A. Immediate decontamination on arrival at hospital
   • Remove all contaminated clothing and thoroughly wash skin and eyes as necessary.
B. Assessment of patient
   • Examine mucous membrane, eyes, and skin for signs of corrosive injury.
   • Check lung sounds, peak flow, and vital signs: if patient is known to have been heavily exposed or has a cough or difficulty in breathing at rest, perform baseline chest x-ray examination.
   • Take brief medical history, with particular attention to any history of respiratory or cardiovascular disease.
C. Initial treatment
   • *Oxygen*: All patients identified as being at risk (see B earlier) should initially receive 100% oxygen, humidified if possible (unless this is contraindicated by their medical history). Oxygen concentration may subsequently be adjusted to the comfort of the patient.
   • *Bronchodilators*: Salbutamol or terbutaline, used by nebulizer, may help relieve respiratory difficulties.

   Dose: **Salbutamol**: adults—2.5–5 mg increasing to 10 mg if
                                       required
                                       children—2.5 mg increasing to 5 mg if
                                       required
             Terbutaline: adults—2–5 mg increasing to 10 mg if
                                       required
                                       children—up to 200 pg/kg
   • *Corticosteroids* have not been proved to produce improvement in chlorine poisoning but have caused pronounced improvement after smoke inhalation. If patient is exposed less than 4 h previously and at risk of pulmonary injury (as defined in B earlier), steroids should be given.
     Dose: **Methylprednisolone**: adults—2 g IV stat (or equivalent)
                                       children—400 mg IV stat (or equivalent)
   • *Laryngeal edema*: give corticosteroids (dosage as earlier) if patient develops laryngeal edema.
   • *Skin burns* should be treated as thermal burns.
   • *Eye damage* requires ophthalmic referral.
D. Monitoring
   • Monitor respiratory function and arterial blood gases regularly: pulmonary edema may occur up to 24 h after exposure. Patients who are well 24 h after exposure may be discharged.
E. Pulmonary edema
   • If pulmonary edema occurs, give 60% humidified oxygen by face mask or mechanically.
   • If PO_2 still cannot be maintained above 50 mmHg, intubate and give PEEP ventilation.
   • Intravenous fluids should be given with great caution as fluid overload is extremely dangerous in such patients: if this occurs, diuretics such as furosemide are indicated.

Source: Based on Baxter, P.J. et al., *Br. J. Ind. Med.*, 46, 277, 1989.

17.3.1.3.1   Corticosteroids in the Management of Chlorine Lung Injury

Corticosteroids are frequently used in chlorine lung injury even though their clinical value has not been rigorously proven and is not unambiguously supported by the clinical evidence (Russell et al., 2006). The efficacy of corticosteroids in the treatment of lung injury following exposure to chlorine is better supported by animal studies (Russell et al., 2006; De Lange and Muelenbelt, 2011). Intraperitoneal dexamethasone (300 µg/kg/day, 7 days) attenuated the effects of high-level exposure (1500 ppm for 5 min) in the rat RADS model: treatment with dexamethasone reduced airway resistance, methacholine responsiveness, BAL neutrophil levels, and histopathological changes compared with saline-treated control animals (Demnati et al., 1998b). The efficacy of inhaled or systemic corticosteroids was demonstrated in the anesthetized and mechanically ventilated pig model. A single dose of nebulized budesonide (0.1 mg/kg) attenuated the physiological effects and reduced pulmonary edema following exposure of pigs to chlorine (400 ppm for 5 min) if given 30 min after exposure (Wang et al., 2002a). In another study with this model, the effects of inhaled budesonide and inhaled β2-agonist, terbutaline, were assessed when given alone or in combination. In this experiment, each drug was equally effective in attenuating the toxic effects when given alone, but a combined treatment with aerosolized terbutaline (0.1 mg/kg) followed by aerosolized budesonide (0.1 mg/kg) was superior to treatment with each drug alone (Wang et al., 2004). The efficacy of inhaled and intravenous corticosteroids was compared in a modified protocol, in which the exposed pigs were ventilated in prone position (Wang et al., 2005). Inhaled budesonide and intravenous betamethasone were equally effective in further enhancing the improvement that occurred during 23 h of ventilation in prone position compared to animals receiving placebo as additional treatment.

Case reports of individuals and groups of chlorine-exposed patients whose treatment included corticosteroids suggest their safety and possible efficacy. The reports cover a broad range of clinical severity, from moderate irritation and inflammation to ARDS, and some of these will be outlined as examples. An illustrative example is a case of two sisters exposed to chlorine gas at their home during an industrial accident (Chester et al., 1977). The older sister arrived at the emergency room with severe cough, chest pain, and lacrimation. Chest radiography showed bilateral infiltrates in the midpulmonary zones. She received oxygen therapy and was sent home after a few hours of observation, where she was symptomatic and bedridden for several days and remained moribund for the next 4 years with dyspnea and chronic productive cough. The younger sister presented with more severe symptoms (initially hoarseness, burning sensation in the eyes, and later marked dyspnea, productive cough, and bilateral diffuse rales on auscultation) and chest radiography (bilateral diffuse infiltrates in the lower pulmonary zones). She was hospitalized, received oxygen therapy, and started corticosteroid treatment with 100 mg of IM hydrocortisone and an oral dose of 60 mg prednisone, which was continued at a dosage of 60 mg daily for 2 days and then reduced to 45 mg. On the second day, her condition worsened, but started to improve on day 4, and on day 5, as the symptoms abated, the chest radiography became normal, therapy was stopped, and she was discharged. Lung functions, lung mechanics, and gas-exchange studies done along a period of 55 months showed marked differences between the two sisters. The younger sister reached the normal value range within 1 month and continued to improve. The older sister, who did not receive corticosteroids, showed only a marginal improvement throughout the follow-up period. This case may imply that corticosteroid treatment had a role in sparing the younger sister from the persistent inflammation, fibrosis, and impaired respiratory physiology of her older sister. A detailed study reported on a single patient who developed RADS after exposure to a high level of chlorine while working at a water treatment plant (Lemiére et al., 1997). Sequential spirometry and methacholine responsiveness tests showed improvement with corticosteroid therapy (inhaled budesonide, 0.8–1.6 mg/day) from a PC_{20} (the concentration of methacholine causing 20% reduction of FEV_1) of 2.5 mg methacholine/mL initially to 57 mg/mL after 1 month of therapy, worsening (4 mg/mL) 1 month after cessation of therapy, and renewed improvement when therapy was resumed (inhaled budesonide, 0.8 mg/day), with final normalization of methacholine responsiveness (PC_{20}, 24 mg/mL), and improved lung biopsy finding, showing repaired epithelium and decreased inflammation 5 months after the exposure. A young woman who was exposed to chlorine in a swimming pool accident deteriorated to ARDS (Babu et al., 2008). She was treated with methylprednisolone (a bolus of 2 mg/kg initially and 2 mg/kg divided into two doses per day) as a rescue therapy starting on day 10 of intubation. She finally recovered, with slow tapering of steroids (Babu et al., 2008). A combined administration of inhaled budesonide and nebulized sodium bicarbonate, the latter given to neutralize the acids formed by the hydrolysis of chlorine in the respiratory tract, was used as additional therapy to the standard oxygen and inhaled salbutamol in the multi casualty exposure in Turkey, mentioned earlier (Cevik et al., 2009). All patients received humidified oxygen and inhaled salbutamol, but the more severe patients diagnosed with cough and dyspnea, sore throat, chest pain, and pathological pulmonary auscultation findings received an additional treatment of inhaled budesonide and bicarbonate (4 mL of 3.75% solution). Of these patients, about a half was discharged and the others were hospitalized. All patients recovered. Treatment with inhalation of nebulized sodium bicarbonate was practiced, in several occasions of chloride exposure, with beneficial results and without adverse effects (reviewed in Jones et al., 2005). It was used in two patients of the Graniteville accident, and the reporting authors recommended considering it as a possible therapy in future mass events of chlorine exposure (Van Sickle et al., 2009).

17.3.1.4 Experimental and Future Therapies

The prevailing trend in the search for novel treatment measures against chlorine-induced lung injury is a transition from empiric to scientifically based treatment modalities, incorporating the growing knowledge on the mechanisms of chlorine lung-injury pathogenesis. The main issues addressed in this research are oxidative stress and signal transduction pathways in inflammatory and toxic processes. Antioxidants are not included in the current therapy, and β2-agonists and corticosteroids were incorporated on an empirical basis, with little consideration of their mechanisms of action, which are not yet completely clear in the context of chlorine lung injury. Novel experimental therapy studies reported recently used antioxidants, cyclic adenosine monophosphate (cAMP) modulators, and sodium nitrite.

17.3.1.4.1 Antioxidants

The effects of antioxidants on chlorine-induced lung injury were studied in conscious rats, exposed to 184 and 400 ppm of chlorine in ambient air (Leustik et al., 2008). Exposure caused lung injury with a dose-dependent severity, which resolved after 24 h in the low-dose exposed rats but persisted in the high-dose exposed animals. BALF and lung tissue showed a dose-dependent decrease in ascorbate, reduced glutathione, and increase in urate, which were reversible after 24 h in the 184 ppm exposed rats. Systemic pretreatment with an antioxidant mixture of N-acetylcysteine, ascorbate, and the iron chelator deferoxamine, injected 18 h and 1 h before exposure to 184 ppm of chlorine, mitigated the injury, as seen by the higher level of BALF ascorbate, arterial blood oxygenation, and reduced BALF protein compared to saline-injected controls. Postexposure administration of ascorbate and deferoxamine, either systemic or by inhalation, reduced lethality and lung injury in mice exposed to 600 ppm chlorine for 45 min. Treatment started 1 h after the end of exposure and continued every 12 h by injection and every 24 h by inhalation for 3 days (Zarogiannis et al., 2011). Postexposure treatment of rats exposed to 400 ppm of chlorine for 30 min, and given ascorbate and deferoxamine by inhalation 1 and 15 h afterward, mitigated lung hyperplasia and methacholine hyperreactivity, which returned to normal level after 7 days (Fanucchi et al., 2012). Treatment of mice with the antioxidant dimethylurea, either 1 h before or 9 h after exposure to chlorine (100 ppm of chlorine for 5 min), reduced airway hyperreactivity to methacholine, with reduction of markers of inflammation and oxidative stress (McGovern et al., 2010).

17.3.1.4.2 Cyclic Adenosine Monophosphate Modulators

β2-Adrenergic agonists are frequently used in the therapy of chlorine-induced lung injury as bronchodilators, but their value as a therapeutic strategy in the broader context of their pharmacological properties as regulators of cAMP levels has been explored only recently (Hoyle, 2010). Postexposure treatment of chlorine-exposed mice (400 ppm for 30 min) with the long-acting β2-adrenergic agonist arformoterol, administered by intranasal instillation 10 min and every 24 h postexposure, ameliorated airway hyperreactivity and reversed the chlorine-induced reduction in cAMP levels and alveolar fluid clearance but had no effect on the inflammatory responses (Song et al., 2011). Another pharmacological strategy of cAMP level manipulation is inhibition of type 4 phosphodiesterase, which is the main regulator of cAMP turnover in pulmonary inflammatory cells (Hoyle, 2010). The effect of the type 4 phosphodiesterase inhibitor rolipram was studied in mice exposed to chlorine as a possible postexposure rescue therapy (Chang et al., 2012). Intraperitoneal or intranasal administration of rolipram to chlorine-exposed mice inhibited pulmonary edema and airway hyperreactivity. No reduction was seen in chlorine-induced BALF IgM concentration, consistent with a mechanism centered on alveolar fluid clearance enhancement and not epithelial or endothelial repair. Also, minor or no effects were observed on levels of inflammatory markers. These studies indicate that both approaches are feasible as postexposure rescue therapies with different advantages and limitations. The major advantage of β2-agonists is their long history of use in lung injury. The gradual decrease in their efficacy in chronic use due to desensitization is a disadvantage. This problem does not occur with phosphodiesterase inhibitors, but they have gastrointestinal side effects (Hoyle, 2010). Both limitations may not be detrimental in the context of chlorine lung injury in view of the generally limited time period of administration and the local mode of delivery.

17.3.1.4.3 Sodium Nitrite

In recent years, sodium nitrite has obtained a special attention as a therapeutic agent in conditions involving ischemia–reperfusion, hypoxic hypertension, and other conditions due to its vasodilator, anti-inflammatory, and cytoprotective effects. These are carried out by direct signaling or mediated by reduction to NO, thus serving as an alternative source for this messenger under hypoxic conditions (Weitzberg et al., 2010; Zuckerbraun et al., 2011). The effect of sodium nitrite was studied in the rat model of chlorine inhalation exposure (Yadav et al., 2011). Rats were exposed to 400 ppm of chlorine for 30 min and received sodium nitrite (1 mg/kg, IP) or saline at 10 min and 2, 4, and 6 h postexposure. Nitrite decreased the early rise in BAL protein and lung water content (6 h postexposure), decreased apoptosis of airway epithelial cells, and increased tissue repair, but inflammation and lipid peroxidation were not affected. A later study with the same model, testing different doses and modes of delivery (Samal et al., 2012), showed that IM administration was as effective as IP in reducing edema, lung hyperreactivity, and inflammation at doses of 0.5–1.0 mg/kg. The efficacy of sodium nitrite in the IM mode of delivery makes it suitable for clinical use at the prehospital environment. The material is readily available and chemically stable and may have a good safety profile in humans, as the relevant doses, which are used in studies of ischemic diseases, are far lower than the dose used for cyanide poisoning (~300 mg/kg IV infusion for adults). As a different pattern of dose dependencies was seen in this study in the efficacy of sodium nitrite against edema and inflammation, more research is required in order to define the precise therapeutic window, as well as its efficacy in combination with other medications like ascorbate and β2-agonists (Samal et al., 2012).

Translation of novel therapies to chlorine exposure from animal studies to human medical practice is problematic, as clinical trials are not feasible due to the rarity of chlorine exposure and ethical restrictions on intentional human exposure. The current conventional therapies are lifesaving, but their efficacy in preventing long-term complications is not proven. Another obstacle lies in knowledge gaps in the mechanism of chlorine-induced lung injury. Translation from the experimental animal studies to clinical practice requires further research to include detailed pathophysiology of chlorine-induced injury, the mechanism of action (and pharmacokinetics) of candidate countermeasures in animal models with relevance to humans, as well as assessment of safety in humans. The case of a promising candidate like sodium nitrite IM is illustrative of the research and development path awaiting any novel therapy, including often-used practices like corticosteroids, β2-agonists, and antioxidants in novel therapeutic contexts. A precedent to a pathway of approval of countermeasures that cannot be tested clinically all the way because of ethical restraints is the FDA *Animal Rule* from 2002. This pathway was established to enable approval of medical countermeasures for CBRN threats, which cannot be assessed in phase III trials (Abersold, 2012).

17.3.2   Phosgene

17.3.2.1   General Properties and Exposure Assessment

Phosgene is a colorless gas at room temperature and a fuming liquid below 8°C. The gas has a weak odor of mown hay, but this is not a reliable warning as the human odor threshold (0.4–1.5 ppm) is five times higher than the OSHA occupational exposure level (OEL, 0.1 ppm or 0.4 mg/m^3, averaged over 8 h) (Glass et al., 2009). It is also 3.5 times heavier than air, so when released, it tends to concentrate at low places. Phosgene was first synthesized in 1812 by passing carbon monoxide and chlorine over charcoal, a process used today as well in production of the material. Due to its high chemical reactivity as an acylating agent, phosgene is widely used and produced in large amounts by the chemical industry (Glass et al., 2009). As the greater part of the immense production volume is done on site by the consumers, only small quantities are shipped as compressed or liquefied gas (Pauluhn et al., 2007). Exposure risk may result from industrial accidents, which are uncommon (Noltkamper and Burgher, 2012). Another source of exposure to phosgene is decomposition of halogenated hydrocarbons. Some cases of lung injury were caused by occupational or domestic exposure to phosgene and other products of thermal or photochemical decomposition of chlorinated hydrocarbons or freons in poorly ventilated sites (Sjögren et al., 1991; Snyder et al., 1992; Wyatt and Allister, 1995). Phosgene was used by both Germany and the Allies as a deadly chemical warfare agent in World War I (Russell et al., 2006; Tourinsky and Sciuto, 2008). Today, it has a minor, if any, military significance, but it is regarded as a highly credible chemical terrorism threat agent due to its toxicity and availability (CDC, 2012).

Phosgene is less soluble in water than chlorine. In contact with water, it hydrolyzes to carbon monoxide and hydrochloric acid, but the rate of this reaction is insufficient to have any importance in toxicity. Therefore, phosgene is not scrubbed by the lining fluid of the upper airways and penetrates into the deep airways where the toxic processes take place (Diller, 1985a; Pauluhn et al., 2007).

17.3.2.2   Clinical Manifestations and Pathogenesis

The main feature of phosgene-induced lung injury is noncardiogenic pulmonary edema, damage to the lower airways and alveolar epithelial cells, and inflammation, resulting in hypoxemic respiratory failure and lethal hemodynamic instability (Diller, 1985a; Pauluhn et al., 2007).

The concentration–effect relationships for humans are shown in Table 17.2.

TABLE 17.2
Effects of Phosgene Exposure in Humans

Effect	Concentration or Dose
Perception of odor	>0.4 ppm
Recognition of odor	>1.5 ppm
Irritation of eyes, nose, throat, and bronchi	>3 ppm
Beginning of lung damage	>30 ppm
Clinical pulmonary edema	>150 ppm
LCt_{01}	>300 ppm min(~1200 mg min/m^3)
LCt_{50}	~500 ppm min(~2000 mg min/m^3)
LCt_{100}	~1200 ppm min(~5200 mg min/m^3)

Sources: Adapted from Diller, W.F., *Toxicol. Ind. Health*, 1, 7, 1985a; Diller, W.F., *Toxicol. Ind. Health*, 1, 93, 1985b.

The course of phosgene poisoning can be divided into three phases (Coman et al., 1947; Diller, 1985a; Pauluhn et al., 2007; Grainge and Rice, 2010): The initial and immediate response to exposure of >3 ppm is a protective vagal reflex, characterized by slowdown of respiration and heart rate, which may sometimes be accompanied by an irritation response to the hydrolysis product HCl. This phase is followed by a clinical latency phase of 6–48 h. Its duration depends on the exposure dose. During this phase, the edematous processes start taking place, and the injury progresses without being noticeable. The clinical edema phase follows when lung injury becomes extensive enough to be clinically apparent. The initial symptoms are cough, dyspnea, tachypnea, and respiratory distress. This phase is accompanied by inflammation, for example, neutrophil infiltration, and pulmonary fibrosis. Exposure to >200 ppm results in apnea for several seconds, epithelial desquamation, inflammatory bronchiolar changes, and death from cardiopulmonary arrest. Animal model studies have shown that the initial pulmonary effect of phosgene is depletion of pulmonary surfactants, leading to increased permeability of the blood–air barrier and influx of fluid and proteins into the interstitial and alveolar spaces (Frosolono and Curie, 1985; Pauluhn et al., 2007). This edematous process was seen in animals—rats, mice, and guinea pigs—by an early rise in BAL protein levels (1–4 h) after exposure to sublethal doses, before any injury is detected histologically (Sciuto, 1998; Duniho et al., 2002). Oxidative stress evident by lipid peroxidation and increased levels of glutathione in BALF were also observed at early stages in rodents (Sciuto, 1998). Neutrophil and other inflammatory cell infiltration, epithelial cell damage, and deposition of fibrin and collagen were observed later, followed by cellular proliferation and tissue repair (Duniho et al., 2002). These changes subsided within 2–4 weeks (Pauluhn, 2006a–c). Similar changes were also observed in phosgene-exposed dogs, but the immediate respiratory pattern did not show the vagal respiratory depression reflex observed in rodents (Pauluhn, 2006a,c). A porcine model of lethal exposure to phosgene was developed in order to provide a clinically relevant large animal model (Brown et al., 2002). Anesthetized pigs were exposed for 10 min to phosgene at a Ct of 2443 mg min/m^3, which caused the death of 4/5 animals within 24 h. Physiological measurements showed increases in blood pH, pulmonary artery pressure, and pulmonary vascular resistance and decreases in blood oxygenation (arterial PO$_2$) and lung compliance within 6–15 h postexposure until death or euthanasia. Postmortem histological examination of the lungs showed extensive inflammation, cell necrosis, and alveolar edema (Brown et al., 2002). Other experimental models and pathophysiological processes will be discussed in the context of experimental therapies.

17.3.2.3 Current Therapies

Currently, there are no specific antidotes to phosgene, and the medical treatment is supportive and symptomatic (Russell et al., 2006; Tourinsky and Sciuto, 2008; Grainge and Rice, 2010). A major difficulty in the management of phosgene poisoning is the occurrence of an asymptomatic latent period, which follows an initial phase of nonspecific minor symptoms. As outlined previously, this latent period is very active in the buildup of pulmonary edema and other pathophysiological features, so that the symptomatic phase takes place when the pulmonary injury is progressive. Therefore, asymptomatic victims should be referred to a medical facility for evaluation and follow-up, patients with respiratory complaints should be admitted to intensive care unit, and those with bronchorestriction or signs of pulmonary edema should be watched for risk of respiratory failure (ATSDR, 2001d). Monitoring for lung edema development by means like chest radiography is recommended (Diller, 1985b; ATSDR, 2001d), and as the duration of the latent period is inversely proportional to the exposure dose, patients exposed to a high dose should have chest radiography from as early as 2 h postexposure (Diller, 1985b).

Patients suspected of occupational exposure to phosgene should be managed promptly according to their clinical condition and diagnosis. A few case reports from the literature are outlined in the following to illustrate this point. A mild symptomatic patient was hospitalized for a 24 h observation following occupational exposure to phosgene. He initially had lacrimation, nausea, and cough with burning sensation in his mouth and throat, and subsequently, he experienced dyspnea and chest pain. He made a slow recovery in the following 2 weeks, suffering from lethargy and exertional dyspnea (Wyatt and Allister, 1995). Another patient, presenting with cough, dyspnea, infiltrates in chest x-rays, and diffuse wheezing and rales, was treated with supplemental oxygen and albuterol and released after 48 h but had subsequent RADS that was responsive to albuterol for at least a year after the incident (Snyder et al., 1992). A patient exposed to a *massive* dose of phosgene felt initially a severe burning sensation of the eyes and cough, which resolved after removal from the exposure site and stripping off his clothes, but the symptoms resumed and worsened in spite of supplementary oxygen treatment and deteriorated to overt pulmonary edema within 6 h. In the hospital, he was treated with supplementary oxygen, mechanical ventilation, antibiotics, hydrocortisone, inhaled isoproterenol, and intravenous aminophylline and recovered after 5 days (Everett and Overholt, 1968). The authors recommended the use of aminophylline as a smooth muscle relaxant and a respiratory center stimulant, as well as steroids and prophylactic antibiotics (Everett and Overholt, 1968). Another severely exposed patient was a welder exposed to phosgene, generated as a photochemical decomposition product of trichloroethylene while working in a poorly ventilated environment (Sjögren et al., 1991). At night, he was admitted to the ICU with severe dyspnea and clinical and chest radiographic evidence of pulmonary edema. He was treated with supplementary oxygen and intravenous diuretics, antibiotics, and prednisolone, starting with 80 mg/day. After transfer to an ordinary ward, he returned to the ICU due to respiratory failure and was mechanically ventilated for 5 days. BAL taken at this phase showed marked neutrophilia (51%, compared to the normal value of 1%). As the patient's condition improved, the prednisolone dosage was reduced gradually down to 30 mg/day upon leaving the hospital after ~50 days of hospitalization (Sjögren et al., 1991). As these examples show, the treatment included supplementary oxygen, corticosteroids, bronchodilators, and mechanical ventilation if needed. The scarce reports in the literature on occupational exposure to phosgene show that most of the treated patients recovered (some with sequelae) and there were some deaths from cardiac complications (Glass et al., 2009).

An exposure dose-oriented management strategy was suggested by Diller (1985b), using different monitoring systems in the industrial setting. According to this strategy, exposure to doses smaller than 25 ppm min (100 mg min/m^3) may be regarded harmless, and exposed victims without any signs, symptoms, or complaints may be discharged immediately. After exposures to doses of 50–150 ppm min (200–600 mg min/m^3), subclinical pulmonary edema with inflammatory changes in distal bronchioles may be expected. Inhaled corticosteroid administration and medical supervision for at least 8 h are recommended, and patients may be discharged if chest radiography shows no signs of pulmonary edema. If chest radiography is not available, this supervision period should be extended to 24 h. After exposure to doses >150 ppm min, pulmonary edema must be expected, and such patients should be given large doses of corticosteroids (inhaled and IV) and referred to an ICU for definitive care. Such a dose-oriented strategy is most suitable for industrial facilities where personal dosimetry devices for phosgene are used. Environmental monitoring and model-based hazard prediction by HAZMAT professionals are highly helpful and important in exposure assessment and triage of exposed victims and should be implemented as an essential element in the planning and management of large-scale industrial accidents or chemical terror events.

Decontamination is needed only for victims with skin or eye irritation or contaminated with liquid phosgene when the ambient temperature on site is below 8°C. Supplemental oxygen therapy is recommended for symptomatic phosgene-poisoned victims, but recent studies have shown that it can be deferred until circumstances are more favorable and the fraction of inspired oxygen concentration (FiO$_2$) can be minimized to normalize oxygen saturation and avoid oxygen toxicity. This in turn may help in avoiding excessive use of oxygen, which may be limiting under such conditions (Grainge et al., 2010a; Grainge and Rice, 2010). More detailed recommendations for medical care of phosgene inhalation, based on large animal studies (Grainge and Rice, 2010), are listed in Table 17.3, in the ATSDR guidelines (ATSDR 2001d) and at the end of this section.

TABLE 17.3
Management of Phosgene Inhalation Injury

Small-Scale Confirmed Exposure	Large-Scale Unconfirmed Exposure
Patients should be rested and observed (repeated physical examination and SaO$_2$), and chest x-rays should be performed some 12–24 h after exposure or earlier if clinically indicated and repeated if appropriate.	Patients should be rested and observed (repeated physical examination and SaO$_2$).
An intravenous bolus of high-dose corticosteroid (e.g., methylprednisolone 1 g) may be considered if presentation is <6 h (although there are no experimental data to support this recommendation[a]).	An intravenous bolus of high-dose corticosteroid (e.g., methylprednisolone 1 g) may be considered if presentation is <6 h, although there are no experimental data to support this recommendation, and resources allow[a].
Nebulized N-acetylcysteine 1–2 g (5–10 mL of 20% solution) may be considered (although there is no substantial evidence of benefit and there is a possibility of adverse effects[b]).	If SaO$_2$ falls below 94%, the patients should receive the lowest concentration of supplemental oxygen to maintain SaO$_2$ in the normal range.
If SaO$_2$ falls below 94%, the patients should receive the lowest concentration of supplemental oxygen to maintain SaO$_2$ in the normal range.	Once patients require oxygen, nebulized β-agonist (e.g., salbutamol, 5 mg by nebulizer every 4 h) may reduce lung inflammation (although delayed administration has not been tested formally).
Once patients require oxygen, nebulized β-agonist (e.g., salbutamol, 5 mg by nebulizer every 4 h) may reduce lung inflammation (although delayed administration has not been tested formally).	Ventilation should be initiated either electively once symptoms persist (especially if there is a short latent period, indicating the likelihood of more significant injury) or delayed until required. Ventilation should be with high PEEP, ARDSNet recommended ventilation strategy.
Consider elective intubation early using the ARDSNet protective ventilation strategy as it lessens injury and significantly improve survival.	

Source: Adapted from Grainge, C. and Rice, P., *Clin. Toxicol.*, 48, 245, 2010.

[a] The evidence shows no benefit from nebulized steroids even if administered 1 h after exposure or high-dose corticosteroids if administered intravenously ≥6 h after exposure.

[b] The only evidence comes from small animal or isolated lung studies, and these suggest early administration (not later than 50 min postexposure), making this type of intervention suitable only for small-scale and confirmed exposure events (Grainge and Rice, 2010).

17.3.2.4 Experimental and Future Therapies

Experimental studies in the management of phosgene poisoning, done in the years 1989–2010 (reviewed by Grainge and Rice, 2010), employed several model systems. These were isolated per-fused rabbit lungs (Sciuto and Hurt, 2004); small animals, rats and mice (Sciuto and Hurt, 2004; Pauluhn et al., 2007); large animals, awake dogs (Bruner et al., 1947; Coman et al., 1947; Pauluhn, 2006c); and anesthetized pigs (Brown et al., 2002). The isolated lungs and small animal models were most useful in understanding the pathogenesis of phosgene-induced lung injury and discovery of pharmacological treatment options, while the large animals were used to simulate realistic clini-cal management and evaluation of supportive practices like supplemental oxygen and mechanical ventilation (Brown et al., 2002; Grainge and Rice, 2010; Jugg et al., 2011).

17.3.2.4.1 Studies in Perfused Lungs and Small Animals

17.3.2.4.1.1 Modulators of Cyclic AMP In the rabbit lung model, rabbits were exposed to a cumulative dose of 1500 ppm min of phosgene, after which they were sacrificed, their lungs isolated, perfused with physiological solution, and mechanically ventilated for a period of 60 or 150 min during which injury-related physiological and biochemical measurements and observations were made (Sciuto and Hurt, 2004). The observed effects of phosgene exposure were lung weight gain and fluid flux increase, albumin leakage, bronchoconstriction, oxidative stress including lipid peroxidation and depletion of reduced glutathione, and changes in toxicity-related mediators. These included increased leukotriene $C_4/D_4/E_4$ and decreased cAMP levels (Sciuto and Hurt, 2004). Treatment of the animals with dibutyryl-AMP, the phosphodiesterase inhibitor aminophylline, and the β2-agonist terbutaline, either before or after exposure to phosgene, ameliorated the injury parameters measured in the lung-perfusion studies either immediately or 4 h after exposure (Kennedy et al., 1989). Perfusion of phosgene-exposed lungs with aminophylline (Sciuto et al., 1997) or the β2-agonist isoproterenol (Sciuto et al., 1998) ameliorated the edema, leukotriene production, oxidative stress, and the phosgene-induced reduction in cAMP levels. These experiments demonstrated the role of the reduction of cAMP in the pathogenesis of phosgene-induced lung injury and the possible protective utility of drugs that elevate cAMP levels and promote its beneficial effects of smooth muscle relaxation, inhibition of inflammatory mediator production, and maintenance of epithelial barrier function and redox balance (Sciuto and Hurt, 2004; Grainge and Rice, 2010; Hoyle, 2010). Postexposure isoproterenol was effective when given IT, IV, or in combination of IT and IV, indicating that it can be administered either locally, systemically, or by both routes (Sciuto et al., 1998).

17.3.2.4.1.2 Ibuprofen   The nonsteroidal anti-inflammatory drug (NSAID) ibuprofen was shown to protect rabbit lungs from phosgene-induced injury (2000 ppm min), if given before or after exposure. The protective effects were seen if the perfused lung studies were done either immediately or 4 h after exposure (Kennedy et al., 1990). The protective action of ibuprofen was shown to be as an iron-chelating antioxidant (Kennedy et al., 1990). Postexposure treatment of mice exposed to lethal dose (640 ppm min) with various doses of ibuprofen enhanced survival in a dose-dependent manner, with reduction in lung edema, lipid peroxidation, and depletion of reduced glutathione (Sciuto, 1997).

17.3.2.4.1.3 N-Acetylcysteine   An IT bolus of N-acetylcysteine (40 mg/kg) given to lungs from phosgene-exposed rabbits (1500 ppm min) at the start of perfusion (40–60 min from start of exposure) protected the lungs, as seen by decreases in lung weight, leukotrienes, lipid peroxidation, and depletion of reduced glutathione compared to untreated exposed lungs (Sciuto et al., 1995; Sciuto and Hurt, 2004). A recent study showed that IP injections of N-acetylcysteine (50–200 mg/kg) after exposure to phosgene in rats mitigated the effects of poisoning in a dose-dependent manner. Further studies addressing the mechanism of protection showed a parallel augmentation of the expression of the glutathione pathway by the transcription regulator nuclear factor erythroid-related factor 2 (Nrf2), whose expression was upregulated in the lungs of rats exposed to phosgene and treated with N-acetylcysteine (Ji et al., 2010).

17.3.2.4.1.4 Ethyl Pyruvate and Inhaled Nitric Oxide   The anti-inflammatory and antioxidant drug ethyl pyruvate was shown to protect rats from phosgene-induced lung edema (Chen et al., 2013). In this study, rats were exposed to phosgene (400 ppm for 1 min) and the animals in the treatment group received IP 40 mg/kg ethyl pyruvate immediately after exposure. Lung wet-to-dry weight ratios, tissue, and BALF levels of nitrate/NO and prostaglandin E_2 (PGE_2) were reduced in ethyl pyruvate compared with saline-treated rats. Molecular analysis has shown that the expression of iNOS and cyclooxygenase-2, the enzyme converting arachidonic acid to PGE_2, increased in phosgene-exposed lungs, and this increase was abolished by ethyl pyruvate. Furthermore, the phosgene-induced increases and their inhibition by ethyl pyruvate were shown to be mediated by the MAP kinase pathway. Different results were obtained in a rat study employing a different exposure protocol, with $C \times t$ values ranging from 880 to 900 mg min/m³ (220–225 ppm min) and exposure times of 20 and 30 min (Li et al., 2011). Treatment of exposed rats with inhaled NO aggravated protein leakage, and ethyl pyruvate (80 mg/kg, IP) did not mitigate the edema by either pre- or postexposure administration. The discrepancy between this and the Chen et al. study could be attributed to the different exposure protocols—in the short exposures, the rats are protected by a reflex response that limits phosgene or any irritant gas intake and reduces the effective dose, which is overcome in the longer exposure durations (Li et al., 2011). Thus, the animals could be protected by ethyl pyruvate after a short but not a long exposure protocol. These results emphasize the caution needed in choice of animal model, exposure conditions, and interpretation of the results.

17.3.2.4.1.5 Inhibition of Neutrophil Recruitment and Activation The role of neutrophils in the pathogenesis of phosgene-induced lung injury and their potential as a target for therapy were demonstrated by the protective effects of various agents that reduce neutrophil influx into phosgene-exposed lungs. Pretreatment of rats with cyclophosphamide for neutrophil depletion, the 5-lipooxygenase inhibitor AA861, and the microtubule-disrupting agent colchicine, an inhibitor of neutrophil chemotactic movement, protected the animals from lung injury by phosgene (0.5 ppm for 60 min), as indicated by decreases in BAL neutrophils, BAL protein, and lipid peroxidation compared to untreated controls (Ghio et al., 1991). Colchicine (1.0 mg/kg, IP) was also protective when given 30 min postexposure. All four treatment protocols decreased mortality in phosgene-exposed mice (Ghio et al., 1991). Pretreatment of rats with colchicine (1.0 mg/kg, IP) reduced pulmonary hyperreactivity to acetylcholine, BAL protein, and neutrophil influx, resulting from exposure to phosgene (1 ppm for 60 min) (Ghio et al., 2005). Colchicine is used clinically as an anti-inflammatory drug but at a low dose due to its systemic toxicity. It is not known if clinically accustomed doses have therapeutic values in phosgene-induced lung injury in humans (Grainge and Rice, 2010). The phosphodiesterase and leukocyte activation inhibitor pentoxifylline did not reduce pulmonary edema in rats exposed to phosgene (80 mg min/m³ or 320 ppm for 20 min), treated with the drug (10–120 mg/kg, IP, at 15 min before, 45 and 105 min after exposure), and examined 5 h after the start of exposure (Sciuto et al., 1996). However, it did protect rats exposed to phosgene in a different exposure protocol (400 ppm for 1 min), treated with pentoxifylline (100 mg/kg, IP) before or after exposure, and observed for 48 h for lung pathology, BAL neutrophil counts, and lung tissue myeloperoxidase (MPO, a biomarker for activated neutrophils), and intercellular cell adhesion molecule-1 (ICAM-1) expression. BAL neutrophils and tissue expression of MPO and ICAM-1 increased from 3 to 48 h in exposed and untreated but not in pre- and postexposure treated rats, which had also reduced lung injury (Zhang et al., 2010).

17.3.2.4.2 Studies in Large Animal Models

17.3.2.4.2.1 Supplemental Oxygen Oxygen therapy is a cornerstone in the management of phosgene poisoning since World War I, where it has been the sole therapy (Grainge, 2004). However, this raised the issue of hyperoxic lung injury, mechanical stress due to positive pressure, and hemodynamic effects, which may affect survival after phosgene poisoning (Diller, 1985b; Grainge and Rice, 2010). In a study to assess the effects of oxygen therapy on survival after phosgene exposure, three groups of dogs were exposed to LCt_{60}–LCt_{80} of phosgene and were transferred afterward to ambient air or 95%, 80%, or 40% oxygen-enriched atmospheres for 72 h and later monitored for survival up to 12 days (Bruner et al., 1947). In the 95% oxygen treatment group, survival was higher during the first 72 h but afterward became lower than in the air-breathing animals. Treatment with 80% oxygen had a marginal effect on early survival, but late survival was markedly reduced. Treatment with 40% oxygen affected neither early nor late survival of phosgene-exposed dogs. These experiments showed that toxicity of oxygen can be imposed on that of phosgene, especially in the case of 95% oxygen treatment, which had some pulmonary toxicity in dogs when given alone (Bruner et al., 1947; Coman et al., 1947). A study with dogs was conducted in order to assess the effect of basic and additional therapies on the early postexposure cardiovascular and respiratory function (Mautone et al., 1985). The animals were exposed to a dosage of 94 ppm for 20 min or 188 ppm min, which is roughly $2 \times LCt_{50}$. The basic treatment, designed to simulate a realistic field pattern, included supplemental oxygen (FiO_2 1.0 for 30 min), $NaHCO_3$ (3 mEq/kg, IV), and a single dose of hydrocortisone (40 mg/kg, PO). The additional treatments, given at 30 min postexposure, included aminophylline, high- and low-dose PGE$_1$, and atropine. After a 2 h observation period, all the animals were euthanized and their lungs were processed for histology. The immediate effects of phosgene exposure were hypoxemia, hypercapnia, acidemia, tachypnea, hypopnea, systemic and pulmonary hypotension, and bradycardia. The most prominent effects of the basic therapy, measured at 30 min postexposure, were a prominent and significant increase in PaO$_2$ relative to air-breathing controls, moderate increase in PaCO$_2$, and correction of blood base deficit, attributed to the NaHCO$_3$ infusion, stressing the beneficial effect of supplemental oxygen in the presence of a pulmonary shunt. The authors concluded that supplemental oxygen (and NaHCO$_3$) should be an essential component in the immediate postexposure treatment but suggested that FiO$_2$ of 0.4–0.5 rather than 1.0 is adequate. A study using the anesthetized pig model was designed to see if there is a minimal effective and safe oxygen dosage and if treatment may be delayed until the symptomatic phase (Grainge et al., 2010a). Anesthetized pigs, exposed to phosgene at a $C \times t$ of 2500 mg min/m^3 and artificially ventilated (conventional protocol—tidal volume of 10 mL/kg, PEEP of 3 cm H$_2$O, frequency of 20 breaths/min), were divided into groups differing in FiO$_2$ and timing: (1) FiO$_2$ 0.30; (2) FiO$_2$ 0.80 immediately following exposure; (3) FiO$_2$ 0.80, at first 6 h postexposure; (4) FiO$_2$ 0.30 at 30 min postexposure and increasing to 0.40 at 6 h postexposure; and (5) FiO$_2$ 0.30 at 30 min postexposure and increasing to 0.40 at 12 h postexposure. Mortality rates at 24 h of monitoring were 70% in the FiO$_2$ 0.30 group, 20% in the FiO$_2$ 0.80 group, and 0% in the delayed and lower oxygen groups. Beneficial effects in lung wet weight/body weight ratio, blood oxygenation, and pulmonary shunt fraction (Q_s/Q_t) were seen in all supplemental oxygen groups, and the low-level delayed oxygen (FiO$_2$ 0.40 at 12 h) showed a nonsignificant ($p = 0.052$) improvement in histology score compared to the FiO$_2$ 0.80 group. Logistic regression modeling showed that the lung wet weight/body weight ratio and the pulmonary shunt fraction were associated with reduced survival. This study suggested that supplementary oxygen dosage can be optimized to the minimum required for adequate blood oxygenation and treatment can be postponed until the appearance of clinical signs of hypoxia, thus having a desirable impact on oxygen economy when medical resources are limiting.

17.3.2.4.2.2 Protective Ventilation　The ARDSNet lung-protective ventilation protocol (Brower et al., 2000) was evaluated in the anesthetized pig model of phosgene lung injury (Grainge and Rice, 2010). Pigs exposed to phosgene at a dosage of 2000 mg min/m^3 (LCt$_{70}$) were divided at 6 h postexposure, the time when hypoxic symptom appearance is expected, into three groups: One group was ventilated conventionally (FiO$_2$ 0.24, tidal volume of 10 mL/kg, PEEP of 3 cm H$_2$O, frequency of 20 breaths/min), another group received protective ventilation (FiO$_2$ 0.40, tidal volume of 8 mL/kg, PEEP of 8 cm H$_2$O, frequency of 20 breaths/min), and the third group received a different protective protocol, with a tidal volume of 6 mL/kg and frequency of 25 breaths/min. In the conventionally ventilated animals, only 3 out of 10 survived the 24 h experiment, while in the protective ventilation groups, all animals survived. The protective ventilation protocol was better than the conventional protocol not only in survival but also in other outcome parameters of blood oxygenation, lung wet weight/body weight ratio, and histopathology.

17.3.2.4.2.3 Corticosteroids　Mautone et al. (1985) included oral hydrocortisone in the basic treatment protocol employed in their dog study. They did not demonstrate any beneficial effect of corticosteroids, and this can be accounted for in their study design: as they wanted to simulate a characteristic treatment regimen in the field, they included steroids as part of the protocol with no intention to check it as a controlled experimental variable. Furthermore, they admitted that they did not expect to see any effect of steroid treatment at the early time points (2 h postexposure) of their study. The effect of intravenous methylprednisolone and nebulized budesonide was studied in the anesthetized pig model (Grainge and Rice, 2010). The pigs were exposed to phosgene at a dosage of 2000 mg min/m^3 and subsequently treated with IV glucose saline (20 mL), methylprednisolone (12.5 mL/kg, IV) at 6 h postexposure, inhaled glucose saline (2 mL), and inhaled budesonide (2 mL of 0.5 mg/mL solution) starting at 1 h and continuing at 6, 12, and 18 h postexposure. There were no differences in survival at 24 h between the treatment and control groups, and there was no improvement in other outcomes, including inflammation, lung weight, or pulmonary shunt fraction. Even though the result was *negative*, the authors did not preclude the use of corticosteroids at higher doses, at earlier time points after exposure, and not alone.

17.3.2.4.2.4 Modulation of Cyclic AMP Aminophylline was tested in the study of Mautone et al. (1985). Given in the clinical dosage (5 mg/kg IV for 20 min, followed by 2 mg/kg IV for 70 min) starting at 30 min postexposure, after the basic treatment (supplemental oxygen, sodium bicarbonate, and hydrocortisone), it was shown to reverse the phosgene-induced bradycardia and systemic hypertension and therefore recommended as a useful therapeutic agent. Although aminophylline was chosen for the study by virtue of its smooth muscle relaxation activity, no significant effect was seen on lung mechanics parameters in this study (Mautone et al., 1985). As a drug requiring intravenous administration soon after exposure, it should be considered for use in small-scale rather than large, multiple-casualty events, including TMCE (Grainge and Rice, 2010). When studied in the anesthetized pig model as a sole pharmacological intervention, inhaled salbutamol was not only poorly effective but also detrimental (Grainge et al., 2009). Anesthetized pigs were exposed to phosgene (dosage of 1978 mg min/m^3), mechanically ventilated in a conventional protocol (FiO$_2$ 0.21, tidal volume 10 mL/kg, PEEP 3 cm H$_2$O, frequency 20 breaths/min), and treated with nebulized salbutamol (2.5 mg in 6 mL saline) or saline alone at 1, 5, 9, 13, 17, and 21 h postexposure. No difference in mortality was observed between the treatment and control groups up to 24 h (5/5 and 4/5, respectively). The physiological parameters of blood oxygenation (PaO$_2$) and shunt fraction in the treatment group were inferior compared to the control group. However, mitigation of lung inflammation was seen in the salbutamol-treated animals, as BAL neutrophil counts were lower (12%) compared to the control (24%). No differences were seen in other parameters, including lung edema. As in the case of corticosteroids, the authors did not preclude the use of inhaled salbutamol and recommended to use it as a means to reduce inflammation once oxygen therapy has been initiated (Grainge et al., 2009; Grainge and Rice, 2010).

17.3.2.4.2.5 Antioxidants Thus far, all studies with antioxidants in phosgene poisoning were done with small animals or isolated lung models. The only large animal study reported investigated the effect of furosemide in the anesthetized pig model (Grainge et al., 2010b). Furosemide, a drug with multiple activities, was selected for this study due to its antioxidant properties. Treatment with inhaled furosemide (4 mL of a 10 mg/mL solution, given at 1, 3, 5, 7, 9, 12, 16, and 20 h postexposure) did not show a significant effect on survival compared to saline-treated controls and did not have any effect on physiological parameters, lung edema or inflammation, apart from a significant worsening of blood oxygenation between 19 and 24 h postexposure. The authors recommended avoiding the use of furosemide in the treatment of phosgene exposure. The important lesson from *negative* results like this is that they direct the caregivers to refrain from using ineffective and/or potentially detrimental measures in patients where the clinical picture is not specific and the identity of the toxicant may not be certain at the early phases of the event.

17.3.2.4.2.6 Recommended Treatment Protocols On the basis of the studies reviewed in their recent article (Grainge and Rice, 2010), Grainge and Rice recommended on two management protocols, one for small-scale events, with one or a few victims, and one for a large-scale event, either a mass casualty industrial accident or a terror scenario (Table 17.3). The protocols are similar in their approach, which is a revised symptomatic supportive strategy, but the mass casualty recommendations are adapted to conditions of resource limitations. The initial action is putting the exposed victim to rest and close observation (repeated physical examination and SaO$_2$), with the addition of chest radiography under the more favorable conditions of confirmed exposure and a small number of patients. Symptom-oriented interventions should be applied promptly, like supplemental oxygen in response to hypoxemia, with care to avoid oxygen toxicity and inflammation by employing minimally required FiO$_2$ and optional inhaled salbutamol. N-acetylcysteine is also possible under favorable conditions but not in TMCE due to the early time required for protective treatment (50 min postexposure), which is impractical for the mass casualty scenario. This is not the case for high-dose methylprednisolone, which is not precluded within the time frame of <6 h postexposure and therefore recommended in both protocols. The elective use of lung-protective mechanical ventilation is recommended in the small-scale scenarios but should be used more restrictively in the large-scale scenarios.

17.3.3 SMOKE INHALATION INJURY

17.3.3.1 General Properties and Exposure Assessment

Smoke inhalation is one of the most common causes of chemically induced lung damage, and in recent years, lung injury and the following pneumonia are the major causes of death in fire victims (Mlcak et al., 2007). Fire smoke is the product of incomplete combustion of inflammable structure materials—wood, fabrics, and plastics—and is a mixture of gaseous toxicants and particles, whose composition is variable with the types of burning materials and other factors, but some general characteristics can be stated. The gaseous phase includes systemic toxicants, of which the most prominent are carbon monoxide, cyanides, hydrogen sulfide, and respiratory irritants that include the water-soluble compounds ammonia, sulfur dioxide, and hydrogen chloride and the lipid-soluble compounds acrolein, other aldehydes, phosgene, and oxides of nitrogen (Rehberg et al., 2009; Toon et al., 2010). As discussed earlier, the water-soluble components are scavenged by the lining fluids of the proximal airways, while the less soluble components are carried further to the distal regions of the lungs. The particulates are soot particles of sizes ranging between 0.1 and 10 μm; of these, particles under 5 μm in size penetrate below the glottis and those under 1 μm reach the alveoli (Demling, 2008; Rehberg et al., 2009; Toon et al., 2010). The particles carry toxic materials like adsorbed gases and adhering heavy metals, which are the major contributors to lung injury (Demling, 2008; Rehberg et al., 2009; Toon et al., 2010). Filtered smoke, devoid of particles larger than 0.3 μm in diameter, did not cause injury like unfiltered smoke, in sheep (Lalonde et al., 1994). Studies with animals exposed to artificial smoke prepared from carbon particles laced with different toxicants have shown toxicity patterns characteristic of the toxicant type (Hales et al., 1988). Lung injury frequently develops as a secondary complication of burns, even in the absence of direct pulmonary exposure. The combined occurrence of the two insults intensifies the pulmonary injury to a higher severity and lethality than with each alone (Mlcak et al., 2007). The prevalent risk factors for lung injury and pneumonia in fire victims are exposure in a closed space, presence of facial burns, high percentage of burned body surface, and age (Shirani et al., 1986). Burns and smoke inhalation are among the highest risk factors for ARDS (Gajic et al., 2011).

17.3.3.2 Clinical Manifestations and Pathogenesis

The pathogenesis of smoke inhalation injury is complex, as it involves three mechanisms—thermal injury, local chemical injury, and systemic intoxication by CO and/or cyanides. There is also a differential involvement of the anatomical regions of the respiratory tract. The upper regions, above the vocal cords, may be subject to thermal injury, while the trachea and bronchi are the main parts affected by chemical injury. The smallest airways and alveoli may become involved in the most severe cases or secondary to airway injury or burns (Cancio, 2005; Demling, 2008; Rehberg et al., 2009). Inhalation of hot air, even at temperatures as high as 300°C, damages the larynx and pharynx. The laryngeal edema that develops may cause death from obstructive asphyxia within a few hours (Moritz et al., 1945). In experimental respiratory exposure to hot air through a translaryngeal cannula in dogs, the thermal injury did not extend the trachea due to the fast cooling of the air, but deeper damage was observed if the animals were exposed to steam, which has a larger heat capacity than air (Moritz et al., 1945). Exposure of the tracheobronchial tree to toxic chemicals in smoke from different combustible materials results in an epithelial necrosis and sloughing, peribronchial edema, and inflammation (Hubbard et al., 1991; Cancio, 2005; Demling, 2008; Rehberg et al., 2009). The accumulation of cell debris, fibrin, mucous, and inflammatory cells in the airways results in the formation of casts (Cox et al., 2003, 2008; Demling, 2008; Rehberg et al., 2009). The airway obstruction and alveolar injury are important causative factors in the severe respiratory dysfunction and hypoxemia that deteriorate to ARDS (Cancio, 2005; Demling, 2008; Rehberg et al., 2009). A common complication of the pulmonary injury in smoke inhalation and burn patients is bronchopneumonia. The loss of mucociliary action, deposition of cast material, accumulation of edema fluid, and generalized immunosuppression provide favorable conditions for colonization of common pathogenic bacteria like *Pseudomonas, Staphylococcus,* or *Klebsiella* (Foley et al., 1968).

The pathophysiology of lung injury in smoke inhalation has been studied in several animal models, mostly in sheep (Shimazu et al., 1987; Kimura et al., 1988; Soejima et al., 2001a; Murakami et al., 2002). The extent of injury is dependent on the smoke dose, which is controlled by the number of smoke breaths applied or exposure duration (Shimazu et al., 1987; Kimura et al., 1988; Hubbard et al., 1991). The early effects of smoke inhalation are caused by a neurogenic inflammatory response (Rehberg et al., 2009) and depletion of lung surfactant (Nieman et al., 1980). Surfactant depletion causes alveolar instability and atelectasis (Steinberg et al., 2005). The neurogenic inflammatory response is mediated by neuropeptides like substance P (Wong et al., 2004; Li et al., 2008) and the calcitonin gene–related peptide (Lange et al., 2009). The inflammatory effects are bronchoconstriction, mucus secretion, activation and infiltration of inflammatory cells, and pulmonary edema. The neuroinflammatory response may have a role in the pathogenesis of smoke inhalation injury, as the pathophysiological effects are ameliorated by the respective neuropeptide receptor antagonists (Wong et al., 2004; Li et al., 2008; Lange et al., 2009). After smoke inhalation alone (Murakami and Traber, 2003) or combined with burns (Enkhbaatar and Traber, 2004; Morita et al., 2011), blood flow to the airways via the bronchial circulation increases 10-fold than normally. The airway hyperemia, together with increased microvascular permeability, causes an increase in pulmonary lymph flow, pulmonary edema, and airway cast formation (Murakami and Traber, 2003; Enkhbaatar and Traber, 2004; Morita et al., 2011).

Other actors in the pathophysiology of smoke inhalation injury and its complications are NO, poly-(ADP-ribose) polymerase (PARP), the coagulation cascade, VILI, and activated neutrophils (Murakami and Traber, 2003; Enkhbaatar and Traber, 2004). The importance of NO was demonstrated in the pathogenesis of smoke inhalation (Soejima et al., 2000) and burn–smoke injury in sheep (Soejima et al., 2001a,b; Enkhbaatar et al., 2003a,b), as inhibitors of nitric oxide synthases (NOS) attenuated ARDS development and other pathophysiological features in these models. The effects of NO and its metabolites are the following: (1) increased pulmonary blood flow (Soejima et al., 2000; Enkhbaatar et al., 2003a,b); (2) loss of the hypoxic pulmonary vasoconstriction response, which leads to increased ventilation–perfusion mismatch and pulmonary shunt, and (3) formation of peroxynitrite anion ($ONOO^-$) by reaction with oxygen radicals (Murakami and Traber, 2003; Enkhbaatar and Traber, 2004; Szabó et al., 2007). Peroxynitrite causes modification of biological molecules, for example, lipids and proteins, by nitration, leading to reported functional impairment in different cell systems (Szabó et al., 2007). Its reaction with DNA causes single-strand breaks, which signal the activation of the DNA repair-associated enzyme PARP, which tags nuclear proteins with poly-(ADP-ribose) chains, using NAD^+ as substrate (Jagtap and Szabo, 2005). Extensive PARP activity and NAD^+ consumption interfere with cellular energy metabolism and result in ATP depletion and necrotic cell death (Jagtap and Szabo, 2005). The necrotic cells invoke inflammatory responses with increasing levels of inflammatory mediators and also lead to airway obstruction by deposited cell debris (Murakami and Traber, 2003; Enkhbaatar and Traber, 2004). Activation of the coagulation cascade and concomitant inhibition of fibrinolysis leads to fibrin deposition as airway casts from exudated plasma and alveolar fibrin deposition by activation of the intrinsic tissue factor VIIa pathway (Tuinman et al., 2012; Glas et al., 2013). VILI results from overdistention of uninjured alveoli during the high tidal volume mechanical ventilation required for adequate oxygenation and ventilation of the injured and obstructed lung. The injured alveolar tissue responds by production and release of inflammatory mediators, most importantly the chemoattractant IL-8, which attracts activated neutrophils from the circulation (Murakami and Traber, 2003). Activated neutrophils and their products—ROS, inflammatory mediators, and proteases—contribute to tissue injury, airway obstruction, and pulmonary pathophysiology of smoke or burn–smoke injury (Murakami and Traber, 2003). The agents and mechanisms under study as targets for novel therapies of smoke and burn inhalation injury (Bartley et al., 2008) will be discussed in the following sections.

17.3.3.3 Current Therapies

17.3.3.3.1 Diagnosis and Initial Patient Management

Smoke inhalation exposure may occur with or without mild surface skin burns (Mallory and Brickley, 1943), but in general, smoke inhalation injury is part of a complex trauma injury, which may include surface burns, thermal upper airway injury, and carbon monoxide/cyanide poisoning (Demling, 2008; Rehberg et al., 2009; Toon et al., 2010; Dries and Endorf, 2013). The initial management of severe burn patients may include supplemental oxygen—normobaric and occasionally hyperbaric—for elimination of blood carboxyhemoglobin, administration of cyanide antidotes if cyanide poisoning is suspected, airway clearance, and fluid resuscitation (Ipaktchi and Arbabi, 2006; Latenser, 2009; Dries and Endorf, 2013). The extensive fluid resuscitation with crystalloid solutions raised concerns regarding pulmonary edema, but this was not observed in clinical and animal model studies (Ipaktchi and Arbabi, 2006; Latenser, 2009). As thermal or chemical upper airway injury may be life threatening, most of the fire victims with suspected inhalation injury have to be intubated (Cancio, 2009; Dries and Endorf, 2013). Some use bronchoscopy to remove airway obstructions, like casts, even if the patient is conscious and has a mild injury (Cancio, 2009; Dries and Endorf, 2013). The most prevalent clinical indications of inhalation injury in fire or smoke exposure victims are a history of exposure in a closed space, facial burns, singed nasal hairs, and soot in the proximal airways. The more indicative signs of stridor and voice change are less common (Cancio, 2009; Dries and Endorf, 2013). As these signs are not reliable indicators of inhalation injury, diagnosis is facilitated by fiber-optic bronchoscopy (Pruitt et al., 1990). Bronchoscopic examination detects inflammatory changes—mucosal erythema, edema, ulceration, and submucosal hemorrhage, as well as the presence of carbon particles in the tracheobronchial tree (Shirani et al., 1986; Masanés et al., 1995). The need for accurate diagnosis of inhalation injury on admission is important, on one hand, as some patients may remain asymptomatic for 72 h until the abrupt appearance of ARDS (Masanés et al., 1995), while others could undergo unnecessary intubation if their upper airway condition was not assessed by fiber-optic bronchoscopy or laryngoscopy (Muehlberger et al., 1998).

17.3.3.3.2 Mechanical Ventilation

The optimal mechanical ventilation strategy for smoke inhalation injury is not yet settled, as there is no ideal technique suitable for smoke inhalation injury (Mlcak et al., 2007; Cancio, 2009; Latenser, 2009; Rehberg et al., 2009; Dries and Endorf, 2013). The risk of VILI, which is a significant risk factor for ARDS in mechanically ventilated patients, was addressed by the ACCP consensus conference in 1993 (Slutsky, 1993) that developed guidelines for ventilation in ARDS (Box 17.2), applicable to inhalation injury as well (Mlcak et al., 2007; Latenser, 2009; Dries and Endorf, 2013). A strategy for inhalation injury compatible with these guidelines and the results of the ARDSNet study on low tidal volume (Brower et al., 2000) begins with the tidal volume of 6–8 mL/kg predicted body weight. If the patient's airways become obstructed by fibrin casts, with an acute fall in PaO_2 and rise in $PaCO_2$, the clinician should provide aggressive airway toilet, and if the patient's condition continues to worsen, ventilation with higher tidal volumes, up to 8–10 mL/kg, should be considered (Mlcak et al., 2007). In general, the ARDSNet strategy may not be the optimal one for smoke inhalation because of the need to overcome airway obstruction. This is more prominent in inhalation injury than in the more common forms of clinical ARDS, where the main injury is alveolar edema (Cancio, 2009;

Rehberg et al., 2009). In order to maintain the required balance of adequate oxygenation and lung protection in inhalation injury, alternative ventilation strategies were evaluated in animal model studies with favorable results and are employed by some burn centers. Two of these—high-frequency percussive ventilation (HFPV) and AV-ECCO$_2$R—will be discussed in the following.

17.3.3.3.3 High-Frequency Percussive Ventilation

HFPV is a variant of HFV in which high-frequency subtidal ventilation (50–900 cycles/min) and lower-frequency (2–40 cycles/min) ventilation cycles are given simultaneously (Cancio, 2005, 2009). The typical waveform consists of a series of subtidal oscillating volumes overlaid on an inspiratory pressure wave that rises until a plateau is reached. Exhalation is passive, and the high-frequency subtidal waves occur during the end-expiratory phase as well. The low-frequency component provides the *volumetric* portion of the system, facilitating the mass movement of gas. The high-frequency component provides the *diffusive* element, facilitating alveolar gas exchange. The advantages of this pattern of ventilation are (1) lung protection by efficient gas delivery at peak pressures that increase alveolar recruitment without causing barotrauma, (2) clearance of obstructing secretions and debris, and (3) allowing spontaneous breathing through the inspiratory and expiratory phases, thus improving the patient–machine synchrony and activation of the respiratory muscles (Cancio, 2009). The technology is implemented in one system, the Volumetric Diffusive Respirator (VDR-4®, Percussionaire Corp., Sandpoint, Idaho). The key component is a sliding venturi that generates the low-pressure high-frequency wave (Rehberg et al., 2009; Kunugiyama and Schulman, 2012). The U.S. Army Institute of Surgical Research (USAISR) has tested this system as a ventilation strategy for burn and inhalation-injured patients requiring mechanical ventilation, initially as a rescue therapy for refractory hypoxemia and later as a prophylactic measure in patients with proven lung injury and high risk for ARDS or pneumonia (Cioffi et al., 1989, 1991). The mortality and occurrence of pneumonia in the HFPV-treated patients were significantly lower than the projected risk (Cioffi et al., 1991). Furthermore, the advantage of HFPV was demonstrated in a comparative study in smoke-exposed baboons, comparing conventional positive-pressure ventilation (tidal volume of 12 mL/kg), HFPV, and HFOV. HFOV was inferior in comparison to the other two techniques in obtaining physiological endpoints, but HFPV was superior in lung protection (Cioffi et al., 1993). A retrospective study comparing 92 burn patients with inhalation injury treated with HFPV to 132 similar patients treated with conventional ventilation has shown an overall reduced mortality in the HFPV-treated patients compared to the conventionally ventilated patients (28% vs. 56% in the respective groups), with a significant difference between the treatment groups in a subpopulation with ≤40% total body surface area burns (15% vs. 32%, $p = 0.02$ in the HFPV and conventional groups) (Hall et al., 2007). A randomized, controlled study comparing HFPV to the ARDSNet low tidal volume strategy in burn patients requiring intubation and mechanical ventilation was performed at the USAISR (Chung et al., 2010). For this study, 62 patients were randomized to HFPV ($n = 31$) and ARDSNet strategy ($n = 31$) in a period of 3 years (2006–2009). In each respective group, 39% and 35% had a confirmed diagnosis of lung injury. No difference was found between the groups in the primary outcome of 28-day number of ventilator-free days and in the secondary outcome of mortality (6 deaths in each group) and plasma IL-8 and IL-6 levels. The most prominent difference between the groups was the requirement for rescue therapy for inability to reach the predetermined oxygenation or ventilation goals—only two patients needed rescue therapy in the HFPV group compared to nine in the ARDSNet group (five for oxygenation and four for ventilation insufficiency). The better oxygenation in the HFOV group was also evident in the higher PaO$_2$/FiO$_2$ values obtained on days 0–6. In fact, the study was terminated for safety concerns because of the 11 patients requiring rescue therapy as a whole. The trial showed that HFPV is more efficacious than the ARDSNet strategy for ventilatory support in burn and inhalation lung injury. The VDR-4® has been used extensively at the USAISR Burn Center for ventilation in burn and other respiratory dysfunctions (Cancio, 2005, 2009) and was included in the standard equipment for respiratory support of burn casualties transported by air from Iraq and Afghanistan to Germany and the United States during the 2003–2007 military operations (Renz et al., 2008).

17.3.3.3.4   Arteriovenous Extracorporeal Carbon Dioxide
Removal–Assisted Mechanical Ventilation

Extracorporeal removal of carbon dioxide by a cardiopulmonary bypass as a means to reduce the mechanical ventilation load and afford *lung rest* has been tested in animal models of smoke inhalation (Brunston et al., 1997a,b) and burn–smoke lung injury (Alpard et al., 1999; Zwischenberger et al., 2001) in sheep. The system used in these studies was based on a carotid–jugular AV cardiopulmonary bypass, through a commercially available membrane oxygenator (Brunston et al., 1997a,b). A typical experiment was carried out in the following steps (Alpard et al., 1999): (1) surgical preparation—installation of catheters and instrument accesses—and 7-day recovery; (2) injury by thermal burns (third-degree burns, 40% of total surface area) and insufflation of cotton smoke through a mechanical ventilator line; (3) fluid resuscitation and ventilation at a tidal volume of 15 mL/kg body weight (equivalent to 10 mL/kg for adult humans), PEEP of 5 cm H_2O, and FiO_2 of 1.0 to eliminate carboxyhemoglobin and to inflict pulmonary stress typical of aggressive ventilation (at carboxyhemoglobin level of <10%, FiO_2 was reduced and adjusted to maintain PaO_2 >60 mmHg; from then on, the animals received ventilatory support to maintain this level of PaO_2, $PaCO_2$ <40 mmHg, blood pH 7.35–7.45, and frequent airway suction to remove secretions); and (4) as the animals developed ARDS (PaO_2/FiO_2 < 200 mmHg) within 24–48 h from injury, they were connected to the AV-$ECCO_2R$ circuit, and ventilator settings were promptly reduced to achieve the desired oxygenation and ventilation goals and minimize pulmonary stress. Tidal volume was initially reduced in steps of 20% until peak inspiratory pressure became <30 cm H_2O and was further reduced until proper oxygenation could be obtained at a minimal minute volume of 2–3 L/min. FiO_2 was likewise adjusted, and when PaO_2 > 60 mmHg could be obtained at FiO_2 of 0.21, ventilation was stopped. In a typical experiment, removal of >95% while shunting 11%–13% of cardiac output through the AV bypass increased oxygenation from PaO_2/FiO_2 of 152 mmHg at the start of AV-$ECCO_2R$ to 300 mmHg after 72 h of ventilation (Alpard et al., 1999). A controlled study, in which 18 sheep with burn and smoke injury were randomized to ventilation with AV-$ECCO_2R$ or ventilation only after establishment of ARDS, showed better survival following weaning from ventilation in the AV-$ECCO_2R$ group (Zwischenberger et al., 2001). The superiority of the AV-$ECCO_2R$-assisted low tidal volume strategy on HFPV and low tidal volume ventilation alone was demonstrated in a randomized study with sheep exposed to a harsher smoke–burn protocol, causing mortality of 100% within 120 h postexposure (Schmalstieg et al., 2007). Survival at 96 h postinjury (72 h post-ARDS criteria) was highest in the AV-$ECCO_2R$ group (71%), compared with the HFPV (55%) and low tidal volume (33%). The relative risk for death in the low tidal volume group was 4.64 ± 3.75 ($p = 0.05$) versus the AV-$ECCO_2R$ group and 2.68 ± 1.69 ($p = 0.12$) versus the HFPV group. This study, together with the clinical studies comparing HFPV to the ARDSNet strategy, points at the disadvantages of the ARDSNet strategy for some, if not all, victims of burns and smoke inhalation and the importance of alternative ventilator support strategies.

17.3.3.4   Experimental Pharmacological and Adjunct Therapies

The supportive management of smoke and burn lung injury and its complications is difficult because of the severe, multifactorial insult and the potentially injurious fluid and ventilation resuscitation measures in use (Ipaktchi and Arbabi, 2006; Latenser, 2009). Pharmacological interventions are under study as adjuncts to conventional management in animal models of smoke, burn–smoke, and smoke–sepsis and in a number of clinical studies. The multitude of intervention strategies studied reflects the complex pathophysiology of the injuries. Among the promising results were experiments with inhaled and systemic anticoagulants, NOS inhibitors, antioxidants and some NSAIDs, leukotriene inhibitors, and continuous nebulized albuterol, while some other strategies (parenteral heparin, the NSAIDs indomethacin and ibuprofen) were categorized as mixed, together with antiadhesion molecules (anti E- and P-selectins) and inhaled surfactants (Bartley et al., 2008). Therapies showing no benefit or adverse outcomes were glucocorticoids, manganese superoxide dismutase, or allopurinol. The studies in strategies targeting coagulopathy and inhibitors of the

NO–peroxynitrite–PARP pathways have been expanded as new drugs are being discovered and studied in a range of inflammatory diseases, including ARDS (Jagtap and Szabo, 2005; Szabó et al., 2007; Tuinman et al., 2012; Glas et al., 2013). Other promising future therapies include β-agonists, modulation of pulmonary blood flow by NOS inhibitors or pulmonary artery ablation, anticoagulants, and novel inflammation-targeting agents (thromboxane synthase inhibitor, γ-tocopherol, and muscarinic antagonists) (Dreis and Endorf, 2013). The main developments in these and other important studies in smoke inhalation injury will be discussed in the following.

17.3.3.4.1 Anticoagulants and Fibrinolytic Agents

The prominent coagulopathy—both increased coagulation and decline of fibrinolysis—in the pathophysiology of inhalation lung injury prompted examination of anticoagulant and fibrinolytic interventions. Studies were carried out with ovine models of smoke inhalation (48 breaths of smoke chilled to <40°C) and burn–smoke similar to those described earlier (Enkhbaatar et al., 2004, 2008) and cotton smoke with sepsis, in which the animals received, after smoke inhalation, airway instillation of 2–5 × 10^{11} cfu of viable *Pseudomonas aeruginosa* bacteria (Murakami et al., 2002). The smoke–sepsis injury is more severe, as the blood oxygenation level of ARDS criteria is reached within 3 h postinjury (Murakami et al., 2002) compared to 24–30 h in the burn–smoke model. The anticoagulant drugs heparin, human recombinant antithrombin (hrAT), and human recombinant activated protein C (hrAPC), which have been studied in ALI/ARDS and sepsis (Glas et al., 2013), were tested as adjunct treatments to resuscitation and ventilation in the ovine models described previously. Combined administration of aerosolized heparin (10,000 units per animal) and IV rhAT (0.34 mg/kg/h) starting 1 h postinjury and continued every 4 h until the end of the study (48 h) was tested in the burn–smoke ovine model. The drugs attenuated all pathophysiological changes of lung injury and hemodynamics compared to controls and elevated plasma levels of AT, which decreased markedly in the injured controls, without any evidence of systemic bleeding (Enkhbaatar et al., 2008). Nebulized heparin (10,000 units every 4 h for 24 h) attenuated lung injury in the ovine smoke–sepsis model, with a prominent effect on pulmonary edema and airway obstruction, while systemic high-dose heparin was not effective in this model (Murakami et al., 2003). Combined treatment with rhAPC and ceftazidime in this model was more efficacious than each treatment alone (Maybauer et al., 2012). The onset of ARDS was prevented in the combined treated animals (PaO_2/FiO_2 remained >300 mmHg), while the effect of rhAPC alone on blood oxygenation decrease was more modest (Maybauer et al., 2012). The additional contribution of ceftazidime to the attenuation of lung and systemic pathology is not clear, and it was attributable to anti-inflammatory activity rather than increased clearance of bacteria (Maybauer et al., 2012). The effect of enhanced fibrinolysis therapy with aerosolized tissue plasminogen activator (tPA) was studied in a burn–smoke ovine model (Enkhbaatar et al., 2004). Aerosol administration of recombinant human tPA (2 mg every 4 h, starting 4 h postinjury) attenuated the pathophysiological features of lung injury, notably affecting the clearance of airway casts, which was shown to be a crucial element in the management of ARDS resulting from burn–smoke injury. Noteworthy in this study is the late start of tPA administration, intended to reflect a typical timing of treatment in real scenarios, and the advantage of clearing already formed clots at this timing. The efficacy of aerosolized tPA was dose dependent (Enkhbaatar et al., 2004). Two small-scale human trials of nebulized heparin were done in the context of ALI from various underlying etiologies, demonstrating safety and indicating that further research is plausible (Tuinman et al., 2012). A retrospective single center study with adult inhalation injury patients (Miller et al., 2009; Tuiman et al., 2012) showed reduced mortality and clinical and pathological improvement after 7 days of treatment with nebulized heparin, N-acetylcysteine, and albuterol as adjunct to mechanical ventilation. The role of N-acetylcysteine in this drug combination is not only as an antioxidant but also as a mucolytic agent, breaking down mucins by reduction of disulfide bonds (Mlcak et al., 2007). A recommended 7-day treatment protocol is administration of 5,000–10,000 units of heparin and 3 mL of saline every 4 h, alternating with 3 mL of 20% NAC together with a bronchodilator every 4 h so that a patient is receiving an aerosolized treatment every 2 h. The bronchodilator is needed to prevent bronchoconstriction in response to irritation by N-acetylcysteine (Mlcak et al., 2007).

17.3.3.4.2 Targeting of NOS, Peroxynitrite, and PARP Pathways

The dysregulatory, inflammatory, and cytotoxic pathways of NOS, peroxynitrite, and PARP are pivotal pathophysiologic factors and therapeutic targets in many inflammatory and degenerative diseases (Jagtap and Szabó, 2005; Szabó et al., 2007). Studies on the effects of drugs that target these pathways in models of smoke inhalation with or without burn and sepsis shed light on the central role of these pathways in the pathophysiology of inhalation injury (Murakami and Traber, 2003; Enkhbaatar and Traber, 2004) and, more importantly, explored possible adjunct therapy strategies. The main studies on NOS, PARP, and peroxynitrite-targeting interventions in smoke inhalation, burn–smoke, and smoke–sepsis will be reviewed in the following.

17.3.3.4.3 NOS Inhibitors

NO, a mediator in many physiological and pathological processes, is produced from arginine by three isoforms of NOS: the inducible NOS (iNOS or NOS2), which is induced in inflammatory and other cells in inflammation, and two constitutive isoforms—neuronal NOS (nNOS or NOS1) and endothelial NOS (eNOS or NOS3)—which are expressed mainly in their respective cells and found in lung tissue (Hauser and Radermacher, 2010). The effect of the selective iNOS inhibitor and peroxynitrite scavenger mercaptoethyl guanine was tested in the smoke inhalation ovine model (Soejima et al., 2000). Administration of mercaptoethyl guanine decreased pulmonary blood flow and attenuated the decrease in blood oxygenation, pulmonary shunt, pulmonary edema, and oxidative stress compared to untreated controls. The same treatment was found to attenuate pulmonary lymph flow, pulmonary vascular permeability, edema, and oxidative stress in the burn–smoke model, with reduction of NO_x (NO_2^-/NO_3^-) formation in plasma, lung lymph, and burned tissues, and reduced 3-nitrotyrosine, an indicator of protein nitration, in lung tissue (Soejima et al., 2001b). As the poor selectivity of mercaptoethyl guanine toward iNOS raised safety concerns in clinical use, already known for nonselective iNOS inhibitors, a more selective iNOS inhibitor (BBS-2) was tested in the burn–smoke (Enkhbaatar et al., 2003a) and burn–sepsis (Enkhbaatar et al., 2006) models. Continuous infusion of the compound attenuated the lung injury only partially, suggesting the involvement of other isoforms in the pathogenesis of lung injury in burn–smoke and burn–sepsis and the need to target these isoforms as well. Continuous infusion of different selective nNOS inhibitors attenuated lung injury concomitantly with reduction of plasma NO_x and lung tissue 3-nitrityrosine in the burn–smoke and in the smoke–sepsis models (Enkhbaatar et al., 2003b, 2009). Detailed molecular and time-course studies established the pivotal role of nNOS as the main generator of NO and the NO-related pathology in the early phase of the smoke–burn and smoke–sepsis injuries, to be followed by iNOS, which becomes prominent later (Hauser and Radermacher, 2010; Lange et al., 2010; Saunders et al., 2010). Concomitant infusion of an nNOS inhibitor and an iNOS inhibitor at reduced doses was more effective in attenuating lung injury in burn–smoke ovine model compared to the lesser attenuation observed in the previous single-inhibitor studies (Lange et al., 2011a). Another recently tested strategy is direct infusion into the bronchial circulation. Direct infusion of an nNOS inhibitor at an extremely reduced dose into the bronchial artery attenuated the lung injury comparably to higher systemic doses, in ovine burn–smoke model, thus providing a safe and an effective dosing technique that may be practical in the future (Hamahata et al., 2011).

17.3.3.4.4 Inhibitors of PARP and Catalytic Decomposition of Peroxynitrite

The complexity and safety issues raised in the strategies based on inhibition of NOS can be circumvented by targeting the downstream elements of the pathway—peroxynitrite and PARP. The pathophysiological features of burn–smoke injury in sheep were significantly attenuated by administration of a PARP inhibitor (Shimoda et al., 2003). Metalloporphyrin compounds that decompose peroxynitrite catalytically and are also considered as possible compounds in the treatment of ARDS have been evaluated in animal and human studies of inflammatory and degenerative diseases (Szabo et al., 2007). Administration of an Fe-metalloporphyrin compound attenuated significantly the burn–smoke lung injury in sheep (Lange et al., 2011b). The attenuation of pathophysiological and inflammatory parameters was concomitant with the reduction in peroxynitrite, as measured by lung tissue 3-nitrotyrosine, poly-(ADP-ribose), thus stressing the role of peroxynitrite as a possible therapeutic target.

17.3.3.4.5 Modulation of Bronchial Circulation

The therapeutic potential of bronchial circulation ablation was demonstrated in two recent studies using the ovine smoke–burn model (Hamahata et al., 2009, 2010). First, bronchial artery ablation was demonstrated to increase ventilator-free survival in sheep after burn–smoke injury. Ablation was performed 72 h before injury by injection of 70% ethanol into the bronchial artery to induce sclerosis. All animals survived until the end of the study, 96 h postinjury. Animals in the ablated group were successfully weaned from ventilation, whereas in the nonablated controls, animals met euthanasia criteria within 72 h (Hamahata et al., 2009). The effect of postinjury ablation was studied in animals undergoing bronchial artery ablation 1 h after burn and smoke treatment and observed for pathophysiological effects for 24 h. The ablation attenuated the decline in blood oxygenation, pulmonary shunting, airway pressure, lung lymph flow, and vascular permeability, as well as edema, myeloperoxidase (MPO), and airway cast formation, which was reduced in the bronchi but not in the trachea. These studies point to the therapeutic potential of bronchial circulation limitation, preferably by pharmacological means, in the treatment of this life-threatening injury.

17.4 SUMMARY

Toxic inhalation injury, resulting from exposure of the respiratory tract to toxic or irritant chemicals, is not common but may occur under a broad spectrum of circumstances. These are mainly associated with the industrial domain, where toxic chemicals are used as precursors, and are produced, processed, stored, and transported in huge quantities. Other, less common circumstances that have become part of our lives in the last 100 years include chemical warfare and terror threats (Markel et al., 2008). Exposure events to inhalational toxicants involve, in many cases, single or small groups of individuals exposed at home, recreation, or work. The known TMCE is a rarity, but as learned from major events like the Bhopal disaster of 1985 and the chlorine car derailment in North Carolina in 2005, they may pose a devastating public health threat if not addressed properly.

Primary medical care of hazardous material victims is an inseparable part of the rescue operation in the event theater, which may be contaminated with active agents. Contamination avoidance is an important element in these operations. This includes the use of proper PPE suited to the hazard type and level, and decontamination of those patients known or suspected of contact with liquid hazardous materials. If they are not decontaminated before entering medical facilities, they may pose exposure hazard to the health-care providers. PPE may impair the performance capability of the user, so proper training in using it and performing in protective posture is mandatory. The procedures for contamination avoidance during medical care of hazardous material casualties are outlined in the text, but it is requisite to refer to published authoritative guideline manuals and obtain qualified training and instruction.

The pathophysiology of toxic inhalation injury depends on the type of material and its properties. The water solubility of the agents defines its major site of injury, whether the central airways or the peripheral airways and alveoli. Airway injury is characterized by epithelial degeneration, peribronchial edema, and obstruction due to mucous secretion and fibrin. Alveolar injury is characterized mainly by increased vascular permeability, decreased fluid clearance, and alveolar flooding. The gross pathophysiological changes are impaired lung mechanics and gas exchange. The common elements to all types of lung injuries are tissue damage and the responses of neurogenic irritation and inflammation, which are associated with oxidative and nitrosative stress and coagulopathy. Severe inflammation may lead to tissue remodeling, fibrosis, and long-term sequelae. This chapter discusses in detail the pathophysiology and management in three selected toxicants: chlorine, which affects mainly the airways; phosgene, which leads to alveolar injury similar to clinical ARDS due to sepsis, pneumonia, and other etiologies; and smoke inhalation. Smoke inhalation is the most severe as it affects both airway and alveoli and is complicated by burns and pneumonia/sepsis.

Current management of inhalational lung injury is supportive and symptomatic. Mild to moderate casualties, presenting with respiratory symptoms of cough, dyspnea, and airway constriction as the most representative, are generally treated with bronchodilators, corticosteroids, and supplementary oxygen. If they do not improve immediately, they should be monitored by physical examination, blood gas analysis, and chest radiography for the development of pulmonary edema and respiratory failure. In a more severe injury to the airways and alveoli, ALI may develop and may progress to ARDS. Care of ARDS is based on mechanical ventilation and supportive measures. The aggressive mechanical ventilation needed to obtain the required oxygenation and ventilation goals in the ALI/ARDS patients is injurious and stressful and exacerbates the inflammatory response. To balance the respiratory support needs and lung protection, several lung-protective ventilator strategies are available; of these, the most prevalent is low tidal volume ventilation, shown in the large-scale ARDSNet trial to significantly increase survival and decrease ventilator-induced injury in ARDS patients. Other discussed lung-protective ventilation strategies are HFOV and HFPV. Other auxiliary techniques discussed in the text are high PEEP, inhaled vasodilators like NO, prone positioning, and ECMO. They are accepted as rescue therapies for refractory hypoxemia and their routine use is debated. Some of these techniques have been studied in model systems of toxic inhalation injury, with some success. A few more strategies were not discussed here and are reviewed in the cited literature.

Pharmacological therapeutic agents, targeting pathophysiological processes, are under study as prophylactic measures, to prevent the deterioration of lung injury to ARDS, or as adjuncts to ventilatory support. The main study areas are anti-inflammatory agents—corticosteroids and NSAIDs, antioxidants, and anticoagulants. Unfortunately, large-scale clinical studies have ended mostly with poor results, even when animal models and small-scale clinical results were promising. As randomized, controlled, clinical studies with exposure to hazardous materials are impossible, the clinical information comes from case reports of individuals and cohorts in the less common multicasualty incidents. Clinically relevant large animal models—mainly pigs and sheep—afford studying injury management as close as possible to the clinical setting, based on the similarity of their disease pathology and response to therapy, to that of humans. Detailed knowledge of the pathophysiological and molecular injury mechanisms will enable enhancing the discovery of drugs and proper delivery methods. The best timing to initiate these treatment modalities is still obscure. The challenge is even greater since in most cases, we need to translate the data from the bench to the clinic in the absence of controlled human studies.

QUESTIONS AND ANSWERS

1. State the basic differences in lung-injury characteristics following exposure to chlorine and phosgene.

    Answer: The respiratory tract is the main organ system affected by exposure to chlorine (chlorine = *central* lung damage). Depending on the dose, the acute effects range from a noxious odor and local irritation to chest pain, dyspnea, cough, airway obstruction, and inflammation and in extreme cases pulmonary edema and ARDS.

    The main feature of phosgene-induced lung injury is noncardiogenic pulmonary edema, damage to the lower airways and alveolar epithelial cells, and inflammation, resulting in hypoxemic respiratory failure and lethal hemodynamic instability (phosgene = *peripheral* lung damage).

2. What should be included in the medical aid given for each type of injury?

Answer: The basic initial therapy for chlorine-induced lung damage is supportive and symptomatic and includes the use of supplemental oxygen to obtain proper blood oxygenation and inhaled β2-agonists for bronchospasm.

In phosgene-induced lung damage, the medical treatment is similar, with special emphasis and alertness to the possibility of pulmonary edema appearing after a latent period.

3. What can be done in order to prevent deterioration of the primary injury in each of these agents?

Answer: In case of chlorine exposure, patients should be admitted and transferred to the ICU when there is suspicion of significant exposure or if clinical signs and symptoms such as chest pain, bronchospasm, and abnormal vital signs are prominent. Inhaled or systemic corticosteroids are given in these cases or when further deterioration in the next few hours is imminent, sometimes with the addition of IV aminophylline and diuretics in cases of pulmonary edema.

Following phosgene exposure, asymptomatic victims should be referred to a medical facility for evaluation and follow-up, patients with respiratory complaints should be admitted to intensive care unit, and those with bronchoconstriction or signs of pulmonary edema should be watched for risk of respiratory failure. Monitoring for lung edema development by means like chest radiography is recommended, and patients exposed to a high dose should have chest radiography from as early as 2 h postexposure. This latent period is very active in the buildup of pulmonary edema. As with chlorine, these patients should receive systemic corticosteroids, aminophylline, and diuretics.

REFERENCES

Abersold, P., 2012. FDA experience with medical countermeasures under the Animal rule. *Adv. Prev. Med.*, **2012**, article ID 50751, doi: 10.1155/201/50751.

Abroug, F., Ouanes-Bebes, L., Dachraoui, F., Ouanes, I., and Brochard, L., 2011. An updated study-level meta analysis of randomized controlled trials on proning in ARDS and acute lung injury. *Crit. Care*, **15**, R6.

Adhikari, N.K.J., Burns, K.E.A., Friedrich, J.O., Granton, J.T., Cook, D.J., and Meade, M.O., 2007. Effect of nitric oxide on oxygenation and mortality in acute lung injury: Systemic review and meta-analysis. *BMJ*, **334**, 779.

Afshari, A., Brok, J., Møller, A.M., and Wetterslev, J., 2011. Inhaled nitric oxide for acute respiratory distress syndrome and acute lung injury in adults and children: A systemic review with meta-analysis and trial sequential analysis. *Anesth. Analg.*, **112**, 1411–1421.

Agabiti, N., Ancona, C., Forastiere, F., Di Napoli, A., Lo Presti, E., Corbo, G.M., D'Orsi, F., and Perucci, C.A., 2001. Short term respiratory effects of acute exposure to chlorine due to a swimming pool accident. *Occup. Environ. Med.*, **58**, 399–494.

Agency for Toxic Substances and Disease Registry (ATSDR), 2001a. Emergency medical services response to hazardous material incidents, section II, in *Managing Hazardous Material Incidents*, Vol. I: *Emergency Medical Services: A Planning Guide for the Management of Contaminated Patients*. Atlanta, GA. pp. 13–44. http://www.atsdr.cdc.gov/MHMI/mhmi-v1–2.pdf (last accessed on May 20, 2014).

Agency for Toxic Substances and Disease Registry (ATSDR), 2001b. *Managing Hazardous Material Incidents*, Vol. III: *Medical Management Guidelines for Acute Toxic Exposures*. Atlanta, GA. http://www.atsdr.cdc.gov/MHMI/mhmi-v3.pdf (last accessed on May 20, 2014).

Agency for Toxic Substances and Disease Registry (ATSDR), 2001c. Chlorine (Cl_2). http://www.atsdr.cdc.gov/MHMI/mmg172.pdf (last accessed on May 20, 2014).

Agency for Toxic Substances and Disease Registry (ATSDR), 2001d. Phosgene ($COCl_2$). http://www.atsdr.cdc.gov/MHMI/mmg176.pdf (last accessed on May 20, 2014).

Agency for Toxic Substances and Disease Registry (ATSDR), 2006. Hazardous Substances Emergency Events Surveillance (HSEES) annual report 2006. http://www.atsdr.cdc.gov/hs/hsees/annual2006.pdf (last accessed on May 20, 2014).

Agency for Toxic Substances and Disease Registry (ATSDR), 2009. Hazardous Substances Emergency Events Surveillance (HSEES) annual report 2009. http://www.atsdr.cdc.gov/HS/HSEES/HSEES%202009%20report%20final%2008%2017%2011_9_2012.pdf (last accessed on May 20, 2014).

Aguilar, S., Scotton, C.J., McNulty, K., Nye, E., Stamp, G., Laurent, G., Bonnet, D., and James, S.M., 2009. Bone marrow stem cells expressing keratinocyte growth factor via an inducible lentivirus protects against bleomycin-induced pulmonary fibrosis. *PLoS One*, **4**, e8013.

Alberts, W.M. and do Pico, G.A., 1996. Reactive airway dysfunction syndrome. *Chest*, **109**, 1618–1626.

Alpard, S.K., Zwischenberger, J.B., Tao, W., Deyo, D.J., and Bidani, A., 1999. Reduced ventilator pressure and improved P/F ratio during percutaneous arteriovenous carbon dioxide removal for severe respiratory failure. *Ann. Surg.*, **230**, 215–224.

Amato, M.B.P., Barabs, C.S.V., Medeiros, D.M. et al., 1998. Effect of a protective ventilation strategy on mortality in the acute respiratory distress syndrome. *N. Engl. J. Med.*, **338**, 347–354.

Andersson, U. and Tracey, K.J., 2012. Neural reflexes in inflammation and immunity. *J. Exp. Med.*, **209**, 1057–1068.

Babu, R.V., Cardenas, V., and Sharma, G., 2008. Acute respiratory distress syndrome from chlorine inhalation during a swimming pool accident: Case report and review of the literature. *J. Intensive Care Med.*, **23**, 275–280.

Baker, D., 2005. The problem of secondary contamination following chemical agent release. *Crit. Care*, **9**, 323–324.

Baker, D.J., 1999. Management of respiratory failure in toxic disasters. *Resuscitation*, **42**, 125–131.

Barrett, A.M. and Adams, P.J., 2011. Chlorine truck attack consequences and mitigation. *Risk Analysis*, **31**, 1243–1259.

Bartley, A.C., Edgar, D.W., and Wood, F.M., 2008. Pharmaco-management of inhalation injuries for burn survivors. *Drug Des. Dev. Ther.*, **2**, 9–16.

Bassford, C.R., Thickett, D.R., and Perkins, G.D., 2012. The rise and fall of β-agonists in the treatment of ARDS. *Crit. Care*, **16**, 208.

Batchinsky, A.I., Martini, D.K., Jordan, B.S., Dick, E.J., Fudge, J., Baird, C.A., Hardin, D.E., and Cancio, C.L., 2006. Acute respiratory syndrome secondary to inhalation of chlorine gas in sheep. *J. Trauma*, **60**, 944–957.

Baxter, P.J., Davies, P.C., and Murray, V., 1989. Medical planning for toxic releases into the community: The example of chlorine gas. *Br. J. Ind. Med.*, **46**, 277–285.

Beers, M.F. and Morrisey, E.E., 2011. The three R's of lung health and disease: Repair, remodeling, and regeneration. *J. Clin. Invest.*, **121**, 2065–2073.

Bein, T., Weber, F., Prasser, C., Pfeifer, M., Schmid, F.-X., Butz, B., Birnbaum, D., Taeger, K., and Schlitt, H., 2006. A new pumpless extracorporeal interventional lung assist in critical hypoxemia/hypercapnia. *Crit. Care Med.*, **34**, 1372–1377.

Bein, T., Weber-Carstens, S., Goldman, A. et al., 2013. Lower tidal volume strategy (~3 ml/kg) combined with extracorporeal CO_2 removal versus 'conventional' protective ventilation (6 ml/kg) in severe ARDS. The prospective randomized Xtravent-study. *Intensive Care Med.*, **39**(5), 847–856, Published online 10 January 2013, doi: 10.1007/s00134-012-2787-6.

Bein, T., Zimmermann, M., Hergeth, K., Ramming, M., Rupprecht, L., Schlitt, H.J., and Slutsky, A.S., 2009. Pumpless extracorporeal removal of carbon dioxide combined with ventilation using low tidal volume and high positive end-expiratory pressure in a patient with severe acute respiratory syndrome. *Anaesthesia*, **64**, 95–198.

Ben-Abraham, R., Gur, I., Vater, Y., and Weinbroum, A.A., 2003. Intraosseous emergency access by physicians wearing full protective gear. *Acad. Emerg. Med.*, **10**, 1407–1410.

Berkenstadt, H., Ziv, A., Barsuk, D., Levine, I., Cohen, A., and Vardi, A., 2003. The use of advanced simulation in the training of anesthesiologists to treat chemical warfare casualties. *Anesth. Analg.*, **96**(6), 1739–1742.

Bernard, G.R., Artigas, A., Brigham, K.L., Carlet, C., Falke, K., Hudson, L., Lamy, M., Legall, J.R., Morris, A., and Spragg, R., the Consensus Committee, 1994. The American-European consensus conference on ARDS: Definition, mechanisms, relevant outcomes, and clinical trial coordination. *Am. J. Respir. Crit. Care Med.*, **149**, 818–824.

Bessac, B.F. and Jordt, S.-E., 2008. Breathtaking TRP channels: TRPA1 and TRPV1 in airway chemosensation and reflex control. *Physiology*, **23**, 360–370.

Bessac, B.F. and Jordt, S.-E., 2010. Sensory detection and response to toxic gases: Mechanisms, health effects, and countermeasures. *Proc. Am. Thorac. Soc.*, **7**, 269–277.

Bonetto, G., Corradi, M., Carraro, S., Zanconato, S., Alinovi, R., Folesani, G., Da Dalt, L., Mutti, A., and Baraldi, E., 2006. Longitudinal monitoring of lung injury in children after acute chlorine exposure in a swimming pool. *Am. J. Respir. Crit. Care Med.*, **174**, 545–549.

Branson, R.D., Johanningman, J.A., Daugherty, E.L., and Rubinson, L., 2008. Surge capacity mechanical ventilation. *Respir. Care*, **53**, 78–90.

Briel, M., Meade, M., Mercat, A. et al., 2010. Higher vs. lower positive-end-expiratory pressure in patients with acute lung injury and acute respiratory distress syndrome: Systemic review and meta-analysis. *JAMA*, **303**, 865–873.

Brinker, A., Stratling, W.M., and Schumacher, J., 2008. Evaluation of bag-valve-mask ventilator in simulated toxic environments. *Anaesthesia*, **63**, 1234–1237.

Brower, R.G. and Fessler, H.E., 2011. Another "negative trial" of surfactant. Time to bury the idea? *Am. J. Respir. Care Med.*, **183**, 966–967.

Brower, R.G., Lanken, P.N., MacIntyre, N. et al., The National Heart Lung and Blood Institute ARDS Clinical Trials Network, 2004. Higher versus lower positive end-expiratory pressures in patients with the acute respiratory distress syndrome. *N. Engl. J. Med.*, **351**, 327–336.

Brower, R.G., Matthay, M.A., Morris, A. et al., The Acute Respiratory Distress Network, 2000. Ventilation with lower tidal volumes as compared with traditional tidal volumes for acute lung injury and acute respiratory distress syndrome. *N. Engl. J. Med.*, **342**, 1301–1308.

Brown, R.F.R., Jugg, B.J.A., Harban, F.M.J., Adhley, Z., Kenward, C.E., Platt, J., Hill, A., Rice, P., and Watkins, P.E., 2002. Pathophysiological responses following phosgene exposure in the anaesthetized pig. *J. Appl. Toxicol.*, **22**, 263–269.

Bruner, H.D., Boche, R.D., Chapple, C.C., Gibbon, M.H., and McCarthy, M.D., 1947. Studies on experimental phosgene poisoning. III. Oxygen therapy in phosgene-poisoned dogs and rats. *J. Clin. Invest.*, **26**, 923–944.

Brunston, R.L., Tao, W., Bidani, A., Alpard, S.K., Traber, D., and Zwischenberger, J.B., 1997b. Prolonged hemodynamic stability during arteriovenous carbon dioxide removal for severe respiratory failure. *J. Thorac. Cardiovasc. Surg.*, **114**, 1107–1114.

Brunston, R.L., Zwischenberger, J.B., Tao, W., Cardenas, V.J., Traber, D.L., and Bidani, A., 1997a. Total arteriovenous CO_2 removal: Simplifying extracorporeal support for respiratory failure. *Am. Thorac. Surg.*, **64**, 1599–1605.

Byers, M., Russell, M., and Lockey, D.J., 2008. Clinical care in the "Hot Zone". *Emerg. Med. J.*, **25**, 108–112.

Caironi, P., Cressoni, M., Chiumello, D. et al., 2010. Lung opening and closing during ventilation of acute respiratory distress syndrome. *Am. J. Respir. Crit. Care Med.*, **181**, 578–586.

Cancio, L.C., 2005. Current concepts in the pathophysiology and treatment of inhalation injury. *Trauma*, **7**, 19–35.

Cancio, L.C., 2009. Airway management and smoke inhalation injury in burn patients. *Clin. Plast. Surg.*, **36**, 555–567.

Castle, N., Pillay, Y., and Spencer, N., 2011. Insertion of six different supraglottic airway devices whilst wearing chemical, biological, radiation, nuclear-personal protective equipment: A manikin study. *Anaesthesia*, **66**, 983–988.

CDC, 2012. Emergency preparedness and response; chemical agents (by category) http//emergency.cdc.gov/agent/agent listchem-category.asp (last accessed on May 20, 2014).

Cepkova, M. and Matthay, M.A., 2006. Pharmacotherapy of acute lung injury and acute respiratory distress syndrome. *J. Intensive Care Med.*, **21**, 119–142.

Cevik, Y., Onay, M., Akmaz, L., and Sezigan, S., 2009. Mass casualties from acute inhalation of chlorine gas. *South. Med. J.*, **102**, 1209–1213.

Chabot, F., Mitchell, J.A., Gutterodge, J.M.C., and Evans, T.W., 1998. Reactive oxygen species in acute lung injury. *Eur. Respir. J.*, **11**, 745–757.

Chan, K.P.W., Stewart, T.E., and Mehta, S., 2007. High frequency oscillatory ventilation for patients with ARDS. *Chest*, **131**, 1907–1916.

Chang, W., Chen, J., Schlueter, C.F., Rando, R.J., Parthak, Y.V., and Hoyle, G.W., 2012. Inhibition of chlorine-induced lung injury by the type 4 phosphodiesterase inhibitor rolipram. *Toxicol. Appl. Pharmacol.*, **263**, 251–258.

Chen, H.-L., Bai, H., Xi, M.-M., Liu, R., Qin, X.-J., Liang, X., Zhang, W., Zhang, X.-D., Li, W.-L., and Hai, C.-X., 2013. Ethyl pyruvate protects rats from phosgene-induced pulmonary edema by inhibiting cyclooxygenase 2 and inducible nitric oxide synthase expression. *J. Appl. Toxicol.*, **33**, 71–77.

Chester, E.H., Kaimal, J., Payne, C.B., and Kohn, P.M., 1977. Pulmonary injury following exposure to chlorine gas. Possible beneficial effects of steroid treatment. *Chest*, **72**, 247–250.

Chow, C.-W., Herrera Abreu, M.T., Suzuki, T., and Downey, G.P., 2003. Oxidative stress and acute lung injury. *Am. J. Respir. Cell Mol. Biol.*, **29**, 427–431.

Chung, K.K., Wolf, S.E., Renz, E.M. et al., 2010. High-frequency percussive ventilation and low tidal volume ventilation in burns: A randomized controlled trial. *Crit. Care Med.*, **38**, 2970–1977.

Cioffi, W.G., de Lemos, R.A., Coalson, J.J., Gerstman, D.A., and Pruitt, B.F. Jr., 1993. Decreased pulmonary damage in primate with inhalation injury treated with high-frequency ventilation. *Ann. Surg.*, **218**, 328–337.

Cioffi, W.G., Graves, T.A., McManus, W.F., and Pruitt, B.F. Jr., 1989. High-frequency percussive ventilation in patients with inhalation injury. *J. Trauma*, **29**, 350–354.

Cioffi, W.G. Jr., Rue, L.W. III, Graves, T.A., McManus, W.F., Mason, A.D. Jr., and Pruitt, B.F. Jr., 1991. Prophylactic use of high-frequency percussive ventilation in patients with inhalation injury. *Ann. Surg.*, **213**, 575–580.

Coats, M.J., Jundi, A.S., and James, M.R., 2000. Chemical protective clothing; a study into the ability to perform lifesaving procedures. *J. Accid. Emerg. Med.*, **17**, 115–118.

Collins, S.R. and Blank, R.S., 2011. Approaches to refractory hypoxemia in acute respiratory distress syndrome: Current understanding, evidence and debate. *Respir. Care*, **56**, 1573–1578.

Coman, D.R., Bruner, H.L., Horn, R.C., Friedman, M., Boche, R.D., McCarthy, M.D., Gibbon, M.H., and Schultz, J.S., 1947. Studies on experimental phosgene poisoning. I. The pathologic anatomy of phosgene poisoning, with special reference to the early and late phases. *Am. J. Pathol.*, **23**, 1037–1073.

Cove, M.E., MacLaren, G., Federspiel, W.J., and Kellum, J., 2012. Bench to bedside review: Extracorporeal carbon dioxide removal, past, present and future. *Crit. Care*, **16**, 232.

Cox, R.A., Burke, A.S., Soejima, K.S., Murakami, K., Katahira, J., Traber, L.D., Herndon, D., Schmalstieg, F.C., Traber, D.L., and Hawkins, H.K., 2003. Airway obstruction in sheep with burn and smoke inhalation injuries. *Am. J. Respir. Cell Mol. Biol.*, **29**, 295–302.

Cox, R.A., Mlcack, R.P., Chinkes, D.L., Jacob, S., Enkhbaatar, P., Jaso, J., Parish, L.P., Traber, D.L., Jeschke, M.G., Herndon, D.N., and Hawkins, H.K., 2008. Upper airway mucus deposition in lung tissue of burn trauma victims. *Shock*, **29**, 356–361.

Craig, T.R., Duffy, M.J., Shyamssundar, M., McDowell, C., O'Kane, C.M., Elborn, J.S., and McAuley, D.F., 2011. A randomized clinical trial of hydroxymethylglutaryl coenzyme A reductase inhibition for acute lung injury (the HARP study). *Am. J. Respir. Crit. Care Med.*, **183**, 620–626.

Crimi, E., Sica, V., Williams-Ignarro, S., Zhang, H., Slutsky, A.S., Ignarro, L.J., and Napoli, C., 2006. The role of oxidative stress in adult critical care. *Free Rad. Biol. Med.*, **40**, 398–406.

D'Allesio, A., Tsushima, K., Aggrawal, NR., West, E.E., Willett, M.H., Britos, M.F., Brower, R.G., Tuder, R.M., Mc Dyer, J.F., and King, L.S., 2009. CD4+CD25+Foxp3+ Tregs resolve experimental lung injury in mice and are present in humans with acute lung injury. *J. Clin. Invest.*, **119**, 2898–2912.

Davies, A., Jones, D., Bailey, M. et al., Australia and New Zealand Extracorporeal Membrane Oxygenation (ANZ ECMO) Influenza Investigators, 2009. Extracorporeal membrane oxygenation for 2009 influenza A (H1N1) acute respiratory distress syndrome. *JAMA*, **302**, 1888–1895.

De Jonge, W.J. and Ulloa, L., 2007. The alpha 7 nicotinic acetylcholine receptor as a pharmacological target for inflammation. *Br. J. Pharmacol.*, **151**, 915–929.

De Lange, D.W. and Meulenbelt, J., 2011. Do corticosteroids have a role in prevention or reducing acute toxic lung injury caused by inhalation of chemical agents? *Clin. Toxicol.*, **46**, 61–67.

Demling, R.H., 2008. Smoke inhalation lung injury. An update. *Eplasty*, **8**, e27.

Demnati, R., Fraser, R., Ghezzo, H., Martin, J.G., Plaa, G., and Malo, J.-H., 1998a. Time-course of functional and pathological changes after a single high acute inhalation of chlorine in rats. *Eur. Respir. J.*, **11**, 922–928.

Demnati, R., Fraser, R., Martin, J.G., Plaa, G., and Malo, J.-H., 1998b. Effects of dexamethasone on functional and pathological changes in rat bronchi caused by high acute exposure to chlorine. *Toxicol. Sci.*, **45**, 242–246.

De Prost, N., Ricard, J.-D., Saumon, G., and Dreyfuss, D., 2011. Ventilator-induced lung injury: Historical perspective and clinical implications. *Ann. Intensive Care*, **1**, 28.

Derdak, S., Mehta, S., Stewart, T.E., Smith, T., Rogers, M., Buchman, T.G., Carlin, B., Lowson, L., and Granton, J., the Multicenter Oscillatory Ventilation for Acute Respiratory Distress Syndrome Trail (MOAT) Study Investigators, 2002. High-frequency oscillatory ventilation for acute respiratory distress syndrome in adults. A randomized controlled trial. *Am. J. Respir. Care Med.*, **166**, 801–808.

Deschamps, D., Soler, P., Rosenberg, N., Baud, F., and Gervais, P., 1994. Persistent asthma after inhalation of a mixture of sodium hypochlorite and hydrochloric acid. *Chest*, **105**, 1895–1896.

Devaney, J.D., Conteras, M., and Laffey, J.G., 2011. Clinical review: Gene-based therapies for ALI/ARDS: Where are we now? *Crit. Care*, **15**, 224.

Devereaux, A.V., Dichter, J.R., Christian, M.D. et al., 2008. Definitive care for the critically ill during a disaster: A framework for allocation of scarce resources in mass critical care. From a Task force for Mass critical Care summit meeting, January 26–27, 2007, Chicago, IL. *Chest*, **133**, 51S–66S.

Dhara, V.R. and Dhara, R., 2002. The Union Carbide disaster in Bhopal: A review of health effects. *Arch. Environ. Health*, **57**, 391–404.

Diller, W.F., 1985a. Pathogenesis of phosgene poisoning. *Toxicol. Ind. Health*, **1**, 7–15.

Diller, W.F., 1985b. Therapeutic strategies in phosgene poisoning. *Toxicol. Ind. Health*, **1**, 93–99.

Dompeling, E., Jöbsis, Q., Vandevijver, N.M.A., Wesseling, G., and Hendriks, H., 2004. Chronic bronchiolitis in a 5-yr-old child after exposure to sulfur mustard gas. *Eur. Respir. J.*, **23**, 343–346.

Dries, D.L. and Endorf, F.W., 2013. Inhalation injury: Epidemiology, pathology, treatment strategies. *Scand. J. Trauma Resusc. Emerg. Med.*, **21**, 31.

Duniho, S.M., Martin, J., Forster, J.S., Cascio, M.B., Moran, T.S., Carpin, L.B., and Sciuto, A.M., 2002. Acute changes in lung histopathology and bronchoalveolar lavage parameters in mice exposed to the chocking agent gas phosgene. *Toxicol. Pathol.*, **30**, 339–349.

Dushiantan, A., Grocott, M.P., Postle, A.D., and Cusak, R., 2011. Acute respiratory distress syndrome and acute lung injury. *Postgrad. Med. J.*, **87**, 612–622.

Eisenkraft, A., Gilat, E., Chapman, S., Baranes, S., Egoz, I., and Levy, A., 2007. Efficacy of the bone injection gun in the treatment of organophosphate poisoning. *Biopharm. Drug Dispos.*, **28**(3), 145–150.

Enkhbaatar, P., Esechie, A., Wang, J. et al., 2008. Combined anticoagulants ameliorate acute lung injury in sheep after burn and smoke inhalation. *Clin. Sci.*, **114**, 321–329.

Enkhbaatar, P., Lange, M., Nakano, Y., Hamahata, A., Jonkam, C., Wang, J., Jaroch, S., Traber, L., Herndon, D., and Traber, D., 2009. Role of neuronal nitric oxide synthase in ovine sepsis model. *Shock*, **32**, 253–247.

Enkhbaatar, P., Murakami, K., Cox, R. et al., 2004. Aerosolized tissue plasminogen activator improves pulmonary function in sheep with burn and smoke inhalation. *Shock*, **22**, 70–75.

Enkhbaatar, P., Murakami, K., Shimoda, K. et al., 2003a. The inducible nitric oxide synthase inhibitor BBS-2 prevents acute lung injury in sheep after burn and smoke inhalation injury. *Am. J. Respir. Crit. Care Med.*, **167**, 2012–1026.

Enkhbaatar, P., Murakami, K., Shimoda, K. et al., 2003b. Inhibition of neuronal nitric oxide synthase by 7-nitroindazole attenuates acute lung injury in an ovine model. *Am. J. Physiol. Regul. Integr. Comp. Physiol.*, **285**, R366–R372.

Enkhbaatar, P., Murakami, K., Traber, L.D. et al., 2006. The inhibition of inducible nitric oxide synthase in ovine sepsis model. *Shock*, **25**, 522–527.

Enkhbataar, P. and Traber, D.L., 2004. Pathophysiology of acute lung injury in combined burn and smoke inhalation injury. *Clin. Sci.*, **107**, 137–143.

Esan, A., Hess, D.R., Raoof, S., George, L., and Sessler, C.N., 2010. Severe hypoxemic respiratory failure. Part I—Ventilator strategies. *Chest*, **137**, 1203–1216.

Everett, E.D. and Overholt, E.L., 1968. Phosgene poisoning. *JAMA*, **205**, 243–245.

Fang, X., Neyrinck, A.P., Matthay, M.A., and Lee, J.W., 2010. Allogeneic human mesenchymal stem cells restore epithelial protein permeability in cultured human alveolar type II cells by secretion of angiopoietin-1. *J. Biol. Chem.*, **285**, 26211–26222.

Fanucchi, M., Bracher, A., Doran, S.F., Squadrito, G.L., Fernandez, S., Poslewaith, E.M., Bowen, L., and Matalon, S., 2012. Post-exposure antioxidant treatment in rats decreases airway hyperplasia and hyperreactivity due to chlorine inhalation. *Am. J. Respir. Cell Mol. Biol.*, **46**, 599–606.

Ferguson, N.D., Cook, D.J., Guyatt, G.H. et al., for the OSCILLATE Trial Investigators and the Canadian Critical Care Trials Group, 2013. High-frequency oscillation in early acute respiratory distress syndrome. *N. Engl. J. Med.*, **368**, 795–805.

Ferguson, N.D., Fan, E., Camporota, L. et al., 2012. The Berlin definition of ARDS: An expanded rationale, justification, and supplementary material. *Intensive Care Med.*, **38**, 1573–1582.

Ferreira, F.L., Peres Bota, D., Bross, A., Melot, C., and Vincent, J.L., 2001. Serial evaluation of the SOFA score to predict outcome in critically ill patients. *JAMA*, **286**, 1754–1758.

Fessler, H.F. and Talmor, D., 2010. Should prone positioning be routinely used for lung protection during mechanical ventilation? *Respir. Care*, **55**, 88–96.

Fisher, A.B. and Beers, M.F., 2008. Hyperoxia and acute lung injury. *Am. J. Physiol. Lung cell Mol. Biol.*, **295**, L1066.

Flaishon, R., Sotman, A., Ben-Abraham, R., Rudick, V., Varssano, D., and Weinbaum, A., 2004a. Antichemical protective gear prolongs time to successful airway management: A randomized crossover study in humans. *Anaesthesiology*, **100**, 260–266.

Flaishon, R., Sotman, A., Ben-Abraham, R., Rudick, V., Varssano, D., and Weinbaum, A., 2004b. Laryngeal mask airway insertion by anaesthetists and nonanaethetists wearing protective gear: A prospective, randomized, crossover study in humans. *Anaesthesiology*, **100**, 267–273.

Foley, F.D., Moncrief, J.A., and Mason, A.D. Jr., 1968. Pathology of the lung in fatally burned patients. *Ann. Surg.*, **167**, 267–273.

Forel, J.-M., Roch, A., Marin, V., Michelet, P., Demory, D., Blache, J.-H., Perrin, G., Gainnier, M., Bongrand, P., and Papazian, L., 2006. Neuromuscular blocking agents decrease inflammatory response in patients presenting with acute respiratory distress syndrome. *Crit. Care Med.*, **34**, 2749–2757.

Francis, H.C., Prys-Picard, C.O., Fishwick, D., Stenton, C., Burgs, P.S., Bradshaw, L., Ayres, J.G., Campbell, S.M., and Niven, R.McL., 2007. Defining and investigating occupational asthma: A consensus approach. *Occup. Environ. Med.*, **64**, 361–356.

Frank, A.J. and Thompson, B.T., 2010. Pharmacological treatments for acute respiratory distress syndrome. *Curr. Opin. Crit. Care*, **16**, 62–68.

Frosolono, M.F. and Curie, W.D., 1985. Response of the pulmonary surfactant system to phosgene. *Toxicol. Ind. Health*, **1**, 29–35.

Frostell, C., Fratacci, M.-D., Wain, J.C., Jones, R., and Zapol, W.M., 1991. Inhaled nitric oxide. A selective pulmonary vasodilator reversing hypoxic pulmonary vasoconstriction. *Circulation*, **83**, 2038–2047.

Frostell, C.G., Blomqvist, H., Hendenstierna, G., Lundberg, J., and Zapol, W.M., 1993. Inhaled nitric oxide selectively reverses human hypoxic pulmonary vasoconstriction without causing systemic vasodilation. *Anaesthesiology*, **78**, 427–435.

Gainnier, M., Roch, A., Forel, J.-M., Thirion, X., Arnal, J.-M., Donati, S., and Papazian, L., 2004. Effect of neuromuscular blocking agents on gas exchange in patients presenting with acute respiratory distress syndrome. *Crit. Care Med.*, **32**, 113–119.

Gajic, O., Dabbagh, O., Park, P.K. et al., on behalf of the U.S. Critical Illness and Injury Trials Group: Ling Injury Prevention Study Investigators (USCIITG-LIPS), 2011. Early identification of patients at risk of acute lung injury. Evaluation of lung injury prediction score in a multicenter cohort study. *Am. J. Respir. Crit. Care Med.*, **183**, 462–470.

Gao Smith, F., Perkins, G.D., Gates, S., Young, D., McAuley, D.F., Tunnicliffe, W., Khan, Z., and Lamb, S.E., for the BALTI-2 Study Investigators, 2012. Effects of intravenous β-2 agonist treatment on clinical outcomes in acute respiratory distress syndrome (BALTI-2): A multicenter, randomized controlled trial. *Lancet*, **379**, 229–235.

Garner, A., Laurence, H., and Lee, A., 2004. Practicality of performing medical procedures in chemical protective ensembles. *Emerg. Med. Austral.*, **16**, 108–113.

Gattinoni, L., Caironi, P., Cressoni, M., Ciumello, D., Ranieri, V.M., Quintel, M., Russo, S., Patroniti, N., Cornejo, R., and Bugedo, D., 2006. Lung recruitment in patients with acute respiratory distress syndrome. *N. Engl. J. Med.*, **354**, 2775–1786.

Gattinoni, L., Carlesso, E., and Langer, T., 2012. Toward ultraprotective mechanical ventilation. *Curr. Opin. Anaesthesiol.*, **25**, 141–147.

Gattinoni, L., Kolobow, T., Tomlinson, T., Iapichino, G., Samaja, M., White, D., and Pierce, J., 1978. Low-frequency positive pressure ventilation with extracorporeal carbon dioxide removal (LFPPV-ECCO$_2$R): An experimental study. *Anesth. Analg.*, **57**, 470–477.

Gattinoni, L., Protti, A., Caironi, P., and Carlesso, E., 2010. Ventilator-induced lung injury: The anatomical and physiological framework. *Crit. Care Med.*, **38** (Suppl.) s539–s548.

Gattinoni, L., Tognoni, G., Pesenti, A. et al., for the Prone-Supine Study Group, 2001. Effect of prone positioning on the survival of patients with acute respiratory failure. *N. Engl. J. Med.*, **345**, 568–573.

Ghio, A.J., Kennedy, T.S., Hatch, G.S., and Tepper, J.S., 1991. Reduction of neutrophil influx diminishes lung injury and mortality following phosgene inhalation. *J. Appl. Physiol.*, **71**, 657–665.

Ghio, A.J., Lehman, J.R., Winsett, D.W., Richards, J.H., and Costa, D.L., 2005. Colchicine decreases airway hyperreactivity after phosgene exposure. *Inhal. Toxicol.*, **17**, 227–285.

Gill, A.L. and Bell, C.N.A., 2004. Hyperbaric oxygen: It's uses, mechanisms of action and outcomes. *Q. J. Med.*, **97**, 385–395.

Glas, G.J., van der Sluis, K.F., Schultz, M.J., Hofstra, J.-J.H., van der Poll, T., and Levi, M., 2013. Bronchoalveolar hemostasis in lung injury and acute respiratory distress syndrome. *J. Thromb. Haemost.*, **11**, 17–25.

Glass, D., McClanahan, M., Koller, L., and Adeshina, F., 2009. Provisional Advisory Levels (PALs) for phosgene. *Inhal. Toxicol.*, **21**(S3), 73–79.

Graham, D.L., Laman, D., Theodore, J., and Robin, E.D., 1997. Acute cyanide poisoning complicated by lactic acidosis and pulmonary edema. *Arch. Intern. Med.*, **137**, 1051–1055.

Grainge, C., 2004. Breath of life: The evolution of oxygen therapy. *J. R. Soc. Med.*, **97**, 489–493.

Grainge, C., Brown, R., Jugg, B.J., Smith, A.J., Mann, T.M., Jenner, J., Rice, P., and Parkhouse, D.A., 2009. Early treatment with nebulized salbutamol worsens physiological measures and does not improve survival following phosgene-induced acute lung injury. *J. R. Army Med. Corps*, **155**, 205–109.

Grainge, C., Brown, R., Jugg, B.J., Smith, A.J., Mann, T.M., Jenner, J., Rice, P., and Parkhouse, D.A., 2010b. Furosemide in the treatment of phosgene induced acute lung injury. *J. R. Army Med. Corps*, **156**, 245–250.

Grainge, C., Jugg, B.J., Smith, A.J., Brown, R.F.R., Jenner, J., Parkhouse, A.A., and Rice, P., 2010a. Delayed low-dose supplemental oxygen improves survival following phosgene-induced acute lung injury. *Inhal. Toxicol.*, **22**, 552–560.

Grainge, C. and Rice, P., 2010. Management of phosgene-induced acute lung injury. *Clin. Toxicol.*, **48**, 245–250.

Griffiths, M.J.D. and Evans, T.W., 2005. Inhaled nitric oxide therapy in adults. *N. Engl. J. Med.*, **353**, 2683–2965.

Grommes, J., Vijayan, S., Drechsler, M., Hartwig, H., Mörgelin, M., Dembinski, R., Jacobs, M., Koeppel, T.A., Binnebösel, M., Weber, C., and Soehnlein, O., 2012. Simvastatin reduces endotoxin-induced acute lung injury by decreasing neutrophil recruitment and radical formation. *PLoS One*, **7**, e38917.

Guidotti, T.L., 2010. Hydrogen sulfide: Advances in understanding human toxicity. *Int. J. Toxicol.*, **29**, 569–581.

Gunnarsson, M., Walther, S.M., Seidel, T., Bloom, G.D., and Lennquist, S., 1998. Exposure to chlorine gas: Effects on pulmonary function and morphology in anaesthetized and mechanically ventilated pigs. *J. Appl. Toxicol.*, **18**, 249–255.

Gupta, N., Su, X., Popov, B., Lee, J.W., Serikov, V., and Matthay, M.A., 2007. Intrapulmonary delivery of bone marrow-derived mesenchymal cells improves survival and attenuate endotoxin-induced acute lung injury in mice. *J. Immunol.*, **179**, 1855–1963.

Hales, C.A., Berkin, P.W., Jung, W., Trautman, E., Lamborghini, D., Herrig, N., and Burke, J., 1988. Synthetic smoke with acrolein but not HCl produces pulmonary edema. *J. Appl. Physiol.*, **64**, 1121–1133.

Hall, J.J., Hunt, J.L., Arnoldo, B.D., and Purdue, G.F., 2007. Use of high-frequency percussive ventilation in inhalation injuries. *J. Burn Care Res.*, **28**, 396–400.

Hamahata, A., Enkhbaatar, P., Hiroyuki, S., Nozaki, M., and Traber, D.L., 2009. Effect of ablated bronchial blood flow on survival rate and pulmonary function after burn and smoke inhalation in sheep. *Burns*, **35**, 802–810.

Hamahata, A., Enkhbaatar, P., Hiroyuki, S., Nozaki, M., and Traber, D.L., 2010. Sclerosis therapy of bronchial artery attenuates acute lung injury induced by burn and smoke inhalation injury in an ovine model. *Burns*, **36**, 1042–1049.

Hamahata, A., Enkhbaatar, P., Lange, M., Cox, R.A., Hawkins, H.K., Sakurai, H., Traber, L.D., and Traber, D.L., 2011. Direct delivery of low-dose 7-nitroimidazole into the bronchial artery attenuates pulmonary pathophysiology after smoke inhalation and burn injury in an ovine model. *Shock*, **36**, 575–579.

Hart, G.B., Strauss, M.B., Lennon, P.A., and Whitcraft, D.D. III, 1985. Treatment of smoke inhalation by hyperbaric oxygen. *J. Emerg. Med.*, **3**, 211–215.

Hauser, B. and Radermacher, P., 2010. Right man, right time, right place?—On time course of the mediator orchestra in septic shock. *Crit. Care*, **14**, 190.

Hayes, M., Curley, G., Ansari, B., and Laffey, J.G., 2012. Clinical review: Stem cell therapies for acute lung injury/acute respiratory distress syndrome—Hope or hype? *Crit. Care*, **16**, 205.

Hick, J.L., Hanfling, D., Burstein, J.L., Markham, J., Mscintyre, A.G., and Barberra, J.A., 2003. Protective equipment for health care facility decontamination personnel: Regulation, risks, and recommendations. *Ann. Emerg. Med.*, **42**, 370–380.

Hilmas, E.H. and Hilmas, C.J., 2009. Medical management and chemical toxicity in pediatrics. Chapter 61, in *Handbook of Toxicology of Chemical Warfare Agents*, Gupta, R.C., ed., Academic Press, London, U.K., pp. 919–950.

Ho, M.-P., Yang, C.-C., Cheung, W.-K., Liu, C.-M., and Tsai, K.-C., 2010. Chlorine gas exposure manifesting acute lung injury. *J. Intern. Med. Taiwan*, **21**, 210–215.

Hoesel, L.M., Flierel, M.A., Niederbichler, A.D. et al., 2008. Ability of antioxidant liposomes to prevent acute and progressive pulmonary injury. *Antioxidant. Redox Signal.*, **10**, 973–981.

Hoyle, G.W., 2010. Mitigation of chlorine lung injury by increasing cyclic AMP levels. *Proc. Am. Thorac. Soc.*, **7**, 284–289.

Hubbard, G.B., Langlinais, P.C., Shimazu, T., Okerberg, C.V., Mason, A.D. Jr., and Pruitt, B.A., Jr., 1991. The morphology of smoke inhalation injury in sheep. *J. Trauma*, **31**, 1477–1486.

Imai, Y., Nakagawa, S., Ito, Y., Kawano, T., Slutsky, A.S., and Miyasaka, K., 2001. Comparison of lung protection strategies using conventional and high-frequency oscillatory ventilation. *J. Appl. Physiol.*, **91**, 1836–1844.

Ipaktchi, K. and Arbabi, S., 2006. Advances in burn critical care. *Crit. Care Med.*, **34**, S239–S243.

Jackson, R.M., 1985. Pulmonary oxygen toxicity. *Chest*, **98**, 900–905.

Jacobson, J.R., Barnard, J.W., Grigoryev, D.N., Ma, S.-F., Tuder, R.M., and Garcia, J.G.N., 2005. Simvastatin attenuates vascular leak and inflammation in murine inflammatory lung injury. *Am. J. Physiol. Lung Cell Mol. Physiol.*, **288**, L1026–L1032.

Jagtap, P. and Szabo, C., 2005. Poly (ADP-ribose) polymerase and the therapeutic effect of its inhibitors. *Nat. Rev. Drug Discov.*, **4**, 421–440.

Ji, L., Zhang, X.D., Chen, H.L., Bai, H., Wang, X., Zhao, H.L., Liang, X., and Hai, C.X., 2010. N-acetylcysteine attenuates phosgene-induced lung injury via up-regulation of Nrf2 expression. *Inhal. Toxicol.*, **22**, 535–542.

Jones, R., Wills, B., and Kang, K., 2010. Chlorine gas: An evolving hazardous material threat and unconventional weapon. *West J. Emerg. Med.*, **11**, 151–156.

Jugg, B.J.A., Smith, A.J., Ruddal, S.J., and Rice, P., 2011. The injured lung: Clinical issues and experimental models. *Philos. Trans. R. Soc. B*, **366**, 309–309.

Kales, S.N. and Christiani, D.C., 2004. Acute chemical emergencies. *N. Engl. J. Med.*, **350**, 800–808.

Kallet, R.H. and Matthay, M.A., 2013. Hyperoxic lung injury. *Respir. Care*, **58**, 123–141.

Kenar, L. and Karayilanoglu, T., 2004. Prehospital management and medical intervention after a chemical attack. *Emerg. Med. J.*, **21**, 84–88.

Kennedy, T.P., Michael, J.R., Hoidal, J.R., Hasty, D., Sciuto, A.M., Hopkins, C., Lazar, R., Bysani, G.K., Tolley, E., and Gurtner, G.H., 1989. Dibutyryl cAMP, aminophylline, and beta-adrenergic agonists protect against pulmonary edema caused by phosgene. *J. Appl. Physiol.*, **67**, 2542–2552.

Kennedy, T.P., Rao, N.V., Noah, W., Michael, J.R., Jafri, M.H. Jr., Gurtner, G.H., and Hoidal, J.R., 1990. Ibuprofen prevents oxidant lung injury and in-vitro lipid peroxidation by chelating iron. *J. Clin. Invest.*, **86**, 1565–1573.

Kimura, R., Traber, L.D., Herndon, D.N., Linares, H.A., Lubbesmeyer, H.J., and Traber, D.L., 1988. Increasing duration of smoke exposure induces more severe lung injury in sheep. *J. Appl. Physiol.*, **64**, 1107–1113.

Kox, M., Pompe, J.C., Peters, E., Vaneker, M., van der Laak, J.W., van der Hoven, J.G., Scheffer, J.G., Hoedemaekers, C.W., and Pikkers, P., 2011. α7 nicotinic acetylcholine receptor agonist GTS-21 attenuates ventilator-induced tumor necrosis factor-α production and lung injury. *Br. J. Anaest.*, **107**, 559–566.

Kunugiyama, S.K. and Schulman, C.S., 2012. High-frequency percussive ventilation using the VDR-4 ventilator: An effective strategy for patients with refractory hypoxemia. *AACN Adv. Crit. Care*, **23**(4), 370–380.

Lalonde, C., Demling, R., Brain, J., and Blanchard, J., 1994. Smoke inhalation injury in sheep is caused by the particle phase, not the gas phase. *J. Appl. Physiol.*, **77**, 15–22.

Lang, J.D., McArdle, P.J., O'Reilly, P.J., and Matalon, S., 2002. Oxidant–antioxidant balance in acute lung injury. *Chest*, **122**, 314S–320S.

Lange, M., Connelly, R., Traber, D.L. et al., 2010. Time course of nitric oxide synthases, nitrosative stress, and poly (ADP ribosylation) in an ovine sepsis model. *Crit. Care*, **14**, R129.

Lange, M., Enkhbaatar, P., Traber, D.L., Cox, R.A., Jacob, S., Mathew, B.P., Hamahata, A., Traber, D.L., Herndon, D.N., and Hawkins, H.K., 2009. Role of calcitonin gene-related peptide (CGRP) in ovine burn and smoke inhalation injury. *J. Appl. Physiol.*, **107**, 176–184.

Lange, M., Hamahata, A., Enkhbaatar, P., Cox, R.A., Nakano, Y., Westphal, M., Traber, L.D., Herndon, D.N., and Traber, D.L., 2011a. Beneficial effects of concomitant neuronal and inducible nitric oxide synthase inhibition in ovine burn and inhalation injury. *Shock*, **35**, 626–631.

Lange, M., Szabo, C., Enkhbaatar, P. et al., 2011b. Beneficial pulmonary effects of a metalloporphyrinic peroxynitrite decomposition catalyst in burn and smoke inhalation injury. *Am. J. Physiol. Lung Cell Mol. Physiol.*, **300**, L167–L175.

Latenser, B.A., 2009. Critical care of the burn patient: The first 48 hours. *Crit. Care Med.*, **37**, 2819–2826.

Lee, J.W., Fang, X., Gupta, N., Sedikov, V., and Matthay, M.A., 2009. Allogeneic human mesenchymal stem cells for treatment of *E. coli* endotoxin-induced acute lung injury in ex vivo perfused human lung. *Proc. Natl. Acad. Sci. USA*, **106**, 16357–16362.

Lee, J.W., Fang, X., Krasnodembskaya, A., Howard, J.P., and Matthay, M.A., 2011. Concise review: Mesenchymal stem cells for acute lung injury: Role of paracrine soluble factors. *Stem Cells*, **29**, 913–919.

Lehavi, O., Leiba, A., Dahan, Y., Schwartz, D., Benin-Goren, O., Ben-Yehuda, Y., Wess, G., Levi, Y., and Bar-Dayan, Y., 2008. Lessons learned from chlorine intoxications in swimming pools: The challenge of pediatric mass toxicological events. *Prehospital Disast. Med.*, **23**, 90–95.

Lemiére, C., Malo, J.-L., and Boutet, M., 1997. Reactive airways dysfunction syndrome due to chlorine: Sequential bronchial biopsies and functional assessment. *Eur. Respir. J.*, **10**, 241–244.

Leustik, M., Doran, S., Bracher, A., Williams, S., Squadrito, G.L., Schoesb, T.R., Postlethwaith, E., and Matalon, S., 2008. Mitigation of chlorine-induced lung injury by low-molecular weight antioxidants. *Am. J. Physiol. Lung Cell Mol. Physiol.*, **295**, L733–L743.

Levitt, J.E. and Matthay, M.A., 2012. Clinical review: Early treatment of acute lung injury—Paradigm shift toward prevention and treatment prior to respiratory failure. *Crit. Care*, **16**, 223.

Li, P.-C., Chen, W.-C., Chang, L.-C., and Lin, S.-C., 2008. Substance P acts via the neurokinin receptor 1 to elicit bronchoconstriction, oxidative stress, and upregulated ICAM-1 expression after oil smoke exposure. *Am. J. Physiol. Lung Cell Mol. Physiol.*, **294**, L912–L920.

Li, W.-L., Hai, C-.X., and Pauluhn, J., 2011. Inhaled nitric oxide aggravates phosgene model of acute lung injury. *Inhal. Toxicol.*, **23**, 842–852.

Lindsay, C.D., 2011. Novel therapeutic strategies for acute lung injury induced by lung damaging agents: The potential role of growth factors as treatment options. *Hum. Exp. Toxicol.*, **30**, 701–724.

Litell, J.M., Gong, M.N., Talmor, D., and Gajic, O., 2011. Acute lung injury: Prevention may be the best medicine. *Respir. Care*, **56**, 1546–1554.

Liu, L.L., Aldrich, J.M., Shimabukuru, D.W., Sullivan, K.R., Taylor, J.M., Thornton, K.C., and Gropper, M.A., 2010. Rescue therapies for acute hypoxemic respiratory failure. *Anest. Analg.*, **111**, 693–702.

Livigni, S., Maio, M., Ferretti, E., Longobardo, A., Potenza, R., Rivalta, L., Selvaggi, P., Vergano, M., and Bertolini, G., 2006. Efficacy and safety of a low-flow veno-venous carbon dioxide removal device: Results of an experimental study in adult sheep. *Crit. Care*, **10**, R151.

Macintyre, A.G., Christopher, W., Eitzen, E. Jr, Gum, R., Weir, S., DeAtley, C., Tonat, K., and Barberra, J., 2000. Weapons of mass destruction events with contaminated casualties. Effective planning for health care facilities. *JAMA*, **283**, 242–249.

MacLaren, G., Combes, A., and Bartlett, R.H., 2012. Contemporary extracorporeal membrane oxygenation for adult respiratory failure: Life support in the new era. *Intensive Care Med.*, **38**, 210–220.

Mallory, T.B. and Brickley, W.J., 1943. Symposium on the management of the coconut Grove burns at the Massachusetts General Hospital. Pathology: With special reference to pulmonary lesions. *Ann. Surg.*, **17**, 865–884.

Mapp, C.E., Pozzato, V., Pavoni, V., and Gritti, G., 2000. Severe asthma and ARDS triggered by acute short-term exposure to commonly used cleaning detergents. *Eur. Respir. J.*, **16**, 570–572.

Marik, P.E., Meduri, G.U., Rocco, P.M., and Annane, D., 2011. Glucocorticoid treatment in acute lung injury and acute respiratory distress syndrome. *Crit. Care Clin.*, **27**, 589–607.

Marik, P.E., Pastores, S.M., Annane, D. et al., 2008. Recommendations for the diagnosis and management of corticosteroid insufficiency in critically ill adult patients: Consensus statements from an international task force by the American College for Critical Care Medicine. *Crit. Care Med.*, **36**, 1937–1949.

Marini, J.J., 2010. Prone positioning in ARDS: Defining the target. *Intensive Care Med.*, **36**, 559–561.

Markel, G., Krivoy, A., Rotman, E., Schein, O., Shrot, S., Brosh-Nissimov, T., Dushnitsky, T., and Eisenkraft, A., 2008. Medical management of toxicological mass casualty events. *Isr. Med. Assoc. J.*, **10**(11), 761–766.

Martin, J.G., Campbell, H.R., Ijima, H., Gautrin, D., Malo, J.-L., Eidelman, D.H., Qutayba, H., and Maghni, K., 2003. Chlorine-induced injury to the airway in mice. *Am. J. Respir. Crit. Care Med.*, **168**, 568–574.

Masanés, M.-J., Legendre, C., Lioret, N., Saizy, R., and Lebeau, B., 1995. Using bronchoscopy and biopsy to diagnose early inhalation injury. Macroscopic and histologic findings. *Chest*, **107**, 1365–1367.

Matthay, M.A., Brower, R.G., and Carson, M.D., the National Heart, Lung and Blood Institute Acute Respiratory Distress Syndrome (ARDS) Clinical Trials Network, 2011. Randomized, placebo-controlled clinical trial of an aerosolized β2-agonist for treatment of acute lung injury. *Am. J. Respir. Crit. Care Med.*, **184**, 561–568.

Mattahy, M.A., Ware, L.B., and Zimmermann, G.A., 2012. The acute respiratory distress syndrome. *J. Clin. Invest.*, **122**, 2731–2740.

Mautone, A.J., Katz, J., and Scarpelli, E.M., 1985. Acute responses to phosgene inhalation and selected corrective measures. *Toxicol. Ind. Health*, **1**, 37–57.

Maybauer, M.O., Maybauer, D.M., Fraser, J.F. et al., 2012. Combined recombinant human activated protein C and ceftazidime prevent the onset of acute respiratory distress syndrome in severe sepsis. *Shock*, **37**, 170–176.

McClintock, S.D., Hoesel, L.M., Das, S.K., Till, G.O., Neff, T., Kunkel, R.G., Smith, M.G., and Ward, P.A., 2006. Attenuation of half sulfur mustard gas-induced acute lung injury in rats. *J. Appl. Toxicol.*, **26**, 126–131.

McGovern, T.K., Powell, W.S., Day, B.J., White, C.W., Govindaraju, K., Karmouty-Quintana, H., Lavoie, N., Tan, J.J., and Martin, J.G., 2010. Dimethylurea protects against chlorine induced changes in airway function in a murine model of irritant induced asthma. *Respir. Res.*, **11**, 138.

Meade, M.O., Cook, D.J., Guyatt, G.D. et al., for the Lung Ventilation Study Investigators, 2008. Ventilation strategy using low tidal volumes, recruitment maneuvers, and high positive end-expiratory pressure for acute lung injury and acute respiratory distress syndrome: A randomized controlled trial. *JAMA*, **299**, 637–645.

Meduri, G.U., 1996. The role of host defense response in the progression and outcome of ARDS: Pathophysiological correlations and response to glucocorticoid treatment. *Eur. Respir. J.*, **9**, 2650–2670.

Meduri, G.U., Golden, E., Freier, A.X., Taylor, E., Zaman, M., Carson, S.J., Gibson, M., and Umberger, R., 2007. Methylprednisolone infusion in early severe ARDS: Results of a randomized controlled trial. *Chest*, **131**, 954–963.

Meduri, G.U., Headley, A.S., Golden, E., Carson, S.J., Umberger, R.A., Kelso, T., and Toiley, E.A., 1998. Effect of prolonged methylprednisolone therapy in unresolving acute respiratory distress syndrome. A randomized controlled trial. *JAMA*, **280**, 159–165.

Meduri, G.U., Marik, P.E., Chrousos, G.P., Pastores, S.M., Arlt, W., Beishuisen, A., Bokhari, F., Zaloga, G., and Annane, D., 2008. Steroid treatment in ARDS: A critical appraisal of the ARDS network trial and the recent literature. *Intensive Care Med.*, **34**, 61–69.

Meduri, U.G., Annane, D., Chrousos, G.P., Marik, P.E., and Sinclair, S.E., 2009. Activation and regulation of systemic inflammation in ARDS. *Chest*, **136**, 1631–1643.

Mei, S.H.J., McCarter, S.D., Deng, Y., Parker, C.H., Liles, W.C., and Stewar, D.J., 2007. Prevention of LPS-induced acute lung injury in mice by mesenchymal stem cells overexpressing angiopoietin 1. *PLoS Med.*, **4**, e269.

Mercat, A., Richard, J.-C.M., Vielle, B. et al., for the Expiratory Pressure (Express) Study Group, 2008. Positive end-expiratory pressure setting in adults with acute lung injury and acute respiratory distress syndrome: A randomized controlled trial. *JAMA*, **299**, 646–654.

Miller, A.C., Rivero, A., Ziad, S., Smith, D.J., and Elamin, E.M., 2009. Influence of nebulized unfractionated heparin and N-acetylcysteine in acute lung injury after smoke inhalation injury. *J. Burn Care Res.*, **30**, 249–256.

Mishra, P.K., Samarth, R.M., Pathak, N., Jain, S.K., Banerjee, S., and Maudar, K.K., 2009. Bhopal gas tragedy: Review of clinical and experimental findings after 25 years. *Int. J. Occup. Med. Environ. Health*, **22**, 193–202.

Mlcak, R.P., Suman, O.E., and Herndon, D.N., 2007. Respiratory management of inhalation injury. *Burns*, **33**, 2–13.

Mohan, A., Kumar, N.S., Rao, M.H., Bollineni, S., and Manohar, I.C., 2010. Acute exposure to chlorine gas: Clinical presentation, pulmonary functions and outcomes. *Indian J. Chest Dis. Allied Sci.*, **52**, 149–152.

Moloney, E.D. and Evans, T.W., 2003. Pathophysiology and pharmacological treatment of pulmonary hypertension in acute respiratory distress syndrome. *Eur. Resp. J.*, **21**, 720–727.

Morita, M., Enkhbataar, P., Maybauer, D.M., Maybauer, M.O., Westphal, M., Murakami, K., Hawkins, H.K., Cox, R.A., Traber, L.D., and Traber, D.L., 2011. Impact of bronchial circulation on bronchial exudation following combined burn and smoke injury in sheep. *Burns*, **37**, 465–473.

Moritz, A.R., Henriques, F.C., and McLean, R., 1945. The effects of inhaled heat on the air passages and lungs. *Am. J. Pathol.*, **21**, 311–331.

Morris, J.B., Wilkie, W.S., and Shusterman, D.J., 2005. Acute respiratory responses of the mouse to chlorine. *Toxicol. Sci.*, **83**, 380–387.

Muehlberger, T., Kunar, D., Munster, A., and Couch, M.C., 1998. Efficacy of fiberoptic laryngoscopy in the diagnosis of inhalation injury. *Arch. Otolaryngol. Head Neck Surg.*, **124**, 1003–1007.

Müller, H.C., Hellwig, K., Rosseau, S. et al., 2010. Simvastatin attenuates ventilator-induced lung injury in mice. *Crit. Care*, **14**, R142.

Murakami, K., McGuire, R., Cox, R.A. et al., 2002. Heparin nebulization attenuates acute lung injury in sepsis following smoke inhalation in sheep. *Shock*, **18**, 236–241.

Murakami, K. and Traber, D.L., 2003. Pathophysiological basis of smoke inhalation injury. *News Physiol.*, **18**, 125–129.

Nieman, G.F., Clark, W.R. Jr., Wax, S.D., and Webb, S.R., 1980. The effect of smoke inhalation on pulmonary surfactant. *Ann. Surg.*, **191**, 171–181.

Noah, M.A., Peek, G.J., Finney, S. et al., 2011. Referral to an extracorporeal membrane oxygenation center and mortality among patients with severe 2009 influenza A (H1N1). *JAMA*, **306**, 1659–1668.

Nodelman, V. and Ultman, J.S., 1999. Longitudinal distribution of chlorine absorption in human airways: Comparison of nasal and oral quiet breathing. *J. Appl. Physiol.*, **86**, 1984–1993.

Noltkamper, D. and Burgher, S.W., 2012. Phosgene toxicity. *Medscape reference*, http//emedicine.medscape.com/article/820649-treatment; http//emedicine.medscape.com/article/820649-medication (last accessed on May 20, 2014).

Ogura, H., Cioffi, W.G. Jr., Jordan, B.S., Okerberg, C.V., Johnson, A.A., Mason, A.D., and Pruitt, B.A. Jr., 1994a. The effect of inhaled nitric oxide on smoke inhalation injury in an ovine model. *J. Trauma*, **37**, 294–302.

Ogura, H., Saitoh, D., Johnson, A.A., Mason, A.D., Pruitt, B.A. Jr., and Cioffi, W.G. Jr., 1994b. The effect of inhaled nitric oxide on pulmonary ventilation–perfusion matching following smoke inhalation injury. *J. Trauma*, **37**, 893–898.

Ohbu, S., Yamashina, A., Takasu, N., Yamaguchi, T., Murai, T., Nakano, K., Matsui, Y., Mikami, R., Sakurai, K., and Hinohara, S., 1997. Sarin poisoning on Tokyo subway. *South. Med. J.*, **96**, 587–593.

Okolie, N.P. and Osagie, A.U., 2000. Differential effects of chronic cyanide intoxication on heart, lung and pancreatic tissues. *Food Chem. Toxicol.*, **38**, 543–548.

Okudera, H., 2002. Clinical features on nerve gas terrorism in Matsumoto. *J. Clin. Neurosci.*, **9**, 17–21.

O'Neal, H.R. Jr., Koyama, T., Koehler, E.A.S., Siew, E., Curtis, B.R., Fremont, R.D., May, A.K., Bernard, G.R., and Ware, L.B., 2011. Prehospital statin and aspirin use and the prevalence of severe sepsis and ALI/ARDS. *Crit. Care Med.*, **39**, 1343–1350.

OSHA, 2005. OSHA/NIOSH interim guidance. Chemical–Biological–Radiological (CBRN) personal protective equipment selection matrix for emergency responders. http//osha.gov/SLTC/emergencypreparedness/cbrnmatrix/index (last accessed on May 20, 2014).

Papazian, L., Forel, J.-M., Gacouine, A. et al., for the ACURASYS Study Investigation, 2010. Neuromuscular blockers in early acute respiratory distress syndrome. *N. Engl. J. Med.*, **363**, 1197–1116.

Pauluhn, J., 2006a. Acute nose-only exposure of rats to phosgene. Part I. Concentration × time dependence of LC_{50}s, nonlethal-threshold concentrations, and analysis of breathing patterns. *Inhal. Toxicol.*, **18**, 423–435.

Pauluhn, J., 2006b. Acute nose-only exposure of rats to phosgene. Part II. Concentration × time dependence on changes in bronchoalveolar lavage during a follow-up period of three months. *Inhal. Toxicol.*, **18**, 595–607.

Pauluhn, J., 2006c. Acute head-only exposure of rats to phosgene. Part III. Comparison of indicators of lung injury in dogs and rats. *Inhal. Toxicol.*, **18**, 609–621.

Pauluhn, J., Carson, A., Costa, D.L., Gordon, T., Kondavanti, U., Last, J.A., Matthay, M.A., Pinkerton, K.E., and Sciuto, A.M., 2007. Workshop summary: Phosgene-induced pulmonary toxicity revisited: Appraisal of early and late markers of pulmonary injury from animal models with emphasis on human significance. *Inhal. Toxicol.*, **19**, 789–810.

Peek, G.J., Mugford, M., Tiruviopati, R. et al., for the CESAR Trial Collaboration, 2009. Efficacy and economic assessment of conventional ventilator support versus extracorporeal membrane oxygenation for severe adult respiratory failure (CESAR): A multicentre randomized controlled study. *Lancet*, **374**, 1351–1363.

Pelosi, L., Brazzi, L., and Gattinoni, L., 2002. Prone position in acute respiratory distress syndrome. *Eur. Respir. J.*, **20**, 1017–1028.

Perkins, G.D., Gao, F., and Thickett, D.R., 2008. In vitro and in vivo effects of salbutamol on alveolar epithelial repair in acute lung injury. *Thorax*, **63**, 215–220.

Perkins, G.D., McAuley, D.F., Thickett, D.R., and Gao, F., 2006. The beta-agonist lung injury trial (BALTI): A randomized placebo-controlled clinical trial. *Am. J. Respir. Crit. Care Med.*, **173**, 281–287.

Perkins, G.D., Nathani, N., McAuley, D.F., and Gao, F., 2007. In vitro and in vivo effects of salbutamol on neutrophil function in acute lung injury. *Thorax*, **62**, 36–42.

Phua, J., Badia, J.R., Adhikari, N.K.J. et al., 2009. Has mortality from acute respiratory distress syndrome decreased over time? A systematic review. *Am. J. Respir. Crit. Care Med.*, **179**, 220–227.

Pipeling, M.R. and Fan, E., 2010. Therapies for refractory hypoxemia in acute respiratory distress syndrome. *JAMA*, **304**, 2521–2527.

Pruitt, B.A. Jr., Cioffi, W.G., Shimazu, M., Ikeuchi, H., and Mason, A.D. Jr., 1990. Evaluation and management of patients with inhalation injury. *J. Trauma*, **30**, S63–S69.

Rancourt, R.C., Veress, L.A., Guo, X.L., Jones, T.N., Hendry-Hofer, T.B., and White, C.W., 2012. Airway tissue factor-dependent coagulation activity in response to sulfur mustard analog 2-chloroethyl ethyl sulfide. *Am. J. Physiol. Lung Cell Mol. Physiol.*, **302**, L82–L92.

Ranieri, V.M., Sutter, P.M., Tortorella, C., De Tullio, R., Dayer, J.-M., Brienza, A., Bruno, F., and Slutsky, A.S., 1999. Effect of mechanical ventilation on inflammatory mediators in patients with acute respiratory distress syndrome. A randomized controlled trial. *JAMA*, **281**, 54–61.

Raoof, S., Goulet, K., Esan, A., Hess, D.R., and Sessler, C.N., 2010. Severe hypoxemic respiratory failure: Part 2—Nonventilatory strategies. *Chest*, **137**, 1437–1445.

Reade, M.C. and Milbrandt, E.B., 2007. Is there evidence to support a phase II trial of inhaled corticosteroids in the treatment of incipient or persistent ARDS? *Crit. Care Resusc.*, **9**, 276–285.

Rehberg, S., Maybauer, M.O., Enkhbaatar, P., Maybauer, D.M., Yamamoto, Y., and Traber, D.T., 2009. Pathophysiology, management and treatment of smoke inhalation injury. *Expert Rev. Respir. Med.*, **3**, 283–297.

Ren, G., Chen, X., Dong, F., Ren, X., Zhang, Y., and Shi, Y., 2012. Concise review: Mesenchymal stem cells and translational medicine: Emerging issues. *Stem Cells Trans. Med.*, **1**, 51–58.

Renz, E.M., Cancio, L.C., Barillo, D.J. et al., 2008. Long range transport of war-related burn casualties. *J. Trauma*, **64**, S136–S145.

Ronchi, C.F., dos Anjos Ferreira, A., Campos, F.J., Kurokawa, C.S., Carpi, M.F., de Moraes, M.A., Bonatto, R.C., Defaveri, J., Yeum, K.-J., and Fioretto, J.R., 2011. High frequency oscillatory ventilation attenuates oxidative lung injury in a rabbit model of acute lung injury. *Exp. Biol. Med.*, **236**, 1188–1196.

Rossaint, R., Falke, K.J., López, F., Slama, K., Pison, U., and Zapol, W.M., 1993. Inhaled nitric oxide for the adult respiratory distress syndrome. *N. Engl. J. Med.*, **328**, 399–405.

Rubinson, L., Branson, R., Pesik, N., and Talmor, D., 2006. Positive-pressure ventilation equipment for mass casualty respiratory failure. *Biosecur. Bioterror*, **4**, 183–194.

Russell, D., Blaine, P.G., and Rice, P., 2006. Clinical management of casualties exposed to lung damaging agents: A critical review. *Emerg. Med. J.*, **23**, 421–424.

Samal, A.A., Honavar, J., Brandon, A. et al., 2012. Administration of nitrite after chlorine gas exposure prevents lung injury: Effect of administration modality. *Free Rad. Biol. Med.*, **53**, 1431–1439.

Saunders, F.D., Westphal, M., Enkhbaatar, P. et al., 2010. Molecular biological effects of selective neuronal nitric oxide synthase inhibition in ovine lung injury. *Am. J. Physiol. Lung Cell Mol,. Physiol.*, **298**, L427–L436.

Schmalstieg, F.C., Keeny, S.E., Rudloff, H.E., Palkowetz, K.H., Cevallos, M., Zhou, X., Cox, R.A., Hawkins, H.K., Traber, D.L., and Zwischenberger, J.B., 2007. Arteriovenous CO_2 removal improves survival compared to high percussive and low tidal volume ventilation in smoke/burn sheep acute respiratory distress syndrome model. *Ann. Surg.*, **246**, 512–523.

Schwartz, D.A., Smith, D.D., and Lakshminarayan, S., 1990. The pulmonary sequelae associated with accidental inhalation of chlorine gas. *Chest*, **97**, 820–825.

Sciuto, A.M., 1997. Ibuprofen treatment enhances the survival of mice following exposure to phosgene. *Inhal. Toxicol.*, **9**, 389–403.

Sciuto, A.M., 1998. Assessment of early acute lung injury in rodents exposed to phosgene. *Arch. Toxicol.*, **72**, 283–288.

Sciuto, A.M. and Hurt, H.H., 2004. Therapeutic treatment of phosgene-induced lung injury. *Inhal. Toxicol.*, **16**, 565–580.

Sciuto, A.M., Stotts, R.R., and Hurt, H.H., 1996. Efficacy of ibuprofen and pentoxifylline in the treatment of phosgene-induced acute lung injury. *J. Appl. Toxicol.*, **16**, 381–384.

Sciuto, A.M., Strickland, P.T., and Gurtner, G.H., 1998. Post-exposure treatment with isoproterenol attenuates pulmonary edema in phosgene-exposed rabbits. *J. Appl. Toxicol.*, **18**, 321–329.

Sciuto, A.M., Strickland, P.T., Kennedy, T.P., and Gurtner, G.H., 1995. Protective effects of N-acetylcysteine treatment after phosgene exposure in rabbits. *Am. J. Respir. Crit. Care Med.*, **151**, 768–772.

Sciuto, A.M., Strickland, P.T., Kennedy, T.P., and Gurtner, G.H., 1997. Postexposure treatment with aminophylline protects against phosgene-induced acute lung injury. *Exp. Lung Res.*, **23**, 317–332.

Segal, E. and Lang, S., 2012. Chlorine gas toxicity. Medscape reference, http://emedicine.medscape.com/article/820779-treatment; http://emedicine.medscape.com/article/820779-medicatiom (last accessed on May 20, 2014).

Sessler, C.N. and Gay, P.C., 2010. Are corticosteroids useful in late-stage acute respiratory syndrome? *Respir. Care*, **55**, 43–52.

Shakeri, M.S., Dick, F.D., and Ayres, J.G., 2008. Which agents cause reactive airways dysfunction syndrome (RADS)? A systematic review. *Occup. Med.*, **58**, 205–211.

Shimazu, T., Yukioka, T., Hubbard, G.B., Langlinais, P.C., Mason, A.D. Jr., and Pruitt, B.A. Jr., 1987. A dose–responsive model of smoke inhalation injury. Severity-related alteration in cardiopulmonary function. *Ann. Surg.*, **206**, 89–98.

Shimoda, K., Murakami, K., Enkhbaatar, P. et al., 2003. Effect of poly (ADP ribose) synthase inhibition on burn and smoke inhalation injury in sheep. *Am. J. Physiol. Lung Cell Mol. Physiol.*, **285**, L240–L249.

Shirani, K.Z., Pruitt, B.A. Jr., and Mason, A.D. Jr., 1986. The influence of inhalation injury and pneumonia on burn mortality. *Ann. Surg.*, **205**, 82–87.

Shyamsundar, M., McKeown, S.T.W., O'Kane, C.M. et al., 2009. Simvastatin decreases lipopolysaccharide-induced pulmonary inflammation in healthy volunteers. *Am. J. Respir. Crit. Care Med.*, **179**, 1107–1114.

Sinclair, K., Yerkovich, S.T., and Chambers, D.C., 2013. Mesenchymal stem cells and the lung. *Respirology*, **18**, 397–411.

Singla, S. and Jackson, J.R., 2012. Statins as a novel therapeutic strategy in acute lung injury. *Pulm. Circ.*, **2**, 397–406.

Siobal, M.S. and Hess, D.R., 2010. Are inhaled vasodilators useful in acute lung injury and acute respiratory distress syndrome? *Respir. Care*, **55**, 144–157.

Sjögren, B., Plato, N., Alexandersson, R., Eklund, A., and Falkenberg, C., 1991. Pulmonary reaction caused by welding-induced decomposed trichloroethylene. *Chest*, **99**, 237–238.

Slutsky, A.S., 1993. ACCP consensus conference: Mechanical ventilation. *Chest*, **104**, 1833–1859.

Snyder, R.W., Mishel, H.S., and Christensen, G.C., 1992. Pulmonary toxicity following exposure to methylene chloride and its combustion product phosgene. *Chest*, **101**, 860–861.

Soejima, K., McGuire, R., Snyder, N., Uchida, T., Szabo, C., Salzman, A., Traber, L.D., and Traber, D.L., 2000. The effect of inducible nitric oxide synthase (iNOS) inhibition on smoke inhalation injury in sheep. *Shock*, **13**, 261–266.

Soejima, K., Schmalstieg, F.C., Sakurai, H., Traber, L.D., and Traber, D.L., 2001a. Pathophysiological analysis of combined burn and smoke inhalation injuries in sheep. *Am. J. Physiol. Lung Cell Mol. Physiol.*, **280**, L1233–L1241.

Soejima, K., Traber, L.D., Schmalstieg, F.C., Hawkins, H., Jodoin, J.M., Szabo, C., Szabo, E., Varig, L., Salzman, A., and Traber, D.L., 2001b. Role of nitric oxide in vascular permeability after combined burns and smoke inhalation injury. *Am. J. Crit. Care Med.*, **163**, 745–752.

Song, W., Wei, S., Liu, G., Yu, Z., Estell, K., Yadav, A.K., Schweibert, L.M., and Matalon, S., 2011. Postexposure administration of a β_2-agonist decreases chlorine-induced airway hyperreactivity in mice. *Am. J. Respir. Cell Mol. Biol.*, **45**, 88–94.

Spragg, R.G., Bernard, G.R., Checkley, W. et al., 2010. Beyond mortality: Future clinical research in acute lung injury. *Am. J. Respir. Crit. Care Med.*, **181**, 1121–1127.

Squadrito, G.L., Postlethwait, E.M., and Matalon, S., 2010. Elucidating mechanisms of chlorine toxicity: Reaction kinetics, thermodynamics and physiological implications. *Am. J. Physiol. Lung Cell Mol. Physiol.*, **299**, L289–L300.

Steinberg, J.M., Schiller, H.J., Tsvaygenbaum, B., Mahoney, G.K., DiRocco, J.D., Gatto, L.A., and Nieman, G.F., 2005. Wood smoke inhalation causes alveolar instability in a dose-dependent fashion. *Respir. Care*, **50**, 1062–1070.

Steinberg, K.P., Hudson, L.D., Goodman, R.B. et al., The National Heart, Lung and Blood Institute Acute Respiratory Distress (ARDS) Clinical Trials Network, 2006. Efficacy and safety of corticosteroids for persistent acute respiratory distress syndrome. *N. Engl. J. Med.*, **354**, 1671–2684.

Su, X., Lee, J.W., Matthay, Z.A., Mednick, G., Uchida, T., Fang, X., Gupta, N., and Matthay, M.A., 2007 Activation of the $\alpha7$ nAChR reduces acid-induced acute lung injury in mice and rats. *Am. J. Respir. Cell Mol. Biol.*, **37**, 186–192.

Su, X., Matthay, M.A., and Malik, A.B., 2010. Requisite role of the cholinergic $\alpha7$ nicotinic acetylcholine receptor pathway in suppression of gram-negative sepsis induced acute lung inflammatory injury. *J. Immunol.*, **184**, 401–410.

Sud, S., Friedrich, J.O., Taccone, P. et al., 2010b. Prone ventilation reduces mortality in patients with acute respiratory failure and severe hypoxemia: Systematic review and meta-analysis. *Intensive Care Med.*, **36**, 585–599.

Sud, S., Sud, M., Friedrich, J.O., Meade, M.O., Ferguson, N.D., Wunsch, H., and Adhikari, N.K.J., 2010a. High frequency oscillation in patients with acute lung injury and acute respiratory distress syndrome (ARDS): Systematic review and meta-analysis. *BMJ*, **340**, c2327.

Sun, J., Han, Z.-B., Liao, W., Yang, S.G., Yang, Z., Yu, J., Meng, L., Wu, R., and Han, Z.C., 2011. Intrapulmonary delivery of human umbilical cord mesenchymal cells attenuate acute lung injury by expanding $CD4^+CD25^+$ Forkhead Boxp3 $(FOXP3)^+$ regulatory T cells and balancing anti- and pro-inflammatory factors. *Cell Physiol. Biochem.*, **27**, 587–596.

Suntres, Z.E., 2011. Liposomal antioxidants for protection against oxidant-induced damage. *J. Toxicol.*, **2011**, Article ID 152474, doi:10.1155/2011/152474.

Szabó, C., Ischiropoulos, H., and Redi, R., 2007. Peroxynitrite: Biochemistry, pathophysiology, and development of therapeutics. *Nat. Rev. Drug Discov.*, **6**, 662–680.

Taccone, P., Pesanti, A., Latini, R. et al., for the Prone-Supine II Study Group, 2009. Prone positioning in patients with moderate and severe acute respiratory distress syndrome. A randomized controlled trial. *JAMA*, **302**, 1977–1984.

Talmor, D., 2008. Airway management during a mass casualty event. *Resp. Care*, **53**, 226–230.

Tang, B.M.P., Craig, J.C., Eslick, G.D., Seppelt, I., and McLean, A.S., 2009. Use of corticosteroids in acute lung injury and acute respiratory distress syndrome: A systematic review and meta-analysis. *Crit. Care Med.*, **37**, 1594–1603.

Terragni, P.P., Del Sorbo, L., Mascia, L., Urbino, R., Martin, E.L., Biroco, A., Faggiano, C., Quintel, M., Gattinoni, L., and Ranieri, V.M., 2009. Tidal volume lower than 6 ml/kg enhances lung protection. Role of extracorporeal carbon dioxide removal. *Anesthesiology*, **11**, 826–835.

Terragni, P.P., Rosboch, G., Tealdi, A. et al., 2007. Tidal hyperinflation during low tidal volume ventilation in acute respiratory distress syndrome. *Am. J. Respir. Crit. Care Med.*, **175**, 160–166.

Thom, S.R., Mendiguren, I., and Fisher, D., 2001. Smoke inhalation induced alveolar lung injury is inhibited by hyperbaric oxygen. *Undersea Hperb. Med.*, **28**, 175–179.

Toon, M.H., Maybauer, M.O., Greenwood, J.E., Maybauer, D.M., and Fraser, J.F., 2010. Management of acute smoke inhalation. *Crit. Care Resusc.*, **12**, 53–61.

Tourinsky, S.D. and Sciuto, A.M., 2008. Toxic inhalation injury and toxic industrial chemicals. Chapter 10, in: *Medical Aspects of Chemical Warfare*. Tourinsky, S.D. (senior editor), Textbook of Military Medicine. Office of the Surgeon General, US Army, Borden Institute, Walter Reed Army Medical Center, Washington, DC, pp. 339–370.

Tracey, K.J., 2002. The inflammatory reflex. *Nature*, **420**, 853–859.

Tuck, S.A., Ramos-Barbón, D., Campbell, H., McGovern, T., Karmouty-Quintana, H., and Martin, J.G., 2008. Time course of airway remodeling after an acute chlorine gas exposure in mice. *Respir. Res.*, **9**, 61.

Tuinman, P.R., Dixon, B., Levi, M., Juffermans, N.P., and Schultz, M.J., 2012. Nebulized anticoagulants for acute lung injury—A systemic review of preclinical and clinical investigations. *Crit. Care*, **16**, R70.

US Army Medical Research Institute of Chemical Defense, 2007. *Medical Management of Chemical Casualties Handbook*, 4th edn., Aberdeen Proving Ground, Aberdeen, MD.

US National Library of Medicine, Haz-Map®. http://hazmap.nlm.nih.gov/index-php. (Accessed on March 2013).

Van Sickle, W., Wenck, M.A., Beltflower, A.B. et al., 2009. Acute health effects after exposure to chlorine gas released from train derailment. *Am. J. Emerg. Med.*, **27**, 1–7.

Villar, J., Kacmarek, R.M., Pérez-Méndez, L., and Aguirre-Jaime, A., for the ARIES Network, 2006. A high positive end-expiratory pressure, low tidal volume ventilatory strategy improves outcome in persistent acute respiratory distress syndrome: A randomized, controlled trial. *Crit. Care Med.*, **34**, 1311–1318.

Vincent, J.-L., Moreno, R., Takala, L., Willats, S., De Mendoca, A., Bruining, H., Reinhardt, C.K., Suter, P.M., and Thijs, L.G., 1996. The SOFA (Sepsis-related Organ Failure Assessment) score to describe organ dysfunction/failure. *Intensive Care Med.*, **22**, 707–710.

Wang, J., Abu-Zidan, F.M., and Walther, S.M., 2002b. Effects of prone and supine posture on cardiopulmonary function after experimental chlorine gas lung injury. *Acta Anaesthesiol. Scand.*, **46**, 1094–1102.

Wang, J., Winskog, C., Edston, E., and Walther, S.M., 2005. Inhaled and intravenous corticosteroids both attenuate chlorine gas-induced lung injury in pigs. *Acta Anaesthesiol. Scand.*, **49**, 183–190.

Wang, J., Zhang, L., and Walther, S.M., 2002a. Inhaled budesonide in experimental chlorine gas injury: Influence of time interval between injury and treatment. *Intensive Care Med.*, **28**, 352–357.

Wang, J., Zhang, L., and Walther, S.M., 2004. Administration of aerosolized terbutaline and budesonide reduces chlorine-gas induced lung injury in pigs. *J. Trauma*, **56**, 850–862.

Ware, L.B. and Matthay, M.A., 2000. The acute respiratory distress syndrome. *N. Engl. J. Med.*, **342**, 1334–1343.

Watson, A., Opresko, D., Young, R., and Hauschild, V., 2006. Development and application of acute exposure guideline levels (AEGLs) for chemical warfare nerve and sulfur mustard agents. *J. Toxicol. Environ. Health B: Crit. Rev.*, **9**(3), 173–263.

Weinberger, B., Laskin, J.D., Sunil, V.R., Sinko, P.J., Heck, D.E., and Laskin, D.L., 2011. Sulfur mustard-induced pulmonary injury: Therapeutic approaches to mitigating toxicity. *Pulm. Pharmacol. Ther.*, **24**, 92–99.

Weiselberg, R. and Nelson, L.S., 2011. A toxic swimming pool hazard. *Emerg. Med.*, **43**(4), 19–21.

Weitzberg, E., Hazel, M., and Lundberg, J.O., 2010. Nitrate–nitrite–nitric oxide pathway: Implications for anaesthesiology and intensive care. *Anaesthesiology*, **113**, 1460–1475.

Wenk, W.A., Van Sickle, D., Dronciuk, D., Belflower, A., Youngblood, C., Whisnant, M.D., Taylor, R., Rudnick, V., and Gibson, J.J., 2007. Rapid assessment of exposure to chlorine released from a train derailment and resulting health impact. *Public Health Rep.*, **122**, 784–792.

Wheeler, A.P. and Bernard, G.R., 2007. Acute lung injury and acute respiratory distress syndrome: A clinical review. *Lancet*, **369**, 1553–1565.

White, C.W. and Martin, J.G., 2010. Chlorine gas inhalation. Human clinical evidence of toxicity and experience in animal models. *Proc. Am. Thorac. Soc.*, **7**, 257–263.

Wigenstam, E., Rocksén, D., Ekstrnd-Hammerström, B., and Bucht, A., 2009. Treatment with dexamethasone or liposome-encapsulated vitamin E provides beneficial effects after chemical-induced injury. *Inhal. Toxicol.*, **21**, 958–964.

Wong, S.S., Sun, N.N., Lantz, R.C., and Witten, M.R., 2004. Substance P and neutral endopeptidase in development of acute respiratory distress syndrome following fire smoke inhalation. *Am. J. Physiol. Lung Cell Mol. Physiol.*, **287**, L859–L866.

Wyatt, J.P. and Allister, C.A., 1995. Occupational phosgene poisoning: A case report and review. *J. Accid. Emerg. Med.*, **12**, 212–213.

Yadav, A.K., Bracher, A., Doran, S.F., Leustik, M., Squardito, G.L., Postlethwaith, E.M., and Matalon, S., 2010. Mechanisms and modifications of chlorine-induced lung injury in animals. *Proc. Am. Thorac. Soc.*, **7**, 278–283.

Yadav, A.K., Doran, S.F., Samal, A.A. et al., 2011. Mitigation of chlorine gas lung injury in rats by postexposure administration of sodium nitrite. *Am. J. Physiol. Lung Cell Mol. Physiol.*, **300**, L362–L369.

Young, D., Lamb, S.E., Shah, S., MacKenzie, I., Tunnicliffe, W., Lall, R., Rowan, K., and Cuthbertson, B.H., for the OSCAR Study Group, 2013. High-frequency oscillation for acute respiratory distress syndrome. *N. Engl. J. Med.*, **368**, 806–813.

Zambon, M. and Vincent, J.-L., 2008. Mortality rates for patients with acute lung injury/ARDS have decreased over time. *Chest*, **133**, 1120–1127.

Zarogiannis, S.G., Jurkuveniate, A., Fernandez, S., Doran, S.F., Yadav, A.K., Squardito, G.L., Postlethwaith, E.M., Bowen, L., and Matalon, S., 2011. Ascorbate and deferoxamine administration after chlorine exposure decrease mortality and lung injury in mice. *Am. J. Respir. Cell Mol. Biol.*, **45**, 386–895.

Zhang, X.-D., Hong, J.-F., Qin, X.-J., Li, W.-L., Chen, H.-L., Liu, R., Liang, X., and Hai, C.-X., 2010. Pentoxifylline inhibits intercellular adhesion molecule-1 (ICAM-1) and lung injury in experimental phosgene-exposure rats. *Inhal. Toxicol.*, **22**, 889–895.

Zhu, Y.-G., Qu, J.-M., Zhang, J., Jiang, H.-N., and Xu, J.-F., 2011. Novel interventional approaches for ALI/ARDS: Cell based gene therapy. *Med. Inflamm.*, **2011**, Article ID 560194, doi: 10.1155/2011/560/560194.

Zuckerbraun, B.S., George, P., and Gatwin, M.T., 2011. Nitrite in pulmonary arterial hypertension: Therapeutic avenues in the setting of dysregulated arginine/nitric oxide synthase signaling. *Cardiovasc. Res.*, **89**, 542–552.

Zwischenberger, J.B., Alpard, S.K., Tao, W., Deyo, D.J., and Bidani, A., 2001. Percutaneous extracorporeal arteriovenous carbon dioxide removal improves survival in respiratory distress syndrome: A prospective randomized outcome study in adult sheep. *J. Thorac. Cardiovasc. Surg.*, **121**, 542–551.

18 Transcriptomic Responses to Crystalline Silica Exposure*

Pius Joseph, Christina Umbright, and Rajendran Sellamuthu

CONTENTS

18.1 INTRODUCTION

18.1.1 Exposure to Crystalline Silica

Silica, because of its abundance in Earth's crust, presents numerous opportunities for human exposure. Between the two forms (amorphous and crystalline) of silica† present, crystalline silica assumes a major importance with respect to potential harmful human health effects. Even though

---

* The findings and conclusions in this report are those of the author(s) and do not necessarily represent the views of the National Institute for Occupational Safety and Health.
† Based on the bonding geometry of silica (SiO_2), five different crystalline silica structures have been characterized—quartz, cristobalite, coesite, tridymite, and stishovite (Mandel and Mandel, 1996). In this chapter, the term *crystalline silica* refers to any one of the crystalline silica structures.

incidental exposure to crystalline silica, mainly through dust, takes place routinely, occupational exposures are of major concern from a health effects standpoint. Any occupation that involves movement of the earth's surface, especially mining, construction, silica milling, stonecutting, and sandblasting is considered as a major source for exposure to crystalline silica. While small quantities of crystalline silica may enter the human body orally, the major route for occupational exposure to crystalline silica is through inhalation. It has been estimated that approximately two million workers in the United States and millions more worldwide are occupationally exposed to crystalline silica annually (Sanderson, 2006). In several cases, occupational exposure to crystalline silica takes place at levels much higher than the National Institute for Occupational Safety and Health (NIOSH) recommended exposure limit (REL) of 0.05 mg/m^3 (Linch et al., 1998).

18.1.2   HEALTH EFFECTS OF CRYSTALLINE SILICA EXPOSURE

A large number of in vitro cell culture (Ding et al., 1999; Gwinn et al., 2009) and in vivo animal (Porter et al., 2001, 2002, 2004) studies conducted in the past have demonstrated the toxicity potential of crystalline silica. It has been reported that occupational exposure to crystalline silica is associated with the development of autoimmune diseases, rheumatoid arthritis, chronic renal diseases, and lupus (NIOSH, 2002). Based on overwhelming evidence, the International Agency for Research on Cancer (IARC) has classified crystalline silica as a type I human carcinogen (IARC, 1997; Steenland and Sanderson, 2001). However, the pulmonary effect associated with crystalline silica exposure, which is of greatest concern, is silicosis, an irreversible but preventable interstitial lung disease characterized by alveolar proteinosis and diffuse fibrosis resulting in progressively restrictive lung function and death (Castranova and Vallyathan, 2000). Epidemiologic studies have demonstrated that workers have a significant risk of developing chronic silicosis when they are exposed to respirable crystalline silica over a working lifetime at the current Occupational Safety and Health Administration permissible exposure limit (PEL), the Mine Safety and Health Administration (MSHA) PEL, or the NIOSH REL (Hnizdo and Sluis-Cremer, 1993; Steenland and Brown, 1995; Kreiss and Zhen, 1996). The number of cases of silicosis and silicosis-related diseases in the United States is currently unknown. However, approximately 200 deaths were attributed to silicosis in 2005 in the United States (NIOSH, 2006). Currently, silicosis is diagnosed by chest x-ray and pulmonary function tests. Both of these tests, unfortunately, detect silicosis based on structural and/or functional impairment of the lungs most likely representing an advanced stage of the disease. As such, the current diagnostic tools may not be helpful to prevent deaths associated with this incurable disease. On the other hand, the preventable nature of silicosis strongly supports the argument to develop tests sensitive enough to predict onset of the disease at an early preventable stage. In fact, NIOSH has recommended developing a highly sensitive and noninvasive or minimally invasive test to detect silicosis at an early preventable stage (NIOSH, 2002).

18.1.3   TRANSCRIPTOME: A SENSITIVE AND MECHANISTICALLY RELEVANT TARGET FOR TOXICITY

Transcriptome refers to the entire set of transcripts or mRNAs present in a cell or an organism. Transcriptomics, also referred to as expression profiling, is the determination of the expression levels of all of the transcripts or mRNAs present in a cell or an organism at a given time by employing techniques such as DNA microarray, next-generation RNA sequencing, subtractive hybridization, differential display, and serial analysis of gene expression. Transcriptome, unlike genome, is highly dynamic in nature and often responds sensitively to exposure of a cell or an organism to a toxic agent. Therefore, expression profiling of the entire transcriptome in a cell or an organism following its exposure to a toxic agent under controlled experimental conditions may be considered as a sensitive indicator of toxicity resulting from exposure to the toxic agent. The differentially expressed genes and/or their products, following appropriate validation, may be employed as biomarkers for exposure and toxicity of the agent being investigated. Similarly, diligent bioinformatics analysis of the significantly differentially expressed transcripts detected in a cell or an organism in response to its exposure to a toxic agent has been shown to provide valuable insight into the mechanism(s) underlying the toxicity of the agent. Therefore, global gene expression profiling, as supported by the results of several studies (Waring et al., 2001; Hamadeh et al., 2002; Amin et al., 2004; Heinloth et al., 2004), is often considered as a sensitive and mechanistically relevant approach to detect and understand toxicity associated with exposure to toxic agents.

18.2 TRANSCRIPTOMIC RESPONSES TO CRYSTALLINE SILICA EXPOSURE

Crystalline silica particles, following their entry into the lungs, are phagocytosed by the alveolar macrophages (AMs) for elimination. In the absence of efficient elimination from the lungs, such as under conditions of excessive crystalline silica exposure, interaction with the inhaled silica particles activates the AMs resulting in AM death and release of the silica particles and various signaling molecules within the lungs. The crystalline silica particles as well as the signaling molecules released into the lungs may interact with the alveolar epithelium to initiate a cascade of pulmonary and extra-pulmonary events. The net result is the recruitment of inflammatory cells into the lungs, release of toxic ROS and reactive nitrogen species (RNS), the induction of inflammation, DNA damage, apoptosis, and fibrosis leading to the development of diseases, for example, silicosis and cancer. Normal functioning of cells/tissues/organs is regulated by the expression of a large number of genes that are organized as specific biological functions, pathways, and networks. The crystalline silica particles as well as the various signaling molecules released from activated AMs may cause alterations in the expression of one or several genes to result in the functional disruption of the corresponding biological functions, pathways and networks that are vital for normal cell/tissue/organ function. In addition, the transcriptome may be affected as a secondary effect of the pulmonary toxicity resulting from the inhaled crystalline silica particles. Therefore, global gene expression profiling of biological samples that are exposed to crystalline silica and functional analysis of the genes differentially expressed in response to the exposure may provide valuable information with respect to the toxicity potential of silica as well as the underlying molecular mechanisms of crystalline silica–induced pulmonary and extra-pulmonary toxicity.

18.2.1 TRANSCRIPTOMIC RESPONSE TO CRYSTALLINE SILICA EXPOSURE CORRESPONDS TO SILICA-INDUCED TOXICITY

A relationship between crystalline silica–induced toxicity and global gene expression changes taking place in biological samples has been investigated using cell culture and animal models. It has been fairly well established that crystalline silica, compared to amorphous silica, is more biologically active and pathogenic (Warheit et al., 1995; Johnston et al., 2000; Fubini et al., 2001). Whether global gene expression changes would account for the reported differences in the biological activity and pathogenicity reported between amorphous and crystalline silica was investigated in a human bronchial epithelial cell line (BEAS2B) and primary human bronchial epithelial cells (NHBE) (Perkins et al., 2012). The effect of crystalline silica on the transcriptome, compared with amorphous silica, was more profound in both cell types. More genes were significantly differentially expressed in both cell types exposed to crystalline silica compared with amorphous silica. Furthermore, the fold changes in gene expression levels were significantly higher in the cells exposed to crystalline silica compared with those exposed to amorphous silica. Therefore, the effect of silica particles (amorphous and crystalline) on the transcriptome appeared to be related to their biological activities and pathogenic potentials. In a study reported by Sellamuthu et al. (2011a), human lung epithelial cells, A549, were treated with Min-U-Sil 5 crystalline silica at final concentrations of 15, 30, 60, 120, and 240 µg/cm² cultured area for 6 h or 60 µg/cm² cultured area for time intervals of 2, 6, and 24 h. At the end of the silica exposure period, cytotoxicity was determined by assaying lactate dehydrogenase (LDH) activity in the cell culture medium. Simultaneously, total RNA was isolated from the control and silica-treated cell culture samples to determine global gene expression profiles using Human HT 12-v3-Beadchip Arrays (Illumina, Inc., San Diego, CA). The crystalline silica–induced

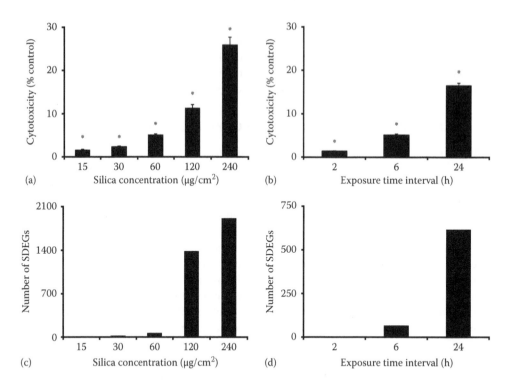

FIGURE 18.1 Cytotoxicity and differential gene expression profile in crystalline silica–exposed A549 cells. Exponentially growing human lung epithelial cells (A549) were treated with Min-U-Sil 5 crystalline silica particles at concentrations of 15, 30, 60, 120, and 240 µg/cm² for 6 h (a, c) or 60 µg/cm² for time intervals of 2, 6, or 24 h (b, d). Cytotoxicity and global gene expression profile were determined by LDH assay and microarray analysis, respectively, as described in our original publication (Sellamuthu et al., 2011a). Cytotoxicity and the number of SDEGs (FDR p value <0.05) were calculated with respect to the corresponding control. *Statistically significant, p < 0.05 (n = 5). (Reproduced from Sellamuthu, R. et al., *Inhal. Toxicol.*, 23(14), 927, 2011a. With permission.)

cytotoxicity and the number of significantly differentially expressed genes (SDEGs)* in the cells correlated very well (Figure 18.1a–d) and exhibited correlation coefficients (r² values) of 0.89 and 0.98, respectively, for the silica concentration–response and time-course study. Bioinformatics analysis of the SDEGs in the A549 cells identified the biological functions that were significantly enriched in response to crystalline silica–induced cytotoxicity. As presented in Figure 18.2a and b, the biological functions significantly enriched in response to silica exposure in the A549 cells were highly relevant to the already established mechanisms of crystalline silica–induced pulmonary toxicity. It is worth mentioning that the enrichment of the biological functions, represented by the number of SDEGs belonging to each of the biological function categories, in the silica-exposed A549 cells exhibited a similar quantitative response to the concentration and duration of crystalline silica exposure as was seen in the case of the crystalline silica–induced cytotoxicity. A similar positive relationship has been reported between cytotoxicity and the number of SDEGs detected in human bronchial epithelial cells, BEAS2B cells, treated with increasing concentrations of crystalline silica (Perkins et al., 2012).

Further evidence to support the existence of a relationship between crystalline silica–induced pulmonary toxicity and global gene expression changes is obtained from the results of animal experiments. In a study conducted in our laboratory, rats were exposed to Min-U-Sil 5 crystalline

---

* Genes whose expressions are significantly different between control and toxicant-exposed samples are considered as SDEGs. The criteria commonly employed to select significantly differentially expressed genes include a fold change in expression, p value [either simple or false discovery rate (FDR) p value], or a combination of the two.

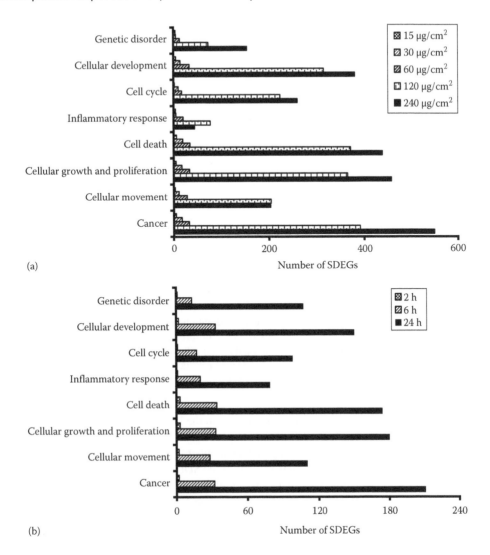

(a)

(b)

FIGURE 18.2 Effect of crystalline silica concentration and duration of exposure on the major biological functions perturbed by crystalline silica exposure in A549 cells. Exponentially growing human lung epithelial cells (A549) were treated with Min-U-Sil 5 crystalline silica particles at concentrations of 15, 30, 60, 120, and 240 g/cm^2 for 6 h (a) or 60 μg/cm^2 for time intervals of 2, 6, or 24 h (b). Global gene expression profiles in the cells were determined by microarray analysis, and the genes significantly differentially expressed in response to crystalline silica exposure were subjected to bioinformatics analysis as described in detail in our original publication (Sellamuthu et al., 2011a). The number of SDEGs involved in the top eight ranking IPA biological processes that were significantly enriched in response to crystalline silica exposure is presented. Data represent the mean of five independent experiments. (Reproduced from Sellamuthu, R. et al., *Inhal. Toxicol.*, 23(14), 927, 2011a. With permission.)

silica by inhalation (15 mg/m^3, 6 h/day, 5 days). Pulmonary damage and global gene expression profiles were determined in the lungs at post–silica exposure time intervals of 0, 1, 2, 4, 8, 16, and 32 weeks (Sellamuthu et al., 2011a,b, 2012, 2013a). Determination of pulmonary damage was performed on the basis of LDH activity and concentrations of albumin, total protein, and the pro-inflammatory cytokine, macrophage chemo-attractant protein 1 (MCP1), in the bronchoalveolar lavage fluid (BALF). Global gene expression profiles were determined in the lungs by employing RatRef-12V1.0 Expression Beadchip Arrays (Illumina, Inc., San Diego, CA). A steady progression of pulmonary toxicity was seen during the post–silica exposure time intervals in the crystalline

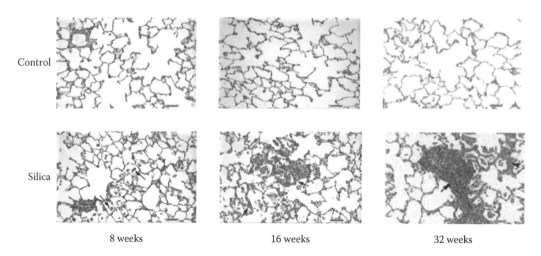

Control

Silica

8 weeks   16 weeks   32 weeks

FIGURE 18.3 (**See color insert**.) Progression of lung damage in crystalline silica–exposed rats. Rats were exposed to Min-U-Sil 5 crystalline silica (15 mg/m^3, 6 h/day, 5 days) or air (control) as described in detail in our original publications (Sellamuthu et al., 2011b, 2012). Lung sections prepared at the post–silica exposure time intervals of 0, 1, 2, 4, 8, 16, and 32 weeks were stained with hematoxylin and eosin. Only photomicrographs of lung sections of 8, 16, and 32 weeks post–silica exposure time intervals are presented, and those of earlier time intervals can be found in our original publication (Sellamuthu et al., 2011b). Arrows indicate AMs in alveolar space (8 weeks) and type II pneumocyte hyperplasia (16 and 32 weeks). Magnification = 20×. (Modified from Sellamuthu, R. et al., *Toxicol. Sci.*, 122(2), 253, 2011b; Sellamuthu, R. et al., *Inhal. Toxicol.*, 24(9), 570, 2012. With permission.)

silica–exposed rats on the basis of histological changes in the lungs compared to the corresponding time-matched controls (Figure 18.3). A very strong positive correlation was seen between the progression of silica-induced pulmonary toxicity and the number of SDEGs detected in the lungs of the rats (Figure 18.4 and Table 18.1). Functional analysis of the SDEGs in the lungs of the silica-exposed rats identified the biological functions that were significantly enriched in response to crystalline silica–induced pulmonary toxicity. The biological functions that were significantly enriched in the lungs of the silica-exposed rats were mechanistically relevant to the already known pulmonary effects of crystalline silica exposure (Figure 18.5). In addition, the number of SDEGs present in each of the significantly enriched biological functions steadily increased similar to the progression of pulmonary toxicity seen in the rats during the post–silica exposure time intervals. The strong correlation between crystalline silica–induced toxicity and the number of SDEGs as well as the enrichment of biological functions seen in the cell culture as well as animal tissue samples suggested the potential application of transcriptomics as a relevant approach to detect and study crystalline silica–induced toxicity.

18.2.2 Crystalline Silica and Oxidative Stress

The finding that silicotic lungs are in a state of oxidative stress (Vallyathan and Shi, 1997) has led to the belief that oxidant-mediated lung damage may play a role in the development of toxicity and diseases associated with crystalline silica exposure. The observations that freshly fractured crystalline silica is more toxic to cells (Ding et al., 1999; Gwinn et al., 2009) and animals (Vallyathan et al., 1995) compared to aged crystalline silica further support for the involvement of oxidative stress, through the generation of reactive oxygen species (ROS), in crystalline silica toxicity. Generation of toxic ROS taking place in biological samples in response to their exposure to crystalline silica may be the direct effect of silica particles (Vallyathan et al., 1995), and/or it may be mediated by silica particles indirectly through cellular processes (Vallyathan et al.,1992). It is known that crystalline

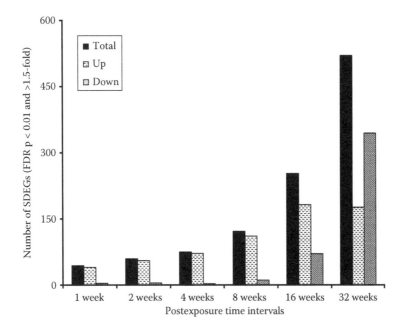

FIGURE 18.4 Differential gene expression profile in the lungs of crystalline silica–exposed rats. Rats were exposed to Min-U-Sil 5 crystalline silica (15 mg/m³, 6 h/day, 5 days), and the number of SDEGs (>1.5-fold change and <0.01 FDR p value) was determined by microarray analysis as described in detail in our original publications (Sellamuthu et al., 2012, 2013a). The number of genes significantly differentially expressed (total), overexpressed (up), and underexpressed (down) in the silica-exposed rat lungs compared with the corresponding time-matched controls is presented for the postexposure time intervals presented on the X-axis. Data represent the mean of eight crystalline silica–exposed rats compared with four corresponding time-matched control rats per time point. (Modified from Sellamuthu, R. et al., *J. Appl. Toxicol.*, 33(4), 301, 2013; Sellamuthu, R. et al., *Inhal. Toxicol.*, 24(9), 570, 2012. With permission.)

TABLE 18.1

Correlation Coefficients (r^2 Values) for the Relationship between Crystalline Silica–Induced Pulmonary Toxicity (BALF LDH, PMN, and MCP-1) and the Number of SDEGs in the Lungs of the Rats

	BALF LDH	BALF PMN	BALF MCP-1
Lung SDEGs	0.776	0.879	0.927

Source: Modified from Sellamuthu, R. et al., *J. Appl. Toxicol.*, 33(4), 301, 2013a. With permission.

Note: The toxicity measurements and the number of SDEGs in the lungs of the silica-exposed rats at the postexposure time intervals of 0, 1, 2, 4, 8, 16, and 32 weeks following a 1-week exposure to crystalline silica (15 mg/m³, 6 h/day) were done as described in our original publications (Sellamuthu et al., 2012, 2013a).

---

silica–induced oxidative stress rises even after the termination of silica exposure and lungs have cleared most of the deposited silica (Fubini and Hubbard, 2003; Rimal et al., 2005).

Transcriptome analysis has provided support for the generation of ROS in biological samples in response to crystalline silica exposure. A significant and crystalline silica concentration-dependent overexpression of several oxidative stress–responsive genes belonging to the nuclear factor kappa B and AP-1 family has been observed in the A549 cells (Table 18.2). Superoxide anion, an ROS

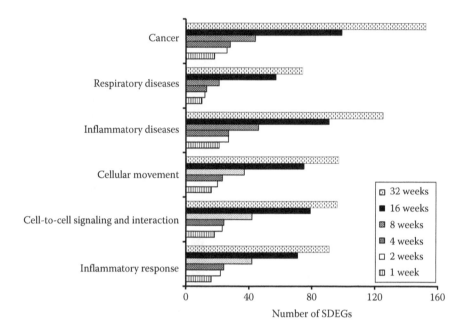

FIGURE 18.5 Enrichment of top ranking IPA biological functions in crystalline silica–exposed rat lungs. Rats were exposed to Min-U-Sil 5 crystalline silica (15 mg/m³, 6 h/day, 5 days), and the number of SDEGs in their lungs was determined by microarray analysis as described in detail in our original publications (Sellamuthu et al., 2012, 2013a). The microarray data were subjected to bioinformatics analysis, and the number of SDEGs in the crystalline silica–exposed rat lungs belonging to the six top ranking IPA biological functions at each of the postexposure time interval is presented. Data represent the group mean of eight crystalline silica–exposed and four time-matched control rats per time point. (Modified from Sellamuthu, R. et al., *J. Appl. Toxicol.*, 33(4), 301, 2013a; Sellamuthu, R. et al., *Inhal. Toxicol.*, 24(9), 570, 2012. With permission.)

generated in response to crystalline silica exposure, undergoes dismutation catalyzed by superoxide dismutase to generate hydrogen peroxide (H_2O_2) (Liochev and Fridovich, 2007). H_2O_2 is also generated during the metabolism of spermine catalyzed by spermine oxidase (Murray-Stewart et al., 2008). The toxic and reactive H_2O_2 generated is detoxified, predominantly by catalase, in order to prevent its interaction with intracellular target(s) potentially leading to toxicity. The net result of the significant and crystalline silica concentration-dependent overexpression of SOD* and SMOX with simultaneous decrease in the expression of catalase, as noticed in the crystalline silica–exposed A549 cells (Table 18.2), is the cellular accumulation of toxic and reactive H_2O_2 contributing to crystalline silica–induced oxidative stress and toxicity. A significant overexpression of the SOD gene has also been reported in a human bronchial epithelial cell line (BEAS2B) in response to their exposure to crystalline silica (Perkins et al., 2012).

In laboratory animals, inhaled crystalline silica particles interact with AMs and alveolar epithelium to result in the generation of ROS (Porter et al., 2002). Several genes involved in the generation of ROS as well as in the cellular response to oxidative stress, viz, SOD2, HMOX1, MT1A, NCF1, LCN2, ARG1, LPO, and NOXO1, were significantly overexpressed in the lungs of the crystalline silica–exposed rats suggesting the involvement of oxidative stress in the induction of pulmonary toxicity (Table 18.3). The NOXO1 gene codes a protein that is an activator of the superoxide-generating gene NOX1 (Banfi et al., 2003), and its significant overexpression in the crystalline silica–exposed rat lungs may, therefore, imply the generation of superoxide anion capable of resulting in oxidative stress. As seen in the A549 cells exposed to crystalline silica, the significant overexpression of SOD2 with no corresponding increase in the expression of H_2O_2

* Expanded names of the genes abbreviated in the text can be found in Tables 18.2, 18.3, or 18.6.

TABLE 18.2
Differentially Expressed Genes in A549 Cells Exposed to Crystalline Silica

Gene		Crystalline Silica (µg/cm²)					Time Interval (h)		
		15	30	60	120	240	2	6	24
Reactive oxygen species									
SOD2	Superoxide dismutase 2	1.08	1.11	1.21[a]	1.98[a]	2.50[a]	−1.05	1.21[a]	4.52[a]
CAT	Catalase	1.02	−1.05	−1.07	−1.16[a]	−1.14[a]	−1.08	−1.07	−1.24
SMOX	Spermine oxidase	1.33	1.39[a]	1.55[a]	2.99[a]	3.52[a]	1.33	1.55[a]	2.55[a]
CYR61	Cysteine-rich angiogenic inducer 61	1.12	1.59[a]	1.90[a]	2.47[a]	1.57[a]	2.11[a]	1.90[a]	1.55[a]
Antioxidant/oxidative stress									
NFE2L2	Nuclear factor erythroid-derived 2-like 2	1.03	1.05	1.15	1.83[a]	1.61[a]	1.28[a]	1.15	−1.05
NFKB1	Nuclear factor of kappa light polypeptide gene enhancer in B-cells 1	−1.04	1.18	1.31[a]	2.00[a]	2.03[a]	1.11	1.31[a]	1.50[a]
NFKB2	Nuclear factor of kappa light polypeptide gene enhancer in B-cells 2	1.04	1.02	1.05	1.12[a]	1.11[a]	1.05	1.05	1.09[a]
NFKBIZ	Nuclear factor of kappa light polypeptide gene enhancer in B-cells inhibitor, zeta	1.15	1.21	1.23	2.51[a]	3.31[a]	1.18	1.23	2.86[a]
DUSP1	Dual specificity phosphatase 1	1.24	1.22[a]	1.49[a]	2.47[a]	3.00[a]	1.56	1.49[a]	1.55[a]
DUSP5	Dual specificity phosphatase 5	1.92[a]	1.79[a]	2.48[a]	5.74[a]	8.02[a]	2.84[a]	2.48[a]	2.98[a]
FOS	v-fos FBJ murine osteosarcoma viral oncogene	1.24	1.36	1.42	3.93[a]	20.34[a]	2.06[a]	1.42	1.24[a]
JUNB	homolog jun B proto-oncogene	1.02	−1.00	1.07	1.46[a]	1.72[a]	1.20	1.07	−1.03
c-JUN	jun oncogene	1.27	1.46	1.87[a]	5.46[a]	10.64[a]	2.17[a]	1.87[a]	2.40[a]
STC1	Stanniocalcin 1	1.63[a]	1.75[a]	2.37[a]	4.67[a]	5.40[a]	1.90[a]	2.37[a]	2.75[a]
STC2	Stanniocalcin 2	1.22	1.31[a]	1.55[a]	2.06[a]	2.03[a]	1.02	1.55[a]	3.35[a]
Inflammation									
IRF1	Interferon regulatory factor 1	1.10	1.06	1.14	1.54[a]	1.33[a]	−1.04	1.54[a]	1.15
RELA	v-rel reticuloendotheliosis viral oncogene homolog A	1.02	1.06	1.11[a]	1.22[a]	1.12[a]	1.05	1.11[a]	1.13[a]
RELB	v-rel reticuloendotheliosis viral oncogene homolog B	1.20[a]	1.24[a]	1.51[a]	1.92[a]	1.73[a]	−1.01	1.51[a]	1.92[a]

(continued)

TABLE 18.2 (continued)
Differentially Expressed Genes in A549 Cells Exposed to Crystalline Silica

Gene		Crystalline Silica (µg/cm²)					Time Interval (h)		
		15	30	60	120	240	2	6	24
Inflammation									
IL1A	Interleukin 1 alpha	1.05	1.02	1.05	1.86[a]	2.62[a]	ND	1.05	1.22[a]
IL1B	Interleukin 1 beta	ND	−1.00	1.02	1.28[a]	1.45[a]	ND	1.28[a]	1.6[a]
IL6	Interleukin 6	1.12	1.24	1.58[a]	6.17[a]	8.51[a]	1.43[a]	1.58[a]	6.14[a]
IL8	Interleukin 8	3.16[a]	3.84[a]	7.13[a]	35.72[a]	41.60	9.77[a]	7.13[a]	18.21[a]
IL11	Interleukin 11	1.08	1.23	1.47	3.75[a]	4.92[a]	1.89[a]	1.47	1.49[a]
IRAK2	Interleukin-1 receptor-associated kinase 2	1.63[a]	1.77[a]	1.89[a]	5.56[a]	7.15[a]	1.66[a]	1.89[a]	3.54[a]
PTGS2	Prostaglandin-endoperoxide synthase 2	1.51	1.55[a]	2.21[a]	11.13[a]	15.09[a]	2.38[a]	2.21[a]	3.91[a]
CCL2/ MCP1	Chemokine (C–C motif) ligand 2	1.68	1.77[a]	1.67[a]	3.30[a]	3.57[a]	1.38	1.67[a]	3.23[a]
CCL20/ MIP3α	Chemokine (C–C motif) ligand 20	1.06	1.06	1.23[a]	3.20[a]	4.16[a]	1.13	1.23[a]	2.39[a]
CXCL1/ GRO1	Chemokine (C–X–C motif) ligand 1	ND	−1.00	1.04	1.41[a]	1.50[a]	ND	1.04	1.60[a]
CXCL2/ MIP-2	Chemokine (C–X–C motif) ligand 2	1.24	1.12	1.29	3.81[a]	3.58[a]	1.38	1.29	3.23[a]
CXCL5	Chemokine (C–X–C motif) ligand 5	1.24	1.17	1.57[a]	2.70[a]	4.16[a]	1.13	1.57[a]	2.39[a]
CXCL8/ IL8	Interleukin 8	3.16[a]	3.84[a]	7.13[a]	35.72[a]	41.60	9.77[a]	7.13[a]	18.21[a]
PLAU	Plasminogen activator urokinase	1.23	1.44	1.75[a]	2.44[a]	1.83[a]	1.48[a]	1.75[a]	1.53[a]
ITGA2	Integrin alpha 2 (CD49B, alpha 2 subunit of VLA-2 receptor)	−1.05	1.11	1.26	1.74[a]	1.57[a]	1.03	1.26	1.81[a]
MMP10	Matrix metallopeptidase 10	ND	1.03	1.04	1.87[a]	3.82[a]	ND	1.04	1.71[a]
CEBPB	CCAAT/enhancer-binding protein	1.13	1.23[a]	1.42[a]	2.78[a]	3.62[a]	1.86[a]	1.42[a]	2.08[a]
Apoptosis									
NFE2L2	Nuclear factor erythroid-derived 2-like 2	1.03	1.05	1.15	1.83[a]	1.61[a]	1.28[a]	1.15	−1.05
NFKB1	Nuclear factor of kappa light polypeptide gene enhancer in B-cells 1	−1.04	1.18	1.31[a]	2.00[a]	2.03[a]	1.11	1.31[a]	1.50[a]
NFKB2	Nuclear factor of kappa light polypeptide gene enhancer in B-cells 2 (p49/p100)	1.04	1.02	1.05	1.12[a]	1.11[a]	1.05	1.05	1.09[a]

TABLE 18.2 (continued)

Differentially Expressed Genes in A549 Cells Exposed to Crystalline Silica

Gene		Crystalline Silica (µg/cm²)					Time Interval (h)		
		15	30	60	120	240	2	6	24
NFKBIA	Nuclear factor of kappa light polypeptide gene enhancer in B-cells inhibitor alpha	1.10	1.32[a]	1.45[a]	2.93[a]	4.48[a]	1.35	1.45[a]	1.94[a]
NFKBIZ	Nuclear factor of kappa light polypeptide gene enhancer in B-cells inhibitor, zeta	1.15	1.21	1.23	2.51[a]	3.31[a]	1.18	1.23	2.86[a]
GDF15	Growth differentiation factor 15	1.65[a]	1.50[a]	1.75[a]	5.33[a]	9.68[a]	1.74	1.75[a]	2.38[a]
GADD45α	Growth arrest and DNA-damage-inducible 45 alpha	1.19	1.46[a]	1.62[a]	3.08[a]	3.81[a]	1.59[a]	1.62[a]	1.79[a]
GADD34	Growth arrest and DNA-damage-inducible 34	1.86[a]	2.10[a]	2.82[a]	5.88[a]	10.93[a]	2.72[a]	2.82[a]	3.56[a]
EGR1	Early growth response 1	1.91[a]	2.05[a]	2.51[a]	13.65[a]	36.34[a]	6.46[a]	2.51[a]	2.49[a]
BIRC3	Baculoviral IAP repeat-containing 3	1.27	1.58[a]	2.01[a]	5.78[a]	8.51[a]	1.59[a]	2.01[a]	2.43[a]
c-FOS	v-fos FBJ murine osteosarcoma viral oncogene homolog	1.24	1.36	1.42	3.93[a]	20.34[a]	2.06[a]	1.42	1.24[a]
FOSB	FBJ murine osteosarcoma viral oncogene homolog B	3.58[a]	3.87[a]	4.30[a]	11.64[a]	34.73[a]	9.83[a]	4.30[a]	3.74[a]
FOSL1	FOS-like antigen 1	1.10	1.31[a]	1.63[a]	2.93[a]	2.33[a]	1.39[a]	1.63[a]	1.44[a]
CYR61	Cysteine-rich angiogenic inducer 61	1.12	1.59[a]	1.90[a]	2.47[a]	1.57[a]	2.11[a]	1.90[a]	1.55[a]
JUND	jun D proto-oncogene	−1.03	1.10	1.20	2.26[a]	3.26[a]	1.17	1.20	−1.01
OKL-38	Pregnancy-induced growth inhibitor	−1.18	1.10	1.16	1.79[a]	2.49[a]	1.32	1.16	−1.16
Cancer									
EFNA1	Ephrin-A1 v-myc myelocytomatosis	1.31	1.32	1.45	2.72[a]	2.93[a]	1.10	1.45	1.88[a]
MYC	viral oncogene homolog (avian)	1.14	1.22	1.32	1.69[a]	2.16[a]	1.23	1.32	1.16
CEBPB	CCAAT/enhancer-binding protein	1.13	1.23[a]	1.42[a]	2.78[a]	3.62[a]	1.86[a]	1.42[a]	2.08[a]
TRIB1	Tribbles homolog 1 (Drosophila)	1.13	1.09	1.22	2.57[a]	3.90[a]	1.34	1.22	1.55[a]
PTGS2	Prostaglandin-endoperoxide synthase 2	1.51	1.55[a]	2.21[a]	11.13[a]	15.09[a]	2.38[a]	2.21[a]	3.91[a]
ZFP36	Zinc finger protein 36	1.36	1.44	1.58[a]	6.43[a]	10.65[a]	2.16[a]	1.58[a]	1.10

(continued)

TABLE 18.2 (continued)
Differentially Expressed Genes in A549 Cells Exposed to Crystalline Silica

Gene		Crystalline Silica (µg/cm²)					Time Interval (h)		
		15	30	60	120	240	2	6	24
Cancer									
KLF2	Kruppel-like factor 2	1.33	1.67	1.87[a]	2.38[a]	1.90[a]	1.88[a]	1.87[a]	1.18[a]
KLF6	Kruppel-like factor 6	1.38[a]	1.61[a]	1.97[a]	5.14[a]	6.84[a]	2.03[a]	1.97[a]	2.38[a]
KLF11	Kruppel-like factor 11	1.08	1.13	1.21	1.60[a]	1.65[a]	1.14	1.21	1.58[a]
KLF13	Kruppel-like factor 13	1.14	1.22	1.38[a]	1.71[a]	1.32[a]	1.01	1.38[a]	1.15
ETS1	v-ets erythroblastosis virus E26 oncogene homolog 1	1.36	1.44[a]	1.77[a]	3.50[a]	3.58[a]	1.35[a]	1.77[a]	2.06[a]
DDIT3	DNA-damage-inducible transcript 3	1.36[a]	1.48[a]	1.63[a]	3.11[a]	4.15[a]	1.25	1.63[a]	2.17[a]
FGF2	Fibroblast growth factor 2	1.00	1.00	1.10	1.25[a]	1.27[a]	1.07	1.10	1.07
Cellular growth and proliferation									
EGR1	Early growth response 1	1.91[a]	2.05[a]	2.51[a]	13.65[a]	36.34[a]	6.46[a]	2.51[a]	2.49[a]
RELB	v-rel reticuloendotheliosis viral oncogene homolog B	1.20[a]	1.24[a]	1.51[a]	1.92[a]	1.73[a]	−1.01	1.51[a]	1.92[a]
IL8	Interleukin 8	3.16[a]	3.84[a]	7.13[a]	35.72[a]	41.60[a]	9.77[a]	7.13[a]	18.21[a]
DUSP5	Dual specificity phosphatase 5	1.92[a]	1.79[a]	2.48[a]	5.74[a]	8.02[a]	2.84[a]	2.48[a]	2.98[a]
FST	Follistatin	1.41[a]	1.62[a]	2.20[a]	4.57[a]	4.60[a]	1.25	2.20[a]	2.48[a]
ETS1	v-ets erythroblastosis virus E26 oncogene homolog 1	1.36	1.44[a]	1.77[a]	3.50[a]	3.58[a]	1.35[a]	1.77[a]	2.06[a]
Cell cycle									
EGR1	Early growth response 1	1.91[a]	2.05[a]	2.51[a]	13.65[a]	36.34[a]	6.46[a]	2.51[a]	2.49[a]
ETS1	v-ets erythroblastosis virus E26 oncogene homolog 1	1.36	1.44[a]	1.77[a]	3.50[a]	3.58[a]	1.35[a]	1.77[a]	2.06[a]
STC1	Stanniocalcin 1	1.63[a]	1.75[a]	2.37[a]	4.67[a]	5.40[a]	1.90[a]	2.37[a]	2.75[a]
c-JUN	jun oncogene	1.27	1.46	1.87[a]	5.46[a]	10.64[a]	2.17[a]	1.87[a]	2.40[a]
NFKBIA	Nuclear factor of kappa light polypeptide gene enhancer in B-cells inhibitor alpha	1.10	1.32[a]	1.45[a]	2.93[a]	4.48[a]	1.35	1.45[a]	1.94[a]
PTGS2	Prostaglandin-endoperoxide synthase 2	1.51	1.55[a]	2.21[a]	11.13[a]	15.09[a]	2.38[a]	2.21[a]	3.91[a]
Cellular development									
CSF2	Colony stimulating factor 2 (granulocyte-macrophage)	ND	1.07	1.13	2.48[a]	3.81[a]	1.08	1.13	2.04[a]
CEBPB	CCAAT/enhancer-binding protein (C/EBP)	1.13	1.23[a]	1.42[a]	2.78[a]	3.62[a]	1.86[a]	1.42[a]	2.08[a]

TABLE 18.2 (continued)
Differentially Expressed Genes in A549 Cells Exposed to Crystalline Silica

Gene		Crystalline Silica (µg/cm²)					Time Interval (h)		
		15	30	60	120	240	2	6	24
EGR1	Early growth response 1	1.91[a]	2.05[a]	2.51[a]	13.65[a]	36.34[a]	6.46[a]	2.51[a]	2.49[a]
FOSL1	FOS-like antigen 1	1.10	1.31[a]	1.63[a]	2.93[a]	2.33[a]	1.39[a]	1.63[a]	1.44[a]
c-JUN	jun oncogene	1.27	1.46	1.87[a]	5.46[a]	10.64[a]	2.17[a]	1.87[a]	2.40[a]
FOXO1	Forkhead box O1	1.17	1.21[a]	1.41[a]	2.06[a]	1.76[a]	1.16	1.41[a]	1.33[a]

Source: Reproduced from Sellamuthu, R. et al., *Inhal. Toxicol.*, 23(14), 927, 2011a. With permission.

Notes: Exponentially growing human lung epithelial cells (A549) were treated with Min-U-Sil 5 crystalline silica particles at concentrations of 15, 30, 60, 120, and 240 µg/cm² for 6 h or 60 µg/cm² for time intervals of 2, 6, or 24 h. Global gene expression profiles in the cells were determined by microarray analysis, and the genes significantly differentially expressed in response to silica exposure were subjected to bioinformatics analysis as described in detail in our original publication (Sellamuthu et al., 2011a). A partial list of the SDEGs involved in IPA biological processes and canonical pathways that are relevant to crystalline silica toxicity and their fold changes in expressions compared to the control cells is presented.

Data represent the fold change in the expression of the individual genes and are mean of five independent microarray experiments. Some of the genes are listed under multiple categories since they are involved in multiple functions.

ND—Gene expression not detected.

[a] Statistically significant (FDR p value < 0.05) change in the expression compared to the corresponding control.

detoxifying genes, catalase and glutathione peroxidase/reductase (Gaetani et al., 1994), seen in the rat lungs should favor the excessive generation and tissue accumulation of reactive and toxic H_2O_2, contributing to oxidative stress and pulmonary toxicity in the crystalline silica–exposed rat lungs (Sellamuthu et al., 2013a). This argument is further supported by the significant overexpression LPO, an H_2O_2-responsive gene (Davies et al., 2008), seen in the crystalline silica–exposed rat lungs (Langley et al., 2011; Sellamuthu et al., 2013a).

18.2.3 CRYSTALLINE SILICA AND DNA DAMAGE

Direct interaction between crystalline silica and DNA resulting in structural changes in the DNA phosphate backbone has been demonstrated (Mao et al., 1994). Similarly, the ability of crystalline silica to result in DNA strand break has been reported (Shi et al., 1994). Several lines of evidence suggest a major role played by ROS in the DNA damage caused by crystalline silica. Chemical etching of freshly fractured silica particles with hydrofluoric acid resulted in diminished DNA damage mainly due to the removal of metal ion impurities and reactive centers created by fracturing (Daniel et al., 1993). The observations that crystalline silica–induced DNA damage is significantly lower in the absence of molecular oxygen and that catalase and scavengers of ROS are capable of blocking crystalline silica–induced DNA damage further support the involvement of ROS in crystalline silica–induced DNA damage. The DNA damage resulting from exposure to crystalline silica 12 may contribute to apoptosis and carcinogenesis, both of which are attributed to play significant roles in the pulmonary toxicity induced by crystalline silica. Results of transcriptomics analysis further support the DNA damage resulting from exposure of cells to crystalline silica. In a study conducted by Gwinn et al. (2009), human bronchial epithelial cells (BEAS2B) and lung cancer cells (H460 and H1299) were exposed to crystalline silica, and expression levels of several genes involved in DNA damage signaling pathways were identified. The genes whose expression levels were significantly upregulated in response to crystalline silica exposure in the cells included ataxia telangiectasia mutated (ATM), sestrin 1 (SESN1), mouse double minute 2 (MDM2), cell division cycle 25 homolog A (CDC25), B-cell lymphoma 6

TABLE 18.3

Fold Change in the Expression of a Selected List of Significantly Differentially Expressed Genes in the Lungs of Crystalline Silica–Exposed Rats

Gene	Fold Change in Expression (Postexposure Time Intervals)					
	1 Week	2 Weeks	4 Weeks	8 Weeks	16 Weeks	32 Weeks
Antioxidants and oxidative stress						
Superoxide dismutase 2 (SOD2)	1.76[a]	1.85[a]	1.98[a]	2.47[a]	2.44[a]	1.53[a]
Heme oxygenase 1 (HMOX1)	1.40[a]	1.39[a]	1.46[a]	1.58[a]	2.05[a]	1.63[a]
Metallothionein 1a (MT1A)	1.81[a]	1.77[a]	1.65[a]	2.11[a]	2.44[a]	2.03[a]
NADPH oxidase organizer 1 (NOXO1)	2.54[a]	2.07[a]	2.50[a]	3.24[a]	3.16[a]	1.70[a]
Lipocalin 2 (LCN2)	3.25[a]	3.34[a]	3.58[a]	5.53[a]	5.96[a]	4.20[a]
Arginase 1(ARG1)	1.59[a]	1.29[a]	1.38[a]	1.93[a]	2.28[a]	1.55[a]
Lactoperoxidase (LPO)	1.09[a]	1.02	1.03	1.23[a]	2.09[a]	4.08[a]
Cancer						
Lipocalin 2 (LCN2)	3.25[a]	3.34[a]	3.54[a]	5.53[a]	5.96[a]	4.20[a]
Chitinase 3-like 1 (CHI3L1)	1.89[a]	2.15[a]	2.17[a]	2.75[a]	3.34[a]	2.95[a]
Secreted phosphoprotein 1 (SPP1)	1.34	1.01	1.12	1.85[a]	6.27[a]	8.61[a]
Inflammation						
Chemokine (C–C motif) ligand 2 (CCl2)	2.36[a]	1.99[a]	2.30[a]	4.28[a]	6.22[a]	2.68[a]
Chemokine (C–C motif) ligand 3 (CCl3)	1.84[a]	1.53[a]	1.58[a]	2.24[a]	2.25[a]	1.35[a]
Chemokine (C–C motif) ligand 4 (CCl4)	1.21[a]	1.12	1.23[a]	1.50[a]	1.50[a]	1.24[a]
Chemokine (C–C motif) ligand 7 (CCl7)	1.39[a]	1.29[a]	1.35[a]	2.37[a]	4.32[a]	1.96[a]
Chemokine (C–X–C motif) ligand 1 (CXCl1)	3.05[a]	2.26[a]	2.76[a]	3.20[a]	2.71[a]	ND
Chemokine (C–X–C motif) ligand 2 (CXCl2)	1.46[a]	1.12	1.22[a]	1.57[a]	1.42[a]	1.15[a]
Chemokine (C–X–C motif) ligand 5 (CXCl5)	2.52[a]	2.06[a]	3.39[a]	4.32[a]	3.96[a]	1.69[a]
Chemokine (C–X–C motif) ligand 9 (CXCl9)	1.09	1.29[a]	1.31[a]	2.08[a]	2.91[a]	3.17[a]
Chemokine (C–X–C motif) ligand 10 (CXCl10)	1.00	1.05	1.04	1.04	1.52[a]	1.59[a]
Chemokine (C–X–C motif) ligand 11 (CXCl11)	1.15	1.04	1.05	1.40[a]	2.63[a]	2.61[a]
Interleukin 1 beta (IL1β)	1.21[a]	1.16	1.22[a]	1.48[a]	1.73[a]	1.92[a]
Interleukin 1 receptor antagonist, transcript variant 2 (IL1R2)	1.45[a]	1.21[a]	1.21[a]	2.27[a]	2.80[a]	1.82[a]
Resistin-like alpha (RETNLA)	2.46[a]	2.52[a]	2.42[a]	3.57[a]	8.46[a]	11.04[a]
S100 calcium-binding protein A8 (S100A8)	1.26[a]	1.16	1.42[a]	2.50[a]	3.59[a]	2.94[a]
Triggering receptor expressed on myeloid cells 1 (TREM1)	1.30[a]	1.08	1.10	1.46[a]	1.72[a]	1.09
Triggering receptor expressed on myeloid cells 2 (TREM2)	1.50[a]	1.28[a]	1.27	1.63[a]	1.64[a]	1.78[a]
Lipocalin 2 (LCN2)	3.24[a]	3.34[a]	3.58[a]	5.53[a]	5.96[a]	4.20[a]
Chitinase 3–like 1 (CHI3L1)	1.89[a]	2.15[a]	2.17[a]	2.75[a]	3.34[a]	2.95[a]
Secreted phosphoprotein 1 (SPP1)	1.34	1.01	1.12	1.85[a]	6.27[a]	8.61[a]
Arachidonate 15-lipoxygenase (ALOX15)	−1.44[a]	−1.40[a]	−1.30[a]	−1.89[a]	−1.71[a]	−1.50[a]
Tissue remodeling/fibrosis						
Matrix metallopeptidase 8 (MMP8)	1.09	1.13[a]	1.22[a]	1.33[a]	1.50[a]	1.13[a]
Matrix metallopeptidase 12 (MMP12)	3.78[a]	3.49[a]	3.87[a]	4.34[a]	4.96[a]	4.33[a]
Secreted phosphoprotein 1 (SPP1)	1.34	1.01	1.12	1.85[a]	6.27[a]	8.61[a]
Haptoglobin (HP)	1.49	1.56[a]	1.78[a]	2.11[a]	2.42[a]	2.51[a]
Arginase 1 (ARG1)	1.59[a]	1.29[a]	1.38[a]	1.93[a]	2.28[a]	1.55[a]
Chemokine (C–X–C motif) ligand 9 (CXCL9)	1.09	1.29[a]	1.31[a]	2.08[a]	2.91[a]	3.17[a]

TABLE 18.3 (continued)

Fold Change in the Expression of a Selected List of Significantly Differentially Expressed Genes in the Lungs of Crystalline Silica–Exposed Rats

Gene	Fold Change in Expression (Postexposure Time Intervals)					
	1 Week	2 Weeks	4 Weeks	8 Weeks	16 Weeks	32 Weeks
Chemokine (C–C motif) ligand 2 (CCl2)	2.36[a]	1.99[a]	2.30[a]	4.28[a]	6.22[a]	2.68[a]
Chemokine (C–C motif) ligand 7 (CCl7)	1.39[a]	1.29[a]	1.35[a]	2.37[a]	4.32[a]	1.96[a]
Complement component 2 (C2)	1.21[a]	1.02	1.05	1.19[a]	1.5[a]	1.77[a]
Complement component 3 (C3)	2.50[a]	2.14[a]	2.26[a]	2.88[a]	3.02[a]	ND
Complement component 4–binding protein, alpha (C4BPA)	2.14[a]	2.19[a]	2.23[a]	2.90[a]	3.55[a]	3.07[a]
Complement component 5 (C5)	1.19[a]	1.24[a]	1.31[a]	1.533[a]	1.62[a]	ND
Complement factor B (CFB)	1.32[a]	1.18[a]	1.22[a]	1.37[a]	1.5[a]	1.18[a]
Complement factor 1 (CFI)	1.28	1.47[a]	1.65[a]	2.09[a]	4.18[a]	4.88[a]

Sources: Modified from Sellamuthu, R. et al., *J. Appl. Toxicol.*, 33(4), 301, 2013a; Sellamuthu, R. et al., *Inhal. Toxicol.*, 24(9), 570, 2012. With permission.

Notes: Rats were exposed to Min-U-Sil 5 crystalline silica (15 mg/m³, 6 h/day, 5 days), and the number of SDEGs in their lungs was determined by microarray analysis as described in detail in our original publications (Sellamuthu et al., 2012, 2013a). The microarray data were subjected to bioinformatics analysis, and the fold changes in the expressions of a selected list of genes belonging to IPA biological processes that are relevant to mechanism(s) of crystalline silica–induced pulmonary toxicity are presented for each of the post–silica exposure time intervals. Data represent the group mean of eight crystalline silica–exposed and four time-matched control rats per time point.

Data represent the group mean of the silica-exposed rats (n = 8) compared with the corresponding time-matched controls (n = 4) and are obtained from the microarray analysis results. Some of the genes are listed under more than one category since bioinformatics analysis identified their involvement in multiple categories.

[a] Statistically significantly different (FDR p < 0.01) compared to the time-matched control samples.

(BCL6), BCL-2-associated X gene (BAX), proliferating cell nuclear antigen (PCNA), GADD45, and excision repair cross-complementing 3 (ERCC3). Similarly, Sellamuthu et al. (2011a) detected significant overexpression of GADD45α and GADD34 in A549 cells in response to silica exposure (Table 18.2).

18.2.4 Crystalline Silica and Apoptosis

Crystalline silica particles, following their entry into the respiratory system, are engulfed by AMs for clearance from and detoxification of the system. The interaction between crystalline silica particles and AM may result in a cascade of cellular events resulting in the activation of AMs culminating in their death and the release of silica particles into the lungs. Apoptosis plays a major role in crystalline silica–induced death of the activated AMs and neutrophils and in turn may contribute significantly to the pulmonary effects of crystalline silica particles. The observation that apoptosis-deficient FasL⁻/⁻ *gld* mice did not develop silicosis (Borges et al., 2002) suggests the crucial role played by apoptosis in the adverse health effects, especially silicosis, associated with exposure to crystalline silica.

Induction of apoptosis, in response to crystalline silica exposure, has been demonstrated in in vitro cell culture (Hamilton et al., 2000) and in vivo animal (Srivastava et al., 2002) models. The involvement of ROS and RNS (Srivastava et al., 2002; Santarelli et al., 2004) as well as scavenger receptors (Hamilton et al., 2000; Fubini and Hubbard, 2003) in silica-induced apoptosis has been demonstrated. Srivastava et al. (2002) have demonstrated the involvement of interleukin 1 beta (IL1β) and nitric oxide (NO) in crystalline silica–induced apoptosis. Crystalline silica–induced apoptosis in the IC-21 macrophage cell line was inhibited by anti-IL1β antibody and the NO synthase (NOS) inhibitor, N(G)-nitro-L-arginine-methyl ester, suggesting the release of IL1β-mediated NO in silica-induced apoptosis. These findings obtained from cell culture experiments were further supported by the results of animal studies (Srivastava et al., 2002). Exposure of IL1β and inducible NOS (iNOS) knockout mice to crystalline silica resulted in significantly reduced apoptosis, inflammation, and silicotic lesions compared to the wild-type mice. These results, in addition to demonstrating an association between apoptosis and inflammation, supported a potential role for IL1β-dependent NO-mediated apoptosis in silicosis.

An important role for apoptosis in crystalline silica–induced autoimmune diseases has been suggested based on the results of a study conducted in the autoimmune-prone New Zealand mixed (NZM) mouse (Brown et al., 2005). Significant DNA fragmentation noticed in the bone marrow–derived macrophages of the crystalline silica–administered NZM mice suggested apoptosis induction. Increased levels of anti-histone autoantibodies, high proteinuria, and glomerulonephritis seen in the NZM mice suggested the autoimmune effects of crystalline silica. Rottlerin, a protein kinase C (PKC) delta inhibitor, blocked crystalline silica–induced apoptosis and autoimmune effects in the mice. These results, in addition to demonstrating the involvement of PKC in crystalline silica–induced apoptosis, suggested the involvement of apoptosis in the development of autoimmune diseases in response to crystalline silica exposure.

The studies conducted in our laboratory (Sellamuthu et al., 2011a, 2012, 2013a), by employing A549 cells and rats, suggested the involvement of several genes in crystalline silica–induced apoptosis (Tables 18.2 and 18.3). In addition, bioinformatics analysis of the SDEGs in the crystalline silica–exposed A549 cells and rat lungs provided insights into the mechanisms potentially underlying silica-induced apoptosis. DNA damage plays an important role in apoptosis. The ability of silica to interact with DNA and, therefore, to result in DNA damage has been previously demonstrated (Mao et al., 1994). Even though it has been fairly well established that crystalline silica exposure results in apoptosis indirectly through the generation of ROS (Santarelli et al., 2004), the apoptotic implication of direct DNA damage induced by crystalline silica, if any, is not well understood. Growth arrest and DNA damage (GADD) genes are a family of genes that are found significantly overexpressed in cells undergoing apoptosis (Hollander et al., 1997). GADD proteins interact with a diverse array of proteins to facilitate apoptosis. For example, GADD45α, a p53-regulated gene, interacts with other *p53*-regulated genes that play an important role in apoptosis (Sarkar et al., 2002). The significant and crystalline silica concentration-dependent overexpression of GADD34 and GADD45α—two important members of the GADD family of genes seen in the crystalline silica–exposed A549 cells (Table 18.2)—suggests their involvement, by yet-to-be-identified mechanism(s), in crystalline silica–induced apoptosis. EGR1 is a transcription factor that is characterized by rapid and transient upregulation of expression in response to stress (Yu et al., 2007). Expression of the EGR1 mRNA was significantly upregulated in response to crystalline silica exposure and toxicity in A549 cells (Table 18.2). EGR1, similar to the GADD family of genes, plays a prominent role in apoptosis through its interaction with the p53 family of genes (Yu et al., 2007).

18.2.5 Crystalline Silica and Inflammation

Crystalline silica exposure results in the induction of inflammation (Chen et al., 1999; Fubini and Hubbard, 2003; Porter et al., 2004), and a central role for inflammation in the pulmonary effects associated with silica exposure has been established (Castranova, 2004). Pulmonary inflammation has been suggested as a major factor contributing to silicosis, the most devastating health effect associated with exposure to respirable crystalline silica. Significant increases in AMs and polymorphonuclear leukocytes (PMNs) and pro-inflammatory chemokines, MCP1 and macrophage inflammatory protein-2 (MIP2), in the lung samples of rats following their inhalation exposure to crystalline silica are strong indicators of silica-induced pulmonary inflammation

(Sellamuthu et al., 2011b, 2012). Crystalline silica particles, following their entry into the lungs, are taken up by AMs, primarily for their elimination from the lungs. In this process, the silica particles may activate AMs resulting in their death and release of the silica particles along with various signaling molecules. The crystalline silica particles as well as the released signaling molecules, in turn, interact with additional AMs and alveolar epithelial cells, further resulting in AM activation and secretion of additional signaling molecules. Many of the signaling molecules thus released in response to the crystalline silica–mediated activation of AMs and epithelial cells are inflammatory in nature and may result in the recruitment of additional inflammatory cells (i.e., neutrophils, macrophages, lymphocytes, etc.) into the lungs resulting in pulmonary inflammation. The significant increase in the number of neutrophils seen in the blood of crystalline silica–exposed rats may suggest the release of the inflammatory mediators generated in the lungs into systemic circulation and the induction of systemic inflammation (Sellamuthu et al., 2011b, 2012). The crystalline silica–induced pulmonary inflammation in the rats, similar to the trend exhibited by the various parameters of crystalline silica–induced pulmonary toxicity, exhibited a steady progression during the postexposure time intervals analyzed (Sellamuthu et al., 2011b, 2012). These findings strongly support the previous suggestion that inflammation plays a central role in crystalline silica toxicity and the associated adverse health effects.

Microarray analysis of global gene expression profiles in the lungs of the crystalline silica–exposed rats supported the induction and progression of pulmonary inflammation and toxicity seen in the crystalline silica–exposed rats. Inflammatory response, inflammatory diseases, and cellular movement were three of the top ranking Ingenuity Pathway Analysis (IPA) biological functions identified as being significantly enriched by crystalline silica exposure in the rat lungs (Figure 18.5; Sellamuthu et al., 2013a). In addition, multiple canonical pathways and molecular networks involved in the induction of inflammation were significantly and progressively enriched during the silica postexposure time intervals (Figures 18.5 and 18.6). Interestingly, the number of inflammation-related biological functions, pathways, and networks that were significantly affected by crystalline silica exposure in the lungs also steadily increased (Figure 18.5) along with the progression of crystalline silica–induced pulmonary toxicity in the rats (Figure 18.3) suggesting a possible relationship between crystalline silica–induced differential expression of genes involved in the inflammation and the toxicity progression seen in the rat lungs. Bioinformatics analysis of the SDEGs provided insights into the molecular mechanisms underlying the progression of crystalline silica–induced pulmonary inflammation and toxicity in the rats. Several inflammatory response genes that encode inflammatory cytokines/chemokines were significantly overexpressed in crystalline silica–exposed rat lungs, and the magnitude of their overexpression steadily increased during the postexposure time intervals (Table 18.3) along with the progression of pulmonary toxicity induced by crystalline silica in the rats (Figure 18.3). Many of these pro-inflammatory cytokines/chemokines function as chemoattractants and recruit inflammatory cells, especially PMNs, into the lungs (Olson and Ley, 2002) in response to pulmonary damage. This may, therefore, account, at least in part, for the significant increase in the number of PMNs detected in the lungs resulting in the induction and progression of pulmonary inflammation and toxicity in the crystalline silica–exposed rats (Sellamuthu et al., 2011b, 2012). In addition to the genes encoding inflammatory cytokines/chemokines, significant overexpression of other genes known to play prominent roles in the induction of inflammation such as S100A8 (Ryckman et al., 2003), RETNLA (Holcomb et al., 2000), TREM1 and TREM2 (Ford and McVicar, 2009), LCN2 (Zhang et al., 2008), CHI3L1 (Eurich et al., 2009), SPP1 (Sabo-Attwood et al., 2011), and several members of the complement system (Li et al., 2007) and acute phase response (Whicher et al., 1999) was found in the crystalline silica–exposed rat lungs (Table 18.3). It is noteworthy that overexpression of these inflammatory response genes steadily increased along with the progression of crystalline silica–induced pulmonary inflammation and toxicity in the rats during the postexposure time intervals analyzed, further supporting their involvement in the progression of pulmonary inflammation and toxicity. Lipoxins play an important role in the resolution of pulmonary inflammation (Chan and

Moore, 2010). Lipoxins are the products of arachidonic acid metabolism catalyzed by 15-lipoxy-genase (Alox15) (Kronke et al., 2009). An anti-inflammatory role has been attributed to lipoxins mainly because of their ability to inhibit chemotaxis, adhere and transmigrate neutrophils, and antagonize the pro-inflammatory effects of leukotrienes (Colgan et al., 1993; Scalia et al., 1997; Godson and Brady, 2000). ALOX-15 expression was significantly lower in the lungs of the silica-exposed rats compared with the time-matched controls (Table 18.3). Therefore, it is reasonable to assume that, in addition to the significant overexpression of multiple pro-inflammatory genes, the significant downregulation of ALOX-15 gene expression may have contributed to the unresolved pulmonary inflammation in the crystalline silica–exposed rats. Measurement of lipoxins in the crystalline silica–exposed rats would be beneficial to further establish the role of lipoxins in the resolution of crystalline silica–induced pulmonary inflammation.

The role of aryl hydrocarbon receptor (AhR) in the pulmonary effects of crystalline silica, especially in inflammation and fibrosis, was investigated recently (Beamer et al., 2012). AhR plays a major role in the immune system and is best known as the receptor for 2,3,7,8-tetra-chlorodibenzo-p-dioxin (TCDD). Activation of AhR by TCDD results in a multitude of toxic endpoints, including, profound immunosuppression (Marshall and Kerkvliet, 2010). AhR$^{-/-}$ mice were hypersensitive to crystalline silica–induced inflammation and had increased levels of inflammatory cytokines and chemokines in their BALF. Furthermore, macrophages derived from AhR$^{-/-}$ mice secreted enhanced amounts of cytokines and chemokines in response to crys-talline silica exposure compared to the wild-type C57Bl/6 mice. Analysis of gene expression in macrophages revealed that AhR$^{-/-}$ mice exhibited increased levels of pro-IL1β, IL-6, and BCL-2 and decreased levels of STAT2, STAT5A, and serpin B2 (Pai-2) in response to crystal-line silica exposure. Based on their findings, Beamer and colleagues (2012) have concluded that AhR functions as a negative regulator of crystalline silica–induced inflammation. The authors, furthermore, reported that AhR was not involved in the fibrogenic response induced by crystalline silica in mice. Global gene expression profiling and bioinformatics analysis of the gene expression profile of the crystalline silica–exposed A549 cells, in addition to support-ing the involvement of inflammation in the pulmonary effects of crystalline silica, provided insights into the molecular mechanisms underlying crystalline silica–induced inflammation. Inflammation response ranked very high among the functional categories of genes whose expression levels were significantly affected by crystalline silica exposure in the A549 cells (Sellamuthu et al., 2011a). Transcripts for a number of pro-inflammatory interleukins, in agree-ment with the results of previous studies (Rao et al., 2004; Herseth et al., 2008), and several members of the CXCL family of pro-inflammatory cytokines were significantly overexpressed in the crystalline silica–exposed A549 cells; and their overexpression levels, as in the case of crystalline silica–induced cytotoxicity, were dependent on the concentration and duration of crystalline silica exposure (Sellamuthu et al., 2011a). A definite role for the CXCL family of pro-inflammatory cytokines as mediators of crystalline silica–induced pulmonary inflam-mation has been demonstrated in rats (Borm and Driscoll, 1996) and mice (Chao et al., 2001). In summary, the results of the transcriptomics studies conducted in in vitro cell culture and in vivo animal models of crystalline silica toxicity, in addition to supporting the prominent role played by inflammation in crystalline silica–induced pulmonary toxicity, have provided molec-ular insights into the mechanisms underlying the initiation and progression of silica-induced pulmonary inflammation and toxicity.

18.2.6 CRYSTALLINE SILICA AND PULMONARY FIBROSIS

Pulmonary fibrosis is a major component of silicosis (Ng and Chan, 1991), the most serious health outcome of exposure to respirable crystalline silica. Crystalline silica–induced fibrosis includes the release of fibrogenic factors from AMs, proliferation of the fibroblasts, and increased produc-tion of collagen by pulmonary fibroblasts. Positive collagen staining of lung tissues with Masson's trichrome stain and increased levels of hydroxyproline, a component of collagen, in the lungs are often considered markers for pulmonary fibrosis. Inhalation exposure of rats to crystalline silica for 1 week (15 mg/m^3, 6 h/day, 5 days) resulted in pulmonary fibrosis that was detectable histologically by Masson's trichrome stain at the 32-week post–silica exposure time interval (Sellamuthu et al., 2012). A further progression of pulmonary fibrosis was seen at the 44-week post–silica exposure time interval in the same rat model (Sellamuthu et al., 2013b). Crystalline silica–induced pulmonary fibrosis has also been reported by other investigators (Porter et al., 2001). Global gene expression profiling and bioinformatics analysis of the gene expression data identified significant differential expression of several genes involved in tissue remodeling and fibrosis in the lungs of the crystalline silica–exposed rats as compared to the time-matched control rats (Table 18.3). Many of the fibrosis-related genes whose expression levels were significantly different following inhalation exposure to crystalline silica in our rat model (Table 18.3) were also significantly differentially expressed in the lungs of acute and chronic rat silicosis models developed by Langley et al. (2011) as well as in human fibrotic diseases (Nau et al., 1997; Pardo et al., 2005). MMPs are a family of proteins that participate in many homeostatic biological processes as well as in pathological processes including fibrotic lung diseases (Nagase and Woessner, 1999). MMPs have been implicated in airway remodeling and granuloma formation because of their involvement in extracellular matrix degradation (Scabilloni et al., 2005). Of the *MMPs* that were significantly overexpressed in the crystalline silica–exposed rat lungs, *MMP12* overexpression was most significant (Table 18.3). A definite role for MMP12 in the induction of pulmonary fibrosis has been demonstrated previously in mice carrying a targeted deletion of the MMP12 gene (Matute-Bello et al., 2007). Osteopontin, one of the key components of extracellular matrix, mediates the migration, adhesion, and proliferation of fibroblasts culminating in pulmonary fibrosis (Takahashi et al., 2001). The profibrotic gene SPP1, which codes osteopontin protein, was significantly overexpressed in the crystalline silica–exposed rat lungs, especially at the late post–silica exposure time intervals (Table 18.3). A definite role for SPP1 in fibrosis is suggested based on decreased expression levels of collagen type 1 in SPP1$^{-/-}$ mice (Berman et al., 2004). The ARG1 gene, which was significantly and progressively overexpressed in the silica-exposed lung samples (Table 18.3), has been associated with bleomycin-induced pulmonary fibrosis in mice (Endo et al., 2003). The significant overexpression of the profibrotic chemokines CCl2 (Mercer et al., 2009) and CCl7 (Moore and Hogaboam, 2008) observed in the crystalline silica–exposed rat lungs (Table 18.3; Langely et al., 2011) may indicate their involvement in silica-induced pulmonary fibrosis in the rats. This view is further supported by the significant overexpression of these chemokines in tuberculosis, a human fibrotic disease (Nau et al., 1997). The involvement of RETNLA in the induction of pulmonary fibrosis by promoting the differentiation of myoblasts that mediate collagen deposition has been suggested (Liu et al., 2004). The RETNLA gene was highly overexpressed in the silica-exposed rat lungs (Table 18.3). Since a definite relationship is known to exist between unresolved pulmonary inflammation and fibrosis (Reynolds, 2005), it is reasonable to assume that the significant overexpression of the proinflammatory genes presented in Table 18.3 facilitated the unresolved pulmonary inflammation observed in the rat lungs contributing to crystalline silica–induced pulmonary fibrosis. The magnitude of overexpression for all of these genes known to be involved in tissue remodeling and fibrosis steadily increased in parallel with the progression of crystalline silica–induced pulmonary toxicity in the rats suggesting their role in crystalline silica–induced pulmonary fibrosis and toxicity. An interesting finding of the gene expression data, with respect to crystalline silica–induced pulmonary fibrosis, was the superior sensitivity of the changes in the expression levels of the marker genes of fibrosis compared with a conventional approach, Masson's trichrome staining of lung tissue. The findings by Porter et al. (2001) have suggested a slightly superior sensitivity of Masson's trichrome staining as an indicator for crystalline silica–induced pulmonary fibrosis compared with a biochemical estimation of hydroxyproline in rat lung samples. In our rat model, the earliest indication of crystalline silica–induced pulmonary fibrosis, as detectable by positive trichrome staining of lung tissues, was detected at the 32-week post–silica exposure time interval (Sellamuthu et al., 2012).

However, as presented in Table 18.3, significant overexpression of several genes involved in tissue remodeling and fibrosis was detectable in the lungs of crystalline silica–exposed rats as early as 1 week following the termination of crystalline silica exposure. Overexpression of these genes steadily increased during the post–silica exposure time intervals. The post–silica exposure time interval with the highest overexpression of the fibrosis-related genes detected matched the onset of fibrosis, as revealed by the results of Masson's trichrome staining of the lung tissues in the rats (Sellamuthu et al., 2012).

18.2.7 CRYSTALLINE SILICA AND CANCER

One of the human disease research areas that has benefited immensely by the advancements in genomics, especially transcriptomics, is cancer research. Cancer, in general, is considered to be the outcome of an imbalance in cell division regulated by abnormal gene expression patterns. In general, oncogenes promote and tumor suppressor genes suppress cell growth and division. Loss of control of cell division, by the activation of oncogenes and/or the inactivation of tumor suppressor genes, may facilitate uncontrolled cell division resulting in cancer. There is overwhelming evidence in literature, which include those obtained from in vitro cell culture studies, in vivo animal studies, and epidemiologic studies, supporting the carcinogenic potential of crystalline silica. Initial evidence of neoplastic transformation induced by crystalline silica was obtained in cultures of Syrian hamster embryo cells (Hesterberg and Barrett, 1984). Subsequently, Saffiotti and Ahmed (1995) demonstrated that crystalline silica is capable of morphologically transforming the mouse embryo cell line BALB/3T3/A31-1-1. However, the report that crystalline silica is not carcinogenic in hamsters and mice (Saffiotti, 2005) questions the relevance of the cell transformation data obtained in the cell lines developed in these species with respect to the carcinogenic potential of crystalline silica. In a study conducted by Williams et al. (1996), crystalline silica transformed fetal rat lung epithelial cells retaining markers of alveolar type II pneumocytes. Similarly, Saffiotti (1998) demonstrated crystalline silica–induced cell transformation in AE6 cell lines developed from rat primary cultures of alveolar type II pneumocytes. In both these studies, involving rat lung cell lines, the transformed cells were tumorigenic in immune-deficient mice confirming the carcinogenic potential of crystalline silica. Intratracheal administration of crystalline silica resulted in adenocarcinomas, epidermoid carcinomas, undifferentiated carcinomas, mixed carcinomas, and adenomas in the lungs of F344 rats (Saffiotti et al., 1996), further confirming the carcinogenic potential of crystalline silica. There is also substantial epidemiologic evidence for human carcinogenesis by crystalline silica (IARC, 1997). In spite of the identification of crystalline silica as an animal and human carcinogen (IARC, 1997) and the unequivocal role played by the transcriptome in cancer (Stadler and Come, 2009; Abba et al., 2010), the involvement of gene expression changes, if any, in crystalline silica–induced cancer is not well understood. Therefore, expression profiling of the whole genome in cells transformed with crystalline silica or tumor samples developed in response to crystalline silica exposure in animals and/or humans may provide valuable insight into the mechanisms underlying cancer resulting from exposure to crystalline silica. Similarly, the expression levels of these genes, following appropriate validation, may be employed as biomarkers for detecting crystalline silica–induced cancer. To date, neither transcriptomics studies using either cells transformed with crystalline silica or tumor samples developed in animals and/or humans in response to their exposure to crystalline silica have been reported. Nevertheless, the involvement of a large number of genes has been implicated in crystalline silica–induced cancer based on the associations seen between differentially expressed genes and crystalline silica exposure and/or toxicity in cell culture and animal tissue samples. Enhanced cell proliferation is the characteristic of a carcinogenic response, and, therefore, the overexpression of genes regulating the cellular process of proliferation has been considered for their involvement in cancer. Activator protein 1 (AP1) is a transcription factor consisting of homo- or heterodimers of the FOS and JUN genes, and a definite role for AP1, through transcriptional activation of its downstream target genes, in cellular proliferation has been established (Ryseck et al., 1988). The significant overexpression of AP1 and its downstream target genes has been reported in cells and lung samples in response to their exposure to crystalline silica (Ding et al., 1999; Sellamuthu et al., 2011a; Perkins et al., 2012). The Kruppel-like factors (KLFs) are a group of transcription factors whose involvement in cellular proliferation and cancer has been recognized for some time (Ghaleb and Yang, 2008; Mori et al., 2009; Nakamura et al., 2009). Several KLF genes were significantly over-expressed in crystalline silica–exposed A549 human lung epithelial cells (Table 18.2). Expression of the tumor suppressor gene, p53, was significantly lower in the lungs of mice exposed to crystal-line silica (Ishihara et al., 2002). However, in spite of the major role played by p53 in cancer in general, the report that crystalline silica is not carcinogenic in mice (Saffiotti, 2005) questions the role of p53 in silica-induced carcinogenesis. The involvement of oxidative stress, inflammation, DNA damage, and fibrosis in carcinogenesis has been well established. As presented in Tables 18.2, 18.3, and 18.6 and described in the corresponding sections elsewhere in this chapter, several genes involved in oxidative stress, DNA damage, inflammation, and fibrosis were significantly differentially expressed in cells and animal tissues in response to crystalline silica exposure and/ or toxicity. Additional transcriptomics studies, as mentioned earlier, employing crystalline silica–transformed cells and/or tumors developed in animals and/or in humans in response to crystal-line silica exposure are required to determine precisely the transcriptional changes of specific genes involved in crystalline silica–induced cancer. Furthermore, whole transcriptome studies may potentially explain the species differences seen among rats, mice, and hamsters with respect to their responses to crystalline silica–induced lung cancer.

18.2.8 Novel Mechanisms of Crystalline Silica–Induced Pulmonary Toxicity

A unique feature of global gene expression profiling is its potential to screen all the cellular targets and processes for their potential involvement in the response of a cell or an organism to exposure to a toxic agent. Therefore, whole-genome expression profiling may facilitate the identi-fication of novel target(s) and mechanism(s) of toxicity that may not be achieved by conventional toxicity studies. The involvement of the novel gene target(s) and/or mechanism(s) in the toxicity of the agent being investigated can be further studied and confirmed by employing cell culture and/or animal models that are transgenic for the novel target gene(s) identified. The transcriptomics studies conducted in the recent past employing RNA isolated from cells and animal tissues that were exposed to crystalline silica, in addition to confirming many of the previously identified targets and mechanisms of crystalline silica toxicity, facilitated the identification of several novel toxicity targets and mechanisms potentially involved in crystalline silica toxicity. As presented in Table 18.4, significant overexpression of several members of the solute carrier (SLC) fam-ily of genes was found in the lungs of the crystalline silica–exposed rats at all postexposure time intervals. In parallel with the progression of pulmonary toxicity noticed in the crystalline silica–exposed rats (Figure 18.3), all of the SLC genes listed in Table 18.4 exhibited a steady increase in their overexpression levels in the lungs of the silica-exposed rats compared with the corresponding time-matched controls suggesting their involvement in the initiation as well as the progression of crystalline silica–induced pulmonary toxicity. The SLC gene that was most significantly overexpressed in the lungs of the crystalline silica–exposed rats was SLC26A4, and several lines of evidence suggest a potential role for this gene in crystalline silica–induced pul-monary toxicity. The SLC26A4 gene codes the protein pendrin, which is responsible for exces-sive mucus production by airway epithelial cells (Nakao et al., 2008). A relationship is known to exist between excessive mucus production by airway epithelial cells and morbidity and mortality from certain respiratory diseases (Rogers, 2004; Rose and Voynow, 2006). The steady increase in the overexpression of the SLC26A4 gene noticed in the crystalline silica–exposed rats may also

TABLE 18.4

Differential Expressions of Solute Carrier (SLC) Family of Genes in Rat Lungs

	Fold Change in Expression (Postexposure Time Intervals)					
Gene	1 Week	2 Weeks	4 Weeks	8 Weeks	16 Weeks	32 Weeks
Solute carrier family 26, member 4 (SLC26A4)	4.46[a]	3.67[a]	4.10[a]	6.51[a]	9.93[a]	6.69[a]
Solute carrier family 13, member 2 (SLC13A2)	1.74[a]	1.51[a]	1.72[a]	2.21[a]	2.50[a]	1.67[a]
Solute carrier family 7, member 7 (SLC7A7)	1.50[a]	1.43[a]	1.48[a]	1.86[a]	2.33[a]	3.13[a]
Solute carrier family 16, member 3 (SLC16A3)	1.43[a]	1.23[a]	1.22[a]	1.72[a]	2.02[a]	2.04[a]
Solute carrier family 16, member 11 (SLC16A11)	1.43[a]	1.58[a]	1.59[a]	1.81[a]	1.88[a]	1.89[a]

Sources: Modified from Sellamuthu, R. et al., *Inhal. Toxicol.*, 24(9), 570, 2012; Sellamuthu, R. et al., *J. Appl. Toxicol.*, 33(4), 301, 2013a. With permission.

Notes: Rats were exposed to Min-U-Sil 5 crystalline silica (15 mg/m^3, 6 h/day, 5 days), and the expressions of the SLC genes listed in the table were determined by microarray analysis as described in detail in our original publication (Sellamuthu et al., 2013a). Data represent the group mean of eight crystalline silica–exposed and four time-matched control rats per time point.

[a] Statistically significantly different (FDR, p < 0.01) compared to the time-matched control samples.

account, at least in part, for the progression of pulmonary inflammation noticed in them. It has been reported previously that forced overexpression of the SLC26A4 gene, by yet-to-be-identified mechanism(s), resulted in the activation of the CXCl1 and CXCl2 chemoattractants and facilitated the infiltration of neutrophils into lungs resulting in the induction of pulmonary inflammation (Nakao et al., 2008). In this regard, it is important to note that both CXCl1 and CXCl2 genes were significantly and progressively overexpressed, and a significant increase in the number of infiltrating PMNs and induction of inflammation occurred in rats in response to their exposure to crystalline silica (Sellamuthu et al., 2011b, 2012, 2013a). Collectively, the findings of our study and those reported previously (Nakao et al., 2008) suggest the involvement of the SLC26A4 gene in crystalline silica–induced pulmonary inflammation and toxicity in the rats. Similar to our finding, a significant overexpression of the SLC26A4 transcript has been reported in acute and chronic rat silicosis models (Langley et al., 2011). It is encouraging to know that a transgenic mouse model for the SLC26A4 gene is available (Lu et al., 2011), and future investigations employing this mouse model may facilitate the understanding of, and/or confirm, the role of this gene in crystalline silica–induced pulmonary toxicity.

18.3 PREDICTION OF CRYSTALLINE SILICA EXPOSURE/TOXICITY

18.3.1 Blood Transcriptomics and Crystalline Silica–Induced Pulmonary Toxicity

It has been well recognized that silicosis, the most significant pulmonary health effect associated with crystalline silica exposure in humans, is an incurable, but preventable, fatal disease. The incurable but preventable nature of silicosis makes it absolutely necessary to detect silicosis preclinically, that is, prior to the appearance of clinical symptoms that represent the irreversible stage of the disease. Currently, silicosis is detected by chest x-ray and pulmonary function tests. Both techniques detect silicosis based on structural and/or functional impairment of lungs that are associated with an advanced, and therefore irreversible, stage of the disease. In consideration of the health risks, especially silicosis, and the large number of workers who are occupationally exposed to potentially toxic levels of crystalline silica in the United States and elsewhere, NIOSH has recommended developing highly sensitive and practical (noninvasive or minimally invasive) techniques capable of detecting silicosis preclinically (NIOSH, 2002). Transcriptomics studies conducted in the past several years have demonstrated the earlier appearance of gene expression changes in target organs as markers of toxicity compared to traditional histological and biochemical toxicity markers, suggesting their superior sensitivity as markers of target organ toxicity (Heinloth et al., 2004; Luhe et al., 2005; McBurney et al., 2009, 2012). It is not ethical or practical to obtain lung samples to determine either exposure to crystalline silica or a person's probability of developing silicosis, limiting the use of lung gene expression profiling in monitoring workers routinely for their occupational exposure to crystalline silica and the potential health effects. Surrogate biospecimens that can be obtained by noninvasive or minimally invasive techniques are essential for routine monitoring of occupational exposure to toxic agents and the resulting health effects. Since small quantities of blood required for gene expression profiling can be safely obtained by a minimally invasive approach from humans, blood may be a suitable surrogate biospecimen to predict/detect silicosis preclinically, provided gene expression changes taking place in the blood, in response to crystalline silica exposure/toxicity, reflect crystalline silica's effects in lungs, the primary target organ. Recently, a research project was undertaken in our laboratory, in compliance with the NIOSH recommendation, to develop highly sensitive and practical biomarkers to predict/detect silicosis preclinically. It has been reported previously that rats, similar to humans, develop silicosis following exposure to crystalline silica (Porter et al., 2001). For this reason, a rat silicosis model was employed to investigate the potential application of blood gene expression profiling as a highly sensitive and practical surrogate approach to detect and/or to predict crystalline silica exposure and the resulting pulmonary toxicity. Details regarding the study design can be found in our recent publications (Sellamuthu et al., 2011a,b, 2012, 2013a). Stated briefly, approximately 3-month-old healthy male Fischer 344 rats (CDF strain) were exposed to filtered air (control) or an aerosol consisting of respirable crystalline silica particles (15 mg/m^3, 6 h/day for 5 days). Following exposure, groups of control and silica-exposed rats were sacrificed at postexposure time intervals of 0, 1, 2, 4, 8, 16, 32, and 44 weeks. BALF, including the cells within it, lungs, and blood were collected from the rats to determine the effects of pulmonary exposure to crystalline silica as well as to determine global gene expression profiles in the blood. Specifically, we investigated (1) whether global gene expression changes in the blood reflected crystalline silica–induced pulmonary toxicity, (2) whether bioinformatics analysis of the differentially expressed genes in the blood of the crystalline silica–exposed rats could provide insights into the mechanisms underlying the pulmonary toxicity induced by crystalline silica exposure, and (3) whether exposure to a subtoxic concentration of crystalline silica could be detected or predicted using a blood gene expression signature.

18.3.2 TRANSCRIPTOMICS CHANGES IN THE BLOOD REFLECTED CRYSTALLINE SILICA–INDUCED PULMONARY TOXICITY IN RATS

As presented in our original publications (Sellamuthu et al., 2011b, 2012, 2013a), inhalation exposure of rats to crystalline silica resulted in pulmonary toxicity as evidenced from the increased LDH activity and albumin and protein contents in their BALF compared with corresponding time-matched control rats. Lung histological changes such as the accumulation of inflammatory cells, the appearance of type II pneumocyte hyperplasia, and fibrosis further supported the silica-induced pulmonary toxicity in the rats (Figure 18.3). Induction of pulmonary inflammation in the silica-exposed rats compared with the control rats was evidenced from an increase in the number of neutrophils and the level of MCP1 protein in their BALF. The pulmonary toxicity in the crystalline silica–exposed rats steadily progressed during the postexposure time intervals as evidenced from the various pulmonary toxicity parameters determined in the rats. Pulmonary fibrosis, characteristic of silicosis, was detectable in the rats at late postexposure time intervals of 32 and 44 weeks (Sellamuthu et al., 2012, 2013b).

Global gene expression profiles in the blood samples collected from the control and crystalline silica–exposed rats were determined using RatRef-12 V1.0 Expression Beadchip Arrays (Illumina, Inc., San Diego, CA) as described in detail in our original publications (Sellamuthu et al., 2011b, 2012). The number of SDEGs in the blood samples of the crystalline silica–exposed rats, compared

TABLE 18.5

Correlation Coefficients (r^2 Values) for the Relationship between Crystalline Silica–Induced Pulmonary Toxicity (BALF LDH, PMN, and MCP-1) and the Number of SDEGs in the Blood of the Rats

	BALF LDH	BALF PMN	BALF MCP-1
Blood SDEGs	0.831	0.923	0.958

Sources: Modified from Sellamuthu, R. et al., *Toxicol. Sci.*, 122(2), 253, 2011b; Sellamuthu, R. et al., *Inhal. Toxicol.*, 24(9), 570, 2012. With permission.

Note: The toxicity measurements and the number of SDEGs in the blood of the crystalline silica–exposed rats at postexposure time intervals of 0, 1, 2, 4, 8, 16, and 32 weeks following their 1-week exposure to crystalline silica (15 mg/m^3, 6 h/day, 5 days) were done as described in our original publications (Sellamuthu et al., 2011b, 2012).

with the corresponding time-matched control rat samples, was determined at each post–silica exposure time interval. As presented in Table 18.5, the number of SDEGs in the surrogate tissue, blood, correlated well with markers of pulmonary toxicity (BALF LDH activity) and inflammation (BALF PMN count and MCP1 level). The strong correlation seen between the pulmonary toxicity and inflammation parameters and the number of SDEGs in the blood of the silica-exposed rats suggested that the blood gene expression changes, in fact, were indicative of the silica-induced pulmonary toxicity. Furthermore, these results, in agreement with several previous publications (Bushel et al., 2007; Lobenhofer et al., 2008; Huang et al., 2010; Umbright et al., 2010), confirmed the potential value of global gene expression changes in the surrogate biospecimen, blood, as an indicator of target organ toxicity. However, a slightly better correlation noticed between pulmonary toxicity markers and the number of SDEGs in the blood (Table 18.5), compared to that of the lungs (Table 18.1), suggested that the gene expression changes taking place in blood, the surrogate tissue, may be better indicators of target organ toxicity than the gene expression changes taking place in the target organ itself. A similar superior sensitivity of blood gene expression changes, compared with target organ liver gene expression changes, as markers of hepatotoxicity has been reported previously by Lobenhofer et al. (2008).

18.3.3 BIOINFORMATICS ANALYSIS OF BLOOD TRANSCRIPTOME PROVIDED MOLECULAR INSIGHTS INTO THE MECHANISMS OF CRYSTALLINE SILICA–INDUCED PULMONARY TOXICITY

Bioinformatics analysis of the SDEGs in the blood of crystalline silica–exposed rats (Sellamuthu et al., 2011b, 2012), in addition to supporting the previously recognized toxicity and health effects of crystalline silica, provided insights into the molecular mechanisms underlying crystalline silica–induced pulmonary toxicity. A remarkable similarity was noticed with respect to the IPA biological functions, canonical pathways, and molecular networks that were significantly enriched in the lungs (target organ) and blood (surrogate tissue) (Figure 18.6a and b). The top eight IPA biological functions that were significantly enriched in the target organ, lungs, were also significantly enriched in the surrogate tissue, blood, of the crystalline silica–exposed rats (Figure 18.6a). Most of the IPA biological function categories that were significantly enriched in the blood of the silica-exposed rats, viz, respiratory diseases, cell-to-cell signaling and interaction, immune cell trafficking, cellular movement, cancer, and inflammatory response, were functions that are known to be associated with toxicity and negative health effects of crystalline silica exposure. Comparable similarities in the canonical pathways and molecular networks that were significantly enriched in the lungs and the blood of the rats in response to the crystalline silica–induced pulmonary toxicity were also seen (Figure 18.6b). It has been very well established that induction of inflammation plays a central role

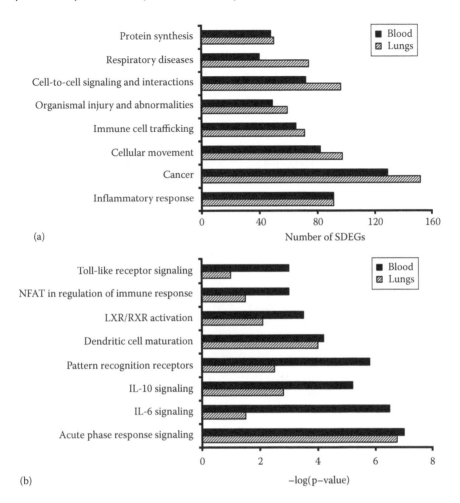

FIGURE 18.6 Enrichment of IPA biological functions and canonical pathways in the lungs and blood of crystalline silica–exposed rats. Rats were exposed to Min-U-Sil 5 crystalline silica (15 mg/m³, 6 h/day, 5 days), and the number of SDEGs at 32-weeks post–silica exposure time interval was determined by microarray analysis as described in detail in our original publications. Bioinformatics analysis of the SDEGs identified in the silica-exposed rat lungs and blood was done using IPA software. The top eight significantly enriched IPA biological functions (a) and canonical pathways (b) in the lungs and blood of the silica-exposed rats are presented to demonstrate the similarity in gene expression profile between lungs and blood of the silica-exposed rats. Data represent the mean of six rats per group. (Reproduced from Sellamuthu, R. et al., *Inhal. Toxicol.*, 24(9), 570, 2012. With permission.)

in the pulmonary effects of crystalline silica exposure in animal models (Castranova, 2004). In fact, the majority of the biological functions, molecular networks, and canonical pathways that were significantly enriched in the blood of the rats in response to crystalline silica–induced pulmonary toxicity were those involved in an inflammatory response (Sellamuthu et al., 2011b, 2012). Given these results, bioinformatics analysis of the differentially expressed genes in the blood appears to be a toxicologically relevant surrogate approach to gain insights into the mechanisms of crystalline silica–induced pulmonary toxicity. A partial list of the genes that were significantly differentially expressed in the blood of the crystalline silica–exposed rats compared to the corresponding time-matched control rats that were involved in biological functions highly relevant to crystalline silica–induced pulmonary toxicity is presented in Table 18.6. The functional significance of the genes listed in Table 18.6 with respect to their relevance to pulmonary toxicity induced by crystalline silica exposure is discussed later. Inhaled crystalline silica particles, either directly or due to their

TABLE 18.6

Fold Change in the Expression of a Selected List of Significantly Differentially Expressed Genes in the Blood of Crystalline Silica–Exposed Rats

Gene		Postexposure Time Intervals (Weeks)						
		0	1	2	4	8	16	32
Reactive oxygen species								
CYBA	Cytochrome b-245, alpha polypeptide	Up	Up[a]	Up[a]	Dn	Up	Up	Up[a]
NCF1	Neutrophil cytosolic factor 1	Up[a]	Up[a]	Up	Dn	Up	Up	Up[a]
NCF2	Neutrophil cytosolic factor 2	Up	Up[a]	Dn	Up	Up[a]	Up[a]	Up[a]
SOD2	Superoxide dismutase 2, mitochondrial	Up[a]	Dn	Dn	Dn	Up	Up	Dn[a]
CAT	Catalase	Dn	Dn[a]	Dn[a]	Up	Dn	Dn	Up[a]
XDH	Xanthine dehydrogenase	Up[a]	Dn	Dn	Dn	Up[a]	Up	ND
NOS2	Nitric oxide synthase, inducible (NOS type II) nitric oxide synthase 2, inducible	Up[a]	Dn	Up	Up	Up[a]	Up[a]	Up[a]
Antioxidant								
NFE2L2	Nuclear factor erythroid-derived 2-like 2	Up[a]	Up	Up	Dn	Up[a]	Up[a]	Up[a]
FOS	FBJ osteosarcoma oncogene	Up	Up	Up	Dn	Up[a]	Up[a]	Up[a]
JUNB	jun B proto-oncogene	Up	Up	Dn	Up	Up[a]	Up[a]	Up[a]
Danger signals or alarmins								
S100A8	S100 calcium-binding protein A8	Up[a]	Up[a]	Up[a]	Up	Up[a]	Up[a]	Up[a]
S100A9	S100 calcium-binding protein A9	Up	Up	Up	Dn	Up[a]	Up[a]	Dn[a]
Inflammation								
NLRP3	NLR family, pyrin domain containing 3	Up	Up	Up	Up	Up[a]	Up[a]	Up[a]
TLR2	Toll-like receptor 2	Up[a]	Up[a]	Up	Up	Up[a]	Up[a]	Up[a]
CLEC4E	C-type lectin domain family 4, member e	Up[a]	Up[a]	Up[a]	Dn	Up[a]	Up[a]	Up[a]
FPR1	Formyl peptide receptor 1	Up[a]	Up[a]	Up	Up	Up[a]	Up[a]	Up[a]
TLR13	Similar to toll-like receptor 13	Up[a]	Up	Up	Dn	Up[a]	Up[a]	ND
TREM1	Triggering receptor expressed on myeloid cells 1	Up[a]	Up[a]	Up[a]	Dn	Up	Up[a]	Up[a]
CD44	Cd44 molecule	Up[a]	Up[a]	Up[a]	Dn	Up[a]	Up[a]	Up[a]
P2RX4	Purinergic receptor P2X, ligand-gated ion channel 4	Up[a]	Up	Up	Up	Up[a]	Up[a]	Dn
MYD88	Myeloid differentiation primary response gene 88	Up[a]	Up	Up	Up	Up[a]	Up[a]	Up[a]
CASP1	Caspase 1	Up[a]	Up	Up	Dn	Up	Up	Dn[a]
IL1β	Interleukin 1 beta	Up[a]	Up	Up[a]	Up	Up[a]	Up[a]	Up[a]
PTAFR	Platelet-activating factor receptor	Up[a]	Up[a]	Up	Up	Up[a]	Up[a]	Up[a]
CXCR2	Interleukin 8 receptor, beta	Up[a]	Up	Up	Up	Up[a]	Up[a]	Up[a]
LGALS3	Lectin, galactoside-binding, soluble, 3	Up[a]	Up[a]	Up[a]	Up	Up[a]	Up[a]	Up[a]
ILR17A	Interleukin 17 receptor A	Up[a]	Up[a]	Up[a]	Dn	Up[a]	Up[a]	Up[a]
MMP8	Matrix metallopeptidase 8	Up[a]	Up[a]	Up	Up[a]	Up[a]	Up[a]	Up[a]
MMP9	Matrix metallopeptidase 9	Up	Up[a]	Up[a]	Up	Up[a]	Up[a]	Up[a]
ALOX5AP	Arachidonate 5-lipoxygenase activating protein	Up[a]	Up[a]	Up[a]	Up	Up[a]	Up[a]	Up[a]
Fibrosis/tissue remodeling								
ALOX5	Arachidonate 5-lipoxygenase	Dn	Up	Up	Up	Up	Up[a]	Up[a]
CCR2	Chemokine (C–C motif) receptor 2	Up	Dn	Dn	Up	Dn	Up[a]	Up[a]
CCL17	Chemokine (C–C motif) ligand 17	Up	Up	Up	Dn	Up	Up[a]	Up[a]
FAS	Fas (TNF receptor superfamily, member 6)	Up	Up	Up	Dn	Up	Up[a]	Up[a]
FOS	FBJ osteosarcoma oncogene	Up	Up	Up	Dn	Up[a]	Up[a]	Up[a]

TABLE 18.6 (continued)

Fold Change in the Expression of a Selected List of Significantly Differentially Expressed Genes in the Blood of Crystalline Silica–Exposed Rats

Gene		Postexposure Time Intervals (Weeks)						
		0	1	2	4	8	16	32
JUNB	jun B proto-oncogene	Up	Up	Dn	Up	Up[a]	Up[a]	Up[a]
MMP8	Matrix metallopeptidase 8	Up[a]	Up[a]	Up	Up[a]	Up[a]	Up[a]	Up[a]
MMP9	Matrix metallopeptidase 9	Up	Up[a]	Up[a]	Up	Up[a]	Up[a]	Up[a]

Sources: Modified from Sellamuthu, R. et al., *Toxicol. Sci.*, 122(2), 253, 2011b; Sellamuthu, R. et al., *Inhal. Toxicol.*, 24(9), 570, 2012. With permission.

Note: Rats were exposed to Min-U-Sil 5 crystalline silica (15 mg/m³, 6 h/day, 5 days), and the number of SDEGs in their blood samples was determined by microarray analysis as described in detail in our original publications (Sellamuthu et al., 2011b, 2012). The microarray data were subjected to bioinformatics analysis and the fold changes in the expressions of a representative list of genes belonging to IPA biological processes that are relevant to mechanism(s) of crystalline silica–induced pulmonary toxicity are presented for each of the post–silica exposure time intervals. Data represent the group mean of eight crystalline silica–exposed and four time-matched control rats per time point; Up—increased; Dn—decreased gene expression in the crystalline silica–exposed rats compared with the time-matched controls.

[a] Statistically significant (FDR $p \leq 0.05$) compared with the time-matched control rats.

interaction with AMs, result in the generation of ROS and RNS. Functional analysis of the SDEGs in the blood of the crystalline silica–exposed rats supported the existing evidence for the involvement of ROS and RNS in the crystalline silica–induced pulmonary toxicity. Compared with the corresponding time-matched control rats, significant differential expression of several genes that code for enzymes involved in the generation and/or detoxification of ROS and RNS was noticed in the blood of the crystalline silica–exposed rats. The NOX family of NADPH oxidases (NCF1, NCF2, CYBA, and CYBB), NOS2, XDH, and SOD2 were significantly overexpressed, while the expression of catalase was significantly downregulated. The NOX family of NADPH oxidases is involved in the generation of toxic superoxide anion. Similarly, through its involvement in purine metabolism, XDH plays a significant role in the generation of ROS. Significant SOD overexpression along with downregulated catalase expression, as noticed in the blood of the crystalline silica–exposed rats, may facilitate excess production and accumulation of toxic H_2O_2. As a result, H_2O_2 may induce oxidative stress–mediated pulmonary toxicity. NOS2, primarily responsible for the generation of RNS, may also facilitate the interaction of superoxide anion with nitric oxide to form toxic peroxynitrite. The significant overexpression of oxidative stress–responsive genes such as NRF2, JUN-B, and FOS, as observed in the blood of the silica-exposed rats, further supported the induction of oxidative stress in the crystalline silica–exposed rats.

The significant induction of pulmonary inflammation seen in our rat model is in agreement with the previously suggested central role for inflammation in crystalline silica–induced pulmonary toxicity (Castranova, 2004). A significant increase in the number of AMs and infiltrating neutrophils as well as an increased level of the pro-inflammatory cytokine, MCP-1, in the BALF suggested the induction of pulmonary inflammation in the rat model employed in our silica study. Microarray analysis of the blood gene expression profile and further functional analysis of the differentially expressed genes in the silica-exposed rats provided evidence for the induction of inflammation and insight into the various molecular events that are relevant to crystalline silica–induced pulmonary inflammation and damage. It has been well documented that, following inhalation exposure, crystalline silica causes cellular damage resulting in necrosis and death of AMs. This is often associated with the release of intracellular molecules, referred to commonly as endogenous danger signals or alarmins, which play a significant role in the inflammatory response. The transcripts for two such alarmins, S100A8 and S100A9, were significantly overexpressed in the blood of the crystalline silica–exposed rats during the postexposure time intervals (Table 18.6). The alarmins are recognized by innate immune receptors such as the pattern recognition receptors (PRRs) that are involved in the induction of inflammation. Transcripts for the PRRs, viz, NLRP3, CLEC4E, TLR2, and FPR1, were significantly overexpressed in the blood of the crystalline silica–exposed rats (Table 18.6), suggesting their activation and involvement in crystalline silica–induced pulmonary inflammation. In addition to the PRRs, transcripts for other receptors of the alarmins-mediated signaling pathway such as TREM1, P2RX4, and CD44 were also significantly overexpressed in the blood of the crystalline silica–exposed rats (Table 18.6), suggesting their involvement in crystalline silica–induced inflammatory response. Transcripts for MyD88, an adaptor molecule for TLR, and CASP1, an adaptor molecule for the NLRP3 inflammasome complex, were also significantly overexpressed in the blood of the crystalline silica–exposed rats (Table 18.6). A definite role for the NLRP3 inflammasome complex in silicosis and asbestosis has been demonstrated previously (Cassel et al., 2008). The net result of the activation of the PRRs and other inflammatory response receptors along with their respective pathways is the release of pro-inflammatory cytokines and chemokines resulting in the induction of inflammation. The transcript for IL1-β, a pro-inflammatory cytokine that plays a major role in crystalline silica–induced pulmonary inflammation and damage (Srivastava et al., 2002), was significantly overexpressed in the blood of the crystalline silica–exposed rats (Table 18.6). Thus, as expected and in agreement with the results of previous studies (Castranova, 2004) and as suggested by the blood gene expression data of our studies (Sellamuthu et al., 2011b, 2012), the pulmonary damage induced by crystalline silica in the rats was associated with significant induction of inflammation.

Silicosis, one of the most devastating health outcomes of crystalline silica exposure, is characterized by the development of fibrosis resulting in progressively restrictive pulmonary function. Type II cell hyperplasia seen in the lungs of the silica-exposed rats at the later postexposure time intervals (Figure 18.3) may be considered an early indicator of fibrosis. Thickening of the alveolar septa and positive staining of the lung sections with Masson's trichrome stain further suggested the induction of pulmonary fibrosis at late post–silica exposure time intervals of 32 and 44 weeks in our rat model (Sellamuthu et al., 2012; Sellamuthu et al., 2013b). Functional analysis of the differentially expressed genes in the blood of the crystalline silica–exposed rats indicated the activation of several cellular processes relevant to type II pneumocyte hyperplasia and fibrosis. Hyperplasia is the result of rapid cell proliferation, and the involvement of the redox-sensitive transcription factor, AP1, in cell proliferation has been very well documented (Canettieri et al., 2009). The individual constituents of the AP1 element, FOS and JUN, were significantly overexpressed in the blood samples of the crystalline silica–exposed rats at the late crystalline silica postexposure time intervals. Many tissue remodeling and fibrosis-related genes, for example, CCl17, CCR2, Fas, MMP8, MMP9, and MyD88, were significantly overexpressed in the blood of the crystalline silica–exposed rats (Table 18.6), especially at the late postexposure time intervals.

18.3.4 Blood Gene Expression Signature Predicted Exposure of Rats to a Subtoxic Concentration of Crystalline Silica

Typically, occupational exposure to crystalline silica among workers takes place at very low concentrations over a prolonged period of time. In addition, there is a latency period between occupational exposure to crystalline silica and the onset of pulmonary diseases, especially silicosis and cancer, among exposed workers. Therefore, the adverse health effects associated with crystalline silica exposure may not be detectable immediately. Similar to any other adverse health effect, the key to preventing silicosis and other health effects associated with occupational exposure to crystalline silica is early detection of the exposure and/or the associated health effects. This is because the preventable stage in silicosis, if there is one, would be the very early stage where the pulmonary effects may still be reversible and, therefore, preventable by the application of appropriate preventative and therapeutic approaches. The observation that blood gene expression changes are superior in sensitivity to biochemical, hematological, and histological changes as markers of target organ toxicity (Bushel et al., 2007; Lobenhofer et al., 2008; Umbright et al., 2010) prompted us to investigate whether exposure to crystalline silica at a very low subtoxic concentration (concentration that does not result in pulmonary toxicity detectable by conventional toxicity detection methods such as biochemical and histological changes) could be detected by employing a blood gene expression signature. Blood gene expression data obtained from rats exposed to crystalline silica at 15 mg/m^3, 6 h/day for 5 days (0-week postexposure time interval) or filtered air (control) were used as the training set data to develop gene expression signatures for crystalline silica exposure and/or toxicity. One of the blood gene expression signatures consisting of seven genes (Sellamuthu et al., 2011b) was tested in rats immediately following exposure to lower concentrations of crystalline silica (1 or 2 mg/m^3, 6 h/day, 5 days). Rats exposed to crystalline silica at 2 mg/m^3, 6 h/day for 5 days displayed mild pulmonary toxicity as evidenced from the observation of a slight, but statistically significant, elevation in BALF parameters of pulmonary toxicity (LDH activity, albumin, and protein content). In contrast, rats exposed to crystalline silica at 1 mg/m^3, 6 h/day for 5 days showed no detectable pulmonary toxicity as evidenced from normal LDH activity and protein and albumin contents in their BALF (Sellamuthu et al., 2011b). The predictive blood gene expression signature developed for crystalline silica exposure and toxicity correctly identified seven out of eight rats (87.5%) that were exposed to crystalline silica at 2 mg/m^3, 6 h/day for 5 days (Sellamuthu et al., 2011b). Six out of eight rats (75%) that were exposed to crystalline silica at 1 mg/m^3, 6 h/day, 5 days and did not result in any detectable pulmonary toxicity were correctly identified as crystalline silica–exposed rats by the predictive blood gene expression signature (Sellamuthu et al., 2011b). These results, therefore, demonstrated the potential application of blood gene expression profiling to detect/predict crystalline silica exposure and the resulting pulmonary toxicity in the rats. A similar study reported earlier by Bushel et al. (2007) developed a blood gene expression signature for hepatotoxicity in rats that was able to correctly predicting the exposure of a group of rats that were administered a subtoxic oral dose of acetaminophen. Taken together, these studies show promising use of blood transcriptomics to predict toxic exposures.

18.3.5   BLOOD TRANSCRIPTOMICS TO MONITOR HUMAN EXPOSURE TO CRYSTALLINE SILICA

The ability of the predictive blood gene expression signature to detect crystalline silica exposure in the absence of pulmonary toxicity detectable by traditional approaches (biochemical and histological toxicity markers) in the rat model employed in our studies raises the question whether a similar approach can be employed to monitor workers for occupational exposure to crystalline silica. For this to happen, the blood gene expression signature(s) for crystalline silica exposure, in addition to being highly sensitive, should be specific to the pulmonary effects associated with crystalline silica exposure. Whether the blood gene expression signature developed in our rat model is specific to crystalline silica exposure and resulting pulmonary toxicity has not been investigated any further. However, it is important to mention that a significant specificity has been demonstrated between target organ toxicity and blood gene expression profile, especially, in the case of agents that cause hepatotoxicity (Miyamoto et al., 2008; Wetmore et al., 2010). In addition, the blood gene expression signature reported by Bushel et al. (2007) for hepatotoxicity prediction in rats is quite different from the signature that we have identified for crystalline silica–induced pulmonary toxicity in spite of the predominance of inflammatory response genes in both signatures. Since most of the genes constituting the crystalline silica–responsive blood gene expression signature developed in our study are involved in an inflammatory response, it is reasonable to question whether a similar gene expression profile may be achieved with any exposure that involves induction of inflammation. Recently, Charlesworth et al. (2010) have identified 342 genes that are significantly differentially expressed in blood samples obtained from a population consisting of 297 current smokers. It has been fairly well established that inflammation plays a major role in many of the pulmonary effects associated with cigarette smoking (Bhalla et al., 2009). In spite of the central role played by inflammation in the pulmonary effects of crystalline silica and cigarette smoke, none of the genes constituting the crystalline silica–responsive blood gene expression signature identified in our study were differentially expressed in the blood of cigarette smokers (Charlesworth et al., 2010). Similarly, none of the blood signature genes identified for the crystalline silica–induced pulmonary toxicity were found differentially expressed in blood under conditions of inflammation induced by diesel exhaust particles (Peretz et al., 2007) or endotoxin (Calvano et al., 2005). LDH3 isoenzyme, which is an indicator of pulmonary damage (Drent et al., 1996), is the protein product of LDHC, a member of the blood gene expression signature identified in our rat model. Significant elevation in the blood level of LDH3 has been found associated with the pulmonary damage observed in miners who are exposed to crystalline silica (Cobben et al., 1997; Kuempel et al., 2003). The available evidence, therefore, suggests that it may be possible to develop discriminatory blood gene expression signatures that are specific to the pulmonary effects associated with crystalline silica exposure. This will, of course, require additional investigations. These investigations can be done using appropriate animal models similar to the rat model employed in our study. The validated markers capable of detecting preclinical silicosis should eventually be tested in a population of workers who are occupationally exposed to crystalline silica and, therefore, are at risk of developing silicosis. The relevance of such an approach is justified by the observations that (1) human orthologs of the hepatotoxicity signature genes developed in a rat hepatotoxicity model were able to correctly identify the hepatotoxicity resulting from the ingestion of acetaminophen in individuals (Bushel et al., 2007); and (2) mechanistically relevant gene expression changes were detectable in the blood of individuals who had received a bolus of acetaminophen that did not result in biochemical and clinical changes indicative of liver toxicity (Fannin et al., 2010). Thus, additional investigations involving animal and human studies to determine the potential application of peripheral blood gene expression profiling as a practical approach to monitor human exposure to crystalline silica and the possible detection of preclinical silicosis are required. The outcome of these investigations is expected to have a major impact on the prevention/intervention of silicosis.

QUESTIONS

1. Transcriptomics is the study of _____
   a. Genes
   b. mRNA
   c. tRNA
   d. Proteins
   Answer: b
2. True or false
   As an indicator of crystalline silica–induced pulmonary toxicity, blood gene expression change is more sensitive than lung histological change.
   Answer: True
3. Which technique is least suitable to detect a change in gene expression?
   a. Real-time PCR
   b. Microarray
   c. Western blot
   d. Northern blot
   Answer: c

4. True or false

In gene expression studies, microarray technique offers a higher throughput compared to real-time PCR.

Answer: True

REFERENCES

Abba, M. C., Lacunza, E., Butti, M., and Aldaz, C. M. (2010). Breast cancer biomarker discovery in the functional genomic age: A systematic review of 42 gene expression signatures. *Biomark. Insights* **5**, 103–118.

Amin, R. P., Vickers, A. E., Sistare, F., Thompson, K. L., Roman, R. J., Lawton, M., Kramer, J. et al. (2004). Identification of putative gene based markers of renal toxicity. *Environ. Health Perspect.* **112**, 465–479.

Banfi, B., Clark, R. A., Steger, K., and Krause, K. H. (2003). Two novel proteins activate superoxide generation by the NADPH oxidase NOX1. *J. Biol. Chem.* **278**, 3510–3513.

Beamer, C. A., Seaver, B. P., and Shepherd, D. M. (2012). Aryl hydrocarbon receptor (AhR) regulates silica-induced inflammation but not fibrosis. *Toxicol. Sci.* **126**, 554–568.

Berman, J. S., Serlin, D., Li, X., Whitley, G., Hayes, J., Rishikof, D. C., Ricupero, D. A., Liaw, L., Goetschkes, M., and O'Regan, A. W. (2004). Altered bleomycin-induced lung fibrosis in osteopontin-deficient mice. *Am. J. Physiol. Lung Cell. Mol. Physiol.* **286**, L1311–L1318.

Bhalla, D. K., Hirata, F., Rishi, A. K., and Gairola, C. G. (2009). Cigarette smoke, inflammation, and lung injury: A mechanistic perspective. *J. Toxicol. Environ. Health B, Crit. Rev.* **12**, 45–64.

Borges, V. M., Lopes, M. F., Falcao, H., Leite-Junior, J. H., Rocco, P. R., Davidson, W. F., Linden, R., Zin, W. A., and DosReis, G. A. (2002). Apoptosis underlies immunopathogenic mechanisms in acute silicosis. *Am. J. Respir. Cell. Mol. Biol.* **27**, 78–84.

Borm, P. J. and Driscoll, K. (1996). Particles, inflammation and respiratory tract carcinogenesis. *Toxicol. Lett.* **88**, 109–113.

Brown, J. M., Schwanke, C. M., Pershouse, M. A., Pfau, J. C., and Holian, A. (2005). Effects of rottlerin on silica-exacerbated systemic autoimmune disease in New Zealand mixed mice. *Am. J. Physiol. Lung Cell. Mol. Physiol.* **289**, L990–L998.

Bushel, P. R., Heinloth, A. N., Li, J., Huang, L., Chou, J. W., Boorman, G. A., Malarkey, D. E. et al. (2007). Blood gene expression signatures predict exposure levels. *Proc. Natl. Acad. Sci. USA* **104**, 18211–18216.

Calvano, S. E., Xiao, W., Richards, D. R., Felciano, R. M., Baker, H. V., Cho, R. J., Chen, R. O. et al. (2005). A network-based analysis of systemic inflammation in humans. *Nature* **437**, 1032–1037.

Canettieri, G., Coni, S., Della Guardia, M., Nocerino, V., Antonucci, L., Di Magno, L., Screaton, R., Screpanti, I., Giannini, G., and Gulino, A. (2009). The coactivator CRTC1 promotes cell proliferation and transformation via AP-1. *Proc. Natl. Acad. Sci. USA* **106**, 1445–1450.

Cassel, S. L., Eisenbarth, S. C., Iyer, S. S., Sadler, J. J., Colegio, O. R., Tephly, L. A., Carter, A. B., Rothman, P. B., Flavell, R. A., and Sutterwala, F. S. (2008). The Nalp3 inflammasome is essential for the development of silicosis. *Proc. Natl. Acad. Sci. USA* **105**, 9035–9040.

Castranova, V. (2004). Signaling pathways controlling the production of inflammatory mediators in response to crystalline silica exposure: Role of reactive oxygen/nitrogen species. *Free Radic. Biol. Med.* **37**, 916–925.

Castranova, V. and Vallyathan, V. (2000). Silicosis and coal workers' pneumoconiosis. *Environ. Health Perspect.* **108**(Suppl 4), 675–684.

Chan, M. M. and Moore, A. R. (2010). Resolution of inflammation in murine autoimmune arthritis is disrupted by cyclooxygenase-2 inhibition and restored by prostaglandin E2-mediated lipoxin A4 production. *J. Immunol.* **184**, 6418–6426.

Chao, S. K., Hamilton, R. F., Pfau, J. C., and Holian, A. (2001). Cell surface regulation of silica-induced apoptosis by the SR-A scavenger receptor in a murine lung macrophage cell line (MH-S). *Toxicol. Appl. Pharmacol.* **174**, 10–16.

Charlesworth, J. C., Curran, J. E., Johnson, M. P., Goring, H. H., Dyer, T. D., Diego, V. P., Kent, J. W., Jr. et al. (2010). Transcriptomic epidemiology of smoking: The effect of smoking on gene expression in lymphocytes. *BMC Med. Genomics* **3**, 29.

Chen, F., Demers, L. M., Vallyathan, V., Lu, Y., Castranova, V., and Shi, X. (1999). Involvement of 5′-flanking kappaB-like sites within bcl-x gene in silica-induced Bcl-x expression. *J. Biol. Chem.* **274**, 35591–35595.

Cobben, N. A., Drent, M., Schols, A. M., Lamers, R. J., Wouters, E. F., and Van Dieijen-Visser, M. P. (1997). Serum lactate dehydrogenase and its isoenzyme pattern in ex-coalminers. *Respir. Med.* **91**, 616–623.

Colgan, S. P., Serhan, C. N., Parkos, C. A., Delp-Archer, C., and Madara, J. L. (1993). Lipoxin A4 modulates transmigration of human neutrophils across intestinal epithelial monolayers. *J. Clin. Invest.* **92**, 75–82.

Daniel, L. N., Mao, Y., and Saffiotti, U. (1993). Oxidative DNA damage by crystalline silica. *Free Radic. Biol. Med.* **14**, 463–472.

Davies, M. J., Hawkins, C. L., Pattison, D. I., and Rees, M. D. (2008). Mammalian heme peroxidases: From molecular mechanisms to health implications. *Antioxid. Redox. Signal.* **10**, 1199–1234.

Ding, M., Shi, X., Dong, Z., Chen, F., Lu, Y., Castranova, V., and Vallyathan, V. (1999). Freshly fractured crystalline silica induces activator protein-1 activation through ERKs and p38 MAPK. *J. Biol. Chem.* **274**, 30611–30616.

Drent, M., Cobben, N. A., Henderson, R. F., Wouters, E. F., and van Dieijen-Visser, M. (1996). Usefulness of lactate dehydrogenase and its isoenzymes as indicators of lung damage or inflammation. *Eur. Respir. J.* **9**, 1736–1742.

Endo, M., Oyadomari, S., Terasaki, Y., Takeya, M., Suga, M., Mori, M., and Gotoh, T. (2003). Induction of arginase I and II in bleomycin-induced fibrosis of mouse lung. *Am. J. Physiol. Lung Cell. Mol. Physiol.* **285**, L313–L321.

Eurich, K., Segawa, M., Toei-Shimizu, S., and Mizoguchi, E. (2009). Potential role of chitinase 3-like-1 in inflammation-associated carcinogenic changes of epithelial cells. *World J. Gastroenterol.* **15**, 5249–5259.

Fannin, R. D., Russo, M., O'Connell, T. M., Gerrish, K., Winnike, J. H., Macdonald, J., Newton, J. et al. (2010). Acetaminophen dosing of humans results in blood transcriptome and metabolome changes consistent with impaired oxidative phosphorylation. *Hepatology* **51**, 227–236.

Ford, J. W. and McVicar, D. W. (2009). TREM and TREM-like receptors in inflammation and disease. *Curr. Opin. Immunol.* **21**, 38–46.

Fubini, B., Fenoglio, I., Elias, Z., and Poirot, O. (2001). Variability of biological responses to silicas: Effect of origin, crystallinity, and state of surface on generation of reactive oxygen species and morphological transformation of mammalian cells. *J. Environ. Pathol. Toxicol. Oncol.* **20**(Suppl 1), 95–108.

Fubini, B. and Hubbard, A. (2003). Reactive oxygen species (ROS) and reactive nitrogen species (RNS) generation by silica in inflammation and fibrosis. *Free Radic. Biol. Med.* **34**, 1507–1516.

Gaetani, G. F., Kirkman, H. N., Mangerini, R., and Ferraris, A. M. (1994). Importance of catalase in the disposal of hydrogen peroxide within human erythrocytes. *Blood* **84**, 325–330.

Ghaleb, A. M. and Yang, V. W. (2008). The pathobiology of kruppel-like factors in colorectal cancer. *Curr. Colorectal Cancer Rep.* **4**, 59–64.

Godson, C. and Brady, H. R. (2000). Lipoxins: Novel anti-inflammatory therapeutics? *Curr. Opin. Investig. Drugs* **1**, 380–385.

Gwinn, M. R., Leonard, S. S., Sargent, L. M., Lowry, D. T., McKinstry, K., Meighan, T., Reynolds, S. H., Kashon, M., Castranova, V., and Vallyathan, V. (2009). The role of p53 in silica-induced cellular and molecular responses associated with carcinogenesis. *J. Toxicol. Environ. Health A* **72**, 1509–1519.

Hamadeh, H. K., Bushel, P. R., Jayadev, S., DiSorbo, O., Bennett, L., Li, L., Tennant, R. et al. (2002). Prediction of compound signature using high density gene expression profiling. *Toxicol. Sci.* **67**, 232–240.

Hamilton, R. F., de Villiers, W. J., and Holian, A. (2000). Class A type II scavenger receptor mediates silica-induced apoptosis in Chinese hamster ovary cell line. *Toxicol. Appl. Pharmacol.* **162**, 100–106.

Heinloth, A. N., Irwin, R. D., Boorman, G. A., Nettesheim, P., Fannin, R. D., Sieber, S. O., Snell, M. L. et al. (2004). Gene expression profiling of rat livers reveals indicators of potential adverse effects. *Toxicol. Sci.* **80**, 193–202.

Herseth, J., Refsnes, M., Lag, M., Hetland, G., and Schwarze, P. (2008). IL-1beta as a determinant in silica-induced cytokine responses in monocyte-endothelial cell co-cultures. *Hum. Exp. Toxicol.* **27**, 387–399.

Hesterberg, T. W. and Barrett, J. C. (1984). Dependence of asbestos- and mineral dust-induced transformation of mammalian cells in culture on fiber dimension. *Cancer Res.* **44**, 2170–2180.

Hnizdo, E. and Sluis-Cremer, G. K. (1993). Risk of silicosis in a cohort of white South African gold miners. *Am. J. Ind. Med.* **24**, 447–457.

Holcomb, I. N., Kabakoff, R. C., Chan, B., Baker, T. W., Gurney, A., Henzel, W., Nelson, C. et al. (2000). FIZZ1, a novel cysteine-rich secreted protein associated with pulmonary inflammation, defines a new gene family. *EMBO J.* **19**, 4046–4055.

Hollander, M. C., Zhan, Q., Bae, I., and Fornace, A. J., Jr. (1997). Mammalian GADD34, an apoptosis and DNA damage-inducible gene. *J. Biol. Chem.* **272**, 13731–13737.

Huang, J., Shi, W., Zhang, J., Chou, J. W., Paules, R. S., Gerrish, K., Li, J. et al. (2010). Genomic indicators in the blood predict drug-induced liver injury. *Pharmacogenomics J.* **10**, 267–277.

IARC. (1997). Silica, some silicates, coal dust and *para*-aramid fibrils. *IARC, Monogr. Eval. Carcinog. Risk Hum.* **68**, 1–475.

Ishihara, Y., Iijima, H., Matsunaga, K., Fukushima, T., Nishikawa, T., and Takenoshita, S. (2002). Expression and mutation of p53 gene in the lung of mice intratracheal injected with crystalline silica. *Cancer Lett.* **177**, 125–128.

Johnston, C. J., Driscoll, K. E., Finkelstein, J. N., Baggs, R., O'Reilly, M. A., Carter, J., Gelein, R., and Oberdorster, G. (2000). Pulmonary chemokine and mutagenic responses in rats after subchronic inhalation of amorphous and crystalline silica. *Toxicol. Sci.* **56**, 405–413.

Kreiss, K. and Zhen, B. (1996). Risk of silicosis in a Colorado Mining Community. *Am. J. Ind. Med.* **30**, 529–539.

Kronke, G., Katzenbeisser, J., Uderhardt, S., Zaiss, M. M., Scholtysek, C., Schabbauer, G., Zarbock, A. et al. (2009). 12/15-lipoxygenase counteracts inflammation and tissue damage in arthritis. *J. Immunol.* **183**, 3383–3389.

Kuempel, E. D., Attfield, M. D., Vallyathan, V., Lapp, N. L., Hale, J. M., Smith, R. J., and Castranova, V. (2003). Pulmonary inflammation and crystalline silica in respirable coal mine dust: Dose response. *J. Biosci.* **28**, 61–69.

Langley, R. J., Mishra, N. C., Pena-Philippides, J. C., Rice, B. J., Seagrave, J. C., Singh, S. P., and Sopori, M. L. (2011). Fibrogenic and redox-related but not proinflammatory genes are upregulated in Lewis rat model of chronic silicosis. *J. Toxicol. Environ. Health A* **74**, 1261–1279.

Li, M., Peake, P. W., Charlesworth, J. A., Tracey, D. J., and Moalem-Taylor, G. (2007). Complement activation contributes to leukocyte recruitment and neuropathic pain following peripheral nerve injury in rats. *Eur. J. Neurosci.* **26**, 3486–3500.

Linch, K. D., Miller, W. E., Althouse, R. B., Groce, D. W., and Hale, J. M. (1998). Surveillance of respirable crystalline silica dust using OSHA compliance data (1979–1995). *Am. J. Ind. Med.* **34**, 547–558.

Liochev, S. I. and Fridovich, I. (2007). The effects of superoxide dismutase on H_2O_2 formation. *Free Radic. Biol. Med.* **42**, 1465–1469.

Liu, T., Jin, H., Ullenbruch, M., Hu, B., Hashimoto, N., Moore, B., McKenzie, A., Lukacs, N. W., and Phan, S. H. (2004). Regulation of found in inflammatory zone 1 expression in bleomycin-induced lung fibrosis: Role of IL-4/IL-13 and mediation via STAT-6. *J. Immunol.* **173**, 3425–3431.

Lobenhofer, E. K., Auman, J. T., Blackshear, P. E., Boorman, G. A., Bushel, P. R., Cunningham, M. L., Fostel, J. M. et al. (2008). Gene expression response in target organ and whole blood varies as a function of target organ injury phenotype. *Genome Biol.* **9**, R100.

Lu, Y. C., Wu, C. C., Shen, W. S., Yang, T. H., Yeh, T. H., Chen, P. J., Yu, I. S. et al. (2011). Establishment of a knock-in mouse model with the SLC26A4 c.919-2A > G mutation and characterization of its pathology. *PLoS One* **6**, e22150.

Luhe, A., Suter, L., Ruepp, S., Singer, T., Weiser, T., and Albertini, S. (2005). Toxicogenomics in the pharmaceutical industry: Hollow promises or real benefit? *Mutat. Res.* **575**, 102–115.

Mandel, G. and Mandel, N. (1996) The structure of crystalline SiO_2. In *Silica and Silica-Induced Diseases*, eds. Castranova, V., Vallyathan, V., and Wallace, W. E., pp. 63–78, Boca Raton, FL: CRC Press.

Mao, Y., Daniel, L. N., Whittaker, N., and Saffiotti, U. (1994). DNA binding to crystalline silica characterized by Fourier-transform infrared spectroscopy. *Environ. Health Perspect.* **102**(Suppl 10), 165–171.

Marshall, N. B. and Kerkvliet, N. I. (2010). Dioxin and immune regulation: Emerging role of aryl hydrocarbon receptor in the generation of regulatory T cells. *Ann. N. Y. Acad. Sci.* **1183**, 25–37.

Matute-Bello, G., Wurfel, M. M., Lee, J. S., Park, D. R., Frevert, C. W., Madtes, D. K., Shapiro, S. D., and Martin, T. R. (2007). Essential role of MMP-12 in Fas-induced lung fibrosis. *Am. J. Respir. Cell. Mol. Biol.* **37**, 210–221.

McBurney, R. N., Hines, W. M., Von Tungeln, L. S., Schnackenberg, L. K., Beger, R. D., Moland, C. L., Han, T. et al. (2009). The liver toxicity biomarker study: Phase I design and preliminary results. *Toxicol. Pathol.* **37**, 52–64.

McBurney, R. N., Hines, W. M., VonTungeln, L. S., Schnackenberg, L. K., Beger, R. D., Moland, C. L., Han, T. et al. (2012). The liver toxicity biomarker study phase I: Markers for the effects of tolcapone or entacapone. *Toxicol. Pathol.* **40**, 951–964.

Mercer, P. F., Johns, R. H., Scotton, C. J., Krupiczojc, M. A., Konigshoff, M., Howell, D. C., McAnulty, R. J. et al. (2009). Pulmonary epithelium is a prominent source of proteinase-activated receptor-1-inducible CCL2 in pulmonary fibrosis. *Am. J. Respir. Crit. Care Med.* **179**, 414–425.

Miyamoto, M., Yanai, M., Ookubo, S., Awasaki, N., Takami, K., and Imai, R. (2008). Detection of cell free, liver-specific mRNAs in peripheral blood from rats with hepatotoxicity: A potential toxicological biomarker for safety evaluation. *Toxicol. Sci.* **106**, 538–545.

Moore, B. B. and Hogaboam, C. M. (2008). Murine models of pulmonary fibrosis. *Am. J. Physiol. Lung Cell. Mol. Physiol.* **294**, L152–L160.

Mori, A., Moser, C., Lang, S. A., Hackl, C., Gottfried, E., Kreutz, M., Schlitt, H. J., Geissler, E. K., and Stoeltzing, O. (2009). Up-regulation of Kruppel-like factor 5 in pancreatic cancer is promoted by interleukin-1beta signaling and hypoxia-inducible factor-1alpha. *Mol. Cancer Res.* **7**, 1390–1398.

Murray-Stewart, T., Wang, Y., Goodwin, A., Hacker, A., Meeker, A., and Casero, R. A., Jr. (2008). Nuclear localization of human spermine oxidase isoforms—Possible implications in drug response and disease etiology. *FEBS J.* **275**, 2795–2806.

Nagase, H. and Woessner, J. F., Jr. (1999). Matrix metalloproteinases. *J. Biol. Chem.* **274**, 21491–21494.

Nakamura, Y., Migita, T., Hosoda, F., Okada, N., Gotoh, M., Arai, Y., Fukushima, M. et al. (2009). Kruppel-like factor 12 plays a significant role in poorly differentiated gastric cancer progression. *Int. J. Cancer* **125**, 1859–1867.

Nakao, I., Kanaji, S., Ohta, S., Matsushita, H., Arima, K., Yuyama, N., Yamaya, M. et al. (2008). Identification of pendrin as a common mediator for mucus production in bronchial asthma and chronic obstructive pulmonary disease. *J. Immunol.* **180**, 6262–6269.

Nau, G. J., Guilfoile, P., Chupp, G. L., Berman, J. S., Kim, S. J., Kornfeld, H., and Young, R. A. (1997). A chemoattractant cytokine associated with granulomas in tuberculosis and silicosis. *Proc. Natl. Acad. Sci. USA* **94**, 6414–6419.

Ng, T. P. and Chan, S. L. (1991). Factors associated with massive fibrosis in silicosis. *Thorax* **46**, 229–232.

NIOSH. (2002). Hazard review: Health effects of occupational exposures to respirable crystalline silica. Publication No. 2002-129. Cincinnati, OH: US Department of Health and Human Services (NIOSH). Available at: http://www.cdc.gov/niosh/docs/2002–129/02–129a.html. Accessed on December 3, 2012.

NIOSH. (2006). National Occupational Respiratory Mortality System (NORMS). Atlanta, GA: US Department of Health and Human Services, Public Health Services, Centers for Disease Control and Prevention, National Institute for Occupational Safety and Health, Division of Respiratory Diseases and Surveillance, Surveillance Branch. http://webappa.cdc.gov.ords/norms/html. Accessed on December 3, 2012.

Olson, T. S. and Ley, K. (2002). Chemokines and chemokine receptors in leukocyte trafficking. *Am. J. Physiol. Regul. Integr. Comp. Physiol.* **283**, R7–R28.

Pardo, A., Gibson, K., Cisneros, J., Richards, T. J., Yang, Y., Becerril, C., Yousem, S. et al. (2005). Up-regulation and profibrotic role of osteopontin in human idiopathic pulmonary fibrosis. *PLoS Med.* **2**, e251.

Peretz, A., Peck, E. C., Bammler, T. K., Beyer, R. P., Sullivan, J. H., Trenga, C. A., Srinouanprachnah, S., Farin, F. M., and Kaufman, J. D. (2007). Diesel exhaust inhalation and assessment of peripheral blood mononuclear cell gene transcription effects: An exploratory study of healthy human volunteers. *Inhal. Toxicol.* **19**, 1107–1119.

Perkins, T. N., Shukla, A., Peeters, P. M., Steinbacher, J. L., Landry, C. C., Lathrop, S. A., Steele, C., Reynaert, N. L., Wouters, E. F., and Mossman, B. T. (2012). Differences in gene expression and cytokine production by crystalline vs. amorphous silica in human lung epithelial cells. *Part. Fibre Toxicol.* **9**, 6.

Porter, D. W., Hubbs, A. F., Mercer, R., Robinson, V. A., Ramsey, D., McLaurin, J., Khan, A. et al. (2004). Progression of lung inflammation and damage in rats after cessation of silica inhalation. *Toxicol. Sci.* **79**, 370–380.

Porter, D. W., Ramsey, D., Hubbs, A. F., Battelli, L., Ma, J., Barger, M., Landsittel, D. et al. (2001). Time course of pulmonary response of rats to inhalation of crystalline silica: Histological results and biochemical indices of damage, lipidosis, and fibrosis. *J. Environ. Pathol. Toxicol. Oncol.* **20**(Suppl 1), 1–14.

Porter, D. W., Ye, J., Ma, J., Barger, M., Robinson, V. A., Ramsey, D., McLaurin, J. et al. (2002). Time course of pulmonary response of rats to inhalation of crystalline silica: NF-kappa B activation, inflammation, cytokine production, and damage. *Inhal. Toxicol.* **14**, 349–367.

Rao, K. M., Porter, D. W., Meighan, T., and Castranova, V. (2004). The sources of inflammatory mediators in the lung after silica exposure. *Environ. Health Perspect.* **112**, 1679–1686.

Reynolds, H. Y. (2005). Lung inflammation and fibrosis: an alveolar macrophage-centered perspective from the 1970s to 1980s. *Am. J. Respir. Crit. Care Med.* **171**, 98–102.

Rimal, B., Greenberg, A. K., and Rom, W. N. (2005). Basic pathogenetic mechanisms in silicosis: Current understanding. *Curr. Opin. Pulm. Med.* **11**, 169–173.

Rogers, D. F. (2004). Airway mucus hypersecretion in asthma: An undervalued pathology? *Curr. Opin. Pharmacol.* **4**, 241–250.

Rose, M. C. and Voynow, J. A. (2006). Respiratory tract mucin genes and mucin glycoproteins in health and disease. *Physiol. Rev.* **86**, 245–278.

Ryckman, C., Vandal, K., Rouleau, P., Talbot, M., and Tessier, P. A. (2003). Proinflammatory activities of S100: Proteins S100A8, S100A9, and S100A8/A9 induce neutrophil chemotaxis and adhesion. *J. Immunol.* **170**, 3233–3242.

Ryseck, R. P., Hirai, S. I., Yaniv, M., and Bravo, R. (1988). Transcriptional activation of c-jun during the G0/G1 transition in mouse fibroblasts. *Nature* **334**, 535–537.

Sabo-Attwood, T., Ramos-Nino, M. E., Eugenia-Ariza, M., Macpherson, M. B., Butnor, K. J., Vacek, P. C., McGee, S. P., Clark, J. C., Steele, C., and Mossman, B. T. (2011). Osteopontin modulates inflammation, mucin production, and gene expression signatures after inhalation of asbestos in a murine model of fibrosis. *Am. J. Pathol.* **178**, 1975–1985.

Saffiotti, U. (1998). Respiratory tract carcinogenesis by mineral fibres and dusts: Models and mechanisms. *Monaldi. Arch. Chest Dis.* **53**, 160–167.

Saffiotti, U. (2005). Silicosis and lung cancer: A fifty-year perspective. *Acta Biomed.* **76**(Suppl 2), 30–37.

Saffiotti, U. and Ahmed, N. (1995). Neoplastic transformation by quartz in the BALB/3T3/A31-1-1 cell line and the effects of associated minerals. *Teratog. Carcinog. Mutagen.* **15**, 339–356.

Saffiotti, U., Williams, A. O., Daniel, L. N., Kaighn, M. E., Mao, Y., and Shi, X. (1996). Carcinogenesis by crystalline silica: Animal, cellular and molecular studies. In *Silica and Silica-Induced Diseases*, eds. Castranova, V., Vallyathan, V., and Wallace, W. E., pp. 345–381, Boca Raton, FL: CRC Press.

Sanderson, W. (2006). The U.S. Population-at-risk to occupational respiratory diseases. In *Occupational Respiratory Diseases*, ed. Mercahnt, J. A., pp. 86–102, Washington, DC: Department of Health and Human Services (NIOSH) Publication.

Santarelli, L., Recchioni, R., Moroni, F., Marcheselli, F., and Governa, M. (2004). Crystalline silica induces apoptosis in human endothelial cells in vitro. *Cell Biol. Toxicol.* **20**, 97–108.

Sarkar, D., Su, Z. Z., Lebedeva, I. V., Sauane, M., Gopalkrishnan, R. V., Valerie, K., Dent, P., and Fisher, P. B. (2002). mda-7 (IL-24) Mediates selective apoptosis in human melanoma cells by inducing the coordinated overexpression of the GADD family of genes by means of p38 MAPK. *Proc. Natl. Acad. Sci. USA* **99**, 10054–10059.

Scabilloni, J. F., Wang, L., Antonini, J. M., Roberts, J. R., Castranova, V., and Mercer, R. R. (2005). Matrix metalloproteinase induction in fibrosis and fibrotic nodule formation due to silica inhalation. *Am. J. Physiol. Lung Cell. Mol. Physiol.* **288**, L709–L717.

Scalia, R., Gefen, J., Petasis, N. A., Serhan, C. N., and Lefer, A. M. (1997). Lipoxin A4 stable analogs inhibit leukocyte rolling and adherence in the rat mesenteric microvasculature: Role of P-selectin. *Proc. Natl. Acad. Sci. USA* **94**, 9967–9972.

Sellamuthu, R., Umbright, C., Li, S., Kashon, M., and Joseph, P. (2011a). Mechanisms of crystalline silica-induced pulmonary toxicity revealed by global gene expression profiling. *Inhal. Toxicol.* **23**, 927–937.

Sellamuthu, R., Umbright, C., Roberts, J. R., Chapman, R., Young, S. H., Richardson, D., Cumpston, J. et al. (2012). Transcriptomics analysis of lungs and peripheral blood of crystalline silica-exposed rats. *Inhal. Toxicol.* **24**(9), 570–579.

Sellamuthu, R., Umbright, C., Roberts, J. R., Chapman, R., Young, S. H., Richardson, D., Leonard, H. et al. (2011b). Blood gene expression profiling detects silica exposure and toxicity. *Toxicol. Sci.* **122**, 253–264.

Sellamuthu, R., Umbright, C., Roberts, J. R., Cumpston, A., McKinney, W., Chen, B. T., Frazer, D., Li, S., Kashon, M., and Joseph, P. (2013a). Molecular insights into the progression of crystalline silica-induced pulmonary toxicity in rats. *J. Appl. Toxicol.* **33**(4), 301–312. doi: 10.1002/jat.2733.

Sellamuthu, R., Umbright, C., Roberts, J. R., Young, S. H., Richardson, D., McKinney, W., Chen, B. et al. (2013b). Pulmonary toxicity and global gene expression profile in response to crystalline silica exposure in rats. *Society of Toxicology Annual Meeting*, San Antonio, TX, March 10–14, 2013.

Shi, X., Mao, Y., Daniel, L. N., Saffiotti, U., Dalal, N. S., and Vallyathan, V. (1994). Silica radical induced DNA damage and lipid peroxidation. *Environ. Health Perspect.* **102**(Suppl 10), 149–154.

Srivastava, K. D., Rom, W. N., Jagirdar, J., Yie, T. A., Gordon, T., and Tchou-Wong, K. M. (2002). Crucial role of interleukin-1beta and nitric oxide synthase in silica-induced inflammation and apoptosis in mice. *Am. J. Respir. Crit. Care Med.* **165**, 527–533.

Stadler, Z. K. and Come, S. E. (2009). Review of gene-expression profiling and its clinical use in breast cancer. *Crit. Rev. Oncol. Hematol.* **69**, 1–11.

Steenland, K. and Brown, D. (1995). Silicosis among gold miners: Exposure—Response analyses and risk assessment. *Am. J. Public Health* **85**, 1372–1377.

Steenland, K. and Sanderson, W. (2001). Lung cancer among industrial sand workers exposed to crystalline silica. *Am. J. Epidemiol.* **153**, 695–703.

Takahashi, F., Takahashi, K., Okazaki, T., Maeda, K., Ienaga, H., Maeda, M., Kon, S., Uede, T., and Fukuchi, Y. (2001). Role of osteopontin in the pathogenesis of bleomycin-induced pulmonary fibrosis. *Am. J. Respir. Cell. Mol. Biol.* **24**, 264–271.

Umbright, C., Sellamuthu, R., Li, S., Kashon, M., Luster, M., and Joseph, P. (2010). Blood gene expression markers to detect and distinguish target organ toxicity. *Mol. Cell. Biochem.* **335**, 223–234.

Vallyathan, V., Castranova, V., Pack, D., Leonard, S., Shumaker, J., Hubbs, A. F., Shoemaker, D. A. et al. (1995). Freshly fractured quartz inhalation leads to enhanced lung injury and inflammation. Potential role of free radicals. *Am. J. Respir. Crit. Care Med.* **152**, 1003–1009.

Vallyathan, V., Mega, J. F., Shi, X., and Dalal, N. S. (1992). Enhanced generation of free radicals from phagocytes induced by mineral dusts. *Am. J. Respir. Cell Mol. Biol.* **6**, 404–413.

Vallyathan, V. and Shi, X. (1997). The role of oxygen free radicals in occupational and environmental lung diseases. *Environ. Health Perspect.* **105**(Suppl 1), 165–177.

Warheit, D. B., McHugh, T. A., and Hartsky, M. A. (1995). Differential pulmonary responses in rats inhaling crystalline, colloidal or amorphous silica dusts. *Scand. J. Work Environ. Health* **21**(Suppl 2), 19–21.

Waring, J. F., Ciurlionis, R., Jolly, R. A., Heindel, M., and Ulrich, R. G. (2001). Microarray analysis of hepatotoxins in vitro reveals a correlation between gene expression profiles and mechanisms of toxicity. *Toxicol. Lett.* **120**, 359–368.

Wetmore, B. A., Brees, D. J., Singh, R., Watkins, P. B., Andersen, M. E., Loy, J., and Thomas, R. S. (2010). Quantitative analyses and transcriptomic profiling of circulating messenger RNAs as biomarkers of rat liver injury. *Hepatology* **51**, 2127–2139.

Whicher, J., Biasucci, L., and Rifai, N. (1999). Inflammation, the acute phase response and atherosclerosis. *Clin. Chem. Lab. Med.* **37**, 495–503.

Williams, A. O., Knapton, A. D., Ifon, E. T., and Saffiotti, U. (1996). Transforming growth factor beta expression and transformation of rat lung epithelial cells by crystalline silica (quartz). *Int. J. Cancer* **65**, 639–649.

Yu, J., Baron, V., Mercola, D., Mustelin, T., and Adamson, E. D. (2007). A network of p73, p53 and Egr1 is required for efficient apoptosis in tumor cells. *Cell Death Differ.* **14**, 436–446.

Zhang, J., Wu, Y., Zhang, Y., Leroith, D., Bernlohr, D. A., and Chen, X. (2008). The role of lipocalin 2 in the regulation of inflammation in adipocytes and macrophages. *Mol. Endocrinol.* **22**, 1416–1426.

19 Mechanisms Involved in the Inhalation Toxicity of Phosgene

Wenli Li and Juergen Pauluhn

CONTENTS

19.1 INTRODUCTION

Phosgene (carbonyl chloride) is manufactured from the reaction of carbon monoxide and chlorine gas in the presence of activated charcoal as a catalyst. It is utilized across numerous industries for legitimate chemical synthetic processes. Wherever it is technically feasible, phosgene is processed using an *on-demand production strategy*, that is, phosgenation reactions are executed at low stationary levels so transportation and storage are abandoned at chemical plants that keep abreast of progress in engineering technology. It is an essential high-production-volume intermediate in the manufacture of building blocks of various types of plastics and of products used in everyday life, including flexible foam in upholstered furniture, rigid foam insulation in walls and roofs, and thermoplastic polyurethane used in medical devices and footwear. Phosgene is a raw material used primarily in the production of other important chemicals, such as methylene diphenyl diisocyanate (MDI) and toluene diisocyanate (TDI). Phosgene is also important in the manufacture of coatings, adhesives, sealants, and elastomers used on floors and automotive interiors, and it is used to make polycarbonate plastics, as well as a wide variety of pharmaceuticals, agricultural chemicals, and specialty chemical intermediates. The global consumption of phosgene was about 5 million metric tons in 2006 (Market Research Estimate of Chemical Industry).

Phosgene is a colorless gas, which may form a white cloud. It has a particular smell of new-mown hay. Because phosgene is four times heavier than air, it tends to remain close to the ground and to collect in low-lying areas. Historically, phosgene and diphosgene were used during World War I in military conflicts. In these conflicts, phosgene was often combined with chlorine in liquid-filled shells, so it is difficult to state the number of casualties and deaths attributable solely to phosgene.

Between the world wars, phosgene was assigned the military designation CG and was classified as a nonpersistent agent because of its rapid evaporation. In military publications, it has been referred to as a choking agent, pulmonary agent, or irritant gas. Although phosgene and chlorine have been superseded by agents that are orders of magnitude more potent (Kluge and Szinicz, 2005; Szinicz, 2005), they have the potential to function as a weapon of mass destruction by any group with simple chemical synthetic capabilities or with the means to sabotage an existing industrial phosgene source.

Several mechanistic concepts have been implicated in the pathogenesis of the acute lung injury (ALI) and acute respiratory distress syndrome (ARDS) typical of phosgene. This syndrome is characterized by an increased permeability of the alveolar–capillary barrier resulting in lung edema with protein-rich fluid and ensuing impairment of arterial oxygenation. ALI/ARDS is defined as a lung disease with acute onset, noncardiac, diffuse bilateral pulmonary infiltrates, and hypoxia (Grommes and Soehnlein, 2011). Although conveyed by the terms *ALI/ARDS*, not all lung irritants are expected to cause injury following the same mechanistic paradigm. This is due to their differences in chemical reactivity (e.g., reactivity toward nucleophilic proteins/peptides or to oxidize antioxidants) and physicochemical properties (e.g., water solubility or capability to interact/partition with lipid layers/membranes), which are key determinants for the location-specific deposition of the irritant gas as well as the location-specific susceptibility within the respiratory tract. Each retention site has its own biochemical and (neuro) physiological response to injury and may require site-specific mitigation measures. The potency of irritant, reactive chemicals, such as ozone, chlorine, nitrogen oxides, or phosgene, to cause pulmonary injury may also depend on species-specific differences in the type and local concentration of extracellular antioxidant substances contained in airway lining fluids (Hatch, 1992; Hatch et al., 1986; Slade et al., 1993). This complexity precludes simple static cytotoxicity in vitro assays to deliver any meaningful information on mechanism-based treatment strategies.

With regard to the primary toxic mode of action, phosgene gas elicits pulmonary irritation that produces noncardiogenic pulmonary edema without the involvement of any substantial airway irritation and inflammation. Despite extensive research on pharmacological countermeasures to mitigate phosgene-induced acute lung edemagenesis, most approaches have shown limited benefits in experimental models simulating clinical usage. In essence, oxygen treatment and mechanical ventilation have received most attention in the past and proved to be efficient in mitigating anoxic anoxia, the most pressing sequel of the life-threatening acute pulmonary edema. There is also a lack of understanding as to how the cardiopulmonary system and systemic circulation respond interdependently to the phosgene-induced pulmonary injury (Gibbon et al., 1948; Grainge and Rice, 2010). The focus of this chapter is to present putative biochemical and physiological mechanisms involved in the etiopathology of the phosgene-induced acute lung edema and whether current treatment strategies need to be reconsidered based on new experimental evidence from controlled rodent inhalation bioassays.

19.2 PHOSGENE AND PHOSGENE CONGENER

19.2.1 PHYSICOCHEMICAL INFORMATION AND CHEMICAL REACTIVITY

The physicochemical characteristics at room temperature of phosgene (gas), diphosgene (liquid), and triphosgene (solid) are summarized in Table 19.1. More comprehensive data of phosgene are compiled in AEGL (2002; IPCS, 1998). The chemistry of these phosgene congeners has been reviewed in detail elsewhere (Cotarca et al., 1996; Cotarca and Eckert, 2003; Pauluhn, 2011).

The analytical distinction of phosgene, diphosgene, and triphosgene is complex. Analytical methods commonly use analytical procedures for the determination of phosgene (Hendricks, 1986) rather than the determination of the intact molecule of the congeners of phosgene. More recent published evidence on the inhalation toxicity of di- and triphosgene demonstrates that these congeners are not

TABLE 19.1
Physical Characteristics

Substance	Formula	Molecular Weight[a] (g/mol)	Melting/Boiling Points[a] (°C)	Vapor Saturation Concentration (mg/m³)	Conversion Factor 1 ppm = x mg/m³
Phosgene[a] (carbonyl chloride) Gas, insoluble in water	O‖ Cl Cl	99	−128/7.4	6.4×10^6 (0.16 MPa at 20°C)	4.1
Diphosgene[a] (trichloromethyl chloroformate) Liquid, insoluble in water	O‖ Cl₃CO Cl	198	−57/128	111,000 (1.3 kPa at 20°C); decomposition at 300°C	8.1
Triphosgene[a] (bis[trichloromethyl] carbonate) Solid crystals/high sublimation potential, insoluble in water	O‖ Cl₃CO OCCl₃	297	80/206	2437 (0.15 Torr at room temperature)	12.2

[a] Physical constants of phosgene, diphosgene, and triphosgene are reported by Cotarca and Eckert (2003).

decomposed spontaneously to phosgene, as often surmised; rather, they act as intact molecules in their own way (Pauluhn, 2011).

Apart from the analytical challenges to differentiate the congeners of phosgene from phosgene itself, this issue is complicated further by the fact that triphosgene is present in two phases, namely, as gas and as particulate. Hence, sampling devices optimized for the gas phase of phosgene may not necessarily collect triphosgene with high collection efficiency. The discoloration of phosgene indicator badges (dosimeters) has often been taken as indirect evidence that instant decomposition to phosgene occurs (Damle, 1993). Of note is that colorimetric changes of such dosimeters are not specific to phosgene. Accordingly, based on the physical and toxicological characteristics relative to phosgene, such colorimetric changes cannot be related to any specific exposure intensity to diphosgene or triphosgene. Thus, phosgene badges may also be suitable to show the presence/absence of congeners of phosgene in a dichotomous, nonquantitative manner; however, more quantitative approaches as published for phosgene (Niessner, 2010) are absent. The lack of any specific analytical method in an environment where triphosgene and diphosgene may coexist with phosgene is considered to be a distinctive disadvantage of these phosgene precursors (Cotarca et al., 1996). Apart from real-time FT-IR (Fourier Transform Infrared Spectroscopy) (Pauluhn, 2011), there is no specific analytical method to quantify airborne triphosgene in the presence of phosgene.

19.2.2 Comparative Acute Inhalation Lethal Toxic Potency

More recent publications convey the message that phosgene-free phosgenation methods using the solid congener of phosgene bis(trichloromethyl) carbonate (triphosgene) may reduce the health risks

TABLE 19.2

Indices of Acute Inhalation Toxicity of Phosgene and Di- and Triphosgene Gas/Vapor Phase of Rats Nose-Only Exposed for 240 min

Substance	Gender	LC_{50} (mg/m³)	(ppm)	LC_{50} (mmol)	LC_{01} (mg/m³)	(ppm)	LC_{50}/LC_{01} Ratio
Phosgene[a]	Combined	7.2	1.8	0.07	4.5	1.1	1.6
Diphosgene	Males	13.9	1.7	0.07	8.4	1.0	1.7
Triphosgene	Combined	41.5	3.4	0.14	24.1	2.0	1.7

Source: Pauluhn, J., *Inhal. Toxicol.*, 18, 423, 2006a.

[a] LC_{50} recalculated from studies combining different C × t products (for details, see Pauluhn, 2006a). The recalculated LC_{50} was adjusted to 240 min.

potentially associated with phosgenation processes (Cotarca, 1999). This conclusion is based on the fact that triphosgene is asserted to be safer and more convenient to handle, transport, and store because it is a solid with minimal vapor pressure at ambient temperature. It has also been stated that "The toxicity of both diphosgene and triphosgene is exactly the same as phosgene since both dissociate/depolymerize to phosgene on heating and upon reaction with any nucleophile" (Cotarca and Eckert, 2003). This may be a valid statement in context with specialized chemical reactions; however, it may not reflect those reactions taking place within the lung.

The key attributes of the acute inhalation toxicity of phosgene, diphosgene, and triphosgene using up-to-date inhalation exposure methodologies have recently been published (Pauluhn, 2006a, 2011; Pauluhn and Mohr, 2000). Gas- or vapor-phase-exposed rats showed somewhat similar mortality patterns following exposure to phosgene and diphosgene, while triphosgene caused a biphasic mortality pattern, which is a typical finding of irritant gases causing injury in both the lower respiratory tract and the airways. The respective 4 h LC_{50} data and the time-adjusted LCt_{50} data are summarized in Table 19.2.

The findings obtained with triphosgene support a conclusion that triphosgene reacts within the lung instantly as an intact molecule and not via its putative degradation products, which would have increased rather than decreased its molar acute toxic potency. The mode of action of triphosgene appears to be different from that of diphosgene and phosgene due to its property to elicit more persistence signs of respiratory distress, increased necropsy findings suggestive of prolonged/persistent lung injury, and a unique additional delayed mortality pattern typical of central airway injury (*bronchiolitis obliterans*). Under the conditions of tests, triphosgene vapor atmospheres were generated (approximately 40°C followed by instant in situ dilution with dry air). Dilution ratios of 1:50 were still in the lethal range. Experimental evidence supports the notion that triphosgene crystals appear to contain more volatile/more toxic yet ill-defined by-products. Hence, due to its tendency to sublimate and recrystallize the presence of yet unknown more volatile impurities and the absence of any robust analytical method to quantify airborne triphosgene (and by-products), the hazards and risks associated with triphosgene appear to be markedly more complex and difficult to characterize as compared to phosgene. Although the solid physical state of triphosgene conveys the erroneous and equally insidious perception of insignificant exposure, its principal mode of action differs from that of phosgene.

Despite the analytical challenges to characterize the parent molecule, the data presented in Table 19.2 demonstrate unequivocally that diphosgene and triphosgene are both stable in air and do not decompose spontaneously to phosgene in the lung. Consistent with the range of nucleophilic substitution reactions shown by Cotarca et al. (1996), this reactivity appears to cause instant acylation of nucleophiles by the intact molecule. Injury patterns appear to be contingent on the site(s) of initial retention and type of chemical reaction. The toxicity of phosgene gas is essentially due to its capability to acylate nucleophilic moieties. Similarly, its lipophilicity prevents rapid hydrolysis from taking place in the upper airways (Nash and Pattle, 1971). The acylation is most important and results from the reaction of phosgene with nucleophiles such as amino, hydroxyl, and sulfhydryl groups of macromolecules with denaturation of lipids and proteins, irreversible membrane changes, and disruption of enzymatic functions. Phosgene depletes lung glutathione (GSH), and GSH reductase and superoxide dismutase increase as a result of the lung's response to injury. Cellular glycolysis and oxygen uptake are decreased following exposure to phosgene, and there is a decrease of intracellular ATP and cyclic AMP associated with increased permeability of pulmonary vessels and pulmonary edema.

While phosgene and diphosgene show similar lethal toxic potency, the cause of the seemingly lower molar toxicity of triphosgene appears to be consistent with its biphasic mortality patterns suggesting that the retention of vapor is not restricted to the pulmonary region alone. In other words, some triphosgene may also react within the bronchial airways causing different injury patterns. Minute differences in the physicochemical properties of these congeners of phosgene may change their primary site of initial deposition within the lung; and due to the inhomogeneous distribution and abundance of nucleophilic scavengers in the lining fluids or cell surfaces, the manifestations of toxicity may differ from one congener to the other. Similar biphasic mortality patterns have been observed following single exposures to chlorine (AEGL, 2002; Gary et al., 2010; Zwart et al., 1988).

The armamentarium of currently available analytical techniques needs to be improved to better characterize both triphosgene per se and its putative, yet ill-defined, more volatile by-products. Notably, the *pure* triphosgene used in the reported study contained off-gassing component(s) reactive with the 2-HMP scavenger. This circumstance caused high initial concentrations of ill-defined toxic gases. It cannot be excluded that these off-gassing by-products would have increased the acute lethal toxic potency of triphosgene due to the preconditioning of the test article prior to the exposure of rats. Due to its availability and ease of handling, many research-based inhalation studies with phosgene utilized triphosgene in the absence of methodologies demonstrating unequivocally that complete decomposition of triphosgene to phosgene occurred. Caution is advised to attribute triphosgene-specific long-lasting effects and delayed mortality to phosgene as long this uncertainty cannot be resolved.

19.3   EXPERIMENTAL MODELS

19.3.1   Acute Inhalation Toxicity and Haber's Rule

Numerous studies were designed in large part to determine the mechanism involved in the late clinical manifestation of pulmonary edema and associated late onset of lethality following acute exposure to phosgene. It is believed that the better understanding of the mechanisms determining the concentration- and concentration × exposure duration (C × t)-dependent outcomes, including the involved temporal changes, contribute to improved diagnostic and prognostic tools for designing more efficacious treatments at the yet asymptomatic stage of intoxication. While there is a preponderance of published literature that describes multiple biochemical pathways and effects subsequent to phosgene inhalation exposure, a generally accepted conceptualization of the sequence of the most crucial physiological events involved is still pending.

Exposure to phosgene may progress into a life-threatening pulmonary edema within 1 day post-exposure in both humans and animals (Borak and Diller, 2000; Diller, 1978; Sciuto, 2005). Mechanisms involved in the latency period for pulmonary edema have been studied extensively during the past decades in numerous animal models (Duniho et al., 2002; Pauluhn et al., 2007; Sciuto, 2005). All animal models have in common that the onset of lethality is inversely related to the C × t relationship. This C × t relationship has been shown to be valid for both the indicator of pulmonary edema *total protein in bronchoalveolar lavage (BAL protein) fluid* and lethality (Figure 19.1). The data shown for BAL protein were obtained in rats from exposure durations of 30 to 360 min and were from two different laboratories. The high equivalence of data demonstrates

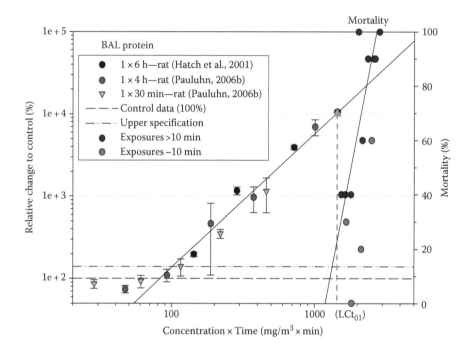

FIGURE 19.1 **(See color insert.)** Comparison of BALF protein and mortality in phosgene-exposed rats at different C × t relationships. BAL-protein data were from whole-body-exposed Fischer 344 rats (Hatch et al., 2001) or nose-only-exposed Wistar rats (Pauluhn, 2006b) to multiple concentrations of phosgene bracketing exposure durations from 30 min to 6 h. Rats were sacrificed approximately 24 h following inhalation exposure. Data represent group means ± SDs. (Data from Pauluhn, J., *Inhal. Toxicol.*, 18, 609, 2006c.) Mortality data were from rat inhalation studies utilizing exposure durations from 30 to 360 min. (From Pauluhn, J., *Inhal. Toxicol.*, 18, 423, 2006a.)

that laboratory-specific exposure modes and procedures have minimal impact when using actual, that is, analytically determined, exposure concentrations (atmosphere collected from the vicinity of the breathing zone of the exposed animals).

Both endpoints reached their climax approximately 1 day postexposure. Mortality occurring beyond 1 day was described as a very rare event (Pauluhn, 2006a). This appears to be a contradiction to some published evidence. However, many research-based inhalation studies used triphosgene to generate phosgene, in the absence of any verifying analytical methodology. Based on the results obtained with triphosgene (see previous section), methodological differences in the generation and characterization of exposure atmospheres need to be appreciated.

Lethality data from 10 min exposures required seemingly higher C × t's to cause similar levels of mortality. Respiratory function measurements during the exposure to phosgene revealed a concentration-dependent, transient depression of the respiratory minute volume (Pauluhn, 2006a). Therefore, especially in rodent species capable of depressing their ventilation as a result of stimulation of sensory nerve endings, Haber's rule does not necessarily apply for very high concentrations at short exposure durations. The relationship of equal C × t products obtained under differing exposure conditions is depicted in Figure 19.2. It shows that reliable comparisons of C × t relationships in rats require a minimal duration of exposure of 30 min. The directed-flow nose-only inhalation chambers used (for details, see Pauluhn, 2006a; Pauluhn and Thiel, 2007) were preequilibrated, which means exposure dose-related phenomena can solely be attributed to ventilation rather than the attainment of inhalation chamber steady-state concentrations.

The comparison made in Figure 19.2 underscores the importance of physiological measurements contemporaneous with inhalation exposure to phosgene to understand and appreciate

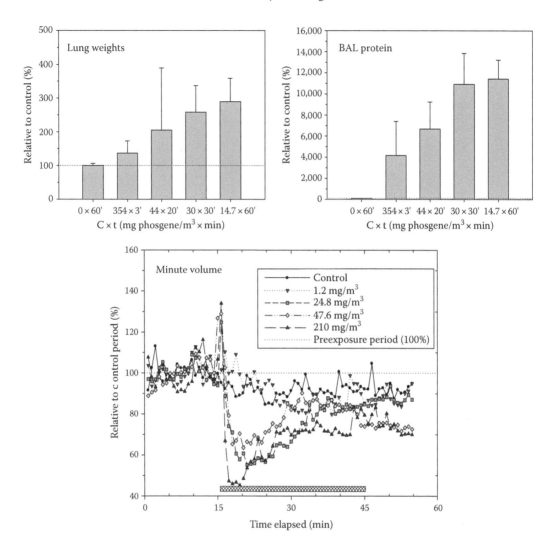

FIGURE 19.2 (**See color insert.**) Wet lung weights and total protein concentrations in BAL from rats exposed for 1 h to air only (control), 354, 44, 30, and 14.7 mg phosgene/m³ at nose-only exposure durations of 3, 20, 30, and 60 min. Rats were sacrificed on postexposure day 1. Data represent group means ± SD (n = 6–8). Data were normalized to the conditioned air nose-only-exposed control group (= 100%). The average respiratory minute volume of 8 rats/group exposed for 30 or 10 min (210 mg/m³) to phosgene is shown in the lower panel. Each exposure was preceded by a 15 min exposure period to air. Data were averaged for time periods of 45 s. Measurements utilized volume-displacement nose-only plethysmographs. (Data modified from Li, W.-L. et al., *Inhal. Toxicol.*, 23, 842, 2011; Pauluhn, J., *Inhal. Toxicol.*, 18, 423, 2006a.)

species-specific differences in ventilation and dosimetry. The ventilation of phosgene-exposed rats depends on complex changes in respiratory rate and tidal volume (TV) as detailed elsewhere (Pauluhn, 2006a). The comparison of increased lung weights and BAL protein with the transient depression in ventilation at continued exposure to phosgene demonstrates the complex interrelationship of the actually inhaled phosgene dose and the ensuing effect. It also demonstrates that exposure durations shorter than 20–30 min duration are biased to generate false-negative findings. In larger species, such as the dog, breath-holding periods are reported to occur at high phosgene concentrations (NDRC, 1946). Collectively, the data summarized in Figure 19.2 demonstrate that the experimental conditions and exposure regimens of phosgene inhalation studies must be accounted for and with an in-depth understanding of the limitations of each experimental approach taken.

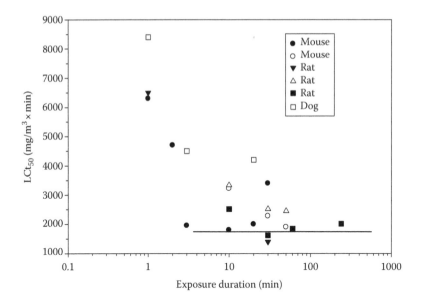

FIGURE 19.3 Dependence of the exposure duration (time)-adjusted median lethal concentration (LCt_{50}) on the exposure duration. This illustration supports the notion that exposure durations <30 min should be judged cautiously due to experimental or animal-related artifacts (i.e., depression of ventilation). Data were from selected studies on mice, rats and dogs using exposure modes with apparent rapid attainment of steady-state.

The literature is replete with studies of short-duration phosgene exposure intended to simulate accidental exposures. These studies show much greater variability at short durations of exposure (<30 min) and, therefore, must be interpreted cautiously especially from a dosimetric perspective (Figure 19.3). Dosimetry variability may stem from four sources: (1) incomplete attainment of steady state within the phosgene delivery system, (2) mismatch of nominal and actual exposure concentrations, (3) irritation-related changes in breathing patterns and ventilation (reflex bradypnea), and (4) susceptibility to stress (perception)-related changes in ventilation as often experienced in larger animal species. The pulmonary toxicity of phosgene gas depends on its retained dose in the lower airways. The major mechanism of retention is dependent on diffusion and the respiratory minute volume (which is the product of respiratory rate and TV). Any reflexively related change in this relationship may affect the retained dose of phosgene at its target site in the lung.

Apart from reflex-induced changes in respiration, exposure methodologies, the purity of phosgene (presence of sensory irritant HCl as an impurity or degradation product), and any resultant psychological stresses can alter ventilation and thus the inhaled toxic dose. In summary, it seems that most reported species-specific differences are related to differences in the species-specific mode of breathing or capability to elicit breath-holding periods, which occur at lethal concentration (high) × time (short) relationships. Accordingly, changes in the dosimetric uptake of inhaled phosgene make extrapolations from short-term to longer-term exposure (and vice versa) error prone and therefore call for experimental data from exposure durations at which stable ventilation can be assumed in order to validate or refute the assumptions made for any $C^n \times t$–dependent effects of phosgene (Pauluhn, 2006a; Pauluhn et al., 2007).

There are few state-of-the-art studies addressing the systematic analysis of $C^n \times t = k$ relationships of interest for emergency response planning. Most studies focused on the C × t relationship of lethal endpoints (Pauluhn, 2006a; Zwart et al., 1990). When fitting the empirical data from the entire concentration–mortality curve of 10, 30, 60, and 240 min exposure periods, the *toxic load exponent (n)* was n = 0.9–1.1, depending on the mathematical procedure applied (Figure 19.4). As anticipated, the LC_{50} at 10 min was overpredicted relative to the empirical data when giving more weight to the more consistent results at longer exposure durations (see also Pauluhn, 2006a; Pauluhn et al., 2007).

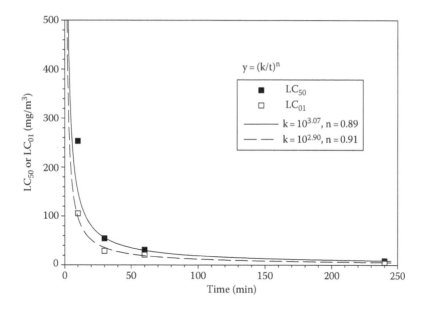

FIGURE 19.4 Hyperbolic relationship of the mathematically derived median lethal concentration (LC_{50}) and the nonlethal threshold concentration (LC_{01}).

This analysis supports a concept that empirically determined short-term exposure values for phosgene may be markedly confounded by concentration-dependent decreased ventilation and inhaled dose. This interpretation is supported by Haber's rule (Flury, 1921; Haber, 1924), which is focused on two factors—the *concentration in inspired air* (C) and the *duration* (t) during which this concentration is inhaled—to yield an identical intensity of biological response (Witschi, 1997, 1999). This rule implicitly assumes that the actual minute volume inhaled by the animal during the exposure period (t) is constant. This negligence and oversimplification is likely why many empirical data obtained with irritant hazardous substances fail to comply with Haber's rule, which has a *toxic load exponent* of n = 1. On the whole, it appears as if this exponent corrects mainly for the substance-specific changes in rodent-specific ventilation and inhaled dose accordingly.

19.3.2 ACUTE INHALATION TOXICITY AND PATHOPHYSIOLOGY

Mechanistic inhalation studies with phosgene in rats were targeted to be in the range of the LCt_{01} (see Figure 19.1) using a C × t product of 1050 mg/m^3 × min ($\approx$35 mg/m^3 × 30 min). The comparison of time-course changes of wet lung weights and BAL protein from a range of postexposure tissue-collection time points demonstrates that both endpoints reach the maximum on the first postexposure day (for further details, see Pauluhn, 2006b) and are useful for C × t dose–response analyses (Figures 19.5 and 19.6). The sensitivity of these edema-related endpoints differs markedly as suggested by the threefold increase in lung weights at a 100-fold increase in BAL protein. In addition to increased lung weights and BAL protein, lung edema formation is paralleled by a proportionally increased *enhanced pause* (Penh) as illustrated in Figure 19.6. This is believed to be caused by elevated lung water content and associated changes in the elastic recoil of lung tissue. At 882 mg/m^3 × min, the C × t of phosgene seems to be high enough to increasingly damage alveolar macrophages (Figure 19.6).

Early (NDRC, 1946) as well as recent inhalation studies on rats with phosgene (Pauluhn, 2006a) demonstrate that phosgene stimulates vagal C-fibers, which represent the majority of vagal afferents innervating the lower airways, with subsequent effects on the control of spontaneous breathing. The most characteristic change of such effect is typified by a reflexively induced prolonged apnea period or *breath-holding period*. In rodents, the stimulation of this reflex results in a transient reduction of the respiratory minute ventilation volume and increased apnea time (AT). Likewise, dogs exposed

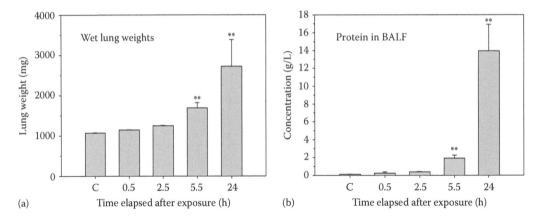

FIGURE 19.5 Time-course changes of wet lung weights (a) and protein in BALF (b) of rats exposed 1050 mg phosgene/m³ × min (30 min exposure duration). Air-exposed control rats (C) were analyzed 24 h postexposure. Rats were sacrificed on postexposure day 1. Asterisks denote statistical significance to nose-only air-exposed control (**P < .01). Data represent group means ± SD (n = 8).

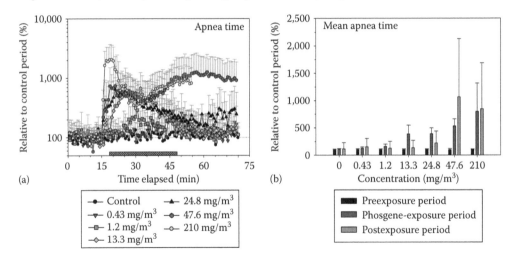

FIGURE 19.6 **(See color insert.)** Average duration of apnea periods (changes in apnea time relative to exposure duration (a) and the mean apnea times during the respective preexposure, exposure, and postexposure (b) segment) between individual breathing cycles of rats exposed nose-only to phosgene in volume-displacement plethysmographs. All data were normalized to the pre–phosgene exposure data collection period of 15 min (baseline data = 100%). With the exception of the 210 mg/m³ group (exposure duration 10 min), all rats were exposed to phosgene for 30 min (hashed bar), followed by an air-only exposure period of 30 min. The averaged values for each time period are given in the bar chart. Data represent means ± SD (n = 4). (Data modified from Pauluhn, J., *Inhal. Toxicol.*, 18, 423, 2006a.)

at concentrations higher than 7000 mg/m³ showed instant cessation of respiration with the lungs in the deflated phase. When the vagus nerve is cut prior to exposure, no reflex inhibition of breathing was observed (NDRC, 1946). The nociceptive role of these reflexes is to sense the onset of any pathophysiological condition within the lower respiratory tract. This also implies that these C-fiber nerve endings are likely to be dosed more intensively than any other tissue. Likewise, the afferent activity arising from these vagal nerve fibers appears to play an important role in regulating the cardiopulmonary function under both normal and abnormal physiological conditions.

The occurrence and prolongation of apnea periods can be measured by the composite lung function endpoint *enhanced pause* (Penh). As detailed elsewhere, Penh is likely more related to changes in breathing control whether caused by stimulation of receptors or changes in the mechanical properties of lung parenchyma (Pauluhn, 2004). Opposite to measurements of respiratory rate and TV in volume-displacement plethysmographs (see Figure 19.2), which require restraint, the composite endpoint, Penh is proposed to be a more attractive endpoint utilizing whole-body plethysmography without any restraint or complex interventions (Hamelmann et al., 1997). Only box pressure is measured and the dimensionless parameter, Penh (among other breathing parameters), is derived from the breathing pattern and the breath structure measured as the box pressure decay during expiration and the ratio of inspiratory and expiratory pressure maxima during tidal breathing. Several observations led to the consideration of Penh as a surrogate of pulmonary or even airway resistance, although the transpulmonary pressure, the driving force of breathing and subsequent flow and volume changes, is not measured in barometric plethysmographs. Several factors can affect the various parameters involved in the calculation of Penh, and its indiscriminate use is subject to misleading interpretations of response. As almost all respiratory mechanics variables show qualitative correlations (Mitzner et al., 2003; Mitzner and Tankersley, 1998), it appears to be difficult to causally relate changes in Penh to any specific pathophysiological event, such as airway constriction. It is also important to recall that nasal airway resistance is the largest component of the total airway resistance in obligate nasal breathing rodents (Lung, 1987).

Measurement of Penh by unrestrained plethysmography does not provide a direct assessment of any specific physiologic variable. Penh is derived during spontaneous tidal breathing from the dimensionless relationship combining peak inspiratory (PIF) and expiratory (PEF) flows, the expiratory time (ET), and relaxation time (Rt; defined as time to expire 65% of the inhaled volume) as follows (Hamelmann et al., 1997):

$$\text{Penh} = \frac{\text{PEF}}{\text{PIF}} \times \left(\frac{\text{ET}}{\text{Rt}} - 1 \right)$$

Rt is commonly defined as the time from start of expiration to the return to 35% of TV. This is the time it takes the animal to expire 65% of its TV from the start of expiration. More details are described elsewhere (Drorbough and Fenn, 1955; Epstein and Epstein 1978; Hamelmann et al., 1997; Kimmel et al., 2002). The *pause* is given by ET/Rt − 1. In phosgene-exposed rats, this pause has been shown to occur between end of expiration and inspiration of the next breath (Pauluhn, 2006a) and is a typical characteristic of receptor stimulation in the lower respiratory tract (likely to be caused by stimulation of juxta-alveolar J-reflexes; Paintal, 1969, 1981; Lee and Widdicombe, 2001). These *apnea periods* are typical of pulmonary (alveolar) vagal C-fiber stimulation. It seems appropriate to assume that any change in pulmonary parenchymal compliance due to increased plasma exudation may be reflected best by prolonged AT and Penh accordingly (Li and Pauluhn, 2010; Pauluhn, 2004, 2006a). If the elastic resistance of the lung increases due to disease, the respiratory pattern is altered accordingly to minimize the mechanical work to overcome increased elastic tissue resistance. However, caution is advised to attribute such changes in breathing control to specific pathophysiological effects. This interpretation is not at variance with other authors (Hantos and Brusasco, 2002). Also Peták et al. (2001) concluded that measurements of the airway and tissue mechanical properties during hyperoxia (mice) are not reflected by those in Penh, and the relationships with other plethysmographic indexes are not easily interpretable either. Despite these shortcomings, Penh appears to integrate several physiological endpoints in a wholly noninvasive and nondisturbing manner so that nonspecific functional changes can readily be identified in studies where incremental rather than absolute changes are the primary focus. However, caution is advised in linking these changes to any specific pathophysiological effects.

While the AT of rats exposed at 25 mg/m³ × 30 min (750 mg/m³ × min) showed evidence of recovery after cessation of exposure to phosgene, C × t relationships at and above 48 mg/m³ × 30 min (1440 mg/m³ × min) remained elevated. These C × t relationships to vagal C-fiber stimulation are in remarkable accord to the time-adjusted median lethal concentration (LCt_{50}) and nonlethal threshold concentration (LCt_{01}) of 1741 and 1075 mg/m³ × min, respectively (Figure 19.1). The prolonged apnea period (Figures 19.6 and 19.8) and bradycardia (Figure 19.9) persisted for more than 20 h post–phosgene exposure, despite unequivocal evidence of a time-dependently increased pulmonary edema (Figure 19.5) and hemoconcentration (Figure 19.10).

Rat studies have shown a clear association of increased Penh with increased lung water content, evidenced by increased lung weights and BAL protein (Figure 19.7). There appears to be no clear association of changes in total BAL cell counts with those suggestive of adversely affected fluid dynamics of the lung. Time-course analyses of Penh following acute exposure to variable concentrations of phosgene are depicted in Figures 19.7 and 18.8. The changes in breathing patterns are taken as indirect evidence of both C- and C × t–dependent mechanisms. The concentration-dependent response to injury is reflected by increased Penh and occurs concurrent to the exposure to phosgene with increasing C × t (Figure 19.8). These findings support the conclusion that the most critical events appear to occur during the preedema manifestation phase, which is consistent with dysfunctional neurogenic factors that control cardiopulmonary and vascular functions.

The sustained depression cardiopulmonary activity appears to portend that a dysregulated, continued neurogenic overstimulation of pulmonary C-fibers is an essential contributing factor for the lethal acute lung edema to occur (Li et al., 2011). Of note is that the transition from reversibility to persistence occurs just at the C × t threshold dose of phosgene that causes frank edemagenesis and

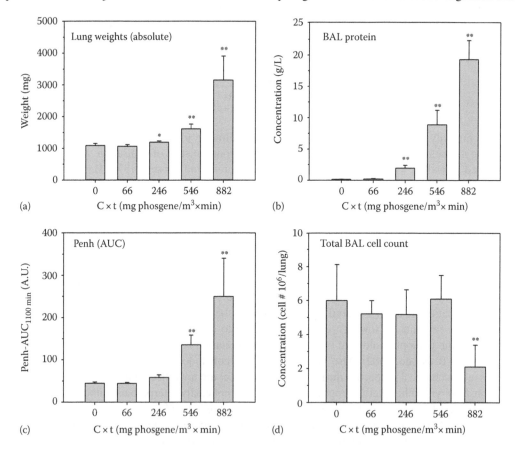

FIGURE 19.7 Comparison of lung weights (a), protein in BALF (b), Penh (AU) (c), and Total Cell counts in BALF (d) of rats exposed for 1 h to 0, 66, 246, 546, and 882 phosgene/m³. Rats were sacrificed on postexposure day. Asterisks denote statistical significance to nose-only air-exposed control (*P < .05, **P < .01). Data represent group means ± SD (n = 8). (Modified from Li, W.-L. et al., *Inhal. Toxicol.*, 23, 842, 2011.)

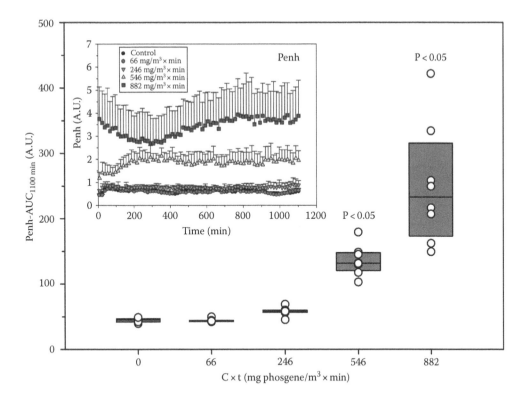

FIGURE 19.8 Measurement of *enhanced pause* (Penh) in barometric plethysmographs of rats exposed for 1 h to 0, 66, 246, 546, and 882 mg phosgene/m^3. Measurements over 20 h started shortly after the phosgene exposure period ceased. Data represent means ± SD of eight rats per group (data partially reproduced from Li et al., 2011). Box–Whisker plots showing individual data points, with boxes (25th to 75th percentile of the data) as well as their median values. Symbols indicate the area under the curve from eight rats/group for the entire data collection period of 20 h (see insert). (Modified from Li, W.-L. et al., *Inhal. Toxicol.*, 23, 842, 2011.)

associated lethality. Hence, continued stimulation of vagal C-fibers appears to make C-fibers unresponsive to overriding autonomic control. Consequently, sympathetic nervous failure and vascular dynamic disturbances both in the lung and systemic circulation are a likely outcome. Likewise, pulmonary vagal reflex stimulation, apnea, bradycardia, and cholinergic symptoms have been described in context with the pathogenesis of phosgene-induced ALI (Bruner et al., 1948a; Diller, 1985a). This stimulatory response to phosgene gas occurs instantly with the onset of exposure and includes both dose rate (i.e., concentration) and C × t–dependent components (Figures 19.6, 19.8, and 19.9). The competitive interaction of factors stimulating concentration-dependent vagal C-fibers is evidently counterbalanced by C × t–dependent inactivation of phosgene. Inactivation of retained phosgene could be attributed to nucleophiles contained in surfactant. The overall outcome is an instant neurogenic stimulation, the magnitude of which depends on the degree to which detoxifying factors are still available. If exhausted, neurogenic stimulation becomes persistent (Li et al., 2011).

Along with the depressed respiration and heart rate, body temperatures were also decreased (Figure 19.10). While normal control rats displayed typical circadian rhythms in body temperature, the rats exposed to phosgene showed a statistically significant, persistent hypothermia unresponsive to circadian rhythms. Due to their small thermal inertia, unlike humans, small laboratory rodents undergo a hypothermic response known as reflex bradypnea when exposed to respiratory tract irritants. Also for other respiratory tract irritants, hypothermia coincided with a reduction in reflexively related ventilation, heart rate, and metabolism that may minimize exposure and toxicity (Gordon, 1990, 1993; Gordon et al., 1988, 2008; Watkinson and Gordon, 1993).

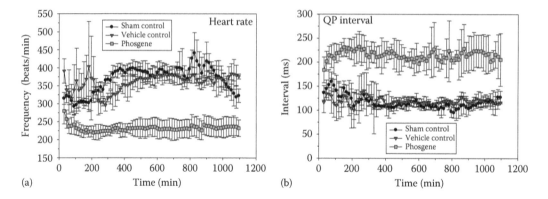

(a) Time (min)   (b) Time (min)

FIGURE 19.9 Time-course analysis of heart rate (a) and QP interval (b) by telemetry in rats exposed to 1050 mg phosgene/m³ × min. Approximately 20 h of data collection began shortly after phosgene exposure ceased. Control groups were either not exposed or received saline by intraperitoneal injection. Data represent the means ± SD (n = 4).

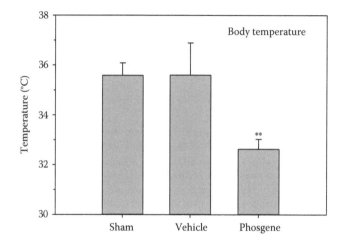

FIGURE 19.10 Body temperature of rats exposed to 1050 mg phosgene/m³ × min relative to nonexposed controls. Data represent the means ± SD (n = 4) over 20 h post–phosgene exposure and were recorded by intraperitoneally implanted telemetry transmitters. Asterisks denote statistical significance to nose-only air-exposed control (**P < .01).

These species-specific differences need to be appreciated when using small-rodent animal models; otherwise, physiological changes secondary to hypothermia can be inappropriately considered adverse. Due to their small thermal inertia, mice are markedly more thermolabile than rats.

Collectively, the combination of prolonged changes in breathing patterns suggestive of vagal involvement (Figure 19.8), the persistent sinus bradycardia after cessation of exposure to phosgene (Figure 19.9), and hypothermia (Figure 19.10) may be considered as pathognomonic for heightened vagal activity. These endpoints were remarkably persistent over a post–phosgene exposure period of approximately 20 h despite the unequivocal evidence of progressive changes in hemodynamics (hemoconcentration, Figure 19.11) and increased lung water content evidenced by increased lung weights and protein in BAL fluid (BALF) (Figure 19.6).

Time-course analyses of hemoglobin in blood, extravasated protein in BALF, and lung weights from rats exposed to either air (control) or 1050 mg phosgene/m³ × min were determined as presented in Figures 19.6 and 19.11. For all endpoints examined, a coherent increase occurred 5–6 h post–phosgene exposure with maximum effects 24 h postexposure. The pathodiagnostic sensitivity

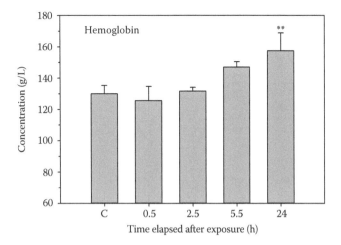

FIGURE 19.11   Time-course changes of hemoglobin from rats exposed to 1050 mg phosgene/m^3 × min (30 min exposure duration). Data represent group means ± SD (n = 8). Asterisks denote statistical significance to nose-only air-exposed control (**P < .01).

of BALF protein was approximately two orders of magnitude higher than that of wet lung weights. In concert, all three endpoints provided unequivocal evidence of redistributed plasma fluid and proteins from the peripheral circulation to the lung.

This series of changes is not peculiar to phosgene poisoning since similar findings have been noted following exposure to other lung irritants (Bruner et al., 1948a; Gordon et al., 2008; Pauluhn, 2004). Within the time periods of heart rate measurement, heart rate variability, that is, the degree of difference in the interbeat intervals of successive heartbeats (an indicator of the balance between the sympathetic and parasympathetic arms of the autonomic nervous system; Farraj et al., 2011), did not occur. In agreement with the sequence of pathophysiological events discussed earlier, the heart rate is also controlled by parasympathetic mechanisms. Any stimulation of this system essentially *prolongs* the time to the next heartbeat, thus slowing the pulse (Huston and Tracey, 2011).

19.3.3   Etiopathological Role of Neutrophilic Granulocytes

The intrapulmonic accumulation of neutrophils is a relatively common finding in certain animal models of increased permeability pulmonary edema and in humans with adult respiratory distress syndrome. The release of toxic oxygen radicals from these cells can result in ALI. Whether these cells mediate the increased permeability in all models of increased permeability pulmonary edema remains controversial (Glauser and Fairman, 1985; Pauluhn et al., 2007). However, neutrophils are considered to play a key role in the progression of the pathogenesis of ALI/ARDS (Abraham, 2003). Conversely, the early acute phase of ALI/ARDS is distinguished by the influx of protein-rich edema fluid into the alveoli as a consequence of increased permeability of the alveolar–capillary barrier and/or pulmonary hypertension.

Consistent with previous findings and published evidence, protein and neutrophilic granulocytes (PMNs) in BAL were among the most sensitive endpoints to probe for phosgene-induced pulmonary effects. It appears that both indicators of an acute inflammatory response and elevation of transmucosal permeability were interdependent (Figure 19.12). Although not directly addressed in detail, the PMNs appeared to be unstimulated and passive and apparently did not pose any danger signals to other cells or tissues as both endpoints followed the same time course of influx and evasion (Pauluhn, 2006b). Exposure to approximately 117 mg/m^3 × min did not cause significant changes exceeding the background (95th percentile), while at about 200 (189–221) mg/m^3 × min a distinctive, significant increase over the time-matched control was observed on the first postexposure day (Pauluhn, 2006b).

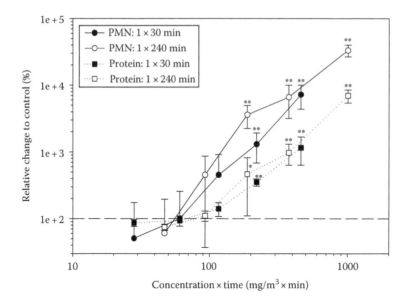

FIGURE 19.12 C × t dependence of phosgene on BAL PMN and protein after exposure to either 30 or 240 min. Rats were sacrificed on postexposure day 1. Data represent group means ± SDs. Asterisks denote statistical significance to nose-only air-exposed controls (*P < 0.05, **P < 0.01). (Data duplicated from Pauluhn, J. et al., *Inhal. Toxicol.*, 19, 789, 2007.)

Most changes in BALF were already indistinguishable from the control on postexposure day 3. The slight left shift of the C × t curve of BAL PMNs appears to be causally related to the methodological factors of cell counting. Taking into account the C × t relationship of the most sensitive endpoints reflective of acute lung edema at the time point of maximum effects (1 day postexposure), 117 mg/m³ × min is considered to be the no-observed-adverse-effect concentration (Figure 19.12). These indicators of pulmonary injury were less pronounced at similar C × t products following the 30 min exposure period when compared to the 240 min exposure period. Based on the initial decrease of the respiratory minute volume observed in rats at similar exposure concentrations (Pauluhn, 2006a), this finding is therefore consistent with the data presented in Figure 19.2 This means the exposure duration-adjusted inhaled dose at 30 min is less than at 240 min.

Tissue injury following exposure to phosgene was shown to be reduced by acute preexposures to phosgene gas (Ghio and Hatch, 1996; Hatch et al., 2001). These authors demonstrated that one day after a 6 h exposure of rats to phosgene, both protein concentrations and percentage neutrophils were increased. Upon reexposure to 480 mg/m³ × min 1 week later, lavage confirmed the development of tolerance with protein concentrations diminished after the second exposure in those rats previously exposed to phosgene, despite having a greater number of BAL neutrophils. The association of the neutrophil incursion with a protective effect was further established in studies employing colchicine and dextran. Colchicine decreased neutrophil influx from occurring after the first exposure and thus diminished the development of tolerance after a second exposure. Intratracheal instillation of dextran produced a neutrophil incursion and subsequently decreased injury after exposure to phosgene. In investigations using both colchicine and dextran, neutrophil influx increased with the development of adaptation. Thus, lung injury after the development of tolerance to phosgene provides a unique paradox in which neutrophils are not associated with injury but rather with a protective effect. Likewise, following exposure to 240 mg/m³ × min, rats were tested at 0, 4, and 24 h postexposure to phosgene by infusing anesthetized animals intravenously with acetylcholine and assessing expiratory resistance and dynamic compliance. Immediately and 4 h postexposure, a significant change in expiratory resistance and dynamic compliance was observed in those animals exposed to phosgene. Rats pretreated with colchicine indicate that colchicine decreased neutrophil influx, protein accumulation, and changes in both expiratory resistance and dynamic compliance after phosgene exposure (Ghio et al., 2005).

Collectively, it can be concluded that tolerance was manifested by a markedly diminished elevation of BAL protein despite a greater number of neutrophils in BAL. One mechanism to explain this finding would be a decreased capacity to generate free radicals and/or silencing of phagocytes to produce and discharge proinflammatory factors. There is mounting experimental circumstantial evidence to suggest that such alterations in the level or activity of antimicrobial peptides contained in surfactant may contribute to deficiencies in host defense (Gilmour and Selgrade, 1993; Pauluhn et al., 2007; Wright, 2003; Yang et al., 1995). The type of adaptation observed in rodents following exposure to phosgene can be associated with modifications in the structure, physiology, and biochemistry of the lung. Such changes may include an increased cell population, decrements in airway receptors, elevation in antioxidant molecules and enzymes, and alterations in mucous secretion and/or capacity to synthesize surfactant.

19.3.4 PULMONARY INFLAMMATION AND NITRIC OXIDE

A beneficial role for nitric oxide (NO) is postulated on the observation that NO inhibits neutrophil migration and cytokine production and that nitric oxide synthase (NOS) inhibitors worsen ALI. There are conflicting conclusions regarding the nature of NO contribution to the pathogenesis of ALI, although inhibitors of NOS activity have been reported to attenuate ALI and reduce the formation of peroxynitrite (Wizemann et al., 1994). NOS accounts for most of the activity of endothelium-derived relaxing factor and is a key mediator in the regulation of blood flow under many physiological conditions. Blockage of NOS, however, may elicit serious side effects in microvascular permeability (Hinder et al., 1997). As an endogenous mediator of vascular tone and host defense, inhaled NO has been claimed to cause vasodilation and lowers pulmonary vascular resistance (Creagh-Brown et al., 2009).

More recent studies have investigated whether phosgene-induced ALI can be diagnosed as increased exhaled NO following exposure of rats to phosgene or mitigated by inhalation exposure to NO gas and treatment with NOS inhibitors. Rats exposed to either air (control) or 1050 mg phosgene/m^3 × min were analyzed for NO and carbon dioxide (CO_2) in the exhaled air 5 and 24 h post–phosgene exposure. NO was minimally, but not clearly time-dependently, increased. L-NAME, an inhibitor of both inducible and constitutive isoforms of NO synthase, has been shown to be effective at mitigating the phosgene-induced lung edema in mice after curative administration of 50 mg/kg bw (Torkunov and Shabanov, 2009). A similar curative dosing regimen using a dose range from 2 to 100 mg L-NAME/kg bw had no conclusive effect on lung weights. Inhaled NO at 1.5 ppm × 6 h or 15 pm × 20 h led to an aggravation of edema formation (Li et al., 2011). Accordingly, concurrent with previously published data (Li et al., 2013), the absence of any robust relationship of exhaled NO (Figure 19.13) with an acute lung edema and hemoconcentration and the unresponsiveness to NOS inhibitors (Li et al., 2011) support the conclusion that NO does not appear to play any direct role in the phosgene-induced lung edemagenesis.

19.4 PUTATIVE MECHANISMS OF PHOSGENE-INDUCED ALI

Multiple hypotheses have been articulated regarding putative modes of action of inhaled phosgene gas:

i. *Indirect cytotoxicity by the release of HCl and acidity-related epithelial (pneumocyte I) injury.* The acidity concept (release of HCl) was abandoned decades ago (Diller, 1985a; Pauluhn et al., 2007). Consistent with its chemical reactivity, the toxicity of phosgene is more likely to be causally linked to instant acylation of nucleophilic moieties in solutes and/or on directly accessible cell surfaces. The C × t–dependent depletion of nucleophilic antioxidants

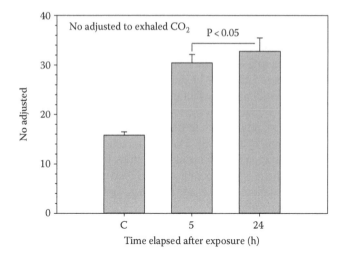

FIGURE 19.13 Dynamic measurement of exhaled nitric oxide in rats 5 and 24 h post–phosgene exposure (1050 mg/m³ × min). Similarly examined sham control rats served as concurrent control. Bars represent means ± SD (n = 3).

and peptides may then lead to disruption of enzymatic functions and membrane barrier functions. Concurrent with this hypothesis is that the prophylactic administration of strong nucleophiles markedly increased the tolerance to phosgene (Diller, 1980; Stevens et al., 1999). The *self-tolerance* to edema formation following pre-phosgene-exposed rats (Ghio and Hatch, 1996; Hatch et al., 2001) does not support the notion that hydrolysis and HCl-related tissue injury occur to any appreciable extent or are the principal causes of lung edema.

ii. *Dysfunctional endothelium.* The permeability and fenestration of the capillary endothelium is possibly affected by changes in (neuro)endocrine/paracrine control. Furthermore, lung microvascular endothelial cells possess an intrinsic capacity to preserve intracellular cyclic adenosine monophosphate (cAMP), which enhances their barrier function (Stevens et al., 1999). Increased cAMP levels are necessary to keep smooth muscle cell tight junctions intact to prevent or impede interstitial and/or alveolar edema. Phosgene exposure lowers lung tissue cAMP concentration. The reconstitution of depleted cAMP levels has been shown to mitigate prophylactically and curatively phosgene-induced lung edema (Sciuto and Hurt, 2004).

iii. *Deterioration of pulmonary surfactant with ensuing high-permeability edema.* Surfactant surely plays a key role in controlling the intra-alveolar surface tension and the fluid balance of the lung. Numerous factors are associated with functional surfactant. For edemagenesis, dysfunctional surfactant and increased surface tension are interrelated and promote atelectasis and reduction in interstitial hydrostatic pressure with consequences both to vascular perfusion and fluid dynamics (Nieman and Bredenberg, 1985). Inhibition of surfactant properties by acylation, inflammatory mediators, and especially plasma-derived proteins has been suggested to present a major step in the impairment of surfactant function progressing into a high-permeability lung edema (Doyle et al., 1999; Parker and Townsley, 2004; Seeger et al., 1985). The *adaptation* or *self-tolerance* to edema formation observed in rats exposed to phosgene (Ghio and Hatch, 1996; Hatch et al., 2001) or other lower respiratory tract irritants (Pauluhn, 2002) for a second time might also be related, at least in part, to an adaptively increased capacity and rate to store and synthesize surfactant and/or nucleophilic scavenger molecules.

iv. *Increased inflammation and oxidative stress.* Tissue injury occurs as an outbalanced relationship of generated oxidants at a stage of depleted antioxidants (Sciuto and Hurt, 2004).

The therapeutic administration of antioxidants by inhalation/instillation did not provide convincing evidence that this mechanism plays any key role for acute edemagenesis but it may be beneficial in mitigating inflammation-related oxidative stress (Pauluhn and Hai, 2011; Sciuto et al., 1995, 2003a,b,c).

v. *Obstruction of the lymphatic system draining the lung.* Increased microvascular permeability by the mode of actions (ii) and (iii) may result in an overflow of lymphatic drainage with subsequent flooding of the alveoli (Parker et al., 1981). The large amount of fluid flowing through the interstitial space toward the lymphatics dilutes interstitial proteins, thus decreasing the microvascular hydrostatic pressure and increasing the perimicrovascular hydrostatic pressure. The triad of increased lymphatic flow, increased perimicrovascular hydrostatic pressure, and decreased perimicrovascular colloid oncotic pressure makes up what is referred to as the *edema safety factors* (Nieman, 1985; Staub, 1974) and is consistent with the delay in phenotypic edema manifestation.

vi. *Hypoxic pulmonary vascular resistance.* Hydrostatic pressure within the pulmonary vascular system due to C-fiber stimulation or end-stage hypoxia-driven increased pulmonary vascular resistance is aggravated further by increased blood viscosity, thrombosis, and disrupted pulmonary capillary endothelium. A similar sequence of lung injury has been shown for other edemagenic noxae, such as oleic acid. Similar to phosgene, oleic acid also caused a decrease in cardiac output with acute permeability pulmonary edema, which was attributed to significant reductions in the plasma volume and marked increases in lung water content (Henning et al., 1986). These authors conclude that the reductions in pulmonary blood flow and right ventricular force were not dependent on neural interaction, ventricular interference, or a decrease in coronary blood flow. Rather, the decrease in ventricular force was explained by poor right ventricular adaption to increased afterload due to increased pulmonary vascular resistance (Sciuto and Hurt, 2004). Despite having a different etiopathology, the oleic acid model mimics many the phosgene model quite well (Ballard-Croft et al., 2012).

vii. Rothlin (1941), for the first time, articulated an additional hypothesis of a vagolytic-like neurogenic etiopathology of edemagenesis and suggested that the phosgene-induced pulmonary edema is brought about by a reflexively mediated vascular response.

It cannot be excluded that such C-fiber-related stimulation may also trigger in rats bradycardia elicited by a baroreflex-activated rise in pulmonary blood pressure following a massive catecholamine discharge from the sympathetic nervous system as a result of high-dose phosgene exposure. This *catecholamine storm* may lead to peripheral vasoconstriction, increased vascular resistance, and redistribution of blood into the circulatory system of the lung. The role of secondary responses to injury in regard to the pulmonary inflammatory response as well as the associated neurogenic changes in microvascular control leading eventually to lung edema needs to be considered (Chen, 1995; Ivanhoe and Meyers, 1964; Sedy et al., 2009). Likewise, cardiovascular-neural dysregulation associated with a translocation of blood from the systemic to the pulmonary circulation has to be accounted for (Baekey et al., 2010; McMullan et al., 2009; Sarnoff et al., 1953). Phosgene-induced pulmonary edema was successfully attenuated by (inadvertent) high-dose pentobarbital anesthesia administered shortly after a lethal C × t of phosgene (unpublished). Such a finding supports the notion that neurogenic mechanisms appear to play a role in the genesis of the phosgene-induced pulmonary edema as deep anesthesia has been demonstrated to be efficacious to mitigate other types of neurogenic lung edema (Sedy et al., 2009).

Although vagotomy and parasympatholytic drugs (atropine) prevented or abolished the neurogenic etiopathology of phosgene, they did not affect pulmonary edemagenesis (Bruner et al., 1948a,b). This comes as no surprise, because these authors exposed dogs to a C × t of 15,000 mg/m^3 × min, which is 15 times higher than the concentration used in previous rat and dog inhalation studies (Li et al., 2011; Pauluhn, 2006c; Pauluhn et al., 2007). There is reason to believe that functional pulmonary endothelium and epithelium are among the most essential prerequisites for any beneficial drug therapy (Berthiaume et al., 1999, 2002; Berthiaume and Matthay, 2007). Bilateral vagotomy does not interrupt all neural connection between parasympathetic ganglia and smooth muscles in lung vessels and airways. Also noncholinergic mechanisms (e.g., tachykinins) can exert multiple actions that can elicit changes in the fluid dynamics in the lungs (Hong et al., 1995). Collectively, with the exception of mechanism (i), all other hypothesized modes of actions appear to be involved in phosgene-induced ALI in one way or another and are phenotypically manifested in dependence on the inhaled dose, dose rate, and time point of analysis.

The sympathetic and parasympathetic nervous systems are not *opposites*; rather, their interactions are complex and a dynamic interaction occurs between them (Olshansky et al., 2008). Vagus nerve afferent activation, originating peripherally, can also modulate efferent sympathetic and parasympathetic function centrally and at the level of the baroreceptor. Efferent vagus nerve activation can have tonic and basal effects that inhibit sympathetic activation and release of catecholamines at the presynaptic level. The related cardiovascular effects include heart rate reduction by inhibition of the sympathetic nervous system. Elimination of the lung's sympathetic tone results in a generalized pulmonary vasoconstriction and shutdown of pulmonary neurogenic control as suggested by Ivanhoe and Meyers (1964). These authors conclude that there is vasomotor preponderance of the venous side over the arterial site of the pulmonary circulation and propose that the pulmonary vasoconstriction is a likely cause of the phosgene-induced acute pulmonary edema, as hypothesized by Sciuto and Hurt (2004). Any stagnant blood flow and pulmonary thromboembolism can be further exacerbated by pulmonary hypoxic vasoconstriction as occurs during life-threatening pulmonary edema. The decrease in heart rate (cardiac output) and systemic vasoconstriction will be an alternate cause of death if the redistribution of the plasma volume into the lung further aggravates the acute edema and anoxic anoxia in a hemodynamic state of accompanying hemoconcentration and increasing blood viscosity. All of these factors further seriously impede with gas exchange and cause an imbalance of lung fluid dynamics.

The simultaneous activation of pro- and anti-inflammatory mechanisms concurrent with pulmonary edemagenesis is analogous to other homoeostatic systems, which act in concert to coordinate hemostasis. Any stagnant blood flow and pulmonary thromboembolism interruption of these pathways leads to excessive inflammation (Huston and Tracey, 2011). Parasympathetic stimulation will inhibit tonic sympathetic activation; this means any elevated sympathetic tone is overridden by intense vagus nerve discharge. This complex interrelationship makes it difficult (in inhalation toxicity studies on rats) to attribute unequivocally any neuroparalytic etiopathology of lung edema (Ivanhoe and Meyers, 1964) to either the parasympathetic or the sympathetic system. Similarly, vagus-mediated neural mechanisms are also known to inhibit proinflammatory cytokine release via the *cholinergic anti-inflammatory pathway* (Bernik et al., 2002; Borovikova et al., 2000). This complex interrelationship between humoral and neurogenic factors makes it increasingly difficult to anticipate the lead pathomechanism of phosgene-induced pulmonary edemagenesis and what appears to be the best strategy of any pharmacological intervention.

In summary, when analyzing the sequence and time course of pathological events in phosgene-induced lung edema, direct all-or-nothing cytotoxic events would have called for different time-course patterns of response to injury. Therefore, it is concluded that an overstimulated sensorimotor vagal reflex involved in the control of cardiopulmonary hemodynamics may eventually lead to a self-perpetuating and self-amplifying acute lung edema within 24 h post–phosgene exposure. As the acute pulmonary edema progresses, the significantly depressed cardiac output further compromises oxygen delivery to tissues, and the pulmonary edema becomes increasingly refractory to pharmacological intervention. Continued bradycardia after exposure to phosgene and other signs typical of excessive parasympathetic tone have also been observed in man (Bruner et al., 1948b). These characteristics may call for a redirection of treatment strategies from anti-inflammatory to reactivation of the mechanisms involved in cardiopulmonary autonomic control and vascular hemodynamics that lead to pulmonary hypertension and edema.

19.5 QUALIFIERS OF TOXICITY AND SPECIES DIFFERENCES

There is a plethora of acute inhalation data of phosgene in experimental animal species (Figure 19.3). In terms of relevance to humans, the strengths and weaknesses of each animal model must be assessed critically with particular attention to the actual concentration of phosgene gas in the exposure atmospheres as well as the applied exposure regimen and mode. The acute toxicity of inhaled phosgene has been reviewed in detail elsewhere (AEGL, 2002; Borak and Diller, 2001; Diller, 1985a,b; Diller and Zante, 1982; Duniho et al., 2002; Pauluhn, 2006a,b,c; Pauluhn et al., 2007; Zwart et al., 1990). Human data are limited to descriptive effects from accidental exposures. Therefore, the human database has not been considered appropriate for the derivation of workplace limits or acute emergency exposure guideline levels (AEGL, 2002), and all exposure limits derived so far have been based on controlled animal exposure studies. Most models address effects associated with acute inhalation exposure, while only one study investigated inhalation toxicity over a 3-month exposure period (Kodavanti et al., 1997).

Hatch et al. (2001) demonstrated that a single exposure of Fischer rats to 4.1 mg/m^3 (1476 mg/m^3 × min; duration of exposure 6 h) caused elevations in BAL protein (18 h following exposure) approximately 100 times control levels. This C × t product did not cause mortality even after repeated exposures when using an exposure regimen of 1 day/week for 4 or 12 weeks (Kodavanti et al., 1997). With regard to the magnitude of effects obtained at 1008 mg/m^3 × min (4.2 mg/m^3 × 240 min) and 1476 mg/m^3 × min (4.1 mg/m^3 × 360 min), the 50% longer exposure period used by Hatch et al. (2001) caused a proportionally higher BAL-protein concentration (Figure 19.1). The diminution of effects after intermittent, subchronic exposure supports the occurrence of the adaptive mechanisms already addressed earlier (Kodavanti et al., 1997). Of note is that the accumulated C × t products from independent single inhalation exposure studies in rats match the borderline NO(A)EL of the 12-week repeated exposure inhalation study from Kodavanti et al. (1997) surprisingly well. This result is consistent with a mode of action depending on noncumulative acute effects. Similar findings have been reported for other pulmonary irritants (Pauluhn, 2002, 2004). Despite the initiation of adaptive responses in intermittent exposure scenarios, no apparent carry-over of effects from one exposure day to another appears to exist. Thus, chronic findings were likely to have been elicited by acute mechanisms, unless the tissue injury-time integral is small enough to allow for sufficient reconstitution without accumulation of injury. This supports the conclusion that the chronic outcome of repeated exposures to phosgene is likely to be contingent upon recurrent occurring acute irritation-related effects (*acute on chronic irritation*), rather than cumulative effects.

19.6 COUNTERMEASURES AND PHARMACOLOGICAL INTERVENTION

The current treatments of phosgene-induced ALI are focused on alleviating hypoxemia by administration of supplemental oxygen when patients develop acute symptoms of respiratory distress as a result of lung edema. Apart from physical mitigation measures (oxygen and mechanical ventilation), any generally accepted rationalized and empirically verified intervention principle has not yet been established for this particular type of ALI. One major obstacle of reaching this objective is the lack of a clear understanding of the involved mechanisms, including their time course of phenotypical manifestation and C × t dependence. Due to the specific physicochemical properties of phosgene, the ALI caused by this chemical is largely confined to the lower airways with typical response to injury for this particular region of the respiratory tract. Previous inhalation of phosgene has shown that nonlethal lung injury is rapidly reversible in the absence of long-term sequelae (Li et al., 2011; Pauluhn, 2006a,b). Moreover, these studies have demonstrated that the occurrence of inflammatory neutrophils shows similar profiles of recruitment and recovery as extravasated proteins. Such findings support the supposition that neutrophilic granulocytes appear to mirror a dysfunctional alveolocapillary barrier rather than being actively involved in edema formation (Pauluhn et al., 2007).

Consistent with the lipophilicity of phosgene and the amphiphilic properties of the surfactant layer lining the alveolus, phosgene gas is preferentially retained in the lower respiratory tract. This surfactant layer contains high concentrations of low- to-high-molecular-weight nucleophilic substrates capable of scavenging phosgene. The capacity of this layer to detoxify phosgene appears to be sufficient to *buffer away* low dose rates of phosgene (Pauluhn et al., 2007). Higher dose rates may overwhelm this capacity and acylation of nucleophilic moieties with subsequent disturbance or deterioration of membrane functions. Similar reactions may take place in the membranes of vagal C-fibers, which represent the majority of vagal afferents innervating the lower airways. Their nociceptive role in detecting the onset of pathophysiological conditions implies that C-fiber nerve endings are likely to be dosed more intensively than any other tissue. The afferent activity arising from these vagal nerve fibers also plays an important role in regulating cardiopulmonary function under both normal and abnormal physiological conditions (Lee, 2009). Hence, the activation of these afferents by phosgene may elicit respiratory and cardiovascular reflex responses alike. Hallmarks of this parasympathetic stimulation are prolonged apneic periods (stimulation of Paintal reflex; Paintal, 1969, 1981; Pauluhn, 2006a) and bradycardia.

Modulation of lung liquid clearance can occur in the injured lung because the alveolar and distal airway epithelia are remarkably resistant to injury. Accordingly, the alveolar epithelium is a key element in the process of edema clearance. In fact, the alveolar epithelial barrier is not only a tight epithelium but is also actively involved in the transport of ions and solutes. Similarly, it is important to note that the resolution of acute lung edema based on beta-adrenoreceptor stimulation requires an intact type II alveolar epithelium to make it possible that this pharmacodynamic principle is operative. In this context, the epithelial type II cells play a critical role despite the fact that these cells comprise only ≈5% of the alveolar surface area, yet they constitute ≈60% of the alveolar cells and about ≈15% of all lung parenchymal cells (the exact characteristics of cells from the alveolar region of normal lungs in different mammalian species have been detailed elsewhere by Pinkerton et al., 1992). The surface area covered by type II cells may markedly increase in response to injury. In order to be effective in keeping the alveolar space free of excess fluid, ions, and serum proteins, this cell is equipped with a number of membrane-bound water channels and ion pumps as well as albumin-binding receptors. Accordingly, the type II cell plays important roles in normal pulmonary function and in the response of the lung to toxic compounds (Castranova et al., 1988; Fehrenbach, 2001). To be effective, however, the transport principles may not already be maximally upregulated and the capacity for further increase of fluid removal must exist. Most importantly, the dose of beta-adrenergic agonists administered must by no means stimulate hemodynamic site effects, which then increase further the extravasation of edema fluid by Starling forces. Collectively, these observations suggest that there could be significant endogenous activation of the sodium transport mechanism in lung injury but this mechanism might not be functional when the alveolar epithelium is severely damaged (Berthiaume, 1998; Berthiaume et al., 1999, 2002).

Thus, for any type of pharmacological manipulation to resolve phosgene-induced lung edema, the ability of the alveolar epithelial barrier to reabsorb alveolar edema fluid within the first 12 h after ALI must be preserved. In areas of the lung, where significant injury to the alveolar epithelium has occurred, reconstitution will be necessary before the resolution of pulmonary edema can be modulated. Finally, in areas of intense alveolar type II cell proliferation, it is possible that there would be intrinsic enhanced alveolar liquid clearance. In fact, some recent studies have shown that the sustained treatment of alveolar type II cells with β-adrenergic agonists can increase the expression of the epithelial sodium channel (ENaC) and resolution of pulmonary edema requires active transepithelial sodium transport. In mild-to-moderate lung injury, alveolar edema fluid clearance is often preserved or even upregulated. Several studies have demonstrated that resolution of alveolar edema depends on active ion transport across the alveolar epithelium. The stimulation of alveolar liquid clearance is related to activation or increased expression of sodium transport molecules such as the ENaC or the Na^+–K^+–ATPase pump (Folkesson et al., 1996; Guidot et al., 2006; Matthay et al., 2002a,b). When severe lung injury occurs, the decrease in alveolar liquid clearance (Ware and Matthay, 2001) may be related to changes in alveolar permeability or to changes in activity or expression of sodium or chloride transport molecules. Multiple pharmacological tools, such as beta-adrenergic agonists, may prove effective in stimulating the resolution of alveolar edema in the injured lung (Berthiaume et al., 1987, 1999, 2002; Frank et al., 2000; Kennedy et al., 1989; Sciuto et al., 1995, 1996).

Glucocorticoids have been shown to upregulate alveolar liquid clearance, although more recently reported systematic studies did not show any benefit from nebulized steroid even when administered 1 h after exposure or high-dose corticosteroid if administered intravenously ≥ 6 h after exposure (Grainge and Rice, 2010). Likewise, anesthetized instrumented pigs, ventilated with intermittent positive pressure ventilation, exposed to phosgene at 2000 mg/m^3 $\times$ min, and treated with intravenous methylprednisolone (12.5 mg/kg) 6 h postexposure or budesonide (2 mL of 0.5 mg/mL solution) at 1, 6, 12, and 18 h postexposure did not show change in mortality, lung edema, or shunt fraction; however, some beneficial effects on cardiac parameters, for example, stroke volume and left ventricular stroke work, were noted (Smith et al., 2009). These and other authors (deLange and Meulenbelt, 2011) concluded that the efficacy of corticosteroids after human exposure to lung-damaging agents such as phosgene is inconclusive as in a number of well-structured controlled studies and the indications for administration of corticosteroids are unclear. Experimental studies on rats provided evidence that corticoids aggravate the toxicity of phosgene (Liu et al., 2014).

19.7 FUTURE DIRECTIONS

A possible explanation for the apparent limited efficacy of anti-inflammatory therapy in phosgene-induced ALI is that by the time the acute edema has become clinically apparent, the disease is already advanced to an extent that it becomes increasingly refractory to treatment. The time course of changes in the control of breathing and cardiopulmonary changes supports the notion that neurogenic factors come into play almost concomitant with exposure and contribute to a significant extent to the time dependence of edema formation. An apparent imbalance of pulmonary and systemic vascular control, concomitant with hemoconcentration and stagnant blood flow, appears to further aggravate phosgene-induced acute lung edema. Depending on the inhaled dose, multiple mechanisms are likely be involved so treatment strategies require adaptation and alignment to the inhaled C $\times$ t relationship. In order to decrease the risk of unwanted side effects associated with each type of countermeasure at an asymptomatic stage, early biomarkers of alveolar injury and loss of pulmonary vascular control are urgently needed for diagnosis, prognosis, and selection of the most appropriate countermeasure.

QUESTIONS

1. *Phosgene (COCl2) gas has been shown to be a lower respiratory tract (alveolar) irritant. What do you think is the most likely time course of mortality?*
   a. Instant death due to hydrolysis and combined toxicity of CO (anoxia) and HCl (lung airway irritation and alveolar edema).
   b. Acute lung edema with time-related aggravation of edema within 24 h postexposure. Therefore, death occurs within 24 h postexposure without delayed mortality.
   c. Instant death due to acute bronchoconstriction.
   d. There is no reason given to anticipate any preferential time period for mortality to occur.
   Answer: b
   Mortality occurs due to acute alveolar irritation and buildup of edema fluid that overwhelms the lymphatic draining system after approximately 10–15 h postexposure. When severe enough, death occurs due to the flooding of the alveoli.
2. *You compare the mortality-based toxic potencies of equal C $\times$ t products of phosgene-exposed rats after 10 and 360 min inhalation exposure. What is the expected outcome?*
   a. The median lethal concentrations (LC$_{50}$s) at 10 and 360 min are equal.
   b. The median lethal concentration (LC$_{50}$) at 10 min provides the lower LC$_{50}$ value.

c. The median lethal concentration (LC_{50}) at 10 min provides the higher LC_{50} value.

d. Short exposure durations of 10 min are experimentally not feasible as the inhalation chamber steady state cannot be attained.

Answer: c

An equal $C \times t$ requires a proportionally higher concentration to be tested at 10 min. This higher concentration prompts a reflexively induced, transient depression in ventilation. Due to the lesser inhaled volume, the LC_{50} at 10 min will be higher.

3. *Phosgene ($COCl_2$) and chlorine (Cl_2) are both potent lung irritants. What appears to be the principal location of irritation-related ALI related to which mechanism?*

a. Both $COCl_2$ and Cl_2 have the same toxic mechanism and location of injury due to their common denominator Cl_2.

b. The principal location of injury of phosgene is the alveolus, while for chlorine, it is both the alveolus and the airways. As both molecules contain 'Cl,' their toxic mechanism is similar.

c. Chlorine gas is more water soluble than phosgene. Therefore, the site(s) of airway injury depends primarily on their water solubility and concentration rather than chemical reactivity or mechanism.

d. Chlorine gas is an oxidant and phosgene is an acylating agent. Therefore, each substance may act in its own way.

Answer: c

The water solubility determines the primary site of injury.

4. *Phosgene ($COCl_2$) has been shown to be a lung irritant. What do you think is the most sensitive endpoint in animal models for acute dose–response analyses?*

a. Protein extravasated from pulmonary capillaries into the alveolus probed by BAL.

b. Histopathology.

c. Blood gas analysis.

d. Determination of proinflammatory cytokines in blood.

Answer: a

Protein in BAL is the most quantifiable endpoint matching the pattern of disease.

5. *Preexposure of animals to irritant $C \times t$ products of phosgene has been shown to increase tolerance to phosgene gas exposures one week later. What is the most likely mechanism causing this tolerance?*

a. More pronounced sensory irritation at the second exposure and instant reduction of ventilation.

b. Plugging of airways due to mucus hypersecretion. Therefore, phosgene cannot penetrate the alveoli and will be exhaled without causing injury.

c. Injury-related adaptive increase of cells producing factors to neutralize phosgene.

d. Induction of anti-inflammatory factors inactivating neutrophilic granulocytes.

Answer: c

Any prior injury increases the tolerance toward phosgene due to either the proliferation of cells or the increased production of surfactant and factors that would scavenge this nucleophile.

6. *You want to simulate high-level potentially lethal accidental exposures to a mixture of chlorine and phosgene gas in a rat inhalation bioassay. Occupational exposure to the gas may be whole body. Which of the following methods would be most appropriate?*

a. Whole-body exposure at durations 1, 3, and 10 min with immediate analysis of BAL.

b. Whole-body exposure at durations 1, 3, and 10 min with postexposure period of at least 14 days.

c. Nose-only exposure at durations 15 and 60 min with immediate analysis of BAL.

d. Nose-only exposure at durations 15 and 60 min with postexposure period of at least 14 days.

Answer: d

Such short exposures require nose-only exposure. Edema formation occurs within 24 h. Therefore, only (d) can deliver information on lethality occurring early and delayed.

REFERENCES

Abraham, E. 2003. Neutrophils and acute lung injury 619. *Crit. Care Med.* 31, S195–S199.

AEGL (Acute Exposure Guideline Levels) for Selected Airborne Chemicals). 2002. Phosgene, Vol. 2. Prepared by the National Research Council of the National Academies. The National Academies Press, Washington, DC. pp. 15–70.

Baekey, D.M., Molkov, Y.I., Paton, J.F., Rybak, I.A., and Dick, T.E. 2010. Effect of baroreceptor stimulation on the respiratory pattern: Insights into respiratory-sympathetic interactions. *Respir. Physiol. Neurobiol.* 174(1–2), 135–145. Epub September 15, 2010.

Ballard-Croft, C., Wang, D., Sumpter, L.R., Zhou, X., and Zwischenberger, J.B. 2012. Large-animal models of acute respiratory distress syndrome. *Ann. Thorac. Surg.* 93, 1331–1339.

Bernik, T.R., Friedman, S.G., Ochani, M., DiRaimo, R., Ulloa, L., Yang, H., Sudan, S., Czura, C.J., Ivanova, S.M., and Tracey, K.J. 2002. Pharmacological stimulation of the cholinergic antiinflammatory pathway. *J. Exp. Med.* 195, 781–788.

Berthiaume, Y. 1998. Mechanisms of edema clearance. In: *Pulmonary Edema*, Weir, E.K. and Reeves, J.T. (eds.). Futura, Armonk, NY, pp. 77–94.

Berthiaume, Y. and Matthay, M.A. 2007. Alveolar edema fluid clearance and acute lung injury. *Respir. Physiol. Neurobiol.* 159, 350–359.

Berthiaume, Y., Folkesson, H.G., and Matthay, M.A. 2002. Lung edema clearance: 20 years of progress—Invited review: Alveolar edema fluid clearance in the injured lung. *J. Appl. Physiol.* 93, 2207–2213.

Berthiaume, Y., Lesur, O., and Dagenais, A. 1999. Treatment of adult respiratory distress syndrome: Plea for rescue therapy of the alveolar epithelium. *Thorax* 54, 150–160.

Berthiaume, Y., Staub, N.C., and Matthay, M.A. 1987. Beta-adrenergic agonists increase lung liquid clearance in anesthetized sheep. *J. Clin. Invest.* 79, 335–343.

Borak, J. and Diller, W.F. 2000. Phosgene exposure: Mechanisms of injury and treatment strategies. *J. Occup. Environ. Med.* 43, 110–119.

Borovikova, L.V., Ivanova, S., Zhang, M., Yang, H., Botchkina, G.I., Watkins, L.R., Wang, H., Abumrad, N., Eaton, J.W., and Tracey, K.J. 2000. Vagus nerve stimulation attenuates the systemic inflammatory response to endotoxin. *Nature* 405, 468–462.

Bruner, H.D., Boche, R.D., Gibbon, M.H., and McCarthy, M.D. 1948a. Electrocardiographic study of heart and effect of vagotomy in phosgene poisoning. *Proc. Soc. Exp. Biol. Med.* 68, 279–281.

Bruner, H.D., Gibbon, M.H., McCarthy, M.D., Boche, R.D., Talbot, T.R., Lockwood, J.S., and Sanders, G.B. 1948b. Studies on experimental phosgene poisoning; infusions in the treatment of experimental phosgene poisoning. *Ann. Intern. Med.* 28, 1125–1131.

Castranova, V., Rabovsky, J., Tucker, J.H., and Miles, P.R. 1988. The alveolar type II epithelial cell: A multi-functional pneumocyte. *Toxicol. Appl. Pharmacol.* 93, 472–473.

Chen, H.I. 1995. Hemodynamic mechanisms of neurogenic edema. *Biol. Signals* 4, 186–192.

Cotarca, L. 1999. Comment on "Chemical Safety. Safe handling of Triphosgene [bis(trichloromethyl)-carbonate"—Letter to the Editor. *Org. Process Res. Dev.* 3, 377.

Cotarca, L. and Eckert, H. 2003. Chapter 2: Phosgenation reagents. In: *Phosgenations: A Handbook*. Wiley-VCH Verlag GmbH, Weinheim, Germany, pp. 3–31.

Cotarca, L., Delogu, P., Nardelli, A., and Sunjic, V. 1996. Bis(trichloromethyl) carbonate in organic synthesis. *Synthesis* 5, 553–576.

Creagh-Brown, B.C., Griffiths, M.J., and Evans, T.W. 2009. Bench-to-bedside review: Inhaled nitric oxide therapy in adults. *Crit. Care* 13(3), 221. Epub May 29, 2009. Review.

Damle, S.B. 1993. Safe handling of diphosgene and triphosgene. *Chem. Eng. News* 71, 4.

deLange, D.W. and Meulenbelt, J. 2011. Do corticosteroids have a role in preventing or reducing acute toxic lung injury caused by inhalation of chemical agents? *Clin. Toxicol. (Phila.)* 49(2), 61–71.

Diller, W.F. 1978. Medical phosgene problems and their possible solution. *J. Occup. Med.* 20, 189–193.

Diller, W.F. 1980. The methenamine misunderstanding in the therapy of phosgene poisoning. *Arch. Toxicol.* 46, 199–206.

Diller, W.F. 1985a. Pathogenesis of phosgene poisoning. *Toxicol. Ind. Health* 1, 7–15.

Diller, W.F. 1985b. Late sequelae after phosgene poisoning: A literature review. *Toxicol. Ind. Health* 1, 129–133.

Diller, W.F. and Zante, R. 1982. Dosis-Wirkungs-Beziehungen bei Phosgen-Einwirkung auf Mensch und Tier. *Zbl. Arbeitsmed.* 32, 360–368.

Doyle, I.R., Nicholas, T.E., and Bersten, A.D. 1999. Partitioning lung and plasma protein: Circulating surfactant proteins as biomarkers of alveolocapillary permeability. *Clin. Exp. Pharmacol. Physiol.* 26, 185–197.

Drorbough, J.E. and Fenn, W.O. 1955. A barometric method for measuring ventilation in newborn infants. *Pediatrics* 16, 81–87.

Duniho, S.M., Martin, J., Forster, J.S., Cascio, M.B., Moran, T.S., Carpin, L.B., and Sciuto, A.M. 2002. Acute pulmonary changes in mice exposed to phosgene. *Toxicol. Pathol.* 30, 339–349.

Epstein, M.A. and Epstein, R.A. 1978. A theoretical analysis of the barometric method for measurement of tidal volume. *Respir. Physiol.* 32, 105–120.

Farraj, A.K., Hazari, M.S., and Cascio, W.E. 2011. The utility of the small rodent electrocardiogram in toxicology. *Toxicol. Sci.* 121, 11–30.

Fehrenbach, H. 2001. Alveolar epithelial type II cell: Defender of alveolus revisited. *Respir. Res.* 2, 33–46.

Flury, F. 1921. Über Kampfgasvergiftungen. I: Über Reizgase. *Z. Gesamte Exp. Med.* 13, 1–15.

Folkesson, H.G., Matthay, M.A., Weström, B.R., Kim, K.J., Karlsson, B.W., and Hastings, R.H. 1996. Alveolar epithelial clearance of protein. *J. Appl. Physiol.* 80, 1431–1445.

Frank, J.A., Wang, Y., Osorio, O., and Matthay, M.A. 2000. β-Adrenergic agonist therapy accelerates the resolution of hydrostatic pulmonary edema in sheep and rats. *J. Appl. Physiol.* 89, 1255–1265.

Gary, W.H., Chang, W., Chen, J., Schlueter, C.F., and Rando, R.J. 2010. Deviations from Haber's law for multiple measures of acute lung injury in chlorine-exposed mice toxicological sciences. *Toxicol. Sci.* 118, 696–703.

Gerriets, J.E., Reiser, K.M., and Last, J.A. 1996 Lung collagen crosslinks in rats with experimentally induced pulmonary fibrosis. *Biochim. Biophys. Acta* 1316, 121–131.

Ghio, A.J. and Hatch, G.E. 1996. Tolerance to phosgene is associated with a neutrophilic influx into the rat lung. *Am. J. Respir. Crit. Care Med.* 153, 1064–1071.

Ghio, A.J., Lehmann, J.R., Winsett, D.W., Richards, J.H., and Costa, D.L. 2005. Colchicine decreases airway hyperreactivity after phosgene exposure. *Inhal. Toxicol.* 17, 277–285.

Gibbon, M.H., Bruner, H.D., Boche, R.D., and Lockwood, J.S. 1948. Studies on experimental phosgene poisoning. II. Pulmonary artery pressure in phosgene-poisoned cats. *J. Thorac. Surg.* 17, 264–273.

Gilmour, M.I. and Selgrade, M.J. 1993. A comparison of the pulmonary defenses against Streptococcal infection in rats and mice following O_3 exposure: Differences in disease susceptibility and neutrophil recruitment. *Toxicol. Appl. Pharmacol.* 123, 211–218.

Glauser, F.L. and Fairman, R.P. 1985. The uncertain role of the neutrophil in increased permeability pulmonary edema. *Chest* 88, 601–607.

Gordon, C.J. 1990. Thermal biology of the laboratory rat. *Physiol. Behav.* 47, 963–991.

Gordon, C.J. 1993. *Temperature Regulation in Laboratory Rodents.* Cambridge University Press, New York.

Gordon, C.J., Mohler, F.S., Watkinson, W.P., and Rezvani, A.H. 1988. Temperature regulation in laboratory mammals following acute toxic insult. *Toxicology* 53, 161–178.

Gordon, C.J., Spencer, P.J., Hotchkiss, J., Miller, D.B., Hinderliter, P.M., and Pauluhn, J. 2008. Thermoregulation and its influence on toxicity assessment. *Toxicology* 244, 87–97.

Grainge, C. and Rice, P. 2010. Management of phosgene-induced lung injury. *Clin. Toxicol.* 48, 497–508.

Grommes, J. and Soehnlein, O. 2011. Contribution of neutrophils to acute lung injury. *Mol. Med.* 17, 293–307.

Guidot, D.M., Folkesson, H.G., Jain, L., Sznajder, J.I., Pittet, J.-F., and Matthay, M.A. 2006. Integrating acute lung injury and regulation of alveolar fluid clearance. *Am. J. Physiol. Lung Cell Mol. Physiol.* 291, L301–L306.

Haber, F. 1924. Zur Geschichte des Gaskrieges (On the history of gas warfare). In: *Fünf Vorträge aus den Jahren 1920–1923* (Five Lectures from the Years 1920–1923). Springer, Berlin, Germany, pp. 76–92.

Hamelmann, E., Schwarze, J., Takeda, K., Oshiba, A., Larsen, G.L., Irvin, C.G., and Gelfand, E.W. 1997. Noninvasive measurement of airway responsiveness in allergic mice using barometric plethysmography. *Am. J. Respir. Crit. Care Med.* 156, 766–775.

Hantos, Z. and Brusasco, V. 2002. Assessment of respiratory mechanics in small animals: The simpler the better? *J. Appl. Physiol.* 93, 1196–1197.

Hatch, G.E. 1992. Chapter 33: Comparative biochemistry of airway lining fluid. In: *Treatise on Pulmonary Toxicology: Comparative Biology of the Normal Lung*, Vol. I, Parent, R.A. (ed.). CRC Press, Boca Raton, FL, pp. 617–632.

Hatch, G.E., Kodavanti, U., Crissman, K., Slade, R., and Costa, D. 2001. An 'injury-time integral' model for extrapolating from acute to chronic effects of phosgene. *Toxicol. Ind. Health* 17, 285–293.

Hatch, G.E., Slade, R., Stead, A.G., and Graham, J.A. 1986. Species comparison of acute inhalation toxicity of ozone and phosgene. *J. Toxicol. Environ. Health* 19, 43–53.

Hendricks, W. 1986. OSHA method 61, Sampling and analytical methods–phosgene, https://www.osha.gov/dts/sltc/methods/organic/org061/org061.html (Accessed on July 8, 2014).

Henning, R.J., Heyman, V., Alcover, I., and Romeo, S. 1986. Cardiopulmonary effects of oleic acid-induced pulmonary edema and mechanical ventilation. *Anesth. Analg.* 65, 925–932.

Hinder, F., Booke, M., Traber, L.D., and Traber, D.L. 1997. Nitric oxide and endothelial permeability. *J. Appl. Physiol.* 83(6), 1941–1946.

Hong, J.-L., Rodger, I.W., and Lee, L.-Y. 1995. Cigarette smoke-induced bronchoconstriction: Cholinergic mechanisms, tachykinins, and cyclooxygenase products. *J. Appl. Physiol.* 78, 2260–2266.

Huston, J.M. and Tracey, K.J. 2011. The pulse of inflammation: Heart rate variability, the cholinergic anti-inflammatory pathway and implications for therapy. *J. Intern. Med.* 269, 45–53.

IPCS. 1998. Phosgene: Health and safety guide no. 106. International Program on Chemical Safety (IPCS), World Health Organization, Geneva, http://inchem.org.documents/hsg/hsg/hsg106.htm. (Accessed on April 9, 2014).

Ivanhoe, F. and Meyers, F.H. 1964. Phogene poisoning as an example of neuroparalytic acute pulmonary edema: The sympathetic vasomotor reflex involved. *Chest* 46, 211–218.

Kennedy, T.P., Michael, R., Hoidal, J.R., Hasty, D., Sciuto, A.M., Hopkins, C., Lazar, R., Bysani, G.K., Tolley, E., and Gurtner, G.H. 1989. Dibutyl cAMP, aminophylline, and β-adrenergic agonists protect against pulmonary edema caused by phosgene. *J. Appl. Physiol.* 67, 2542–2552.

Kimmel, E.C., Whitehead, G.S., Reboulet, J.E., Carpenter, R.L. 2002. Carbon dioxide accumulation during small animal, whole body plethysmography: Effects on ventilation, indices of airway function, and aerosol deposition. *J. Aerosol. Med.* 15, 37–49.

Kluge, S. and Szinicz, L. 2005. Acute Exposure Guideline levels (AEGLs), toxicity and properties of selected chemical and biological agents. *Toxicology* 214, 268–270.

Kodavanti, U.P., Costa, D.L., Giri, S.N., Starcher, B., and Hatch, G.E. 1997. Pulmonary structural and extracellular matrix alterations in Fischer 344 rats following subchronic exposure. *Fundam. Appl. Toxicol.* 37, 54–63.

Lee, L.-Y. 2009. Respiratory sensations evoked by activation of bronchopulmonary C-fibers. *Respir. Physiol. Neurobiol.* 167, 26–35.

Lee, L.-Y. and Widdicombe, J.G. 2001. Modulation of airway sensitivity to inhaled irritants: Role of inflammatory mediators. *Environ. Health Perspect.* 109(suppl 4), 585–589.

Li, W., Liu, F., Wang, C., Truebel, H., and Pauluhn, J. 2013. Novel insights into phosgene-induced acute lung injury in rats: Role of dysregulated cardiopulmonary reflexes and nitric oxide in lung edema pathogenesis. *Toxicol. Sci.*, 131, 612–628.

Li, W.-L., Hai, C.X., and Pauluhn, J. 2011. Inhaled nitric oxide aggravates phosgene model of acute lung injury. *Inhal. Toxicol.* 23, 842–851.

Li, W.L. and Pauluhn, J. 2010. Comparative assessment of the sensory irritation potency in mice and rats nose-only exposed to ammonia in dry and humidified atmospheres. *Toxicology* 276, 135–142.

Liu, F., Pauluhn, J., Trübel, H., and Wang, C. 2014. Single high-dose dexamethasone and sodium salicylate failed to attenuate phosgene-induced acute lung injury in rats. *Toxicology* 315, 17–23.

Lung, M.A. 1987. Effects of lung inflation on nasal airway resistance in the anesthetized rat. *J. Appl. Physiol.* 63, 1339–1343.

Matthay, M.A., Clerici, C., and Saumon, G. 2002a. Lung edema clearance: 20 years of progress. Invited reviews: Active fluid clearance from the distal air spaces of the lung. *J. Appl. Physiol.* 93, 1533–1541.

Matthay, M.A., Folkesson, H.G., and Clerici, C. 2002b. Lung epithelial fluid transport and resolution of pulmonary edema. *Physiol. Rev.* 82, 569–600.

McMullan, S., Dick, T.E., Farnham, M.M., and Pilowsky, P.M. 2009. Effects of baroreceptor activation on respiratory variability in rat. *Respir. Physiol. Neurobiol.* 166, 80–86.

Mitzner, W. and Tankersley, C. 1998. Noninvasive measurement of airway responsiveness in allergic mice using barometric plethysmography. *Am. J. Respir. Crit. Care Med.* 158, 340–341.

Mitzner, W., Tankersley, C., Lundblad, L.K., Adler, A., Irvin, C.G., and Bates, J. 2003. Interpreting Penh in mice. *J. Appl. Physiol.* 94, 828–832.

Nash, T. and Pattle, R.E. 1971. The absorption of phosgene by aqueous solutions and its relation to toxicity. *Ann. Occup. Hyg.* 14, 227–333.

NDRC. 1946. Chemical warfare agents, and related chemical problems Parts I–III. Summary technical report of Division 9, NDRC, Vol. 1, Authored by Kirner, W.R., Bush, V., and Conant, J.B. Office of Scientific Research and Development, National Defense Research Committees, Washington, DC.

Nieman, G.F. 1985. Current concepts of lung-fluid balance. *Respir. Care* 30, 1062–1076.

Nieman, G.F. and Bredenberg, C.E. 1985. High surface tension pulmonary edema induced by detergent aerosol. *J. Appl. Physiol.* 58, 129–136.

Niessner, R. 2010. Quantitative determination of phosgene doses by reflectometric badge readout. *Anal. Bioanal. Chem.* 397, 2285–2288.

Olshansky, B., Sabbah, H.N., Hauptman, P.J., and Colucci, W.S. 2008. Parasympathetic nervous system and heart failure: Pathophysiology and potential implications for therapy. *Circulation* 118, 863–871.

Paintal, A.S. 1969. Mechanism of stimulation of type J pulmonary receptors. *J. Physiol.* 203, 511–532.

Paintal, A.S. 1981. Effects of drugs on chemoreceptors, pulmonary and cardiovascular receptors. In: *International Encyclopedia of Pharmacology and Therapeutics*, Section 104, *Respiratory Pharmacology*, Widdicombe, J.G. (ed.). Pergamon Press, Oxford, U.K., pp. 217–239.

Parker, J.C. and Townsley, M.I. 2004. Evaluation of lung injury in rats and mice. *Am. J. Physiol. Lung Cell Mol. Physiol.* 286, L231–L246.

Parker, J.C., Parker, R.E., Granger, D.N., and Taylor, A.E. 1981. Vascular permeability and transvascular fluid and protein transport in the dog lung. *Circ. Res.* 48, 549–561.

Pauluhn, J. 2002. Critical analysis of biomonitoring endpoints for measuring exposure to polymeric diphenyl-4,4′-diisocyanate (MDI) in rats: A comparison of markers of exposure and markers of effect. *Arch. Toxicol.* 76, 13–122.

Pauluhn, J. 2004. Comparative analysis of pulmonary irritation by measurements of Penh and protein in bronchoalveolar lavage fluid in Brown Norway rats and Wistar rats exposed to irritant aerosols. *Inhal. Toxicol.* 16, 159–175.

Pauluhn, J. 2006a. Acute nose-only exposure of rats to phosgene. Part I: Concentration × time dependence of LC_{50}s and non-lethal-threshold concentrations and analysis of breathing patterns. *Inhal. Toxicol.* 18, 423–435.

Pauluhn, J. 2006b. Acute nose-only exposure of rats to phosgene. Part II: Concentration × time dependence of changes in bronchoalveolar lavage during a follow-up period of 3 months. *Inhal. Toxicol.* 18, 595–607.

Pauluhn, J. 2006c. Acute head-only exposure of dogs to phosgene. Part III: Comparison of indicators of lung injury in dogs and rats. *Inhal. Toxicol.* 18, 609–621.

Pauluhn, J. 2011. Acute nose-only inhalation exposure of rats to di- and triphosgene relative to phosgene. *Inhal. Toxicol.* 23, 65–73.

Pauluhn, J. and Hai, C.X. 2011. Attempts to counteract phosgene-induced acute lung injury by instant high-dose aerosol exposure to hexamethylenetetramine, cysteine or glutathione. *Inhal. Toxicol.* 23, 58–64.

Pauluhn, J. and Mohr, U. 2000. Inhalation studies in laboratory animals—Current concepts and alternatives. Review. *Toxicol. Pathol.* 28, 734–753.

Pauluhn, J. and Thiel, A. 2007. A simple approach to validation of directed-flow nose-only inhalation chambers. *J. Appl. Toxicol.* 27, 160–167.

Pauluhn, J., Carson, A., Costa, D.L., Gordon, T., Kodavanti, U., Last, J.A., Matthay, M.A., Pinkerton, K.E., and Sciuto, A.M. 2007. Workshop summary—Phosgene-induced pulmonary toxicity revisited: Appraisal of early and late markers of pulmonary injury from animals models with emphasis on human significance. *Inhal. Toxicol.* 19, 789–810.

Peták, F., Habre, W., Donati, Y.R., Hantos, Z., and Barazzone-Argiroffo, C. 2001. Hyperoxia-induced changes in mouse lung mechanics: Forced oscillations vs. barometric plethysmography. *J. Appl. Physiol.* 90, 2221–2230.

Pinkerton, K.E., Gehr, P., and Crapo, J.D. 1992. Chapter 11: Architecture and cellular composition of the air-blood barrier. In: *Treatise on Pulmonary Toxicology: Comparative Biology of the Normal Lung*, Vol. I, Parent, R.A. (ed.). CRC Press, Boca Raton, FL, pp. 121–128.

Rothlin, E. 1941. Pathogenesis and treatment for phosgene intoxication. *Schweiz. Med. Wochenschr.* 71, 1526–1535. (Cited in EPA 1986).

Sarnoff, S.J., Berglund, E., and Sarnoff, L.C. 1953. Neurohemodynamics of pulmonary edema. III. Estimated changes in pulmonary blood volume accompanying systemic vasoconstriction and vasodilation. *J. Appl. Physiol.* 5, 367–374.

Sciuto, A.M. 1997. Ibuprofen treatment enhances the survival of mice following exposure to phosgene. *Inhal. Toxicol.* 9, 389–403.

Sciuto, A.M. 1998. Assessment of early acute lung injury in rodents exposed to phosgene. *Arch. Toxicol.* 72, 283–288.

Sciuto, A.M. 2005. Chapter 20: Inhalation toxicology of an irritant gas—Historical perspectives, current research, and case studies of phosgene exposure. In: *Inhalation Toxicology*, 2nd edn., Katz, S.A. and Salem, H. (eds.). Taylor and Francis, CRC-Press, New York, pp. 457–483.

Sciuto, A.M. and Hurt, H.H. 2004. Therapeutic treatments of phosgene-induced lung injury. *Inhal. Toxicol.* 16, 565–580.

Sciuto, A.M., Carpin, L.B., Moran, T.S., and Forster, J.S. 2003b. Chronological changes in electrolyte levels in arterial blood and bronchoalveolar lavage fluid in mice after exposure to an edemagenic gas. *Inhal. Toxicol.* 15, 663–674.

Sciuto, A.M., Cascio, M.B., Moran, T.S., and Forster, J.S. 2003a. The fate of antioxidant enzymes in bronchoalveolar lavage fluid over 7 days in mice with acute lung injury. *Inhal. Toxicol.* 15, 675–685.

Sciuto, A.M., Clapp, D.L., Hess, Z.A., and Moran, T.S. 2003c. The temporal profile of cytokines in the bronchoalveolar lavage fluid in mice exposed to the industrial gas phosgene. *Inhal. Toxicol.* 15, 687–700.

Sciuto, A.M., Lee, R.B., Forster, J.S., Cascio, M.B., Clapp, D.L., and Moran, T.S. 2002. Temporal changes in respiratory dynamics in mice exposed to phosgene. *Inhal. Toxicol.* 14, 487–501.

Sciuto, A.M., Phillips, C.S., Orzolek, L.D., Hege, A.I., Moran, T.S., and Dillman, J.F. 2005. Genomic analysis of murine pulmonary tissue following carbonyl chloride inhalation. *Chem. Res. Toxicol.* 18, 1654–1660.

Sciuto, A.M., Strickland, P.T., Kennedy, T.P., and Gurtner, G.H. 1995. Protective effects of *N*-acetylcysteine treatment after phosgene exposure in rabbits. *Am. J. Respir. Crit. Care Med.* 151, 768–772.

Sciuto, A.M., Strickland, P.T., Kennedy, T.P., Guo Y.-L., and Gurtner, G.H. 1996. Intratracheal administration of DBcAMP attenuates edema formation in phosgene-induced acute pulmonary injury. *J. Apppl. Physiol.*, 80, 149–157.

Sedy, J., Zicha, J., Kunes, J., Hejcl, A., and Syková, E. 2009. The role of nitric oxide in the development of neurogenic pulmonary edema in spinal cord-injured rats: The effect of preventive interventions. *Am. J. Physiol. Regul. Integr. Comp. Physiol.* 97(4), R1111–R1117. Epub August 12, 2009.

Seeger, W., Stöhr, G., Wolf, H.R., and Neuhof, H. 1985. Alteration in surfactant function due to protein leakage: Special interaction with fibrin monomer. *J. Appl. Physiol.* 58, 326–338.

Slade, R., Crissman, K., Norwood, J., and Hatch, G. 1993. Comparison of antioxidant substances in bronchoalveolar lavage cells and fluid from humans, guinea pigs, and rats. *Exp. Lung Res.* 19, 469–484.

Smith, A., Brown, R., Jugg, B., Platt, J., Mann, T., Masey, C., Jenner, J., and Rice, P. 2009. The effect of steroid treatment with inhaled budesonide or intravenous methylprednisolone on phosgene-induced acute lung injury in a porcine model. *Mil. Med.* 174(12), 1287–1294.

Staub, N.C. 1974. Pulmonary edema. *Physiol. Rev.* 54, 678–811.

Stevens, T., Creighton, J., and Thompson, W.J. 1999. Control of cAMP in lung endothelial cell phenotypes. Implications for control of barrier function. *Am. J. Physiol.* 277, L119–L126.

Szinicz, L. 2005. History of chemical and biological warfare agents. *Toxicology* 214, 167–181.

Torkunov, P.A. and Shabanov, P.D. 2009. Using NO-synthase inhibitors derived from L-arginine for preventing acute experimental lung edema development in mice. *Eksp. Klin. Farmakol.* 72, 44–46.

US Environmental Protection Agency (EPA). 2005. Toxicological review of phosgene. In support of summary information on the integrated risk information system (IRIS). EPA/635/R-06/001. U.S. Environmental Protection Agency, Washington, DC. Available at http://www.epa.gov/iris/subst/0487.htm. (Accessed on April 9, 2014).

Ware, L.B. and Matthay, M.A. 2001. Alveolar fluid clearance is impaired in the majority of patients with acute lung injury and the acute respiratory distress syndrome. *Am. J. Respir. Crit. Care Med.* 163, 1376–1383.

Watkinson, W.P. and Gordon, C.J. 1993. Caveats regarding the use of the laboratory rat as a model for acute toxicological studies: Modulation of the toxic response via physiological and behavioral mechanisms. *Toxicology* 81, 15–31.

Winternitz, M.C. 1920. *Pathology of War Gas Poisoning.* Yale-University Press, New Haven, CT, pp. 35–66.

Witschi, H. 1997. The story of the man who gave us "Haber's law". *Inhal. Toxicol.* 9, 199–207.

Witschi, H. 1999. Some notes on the history of Haber's law. *Toxicol. Sci.* 50, 164–168.

Wizemann, T.M., Gardner, C.R., Laskin, J.D., Quinones, S., Durham, S.K., Goller, N.L., Ohnishi, S.T., and Laskin, D.L. 1994. Production of nitric oxide and peroxynitrite in the lung during acute endotoxemia. *J. Leukoc. Biol.* 56(6), 759–768.

Wright, J.R. 2003. Pulmonary surfactant. A front line of lung host defense. *J. Clin. Invest.* 111, 1453–1455.

Yang, Y.G., Gilmour, M.I., Lange, R., Burleson, G.R., and Selgrade, M.K. 1995. Effects of acute exposure to phosgene on pulmonary resistance to infection. *Inhal. Toxicol.* 7, 393–404.

Zwart, A. and Woutersen, R.A. 1988. Acute inhalation toxicity of chlorine in rats and mice: Time-concentration-mortality relationships and effects on respiration. *J. Hazard. Mater.* 19, 195–208.

Zwart, A., Arts, J.H.E., Klokman-Houweling, J.M., and Schoen, E.D. 1990. Determination of concentration-time-mortality relationships to replace LC50 values. *Inhal. Toxicol.* 2, 105–117.

20 Chemical Warfare Agents and Nuclear Weapons*

Terry J. Henderson, Nabil M. Elsayed, and Harry Salem

CONTENTS

---

* The views and opinions expressed in this chapter are those of the author and should not be construed as an official Department of the Army position, policy, or decision, unless so designated by other official documentation. Citation of trade names in this chapter does not constitute an official Department of the Army endorsement or approval of the use of such commercial items.

20.1 INTRODUCTION

Chemical warfare agents (CWAs) are chemical substances with toxic properties specifically for producing lethal, incapacitating, or damaging effects to humans when suitably delivered. Since their reintroduction in World War I by Germany, CWAs have remained weapons for use by combatant armies. In recent years, however, these agents have also been used by armies against rebellious noncombatant civilians and by terrorists against innocent, unsuspecting citizens, as was the case in the 1995 sarin nerve agent release in a Tokyo subway station. The potential risk of their deployment by terrorists increased significantly after the attack on the New York World Trade Center on September 11, 2001. Although the United States and most European countries place a greater emphasis on, and allocate most of their antiterrorist resources to, the preparation for potential chemical and biological terrorism, blasts from explosive detonations remain the single most frequent cause of human death and injury, as well as property damage, by terrorist attacks.

Nuclear weapons (NWs) derive their enormous destructive power from the energy produced by nuclear reactions and are not generally included when discussing CWAs. However, with the increased concern of potential nuclear terrorism and the possible deployment of portable NWs or radiological dispersion devices (dirty bombs), it seems appropriate to include NWs and dirty bombs in the discussion of CWAs. Unlike CWAs, NWs and radiological dispersion devices are more destructive to life and property, and their contaminating radiation is highly persistent in the environment long after detonation. NWs were only used twice in times of war by the United States against Japan during World War II.

20.2 BRIEF HISTORY OF CHEMICAL WARFARE

The use of chemical weapons dates back to 1000 BCE, when the Chinese used arsenical smoke in battle. In 600 BCE, Solon of Athens poisoned the water supply of Athens, Greece, during the siege of Kirrha. Later, in 429 BCE, the Spartans used noxious smoke and flame against the Athenian cities during the Peloponneisian War, and poison-tipped arrows were used during the Trojan War. In fact, the English word for toxin, or poison, comes from the Greek word *toxikon*, which in turn is derived from the Greek word *toxon*, or arrow. Leonardo da Vinci also had a minor role in the history of chemical warfare. In 1100, he proposed a powder of arsenic sulfide and verdigris as an antiship weapon.

Modern chemical warfare was launched by Germany during World War I, when on October 27, 1914, in breach of the 1899 International Ban, the German Army fired 105 mm shells filled with the lung irritant, dianisidine chlorosulfate, against British troops near the town of Neuve-Chapelle, France. The attack was unsuccessful, however, because explosives in the shells destroyed the lung irritant. In January 1915, the German Army fired shells containing xylyl bromide against the Russian troops near Bolimów, Poland, but the cold weather prevented chemical vaporization. Then, on April 22, 1915, under the guidance of renowned chemist Fritz Haber, the German Army released 168 tons of chlorine from 5730 cylinders against French and Algerian troops defending Ypres, Belgium. Two days later, the German Army launched another gas attack with 150–200 tons of chlorine gas against the Canadian Army defending Ypres. The results of these 2 days were estimated at 5,000 dead and 10,000 disabled. The British Army responded with their first gas attack on September 25, 1915, deploying chlorine gas at Loos, Belgium. During the course of the war, Germany developed and used phosgene, a more effective gas than chlorine, and also developed diphosgene. In response, the French used hydrogen cyanide, then cyanogen chloride (Smart, 1997). On July 17, 1917, the German Army launched an attack using a new chemical agent developed by Lommel and Steinkopt that caused skin blisters and death upon inhalation at high concentrations. The Germans named the chemical LoSt after the two chemists who developed the agent. However, the British called the new agent mustard gas, although it was an oily liquid, because of its mustard or garlic smell and the yellow cross identification markings that were used on its containment canisters. The French called it Yprite after the location where it was first used. The mustard gas attack resulted in 20,000 British casualties after 6 weeks of use, and the British retaliated by using a combination of high explosives and mustard gas. The British used mustard gas again after the war, employing mustard-filled shells against the Bolshevik troops of the newly formed Soviet Union during the Russian Civil War between 1918 and 1921. At the end of World War I, the total number of casualties was estimated at over 83,000 dead and almost 1.2 million injured. Germany, Britain, and France each suffered 8,000–9,000 fatalities resulting from the gas war, whereas Russia's losses were the heaviest at 56,000 fatalities, because they were late in deploying protective masks. In addition, the American Expeditionary Force suffered 71,345 chemical casualties and 1,462 deaths (Joy, 1997). After the war, mustard gas production continued and was used in the 1920s and 1930s by the British in Afghanistan, the Italians in Ethiopia, and the Japanese in China (Table 20.1). More recently, it was used by the Iraqis against Iran during the Iran–Iraq war, and again, against the Kurds to quell their uprising.

In December 1936, another German scientist, Gerhard Schrader, who was developing new pesticides, made an accidental discovery. Schrader sprayed insects with a solution of one of his new chemicals that he diluted 200,000-fold, which killed them. In the course of his experiments, he was contaminated in a laboratory accident and developed side effects that included pinpoint-constricted pupils, sensitivity to light, shortness of breath, and giddiness. Unlike other gases that had to be inhaled to produce toxicity, this chemical was capable of inducing spasms and death after dermal exposure (U.S. Army Medical Research Institute of Chemical Defense [USAMRICD], 2000). The new chemical was later named tabun (GA) and marked the introduction of the organophosphates as potent chemical nerve agents. Other G-series agents were later developed, including sarin (GB) and soman (GD); see Figure 20.3 for chemical structures. At the end of World War II, the Soviets captured 12,000 tons of tabun and the processing plant that produced them, which was moved to Russia to form the basis of the Soviet chemical weapons program. After the war, both the Soviet Union and the United States developed the more potent V-series agents, including VX, VE, and VR. Table 20.1 presents a chronology of the major state use of chemical weapons in military conflicts.

TABLE 20.1
Chronology of Major Military User States of Chemical Weapons

Period	User	Chemical(s) Used
1000 BCE	Chinese	Arsenical smoke
Siege of Kiirha Greece (600 BCE)	Solon of Athens	Poisoned water supply
Peloponnesian War (429 BCE)	Spartans *vs.* Athens	Burning pitch and sulfur
Peloponnesian War (424 BCE)	Spartans	Burning pitch and sulfur
World War I (1914)	France *vs.* Germany	Tear gas
World War I (April 22, 1915)	Germany *vs.* France	Chlorine gas
World War I (September 25, 1915)	Britain *vs.* Germany	Chlorine gas
World War I (February 26, 1918)	Germany *vs.* United States	Phosgene and chloropicrin
World War I (June 1918)	United States *vs.* Germany, and establishment of the Chemical Warfare Service	
Russian Civil War (1918–1921)	Britain *vs.* Russian Bolsheviks	Adamsite, and mustard gas artillery shells
Spanish Civil War (1922–1927)	Spain *vs.* RIF rebels in Spanish Morocco	
Abyssinia (Ethiopia) Conquest (1935)	Italy *vs.* Ethiopians	Mustard gas by aerial spray
China (1936)	Japan *vs.* China	Mustard gas, phosgene, and hydrogen cyanide
Vietnam, Cambodia, Laos (1962–1970)	United States *vs.* Vietnamese	Defoliants (agents orange, blue, purple, and white), and tear gas
Yemen, civil war (1963–1967)	Egypt *vs.* Yemeni Royalists	Phosgene and mustard gas
Afghanistan	Soviet Union *vs.* Afghanistan	Alleged use of yellow rain (mycotoxins)
Iran–Iraq War (1983–1988)	Iraq *vs.* Iran	Mustard gas, tabun
Iraq Anfal campaign (1987–1988)	Iraq *vs.* Kurds	Hydrogen cyanide, mustard gas

Source: Adapted from Monterey Institute for International Studies, James Martin Center for Nonproliferation Studies, Chronology of state use and biological and chemical weapons control, 2002 available at: http://cns.miis.edu/cbw/pastuse.htm, Accessed May 23, 2014.

20.3 CLASSIFICATION OF CHEMICAL WARFARE AGENTS

CWAs can be classified in two ways: first, operationally, based on their lethality and persistence in the environment as shown in Figure 20.1; second, categorized by classes according to their mechanism of action or their primary target organ. Table 20.2 lists the general characteristics of CWAs that are used either to inflict lethality, incapacitation, or distraction of the enemy in combat, or for civilian crowd control during riots. Also included in the table are the inhalation lethality values for each CWA, expressed as the median lethal exposure (LCt_{50} in mg·min·m^3) in a 70 kg man. Agents used for incapacitation or distractions have been classified as nonlethal, or less than lethal. In general, these include agents used by law enforcement personnel, such as mace, which although ordinarily are nonlethal at effective concentrations, can be lethal at very high concentration or prolonged exposure.

20.3.1 Lethal Chemical War Agents

20.3.1.1 Pulmonary Agents

Pulmonary agents, or choking agents, are chemicals that exert their toxicity almost exclusively by inhalation, although they can also cause eye and skin irritation. These include chlorine [Cl_2] (CI), phosgene [$COCl_2$] (CG), diphosgene [$ClCO_2CCl_3$] (DP), chloropicrin [Cl_3CNO_2] (PS), and toxic pyrolysis products such as nitrogen oxides [NO_x], phosphorus oxides [PO_x], sulfur oxides [SO_x], and perfluoroisobutene [$CF_2C(CF_3)_2$] (PFIB). Inhaled pulmonary agents readily penetrate the respiratory system, reaching the

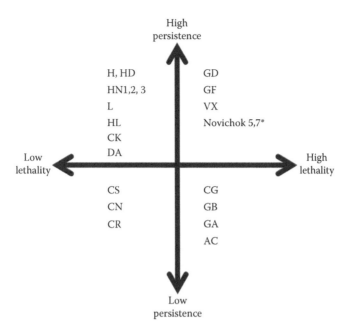

FIGURE 20.1 A system for the classification of CWAs by persistence and lethality. (Modified from Evison, D. and Hinsley, D., *Br. Med. J.*, 324, 332, 2002.) *Novichoks are a class of nerve agents developed by the Soviet Union with very limited available information. These agents are believed to have high lethality and persistence.

deep lung (respiratory bronchioles and alveoli) and the peripheral compartments of the respiratory tree. Choking agents may cause nasopharyngeal, airway, and laryngeal irritation; apnea; chest tightness; capillary and alveolar wall damage; delayed pulmonary edema; and extensive fluid buildup in the lungs that can interfere with oxygen uptake and lead to adult respiratory distress syndrome (ARDS) symptoms or noncardiogenic pulmonary edema (USAMRICD, 2000). Because of their reactivity, pulmonary agents are removed at the alveolar–capillary membrane surface or proximally in the respiratory tract and do not diffuse systemically to cause a significant clinical response.

Chlorine is a lung irritant used during World War I, where it was considered inefficient and was ultimately replaced with other, more potent pulmonary gases. Exposure to chlorine results in both central airway damage and pulmonary edema. Inhalation of 1000 ppm of chlorine will cause death in 2–3 min. The LCt_{50} for chlorine is 6000 ppm.

Phosgene is a more effective pulmonary toxicant than chlorine; it was first prepared by John Davy in 1812 and, as described earlier, was developed as a chemical weapon by Germany during World War I. Phosgene was first used in December 1915 against British troops near Ypres, Belgium. Eighty-eight tons were released from 4000 cylinders, causing 1069 total casualties and 120 deaths. Subsequently, both sides used substantial amounts of phosgene, which accounted for approximately 85% of all chemical weapon deaths by the end of World War I. At present, the United States produces over a billion pounds of phosgene per year for industrial use. Inhalation of phosgene damages the lungs and produces pulmonary edema (Marrs et al., 1996), and its hydrolysis forms HCl, which contributes to its toxicity (Equation 20.1). The phosgene carbonyl group [$>C=O$] can undergo acylation reactions with several functional groups such as amino [$-NH_2$], sulfhydryl [$-SH$], and hydroxyl [$-OH$] groups, rendering them nonfunctional and thus contributing to pathophysiological effects (Marrs et al, 2000; USAMRICD, 2000). No antidote for phosgene intoxication is currently available, and its LCt_{50} is approximately 3200 mg·min/m³, which is twice as toxic as chlorine. The Soviet Union stockpiled phosgene, and Egypt used it against the Yemeni Royalists during the Yemeni Civil War from 1963 to 1967 (Table 20.1). Its medical

TABLE 20.2

Characteristics of Some Lethal Warfare and Nonlethal Riot Control Chemical Agents^a

Category/Name	Code	State (20°C)	Odor	Inhalation Lethality^B	Action	Response
Warfare Agents (Lethal)						
Pulmonary (Choking)						
Chlorine	CL	Gas	Bleach	3,000	Rapid	Lung damage and flooding
Phosgene	CG	Gas	New mown hay green corn	3,200	Immediate	Lung damage and flooding
Diphosgene	DP	Liquid	New mown hay, green corn	3,200	Immediate	Lung damage and flooding
Chloropicrin	PS	Liquid	Stinging pungent	20,000	Rapid	Inhalation
Blister (Vesicants)						
Sulfur mustard	H, HD	Liquid	Garlic	1,500	Delayed	Blisters, tissue burns, blood vessels injury, necrosis
Nitrogen mustard	HN-1	Liquid	Fishy or musty	1,500	Delayed	Blisters, tissue burns, blood vessels injury, necrosis
Nitrogen mustard	HN-2	Liquid	Soapy to fruity	3,000		Blisters, tissue burns, blood vessels injury, necrosis
Nitrogen mustard	HN-3	Liquid	None	1,500	Delayed	Blisters, tissue burns, blood vessels injury, necrosis
Phosgene oxime	CX	Solid or liquid	Sharp, penetrating	3,200 (Est)	Immediate	Violent irritation of mucus membranes of eyes, nose, lung
Lewisite	L	Liquid	Geranium	1,500	Rapid	Blisters, tissue burns, blood vessels injury, necrosis, systemic poisoning
Mustard lewisite	HL	Liquid	Garlic	1,500	Immediate stinging, delayed blistering	Blisters, tissue burns, blood vessels injury, necrosis, systemic poisoning
Phenyldichloroarsine	PD	Liquid	None	2,600	Immediate eye, slower skin effects	Irritation, nausea, vomiting, blisters
Ethyldichloroarsine	ED	Liquid	Fruity, biting; irritating	3,000–5,000	Immediate stinging, delayed blistering	Blisters, death
Methyldichloroarsine	MD	Liquid	None	3,000–5,000	Rapid	Blisters, skin, eyes, respiratory tract damage, systemic poisoning
Blood						
Hydrogen cyanide	AC	Gas or liquid	Bitter almond	2,500–5,000	Very rapid	Interfaces with oxygen uptake, accelerates the rate of breathing
Cyanogen chloride	CK	Gas	Bitter almond	11,000		Chocking, irritating, slow breathing
Arsine	SA	Gas	Garlic	5,000	Delayed	Damage to blood, liver, and kidneys
Nerve (Organophosphates)						
Tabun	GA	Liquid	None	400	Very rapid	Cessation of breathing, death

Sarin	GB	Liquid	None	100	Very rapid	Cessation of breathing, death
Cyclosarin	GF	Liquid	None Sweet, musk, peach, shellac	35	Very rapid	Cessation of breathing, death
Soman	GD	Liquid	Camphor	70	Very rapid	Cessation of breathing, death
VX	VX	Liquid	None	50		
Novichok, 5, 7 (Russian nerve agent)	N/A	N/A	N/A	N/A	N/A	5–10 the potency of VX
Riot Control Agents (Nonlethal)						
Tear (Lacrimators)						
Chloracetophenone	CN	Solid	Apple blossom	14,000	Instantaneous	Lacrimation, respiratory tract irritation
Chloracetophenone and chloroform	CNC	Liquid	Chloroform	11,000	Instantaneous	Lacrimation, respiratory tract irritation
Chloracetophenone in chloroform	CNS	Liquid	Flypaper	11,400	Instantaneous	Lacrimation, vomiting, and chocking
Chloracetophenone/benzene and carbon tetrachloride	CNB	Liquid	Benzene	11,000 (Est)	Instantaneous	Powerful lacrimation
Bromobenzylcyanide	CA	Liquid	Sour fruit	8–11,000 (Est)	Instantaneous	Lacrimation, eye and respiratory tract irritation
o-chlorobenzylmalononitrile	CS	Solid	Pepper	>50,000	Instantaneous	Highly irritating less
Vomiting						
Diphenylchlorarsine	DA	Solid	None	15,000 (Est)	Very rapid	Cold-like symptoms, headache, vomiting, nausea
Adamsite	DM	Solid	None	15,000	Very rapid	Cold-like symptoms, headache, vomiting, nausea
Diphenylcyanoarsine	DC	Solid	Bitter almond/garlic	10,000 (Est)	Rapid	Cold-like symptoms, headache, vomiting, nausea
Incapacitating (Hallucinating)						
3-Quinuclidinyl benzilate	BZ	Solid	None	20,000	Delayed	Tachycardia, dizziness, vomiting, dry mouth, blurred vision, stupor, random activity

Sources: Marrs, T.C. et al., *Chemical Warfare Agents: Toxicology and Treatment*, John Wiley & Sons, New York, 1996; Ellison, H., *Handbook of Chemical and Biological Warfare Agents*, CRC Publishers, Boca Raton, FL, 2000; USAMRICD (U.S. Army Medical Research Institute of Chemical Defense), *Medical Management of Chemical Casualties Handbook*, 3rd edn, Chemical Casualty Care Division. Aberdeen, MD, 2000. Gaylor, D.W., *The use of Haber's law in standard setting and risk assessment*, *Toxicology*, 149, 17–19, 2000.

[a] N/A, not available; Est, estimated for a 70 kg man.

[b] Inhalation lethality expressed as the median lethal exposure (LCt$_{50}$, mg·min/m^3) in a 70 kg man.

management regime includes oxygen therapy and high doses of steroids to prevent pulmonary edema. Treatment of pulmonary edema, if it occurs, is with pulmonary expiratory end pressure (PEEP) to maintain PaO_2 above 60 mmHg.

$$COCl_2 \xrightarrow{H_2O} CO_2 + 2\ HCl \tag{20.1}$$

Chloropicrin is a lung irritant and also a soil fumigant used for its broad biocidal and fungicidal properties, mainly for high-value crops such as strawberries, tomatoes, tobacco, and flowers; it is used to treat lumber as well. The Scottish chemist John Stenhouse first synthesized chloropicrin in 1848, and during World War I, it was classified as a choking agent or sometimes mixed with sulfur mustard to lower its freezing point. Currently, it is classified as a tear agent in the United States. On April 9, 2004, the British Broadcasting Corporation News (BBC, 2004) reported a gas attack using chloropicrin in Sofia, Bulgaria, in which 40 people were injured, one seriously, which brings chloropicrin back into focus as a potential terrorist chemical weapon. Chloropicrin is a colorless to light-green oily liquid with an intense, penetrating odor. Upon decomposition, chloropicrin can produce phosgene, nitrogen oxides, and chlorine compounds. The inhalation median lethal concentration (LC_{50}) for 10 min is 30 ppm, and bronchial or pulmonary lesions are produced at 20 ppm. Dermal and ocular contact with chloropicrin can lead to chemical burns or dermatitis, and prolonged eye exposure can lead to blindness. There is no antidote available for chloropicrin exposure, and in severe cases of respiratory compromise, especially if PaO_2 cannot be maintained above 60 mm Hg, PEEP is recommended to open the alveoli (Ellison, 2000).

20.3.1.2 Cyanide Agents (Blood Agents)

Cyanide agents were earlier referred to as *blood agents* because of their observed cyanide systemic effects in comparison to other agents, such as the blister agents, which were thought to produce only local effects. It has been shown, however, that most chemical agents are equally capable of exerting systemic effects. The blood agents include hydrogen cyanide [HCN] (AC), cyanogen chloride [NCCl] (CK), and arsine [AsH_3] (SA). Liquid cyanide delivered in enclosed munitions rapidly vaporizes and, if inhaled, forms the cyanide anion [CN^-] that is readily distributed to virtually every organ and tissue in the body. During World War I, cyanide was used by the French who deployed almost 4000 tons without much success, possibly because of the small (1–2 lb) caliber munitions used for its delivery. Another potential reason for its ineffectiveness is its high volatility and resulting quick dispersal before building up to a sufficiently high concentration, as inhalation is the primary route for cyanide toxicity. Cyanides bind to the enzyme cytochrome oxidase, blocking oxygen uptake and intracellular oxygen utilization to cause cyanosis and death. Hydrogen cyanide was also used in the United States for execution of criminals in *gas chambers*. Its medical management regime includes oxygen therapy and use of an antidote in a two-step process. In the first step, a methemoglobin-forming drug, such as amyl nitrite or sodium nitrite, is given to cyanotic patients to oxidize hemoglobin iron from Fe(II) to the Fe(III) state, which binds preferentially to cyanide (Equation 20.2). The second step is to provide a sulfur donor such as sodium hyposulfite to convert the toxic cyanide to nontoxic thiocyanate (Equation 20.3). A more effective antidote with fewer side effects is α-ketoglutaric acid. Decontamination of cyanides is conducted under basic conditions by using hypochlorite (Marrs et al., 1996).

$$CNHb \xrightarrow[\text{Sodium nitrite (IV)}]{\text{Amyl nitrite (inhalation)}} CNHb \tag{20.2}$$

$$CN^- \xrightarrow[\text{Sulfur donor}]{\text{Sodium thiosulfate}} SCN^- \tag{20.3}$$

20.3.1.3 Blister Agents (Vesicants)

Blister agents, or vesicants, were the major military threats during World War I. Their most common effects are dermal edema, erythema and blister formation, ocular irritation, conjunctivitis, corneal opacity and eye damage, upper airway sloughing, pulmonary edema, metabolic failure, neutropenia, and sepsis. In addition, blister agents can cause gastrointestinal effects and suppression of bone marrow stem cell production (USAMRICD, 2000). The major vesicants include sulfur mustard [$(ClCH_2CH_2)_2S$] (H), the nitrogen mustards [$(ClCH_2CH_2)_2NCH_2CH_3$, $ClCH_2CH_2)_2NCH_3$, and $ClCH_2CH_2)_3N$] (HN1, HN2, and HN3, respectively), lewisite [$ClCHCHAsCl_2$] (L), mustard lewisite (HL), a mixture of mustard and lewisite, and phosgene oxime [Cl_2CNOH] (CX).

Sulfur mustard, or mustard gas, which is actually a viscous liquid, has a long history of use on the battlefield as listed in Table 20.1. The exact mechanism of action for sulfur mustard injury has not been elucidated to a high degree of certainty, but several hypotheses have been proposed (Papirmeister et al., 1985, 1991; Somani and Babu, 1989). Figure 20.2 presents the current knowledge for the possible mechanism(s) of action of sulfur mustard toxicity (Sidell et al., 1997). It is possible that the application of modern biochemical tools, such as genomics and proteomics, may lead to a more concrete understanding of the process of sulfur mustard injury and repair. In recent years, however, numerous reports have been published suggesting that antioxidant therapy using *N*-acetylcysteine (NAC) and glutathione may potentially have a beneficial effect in the treatment of sulfur mustard injury, or accelerating the process of healing (Elsayed et al., 1992; Gross et al., 1993; Amir et al., 1998; Anderson et al., 2000; Atkins et al., 2000; Kumar et al., 2001; Das et al., 2003; Elsayed and Omaye, 2004). Because an antidote for sulfur mustard has not yet

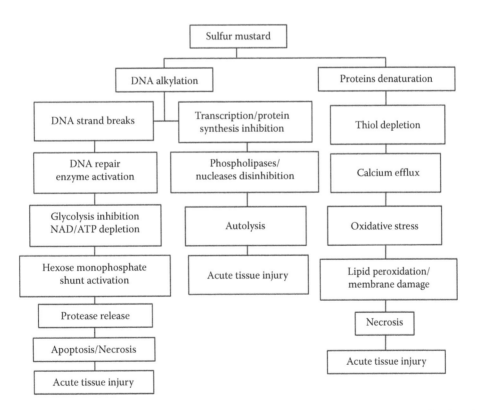

FIGURE 20.2 Diagram summarizing present knowledge of sulfur mustard mechanism(s) of action, leading to acute tissue injury. (Modified from Sidell, F.R. et al., Vesicants, in *Textbook of Military Medicine*, Part I, *Medical Aspects of Chemical and Biological Warfare*, Sidell, F.R., Takafuji, E.T., and Franz, D.R., eds., Walter Reed Army Medical Center, Washington, DC, 1997, pp. 197–228.)

been developed, medical management is limited to decontamination and supportive therapy. The standard decontamination procedure uses 0.5% household bleach (Smith and Dunn, 1991).

The nitrogen mustard class of blister agents contains three members. HN1 and HN3 are more likely to act as efficient vesicants (Mann, 1948), whereas HN2 has been used clinically for many years as a cancer chemotherapeutic agent. Exposure to the vapor of nitrogen mustards for 10 min affects the skin at concentrations as low as 30 ppm, and the eyes at 1 ppm or greater. The LC_{50} for inhalation exposure is 18 ppm, and the median lethal dose (LD_{50}) for skin exposure is 700 mg. Nitrogen mustards are not detoxified by the body, that is, the exposures are cumulative (Ellison, 2000). HN3 and sulfur mustard comprise another group, the *mustard gas group*, because they produce similar vesicating effects (Mann, 1948).

Lewisite (L) was named after Dr. Wilford Lee Lewis who first synthesized the compound in 1918, but its production was too late to join other CWAs used in World War I. Although it is typically used neat, in the absence of other chemicals, some countries such as the Soviet Union combined it with mustard (forming HL) to lower its freezing point for ground dispersal and aerial spraying (USAMRICD, 2000). Inhalation of lewisite has an LCt_{50} of 1500 mg·min/m^3, which is similar to that of mustard. Medical management of the exposure to lewisite and the lewisite/mustard mixture includes intramuscular injection of dimercaprol, often referred to as British anti-lewisite (BAL).

Phosgene oxime (CX) is not a true vesicant; rather, it is an urticant or nettle agent that produces corrosive type lesions of the skin and tissues that may not fully heal for 2 months. Inhaled, CX vapor is extremely irritating and causes almost immediate tissue damage upon contact (USAMRICD, 2000). Exposure causes lung membranes to swell and pulmonary edema formation that can be lethal (Ellison, 2000). Medical management is limited to immediate decontamination of the exposed areas and symptomatic management of the lesions.

20.3.1.4 Nerve Agents (Organophosphates)

Nerve agents are the most toxic class of chemical agents known, and, as illustrated in Figure 20.1, exhibit high lethality with low to high persistence. The chemical structures of a number of representative nerve agents are shown in Figure 20.3. The agents are typically organic esters of phosphoric acid, or organophosphate compounds. Nerve agents were first developed and stockpiled by Germany just before and during World War II but were never deployed in the battlefield. The only known military use of nerve agents was in the Iran–Iraq war during the 1980s. The G-series agents are the most volatile and tend to have high lethality and low persistence. Some G-series agents, however, can be thickened with different substances (commonly an acrylate) to increase their persistence and percutaneous toxicity. G-series agents are colorless to brownish liquids with a consistency ranging from that of water to that of light machine oil, and are hazardous through inhalation, skin, eye, and ingestion exposures. The LC_{50} for G-series agents is estimated to be as low as 1 ppm for a 10 min exposure, and the percutaneous LD_{50} is estimated as 300 mg per individual. V-series agents are less volatile and tend to possess both higher lethality and persistence than their G-series counterparts. These agents can be either solids or liquids at room temperature, and can also be thickened to increase their persistence. The LC_{50} for inhaled V agents is as low as 0.3 ppm for a 10 min exposure, and the LD_{50} for skin exposure is estimated as 100 mg per individual. Finally, and in a class by itself, is agent GV, considered to be an intermediate volatility agent. GV combines attributes of both the G- and V-series agents, that is, they have greater percutaneous toxicity than G-series agents and greater inhalation toxicity than V-series agents, but are not as stable as agents from either of the two series. All nerve agents pose significant additional hazards upon hydrolysis or combustion because they produce hydrogen fluoride, nitrogen oxides, and phosphorous oxides, as well as residual toxic organophosphates (Ellison, 2000).

The newest class of nerve agents is the novichok series developed by the Soviet Union during the 1970s and 1980s. The agents were designed to achieve three major objectives: (1) to be

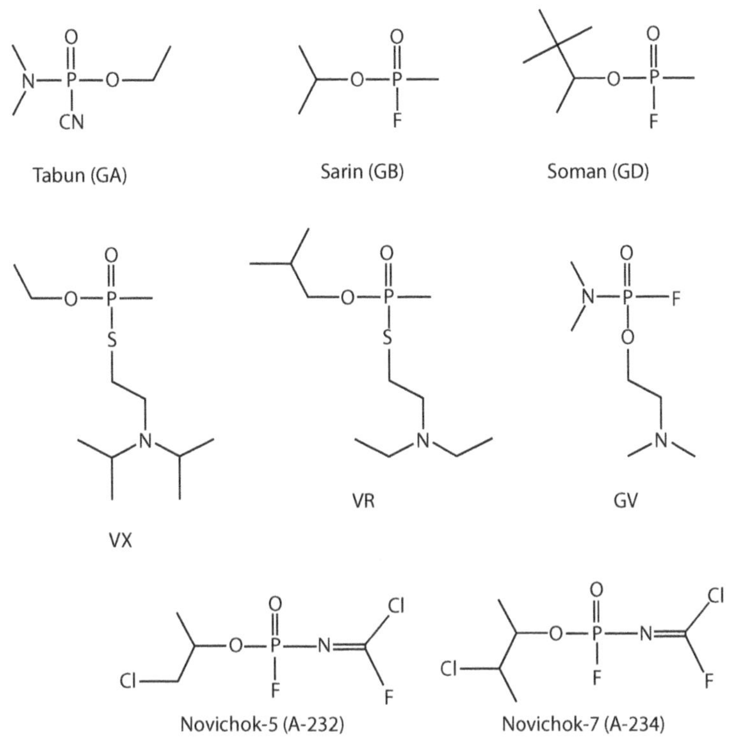

FIGURE 20.3   Chemical structures of nerve agents representative of G-series agents (GA, GB, and GD), V-series agents (VX and VR), agent GV, and Novichok-series agents (A-232 and A-234).

undetectable with standard North Atlantic Treaty Organization (NATO) chemical detection equipment, (2) to defeat NATO chemical protective equipment, and (3) to be significantly safer to handle than G- and V-series nerve agents. Information about the composition, persistence, and human toxicity of this class of agents is very scarce. However, available information does indicate that some novichok agents may be five to eight times as lethal as VX, a highly lethal agent in the US arsenal (Ellison, 2000). The novichok series comprises more than 100 structural analogs, with novichock-5 being the most promising from a military standpoint (Tucker, 2007); see Figure 20.3 for the chemical structure of novichok-5. Some novichok agents are binary agents that can be disguised as harmless agricultural products to evade international inspection and verification regimes. Binary agents exist as nerve agent precursors that are mixed in a munition to produce the agent just before its use. Because the precursors are, in general, significantly less hazardous than the agents themselves, handling binary agents is much safer and easier than authentic nerve agents. Moreover, agent precursors are usually much easier to stabilize than the authentic agents, providing a means to increase nerve agent shelf life.

Nerve agents are cholinesterase inhibitors that disrupt the functions of the nervous system by inhibiting the enzymes butyrylcholinesterase in the plasma and acetylcholinesterase in red blood cells and at cholinergic receptor sites in tissues and organs. These sites include smooth and skeletal muscles, central nervous system, and most exocrine glands (USAMRICD, 2000). The clinical effects of the nerve agents are due to accumulation of excess acetylcholine. A major characteristic associated with exposure to nerve gas vapor is bilateral miosis, often accompanied by sharp or dull pain around the eyes, dim and blurred vision, nausea, and occasional vomiting. Inhalation of nerve agents results in rhinorrhea, bronchoconstriction, and tightness of the chest. The effects on the central nervous system include loss of consciousness, seizure, and apnea. Medical management

includes decontamination, ventilation, administration of antidote, and supportive therapy. The antidotes available include atropine and pralidoxime chloride (2-PAMCl). In 2003, the U.S. Food and
Drug Administration approved the use of pyridostigmine bromide as a pretreatment against the
nerve agent soman (GD).

20.3.2 Nonlethal CWAs

Nonlethal CWAs include riot control agents, more commonly referred to as tear gasses or tear
agents, vomiting agents, and incapacitating agents, which can also be referred to as hallucinating
agents. Tear and vomiting agents are discussed in more detail in Chapter 21 of this book.

20.3.2.1 Riot Control Agents or Tear Agents (Lacrimators)

Nonlethal riot control agents, or tear gases, were the first CWAs deployed during World War I. The
French police used them to control rioters before the war, and the French military used them during
the war with limited success. Use of riot control agents on the battlefield decreased as more lethal
chemicals were developed, and currently, the agents are used mostly by law enforcement agencies.
Tear agents are used in many countries for military training, and were used extensively by the US
military in Vietnam for tunnel denial (TM-8-285, 1956; USAMRICD, 2000).

Tear agents are local irritants that cause transient discomfort and intense eye pain, forcing closure of the eyes that lead to a temporary inability to fight or resist arrest. The agents may also irritate
the respiratory tract, causing breathing difficulty. In high concentration, tear agents can irritate
skin, causing temporary burning as well as itching, nausea, and/or vomiting. These effects are also
transient and persist no more than a few minutes after the exposure is ended (TM-8-285, 1956;
USAMRICD, 2000). In enclosed spaces, very high concentrations of tear agents can become lethal
(Ellison, 2000). Tear agents are mostly solids with low vapor pressure, colorless to yellow in color,
and have a floral to a pepper-like odor. The agents dispersed as aerosols are not persistent; however,
release of large amounts resulting in significant solid or liquid deposition can pose a persistent hazard. Medical management is not required, as casualties of tear agents usually recover within 15 min
after removal from a contaminated environment. High-level exposures can produce dermatitis and
superficial skins burns resembling thermal burns, and are treated similarly. For more detailed information on tear agents, see Chapter 21 of this book.

20.3.2.2 Vomiting Agents

Originally developed as sternutators (sneezing agents), vomiting agents are nonlethal chemicals
with an estimated inhalation LC_{50} of 1000 mg/m^3 for a 10 min exposure and a short persistence
of approximately 30 min. These agents produce a pepper-like irritation of the upper respiratory tract, as well as eye irritation and lacrimation. Additionally, the agents can cause coughing,
sneezing, nausea, vomiting, and a general feeling of malaise, and produce local inflammation of
the upper respiratory tract, nasal accessory sinuses, and eyes. Vomiting agents can be dispersed
as aerosols to produce their effect by inhalation or by direct action on the eyes (TM 8-285, 1956;
Ellison, 2000; USAMRICD, 2000). Agents of this class are crystalline solids that are dispersed
by heat as fine particulate smokes ranging from canary yellow to white that become colorless
after dilution with air. Medical management is not required, and victims usually recover from
exposure within 2 h after removal from the contaminated atmosphere. Duration of the symptoms
can be shortened by vigorous exercise (Ellison, 2000). See Chapter 21 of this book for more
detailed information on vomiting agents.

20.3.2.3 Incapacitating Agents (Hallucinating Agents)

Incapacitating agents are nonlethal CWAs designed to create confusion or hallucinations. These
are anticholinergic agents that act as competitive inhibitors of acetylcholine at postsynaptic and
postjunctional muscarinic sites. 3-Quinuclidinyl benzilate (BZ) is a major example of this class of

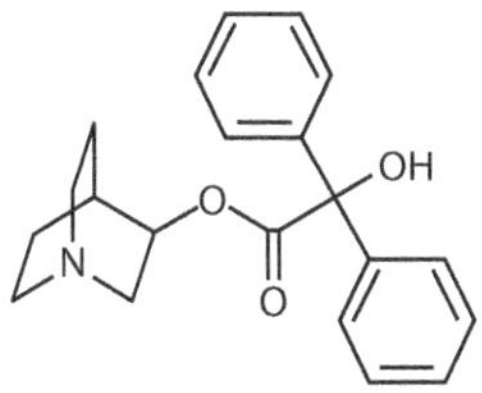

FIGURE 20.4 Chemical structure of 3-quinuclidinyl benzilate (BZ).

agents and is an odorless, nonirritating solid; its chemical structure is illustrated in Figure 20.4. It is extremely persistent in soils, water, and most other surfaces. Exposure to BZ results in tachycardia, dry mouth and mucous membranes, flushing, delirium, and hypertension. Toxicity can occur after exposure to these agents following inhalation, ingestion, or absorption through the skin. Medical management involves administering physostigmine salicylate and supportive care (TM 8-285, 1956; Ellison, 2000; USAMRICD, 2000).

20.4 CWA STOCKPILES IN THE UNITED STATES AND RUSSIAN FEDERATION

As stated earlier, knowledge about modern chemical warfare began during World War I when the Germans used chlorine gas. Many countries, including the United States, prohibited the use of CWAs in theaters of war. During the US Civil War, for example, the US War Department issued General Order 100 on April 24, 1863, proclaiming that "the use of poison in any manner, be it to poison wells, or foods, or arms, is wholly excluded from modern warfare." On July 29, 1899, The Hague Convention (II) with Respect to the Laws and Customs of War on Land was signed, which declared, "It is especially prohibited … to employ poison or poisoned arms." The Geneva protocol of 1925 also prohibited the use of poison in warfare.

After World War I, many governments wished to ban the use of chemical weapons in war because of the horrible means by which people were killed and injured. In 1925, at the League of Nations (the predecessor of the present United Nations), 38 nations signed the Geneva Protocol for "… the prohibition of the use in war of asphyxiating, poisonous, or other gases, and bacteriological methods of warfare"; the protocol has since been signed by over 130 nations. The protocol contained large loopholes, including that which does not prohibit the manufacture and threat of use of chemical weapons, it is vague on the term "other gases," and has no provisions for the punishment of countries that use such weapons illegally. A partial chronology of the major legal frameworks to control the use or prohibit the manufacture of CWAs is presented in Table 20.3.

During World War II, chemical weapons were not used in battle in the European theater of operations, although Italy used CWAs in the invasion of Ethiopia in Africa, and Japan used chemical and biological agents in the invasion of China (Table 20.1). After World War II, it was discovered that Germany had developed a new class of chemicals (nerve agents) that included tabun, sarin, and soman. The Germans possessed large quantities of the agents (20,000–30,000 tons of tabun) that were seized by Allied forces. However, the Russians took over most of their manufacturing plants and moved them to Volgograd, Russia. With the beginning of the Cold War, a new cycle of research and development of CWAs was underway, and large quantities of chemical agents and weapons were stockpiled. The United States and the Soviet Union were the two major producers of CWAs and possessed the largest stockpiles of chemicals. In 1992, the "Convention on the Prohibition of the Development, Production, Stockpiling, and Use of Chemical Weapons and Their Destruction" (Chemical Weapons Convention, CWC) was approved by the United Nations and went into effect in 1997. Tables 20.4 and 20.5 list the CWA stockpiles of the United States and the Russian Federation after both countries agreed to stop CWA production and began destruction of existing stockpiles (Federation of American Scientists, 2000; Harigel, 2003; Russian Munitions Agency, 2003).

TABLE 20.3

Partial Chronological Listing of Major Events and Developments of Legal Frameworks to Control the Use of Chemical Weapons

Date	Framework	Major Provisions/Achievement
April 24, 1863	The US War Department, General Order 100	The use of poison in any manner, be it poison wells, or foods, or arms, is wholly excluded from modern warfare.
July 29, 1899	The Hague Convention (II) with Respect to the Laws and Customs of War on Land	It is especially prohibited ... to employ poison or poisoned arms.
June 17, 1925	Geneva Protocol for the Prohibition of the Use in War of Asphyxiating, poisonous, or Other Gases, and Bacteriological Methods of Warfare	Many countries signed the Protocol but ratified it over 30 years later
April 10, 1972	Convention on the Prohibition of the Development, Production and Stockpiling of Bacteriological (Biological) and Toxin Weapons and on their Destruction	Biological Weapon Convention Signed by the United States, Britain, and the Soviet Union.
January 22, 1975	Geneva Protocol (1925)	The United States ratifies the Geneva Protocol, signed originally by the United States on 17 June 1925.
September 3, 1992	The Convention on the Prohibition of the Development, Production, Stockpiling, and use of Chemical Weapons and their destruction, CWC	The convention was approved by the UN.
April 29, 1997	The Convention on the Prohibition of the Development, Production, Stockpiling, and use of Chemical Weapons and their destruction, CWC	The convention entered into force. As of November 1997, 165 nations had signed the CWC, of which 104 nations had ratified the treaty.

Source: Adapted from The Avalon Project, Laws of war: Laws and customs of war on land (Hague II), Yale Law School, New Haven, CT, 1899, available at: http://www.yale.edu/lawweb/avalon/lawofwar/hague02.htm.

TABLE 20.4

Estimated Stockpiles of Chemical Weapons in the Russian Federation

Chemical Weapons Category 1		Chemical Weapons Category 2		Chemical Weapons Category 3	
Nerve agents (GA, GB, VX)	32.2[a]	Chocking agents (CG)	10[a]	NA	NA
Blister agents (H, L, HL)	7.8[a]	Artillery shells (122 mm) filled with (CG)	3844[b]	Inert chemical munitions, discontinuous and propellant charges	288,300[b]
Total[a]	40[a]		10[a]		0[c]

Source: Adapted from the Russian Munitions Agency, Chemical disarmament, chemical weapons, History of CW development, available at: http://www.munition.gov.ru/eng/hstchw.html, 2003.

[a] Metric tons × 1000.

[b] Units × 1000.

[c] All stockpiles of category 3 chemical weapons have been destroyed as of April 2003.

TABLE 20.5

Estimated Stockpiles of Chemical Weapons in the United States

Unitary Chemicals		Binary Chemicals	
Nerve agents (GA, GB, VX, TGA, TGB)	13352.63[a]	Methylphosphate diflouride (DF)	680.19[a]
		Isopropyl alcohol and isopropylamine (OPA)	
		Ethyl 2-diisopropylaminoethyl methylphosphate (QL)	
Blister agents (H, HD, HT, L)	172469.948[a]		
Total	30599.57[a]		680.19[a]

Source: Adapted from Harigel, G.G., *Chemical and Biological Weapons: Use in Warfare, Impact on Society and Environment*, Nuclear Age Peace Foundation, Santa Barbara, CA, 2003, pp. 1–25, available at: http://www.wagingpeace.org.

[a] Metric tons.

20.5 NUCLEAR WEAPONS

NWs are the most destructive weapons ever produced for use on the battlefield. Although NWs were used only twice during World War II, their acquisition served to deter aggression and coercion by other nations, and the threat of mutually assured destruction served as an instrument of policy and diplomacy. This class of weapons can produce large-scale destruction, such as those experienced by the cities of Hiroshima and Nagasaki in Japan. In addition, the radionuclides produced from NW detonation (fallout) can persist for many years in the environment, causing long-term adverse effects to all living systems and rendering large areas of land *off-limits* for use. With mounting global terrorism since the beginning of the twenty-first century, the concern of potential nuclear terrorist attacks has also increased. Although sophisticated NWs may not be an available option to terrorists for lack of technical expertise or manufacturing facilities, miniature nuclear devices as remnants of the cold war could potentially be manufactured, purchased, or obtained through illegal means. It is also possible for rudimentary electromagnetic weapons to be developed and used by terrorists determined to inflict maximum damage to military personnel or civilian populations. To understand the potential risk from the deployment of NWs or radiological devices, it was necessary to include a review of NWs.

20.5.1 MAJOR TYPES OF NUCLEAR WEAPONS

NWs can be in the form of bombs, missile warheads, artillery shells, and other devices. This review is limited to bombs, because a terrorist nuclear attack will most likely involve a small bomb or explosive radiological dispersal device (RDD). The major types of nuclear bombs include fission bombs, fusion bombs, salted bombs, neutron bombs, and some other devices. We also include deliberate radiological poisoning in this review, such as that of Alexander Litvinenko in 2006, as this has become another option available to terrorists to disperse radiological materials (Goldfarb and Litvinenko, 2007).

20.5.1.1 Fission Bombs

Fission bombs, or atom bombs (A bombs), are weapons that produce energy from the nuclear fission of heavy elements such as uranium or plutonium when bombarded by neutrons, resulting in the liberation of lighter elements together with more neutrons that can trigger chain reactions. In fission weapons, a mass of enriched uranium or plutonium is assembled into a supercritical mass (the amount of material needed to start an exponentially growing nuclear chain reaction), either by firing or shooting a subcritical projectile into a subcritical mass (the gun method), or by using explosives to compress a sphere of subcritical material to many times its original density (the implosion method) (FM 8-10-7, 1993). The latter approach is considered more sophisticated and is the only method that can be used for detonating plutonium weapons.

20.5.1.2 Fusion Bombs

Fusion bombs, or thermonuclear weapons, are generally known as hydrogen bombs (H bombs). These weapons require an extremely high temperature, such as that generated by the detonation of a fission core (an A bomb), to trigger chain reactions that use lighter elements, such as hydrogen or helium, to produce heavier elements and large amounts of energy (Alt et al., 1989; Rhodes, 1995).

20.5.1.3 Salted Bombs

Salted bombs are thermonuclear weapons that apply advanced designs to produce more devastation. One such design is to vary the material used in the shell of a nuclear device. For example, a fusion bomb with an outer shell of neutral (unenriched) uranium, upon detonation, produces intense fast neutrons that can cause the neutral uranium in the shell to fission, despite being unenriched, significantly increasing the explosive yield of the weapon. By using different elements in the NW shell, different fallout durations can be achieved. Using cobalt in the shell, for instance, provides long-term (years) fallout duration when the neutral cobalt is converted by the fusion-emitted neutrons to cobalt-60, a powerful, long-term gamma ray emitter. Other shell types include zinc, an intermediate (months) fallout emitter, and gold, a short-term (days) fallout emitter (Glasstone and Dolan, 1977).

20.5.1.4 Neutron Bombs

Neutron bombs are small, enhanced radiation thermonuclear weapons in which the neutrons generated by the fusion reaction are not absorbed within the weapon body. These weapons use x-ray mirrors and radiation cases made of chromium or nickel rather than the usual uranium or lead found in other NWs to permit neutrons to intentionally escape the weapon. This produces the release of an intense burst of high-energy, destructive neutrons intended to cause major casualties. In contrast to the salted bombs, however, the burst of ionizing radiation occurs only at the time of detonation and is not accompanied by additional enhanced residual radiation (Glasstone and Dolan, 1977; Cohen, 1983). The levels of neutron radiation released by a neutron bomb detonation are able to penetrate thick, protective materials such as armor.

20.5.2 DISTRIBUTION OF ENERGY FROM NUCLEAR DETONATION

Nuclear detonation results in massive release of energy in different forms and magnitudes. The major forms of energy are blast overpressure (BOP); thermal radiation, ionizing radiation (immediate), and residual radiation (delayed fallout); and electromagnetic pulses (EMPs) (Glasstone and Dolan, 1977). The distribution of energy from a moderate size nuclear detonation (less than 10 kilotons [kTs]) is shown in Table 20.6.

20.5.2.1 Blast Overpressure

Between 50% and 60% of the energy produced by a nuclear detonation is due to BOP and the resulting *blast wind*. BOP is the abrupt change in atmospheric pressure above (positive) or below (negative) ambient pressure. Peak overpressure from a 1 KT detonation can reach 200–400 kilopascals (kPa); depending on the distance from the center of the detonation point, different rates of lethality can occur (Table 20.7). In air, the magnitude of the overpressure is proportional to the air density, that is, the higher the air density, the greater the BOP (static overpressure). Another factor that contributes to the destructive power of a nuclear detonation is the drag exerted by the blast wind (dynamic pressure). Both factors contribute to the physical damage produced by a nuclear

TABLE 20.6

Distribution of Total Energy Produced by Nuclear Weapon Detonation

Physical Effect	Contribution (%)	Description
BOP		Abrupt changes in atmospheric pressure (above or below),
Standard Fission/Fusion	50	ambient level representing rapid compression/decompression
Enhanced Radiation Weapon	40	waves causing death or nonlethal injuries
Thermal Radiation		Direct effect of exposure to thermal energy (flash burns)
Standard Fission/Fusion	35	
Enhanced Radiation Weapon	25	Indirect effect of exposure to environmental fires (flame burns)
Initial Radiation		Neutron, gamma rays, and alpha and beta particles emitted
Standard Fission/Fusion	5	within the first minute after detonation (immediate)
Enhanced Radiation Weapon	30	
Residual (fallout) Radiation		Alpha, beta, and gamma rays (delayed)
• Standard Fission/Fusion	10	
• Enhanced Radiation Weapon	5	
EMP	1	Broadband, high-intensity, short-duration burst of
		electromagnetic energy caused by ionized air molecules.

Sources: Adapted from Alt, L.A. et al., Nuclear events and their consequences, in *Textbook of Military Medicine*, Part 1, Vol. 2, *Medical Consequences of Nuclear Warfare*, Walker, R.I. and Cerveny T.J., eds., Office of the Surgeon General, Department of the Army, Washington, DC, 1989, pp. 1–14; FM 8–10-7 Field manual: Health service support in a nuclear, biological and chemical environment, Department of the Army, Washington, DC, 1993.

TABLE 20.7

Range and Lethality of Exposure to Peak Overpressures from a 1 KT Nuclear Detonation

Peak Overpressure, kPa (psi)	Distance from Detonation Point, m (ft)	Lethality (%)
233–294 (34–43)	150 (492)	1
294–415 (43–60)	123 (404)	50
>415 (60)	110 (361)	100

Source: Adapted from FM 8-10-7 Field manual: Health service support in a nuclear, biological and chemical environment, Department of the Army, Washington, DC, 1993.

detonation (Glasstone and Dolan, 1977; FM 8-10-7, 1993; FM 8-9, 1996). The rapid compression/decompression cycle produced by the impacting BOP waves transmits energy through the human body and solid objects that can be extremely destructive. Three major types of injuries can occur from exposure to BOP. First, exposure to the shock waves results in primary blast injury, mostly to the hollow organ systems such as the respiratory, gastrointestinal, and auditory systems. The lungs, however, are the organs most sensitive to damage, and this injury can be fatal. Exposure to flying missiles results in secondary blast injury, and finally, displacement of the body by the incoming BOP waves impacting solid objects results in tertiary blast injury (Rössle, 1950; Schardin, 1950; White et al., 1971; Stuhmiller et al., 1991; Elsayed, 1997). The magnitude and frequency of the shock wave, the position and distance of the body in relation to the incoming wave, and the potential magnification of the BOP wave in enclosures further affect these injuries. A possible mechanism of injury from exposure to the blast was proposed in 1997. In this mechanism, incoming blast waves damage red blood cells and release heme iron, which initiates free radical–mediated reactions that continue to propagate after the initial blast exposure, resulting in delayed death (Gorbounov et al., 1995, 1997; Elsayed et al., 1997).

20.5.2.2 Thermal Radiation

Surface nuclear detonation emits a large amount (between 30% and 40%) of electromagnetic radiation (Table 20.6) outward from the fireball that includes visible, infrared, and ultraviolet lights. Most of the damage from thermal radiation is associated with skin burns and eye injury from the intense light, which can even start fires that are spread by the blast wind to points far from the fireball at the point of detonation (ground zero). Thermal radiation can cause burns, either directly following absorption of the thermal energy by the exposed surface (flash burns), or indirectly by fires in the environment (flame burns) spread by the blast wind. Several factors modulate the energy available to cause flash burns, including the detonation yield, distance from the fireball, height of the burst, weather conditions, and the surrounding environment. The degree of burn injury depends on the flux measured in calories per surface area of exposed skin (cal/cm^2) and duration of the thermal pulse. For example, thermal radiation from a 1 KT detonation would produce second-degree burns on exposed skin at a range of 0.78 km, pulse duration of 0.12 s, and flux of 4.0 cal/cm^2 (Glasstone and Dolan, 1977; FM 8-10-7, 1993; FM 8-9, 1996).

20.5.2.3 Initial Ionizing Radiation

Between 4% and 5% of the energy released from a nuclear detonation is in the form of initial neutrons and gamma rays. Neutrons are released almost entirely from the fission or fusion reactions, whereas initial gamma rays are produced from the same reactions and from the decay of short-lived fission products. Nuclear detonation results in four types of ionizing radiation within the first minute (immediate radiation) following detonation: neutrons and gamma, alpha, and beta radiation.

Neutrons are uncharged particles emitted in the first few seconds of detonation that do not ionize tissue directly and are not a fallout hazard. However, because of their significant mass, they can interact with other atomic nuclei, disrupting atomic structure and potentially causing 20 times more damage than gamma rays. Neutron emission constitutes 5%–20% of the total energy of a NW; however, neutron bombs produce a much higher yield of neutrons to kill people with radiation rather than destroy properties through blast effects.

Gamma rays are emitted during detonation and in fallout residual nuclear radiation. These are short-wavelength, uncharged, high-energy radiation similar to x-rays. Gamma rays are a serious health hazard due to their high energy and penetrability, which can result in whole-body radiation.

Beta particles are high-energy, high-speed electrons or positrons found primarily in fallout residual radiation. The particles have weak penetrability and can travel a short distance in tissues. However, exposure to large quantities of beta particles can produce damage to the basal stratum of the skin (beta burn) similar to thermal skin burns (Glasstone and Dolan, 1977; FM 8-10-7, 1993; FM 8-9, 1996). There are two types of beta decay, β^- and β^+, which, respectively, give rise to electrons and positrons.

Alpha particles consist of two protons and two neutrons bound together in a particle identical to a helium nucleus. The particles are heavier (four times the mass of neutrons) charged particles and a fallout hazard. Because of their size, alpha particles lack penetrability and can be stopped by dead layers of skin or by a military uniform. Alpha particles represent a negligible external hazard but can cause serious internal damage if inhaled or ingested.

20.5.2.4 Residual (Fallout) Radiation

The major hazard from residual nuclear radiation is radioactive fallout and neutron-induced hazards.

20.5.2.4.1 Fission Products

The splitting of heavy uranium or plutonium nuclei in fission reactions produces a large number of fission products (approximately 60 g/KT yield) with different half-lives, and an estimated activity of 1.1×10^{21} Bq (becquerel) (equivalent to 30 million kg of radium) approximately 1 min after detonation. These products have different decay rates (emission of beta and gamma rays), that is, environmental persistence ranging from very short (seconds) to long (months) to very long (years to many years). The fission products with the greatest concern to animals from fallout include strontium-90 ($t_{1/2}$ of 28 years), cesium-137 ($t_{1/2}$ of 30 years), carbon-14 ($t_{1/2}$ of 5800 years), strontium-89 ($t_{1/2}$ of 51 days), and iodine-131 ($t_{1/2}$ of 8 days). Iodine-131 poses the highest risk of internal exposure because it is deposited on the surface of vegetation, eaten by meat animals, then secreted in the milk, and finally ingested by humans. Iodine-131 concentrates preferentially in the thyroid gland and can ultimately destroy it. Strontium-89 and strontium-90 are produced in large amounts in nuclear fallout. Similar to iodine-131, consumption of contaminated fruits, vegetables, and milk is a major health hazard because strontium is readily incorporated in the bones.

20.5.2.4.2 Unfissioned Nuclear Material

Nuclear detonation does not consume all the enriched uranium or plutonium in the fission reactions, and much of it is dispersed unfissioned by the explosion that continues to decay slowly by emitting alpha particles.

20.5.2.4.3 Neutron-Induced Activity

When atomic nuclei exposed to a flux of neutrons capture a neutron, they become radioactive and continue to decay over a long period of time by emitting beta and gamma radiation that can activate other atoms in the environment, thus producing a hazardous area that becomes *off-limits* for use by humans.

20.5.2.5 Electromagnetic Pulses

An EMP is a burst of radiation containing electronic and magnetic components produced by certain types of high-energy explosions, especially nuclear explosions, or from a rapidly fluctuating magnetic field. In the case of a NW detonation, gamma rays (primarily photons) produced from the detonation collide with atoms in the mid-stratosphere, transferring energy to their electrons. The increased energy allows the electrons to escape from their electronic orbitals and become free electrons. This is known as the Compton effect, and the movements of the resulting free electrons produce an electric current referred to as the Compton current. Movement of these electrons starts in a generally downward direction at about 94% the speed of light (Longmire, 1986), but Earth's magnetic field couples with those of the free electrons and changes their motional direction to one perpendicular to the geomagnetic field. The interaction of Earth's magnetic field, and the downward flow of electrons, produces a very large, but very brief, EMP over the affected area. This is referred to as the E1 component of a nuclear EMP and inflicts most of its damage by generating electrical breakdown potential differences (voltages) in electronic devices to be exceeded. The E1 component fluctuates too quickly for lightning protectors to provide protection; and can destroy electronic equipment, wires, antennas, and metal objects; interfere with radiofrequency links and microcircuits; and can disable satellites, computers, and communications equipment. The E2 component, the second component of a nuclear EMP, is generated by the scattered gamma rays and inelastic gamma rays produced by neutrons form a nuclear detonation. The E2 component is an intermediate time pulse lasting from about 1 μs (microsecond) to 1 s. This component has many similarities to the EMPs produced by lightning; however, it is often much less intense than a lightning-derived EMP. Because of the widespread use of lightning protection technologies, the E2 component is the least likely EMP component to cause equipment damage. The final component of a nuclear EMP, the E3 component, is a very slow pulse lasting tens to hundreds of seconds, caused by the perturbation of Earth's magnetic field by the nuclear detonation and the relaxation of the field back to its natural state. The E3 component has similarities to a geomagnetic storm caused by a very severe solar flare. As is the case for geomagnetic storms, the E3 component can induce currents in long electrical conductors that can damage power lines and transformers. The existence of EMPs has been known since the 1940s when NWs were first developed and tested. However, their effect was not fully understood until 1962, when the United States conducted a series of high-altitude atmospheric tests and detonated a nuclear device about 250 miles above the Pacific Ocean in an area about 900 mi from Hawaii. The resulting EMP disrupted radio stations and electric equipment throughout Hawaii (Ricketts et al., 1976). There are no known biological effects from EMPs.

20.5.3 Chemical and Biological Effects of Ionizing Radiation

When an atom absorbs ionizing radiation, it can liberate an atomic particle (typically an electron, proton, or neutron, but sometimes an entire nucleus). Such an event can alter chemical bonds and produce ions, usually in ion pairs, that are especially reactive. This greatly magnifies the chemical and biological damage per unit energy of radiation because chemical bonds will be broken in this process. In terms of living cells, direct interaction of radiation with a cellular molecule could cause irreparable damage and lead to cell malfunction or death. Indirect interaction causing cell damage occurs when radiation interacts with water molecules in the body to create toxic molecules (free radicals) that could, in turn, affect neighboring sensitive molecules. The most radiosensitive organs in the body are those of the hematopoietic and gastrointestinal systems. Direct and indirect cellular radiation damage is a function of specific tissue sensitivity and radiation dose. Although high doses of radiation can cause cell death, lower doses can cause a variety of effects, including delays in the mitotic cycle phases, disruption of cell growth, and changes in permeability and motility. In general, radiosensitivity tends to be inversely related to the degree of cell differentiation. Prediction of radiation damage is very difficult because exposed organs are often unknown. The primary, immediate medical concern is severe acute radiation syndrome (ARS) caused by whole-body irradiation. The LD_{50} for exposed persons within a period of 60 days from radiation exposure ($LD_{50}/60$) is approximately 450 centigray (cGy). Cellular recovery is possible if a sufficient proportion of the stem cell population remains undamaged after irradiation injury. Although complete recovery from radiation can occur, late somatic relapses have a higher probability of occurring because of radiation-induced damage (Glasstone and Dolan, 1977; FM 8-10-7, 1993; FM 8-9, 1996). Table 20.8 lists the relative radiotoxicity, inhalation dose coefficients, and median lethal dose values for some of the more commonly encountered radionuclides.

20.5.4 Acute Radiation Syndrome and Cancers from Ionizing Radiation Exposure

ARS is a constellation of health effects that appear within 24 h of exposure to a large dose of ionizing radiation (10 cGy or greater) over a short period of time. The onset and type of symptoms experienced by ARS patients depends largely on the radiation exposure. Smaller doses result in gastrointestinal effects such as nausea and vomiting, and symptoms related to falling blood counts such as infection and bleeding. Larger doses can result in neurological effects and rapid death. ARS symptoms may persist up to several months after radiation exposure. Treatment is generally supportive with antibiotics, blood products, colony-stimulating factors, and stem cell transplants (Donnelly et al., 2010).

ARS is divided into three main medical presentations: hematopoietic, gastrointestinal, and cerebrovascular. The hematopoietic effects require exposure to the areas of the bone marrow actively forming blood elements (the pelvis and sternum in adults), and is marked by a sharp drop in the number of blood cells. This can result in infections due to low numbers of white blood cells, bleeding from a decreased number of platelets, and anemia due to depressed numbers of red blood cells (Donnelly et al., 2010). These changes can be detected with blood tests after receiving a whole-body dose as low as 25 cGy. The gastrointestinal effects require an absorbed dose of 6–30 Gy to the stomach or intestines (Donnelly et al., 2010). Nausea, vomiting, loss of appetite, and abdominal pain are usually seen within 2 h. The cerebrovascular syndrome typically occurs from exposure to the brain at absorbed doses greater than 30 Gy, although it may occur from doses as low as 10 Gy (Donnelly et al., 2010). It appears with neurological symptoms such as dizziness, headache, or decreased level of consciousness, occurring within a few minutes to a few hours after exposure, and an absence of vomiting. Table 20.9 lists the symptoms of ARS as a function of whole-body absorbed dose.

TABLE 20.8

Relative Radiotoxicity per Unit Activity, Inhalation Dose Coefficients per Unit Intake, and Median Lethal Dose Values for Selected Radionuclides

Radionuclide	Relative Radiotoxicity per Unit Activity[a]	Inhalation Dose Coefficient per Unit Intake[b] (Sv/Bq)	LD50[c] (mg/kg Body Mass)	LD50[c] Included Compounds
^{51}Cr	Moderate	3.20×10^{-11}	13–811	All Cr(IV) compounds
			183–2,365	All Cr(III) compounds
^{54}Mn	High	1.50×10^{-9}	225–3,730	All compounds
^{60}Co	High	1.00×10^{-8}	42–6,170	All compounds
^{63}Ni	Moderate	4.80×10^{-10}	39–9,000	All compounds
^{65}Zn	Moderate	1.60×10^{-9}	237->5,000	All compounds
^{89}Sr	High	6.10×10^{-9}	2350–2900	All compounds
^{90}Sr	High	3.60×10^{-8}		
^{109}Cd	Moderate	6.60×10^{-9}	100–300	All soluble compounds
^{129}I	High	3.60×10^{-8}	3,320–22,000	All compounds
^{131}I	High	7.4×10^{-9}		
^{134}Cs	High	6.60×10^{-9}	1,000–2,500	All soluble compounds
^{137}Cs	High	4.60×10^{-9}		
^{210}Pb	Very high	1.10×10^{-6}	90–300	All soluble inorganic compounds
^{210}Po	Very high	3.30×10^{-6}	0.006–0.015	Chloride only
^{224}Ra	High	3.00×10^{-6}	0.1–0.2	Chloride only
^{226}Ra	Very high	3.50×10^{-6}		
^{228}Th	Very high	3.20×10^{-5}	3,800	All soluble compounds
^{232}Th	Low	4.50×10^{-5}		
^{234}U	Very high	3.50×10^{-6}	114–1,580	All soluble compounds
^{235}U	Low	3.10×10^{-6}		
^{238}U	Low	2.90×10^{-6}		
^{238}Pu	Very high	4.60×10^{-5}	0.005–1.6	All soluble compounds
^{239}Pu	Very high	5.00×10^{-5}		
^{241}Am	Very high	4.20×10^{-5}	0.032–0.262	All soluble compounds
^{243}Cm	Very high	3.10×10^{-5}	0.0014	All soluble compounds

[a] The relative radiotoxicity per unit activity rates the ability of a substance to cause harmful effects because of its radioactivity. Radionuclides have been classified into four risk groups for protection purposes (IAEA, 1973): very high radiotoxicity (group I), high radiotoxicity (group II), moderate radiotoxicity (group III), and low radiotoxicity (group IV).

[b] Inhalation dose coefficients correspond to committed equivalent doses in an organ or tissue per unit intake $(h_T(t))$ or committed effective dose per unit intake $e(t)$, where t is the time period in years over which the dose is integrated (i.e., 50 years for adults). The term dose coefficient per unit intake, measured in Sv/Bq, is also used to mean dose coefficients.

[c] Measures acute toxicity and typically includes toxic effects causing death in a population of animals other than those from radioactivity.

Symptoms very similar to those of ARS may appear months to years after exposure as chronic radiation syndrome when the dose rate is too low to cause the acute form (Reeves and Ainsworth, 1995). Survivors of ARS and chronic radiation syndrome face an increased risk of cancer for the remainder of their lives, because the energy released by ionizing radiation typically damages cellular DNA. Mutagenic events do not occur immediately after irradiation; however, surviving cells appear to acquire a genomic instability that causes an increased rate of mutations in future generations. These cells will progress through multiple stages of neoplastic transformation that may culminate into a tumor after years of incubation. The neoplastic transformation can be divided into three major, independent stages: morphological changes to the cell, cellular immortality (losing

TABLE 20.9
Symptoms of Whole-Body Irradiation from External or Internal Absorption[a]

Phase of Syndrome	Feature	Dose Range (Gy)[b]				
		1–2	2–6	6–8	8–30	>30
Prodrome (immediate)	Nausea and vomiting					
	Incidence	5%–50%	50%–100%	75%–100%	90%–100%	100%
	Onset	2–6 h	1–2 h	10–60 min	<10 min	Minutes
	Duration	<24 h	24–48 h	<48 h	<48 h	NA[c]
	Diarrhea	None	None to mild	Heavy	Heavy	Heavy
	Incidence	—	<10%	>10%	>95%	100%
	Onset	—	3–8 h	1–3 h	<1 h	<1 h
	Headache	Slight	Mild to moderate	Moderate	Severe	Severe
	Incidence	—	50%	80%	80%–90%	100%
	Onset	—	4–24 h	3–4 h	1–2 h	<1 h
	Fever	None	Moderate	Moderate to severe	Severe	Severe
	Incidence	—	10%–100%	100%	100%	100%
	Onset	—	1–3 h	<1 h	<1 h	<1 h
	Central nervous system function	No impairment	Cognitive impairment for 6–20 h	Cognitive impairment for <24 h	Rapid incapacitation	Seizures, tremors, ataxia, and lethargy
Latent period		28–31 days	7–28 days	<7 days	None	None
Illness		Mild-to-moderate leucopenia, fatigue, and weakness	Mild-to-severe leucopenia, purpura, hemorrhage, infections, and epilation after 3 Gy	Severe leucopenia, high fever, diarrhea, vomiting, dizziness, and disorientation, hypotension, and electrolyte imbalance	Nausea, vomiting, severe diarrhea, high fever, electrolyte imbalance, and shock	NA[c]
Mortality	Incidence without Care	0%–5%	5%–100%	95%–100%	100%	100%
	Incidence with Care (%)	0%–5%	5%–50%	50%–100%	100%	100%
	Death	6–8 weeks	4–6 weeks	2–4 weeks	2–14 days	1–2 days

Sources: AFRRI (Armed Forces Radiobiology Research Institute), *Medical Management of Radiological Causalities*, 2nd edn., Armed Forces Radiobiology Research Institute, Bethesda, MD, 2003; Merck Manuals, Radiation exposure and contamination. Available at http://www.merckmanuals.com/professional/injuries_poisonong/radiation_exposure_and_contamination/radiation_exposure_and_contamnation.html, 2012. International Bureau of Weights and Measures, United States National Institute of Standards and Technology (NIST), *The International System of Units (SI)*, NIST Special Publication 330, Taylor, B.N. and Thompson, A., eds., Department of Commerce, National Institute of Standards and Technology, Gaithersburg, MD, 2008.

[a] Whole-body irradiation of up to 1 Gy is unlikely to cause any symptoms.

[b] 1 rad = 1 cGy; 100 rad = 1 Gy.

[c] Not applicable; patients die in <48 h.

normal, life-limiting cell regulatory processes), and adaptations that favor formation of a tumor (Little, 2000). Ionizing radiation increases the risk of certain types of cancer more than others. Leukemia is the most common radiation-induced cancer, with other types, including lung cancer, skin cancer, thyroid cancer, multiple myeloma, breast cancer, and stomach cancer, more prevalent than other cancers.

20.5.5 Radiation Hormesis Hypothesis

It is well established that exposure to ionizing radiation at even modest doses results in biological damage, and the amount of damage is directly proportional to radiological dose. This dose–effect relationship is described by the linear no-threshold (LNT) model for ionizing radiation exposure, which is illustrated in Figure 20.5 for hypothetical data in the region just above background radiation. According to the model, radiation is always considered harmful, so that the sum of several, very small exposures is considered to have the same effect as one larger exposure. The LNT model is used exclusively in radiation protection to quantify radiation exposure and set regulatory limits. There is evidence, however, that very low doses of ionizing radiation just above background levels stimulates activation of repair mechanisms that protect against disease (mostly cancers), and further, these repair mechanisms are not activated in the absence of ionizing radiation. Such evidence has led to the radiation hormesis hypothesis, which dictates that any description of biological damage as a function of radiation dose must include the protective effects from the repair mechanisms. A plot of hypothetical data displaying radiological hormesis is also shown in Figure 20.5. A unique feature to any hormetic response is the zero equivalence point (ZEP), the safety threshold dose between positive and negative biological effects from radiation exposure. The *low dose regime* for a hormesis model occurs between background radiation and the ZEP.

Hormetic models have also been hypothesized for several other types of exposures, including the positive and negative effects of drugs (all drugs are toxic in large quantities) and ethyl alcohol (alcoholic beverages) on heart disease and stroke (Calabrese and Cook, 2006). Hormesis models for dose response are under vigorous debate at present (Kaiser, 2003), and the notion that hormesis is important for chemical risks regulations is not widely accepted (Axelrod et al., 2004). The concept of

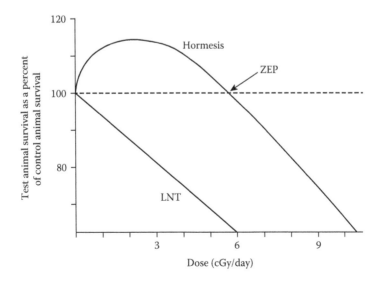

FIGURE 20.5 Plot of hypothetical data for animals exposed to controlled levels of ionizing radiation just above background levels that display a linear LNT and a hormetic (hormesis) response. The ZEP for the hormesis response is indicated.

hormesis has generated the most controversy when applied to ionizing radiation. Consensus reports by the U.S. National Research Council (NRC), the National Council of Radiation Protection and Measurements, and the United Nations Scientific Committee on the Effects of Atomic Radiation (UNSCEAR) cite that there is no clear evidence for radiation hormesis in humans, and in the specific case of the NRC, radiation hormesis is outright rejected. However, a report by the French National Academy of Medicine (Académie nationale de Médecine, 2005) rejected the LNT model for carcinogenic risk at doses less than 20 millisievert (mSv). The academy believes that there may be several dose–effect relationships rather than one, and these relationships may have variables such as target tissue, radiation dose, dose rate, and individual sensitivity factors.

A large body of research with results in support of radiological hormesis derives from animal research. Starting from about 1940, hundreds of studies describing positive biological effects from exposure to low levels of ionizing radiation can be found throughout the literature. For example, rats fed radioactive uranium dust produced more young than controls (Stone, 1942), and other rats exposed to a 2.5 Gy dose of x-rays demonstrated both superovulation and superimplantation (Hahn and Ward, 1967). And when compared to controls, sterility was significantly reduced in mice previously exposed to 2–3 Gy of x-rays (Luning, 1960; Spalding et al., 1981). Other studies found that mice irradiated with a 200 milligray (mGy) dose of x-rays protects them against further x-ray exposure and ozone gas (Miyachi, 2000), and gamma irradiation of mice at 1 mGy/h prevents the development of cancer induced by methylcholanthrene (Sakai et al., 2002). Cells in culture have also been shown to develop protection mechanisms from low-level radiation exposures (Azzam et al., 1994; de Toledo et al., 2006).

Some studies involving human subpopulations exposed to low levels of ionizing radiation have also reported results indicative of radiological hormesis. One of the most cited concerns residents of Kerala, India, and its immediately surrounding area. These residents are exposed to about 8 mSv/h of gamma radiation from the monazite sand in the area, about 80 times the dose rate equivalent in London, England. The sand contains roughly a third of the world's economically recoverable reserves of radioactive thorium. A decade-long study of 69,958 Kerala residents reported no excess cancer risk from their exposure to the terrestrial gamma radiation (Nair et al., 2009). The excess relative risk of cancer, excluding leukemia, was reported at -0.13 Gy^{-1}, suggesting a reduced risk of cancer. It was also found that the leukemia incidence was not significantly related to the high background radiation in the area. Another well-cited case involves the inhabitants of a series of 180 radioactive apartment buildings in Taiwan. In 1982, more than 20,000 tons of steel was accidently contaminated with cobalt-60, much of which was used in the construction of the apartment buildings. Tens of thousands of the buildings' residents were exposed to gamma radiation at levels up to 1000 times the background level; the average excess cumulative dose was 47.7 mSv, with a maximum cumulative dose of 2360 mSv. Many residents lived in the buildings a full decade before the radioactivity was even traced and measured, and some continued living in the buildings for another decade. A report of thousands of the buildings' residents stated that the cancer rates for those living in the radioactive buildings were 3.6% lower than the prevailing Taiwanese rate, more than 20 times lower than expected (Chen et al., 2004, 2007). Additionally, the incidence of congenital malformations was reduced to about 7% of the incidence of the general Taiwanese public.

It should be cautioned that assessing the risk of radiation at low doses below 100 mSv and low dose rates less than 6 mSv/h is very difficult and controversial (Mullenders et al., 2009; Tubiania et al., 2009). This is largely because the baseline cancer rate is very high at about 42%, and fluctuates about 40% due to lifestyle and environmental effects (Parkin et al., 2011; Boice, 2012), obscuring the subtle effects of low-level radiation. Epidemiological studies are often capable of detecting elevated cancer rates as low as 20%–30%, but for low doses below 100 mSv, the predicted elevated risks are only 0.1%–0.4%, and excess cancer cases, if present, cannot be detected due to confounding factors, errors, and biases (Boice, 2010, 2012). Given the absence of direct epidemiological evidence, there is considerable debate as to whether the dose–response behavior for radiation exposure below 100 mSv is hormetic, linear, and indicative of the LNT model, or something else.

20.5.6 THE 1950–2020 LIFE SPAN STUDY AND THE LNT MODEL

Given the controversy over the using the LNT model and radiation hormesis, it is no surprise that using the model for predicting health-related issues for low-level radiation exposures has come under some scrutiny lately (see Cuttler and Pollycove, 2009, for example). This is especially true for data from the atomic bomb survivors of Hiroshima and Nagasaki, Japan. Conducted since 1947 by what is now the Radiation Effects Research Foundation, the Life Span Study (LSS) on the cancer mortality of the Hiroshima–Nagasaki survivors appears to support conclusions that the effects of radiation exposure predicted with the LNT model are grossly overstated and do not reflect real risks to the public. The enormous release of heat from the Hiroshima and Nagasaki bombs killed between 150,000 and 200,000 of the total 429,000 inhabitants of the two cities. The LSS cohort of 86,572 is roughly half of the survivors who were within 2.5 km (about 1.6 mi) of the bomb detonations. The study found 344 excess solid cancer deaths and 87 leukemia deaths among cohort members 40 years after the bomb detonations (Shimizu et al., 1996), much less than 1% of the predicted 10%–30% (Cuttler and Pollycove, 2009). Furthermore, since 36,000 members of the cohort were far enough away from ground zero not to have received severe radiation exposure, the fraction is only 0.7% of the irradiated survivors. Of the cohort, 56% were alive in 1991 and 38,902 had died, indicating that about 1% of them had died from radiation-induced cancer. It is estimated that about 800 will have died from radiation by 2020, the end of the LSS (Lapp, 1995). This is, again, about 1%.

The survivors of the bombings experienced many confounding health risks, such as thermal burns, wounds from blast debris, infection, thirst, starvation, lack of sanitation, shelter, and medical care. And their entire social structure had been destroyed. The excess number of cancer deaths in this population is the basis for estimating the number of excess fatal cancers due to any radiation exposure in the environment. Of the 4489 survivors who received more than 500 mSv, a total of 634 died of cancer, 196 more than expected. The number of expected cancer deaths was determined by fitting a straight line to the excess cancer data, and this line, the LNT model, was extrapolated several orders of magnitude into the low dose range. Expected doses in a nuclear reactor accident would most likely fall into this low dose range. Because cancer originates from a mutated cell and radiation alters cells, radiation protection professionals use the LNT model to predict the excess risk of cancer mortality. Evidence of beneficial effects among these survivors (Kondo, 1993) and radiation hormesis are ignored. Cohen (1990) has pointed out that the LNT model suggests that cancer risk in an animal should be proportional to its mass. Larger, heavier animals have more cells and, therefore, should have a greater incidence of cancer for the same adsorbed dose than smaller, lighter animals. Such proportionality of cancer risk with size has not been observed.

20.6 RADIOLOGICAL DISPERSAL DEVICES (DIRTY BOMBS) AND RADIOLOGICAL EXPLOSIONS

RDDs, or *dirty bombs* as they are also known, are weapons that combine conventional explosives, such as dynamite, trinitrotoluene (TNT), pentaerythritol (PETN), or plastic explosives (C-3 and C-4), with radioactive materials. The radioactivity is dispersed through the initial blast of the conventional explosive, producing airborne radiation and contamination. In general, the conventional explosive detonation associated with the RDD produces greater lethality and damage than the radioactive material. The intent of using dirty weapons is to cause disruption of services and generate public fear and panic rather than heavy casualties, death, and destruction, and the outcome fulfills the objectives of terrorists. Dirty bombs can be small devices that fit into a suitcase, or the devices can be as large as a truck. A dirty bomb, however, is not a NW, because it does not involve nuclear fission reactions or generate destructive blast waves or thermal and ionizing radiation.

Many types of radioactive materials with military, industrial, or medical application could be used to produce dirty weapons by encasing a highly radioactive material in lead and surrounding the device with conventional explosives. Depending on atmospheric conditions, the resulting explosion would disperse radiation over a wide area and could cause potential health problems. For example, weapons-grade uranium and plutonium, or spent nuclear fuel, could be used and would be the deadliest materials. However, they are also the most difficult to obtain and handle by a potential terrorist. In contrast, medical-grade chemicals used mostly in cancer treatment, such as radium and cesium isotopes, could be used with less risk in handling, but also with reduced potency.

20.6.1 HEALTH HAZARDS AND CONTAMINATION FROM RADIOLOGICAL EXPLOSIONS

Three nuclear incidents illustrate the potential health hazards and dissemination of radioactive contamination associated with radiological explosions.

20.6.1.1 The 1986 Chernobyl Nuclear Accident

The 1986 disaster at the Chernobyl Nuclear Power Plant in Ukraine (then officially Ukrainian SSR) remains the largest nuclear disaster in history, and is the first to be rated level 7 on the International Nuclear Event Scale (INES). The disaster began during a systems test on April 26, 1986 at reactor 4 of the Chernobyl plant. Following a sudden and unexpected power surge and emergency shutdown attempt, an exponentially larger spike in power output occurred, which led to a reactor vessel rupture and a series of steam explosions. These events exposed the graphite moderator of the reactor to air, causing its ignition. The resulting fire was very difficult to fight, leading to the release of radioactive materials into the atmosphere (U.S. Naval Academy [USNA], 2001). The accident resulted in a series of explosions that totally destroyed reactor 4, and a wide distribution of radioactive materials that contaminated Northern Ukraine, Belarus, Russia, and Scandinavia. At least 31 were killed directly in the disaster, and 140 developed significant ARS among the plant personnel and rescue workers. The battle to contain the contamination and avert greater catastrophe ultimately involved about 530,000 workers and a cost estimated at 18 billion rubles (about 58 million US dollars). A 31 km (about 19 mi) exclusion zone was established immediately surrounding the site, and farming was prohibited in parts of the radiation fallout area, even 18 years after the disaster.

Repercussions from the Chernobyl nuclear disaster were felt decades after the accident, and will continue to be felt for many years to come. Located about 3 mi from the Chernobyl site, the city of Pripyat was evacuated hours after the reactor explosions and remains completely evacuated to this day. Much of the area around the city and power plant has reverted back to forest. By December 2006, a large concrete sarcophagus containment structure had been erected to seal the reactor and its contents. From 1986 to 2000, 350,400 people were evacuated and resettled from the most severely contaminated areas of Belarus, Russia, and Ukraine. The Union of Concerned Scientists estimates that there will be 50,000 excess cancer cases resulting in 25,000 excess cancer deaths as a consequence of the accident. Some conclude that among the billions of people worldwide exposed to radioactive contamination from the Chernobyl accident, nearly a million premature cancer deaths occurred between 1986 and 2004 (Yablokov et al., 2009). Ukrainian officials estimate the area around the Chernobyl site will not be safe for human life again for another 20,000 years.

20.6.1.2 The 1987 Goiânia Nuclear Incident

The Goiânia incident was a radiological contamination accident that occurred in central Brazil during September 1987. Scrap metal scavengers broke into an abandoned hospital site in the city of Goiânia and stole a radiotherapy source containing about 93 g (3.3 oz) of highly radioactive cesium-137 chloride. The shielding capsule containing the cesium chloride was broken into pieces and passed to friends and relatives. This resulted in four dead, 249 contaminated, 14 overexposed, and 114,000 subjected to continuous monitoring. The four casualties were buried in lead-lined coffins overlaid with concrete. The accident led to an extensive cleanup effort throughout the Goiânia area. Large amounts of contaminated topsoil were removed from several sites, 85 houses were ultimately demolished, and truckloads of personal possessions were destroyed. The accident spread radioactive contamination throughout three districts of central Brazil, and even after cleanup, 7 TBq (tetabecquerel or 10^{12} Bq) of radioactivity still remained unrecovered.

20.6.1.3 The 2011 Fukushima Daiichi Nuclear Disaster

The Fukushima Daiichi nuclear disaster is only the second such disaster to be rated INES level 7. The incident was a series of equipment failures, nuclear meltdowns, and releases of radioactive materials at the Fukushima I Nuclear Power Plant at Okuma, Fukushima, Japan, following the Tohoku earthquake and tsunami on March 11, 2011 (IEEE Spectrum, 2011; New Zealand Herald, 2011). Immediately following the earthquake, all three operating reactors shut down automatically and emergency generators came online to power electronics and cooling systems. The tsunami following the earthquake, however, quickly flooded the low-lying emergency generators, causing their immediate failure. This cut power to the pumps needed to continuously circulate coolant water through the reactors for the several days following shutdown to prevent nuclear meltdowns, and the three reactors promptly overheated. Flooding the reactors with seawater (an emergency procedure) started only after a lengthy delay imposed by the Japanese government, but this was too late to prevent meltdowns. In the hours and days that followed, reactors 1, 2, and 3 experienced full meltdowns (Tanabe, 2011). The intense heat and pressure of the melting reactors caused a chemical reaction between the nuclear fuel metal cladding and the remaining water surrounding them to produce explosive hydrogen gas. As first responders worked diligently to cool and shut down the reactors, several hydrogen-air chemical explosions occurred (Hyer Regional News, 2011; IAEA, 2011). Concerns about the repeated small explosions, the atmospheric venting of radioactive gasses, and the possibility of much larger explosions led to a 20 km (approximately 12 mi) radius evacuation around the plant, with a request for people living 20–30 km (about 12–18 mi) from the plant to remain indoors.

Radioactive contamination from the Fukushima disaster was still significant years after the incident. Substantial amounts of radioactive materials were released into the ground and ocean waters for several months following the disaster, until recirculating units could be put into place for cooling. Measurements by the Japanese government 30–50 km (about 10–30 mi) from the plant revealed cesium-137 levels high enough for concern, leading to a ban of the sale of food grown in the area and a temporarily recommendation to not use tap water for preparing food for infants (Nature News Blog, 2012). In May 2012, the company operating the Fukushima plant reported that at least 900 PBq (petabecquerel or 10^{15} Bq) had been released into the atmosphere during March 2011, equivalent to 17% of that from the Chernobyl disaster (*Kyodo News*, 2012). Approximately 25,000 died from repercussions of the earthquake and tsunami; however, there were no deaths due to radiation exposure. At least six workers exceeded their lifetime limits for ionizing radiation, and more than 300 have received significant radiation doses. The accident led to trace amounts of radiation, including iodine-131, cesium-134, and cesium-137, migrating all over the word, with reports of radioactive fallout detected in New York State, Alaska, Hawaii, Oregon, California, Canada, Australia (*New Scientist*, 2011), Fiji, Malaysia, and New Guinea (CTBTO, 2011). On 16 December 2011, Japanese authorities declared the plant stable, although decades are required to decontaminate the surrounding areas and decommission the plant (BBC, 2011). Radiation was still leaking into the ocean from the reactor site as of October 2012. Fishing in the waters around the site was still prohibited, and levels of cesium-134 and cesium-137 in fish caught in the area were no different than those found immediately after the disaster (Buesseler, 2012).

20.6.2 Preparedness for Radiological Dispersal Device Attacks

According to the International Atomic Energy Agency (IAEA), 370 instances of nuclear smuggling have been reported from 1993 to 2000. In April 2001, customs officers in Uzbekistan intercepted a truck carrying 10 lead-lined boxes filled with strontium-90-irradiated scrap metal. To prepare the public, several national and international organizations have issued information sheets and public guidance in the event of a dirty-bomb detonation (Ford, 1998; CDC, 2003; USNRC, 2003; WHO, 2003; OSHA, 2004).

20.7 DELIBERATE RADIOLOGICAL POISONING

Deliberate radiological poisoning is the use of radiological materials to intentionally contaminate one or more individuals for causing their incapacitation, death, or other damaging effects. The poisonings have only recently come into sharp focus by the media and general public since the Alexander Litvinenko incident in 2006. In contrast to the use of nuclear bombs or RDDs, ingestion by the intended victim is the most common means used to deliver the radiological materials, but inhalation may also be used. Alpha-emitters are the most ideal radionuclides for use in deliberate radiological poisonings. Alpha particles are unable to penetrate very far through matter and are brought to rest by only a few centimeters of air, making them the safest and easiest type of radionuclides to handle. Biological damage from alpha-emitters occurs only after their inhalation, ingestion, or absorption through open wounds. And because they can undergo multiple ionizations within very short distances, alpha-emitters have the potential to cause considerably more biological damage than beta- or gamma-emitters from the same amount of emitted energy. The fact that alpha-emitters are invisible to the human senses, lethal in small (milligram) doses, and easily transported with a minimal chance of detection makes them very attractive as a potential weapon. Although there have been many deliberate radiological poisonings over the last several decades, many are unreported. There have been four high-profile deliberate radiological poisonings since 1957.

20.7.1 THE 1957 NIKOLAI KHOKHLOV POISONING

Nikolai Khokhlov was an officer for the Soviet Committee for State Security (KGB), who in 1953, defected to the United States after refusing to assassinate a Soviet dissident. In 1957, while attending an anti-Soviet conference in Frankfurt, Germany, he drank half a cup of coffee that he described as *not as good as usual*. Hours later, Khokhlov became violently ill and was hospitalized and treated for thallium poisoning. Although physicians believed his death was imminent at times, Khokhlov recovered 1 month later and was released from hospital care. Khokhlov returned to the United States where he ultimately died from a heart attack in 2007. The incident was a failed assassination attempt by the Thirteenth KGB Department (Andrew and Mitrokhin, 2000), and is the first deliberate radiological poisoning ever documented. Many details of the incident remain unclear, including the thallium isotope used in the poisoning.

20.7.2 THE 2003 YURI SHCHEKOCHIKHIN POISONING

Yuri Shchekochikhin was a Russian investigative journalist, writer, and lawmaker in the Russian Duma. He made his name by writing about, and campaigning against, the influence of organized crime and corruption in Russia. On June 17, 2003, while on a business trip to Ryazan, about 200 km (120 mi) from Moscow, Shchekochikhin fell sick with flu-like symptoms. He returned home to Moscow that day with a fever, sore throat, body aches, and a burning sensation over his skin. Shchekochikhin's health rapidly deteriorated over the next several days, and he was finally hospitalized on June 21. During the next 12 days, the journalist's organs failed one by one, and his skin literally peeled off his body. He lost all his hair, and his lungs, liver, kidneys, and, finally, his brain stopped functioning on July 3. Physicians at the Moscow Central Clinical Hospital officially declared his death as the result of Lyell's Syndrome, a severe allergic reaction to medications or infections, but the causative allergen was never identified. Moreover,

Shchekochikhin's clinical test results were classified as *medical secret*, making it impossible for family, colleagues, and the media to access them. Although Shchekochikhin's symptoms over the last 16 days of his life were consistent with ARS, the radiological material or a radionuclide causing the symptoms was never identified. A specialist of the Federal Security Service of the Russian Federation (FSB) reported that Shchekochikhin was mostly likely poisoned with thallium (BBC, 2007).

20.7.3 THE 2004 ROMAN TSEPOV POISONING

Roman Tsepov was a businessman from Saint Petersburg, Russia, and confidant to Vladimir Putin (president and prime minister of the Russian Federation from 1999 to present) during his tenure at the Saint Petersburg City Administration, as well as Putin's personal bodyguard. He fell sick on September 11, 2004 with symptoms of severe food poisoning. Upon his hospitalization, physicians were unable to prevent the poison from affecting his bone marrow and producing symptoms of ARS, which ultimately led to his death on September 24, 2004. A postmortem investigation found clear evidence of radiological poisoning but did not identify the radionuclide (*Sunday Times*, 2006).

20.7.4 THE 2006 ALEXANDER LITVINENKO POISONING

The Alexander Litvinenko incident is the most documented deliberate radiological poisoning to date. Litvinenko was a former officer of the KGB and FSB who fled from court prosecution in Russia and received political asylum in the United Kingdom. On November 1, 2006, he suddenly fell ill and was hospitalized. Litvinenko died 3 weeks later on November 23, 2006, becoming the first confirmed victim of lethal polonium-210-induced ARS (Acton et al., 2007; Goldfarb and Litvinenko, 2007). His symptoms appeared consistent with an administered dose of approximately 2 GBq (gigabecquerel or 10^9 Bq), corresponding to about 10 µg of polonium-210, about 200 times the median lethal dose for ingestion. The polonium was identified in Litvinenko's body only after his death. Physicians and Scotland Yard investigators could not detect polonium earlier because it is not a gamma-emitter. Unlike most common radiation sources, polonium-210 emits only alpha particles, which are invisible to normal radiation detectors. Litvinenko was tested for alpha-emitters using special equipment only hours before his death (Goldfarb and Litvinenko, 2007). In January 2007, British police investigators found a teapot highly contaminated with polonium-210 at the hotel that Litvinenko was known to have visited the day he fell ill. Following the incident, more than 700 people in the United Kingdom were tested for polonium-210 contamination. Of these, more than 100 showed polonium-210 concentrations in their urine indicative of some contamination from the poisoning incident, but fewer than 20 had results indicating committed effective doses of greater than 6 mSv (UNSCEAR, 2011).

20.8 CONCLUSIONS

Almost 100 years have passed since the introduction of CWAs during World War I, and significant advances have been realized in understanding most of their mechanisms of action. In spite of such advances, however, effective antidotes do not exist for several of these agents. For example, sulfur mustard continues to be produced, stockpiled, and even used by some nations, albeit clandestinely, and the agent could be possessed and possibly used by terrorists. There is no antidote available for sulfur mustard, and only decontamination and supportive treatment are recommended for its victims. Another major threat to military and civilian personnel is BOP injury resulting from explosive detonations. In recent years, blast research has become focused in the development and improvement of protective vests that were indispensable in Iraq and elsewhere, as well as the development of more accurate predictive models of blast injury. Current research on chemical and biological terrorism should also include the effects of blast injury. Detonation of an explosive device, with or without the incorporation of chemical or biological materials, has the potential to cause primary blast injury, combined with penetrating wounds from secondary blast injury caused by shattered glass or flying missiles. Moreover, detonation of explosive devices containing chemical or biological agents has the potential of added adverse effects from their agents surviving the explosion. The results of nuclear or radiological detonations can be substantially more complicated. These detonations involve tremendous blast injury from the initial detonation, burn injury from thermal radiation, internal injury from ionizing radiation, as well as long-term effects, such as cancer, also from ionizing radiation. All these adverse effects can be further complicated by inhalation injury from the smoke and combustion gasses in case of a fire, not to mention the psychological stress in response to the unexpected attack. Even deliberate radiological poisonings that commonly target only one individual have real potential for contaminating many others, sometimes at locations at considerable distances from the initial incident. Collectively, these events suggest that awareness and consideration should be given to the complex multiple injuries that can occur simultaneously, and that preparedness and response to potential terrorist attacks should include medical management of combined injuries.

QUESTIONS

1. Why are alpha-emitters ideal for deliberate radiological poisoning?
   Answer: They are hard to detect, easy to handle (low energy), and following inhalation can cause lots of damage.
2. What are the mild signs and symptoms following exposure to radionuclides?
   Answer: nausea, vomiting, diarrhea, headache, and fever.
3. What are the differences between nuclear devices and dirty bombs?
   Answer: Dirty bombs disburse radiological debris by using conventional explosives such as dynamite and plastic explosives, while NWs use nuclear fission (A bombs) or fusion (H bombs) to generate their enormous destructive power, also causing a release of radiological debris in the process.
4. Describe hermetic response and why it is so difficult to prove.
   Answer: Small amounts generate positive rather than detrimental effects, and is contrasted to nonlinear threshold models.

REFERENCES

Académie Nationale de Médecine, Académie des Sciences (National Academy of Medicine, Academy of Sciences), Aurengo, A., Averbeck, A., Bonnin, A., Le Guen, B., Masse, R., Monier, R., Tubiana, M., and Valleron, A.-J., Dose-effect relationships and estimation of the carcinogenic effects of low doses of ionizing radiation, de Vatharie, Florent, Paris, 2005.

Acton, J.M., Rogers, M.B., and Zimmerman, P.D., Beyond the dirty bomb: Re-thinking radiological terror, *Survival*, 49, 151–168, 2007.

AFRRI (Armed Forces Radiobiology Research Institute), *Medical Management of Radiological Causalities*, 2nd edn., Armed Forces Radiobiology Research Institute, Bethesda, MD, 2003.

Alt, L.A., Forcino, C.D., and Walker, R.I., Nuclear events and their consequences, in *Textbook of Military Medicine*, Part 1, Vol. 2, *Medical Consequences of Nuclear Warfare*, Walker, R.I. and Cerveny T.J., eds., Office of the Surgeon General, Department of the Army, Washington, DC, 1989, pp. 1–14.

Amir, A., Chapman, S., Gozes, Y., Sahar, R., and Allon, N., Protection by extracellular glutathione against sulfur mustard induced toxicity in vitro, *Hum. Exp. Toxicol.*, 17, 652–660, 1998.

Anderson, D.R., Byers, S.L., and Vesely, K.R., Treatment of sulfur mustard (HD)-induced lung injury, *J. Appl. Toxicol.*, 20, S129–S132, 2000.

Andrew, C. and Mitrokhin, V., *The Mitrokhin Archive: The KGB in Europe and the West*, Penguin Press, New York, 2000.

Atkins, K.B., Lodhi, I.J., Hurley, L.L., and Hinshaw, D.B., *N*-acetylcysteine and endiothelial cell injury by sulfur mustard, *J. Appl. Toxicol.*, 20, S125–S128, 2000.

Axelrod, D., Burns, K., Davis, D., and von Larebeke, N., "Hormesis"-an inappropriate extrapolation from the specific to the universal, *Int. J. Occup. Med. Environ. Health*, 10, 335–339, 2004.

Azzam, E.I., Raaphorst, G.P., and Mitchel, R.E.J., Radiation-induced adaptive response for protection against micronucleus formation and neoplastic transformation in C3H 10T1/2 mouse embryo cells, *Radiat. Res.*, 138, s28–s31, 1994.

BBC (British Broadcasting Corporation), Forty hurt in Sofia gas accident, 2004. Available at http://news.bbc.co.uk/2/hi/europe/3613979.stm, Accessed June 16, 2014.

BBC (British Broadcasting Corporation), Russia's poisoning "without a poison," 2007. Available at http://news.bbc.co.uk/2/hi/programmes/file_on_/46324241.stm, Accessed July 30, 2007.

BBC (British Broadcasting Corporation), Japan PM says Fukushima nuclear site finally stabilized, 2011. Available at http://www.bbc.co.uk/news/world-asia-16212057, Accessed May 21, 2014.

Boice Jr., J.D., Uncertainties in studies of low statistical power, *J. Radiol. Prot.*, 30, 115–120, 2010.

Boice Jr., J.D., Radiation epidemiology: A perspective on Fukushima, *J. Radiol. Prot.*, 32, N33–N40, 2012.

Buesseler, K.O., Fishing for answers off Fukushima, *Science*, 338, 480–482, 2012.

CDC (Centers for Disease Control and Prevention), *Dirty Bombs. Fact Sheet.* Centers for Disease Control and Prevention, Department of Health and Human Services, Atlanta, GA, 2003.

Calabrese, E.J. and Cook, R., The importance of hormesis to public health, *Environ. Health Perspect.*, 114, 1631–1635, 2006.

Chen, W.L., Luan, Y.C., and Shieh, M.C., Is chronic radiation an effective prophylaxis against cancer? *J. Am. Phys. Surg.*, 9, 6–10, 2004.

Chen, W.L., Luan, Y.C., Shieh, M.C., Chen, S.T., Kung, H.T., Soong, K.L., Yeh, Y.C. et al., Effects of cobalt-60 exposure on health of Taiwan residents suggest new approach needed in radiation protection, *Dose-Response*, 5, 63–75, 2007.

Cohen, B.L., *The Nuclear Energy Option: An Alternative for the 90s*, Plenum Press, New York, 1990.

Cohen, S.T., *The Truth about the Neutron Bomb: The Inventor of the Bomb Speaks Out*, Morrow, New York, 1983.

CTBTO (Comprehensive Nuclear-Test-Ban Treaty Organization), Fukushima-related measurements by the CTBTO – Page 1, 2011. Available at http://www.ctbto.org/press-centre/highlights/2011/fukushima-related-measurements-by-the-ctbto/fukushima-related-measurements-by-the-ctbto-page-1/, Accessed May 21, 2014.

Cuttler, J.M. and Pollycove, M., Nuclear energy and health and the benefits of low-dose radiation hormesis, *Dose-Response*, 7, 52–89, 2009.

Das, S.K., Mukherjee, S., Smith, M.G., and Chatterjee, D., Prophylactic protection by *N*-acetylcysteine against the pulmonary injury induced by 2-chloroethyl ethyl sulfide, a mustard analogue, *J. Biochem. Mol. Toxicol.*, 17, 177–184, 2003.

de Toledo, S.M., Assad, N., Venkatachalam, P., Li, L., Howell, R.W., Spitz, D.R., and Azzam, E.I., Adaptive responses to low-dose/low-dose-rate gamma rays in normal human fibroblasts: The role of growth architecture and oxidative metabolism, *Radiat. Res.*, 166, 849–857, 2006.

Donnelly, E.H., Nemhauser, J.B., Smith, J.M., Kazzi, Z.N., Farfán, E.B., Chang, A.S., and Naeem, S.F., Acute radiation syndrome: Assessment and management, *South. Med. J.*, 103, 541–546, 2010.

Ellison, H., *Handbook of Chemical and Biological Warfare Agents*, CRC Publishers, Boca Raton, FL, 2000.

Elsayed, N.M., Toxicology of blast overpressure, *Toxicology*, 121, 1–15, 1997.

Elsayed, N.M., Gorbunov, N.V., and Kagan, V.E., A proposed biochemical mechanism for blast overpressure induced hemorrhagic injury, *Toxicology*, 121, 81–90, 1997.

Elsayed, N.M. and Omaye, S.T., Biochemical changes in mouse lung after subcutaneous injection of the sulfur mustard 2-chloroethyl 4-chlorobutyl sulfide, *Toxicology*, 199, 195–206, 2004.

Elsayed, N.M., Omaye, S.T., Klain, G.J., and Korte, Jr., D.W., Free radical-mediated lung response to the monofunctional sulfur mustard butyl 2-chloroethy sulfide after subcutaneous injection, *Toxicology*, 72, 153–165, 1992.

Evison, D. and Hinsley, D., Chemical weapons, *Br. Med. J.*, 324, 332–335, 2002.

Federation of American Scientists, *Weapons of Mass Destruction around the World: Chemical Weapons*, Federation of American Scientists, Washington, DC, 2000. Available at www.fas.org.

FM 8-10-7, *Field Manual: Health Service Support in a Nuclear, Biological and Chemical Environment*, Department of the Army, Washington, DC, 1993.

FM 8-9, *Virtual Naval Hospital–NATO Handbook on Medical Aspects of NBC Defensive Operations AMedP-6(B)*, Part I: Nuclear. Departments of the Army, Navy, and Air Force, Washington, DC, 1996.

Ford, J.L., *Radiological Dispersal Devices: Assessing the Transnational Threat*, Strategic Forum, Institute for National Strategic Studies, National Defense University, Washington, DC, 1998.

Gaylor, D.W., The use of Haber's law in standard setting and risk assessment, *Toxicology*, 149, 17–19, 2000.

Glasstone, S. and Dolan, P.J., eds., *The Effects of Nuclear Weapons*, Department of Defense and Energy Research and Development Administration, Washington, DC, 1977.

Goldfarb, A. and Litvinenko, M., *Death of a Dissident: The Poisoning of Alexander Litvinenko and the Return of the KGB*, Free Press, New York, 2007.

Gorbounov, N.V., Elsayed, N.M., Kisin, E.R., Kozlov, A.V., and Kagan, V.E., Air blast overpressure induces oxidative stress in rat lungs: Interplay between hemoglobin, antioxidants, and lipid peroxidation, *Am. J. Physiol.*, 272, L320–L334, 1997.

Gorbounov, N.V., Osipov, A.N., Day, B.W., Zyas-Rivera, B., Kagan, V.E., and Elsayed, N.M., Reduction of ferrylmyoglobin and ferrylhemoglobin by nitric oxide: A protective mechanism against ferryl hemoprotein-induced oxidations, *Biochemistry*, 34, 6689–6699, 1995.

Gross, C.L., Innace, J.K., Hovatter, R.C., Meier, H.L., and Smith, W.J., Biochemical manipulation of intracellular glutathione levels influences cytotoxicity to isolated human lymphocytes by sulfur mustard, *Cell Biol. Toxicol.*, 9, 259–267, 1993.

Hahn, E.W. and Ward, W.F., Increased litter size in the rat x-irradiated during the estrous cycle before mating, *Science*, 157, 956–957, 1967.

Harigel, G.G., *Chemical and Biological Weapons: Use in Warfare, Impact on Society and Environment*, Nuclear Age Peace Foundation, Santa Barbara, CA, pp. 1–25, 2003. Available at www.wagingpeace.org.

Hyer Regional News, Hydrogen explosions Fukushima nuclear power plant: What happened? 2011. Available at http://www.hyer.eu/news/regional-news/hydrogen-in-nuclear-accidents-what-is-the-role-of-the-gas-in-fukushima, Accessed on May 21, 2014.

IAEA (International Atomic Energy Agency), *Safe Handling of Radionuclides*. International Atomic Energy Agency, Vienna, Austria, 1973.

IAEA (International Atomic Energy Agency), Fukushima nuclear accident update log 2011. Available at http://www.iaea.org/newscenter/news/2011/fukushima150311.html, May 08, 2011.

IEEE Spectrum, Explainer: What went wrong in Japan's nuclear reactors. Available at http://spectrum.ieee.org/tech-talk/energy/nuclear/explainer-what-went-wrong-in-japans-nuclear-reactors, April 04, 2011.

International Bureau of Weights and Measures, United States National Institute of Standards and Technology (NIST), *The International System of Units (SI)*, NIST Special Publication 330, Taylor, B.N. and Thompson, A., eds., Department of Commerce, National Institute of Standards and Technology, Gaithersburg, MD, 2008.

Joy, R.J.T., Historical aspects of medical defense against chemical warfare, in *Textbook of Military Medicine*, Part I, *Medical Aspects of Chemical and Biological Warfare*, Sidell, F.R., Takafuji, E.T., and Franz, D.R., eds., Walter Reed Army Medical Center, Washington, DC, 1997, pp. 87–109.

Kaiser, J., Sipping from a poisoned chalice. *Science*, 302, 376–379, 2003.

Kondo, S., *Health Effects of Low-Level Radiation*, Kinki Press, Osaka, Japan (Medical Physics Publishing, USA), 1993.

Kumar, O., Sugendran, K., and Vijayaraghavan, R., Protective effect of various antioxidants on the toxicity of sulphur mustard administered to mice by inhalation or percutaneous routes, *Chem. Biol. Interact.*, 134, 1–12, 2001.

Kyodo News, TEPCO puts radiation release early in Fukushima release at 900 PBq, 2012. Available at http://english, kyodonews.jp/news/2012/05/159960.html, May 24, 2012.

Lapp, R.E., *My Life with Radiation: Hiroshima Plus Fifty Years*, cogito Books, Hexhan, Northumberland, U.K., 1995.

Little, J.B., Ionizing radiation, in *Cancer Medicine*, 6th edn., Kufe, D.W., Pollock, R.E., Weichselbaum, R.R., Bast, Jr., R.C., Gansler, T.S., Holland, J.F., and Frei, III, E., eds., B.C. Decker, Hamilton, Ontario, Canada, 2000.

Longmire, C.L., *Justification and Verification of High-Altitude EMP Theory*, Part 1, LLNL-9323905, Lawrence Livermore National Laboratory, Livermore, CA, 1986.

Lunung, K., Studies of irradiated mouse populations, *Hereditas*, 46, 668–674, 1960.

Marrs, T.C., Maynard, R.L., and Sidell, F.R., *Chemical Warfare Agents: Toxicology and Treatment*, John Wiley & Sons, New York, 1996.

Merck Manuals, Radiation exposure and contamination, 2012. Available at http://www.merckmanuals.com/professional/injuries_poisonong/radiation_exposure_and_contamination/radiation_exposure_and_contamnation.html, March 14, 2013.

Miyachi, Y., Acute mild hypothermia caused by a low-dose of x-irradiation induces a protective effect against mid-lethal doses of x-rays, and a low level concentration of ozone may act as a radiomimetic, *Br. J. Radiol.*, 73, 289–304, 2000.

Mullenders, L., Atkinson, M., Herwig, P., Sabatier, L., and Bouffler, S., Assessing cancer risks of low-dose radiation, *Nat. Rev. Cancer*, 9, 596–604, 2009.

Nair, R.R.K., Balakrishnan, R., Suminori, A., Jayalekshmi, P., Nair, M.K., Gangadharan, P., Koga, T., Morishima, H., Seiichi, N., and Sugahara, T. Background radiation and cancer incidence in Kerala, India-Karanagappally cohort study, *Health Phys.*, 96, 55–66, 2009.

Nature News Blog, World Health Organization weighs in on Fukushima, 2012. Available at http://blogs.nature.com/news/2012/05/world-health-organization-weighs-in-on-fukus hima.html, October 06, 2013.

New Scientist, Fukushima radioactive fallout nears Chernobyl levels, 2011. Available at http://www.newscientist.com/article/dn20285-fukushima-radioactive-fallout-nears-ch ernobyl-levels.html. Accessed on March 19, 2013.

New Zealand Herald, Japan's unfolding disaster "bigger than Chernobyl" 2011. Available at http://www.nzherald.co.nz/world/news/article.cfm?c_id=2&objectid=10716671, Accessed March 19, 2013.

OSHA (Occupational Safety and Health Administration), *Radiological Dispersal Devices (RDD)/Dirty Bombs*, Occupational Safety and Health Administration, Department of Labor, Washington, DC, 2004.

Papirmeister, B., Feister, A.J., Robinson, S.I., and Ford, R.D., *Medical Defense against Mustard Gas: Toxic Mechanisms and Pharmacological Implications*, CRC Press, Boca Raton, FL, 1991.

Papirmeister, B., Gross, C.L., Meir, H.L., Petrali, J.P., and Johnson, J.B., Molecular basis for mustard-induced vesication, *Fundam. Appl. Toxicol.*, 5, S134–S149, 1985.

Parkin, D.M., Boyd, L., and Walker, L.C., 16. The fraction of cancer attributable to lifestyle and environmental factors in the UK in 2010, *Br. J. Cancer*, 105, s77–s81, 2011.

Reeves, G.I. and Ainsworth, E.J., Description of the chronic radiation syndrome in humans irradiated in the former Soviet Union, *Radiat. Res.*, 142, 242–243, 1995.

Rhodes, R., *Dark Sun: The Making of the Hydrogen Bomb*, Simon & Schuster Publishers, New York, 1995.

Ricketts, L.W., Bridges, J.E., and Miletta, J., *EMP Radiation and Protective Techniques*, John Wiley & Sons Publishers, New York, 1976.

Rössle, R., Pathology of blast effects, in *German Aviation Medicine, World War II*, Vol. 2, U.S. Government Printing Office, Washington, DC, 1950, pp. 1260–1273.

Russian Munitions Agency, Chemical disarmament, chemical weapons. History of CW development, 2003. Available at www.munition.gov.ru/eng/hstchw.html, November 12, 2003.

Sakai, K., Iwasaki, T., Hoshi, Y., Nomura, T., Oda, T., Fujita, K., Yamada, T., and Tanooka, H., Suppressive effect of long-term low-dose rate gamma irradiation on chemical carcinogenesis in mice, *Int. Cong. Ser.*, 1236, 487–490, 2002.

Schardin, H., The physical principles of the effects of a detonation, in *German Aviation Medicine, World War II*, Vol. 2, U.S. Government, Washington, DC, 1950.

Shimizu, Y., Pierce, D.A., Preston, D.L., and Mabuchi, K., Studies of the mortality of atomic bomb survivors, Report 12, Part I, Cancer: 1950–1990, *Radiat. Res.*, 146, 1–27, 1996.

Sidell, F.R., Urbanetti, J.S., and Smith, W.J., Vesicants, in *Textbook of Military Medicine*, Part I, *Medical Aspects of Chemical and Biological Warfare*, Sidell, F.R., Takafuji, E.T., and Franz, D.R., eds., Walter Reed Army Medical Center, Washington, DC, 1997, pp. 197–228.

Smart, J.K., History of chemical and biological warfare: An American perspective, in *Textbook of Military Medicine*, Part I, *Medical Aspects of Chemical and Biological Warfare*, Sidell, F.R., Takafuji, E.T., and Franz, D.R., eds., Walter Reed Army Medical Center, Washington, DC, 1997, pp. 9–86.

Smith, W.J. and Dunn, M.A., Medical defense against blistering chemical warfare agents, *Arch. Dermatol.*, 127, 1207–1213, 1991.

Somani, S.M. and Babu, S.R., Toxicodynamics of sulfur mustard, *Int. J. Clin. Pharmacol. Ther. Toxicol.*, 27, 419–435, 1989.

Spalding, J.F., Brooks, M.R., and Tietjen, G.L., Comparative litter reproduction characteristics of mouse populations for 82 generations of x-irradiated male progenators, *Proc. Soc. Exptl. Biol. Med.*, 166, 237–240, 1981.

Stone, R., Health protection activities of the plutonium project, *Proc. Am. Philos. Soc.*, 90, 11–19, 1942.

Stuhmiller, J.H., Phillips, Y.Y, and Richmond, D.R., The physics and mechanisms of primary blast injury, in *Textbook of Military Medicine*, Part 1, Vol. 5, *Conventional Warfare, Ballistic, Blast, and Burn Injuries*, Bellamy, R. and Zajtchuk, R., eds., Office of the Surgeon General, Department of the Army, Washington, DC, 1991, pp. 241–270.

Sunday Times, The Putin bodyguard riddle, 2006. Available at http://www.timesonline.co.uk/article/0,2087-2484298,00.html, April 04, 2013.

Tanabe, F., Analysis of core melt accident in Fukushima Daiichi-unit 1 nuclear reactor, *J. Nucl. Sci. Technol.*, 48, 1135–1139, 2011.

The Avalon Project, *Laws of War: Laws and Customs of War on Land (Hague II)*, Yale Law School, New Haven, CT, 1899. Available at www.yale.edu/lawweb/avalon/lawofwar/hague02.htm.

TM 8-285, *Technical Manual: Treatment of Chemical Warfare Casualties*, Departments of the Army, Navy, and Air Force, Washington, DC, 1956.

Tubiana, M., Feinendegen, L.E., Yang, C., and Kaminski, J.M., The linear no-threshold relationship is inconsistent with radiation biologic and experimental data, *Radiology*, 251, 13–22, 2009.

Tucker, J.B., *War of Nerves: Chemical Warfare from World War I to Al-Qaeda*, Anchor Books, Sioux City, IA, 2007.

UNSCEAR (United Nations Scientific Committee on the Effects of Atomic Radiation), Sources and effects of ionizing radiation: UNSCEAR 2008 report to the general assembly with scientific annexes, Vol. II: Scientific annexes C, D, and E, United Nations, New York, 2011.

USAMRICD (U.S. Army Medical Research Institute of Chemical Defense), *Medical Management of Chemical Casualties Handbook*, 3rd edn., Chemical Casualty Care Division, Aberdeen Proving Ground, MD, 2000.

USNA (U.S. Naval Academy), Explosives. Naval applications of chemistry, Chemistry Department, U.S. Naval Academy, Annapolis, MD, 2001.

USNRC (U.S. Nuclear Regulatory Commission), Fact sheet on dirty bombs, 2003. Available at http://www.nrc.gov/reading-rm/doc-collections/fact-sheets/dirty-bombs.html, April 08, 2013.

White, C.S., Jones, R.K., Damon, E.R., and Richmond, D.R., The biodynamics of airblast, Technical Report, DNA 2738-Tl, Department of Defense, Washington, DC, 1971.

WHO (World Health Organization), Health Protection Guidance in the Event of a Nuclear Weapons Explosion, WHO/RAD Information Sheet, Radiation and Environmental Health Unit, World Health Organization, Geneva, Switzerland, 2003.

Yablokov, A.V., Nesterenko, V.B., and Nesterenko, A.V., *Chernobyl: Consequences of the Catastrophe for People and the Environment*, Wiley-Blackwell, Hoboken, NJ, 2009.

21 Emergency Response Planning Guidelines

Finis Cavender

CONTENTS

21.1 INTRODUCTION

21.1.1 BACKGROUND AND NEED FOR COMMUNITY PROTECTION

In December 1984, the release of 40 tons of methyl isocyanate (MIC) in Bhopal, India, emphatically underscored the need for the development of chemical emergency response plans. The great surprise was that so little was known about the toxicity of MIC. It was primarily used as a captive chemical intermediate in the production of carbaryl, and a toxicological profile had never been developed. The lack of a toxicological base was, in part, due to the following: Its acute toxicity was so great that long-term studies had never been considered. In this tragedy, more than 3,800 residents died and an additional 200,000 suffered adverse health effects (Union Carbide Corporation, 2013). The public in India and even worldwide was enraged that an event of this magnitude could happen, because the local community had no idea that such a dangerous chemical was housed in such large quantities, just across the road from their homes (Rusch, 1993; Cavender and Gephart, 1994; Cavender, 2002, 2006).

Globally, residents living near chemical plants demanded that they be given information on chemicals being produced or used in their communities. They also insisted that solutions be found that would prevent similar catastrophes from happening again. As a result, many chemical companies began developing emergency response plans, most of which included health-based exposure values. However, if each company had their own exposure values, the toxicological basis and rationale for each set of values might vary from one company to the next based on the data available to them. Such variance would be confusing and could lead the public to ask, "Which values should we use?" A better approach was for stakeholder chemical companies to work together in developing a single set of emergency response exposure values for chemicals of interest, eliminating the possibility of a multiplicity of numbers that would also be confusing to emergency response planners and managers (Rusch, 1993; Cavender and Gephart, 1994; Kelly and Cavender, 1998; Cavender, 2002, 2006).

The outcry of the public in the United States resulted in the promulgation of the Superfund Amendments and Reauthorization Act of 1986, which contained provisions entitled Emergency Planning and Community Right-to-Know. This legislation, along with provisions under Title III of the Clean Air Act of 1990, required local communities to set up emergency response plans. How was the public to develop emergency response plans without a single set of exposure values? Without a unified set of exposure values based on acute toxicity data, each planning activity would result in differing numbers for every community across the nation. Thus, it was all the more important that a single set of exposure values be developed for each chemical for emergency response planning purposes.

21.1.2   Were Health-Based Numbers Suitable for Emergency Planning Available in 1986?

In 1986, occupational exposure guidelines and standards such as the American Conference of Governmental Industrial Hygienist's threshold limit values (2005), the American Industrial Hygiene Association (AIHA) Workplace Environmental Exposure Levels (WEELs) (AIHA, 2005b), the National Institute of Occupational Safety and Health's (NIOSH) recommended exposure limits (RELs) (NIOSH, 1992), and the Occupational Safety and Health Administration's permissible exposure limits (Occupational Safety and Health Administration, 1995) were available for a good number of chemicals. However, these values were inappropriate for evaluating brief emergency exposures to the general public. These values were developed for the protection of healthy workers. They were based primarily on repeated dose studies conducted over months and years to simulate the daily exposure of workers over their working lifetime. The levels were set as a time-weighted average over the usual 8 h workday and serve to protect against both acute and chronic health effects. They were not designed to protect children, the elderly, or otherwise compromised individuals, such as individuals with asthma or alcohol dependency.

In addition to these occupational standards, several types of guidelines have been developed for use in emergency situations involving a single exposure to a specific chemical or mixture that may cause adverse health effects. In the 1950s, the National Research Council (NRC) began setting operational emergency numbers as advisory responses to specific exposure scenarios. These emergency numbers were developed principally for the Department of Defense (DOD) and were intended for young robust individuals. For instance, the Navy needed levels for confined spaces such as being submerged in a submarine. In 1964, such exposure levels were termed emergency exposure levels (EELs) (NRC, 1986).

Independently, in 1964, the AIHA Toxicology Committee also introduced the concept of EELs and initially proposed EELs for three chemicals: nitrogen dioxide, 1,1-dimethylhydrazine, and 1,1,1-trichloroethane (AIHA Toxicology Committee, 1964). The EELs were established for a single exposure for periods of 5, 15, 30, or 60 min. They were defined as levels that would not result in irreversible toxicity, impair the ability to perform emergency operations, or impair the ability to escape from the exposure. However, these levels were for healthy workers and were not designed to protect many within the general public. Considered to duplicate the NRC guidelines, no additional AIHA EELs were developed (Jacobson, 1966).

In 1986, the NRC changed from EELs to Emergency Exposure Guideline Levels (EEGLs), which have been developed for a number of chemicals (NRC, 1986). EEGLs were developed primarily for military personnel and were not applicable to the general public. Realizing the need for community numbers, NRC introduced the concept of short-term public emergency guidance levels (SPEGLs), but few SPEGLs were developed (NRC, 1986). The NRC also developed continuous exposure guidance levels (CEGLs) for continuous exposure scenarios, for example, in a submarine (NRC, 1986). In the 1990s, similar levels were needed for the confined quarters of spacecraft. Such levels called spacecraft maximum allowable concentrations (SMACs) have been released periodically in a series of publications (NRC, 1992; SMACs, 2000).

National Institute for Occupational Safety and Health (NIOSH), in the late 1970s, developed a series of values representing exposure levels that were "immediately dangerous to life and health" (IDLHs) (NIOSH, 1994). These values were used for identifying respiratory protection requirements in the NIOSH/OSHA Standards Completion Program. Since 1994, some IDLHs have been reviewed and some documentation is available. One of the problems inherent in using IDLHs is that they were intended for the workplace where protective clothing and equipment are at hand, and they do not take into account the exposure of the more sensitive individuals, such as the elderly, children, or people with asthma, living in the surrounding community. In addition, the rationales and documentations were not peer-reviewed and are not generally available.

Other groups have considered the development of short-term exposure guidelines including, the American National Standards Institute, the Pennsylvania Department of Health, and the National Fire Protection Association. In general, the guidelines from these organizations are directed only to occupational exposure, are no longer being generated, or are merely relative hazard ratings. As such, they do not meet the need for airborne concentrations to be used in planning for emergency situations (AIHA, 2005c; Cavender, 2006).

Although various "emergency numbers" have been recommended by several organizations, none of them have been specifically developed for potential accidental releases to the community with the exception of the few NRC SPEGLs. The process of developing NRC SPEGLs is a thorough process, but it is time consuming. Thus, it became evident that a new set of numbers for emergency planners and managers was needed and that they should be developed rapidly for a wide variety of chemicals.

21.2 DEVELOPMENT OF EMERGENCY RESPONSE PLANNING GUIDELINES

21.2.1 BIRTH OF EMERGENCY RESPONSE PLANNING GUIDELINES

Without suitable numbers in place in 1986, several companies, responding to internal needs, independently undertook the development of these emergency planning guideline numbers. Each had separately reached similar conclusions:

1. The numbers are useful primarily for emergency planning and response.
2. The numbers are suitable for protection from health effects due to short-term exposures.
3. They are not suitable for effects due to repeated exposures nor as ambient air quality guidelines.
4. The numbers are guidelines. They are not absolute levels demarcating safe from hazardous conditions.
5. The numbers do not necessarily indicate levels at which specific actions must be taken.
6. The numbers are only one element of the planning activities needed to develop a program to protect the neighboring community.
7. The selection of chemicals needing emergency planning guidelines should, in general, be based on volatility, toxicity, and releasable quantities.

In discussions among personnel in companies developing these guidelines, it became evident that a consistent approach was needed. Uniform procedures and definitions would provide more consistent guidelines. In addition, sharing guidelines for different chemicals would avoid redundant efforts and increase the number of available guidelines (Rusch, 1993; Cavender and Gephart, 1994; Kelly and Cavender, 1998; Cavender, 2002, 2006; AIHA, 2005c).

21.2.2 ROLE OF THE ORGANIZATION RESOURCES COUNSELORS, INCORPORATED

Recognizing this need, Organization Resources Counselors, Inc. (ORC) established a task force to address the need for reliable, consistent, and well-documented emergency planning guidelines. Through the ORC Emergency Response Planning Guidelines (ERPGs) Task Force, companies willing to play an active role in the development of emergency guidance concentrations, by using a consistent procedure, were able to coordinate their efforts. Members of the ORC ERPG Task Force collectively developed methods for establishing emergency exposure guidance levels and a list of chemicals for which ERPGs were believed to be needed. Participating companies made commitments to develop ERPGs for selected chemicals according to the methods developed by the Task Force. Participation in the ORC Task Force was voluntary (Rusch, 1993; Cavender and Gephart, 1994; Kelly and Cavender, 1998; Cavender, 2002, 2006; AIHA, 2005c).

When the Emergency Response Planning (ERP) Committee was formed, many chemical companies had representatives on the Organizational Resources Counselors (ORC). Chemical companies who needed ERPGs for given chemicals would write a draft document and submit it to the ERP Committee through ORC. This worked well for the first 40 or so documents (Cavender, 2002, 2006; AIHA, 2005c).

21.2.3 ROLE OF THE AMERICAN INDUSTRIAL HYGIENE ASSOCIATION

In looking for an acceptable venue to develop and publish the documents of the ERP Committee, the AIHA seemed to be a logical organization because its membership included both industrial hygienists and toxicologists from stakeholder chemical companies, government agencies, academic institutions, and the public sector. As a result, the ERP Committee was formed in 1987, initially as an ad hoc group under the Workplace Environmental Exposure Level Committee, with the specific charge to develop suitable documents for emergency response planning. In 1988, the ERP Committee began producing ERPGs (Rusch, 1993; Cavender and Gephart, 1994; Cavender, 2002, 2006; AIHA, 2005a,b).

It should be obvious that groups other than chemical companies need ERPGs. For instance, the Department of Energy (DOE) includes many nuclear plants, laboratories, and research facilities. Because DOE and numerous other organizations are not represented on ORC, a method was needed whereby other stakeholders could supply a draft document for a chemical for which they needed an ERPG. As a result, ORC has dropped out of the process and all documents are now submitted directly to the ERP Committee via the Guideline Foundation of AIHA. Submissions have been made by the DOE, the DOD, consortia of producers, and committee members. EPRGs have been integrated in risk management scenarios to protect soldiers in noncombat situations. Documents are considered from any source, and the selection of chemicals is based on interest in the chemical, production volume, and physical/chemical properties such as volatility, odor, reactivity, and solubility (Cavender, 2002, 2006; AIHA, 2005c).

Almost immediately, ERPGs gained worldwide recognition and acceptance and are currently used throughout the world. In 1991, the European Centre for Ecotoxicology and Toxicology of Chemicals (ECETOC) published the concept of Emergency Exposure Indices (EEIs) (ECETOC, 1991). However, only three documents were ever published, mainly because of the wide acceptance of ERPGs by emergency planners (ECETOC, 1991; Woudenberg and von der Torn, 1992; AIHA, 2005c).

Instead of developing additional EEIs, Dutch scientists joined the ERP Committee and have aided in the development of scientifically sound documents.

At the request of the US Environmental Protection Agency and the Agency for Toxic Substances and Disease Registry, NRC convened a Subcommittee on Guidelines for Development of Community Emergency Exposure Levels (CEELs) in 1993. The report of the comprehensive CEEL concept served as a blueprint for risk assessment of short-term exposures to high concentrations of chemical toxicants. This report identified the NRC SPEGLs and the AIHA ERPGs as being standards that might be useful in developing CEELs (NRC, 1993). The NRC's criteria for establishing CEELs were similar to the established methodology used in developing ERPGs. However, no CEEL documents were ever produced (Rucsh, 1993; Cavender and Gephart, 1994; Cavender, 2002, 2006; AIHA, 2005c). In an effort to move forward, representatives from the EPA approached the ERP Committee in 1995 and expressed their interest in developing similar numbers for emergency response planning. After several meetings, EPA spearheaded the development of Acute Exposure Guideline Levels (AEGLs) through the NRC/AEGL committee, which was established in 1996 (NRC, 2001). The ERP Committee welcomed EPA's efforts to develop similar exposure values and helped them establish the AEGL committee. From the outset, the plan was to work independently of each other. The AEGLs have a basis very similar to ERPGs as the ERP Committee has worked with EPA in setting up this program. (Kelly and Cavender, 1998, Cavender, 2002, 2006; AIHA, 2005c; Rusch, 2006).

As of November 2011, the NAC/AEGL committee had developed interim documents for all but 5 of the 329 chemicals on the priority list. No new AEGL documents are planned as of this date. (See the popup entitled *Highlight* on the AEGL website [https://www.aiha.org/]). The committee will proceed to move the interim documents through the final stages of development. When all documents reach final status, the committee will have completed its work (EPA/AEGL, 2012). There is no procedure in place to update the AEGL documents.

This means that the ERP Committee is the only active entity that is producing and/or updating emergency response documents for the protection of the general public.

21.2.4 WHAT ARE THE LEVELS OF CONCERN?

The AIHA ERP Committee has utilized three guidance concentration levels. Each of these levels is defined and briefly discussed in the following.

21.2.4.1 ERPG-3

ERPG-3 is "the maximum airborne concentration below which it is believed that nearly all individuals could be exposed for up to one hour without experiencing or developing life-threatening health effects."

The ERPG-3 level is a worst-case planning level above which there is the possibility that exposure to levels above ERPG-3 will be lethal to some members of the community. This guidance level could be used to determine the maximum releasable quantity of a chemical should an accident occur. This quantity is used in the planning stages to project possible exposure levels in the community should a release occur. Once the releasable quantity is known (size of the storage tanks, etc.), the steps to mitigate the potential for such a release can be established.

21.2.4.2 ERPG-2

ERPG-2 is "the maximum airborne concentration below which it is believed that nearly all individuals could be exposed for up to one hour without experiencing or developing irreversible or other serious health effects or symptoms which could impair an individual's ability to take protective action."

Above ERPG-2, for some members of the community, there may be significant adverse health effects or symptoms, including lung or liver disease, miscarriage, or cancer. On the other hand, this level could impair an individual's ability to take protective action. These effects might include dizziness, severe eye or respiratory tract irritation, central nervous system depression, or muscular weakness.

21.2.4.3　ERPG-1

ERPG-1 is "the maximum airborne concentration below which it is believed that nearly all individuals could be exposed for up to 1 h without experiencing other than mild, transient adverse health effects or without perceiving a clearly defined objectionable odor."

ERPG-1 identifies a level that does not pose a health risk to the community but which may be noticeable because of odor, coughing, discomfort, or irritation. In the event that a small, nonthreatening release has occurred, the community could be notified that they may notice an odor or slight irritation but that concentrations are below those which could cause serious health effects. For some materials, because of their properties, an ERPG-1 may not exist. Such cases would include substances for which sensory perception levels are higher than the ERPG-2 level. In such cases, the ERPG-1 level would be given as "not appropriate." It is also possible that no valid sensory perception data are available for the chemical. In this case, the ERPG-1 level would be given as "insufficient data."

All planning activity is to ensure that the daily operations of the plant or process will not result in lethality even if a catastrophic release occurs. The ERPG-2 is the most significant and useful of the ERPGs in emergency planning. This is the level on which decisions to evacuate a region or to shelter in place are based. After collecting and reviewing all the data that have an impact on the ERPG-2 level, setting ERPG-1 and ERPG-3 levels is straightforward.

21.2.5　What Time Period Is Relevant?

One might consider a range of time periods for these guidelines; however, the decision was made to focus on a single period, namely, 1 h. This decision was based on the availability of toxicity information and on a reasonable estimate for an exposure scenario. Most acute inhalation toxicity studies are for either 1 or 4 h because EPA, the Consumer Product Safety Commission, the Department of Transportation, the Organization of Economic and Community Development, and other government agencies require studies for one or both of these periods. Acute lethality data are rarely available for 10 min, 30 min, or 8 h. Some releases may be shorter or longer than 1 h, depending on volatility and weather conditions, but 1 h should serve the community well in being notified of a release, allow time for suitable rescue operations within the plant, and allow first responders to reach and isolate the site of the release. The committee has resisted the development of values for periods other than 1 h, because, for most exposure scenarios, the 1 h period is sufficient; and when another time period is expected to be important, the emergency planners or managers can develop values specifically for their situation rather than randomly using safety factors or concentration times time ($C \times t$) extrapolations for every chemical. However, when warranted, the ERP Committee has set ERPGs for different time periods (see the 1999 addendum to the hydrogen fluoride ERPG document). The logistics are, in general, against very short periods. It is impossible to report a spill or release, run the air dispersion models, and have the news agencies report the information to the public within 10 min. If the message would be to shelter in place with windows closed and air conditioners turned off to prevent outside air entering the house, the exposure would be over before the public could be alerted to the danger (Kelly and Cavender, 1998; Cavender, 2002; AIHA, 2005c). For emergency managers who need to extrapolate to different time periods, there are two generally recognized approaches. For nonirritants, Haber's rule (concentration × time = constant, or $C \times t = k$) is usually valid over a three- or fourfold range of time (Haber, 1924). A seemingly better method for all chemicals was developed by Dutch workers and is given as ($C^n \times t = k$), where in practice, $n = 3$ in extrapolating to a shorter exposure period or $n = 1$ in extrapolating to a longer exposure period (ten Berge et al., 1986). For the best possible results, n should be derived from the slope of the dose–response curve.

21.2.6 How Are the Numbers Derived?

In developing an ERPG level for a chemical, it is important to emphasize the use of acute or short-term inhalation toxicity data and workplace experience. ERPGs are for a once-in-a-lifetime exposure for 1 h. When evaluating the adverse health effects, both immediate and delayed health effects should be considered. When it is believed that adverse reproductive, developmental, or carcinogenic effects might be caused by a single exposure, these data should be carefully considered in the derivation of the ERPGs (Rucsh, 1993; Cavender and Gephart, 1994; Kelly and Cavender, 1998; Cavender, 2002, 2006; AIHA, 2005c). If carcinogenicity data are used, the mathematical approaches adopted by the NRC are utilized in assessing the risk of developing cancer from a single exposure. (NRC, 1986).

The data for developing an ERPG for any chemical are evaluated on a case-by-case basis because each chemical is likely to have a different dose–response curve and produce significantly different effects. Also, the quantity and quality of data available vary widely. There is no formula for selecting ERPG values and no fixed relationship between the three ERPG values (Rucsh, 1993; Cavender and Gephart, 1994; Kelly and Cavender, 1998; Cavender, 2002, 2006; AIHA, 2005c). For ERPG-3 levels, some relationships for irritant chemicals between the 1-h LC-50, the highest nonlethal level in animals, and the threshold for lethality in humans have been studied. The relationships must be considered carefully as they do not hold for all chemicals (Rusch et al., 2009).

It is very important that the rationales for the ERPGs be documented and published. AIHA publishes a pocket-sized handbook (AIHA, 2005c) that contains the numbers for all ERPG documents. In addition, anyone using these numbers for chemicals of concern should obtain the complete ERPG documents for those chemicals (AIHA, 2005a). These documents provide all the data and rationales used in deriving the ERPGs.

21.2.7 Chemical Selection and Data Requirements

ERPGs are developed for a once-in-a-lifetime exposure for up to 1 h. Thus, acute toxicity data and workplace experience in handling the chemical are important in developing ERPGs. Because most community exposures are anticipated to be via inhalation, inhalation toxicity data are most useful in setting the numbers. A 1 h or 4 h inhalation lethality study (1 h median lethal concentration [LC_{50}] or 4-h LC_{50}) in one or more species, a respiratory depression study in Swiss–Webster mice (RD_{50}), an odor threshold, and workplace exposure or human testing to known concentrations are useful in setting ERPGs. Repeated exposure toxicity data and developmental toxicity data, sensitization data, and carcinogenicity data are also important in setting the final numbers. Toxicity by routes other than inhalation is supportive, and if the only carcinogenicity or developmental studies available are via oral dosing, such studies are carefully considered in setting ERPG-2 levels. Finally, if mechanistic or dose–response data are available, these are applied as appropriate (Rucsh, 1993; Cavender and Gephart, 1994; Kelly and Cavender, 1998; Cavender, 2002, 2006; AIHA, 2005c).

Obviously, the more complete the data set, the easier it is to set ERPG numbers and the greater credibility they will have because there will be greater confidence that the effects noted are due to exposure to the chemical. However, it is important that the data be specific and relevant to the derivation of ERPGs. Consider mercury vapor or toluene di-isocyanate (TDI). Numerous reports are available on the toxicity of these compounds, but most are not useful in deriving ERPGs. It is common knowledge that mercury vapor is toxic, but no one has conducted a 1 h LC_{50} for mercury vapor. Similarly, there are many reports on the sensitizing properties of TDI, yet these data contribute little to the overall acute toxicity profile of TDI. Finally, note that some very common chemicals are extremely reactive, for example, hydrogen fluoride (HF). Studies conducted in standard inhalation chambers destroy the chamber because of the extreme reactivity of HF. Such chemicals are studied in special chambers coated or lined to reduce the damage to the chamber.

The documents are arranged in the following sections in the AIHA 2005 Handbook for ERPGs: and WEELs (AIHA, 2005c):

1. Identification
2. Chemical and Physical Properties
3. Animal Toxicity Data
4. Human Experience
5. Current Occupational Exposure Guidelines
6. Recommended ERPGs and Rationales
7. References

21.2.8 REVIEW PROCESS

For documents that are submitted to the ERP Committee, the author should adhere to the following guidelines:

1. The authoring organization should use a multidisciplinary team, including expertise in industrial hygiene, toxicology, medicine, and other health professions to collect and review data and write a draft ERPG document.
2. The author should identify producers, major users, and industry associations that have a significant interest in the chemical and should request unpublished data and other relevant information from them. Studies of effects in humans or animals at known airborne concentrations are especially useful.
3. A robust literature search should be conducted and should include appropriate on-line databases, including MEDLINE and TOXLINE.
4. The author should make every effort to obtain the original reference for all data because transcriptional errors or significant omissions frequently occur in secondary references.
5. The ERPG document should be drafted by using the format prescribed in the latest edition of the *AIHA Handbook on ERPGs* (AIHA, 2005c).
6. The authoring company or organization should submit the draft ERPG document, marked "Preliminary Draft," to the ERP Committee.
7. Copies of all the referenced literature must accompany the draft document. For some lengthy publications, such as NTP chronic studies, the full document may not be needed. Unpublished data such as internal industry reports are often an enormous help, representing the majority of the toxicity data on a given chemical. Confidential company reports should not be used unless a summary containing some details of the methods, results, and conclusions is provided.
8. Upon receipt of a draft document, the AIHA ERP Committee will assign a primary reviewer and a secondary reviewer who will review the draft in depth, rewriting any sections that do not conform to current practice in developing ERPG documents. The submitting author may be contacted concerning any necessary clarifications or corrections.
9. After this initial review and revision, the document is presented to the full AIHA ERP Committee for a detailed discussion of the data summary, ERPG values, and rationales. A document may be discussed at several meetings before it is approved to ballot.
10. A majority vote of members in attendance is needed before sending the final draft to all members for ballot. Members may vote "yes," "no," or "abstain." "No" votes must be accompanied by a specific explanation. Every attempt to garner unanimous approval is sought for every ERPG document.
11. Before being balloted by the committee, the ERPG numbers, rationales, and references are posted on the AIHA Website (www.aiha.org) for 45 days to garner any private or public comments concerning the chemical or the document.

12. Following approval via the balloting process, all necessary changes are incorporated into the document and it is then sent along with the entire reference package to AIHA headquarters for publication.
13. The ERPG document is filed at AIHA along with copies of all the referenced literature. These will be made available to the public as needed.
14. ERPGs are updated after 7 years but may be reviewed and revised at any time as relevant new data become available.

21.2.9 Policy for Commenting on ERPGs under Review

1. On a quarterly basis, a list of all chemicals for which ERPGs are being reviewed is posted on the AIHA Website (https://www.aiha.org/). The following statement will also be posted: "The following materials are currently being studied for future ERPGs. Information and comments are welcome. If one has any input on candidate or completed ERPGs, he/she should contact the AIHA Department of Scientific and Technical Affairs who will forward all comments to the current Chair of the committee."
2. For chemicals where ERPG levels have been approved to ballot, the ERPG numbers, rationales, and a list of references will be posted on the website for a period of 45 days before balloting. This allows time for public review and comment before publishing the document. All suitable comments will be incorporated into the document before balloting. Comments that do not arrive in 45 days will be incorporated in the next update of the document or if they provide significant new data, the document will be re-reviewed as soon as feasible.
3. The chair will direct the comments to the primary reviewer with a copy to the secondary reviewer and the secretary. The primary and secondary reviewers will develop a response and, with the concurrence of the chair, they will send the response directly to the individual submitting the initial comment. Copies of both comments and responses will be listed in the minutes of the next meeting and will be maintained with the reference packages for the specific ERPG documents. When possible, responses will be made within 30 days after the next committee meeting.
4. If a responsible individual requests the opportunity to attend a committee meeting to discuss a specific document, the chair may, at his discretion, grant permission. In general, the guest is only present for the discussion of the document of interest. The individual should be encouraged but is not required to first submit comments in writing. The chair has the right to limit discussion, as would be necessary to ensure an orderly, productive meeting.
5. All requests to attend meetings and all comments not presented at meetings must be in writing. Written responses, possibly brief, will be given to all written comments.
6. Although the committee may elect to incorporate new information into an ERPG document based on these comments, they are under no obligation to do so.
7. Drafts of documents are rarely provided, even if formally requested. For certain working committees or interested government agencies, a single copy of a draft that is so stamped on every page may be given to a responsible individual. In such documents, the tentative numbers are deleted because they do not reflect the committee's or AIHA's position. The numbers may be communicated verbally, with the caveat that they are only tentative. Draft documents are never published because the draft may contain incomplete or erroneous data that are usually completed or corrected during the review process.
8. ERPG values that have not been approved by ballot by the committee must not be published. These numbers usually change as data are evaluated during the review.

21.3 APPLICATION OF ERPGs FOR EMERGENCY PLANNING

21.3.1 How Are the Numbers Used?

ERPGs are intended for emergency planning. ERPGs can be applied in a variety of mandated or voluntary emergency response planning programs. These programs generally include accident scenarios in which air dispersion models determine concentration isopleths. ERPGs are also used in programs designed to protect the public from transportation incidents. ERPGs are extremely important for compliance with Emergency Planning and Community Right-to-Know legislation. Individuals using ERPGs include

    Air dispersion modelers
    Community action emergency response (CAER) participants
    Fire protection specialists
    Government agencies
    Industrial process safety engineers
    Industrial hygienists and toxicologists
    Local emergency planning coordinators (LEPCs)
    RCRA managers
    Risk assessors and risk managers
    State Emergency Response Commissions (SERCs)
    Transportation safety engineers

ERPGs may be used with air dispersion models together with information such as vapor pressure and storage volumes to provide computerized estimates of the direction and speed at which the released plume or cloud will spread over the neighboring terrain. These models will also provide the concentration within the cloud during the time of its dispersion. These models incorporate the quantity and rate of release, volatility, wind speed and direction, temperature, and other environmental conditions. Such models help emergency planners know whom to alert and where first responders should report should a release occur. From these models, action plans can be developed. The plans may vary for any given emergency depending on such things as population density, type of population (e.g., schools), terrain, weather conditions, and other hazards of the released entity (e.g., flammability).

Many air dispersion models, as related to accidental releases of toxic chemicals, stem from assumptions established in *Technical Guidance for Hazard Analysis—Emergency Planning for Extremely Hazardous Substances*, also known as the Green Book, published in 1987 (US Environmental Protection Agency/US Federal Emergency Management/US Department of Transportation, 1987). This reference provides a basis for technical applications for community exposure limits. This and similar references often specify that ERPGs be used to determine where in the community protective actions are needed (sheltering in place, evacuation, or isolation zones).

Although emergency planners need to know the conditions of the release, the magnitude of a potential release can also be predicted by using these models. This allows plant managers to reevaluate possible "worst-case" situations that might occur as a result of process or human failures. Since 9/11, industrial sabotage and terrorist activities are also considered. To aid in plant design and community planning, the engineers select the size of tank that will ensure that a potential release will never reach an airborne concentration above the ERPG-2 level or whatever level the planning group selects as its action level. This type of information is also useful in transporting chemicals across the country. The quantity within the tankers or tank car is limited based on the predicted concentrations that could result from an accidental release.

21.3.2 LIMITATIONS IN USING ERPG NUMBERS

ERPGs are general reference levels, the best judgment of specialists using the best available data, and are intended to be used as part of an overall emergency planning program. The levels are not to be used as safe limits for repeated exposures, as definitive delineators between safe and unsafe exposure conditions, or as a basis for quantitative risk assessment.

Human responses do not occur at precisely the same exposure level for all individuals but can vary over a wide range of concentrations. The values derived for ERPGs should be applicable to nearly all individuals in the general population; however, in any population, there may be hyper-sensitive individuals who have adverse responses at exposure concentrations far below levels where most individuals would respond. Furthermore, because these ERPG values have been derived as planning and emergency response guidelines, *not* as exposure guidelines, they do not contain the safety factors normally incorporated into exposure guidelines. Instead, they are estimates of the concentrations above which there would be an unacceptable likelihood of observing the defined effects. The estimates are based on the available data that are summarized in the documentation (AIHA, 2005a). In some cases where the data are limited, the uncertainty of these estimates may be large. Users of the ERPG values are strongly encouraged to carefully review the documentation before applying these values.

Using ERPG values to determine the actions to be taken when planning for or responding to a given emergency requires careful evaluation of site-specific or situation-specific factors. These may include how the 1-h ERPG values might relate to exposures of different duration; whether there are populations at special risks (such as the elderly, the very young, or those with existing illnesses); and other factors such as the volatility and vapor density of the chemical, storage quantities, weather conditions, and terrain.

21.3.3 WHAT DOES THE FUTURE HOLD?

As of March 2014, the ERP Committee published ERPG documents for 145 chemicals and updates for 79 of these chemicals. Although other groups are producing emergency response planning documents, there will always be a need for the ERP Committee because the chemical selection process for government agencies or foreign entities will never include many of the chemicals that are of interest to industry in the United States. So, the ERP Committee, as volunteers working steadily on behalf of communities worldwide, will continue to evaluate data and produce these documents to reduce the risk of such exposures. After all, chemicals are an integral part of our lives and the potential of exposure to chemicals manufactured, transported, or stored within our communities should not pose an undue risk. It would be great to say, "A catastrophe such as Bhopal will never happen again!" However, "never" does not have a statistical basis, and so we plan shrewdly, we consider all of the data available, and we strive to reduce or minimize the risk of significant exposure (Kelly and Cavender, 1998; Cavender, 2002, 2006; AIHA, 2005c).

QUESTIONS

1. The chemical that killed thousands in Bhopal, India, which led to the establishment of ERPGs is
   a. TDI
   b. MIC
   c. Carbaryl or sevin
   d. Acrolein
   e. Dimethyl cyanide
   Answer: b

2. The 60 min time period for ERPGs is based on the following:
    a. Many acute inhalation studies were for a duration of 1 h.
    b. Very few 10 min LC_{50}s have been conducted.
    c. For a significant exposure, no one would wait 8 h to escape exposure.
    d. Even in environmental exposures which may cover a wide area, nearly everyone could escape exposure or take protective action within 1 h.
    e. It is easier to estimate 1 h than it is any other time frame.
    Answer: d
3. ERPG-2 is the most critical when considering the size of tanks for chemicals, etc. because
    a. Air dispersion models are more difficult for lethal levels.
    b. While no one wants to have a significant environment release, there are good reasons not to know the actual concentration.
    c. While no one wants to have a significant environment release, levels that do not cause permanent health effects are more acceptable than death or permanent health effects.
    d. Evacuation due to high concentrations can lead to street and highway grid-lock.
    e. While no one wants to have a significant environment release, one cannot be competitive if all reactions have to be conducted in a closed system.
    Answer: c
4. The ERPG-1 level is set so that
    a. No one will know that there was a release.
    b. No one will call the Poison Control Center.
    c. No one will go to the emergency room.
    d. No one will allow their children to play outside.
    e. No one will experience more than mild, transient effects.
    Answer: e
5. ERPGs are used worldwide because
    a. They were developed specifically for environmental response to chemical releases.
    b. They are the only levels for this purpose that are updated at least every 10 years.
    c. They are the numbers that are prescribed by EPA to be used for air dispersion modeling following accidental environmental releases.
    d. All of the above.
    e. None of the above.
    Answer: d
6. The type of data useful for setting ERPG-3 levels are
    a. Inhalation developmental studies.
    b. Epidemiology studies.
    c. Acute eye irritation studies.
    d. All of the above.
    e. None of the above.
    Answer: e

REFERENCES

American Conference of Governmental Industrial Hygienists (ACGIH), *Documentation of Threshold Limit Values and Biological Exposure Indices*, ACGIH Worldwide, Cincinnati, OH, 2005.
American Industrial Hygiene Association (AIHA), *Documentation of Emergency Response Planning Guidelines*, AIHA Press, Fairfax, VA, 2005a.
American Industrial Hygiene Association (AIHA), *Documentation of Workplace Environmental Exposure Levels*, AIHA Press, Fairfax, VA, 2005b.
American Industrial Hygiene Association (AIHA), *The AIHA 2003 Handbook for ERPGs and WEELs*, AIHA Press, Fairfax, VA, 2000, 2005c.

American Industrial Hygiene Association Toxicology Committee, Emergency exposure limits, *Am. Ind. Hyg. Assoc. J.*, 25, 578–586, 1964.

Cavender, F., Emergency response planning guidelines. In *Inhalation Toxicology*, 2nd edn., H. Salem and S. Katz (eds.). Taylor & Francis Group, Boca Raton, FL, 2006, pp. 61–72.

Cavender, F.L., Protecting the community, one guideline at a time, *Synergist (AIHA)*, 13, 29–31, 2002.

Cavender, F.L. and Gephart, L.A., Emergency response planning, *J. Air Waste Manage.*, 11, 111–122, 1994.

EPA/AEGL Environmental Protection Agency/Acute Exposure Guideline Levels. www.epa.gov/oppt/aegl/.

European Centre for Ecotoxicology and Toxicology of Chemicals, Emergency exposure indices for industrial chemicals, Technical Report 43, ECETOC, Brussels, Belgium, March 1991.

Haber, F., *Funf Vortage aus den Jahren 1920–1923: Geschichte des Gaskrieges*. Verlag von Julius Springer, Berlin, Germany, 1924, pp. 76–92.

Jacobson, K.H., AIHA short-term values, *Arch. Environ. Health*, 12, 486–487, 1966.

Kelly, D.P. and Cavender, F.L., Emergency response. In *Encyclopedia of Toxicology*, Vol. 1, Wexler, P. (ed.). Academic Press, San Diego, CA, 1998, pp. 527–531.

National Research Council (NRC), Standard operating procedures for developing acute exposure guideline levels (AEGL's) for hazardous chemicals, National Academy Press, Washington, DC, 2001.

National Research Council (NRC), Commission on Life Sciences, Board on Environmental Studies and Toxicology, Committee on Toxicology: Criteria and methods for preparing emergency exposure guidance level (EEGL), short-term public emergency guidance level (SPEGL), and continuous exposure guidance level (CEGL) documents, National Academy Press, Washington, DC, 1986.

National Research Council (NRC), Commission on Life Sciences, Board on Environmental Studies and Toxicology, Committee on Toxicology: Guidelines for developing community emergency exposure levels (CEEL's) for hazardous substances, National Academy Press, Washington, DC, 1993.

National Research Council (NRC), Committee on Toxicology: Guidelines for developing spacecraft maximum allowable concentrations (SMACs) for space station containments, National Academy Press, Washington, DC, 1992.

NIOSH 1994. Documentation for Immediately Dangerous to Life or Health Concentrations (IDLHs). http://www.cdc.gov/niosh/idlh/intridl4.html.

NIOSH 1994a to NIOSH 1992. NIOSH RELs and general recommendations for safety and health. http://www.aresok.org/npg/nioshdbs/docs/92-100.htm.

Occupational Safety and Health Administration, Permissible exposure limits (PELs), 1910:1000, *Federal Register*, June 29, 1995.

Rusch, G.M., The history and development of emergency response planning guidelines, *J. Hazard. Mater.*, 33, 192–202, 1993.

Rusch, G.M. The development and application of acute exposure guideline levels for hazardous substances. In *Inhalation Toxicology*, 2nd edn., H. Salem and S. Katz (eds.). Taylor & Francis Group, Boca Raton, FL, 2006, pp. 39–60.

Rusch, G.M., Bast, C.B., and Cavender, F.L. 2009. Establishing a point of departure for risk assessment using acute inhalation toxicology data, *Reg. Toxicol. Pharmacol. Reg. Toxicol. Pharmacol.* 54, 247–255.

Subcommittee on Spacecraft Maximum Allowable Concentrations National Research Council; Commission on Life Sciences (CLS), Spacecraft Maximum Allowable Concentrations (SMACs) for selected airborne contaminants 1994–2000, Vol. I–IV. National Academy Press, Washington, DC.

Superfund Amendments and Reauthorization Act, 1986.

ten Berge, W.F., Zwart, A., and Appleman, L.M., Concentration-time mortality response relationships of irritant and systematically acting vapors and gases, *J. Hazard. Mater.*, 13, 301–309, 1986.

Title III, Clean Air Act Amendments, 1990. http://epa.gov/oar/caa/caaa.html

Union Carbide Corporation, Bhopal Information Center, http://www.Bhopal.com/chrono.htm accessed May 19, 2014.

US Environmental Protection Agency/US Federal Emergency Management Agency/US Department of Transportation, Technical guidance for hazard analysis: Emergency planning for extremely hazardous substances. US Government Printing Office, Washington, DC, December 1987.

Woudenberg, F. and von der Torn, P., Emergency exposure limits: A guide to quality assurance, *Good Pract. Regul. Law*, 1, 249–293, 1992.

22 Safety Assessment of Therapeutic Agents Administered by the Respiratory Routes

Shayne C. Gad

CONTENTS

TABLE 22.1

Nasal Delivery Systems

• Liquid Nasal Formulations	
• Instillation and rhinyle catheter	Drops
• Unit dose containers	Squeezed bottle
• Metered dose pump sprays	Airless and preservative free sprays
• Compressed air nebulizers	
• Powder dosage dorms	
• Insufflators	Mono-dose powder inhaler
• Multidose dry powder system	
• Pressurized MDIs (dose inhalers)	
• Nasal gels	

22.1   INTRODUCTION

Drugs and medicinal agents administered by the inhalation route include gaseous and vaporous anesthetics, coronary vasodilators, and aerosols of bronchodilators, corticosteroids, mucolytics, expectorants, antibiotics, and an increasing number of peptides and proteins where there is significant nasal absorption (Cox et al. 1970; Williams 1974; Paterson et al. 1979; Hodson et al. 1981; Lourenco and Cotromanes 1982, Tamulinas and Leach, 2000). Concerns with the environmental effects of chlorofluorocarbons have also led to renewed interest in dry powder inhalers (DPIs), which have additionally shown promise for better tolerance and absorption for some new drugs. Recent advances have also led to new nasal delivery systems, such as in Table 22.1. Excessive inhalation of a drug into the pulmonary system during therapy or manufacturing may result in adverse local and/ or systemic effects. Consequently, safety assessment of medicinal preparations delivered via respiratory routes with respect to local tissue toxicity, systemic toxicity, and the therapeutic-to-toxicity ratio is essential. The data generated are essential for charting the course of evaluation and the development of a potential therapeutic agent, the general course of which is summarized in Figure 22.1.

Currently, respiratory-route drugs represent less than 1% of marketed therapeutics. However, advances in pharmaceutical technology and the introduction of new structural classes of therapeutic entities (particularly peptides and highly reactive small molecules such as the nitrous oxide analogs) have led to renewed interest in the use of respiratory routes of delivery to treat a range of diseases, including both respiratory (asthma, SARS, and cystic fibrosis), central nervous system (CNS) (pain), and systemic (such as diabetes).

22.2   PULMONARY DELIVERY OF THERAPEUTIC
INHALED GASES AND VAPORS

Pulmonary dynamics, the dimension and geometry of the respiratory tract, and the structure of the lungs, together with the solubility and chemical reactivity of therapeutic inhalants, greatly influence the magnitude of penetration, retention, and absorption of inhaled gases, vapors (Dahl 1990), and aerosols (Phalen 2009), and therefore the therapeutic potential of the agents. The quantity of an inhalant effectively retained in the pulmonary system constitutes the inhaled *dose* that leads to both therapeutic and toxic (dose limiting) responses (NAS, 1958).

Highly reactive and soluble gaseous or vaporous drugs react and dissolve readily in the mucosal membrane of the nasopharynx and the upper respiratory tract (URT), thereby exerting pharmacological effects or causing local irritation and/or adverse effects on the ciliated, goblet, brush border columnar, and squamous cells of the epithelium. The dissolved drug is also absorbed into the bloodstream and transported to a target organ where it can exert systemic effects. Less reactive and less soluble gaseous or vaporous drugs are likely to penetrate beyond the URT and reach the

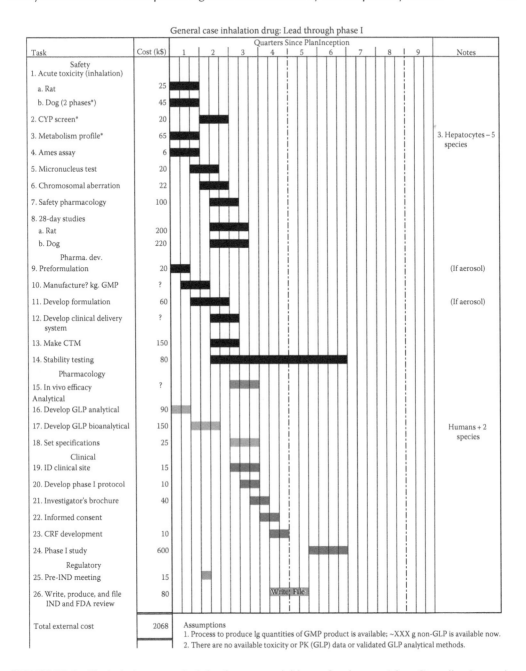

General case inhalation drug: Lead through phase I

Task	Cost (k$)	Quarters Since PlanInception	Notes

Safety
1. Acute toxicity (inhalation)
   a. Rat — 25
   b. Dog (2 phases*) — 45
2. CYP screen* — 20
3. Metabolism profile* — 65 — 3. Hepatocytes – 5 species
4. Ames assay — 6
5. Micronucleus test — 20
6. Chromosomal aberration — 22
7. Safety pharmacology — 100
8. 28-day studies
   a. Rat — 200
   b. Dog — 220

Pharma. dev.
9. Preformulation — 20 — (If aerosol)
10. Manufacture? kg. GMP — ?
11. Develop formulation — 60 — (If aerosol)
12. Develop clinical delivery system — ?
13. Make CTM — 150
14. Stability testing — 80

Pharmacology
15. In vivo efficacy — ?

Analytical
16. Develop GLP analytical — 90
17. Develop GLP bioanalytical — 150 — Humans + 2 species
18. Set specifications — 25

Clinical
19. ID clinical site — 15
20. Develop phase I protocol — 10
21. Investigator's brochure — 40
22. Informed consent —
23. CRF development — 10
24. Phase I study — 600

Regulatory
25. Pre-IND meeting — 15
26. Write, produce, and file IND and FDA review — 80

Total external cost — 2068

Assumptions
1. Process to produce lg quantities of GMP product is available; ~XXX g non-GLP is available now.
2. There are no available toxicity or PK (GLP) data or validated GLP analytical methods.

FIGURE 22.1 Typical pharmaceutical development activities to first-in-man (phase I) studies for respiratory route drugs (Gad 2009). What is presented above are all the tasks involved in developing a potential drug from lead designation to the completion of an initial (FIM—first-in-man) clinical study, usually performed in normal healthy volunteers.

bronchial and alveolar regions causing local and systemic effects. The unabsorbed gases or vapors are then exhaled. For example, ammonia gas generated from a 10% ammonia water may be inhaled for reflex respiratory stimulation purposes (O'Neil, 2006). Ammonia is extremely soluble in water at a concentration of 715 mL of ammonia per mL of water (Phalen 2009) and is readily solubilized in the mucous lining causing URT irritation. By contrast, oxygen is only sparingly soluble in water (0.031 cc of oxygen per mL of water) and capable of penetrating deeply into the alveoli, where gas exchange takes place. Oxygen that binds reversibly with the hemoglobin of erythrocytes is unloaded at the target tissues, while the unbound oxygen is exhaled. Inhalation of properly humidified oxygen is life supporting, but inhalation of unhumidified oxygen may cause a reduction in the mucociliary clearance of secretions and other substances deposited in the trachea of animals (Pavia 1984) and humans (Lichtiger et al. 1975; Gamsu et al. 1976). Gases or vapors of low lipid solubility are also poorly absorbed in the lungs, with much of the inhaled vapor exhaled. Other pharmacological gases and vapors, such as the anesthetics (nitrous oxide, halothane, enflurane, isoflurane, etc.) and the coronary vasodilators (amyl nitrite), likewise affect the epithelium of the respiratory tract and the lungs. The absorbed drugs exert local effects on various types of epithelial cells of the respiratory tract and on type I and II cells and the alveolar macrophages (AMs) in the alveoli. Repeated inhalation of some halogenated hydrocarbon anesthetics will result in the accumulation of the vapors and systemic toxicity (Chenoweth et al. 1972). By contrast, vapors such as the fluorocarbons (FC 11 and FC 12), which are used extensively as propellants for bronchodilator and corticosteroid aerosols, are absorbed rapidly but are not accumulated in the body even upon repeated inhalation (Aviado 1981).

In general, dissolved gases or vapors at a nontoxic concentration are absorbed and metabolized locally by the lungs and systemically by the liver. The unchanged parent drug and its metabolites may be excreted to some extent via exhalation but mainly via the renal system. A dissolved gas or vapor at a toxic concentration, however, is likely to exert local effects such as altering the surface tension of the alveoli linings (surfactant) or disrupting the normal functions of the epithelial cells, the pneumocytes, and the AMs. The disrupted AMs in turn release their intracellular enzymes, potentially causing destruction of the alveolar septa and contributing to histopathologic changes of the respiratory tract and the lungs. Again, the magnitude of the adverse effects is dependent on pulmonary dynamics and the solubilities of the inhalants in the mucous membrane of the URT and in the plasma or lipids of the erythrocytes. The pharmacologic and toxicologic aspects of this are discussed later.

22.3 PULMONARY DELIVERY OF INHALED AEROSOLS

For inhaled aerosols, particle size is a major factor affecting the penetration, deposition, and hence the *dose* and site of pharmacological action (Dautreband 1962 a,b; Agnew 1984). Particle size is expressed in terms of *aerodynamic diameter* (AD), defined as the diameter of a spherical particle of unit density (lg/cm^3) that has the same terminal settling velocity as the particle in question, regardless of its shape and density (Marple and Rubow 1980). The unit for AD is micrometers (μm). A sample of aerosol particles having ADs within a narrow size range is considered to be a monodisperse aerosol, whereas a sample of aerosols with a wide range of ADs is a heterodisperse, or polydisperse, aerosol. The pattern of particle-size distribution usually bell shaped, with smaller and larger particles on both sides of the mean AD. An aerosol sample with a high proportion of particles of similar size has a narrow particle-size distribution or small geometric standard deviation (GSD). An aerosol sample with a GSD of less than two is considered to be a monodisperse aerosol. Therefore, both the AD and GSD of two or less are considered to be optimal for pulmonary penetration and distribution in the respiratory tract and the lungs. For example, in nose breathing, aerosol particles with ADs > 15 μm are likely to be trapped in the nasopharynx (extrathoracic or head region) by filtration and impaction. Particles deposited in the nasopharynx are considered to be *noninhalable* (Lippmann 1970; Miller et al. 1979).

In mouth breathing, only 10%–15% of 15 μm particles penetrate through the larynx to the intrathoracic tracheobronchial (TB) region. Particles reaching the TB region are considered to be *inhalable* (Lippmann 1970; Miller 1979).

In natural nose and mouth breathing, only a negligible proportion of aerosol particles of AD > 10 μm reach the lungs (Swift and Proctor 1982). Aerosol particles of 3–4 μm in AD are considered to be optimal sizes for TB deposition. The mechanisms of deposition are by impaction along the trachea and at bronchial branchings, where the direction of airflow changes, and by gravity settlement in the fine airways in amounts proportional to the particle-settling velocity and the time available for settlement (Hatch and Gross 1964; Heyder et al. 1980). Aerosol particles of 1–2 μm in AD, however, decrease in TB deposition because the particles are too small for effective impaction and sedimentation (Lippmann 1977; Chan and Lippmann 1980; Stahlhofen et al. 1980). Consequently, the majority of the very fine particles are exhaled. However, the deposition of the ultrafine particles of approximately 0.5 μm in AD on the walls of the finest bronchioles and the alveoli increases again due to molecular diffusion processes. Even so, some 90% of the inhaled 0.5 μm particles will still be exhaled during quiet tidal breathing and much more underforced exhalation (Davis et al. 1972; Taulbee 1978). Those fine particles reaching the finest bronchioles and alveoli are considered to be *respirable* (Lippmann 1970).

In general, particles of AD > 10 μm deposit mainly in the URT, whereas particles of 1–5 μm AD, with a GSD of less than 2, are likely to reach the lower respiratory tract (LRT), which includes the TB region and the alveoli, with small oropharyngeal loss.

The proportion of an aerosol sample suitable for inhalation can also be determined on the basis of mass median aerodynamic diameter (MMAD), which is defined as the AD for which 50% (by weight) of an aerosol sample is less than or equal to the stated median AD. For example, a sample with an MMAD of 5 μm means that 50% by weight of that sample has ADs of 5 μm and smaller. The MMAD is, therefore, a good index for determining the proportion of an aerosol sample that is *noninhalable*, *inhalable*, or *respirable*. An aerosol sample with an MMAD of 5 μm and a GSD of less than 2 is considered to be optimal for pulmonary deposition and retention (Task Group on Lung Dynamics 1966).

In addition to AD and GSD, the pulmonary dynamics of a subject also greatly influence the distribution of aerosol particles in various regions of the respiratory tract (Agnew 1984). For example, the velocity of airflow in the respiratory tract significantly influences the pattern of TB deposition. An increase in airflow velocity in the airways (possibly due to exercise or work) increases the effectiveness of particle impaction at the bifurcations of the large airways (Dennis 1961; Hatch and Gross 1964; Parent 1991). As a result, spots impacted with a high concentration of particles (hot spots) are frequently present at the carina and the bifurcations of the airways (Lee and Wang 1977; Bell 1978; Stahlhofen et al. 1981). Furthermore, the depth of each breath (tidal volume) also influences the distribution of aerosols. A small tidal volume permits greater impaction in the proximal conducting airways and less sedimentation in the distal airways.

In general, slow, deep inhalation followed by a period of breath holding increases the deposition of aerosols in the peripheral parts of the lungs, whereas rapid inhalation increases the deposition in the oropharynx and in the large central airways. Thus, the frequency of respiration (the flow velocity) and the depth of breath (tidal volume) influence the pattern of pulmonary penetration and deposition of inhaled aerosols. Therefore, a therapeutic aerosol of ideal size will penetrate deeply into the respiratory tract and the lungs, producing an ideal effect only when the aerosols are inhaled in the correct manner.

22.4 ABSORPTION AND CLEARANCE OF INHALED AEROSOLS

Soluble therapeutic aerosols deposited on the epithelial linings of the respiratory tract are absorbed and metabolized in the same way as soluble gases and vapors. Insoluble medicinal aerosols are few in number. Sodium cromoglycate (SCG) is probably the only insoluble powder to be administered as a prophylactic antiasthmatic (Wanner 1979). Insoluble particles deposited on the ciliated linings of the URT are removed by a mucociliary clearance mechanism. Particles deposited on a terminal airway devoid of ciliated cells may be endocytosed into the epithelial cells. At a toxic concentration, the cells die and the debris is then phagocytized and transported into the interstitial space for removal via the lymph or vascular drainages or reenters the ciliated zone of the airway. Particles deposited on the alveolar walls may be phagocytized by the AMs or neutrophils (PMNs) and transported from the low–surface tension surfactant in the alveolar lining to the high–surface tension bronchial fluid of the ciliate airways for elimination by the mucociliary clearance mechanism (Lauweryns and Baert 1977). The particle sizes optimal for phagocytosis are 2–3 μm, while particles smaller than 0.26 μm are less effective in activating the macrophages (Holma 1967). In any case, AMs can phagocytose only a small fraction of a large number of deposited particles. The nonphagocytosed particles are translocated to the lymphatic system or vascular circulation (as discussed earlier) for elimination (Ferin 1977).

Like the inhaled gases or vapors, soluble and insoluble aerosol particles can directly exert desirable and undesirable local effects at the site of deposition and/or systemic effects after solubilization, absorption, and metabolization.

22.5 PHARMACOTOXICITY OF INHALED AEROSOLS, GASES, AND VAPORS

The inhalation route for administering drugs into the pulmonary system for treatment of respiratory diseases eliminates many bioavailability problems such as plasma binding and *first-pass* metabolism, which are encountered in parenteral or oral administration. Consequently, a small inhalation dose is adequate for achieving the desirable therapeutic dose and response without inducing many undesirable side effects. Furthermore, the direct contact of the drug with the target site ensures rapid action. Nevertheless, the effects from inhaled drug aerosols, gases, and vapors also depend on the pharmacological properties of these substances and the location of their deposition in the respiratory system. For example, the classic experiments on bronchodilation drugs (Dautrebande 1962a,b) showed that fine aerosol particles of isoproterenol penetrate deeply in to the lower respiratory airways (LRAs). In this way, an effective concentration of the drug aerosol can reach the beta-adrenergic receptors of the bronchial smooth muscles. Stimulation of the receptors causes relaxation of the smooth muscle fibers and results in bronchodilation (McFadden 1986; Westfall and Westfall 2011). Such rapid bronchial responses can be produced in both healthy and asthmatic subjects without inducing any adverse cardiac effects. By contrast, the same delivered dose of isoproterenol of large particle sizes deposits mainly along the URT, with a minimal amount reaching the smooth muscles of the LRA. The drug is quickly absorbed into the tracheal and bronchial veins and delivered immediately to the left ventricle of the heart. A high plasma concentration of the drug in the heart causes prominent cardiovascular effects such as tachycardia and hypertension. Other aerosols of beta-adrenergic drugs, such as epinephrine, isoprenaline, terbutaline, and salbutamol, induce bronchodilation effects in animals and humans (Pavia 1984) via inhalation and stimulate ciliary beat frequency and mucous production at the site of deposition in the trachea (Wanner 1981). Such stimulation of mucous production can be detrimental to the breathing process, especially in asthmatics and cystic fibrosis patients. Thus, the TB mucociliary clearance mechanism is also stimulated. By contrast, anticholinergic bronchodilators, such as atropine and ipratropium bromide, cause mucous retention in the lungs (Pavia et al. 1983a,b). Therefore, in pharmacological or safety assessments of inhalant beta-adrenergic bronchial dilatation drugs, aerosols should be of small particle sizes suitable for deposition in the peripheral airways to minimize side effects. However, anticholinergic agents should be of larger particle sizes suitable for deposition in the large airways (Ingram 1977; Hensley et al. 1978).

Other therapeutic aerosols, such as beclomethasone dipropionate, betamethasone valerate, and budesonide corticosteroid (Williams 1974); the carbenicillin and gentamicin antibiotics (Hodson et al. 1981); the 2-mercaptoethane-sulfonate (Pavia et al. 1983b) and *n*-acetylcysteine mucolytics; and even vaccines for the prevention of influenza and tuberculosis (Lourenco and Cotromanes 1982), are active by inhalation and/or oral administration. When these drugs are produced so as to be administered as aerosols, certain particle sizes may be targeted to a specific region or to multiple regions of the pulmonary system depending on the therapeutic target site(s). In any case, when aerosols are delivered as fine particles, the rate of absorption is increased because of an increase in the distribution area per unit mass of the drug. Thus, an effective aerosol dose of corticosteroid for the treatment of asthma and bronchitis is merely a fraction of an oral dose (Williams 1974). An aerosol of disodium cromoglycate (DSG) dry powder, a prophylactic dose for preventing the onset of bronchoconstriction in asthmatic attacks (Cox et al. 1970), is effective mainly by local inhibition of the release of chemical mediators from mast cells in bronchial smooth muscle. Therefore, DSG particle sizes should be approximately 2 μm in AD for the most effective penetration into the bronchial regions (Godfrey et al. 1974; Curry et al. 1975). Likewise, therapeutic aerosols of local anesthetics and surfactants may require appropriate particle sizes to be targeted to a specific region of the pulmonary system.

Other than undesirable pharmacological effects, toxic concentrations of soluble or insoluble aerosol particles may lead to adverse physiological and/or histophathologic responses. For example, irritating aerosols cause dose-related reflex depression of the respiratory rate (Alarie 1966, 1981a), while phagocytosed particles cause chemotaxis of AMs and neutrophils to the site of deposition (Brain 1971). The maximum response usually occurs at 24 h postexposure and returns to normal in approximately 3 days postexposure (Kavet et al. 1978). Furthermore, a toxic quantity of phagocytosed particles may interact with the lysosomal membrane within a macrophage, releasing cytotoxic lysosomal enzymes, proteases, and free radicals that in turn damage the adjacent lung tissue (Hocking and Golde 1979).

In general, a specific category of drug delivered to a specific site of the pulmonary system will exert a specific pharmacological or toxicological action locally or systemically. Therefore, in safety assessments of inhalants, a drug should be delivered to the target sites of the pulmonary system according to the toxicological information required.

Finally, there are many drugs in the categories of amphetamines, anorectics, antihistamines, antipsychotics, tricyclic antidepressants, analgesics and narcotics, and beta-adrenergic blocking agents that are known to accumulate in the lung (Wilson 1982) even though these drugs are not administered via the inhalation route. Therefore, in safety assessments of these drugs, their pulmonary toxicity should also be evaluated. There are also in vitro techniques that are proposed for use in evaluating inhalation effects on respiratory tissues (Agu et al. 2002), which may serve as useful screening tools for sets of potential therapeutics.

22.6 NASAL DELIVERY OF THERAPEUTICS

The biotechnology revolution in therapeutics has led to renewed interest in the nasal route as a means of safely delivering an effective dose of many therapeutics. This, in turn, has led to investigations of new approaches to drug administration by this route, which have proved very rewarding (Smaldone 1997; Sharma et al. 2001; Lawrence 2002; Aldridge 2003). Table 22.2 presents a comparison of this route versus others, which points out the advantages. These advantages have led to the development (and successful marketing) of an impressive number of new drugs with both local (Table 22.3) and systemic (Table 22.4) therapeutic uses. At the same time, this has also led to new guidelines on the subject being promulgated by FDA (CDER 2002) and concerns about end use safety (Kannisto et al. 2002). These guidelines and concerns will directly impact the development of new drugs, such as those in Table 22.5.

The target for nasal administration is the nasal cavity, with a volume (in adults) of only 20 mL but a total surface area (in humans) of ~180 cm². The cavity surface is covered with a 2–4 mm thick nasal mucosa, composed of both respiratory and olfactory components.

There are three separate mechanisms for transmucosal transport of potential drugs and other substances including toxicants:

1. Simple diffusion: a nonsaturable mechanism with no carrier or energy involvement.
2. Facilitated transport: saturable, with carrier involvement but no energy directly expended.
3. Active transport: a saturable mechanism that involves both a carrier and energy expenditure.

TABLE 22.2

Factors Influencing the Selection of the Nasal Route

Major Considerations	Routes of Administration					
	Oral	IV	IM/SC	Transdermal	Nasal	Pulmonary
Delivery interface to blood	Indirect; absorbed through GI system	Direct bolus administration into vein	Indirect; absorbed from muscular/subcutaneous tissue	Indirect; absorbed through relatively impermeable skin	Indirect; absorbed through the highly permeable nasal mucosa	Indirect; but drug delivered to a large highly permeable epithelia
Delivery issues and concerns	Subject to digestive process, first-pass metabolism	Requires administration by healthcare professional	Painful injection, may require administration by healthcare professional	Highly variable, slow delivery; potential for skin reactions	Self-administration. Requirement of high solubility	Requires deep, slow inhalation of small aerosol particles
Patient convenience	High	Low	Low	Moderate	Moderate to high	Moderate to high
Onset of action	Slow	Rapid	Moderate	Slow	Rapid	Moderate to rapid
Delivery of macromolecules	No	Yes	Yes	No	Yes	Yes
Bioavailability	Low to high	Reference standard	Moderate to high	Low	High	Moderate to high
Dose control	Moderate	Good	Moderate	Poor	Moderate	Moderate to good

Sources: Moren, F., Aerosol dosage forms and formulations, in *Aerosols in Medicine*, 2nd edn., Moren, S., Dolovich, M.B., Newhouse, M.T., and Newman, S.P., eds., Elsevier, Amsterdam, the Netherlands, 1993, pp. 329–336; Durham, S.R., *Clin. Exp. Allergy Rev.*, 2, 32, 2002; Greenstone, 2001.

TABLE 22.3

Marketed Nasal Products (for Topical Activity)

Product	Drug	Indication	Manufacturer
Astelin® Nasal Spray	Azelastine hydrochloride	Treatment of seasonal allergic rhinitis	Wallace Laboratories
Beconase® AQ Nasal Spray	Beclomethasone dipropionate monohydrate	Symptomatic treatment of seasonal and perennial allergic rhinitis	Allen and Hanbury's/Glaxo Wellcome Inc
Vancenase® AQ Nasal Spray	Beclomethasone dipropionate monohydrate	Symptomatic treatment of seasonal and perennial allergic rhinitis	Schering Plough Corp
Rhinocort® Nasal Inhaler	Budesonide	Management of symptoms of seasonal and perennial allergic rhinitis and nonallergic perennial rhinitis	Astra USA, Inc
Nasalcrom® Nasal Solution	Cromolyn sodium	Symptomatic prevention and treatment of seasonal and perennial allergic rhinitis	Sandoz. Pharmaceutical Coop
Adrenalin® Chloride	Epinephrine hydrochloride	Nasal decongestant	Parke Davis
Nasalide® Nasal Solution	Flunisolide	Treatment of seasonal and perennial allergic rhinitis	Dura trading Co LTD
Flonase® Nasal Spray	Fluticasone propionate	Symptomatic treatment of seasonal and perennial allergic rhinitis	GlaxoSmithKline, Inc
Atrovent® Nasal Spray	Ipratropium bromide	Symptomatic relief of rhinitis	Boehringer Ingelheim Pharmaceuticals, Inc
Livostin® Nasal Spray	Levocabastine	Treatment of allergic rhinitis	Janssen Research FDN Div/Johnson and Johnson
Privine® Nasal Spray, Nasal Solution and Nasal Drops	Naphazoline hydrochloride	Prompt and prolonged relief of nasal congestion due to common colds, sinusitis	Ciba Consumer Pharmaceuticals
Flunisolide Nasal Solution	Flunisolide	Nasal decongestant	Bausch & Lomb
Afrin® Nasal Spray	Oxymetazoline hydrochloride	Temporary relief of nasal congestion associated with colds, hay fever, and sinusitis	Schering Plough Healthcare Products
Vicks® Sinex® Regular Decongestant Nasal Spray and Ultra Fine Mist	Phenylephrine hydrochloride	Temporary relief of nasal congestion due to colds, hay fever, upper respiratory allergies, or sinusitis	Proctor and Gamble
Vick® Vapor Inhaler *OTC	1-Desoxyephedrine	Nasal decongestant	Proctor and Gamble
Nasonex®	Mometasone	Treatment of seasonal and perennial nasal allergy symptoms	Schering Corp.
Nasacort ®	Triamcinole acetonide	Treatment of seasonal and perennial allergic rhinitis	Aventis

Sources: Moren, F. Aerosol dosage forms and formulations. In: *Aerosols in Medicine*, 2nd edn. (S. Moren, M.B. Dolovich, M.T. Newhouse, and S.P. Newman, eds.). Elsevier, Amsterdam, the Netherlands, pp. 329–336, 1993; Durham, S.R. *Clin. Exp. Allergy Rev.* 2:32–37, 2002; Wills, P. and Greenstone, M. *Cochrane Database Syst. Rev.* CD002996, 2001.

TABLE 22.4

Marketed Nasal Products (for Systemic Activity)

Product	Drug	Indication	Manufacturer
Stadol NS® Nasal Spray	Butorphanol tartrate	Management of pain including migraine headache pain	Bristol Myers Squibb
Miacalcin® Nasal Spray	Calcitonin-salmon	Treatment of hypercalcemia and osteoporosis	Novartis
DDAVP® Nasal Spray	Desmopressin acetate	Diabetes insipidus	Aventis Pharmaceuticals
Migranal® Nasal Spray	Dihydroergotamine mesylate	Treatment of migraine	Novartis
Medihaler-ISO® Spray	Isoproterenol sulfate	Treatment of bronchospasm	3M Pharmaceuticals INC
Nitrolingual® Spray	Nitroglycerin	Prevention of angina pectoris due to coronary artery disease	G Pohl Boskamp GMBH and Co.
Synarel® Nasal Solution	Nafarelin acetate	Central precocious puberty, endometriosis	Roche Laboratories
Nicotrol® Inhalation	Nicotine	Smoking cessation	Pharmacia
Syntocinon® Nasal Spray	Oxytocin	Promote milk ejection in breast-feeding mothers	Novartis
Imitrex® Nasal Spray	Sumatriptan	Migraine	GlaxoSmithKline
Relenza® Powder for Inhalation	Zanamivir	Treatment of uncomplicated acute illness due to influenza A and B	GlaxoSmithKline

22.6.1 Nose to Brain Delivery

- Transport of substances from nasal cavity to CNS has been observed in mid-1900s when it was observed that viruses can move from nose to brain via olfactory pathways.
- A number of studies have also reported the transport of heavy metals from nose to brain via olfactory pathways.
- Studies with tracer materials like potassium ferricyanide, horseradish peroxidase, colloidal gold, and albumin have shown transport of these substances from nose to brain.
- Various low-molecular-weight drugs like estradiol, cephalexin, cocaine, and certain peptides have been shown to reach cerebrospinal fluid, the olfactory bulb, and some parts of brain after nasal administration.

What has increased the utility of the nasal route is the development of strategies for enhancing nasal delivery of drugs? Two major categories or strategies that have increased such utilization are (1) manipulation of formulation (by either coadministration with an enzyme inhibitor or an absorption enhancer or use of a bioadhesive system) and (2) structural modification of the drug molecule (i.e., a pro-drug approach) (Ugwoke et al. 2001). Each of these is discussed in the following text.

22.6.2 Advantage of Formulations

- Increase the permeability of the nasal mucosa by interaction of the formulation components with the nasal membrane in a safe, effective, and reversible manner.
- Increase in drug solubility and protection against enzymatic degradation.
- Increase in the residence time of the drug in the nasal cavity. Commonly used formulations approaches include the following:
  - Liquid formulations
  - Aqueous solutions
  - Synthetic surfactants

TABLE 22.5
Drugs in Development

Product	Indication	Manufacturer	Development
Nasal Nicotine Spray (NNS)	Smoking cessation	Pharmacia Corporation	Phase III trials
formoterol, Oxis Turbuhaler®	Asthma	Astrazeneca PLC	Phase III trials
Zomig (zolmitriptan)®	Migraine therapy	Astrazeneca PLC	NDA filed
Ciches onide	Bronchial-respiratory	Aventis Pharmaceuticals	Phase III trials
FluMist	Vaccine	Aiviron	Approval recommended
Inhaled Insulin	Diabetes therapy	Eli Lilly & Co.	Phase II trials
GW	Bronchial-respiratory	GlaxoSmithKline	Phase II trials
INS 37217 Intranasal	Bronchial-respiratory	Inspire Pharmaceuticals	Phase II trials
Beclomethasone	Bronchial-respiratory	Medeira Pharmaceuticals	Clinical
Salbutamol	Bronchial-respiratory	Medeira Pharmaceuticals	Clinical
VLA-4 antagonist	Bronchial-respiratory	Merck & Co., Inc.	Phase II trials
PT-141	Reproductive system therapy	Palatin Technologies, Inc.	Phase II trials
PA-1806	Bronchial-respiratory	Patho Genisis Corporation	Clinical
Exubera-inhaled insulin	Diabetes therapy	Pfizer	Phase III trials
NNS	Smoking cessation aid	Pharmacia Corporation	Phase III trials
Iloprost	Cardiovascular agent	Schering-Plough Corporation	Phase III trials
Salloutanol	Bronchial-respiratory	Sheffield Pharmaceutical	Phase II trials
Pulmicort	Bronchial-respiratory	Astrazeneca PLC	New labeling approval
Serevent Diskus	Bronchial-respiratory	GlaxoSmithKline	New indication approval
Nicotrol	Smoking cessation aid	Pharmacia	New formation approval
Flunisolide	Severe asthma and lung disease	Bausch & Lomb Pharma	Approved
Cromolyn Sodium	Persistent asthma	Novex Pharma	Approved

- Bile salts
- Phospholipids
- Cyclodextrins
- Micelles
- Liposomes
- Emulsions
- Polymeric microspheres

22.6.2.1 Excipients Used in Formulation

A wide range of excipients and excipient technologies have become available for use with nasally administered products since the 1980s. A broad overview of these follows.

22.6.2.1.1 Surfactants

The effect of surfactant on drug absorption across nasal mucosa has been studied since the 1970s.

In 1981, Hirai et al. compared nonionic, anionic, amphoteric synthetic surfactants and natural anionic surfactants for in vivo nasal absorption of insulin. These authors reported the following:

- A majority of surfactants enhanced insulin absorption relative to the extraction of membrane components.
- Enhancing effects correlate with the extraction efficiency of the membrane components by the surfactants.
- Alkyl glycosides constitute a novel class of sugar-derived surfactants used in cosmetics.

- Maltoside derivatives with an alkyl chain length between 12 and 14 enhance insulin absorption at low surfactant concentrations.
- Mechanism involved the loosening of tight junctions and increasing the paracellular transport.
- Enhancement effect produced by surfactant monomers is related to the ability of the surfactant molecules to penetrate and fluidize the lipid bilayers.
- Carboxymethyl cellulose (CMC), monomers aggregate into micelles, which can solubilize components, particularly cholesterol and phospholipids.

22.6.2.1.2 Bile Salts

- Enhancement of insulin absorption but with milder effects on the biomembrane.
- Additionally, the absorption-promoting effects appear to arise from an inhibitory effect on the enzyme degradation of insulin.
- Bile salts appeared to be most promising and effective absorption enhancers of peptides and proteins and still remain widely used as permeation enhancer.

22.6.2.1.3 Cyclodextrins

- Cyclodextrins are cyclic oligosaccharides containing a minimum of 6-D-glycopyranose units attached by an alpha 1, 4-linkage.
- Cyclodextrins are produced by enzymatic conversion of prehydrolyzed starch.
- Natural cyclodextrins are designated by a Greek letter—α, β, and γ. The β form is the most soluble of the three.
- The ring structure resembles a truncated cone with characteristic cavity volume.
- The internal surface of cavity has slight hydrophobic properties whereas the outer surface is hydrophilic.
- Cyclodextrins form inclusion complexes with lipophilic molecules.
- Generally appear to be less irritating to the nasal mucosa than bile salts and surfactants.
- Large increase in solubility through complex formation and protection from enzymatic degradation improve nasal absorption of lipophilic drugs.
- Very potent absorption enhancers for hydrophilic peptides that are not complexes.
- Shao and Mitra (1992) evaluated several cyclodextrins for their absorption-promoting effect on insulin.

The best results for absorption-promoting effect on insulin were obtained with dimethyl-β-cyclodextrin:

- Promoting order correlated well with the extent of nasal mucosal perturbation.

22.6.2.1.4 Mixed Micelles

- Nasal absorption of insulin in the presence of sodium glycocholate (NaGC) and linoleic acid increased relative to the increase by NaGC and linoleic acid alone.
- Mixed micelles of bile salts and fatty acid appear to have a synergistic effect on the absorption of peptides.
- Maximal nasal absorption enhancement of [D-Arg] kyotorphin has been observed with mixed micelles of NaGC and linoleic acid, effect greater than that with glycocholate alone.

22.6.2.1.5 Formulation and Potential Mucosal Damage

- Improved absorption involves interactions with the mucosal membrane.
- Proposed enhancement mechanisms are as follows:
  - Extraction of membrane components.
  - Penetration and fluidization of membrane.
  - Loosening of tight junctions.
  - Perturbation of nasal mucociliary clearance system.

- Simultaneous transport of environmental toxins.
- Adverse effects have to be of short duration, mild, and rapidly reversible.
- Kinetics of lipid and protein extraction from the membrane are measures of the extent of damage evaluated by measuring the activity of membrane marker enzymes.
- Lactate dehydrogenase: cytosolic enzyme related to intracellular damage.
- 5′-nucleotidase: membrane-bound enzyme, indicator of membrane perturbations.
- Alkaline phosphatase: membrane-bound enzyme, related to membrane damage.
- Ideal characteristics of absorption enhancers include the following:
  - Pharmacological inertness.
  - Nonirritant, nontoxic, and nonallergenic.
  - Effect on nasal mucosa should be transient and completely reversible.
  - Potent in low concentrations.
  - Compatible with other adjuvants.
  - Possesses no offensive odor or taste.
  - Inexpensive and readily available.
  - The factors influencing mucosal damage include the following:
  - Drug administration.
  - Dose.
  - Frequency.
  - Interspecies difference.
  - Sensitivity toward absorption enhancers.
- Clinical signs of nasal irritation studies in rats include the following:
  - In studies for less than 90 days.
  - Struggling, sneezing, salivation, head shaking, and nose rubbing.
- In studies for more than 90-day studies:
  - Histological signs of nasal irritation including inflammation of septal and turbinate mucosal surfaces, epithelial and submucosal infiltration of inflammatory cells, purulent exudates, and mucosal hyperplasia.

Zhang and Jiang (2001) have recently characterized specific approaches for the reduction of the local tissue nasal toxicity of drugs and should be consulted for such.

22.6.2.1.6 Methods to Assess Irritancy and Damage
Erythrocytes

- Used to study the membrane activity of absorption enhancers.

Histology

- Histological studies of nasal membranes.

Intracellular protein release

- Index of cellular damage due to exposure to absorption enhancers.

Tolerability

- These are subjective (double-masked) studies in which individuals report any effects due to the use of enhancers in the formulation.

Cilia function

- Cilia beat frequency is obtained from tissue samples at sacrifice using video capture systems.
- Tissues used for ciliary function studies include chicken embryo trachea, cryopreserved human mucosa taken from sphenoidal sinus, rat nasal mucosa, and recently human nasal epithelial cells.

22.6.2.1.7 Reported Nasal Irritancy Responses

- Local irritation, burning, and stinging upon both acute and/or chronic administration were reported for laureth-9, bile salts, and sodium taurodihydrofusidate.
- Slight nasal itch was reported when dimethyl-cyclodextrin was used for nasal insulin delivery.
- Nasal burning and sinusitis were reported during studies involving nasal insulin delivery with glycocholate and methylcellulose.

22.6.2.2 Delivery Forms

There are a wide range of approaches to clinical drug products and delivery—that is, the actual final form of the drug as administered to patients. The key points on the major classes of these are summarized as follows. It should be noted that if a delivery device is used, it must also be reviewed and approved by the Food and Drug Administration.

22.6.2.2.1 Liquid Nasal Formulation

- These are the most widely used dosage forms in clinical practice.
- These are mainly based on aqueous formulations.
- Humidifying effect is convenient due to the drying of mucous membranes owing to allergic and chronic diseases.
- Major drawback is the limitation on microbiological stability.
- Reduced chemical stability of the drug and short residence time in the nasal cavity are other disadvantages.
- Deposition site and deposition pattern are dependent on delivery device, mode of administration, and physicochemical properties of formulation.
- Preparations depend on whether administered for local or systemic application.
- Patient compliance, cost effectiveness, and risk assessment are concerns.

22.6.2.2.2 Instillation and Rhinyle Catheter

- Catheters are used for delivery to a defined region.
- The combination of an instillation catheter to a Hamilton threaded plunger syringe has been used in order to compare the deposition of drops, nebulizers, and sprays in rhesus monkeys.
- These are used only for experimental studies and not for commercial clinical products.

22.6.2.2.3 Drops

- This is one of the oldest delivery systems.
- Low-cost devices are utilized.
- It is easy to manufacture.
- Disadvantages are related to microbiological and chemical stability.
- Delivered volume cannot be controlled.
- Formulation can be easily contaminated by pipette (delivery device).

22.6.2.2.4 Powder Dosage Forms

- Dry powders are less frequent in nasal drug delivery.
- Major advantages include lack of need for preservatives and improved drug stability.
- It prolongs retention times in nasal region when compared to solutions.
- Addition of bioadhesive excipients results in further decreased clearance rates.
- Nasal powders may increase patient compliance especially for children if smell and taste of drug are otherwise unacceptable.

22.6.2.2.5 Insufflators and Mono-Dose Powder Inhaler

- Many insufflators work with predosed powder doses in capsules.
- The use of gelatin capsules enables the filling and application of different amounts of powder.

- In a mono-dose powder inhaler, pushing a piston results in a precompression of air in chamber.
- The piston pierces a membrane, and the expanding air expels air into the nostrils.

22.6.2.2.6 Pressurized MDIs

- They are manufactured by suspending the drug in liquid propellants with the aid of surfactants.
- Physicochemical compatibility between the drug and propellants must be evaluated.
- Phase separation, precipitation, crystal growth, polymorphism, dispersibility, and adsorption of drug influence drug particle size, dose distribution, and deposition pattern.
- Their advantages include the following:
  - Portability, small size, availability over a wide dose range, dose consistency and accuracy, and protection of contents
- Their disadvantages include the following:
  - Nasal irritation by propellants and depletion of ozone layer by CFCs

22.6.2.2.7 Nasal Gels

- Nasal administration of gels can be achieved by precompression pumps.
- The deposition of gel in the nasal cavity depends on the mode of administration, due to its viscosity and poor spreading properties.
- Nasal gel containing vitamin B_{12} for systemic administration is available in market.

22.6.2.2.8 Patented Nasal Formulations

- West Pharma developed nasal technology (ChiSys) based on the use of chitosan as an absorption enhancer.
- Chitosan is a natural polysaccharide with bioadhesive properties.
- It prolongs the retention time of the formulation in the nasal cavity.
- It may facilitate absorption through promoting paracellular transport.

22.7 METHODS FOR SAFETY ASSESSMENT OF INHALED THERAPEUTICS

Methods for the evaluation of inhalation toxicity should be selected according to the pharmacological and/or the toxicological questions asked, and the design of experiments should specify the delivery route of a drug to the target sites in the pulmonary system. For example, if an immunologic response of the lungs to a drug is in question, then the lymphoid tissues of the lungs should be the major target of evaluation. The following are some of the physiological, biochemical, and pharmacological tests that are applicable for the safety assessment of inhaled medicinal gases, vapors, or aerosols.

URT irritation can occur from the inhalation of a medicinal gas (nitrous oxide), vapor (salicylates), or aerosol (virtually any using a surfactant). For assessing the potential of an inhalant to cause URT irritation, the mouse body plethysmographic technique (Alarie 1966, 1981a,b) has proven to be extremely useful. This technique operates on the principle that respiratory irritants stimulate the sensory nerve endings located at the surface of the respiratory tract from the nose to the alveolar region. The nerve endings in turn stimulate a variety of reflex responses (Alarie 1973; Widdicombe 1974) that result in characteristic changes in inspiratory and expiratory patterns and, most prominently, depression of respiratory rate. Both the potency of irritation and the concentration of the irritant are positively related to the magnitude of respiratory rate depression. The concentration response can be quantitatively expressed in terms of RD_{50}, defined as the concentration (in logarithmic scale) of the drug in the air that causes a 50% decrease in respiratory rate. The criteria for positive URT irritation in intact mice exposed to the drug atmosphere are depression in breathing frequency and a qualitative alteration of the expiratory patterns. Numerous experimental results have shown that the responses of mice correlated almost perfectly with those of humans (Alarie 1980; Alarie and Luo 1984). Thus, this technique is useful for predicting the irritancy of airborne medicinal compounds in humans. From the drug-formulating point of view, an inhalant drug with URT-irritating properties indicates the need for an alternate route of administration. From the industrial hygiene point of view, the recognition of the irritant properties is very important. If a chemical gas, vapor, or aerosol irritates, it has a *warning property*. With an adequate warning property, a worker will avoid inhaling damaging amounts of the airborne toxicant; without a warning property, a worker may unknowingly inhale a harmful amount of the toxicant. However, warning properties are not very reliable for alerting individuals to potential health risks since a person can become tolerant or adapt to smell or irritation properties.

Respiratory tract irritation can alter absorption of a therapeutic agent in at least two ways. If irritation causes cell death and/or loss, such damage will act to improve the extent of absorption. If, however, irritation leads to increased secretion, of mucus, this can serve to act as an increased barrier to absorption and therefore decreased systemic drug availability.

Inhalation of a cardiovascular drug, such as an aerosol of propranolol (a beta-adrenergic receptor agonist), may affect the respiratory cycle of a subject. For evaluating the cardiopulmonary effects of an inhalant, the plethysmograph technique using a mouse or a guinea pig model is useful. The criteria for a positive response in intact mice or guinea pigs are changes in the duration of inspiration and expiration, and the interval between breaths (Schaper 1989).

Pulmonary sensitization may occur from inhalation of drug vapors such as enflurane (Schwettmann and Casterline 1976), and antibiotics such as spiramycin (Davies and Pepys 1975) and tetracycline (Menon and Das 1977). To detect pulmonary sensitization from inhalation of drug and chemical aerosols, the body plethysmographic technique using a guinea pig model has been shown to be useful (Patterson and Kelly 1974; Karol 1988, 1989; Thorne and Karol 1989). The criteria for positive pulmonary sensitization in intact guinea pigs are changes in breathing frequency and their extent, and the time of onset of an airway constrictive response after induction and after a challenge dose of the test drug (Karol et al. 1989).

The mucociliary transport system of the airways can be impaired by respiratory irritants, local analgesics, and anesthetics, and parasympathetic stimulants (Pavia 1984). Any one of these agents will retard the beating frequency of the cilia and the secretion of the serous fluid of the mucous membranes. As a result, the propulsion of the inhaled particles, bacteria, or endogenous debris toward the oral pharynx for expectoration or swallowing will be retarded. Conversely, inhalation of adrenergic agonists increases the activity of the mucociliary transport system and facilitates the elimination of noxious material from the pulmonary system. Laboratory evaluation of the adverse drug effects on mucociliary transport in animal models can be achieved by measuring the velocity of the linear flow of mucus in the trachea of surgically prepared animals (Rylander 1966). Clinically, the transportation of markers placed on the tracheal epithelium of normal human subjects can also be observed using a fiber-optic bronchoscopic technique (Pavia et al. 1980; Mussatto et al. 1988). The criteria of a positive response are changes in the transport time over a given distance of markers placed on the mucus or changes in the rate of mucous secretion (Davis 1976; Johnson et al. 1983, 1987; Webber and Widdicombe 1987). More comprehensive discussion on mucociliary clearance can be found in several reviews (Last 1982; Pavia 1984).

Cytological studies on the bronchial alveolar lavage fluid (BALF) permit the evaluation of the effects of an inhaled drug on the epithelial lining of the respiratory tract. This fluid can be obtained form intact animals or from excised lungs (Henderson 1984, 1988, 1989). Quantitative analyses of fluid constituents such as neutrophils, antibody-forming lymphocytes, and antigen-specific IgG provide information on the cellular and biochemical responses of the lungs to the inhaled agent (Henderson 1984, Henderson et al. 1985, 1987). For example, BALF parameters were found to be unperturbed by the inhalation of halothane (Henderson and Lowrey 1983). The criteria of a positive response are increase in protein content, increase in the number of neutrophils and macrophages for inflammation, increase in the number of lymphocytes and alteration of lymphocyte profiles for immune response, increase in cytoplasmic enzymes (lactate dehydrogenase) for cell lysis (Henderson 1989), and the presence of antigen-specific antibodies for specific immune responses (Bice 1985).

Morphological examination of the cellular structure of the pulmonary system is the foundation of most inhalation toxicity studies. Inhalation of airborne drug vapors or aerosols at harmful concentrations results mainly in local histopathologic changes in the epithelial cells of the airways, of which there are two types: nonciliated and ciliated cells. The nonciliated cells are the Clara cells, which contain secretory granules with smooth endoplasmic reticulum (SER); the secretory granules that lack SER; and the brush cells, which have stubby microvilli and numerous cytoplasmic fibers on their free surfaces. If the concentration gradient of the drug in the lung is high enough to reach the alveoli, the type I alveoli cells will also be affected (Evans 1982). Drugs that affect the lungs via the bloodstream, such as bleomycin (Aso 1976), cause changes to the endothelial cells of the vascular system that result in diffuse damage to the alveoli. The criteria of cellular damage are loss of cilia, swelling, and necrosis and sloughing of cell debris into the airway lumina. Tissues recovering from injuries are characterized by increases in the number of dividing progenitor cells followed by increases in intermediate cells that eventually differentiate into normal surface epithelium.

Pulmonary drug disposition studies are essential in research and development of new inhalant drugs. Inhaled drugs are usually absorbed and metabolized to some extent in the lungs because the lungs, like the liver, contain active enzyme systems. A drug may be metabolized to an inactive compound for excretion or to a highly reactive toxic metabolite(s) that causes pulmonary damage. In most pulmonary disposition studies, a gas or vapor is delivered via whole-body exposure (Paustenbach 1983) or head-only exposure (Hafner 1975). For aerosols, over 90% of a dose administered by mouth breathing is deposited in the oropharynx and swallowed. Consequently, the disposition pattern reflects that of ingestion in combination with a small contribution form pulmonary metabolism. For determining the disposition of inhaled drugs by the pulmonary system alone, a dosimetric endotracheal nebulization technique (Leong 1988) is useful. In this technique, microliter quantities of a radiolabeled drug solution can be nebulized within the trachea using a miniature air-liquid nebulizing nozzle. Alternatively, a small volume of liquid can be dispersed endotracheally using a microsyringe. In either technique, an accurate dose of a labeled drug solution is delivered entirely into the respiratory tract and lungs. Subsequent radioassay of the excreta thus reflects only the pulmonary disposition of the drug without complication from aerosols deposited in the oropharyngeal regions as would be the case if the drug had been delivered by mouth inhalation. For example, in a study of the antiasthmatic drug lodoxamide tromethamine, the urinary metabolites produced by beagle dogs after receiving a dose of the radiolabeled drug via endotracheal nebulization showed a high percentage of the intact drug. However, metabolites produced after oral administrations were mainly nonactive conjugates. The differences were due to the drug's escape from first-pass metabolism in the liver when it was administered through the pulmonary system. The results thus indicated that the drug had to be administered by inhalation to be effective. This crucial information was extremely important in the selection of the most effective route of administration and formulation of this antiasthmatic drug (Leong et al. 1988).

Cardiotoxicity of inhalant drugs should also be evaluated. For example, adverse cardiac effects may be induced by inhaling vapors of fluorocarbons, which are used extensively as propellants in drug aerosols. Inhalations of vapors of anesthetics have also been shown to cause depression of the heart rate and alteration of the rhythm and blood pressure (Merin 1981; Leong and Rop 1989). More importantly, inhalation of antiasthmatic aerosols of beta-receptor agonists delivered in a fluorocarbon propellant has been shown to cause marked tachycardia, electrocardiogram (ECG) changes, and sensitization of the heart to arrhythmia (Aviado 1981; Balazs 1981). Chronic inhalation of drug aerosols can also result in cardiomyopathy (Balazs 1981). For the detection of cardiotoxicity, standard methods of monitoring arterial pressures, heart rate, and ECGs of animals during inhalation of a drug or at frequent intervals during a prolonged treatment period should be useful in safety assessments of inhalant drugs.

Since the inhalation route is just a method for administering drugs, other nonpulmonary effects, such as behavioral effects (Ts'o et al. 1975) and renal and liver toxicity, should also be evaluated. In addition, attention should also be given to drugs that are not administered via the inhalation route, but that accumulate in the lungs where they cause pulmonary damage (Wilson 1982), such as bleomycin.

22.7.1 PARAMETERS OF TOXICITY EVALUATION

Paracelsus stated over 400 years ago that "All substances are poison. The right dose differentiates a poison and a remedy." Thus, in safety assessments of inhaled drugs, the *dose*, or magnitude of inhalation exposure, in relation to the physiological, biochemical, cytological, or morphological response(s) must be determined. Toxicity information is essential to establishing guidelines to prevent the health hazards of acute or chronic overdosage during therapy or of unintentional exposure to the bulk drugs and their formulated products during manufacturing and industrial handling.

22.7.1.1 Inhaled *Dose*

Most drugs are designed for oral or parenteral administration in which the dose is calculated in terms of drug weight in milligrams (mg) divided by the body weight in kilograms (kg):

$$\text{Dose} = \frac{\text{Drug weight (mg)}}{\text{Body weight (kg)}} = \text{mg/kg}$$

For inhalant drugs, the inhaled *dose* has been expressed in many mathematical models (Dahl 1990). However, the practical approach is based on exposure concentration and duration rather than on theoretic concepts. Thus, an inhaled *dose* is expressed in terms of the exposure concentration[©] in milligrams per liter (mg/L) or milligrams per cubic meter (mg/m³) or, less commonly, parts per million (ppm) parts of air, the duration of exposure (*t*) in minutes, the ventilatory parameters including the respiratory rate[®] in the number of breaths per minute and the tidal volume (*Tv*) in liters per breath, and a dimensionless retention factor α (alpha), which is related to the reactivity and the solubility of the drug. The product of these parameters divided by the body weight in kilograms gives the dose:

$$\text{Dose} = \frac{C \cdot t \cdot R \cdot Tv \cdot \alpha}{\text{Body weight}} = \text{mg/kg}$$

In the critical evaluation of the effect of a gas, vapor, or aerosol inhaled into the respiratory tract of an animal, the dosimetric method has been recommended (Oberst 1961). However, due to the complexity of measuring various parameters simultaneously, only a few studies on gaseous drugs or chemicals have employed the dosimetric method (Weston and Karel 1946; Leong and MacFarland 1965; Landy et al. 1983; Stott and McKenna 1984; Dallas et al. 1986, 1989). For studies on liquid or powdery aerosols, modified techniques such as intratracheal instillation (Brain et al. 1976) or endotracheal nebulization (Leong et al. 1988) were used to deliver an exact dose of the test material into the LRT while bypassing the URT and ignoring the ventilatory parameters. These methods deliver a bolus to the lung, which does not mimic the exposure/distribution pattern achieved by actual inhalation.

In routine inhalation studies, it is generally accepted that the respiratory parameters are relatively constant when the animals are similar in age, sex, and body weight. This leaves only *C* and *t* to be the major variables for dose consideration.

$$Dose = C \cdot t = \text{mg/L·min}$$

The product *Ct* is not a true dose because its unit is mg·min/L rather than mg/kg. Nevertheless, *Ct* can be manipulated as though it were a dose, an approximated dose (MacFarland 1976).

TABLE 22.6

Respiratory Parameters for Common Experimental Species and Man

Species	Body Weight (kg)	Lung Volume (mL)	Minute Volume (mL/min)	Alveolar Surface Area (m²)	*Lung Volume%* Surface Area	*Minute Volume%* Lung Volume	*Minute Volume%* Surface Area
Mouse	0.023	0.74	24	0.068	10.9	32.4	353
Rat	0.14	6.3	84	0.39	16.2	13.3	215
Monkey	3.7	184	694	13	14.2	3.77	53
Dog	22.8	1501	2923	90	16.7	1.95	33
Human	75	7000	6000	82	85.4	0.86	73

Source: Altman, P.L. and Dittmer, D.S., *Biological Data Book*, Vol. III, Federation of American Societies for Experimental Biology, Bethesda, MD, 1974.

The respiratory parameters of an animal will dictate the volume of air inhaled and hence the quantity of test material entering the respiratory system. Commonly used parameters for a number of experimental species and man are given in Table 22.6 to illustrate this point and include the alveolar surface area because this represents the target tissue for most inhaled materials. It can be seen that by taking the ratios of these parameters and comparing the two extremes, that is, the mouse and man, that (McNeill 1964) a mouse inhales approximately 30 times its lung volume in 1 min whereas a man at rest inhales approximately the same volume as that of his lung. This can increase with heavy work up to the same ratio as the mouse, but is not sustained for long periods. This means that the dose per unit lung volume is up to 30 times higher in the mouse than man at the same inhaled atmospheric concentration (Touvay and Le Mosquet 2000). The minute volume of the mouse is in contact with five times less alveolar surface area than man, hence the dose per unit area is up to five times greater in the mouse (Akoun 1989). The lung volume in comparison with the alveolar surface area in experimental animals is less than that in humans, meaning that the extent of contact of inhaled gases with the alveolar surface is greater in experimental animals.

While it is possible, and common, to refer to standard respiratory parameters for different species in order to calculate inhaled dose and deposited dose with time, it is usually the case that inhaled materials influence the breathing patterns of test animals. The most common examples of this are irritant vapors, which can reduce the respiratory rate by up to 80%. This phenomenon results from a reflexive pause during the breathing cycle due to stimulation by the inhaled material of the trigeminal nerve endings situated in the nasal passages. The duration of the pause and hence the reduction in the respiratory rate are concentration related, permitting concentration–response relationships to be plotted. This has been investigated extensively by Alarie (1981a) and forms the basis of a test screen for comparing the irritancy of different materials quantitatively and has found application in assessing appropriate exposure limits for human exposure when respiratory irritancy is the predominant cause for concern.

While irritancy resulting from the earlier reflex reaction is one cause of altered respiratory parameters during exposure, there are many others. These include other types of reflex response, such as bronchoconstriction, the narcotic effects of many solvents, the development of toxic signs as exposure progresses, or simply a voluntary reduction in respiratory rate by the test animal due to the unpleasant nature of the inhaled atmosphere. The extent to which these affect breathing patterns and hence inhaled dose can be assessed only by actual measurement.

By simultaneous monitoring of tidal volume and respiratory rate, or minute volume, and the concentration of an inhaled vapor in the bloodstream and the vapor in the exposure atmosphere, pharmacokinetic studies on the $C \cdot t$ relationship have shown that the effective dose was nearly proportional to the exposure concentration for vapors such as 1,1,1-trichloroethane (Dallas et al., 1989), which has a saturable metabolism, found that the steady-state plasma concentrations were dispro-portionately greater at higher exposure concentrations.

Acknowledging the possible existence of deviations, this simplified approach of using C and t for dose determination provides the basis for dose–response assessments in practically all inhalation toxicological studies.

22.7.1.2 Dose–Response Relationship

Often overlooked is that dose response in toxicology has multiple dimensions. The reader should start by recognizing that at least three dimensions are present, and all must be considered to under-stand a biologic system response. As dose increases,

1. Incidence of responders in an exposed population increases (population incidence)
2. Severity of response in affected individuals increases (severity)
3. Time to occurrence of response or of progressive stage of response decreases (lag time)

The oldest principle of dose–response determination in inhalation toxicology is based on Haber's rule, which states that responses to an inhaled toxicant will be the same under conditions where C varies in complementary manner to t (Haber 1924). For example, if $C \cdot t$ elicits a specific magnitude of the same response, that is, $Ct = K$, where K is a constant for the stated magnitude of response. This was first developed for use with war gases, and it holds up reasonably well for shorter exposure to such agents.

This rule holds reasonably well when C or t varies within a narrow range for acute exposure to a gaseous compound (Rinehart and Hatch 1964) and for chronic exposure to an inert particle (Henderson 1991). Excursion of C or t beyond these limits will cause the assumption $Ct = K$ to be incorrect (Adams et al. 1950, 1952; Sidorenko and Pinigin 1976; Andersen et al. 1979; Uemitsu et al. 1985). For example, an animal may be exposed to 1000 ppm of diethyl ether for 420 min or 1400 ppm for 300 min at a constant rate without incurring any anesthesia. However, exposure to 420,000 ppm for 1 min will surely cause anesthesia or even death of the animal. Furthermore, toxicokinetic study of liver enzymes affected by inhalation of carbon tetrachloride (Uemitsu et al. 1985), which has a saturable metabolism in rats, showed that $Ct = K$ does not correctly reflect the *toxicity value* of this compound. Therefore, the limitations of Haber's rule must be recognized when it is used in interpolation or extrapolation of inhalation toxicity data, and such is not recommended for most inhalation situations.

22.7.1.3 Exposure Concentration versus Response

In certain medical situations (e.g., a patient's variable exposure duration to a surgical concentra-tion of an inhalant anesthetic, or the repeated exposures of surgeons and nurses to subanesthetic concentrations of an anesthetic in the operating theater), it is necessary to know the duration of safe exposure to a drug. Duration safety can be assessed by determining a drug's median effective time (Et_{50}) or median lethal time (Lt_{50}). These statistically derived quantities represent the duration of exposure required to affect or kill 50% of a group of animals exposed to a specified concentration of an airborne drug or chemical in the atmosphere.

The graph in Figure 22.2 is the probit plot of cumulative percentage response to the logarithm of exposure duration. It shows the 1000 mg/m³ for 10 h or to 10 mg/m³ for 1000 h, each with a Ct (an approximated dose) of ~10,000 h mg/m³. Similar to concentration–response graphs, the slopes indicate the differences in the mechanism of action and the margins of safe exposure of the three drugs. The ratio of the ET_{50} or LT_{50} of two drugs indicates their relative toxicity, and the ratio of ET_{50} over LT_{50} of the same drug is the therapeutic ratio.

22.7.1.4 Product of Concentration and Duration (Ct) versus Responses

To evaluate inhalation toxicity in situations where workers are exposed to various concentrations and durations of a drug vapor, aerosol, or powder in the work environment during manufacturing or

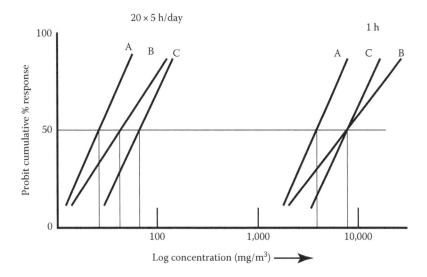

FIGURE 22.2   Dose–response is plotted in terms of the probit of cumulative percentage response to logarithm of the exposure concentrations, where A, B, and C are agents acting by different mechanisms or kinetics.

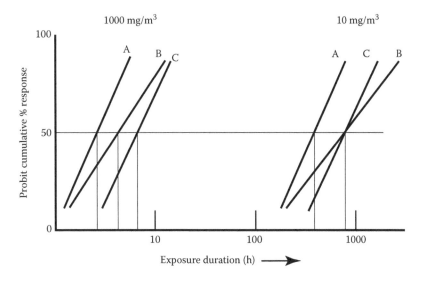

FIGURE 22.3   Dose–response plotted in terms of the probit plot of cumulative percentage response to logarithm of exposure duration.

packaging, a more comprehensive determination of $E(Ct)_{50}$ or $L(Ct)_{50}$ values is used. The $E(Ct)_{50}$ or $L(Ct)_{50}$ values are statistically derived values that represent the magnitude of exposure, expressed as a function of the product of C and t, that is expected to affect or kill approximately 50% of the animals exposed (Figure 22.3). The other curve represents exposures that kill 50% or >50% of each group of animals (Irish and Adams 1940).

The graph in Figure 22.4 illustrates inhalation exposures to a drug using various combinations of C and t that kill 50% of the animals. For example, a 50% mortality occurs when a group of animals is exposed to drug A at a concentration of 1000 mg/m³ for a duration of approximately 2 h, or at a concentration of 100 mg/m³ for a duration of approximately 20 h. Furthermore, the graph also illustrates that the inhalation toxicity of drug A is more than one order of magnitude higher than that of drug B. For example, an exposure to drug A at the concentration of

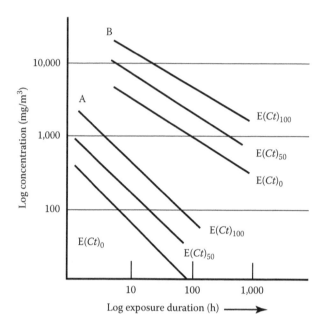

FIGURE 22.4   Dose–response plotted in terms of logarithms of drug exposure concentration and durations.

100 mg/m³ for 100 h kills 100% of the animals, whereas an exposure to drug B at the concentration of 1000 mg/m³ for 100 h does not kill any animals.

22.7.1.5   Units for Exposure Concentration

For therapeutic gases and vapors, exposure concentrations are traditionally expressed in parts per million (ppm). The calculation for the ppm of a gas or vapor in an air sample is based on Avogadro's law, which states that "Equal volumes contain equal numbers of molecules under the same temperature and pressure." In other words, under standard temperature and pressure, 1 g molecular weight (mole) of any gas under a pressure of one atmosphere (equivalent to the height of 760 mm mercury) and a temperature of 273 K has the same number of molecules and occupies the same volume of 22.4 L. However, under ambient conditions, the volume of 22.4 L has to be corrected to a larger volume based on Charles's law, which states that at constant pressure, the volume of gas varies directly with the absolute temperature. Thus, at a room temperature of 25°C, one mole of a gas occupies a volume of 24.5 L.

$$22.4 \text{ L} \times \frac{298 \text{ K}}{273 \text{ K}} = 24.5 \text{ L}$$

Further correction of volume for an atmospheric pressure deviation from one atmosphere may be done by applying Boyle's law, which states that the volume of a gas without change of temperature varies inversely with the pressure applied to it.

$$24.5 \text{ L} \times \frac{758 \text{ mm Hg}}{760 \text{ mm Hg}} = 24.4 \text{ L}$$

In practice, atmospheric pressure in most animal experimental environments usually varies only a few mm Hg, so little or no correction is required.

Using the aforementioned principles, the volume of a vapor generated from a given weight of a liquid can be calculated. For example, 1 mol of water weighs 18 g, while 1 mol of ethanol weighs 46 g. When 1 mol of each liquid is totally vaporized, each will occupy the same volume of 24.5 L at room temperature (25°C) and pressure (760 mm Hg). In an inhalation experiment, if the volume of test liquid and the rate of airflow being mixed in the animal exposure chamber are known, the vapor concentration in the chamber atmosphere can be calculated in parts per million or milligrams per liter. A conversion table published by the US Bureau of Mines enables quick conversion between parts per million and milligrams per liter for compounds with molecular weights up to 300 g (Fieldner et al. 1921; Patty 1958).

For aerosols of nonvolatile liquid and powder pharmaceutical compounds, the concentration of the mist or dust atmosphere must be expressed in terms of milligrams per liter or milligrams per cubic meter (mg/m^3) of air. With advances in biotechnology, many pharmacological testing techniques are based on specific receptor bindings, in which the ratio of the number of molecules to those of the receptors are considered, in which case the exposure concentration may be more appropriately expressed in micromoles per unit volume of air ($\mu mol/m^3$).

22.7.2 RESPIRATORY SAFETY PHARMACOLOGY

Pharmaceuticals differ from industrial and environmental chemicals in that the scope of concern for their adverse safety effects on the respiratory system extends both to reversible functional degradations and to effects on the respiratory systems functionality due to systemically distributed agents administered by routes other than direct respiratory. This is the realm of the relatively new field of safety pharmacology.

As early as 1964, it became apparent that β-adrenergic blocking agents could lead to bronchonstriction (and possible death) in asthmatics (McNeill 1964). Since then, many similar adverse effects have been identified. These known effects of drugs from a variety of pharmacologic/therapeutic classes on the respiratory system are summarized in Tables 22.7, 22.8, and 22.9. Resulting worldwide regulatory requirements (Tables 22.10 and 22.11) require the conduct of prescribed respiratory evaluations prior to drug in humans. The objective of such studies is to evaluate the potential for drugs to cause nonintended pharmacologic or toxicologic effects that influence respiratory function. Changes in respiratory function can result either from alterations in the pumping apparatus that controls the pattern of pulmonary ventilation or from changes in the mechanical properties of the lung that determine the transpulmonary pressures (work) required for lung inflation and deflation.

The respiratory system is responsible for generating and regulating the transpulmonary pressures needed to inflate and deflate the lung. Normal gas exchange between the lung and blood requires breathing patterns that ensure appropriate alveolar ventilation. Ventilatory disorders that alter alveolar ventilation are defined as hypoventilation or hyperventilation syndromes. Hyperventilation results in an increase in the partial pressure of arterial CO_2 above normal limits and can lead to acidosis, pulmonary hypertension, congestive heart failure, headache, and disturbed sleep. Hypoventilation results in a decrease in the partial pressure of arterial CO_2 below normal limits and can lead to alkalosis, syncope, epileptic attacks, reduced cardiac output, and muscle weakness.

Normal ventilation requires that the pumping apparatus provide both adequate total pulmonary ventilation (minute volume) and the appropriate depth (tidal volume) and frequency of breathing. The depth and frequency of breathing required for alveolar ventilation are determined primarily by the anatomic deadspace of the lung. In general, a rapid shallow breathing pattern (tachypnea) is less efficient than a slower deeper breathing pattern that achieves the same minute volume. Thus, any change in minute volume, tidal volume, or the rate of breathing can influence the efficiency of ventilation (Milic-Emili 1982). The inspiratory and expiratory phases of individual breathing have rates of airflow and durations that are distinct and independently controlled (Boggs 1992). Thus, by characterizing changes in the airflow rate and duration of each of these phases, mechanisms responsible for changes in tidal volume or respiratory rate can be identified (Milic-Emili 1982, Indans, 2002). For example, a decrease in airflow during inspiration (the active phase) is generally indicative of a decrease in respiratory drive, while a decrease in airflow during expiration (the passive phase) is generally indicative of an obstructive disorder.

TABLE 22.7
Drugs Known to Cause Pulmonary Disease

Chemotherapeutic
  Cytotoxic
    Azathioprine
    Bleomycin[a]
    Busulfan
    Chlorambucil
    Cyclophosphamide
    Etoposide
    Melphalan
    Mitomycin[a]
    Nitrosoureas
    Procarbazine
    Vinblastine
    Ifosfamide
  Noncytotoxic
    Methotrexate[a]
    Cytosine arabinoside[a]
    Bleomycin[a]
    Procarbazine[a]

Antibiotic
  Amphotericin B[a]
  Nitrofurantoin
    Acute[a]
    Chronic
  Sulfasalazine
  Sulfonamides
  Pentamidine

Anti-inflammatory
  Acetylsalicylic acid[a]
  Gold
  Methotrexate
  Nonsteroidal anti-inflammatory agents
  Penicillamine[a]

Immunosuppressive
  Cyclosporin
  Interleukin-2[a]

Analgesics
  Heroin[a]
  Methadone[a]
  Naloxone[a]
  Ethchlorvynol[a]
  Propoxyphene[a]
  Salicylates[a]

Cardiovascular
  Amiodarone[a]
  Angiotensin-converting enzyme inhibitors
  Anticoagulants
  Beta-blockers[a]
  Dipyridamole
  Fibrinolytic agents[a]
  Protamine[a]
  Tocainide

Inhalants
  Aspirated oil
  Oxygen[a]

Intravenous
  Blood[a]
  Ethanolamine oleate (sodium morrhuate)[a]
  Ethiodized oil (lymphangiogram)
  Talc
  Fat emulsion

Miscellaneous
  Bromocriptine
  Dantrolene
  Hydrochlorothiazide[a]
  Methysergide
  Oral contraceptives
  Tocolytic agents[a]
  Tricyclics[a]
  L-Tryptophan
  Radiation
  Systemic lupus erythematosus (drug induced)[a]
  Complement-mediated leukostasis[a]

Sources: Touvay, C. and Le Mosquet, B., *Therapie* 55, 71, 2000; Akoun, G.M., Natural history of drug-induced pneumonitis, in *Drug Induced Disorders*, Volume 3: *Treatment Induced Respiratory Disorders*, Akoun, G.M. and White, J.P., eds., Elsevier Scientific Publishers B.V., New York, 1989, pp. 3–9; Dorato, M.A., *Drugs Pharm. Sci.*, 62, 345, 1994; Lalej-Bennis, D., *Diabetes Metab.*, 27(3), 372, 2001; Mauderly, J.L., Effects of inhaled toxicants on pulmonary function, in *Concepts in Inhalation Toxicology*, McClellan, R.O. and Henderson, R.F., eds., Hemisphere Publishing Corp, New York, 1989, pp. 347–401; Rosnow, E.C., *Chest*, 102, 239, 1992.

[a] Typically cause acute or subacute respiratory insufficiency.

TABLE 22.8

Drugs Adversely Affect Respiratory Function

Drugs known to cause or aggravate bronchospasm

Vinblastine

Nitrofurantoin (acute)

Acetylsalicylic acid

Nonsteroidal anti-inflammatory agents

Interleukin-2

Beta-Blockers

Dipyridamole

Protamine

Nebulized pentamidine, beclomethasone,
 and propellants

Hydrocortisone

Cocaine

Propafenone

*Agents associated with acute-onset pulmonary
 insufficiency*[a]

Bleomycin plus O_2

Mitocycin

Bleomycin[a]

Procarbazine[a]

Methotrexate[a]

Amphotericin B

Nitrofurantoin (acute)[b]

Acetylsalicylic acid[b]

Interleukin-2[b]

Heroin and other narcotics[b]

Epinephrine[b]

Ethchlorvynol[b]

Fibrinolytic agents

Protamine

Blood products[b]

Fat emulsion

Hydrochlorothiazide

Complement-mediated leukostasis

Hyskon (dextran-70)[b]

Tumor necrosis factor[b]

Intrathecal methotrexate

Tricyclic antidepressants[b]

Amiodarone plus 0_2

Naloxone

• Onset at less than 48 h

Agents associated with pleural effusion

Chemotherapeutic agents

Nitrofurantoin (acute)

Bromocriptine

Dantrolene

Methysergide

L-Tryptophan

Drug inducing systemic lupus erythematosus

Tocolytics

Amiodarone

Esophageal variceal sclerotherapy agents

Interleukin-2

Agents that cause subacute respiratory failure

Chemotherapeutic agents

Nitrofurantoin (chronic)

Amiodarone

L-Tryptophan

Drug inducing systemic lupus erythematosus

Sources: McNeill, R.S., *Lancet*, 21, 1101, 1964; Borison, H.L., *Pharmacol. Ther. B*, 3, 211, 1977; Tattersfield, A.E., *J. Cardiovasc. Pharmacol.*, 8(Suppl. 4), 535, 1986; Illum, L. and Davis, S.S., *Clin. Pharmacokinet.*, 23, 30, 1992; Shao et al. (1992); Shao, Z. and Mitra, A.K., *Pharm. Res.*, 9, 1184, 1992; Fariba et al. (2000).

[a] Associated with hypersensitivity with eosinophilia.

[b] Usually reversible within 48–72 h, implying noncardiac pulmonary edema rather than inflammatory interstitial pneumonitis.

TABLE 22.9

Drugs Known to Influence Ventilatory Control

Depressants	Stimulants
Inhaled anesthetics	Alkaloids
Barbiturates	Nicotine
Benzodiazepines	Lobeline
Diazepam	Piperidine
Temazapam	Xanthine analogs
Chlordiazepoxide	Theophylline
Serotonin analogs	Caffeine
Methoxy-(dimethyl)-tryptamine	Theobromine
Dopamine analogs	Analeptics
Apomorphine	Doxapram
Adenosine analogs	Salicylates
2-Chloroadenosine	Progesterone analogs
R-Phenylisopropyl-adenosine (R-PIA)	Almitrine
N-Ethylcarboxamide (NECA)	Glycine analogs
B-Adrenergic antagonists	Strychnine
Timolol maleate	GABA antagonists
GABA analogs	Picrotoxin
Muscimol	Bicuculline
Baclofen	Serotonin synthesis inhibitors
Opiates	p-Chlorophenylalanine
Morphine	Reserpine
Codeine	
Methadone	
Meperidine	
Phenazocine	
Tranquilizers/analgesics	
Chlorpromazine	
Hydroxyzine	
Rompun (xylazine)	
Nalorphine	

Under ICH (S7A) and FDA guidelines, all new drugs (with limited exception—see Gad, 2009) must be evaluated for pharmacologic safety in three core organ systems (the central nervous system, cardiovascular system, and respiratory system). Table 22.11 presents the required determinations under these regulations for mandated respiratory system evaluations.

Mechanisms of ventilatory disorders can also be characterized as either central or peripheral. Central mechanisms involve the neurologic components of the pumping apparatus that are located in the CNS and include the medullary central pattern generator (CPG) as well as integration centers located in the medulla, pons, hypothalamus, and cortex of the brain that regulate the output of the CPG (Boggs 1992). The major neurologic inputs from the peripheral nervous system that influence the CPG are the arterial chemoreceptors (Boggs 1992). Many drugs stimulate or depress ventilation by selective interaction with the CNS (Eldridge and Millhorn 1981; Mueller 1982; Keats 1985) or arterial chemoreceptors (Heymans 1955, Heymans and Niel 1958).

Defects in the pumping apparatus are classified as hypo- or hyperventilation syndromes and are best evaluated by examining ventilatory parameters in a conscious animal model. The ventilatory

TABLE 22.10
Required Respiratory System Safety Pharmacology Evaluation

Respiratory functions
  Measurement of rate and relative tidal volume in conscious animals

Pulmonary function
  Measurement of rate, tidal volume and lung resistance, and compliance in anaesthetized animals

---

TABLE 22.11
Regulatory Documents Recommending Respiratory Function Testing in Safety Pharmacology Studies

United States	FDA Guideline for the Format and Content of the Nonclinical Pharmacology/Toxicology Section of an Application (Section IID, p. February 12, 1987)
Japan	Ministry of Health and Welfare Guidelines for Safety Pharmacology Studies Required for the Application for Approval to Manufacture (Import) Drugs. Notification YAKUSHIN-YAKU No. January 4, 1991.
Australia	Guidelines for Preparation and Presentation of Applications for Investigational Drugs and Drug Products Under the Clinical Trials Exemption Scheme (STET 12, 15).
Canada	RA5 Exhibit 2, Guidelines for Preparing and Filing Drug Submissions (p. 21).
United Kingdom	Medicines Act 1968, Guidance Notes on Applications for Product Licenses (MAL 2, p. A3F-1).

An example set of data for some standard positive controls is shown in Table 22.13. Note that each of the different agents has a separate pattern of effects.

---

parameters include respiratory rate, tidal volume, minute volume, peak (or mean) inspiratory flow, peak (or mean) expiratory flow, and fractional inspiratory time. Defects in mechanical properties of the lung are classified as obstructive or restrictive disorders and can be evaluated in animal models by performing flow-volume and pressure-volume maneuvers, respectively. The parameters used to detect airway obstruction include peak expiratory flow, forced expiratory flow at 25% and 75% of forced vital capacity, and a timed forced expiratory volume, while the parameters used to detect lung restriction include total lung capacity, inspiratory capacity, functional residual capacity, and compliance. Measurement of dynamic lung resistance and compliance, obtained continuously during tidal breathing, is an alternative method for evaluating obstructive and restrictive disorders, respectively, and is used when the response to drug treatment is expected to be immediate (within minutes post-dose). The species used in the safety pharmacology studies are the same as those generally used in toxicology studies (rats and dogs) since pharmacokinetic and toxicologic/pathologic data are available in these species. These data can be used to help select test measurement intervals and doses and to aid in the interpretation of functional change. The techniques and procedures for measuring respiratory function parameters are well established in guinea pigs, rats, and dogs (Amdur and Mead 1958; King 1966; Mauderly 1974; Diamond and O'Donnell 1977; Murphy 1994).

The key questions in safety pharmacology of the respiratory system are as follows:

• Does the substance affect the mechanisms of respiratory control (central or peripheral) leading to hypoventilation (respiratory depression) or hyperventilation (respiratory stimulation)?
• Does the substance act on a component of the respiratory system to induce, for example, bronchospasm, obstruction, and fibrosis?
• Does the substance induce acute effects, or can we expect chronic effects?
• Are the effects observed dose dependent or independent?

22.7.2.1 Plethysmography

The classic approach to measuring respiratory function in laboratory animals is plethysmography. It has two basic governing principles (Palecek 1969; O'Neil and Raub 1984; Brown and Miller 1987; Boggs 1992).

1. The animal (mice, rat, or dog), anaesthetized or not, restrained or not, is placed in a chamber (single or double) with pneumotachographs.
2. The variations of pressure in chamber(s) at the time of the inspiration and the expiration make it possible to obtain the respiratory flow of the animal.

There are three main types of body plethysmographs: constant volume, constant pressure, and pressure volume. The constant-volume body plethysmograph is a sealed box that detects volume change by the measurement of pressure changes inside the box. While inside the plethysmograph, inhalation of room air (from outside the plethysmograph) by the test animal induces an increase in lung volume (chest expansion) and thus an increase in plethysmograph pressure. On the other hand, exhalation to the atmosphere (outside the plethysmograph) induces a decrease in the plethysmograph pressure. The magnitude of lung volume change can be obtained via measurement of the change in plethysmographic pressure and the appropriate calibration factor. The plethysmograph is calibrated by injecting or withdrawing a predetermined change in box pressure. To avoid an adiabatic artifact, the rate of air injection or withdrawal is kept the same as that of chest expansion, indicated by the same dP/dt (change in pressure over time).

The constant-pressure body plethysmograph is a box with a pneumotachograph port built into its wall. This plethysmograph detects volume change via integration of the flow rate, $\int \Delta \text{Flow}$, which is monitored by the pneumotachograph port. There is an outward flow (air moving from the plethysmograph to the atmosphere) during inspiration and inward flow during expiration. Alternatively, in place of a pneumotachograph, a spirometer can be attached to the constant-pressure plethysmograph to detect volume changes. For the detection of plethysmographic pressure and flow rate, sensitive pressure transducers are usually employed. It is important that the transducer be capable of responding to volume changes in a linear fashion within the volume range studied. The plethysmograph should have negligible leaks, and temperature should not change during all respiratory maneuvers. The plethysmograph should also have linear characteristics with no hysteresis. Dynamic accuracy requires an adequate frequency response. A fast integrated flow plethysmograph, with a flat amplitude response for sinusoidal inputs up to 240 Hz, has been developed for rats, mice, and guinea pigs (Sinnet 1981). Similar plethysmographs can also be provided for use with large mammals.

A third type of pressure–volume plethysmograph has the mixed characteristics of the two types of body box mentioned earlier. For a constant-pressure plethysmograph, the change in volume at first is associated with gas compression or expansion. This fraction of the volume change can be corrected by electronically adding the plethysmographic pressure change to the volume signal. Therefore, the combined pressure–volume plethysmograph has excellent frequency–response characteristics and a wide range of sensitivities (Leigh and Mead 1974).

If volume, flow rate, and pressure changes are detected at the same time, several respiratory variables can be derived simultaneously from the raw signals. The whole-body plethysmograph method can then be used to measure most respiratory variables, such as tidal volume, breathing frequency, minute variables, such as tidal volume, breathing frequency, minute ventilation, compliance, pulmonary resistance, functional residual capacity, pressure–volume characteristics, and maximal expiratory flow–volume curves. Table 22.12 defines the parameters that are typically determined by these methods.

Selection of the proper reference values for the interpretation of findings is essential (Drazen 1984; American Thoracic Society 1991).

TABLE 22.12

Functional Respiratory Responses to Standard Pharmacologic Agents

Parameters	Theophylline 10 mg/kg PO	Pentobarbital 35 mg/kg IP	Diazepam 35 mg/kg IP	Codeine 100 mg/kg IP
F(breaths/min)	+++	---	---	No Change
TV (mL)	No change	No change	No change	-
Ti (s)	--	++	++	+
Te (s)	--	+++	++	-
PIF (mL/s)	++	-	-	-
PEF (mL/s)	++	No change	+	-

Source: Touvay, C. and Le Mosquet, B., *Therapie*, 55, 71, 2000.

Notes: + is an increase and – a decrease, and s are seconds, F is respiratory rate, TV tidal volume, Ti inhalation time or duration, Tc exhalation time, PIF the pulmonary inhalation rate, and PEF the pulmonary exhalation rate.

22.7.2.2 Design of Respiratory Function Safety Studies

The objective of a safety pharmacology evaluation of the respiratory system is to determine whether a drug has the potential to produce a change in respiratory function. Since a complete evaluation of respiratory function must include both the pumping apparatus and the lung, respiratory function safety studies are best designed to evaluate both of these functional components. The total respiratory system is evaluated first by testing for drug-induced changes in ventilatory patterns of intact conscious animals. This is followed by an evaluation of drug-induced effects on the mechanical properties of the lung in anesthetized/paralyzed animals. Together, these evaluations are used to determine (McNeill 1964) whether drug-induced changes in the total respiratory system have occurred and (Touvay and Le Mosquet 2000) whether these changes are related to pulmonary or extra-pulmonary factors.

The time intervals selected for measuring ventilatory patterns following oral administration of a drug should be based on pharmacokinetic data. The times selected generally include the time to reach peak plasma concentration of drug (T_{max}), at least one time before and one after T_{max}, and one time that is approximately 24 h after dosing to evaluate possible delayed effects. If the drug is given as a bolus iv injection, ventilatory parameters are monitored for approximately 5 min predose and continuously for 20–30 min postdose. 1-, 2-, 4-, and 24-h time intervals are also monitored to evaluate possible delayed effects. If administered by inhalation or intravenous infusion, ventilatory parameter would generally be monitored continuously during the exposure period and at 1, 2, 3, and 24 h time intervals after dosing.

The time interval showing the greatest ventilatory change is selected for evaluating lung mechanics. However, if no ventilatory change occurred, the T_{max} would be used. If the mechanical properties of the lung need to be evaluated within 30 min after dosing, then dynamic measurements of compliance and resistance are performed. Measurements include a predose baseline and continuous measurements for up to approximately 1 h postdose. If the mechanical properties of the lung need to be measured at 30 min or longer after dosing, then a single time point is selected, and the pressure–volume and flow–volume maneuvers are performed.

Supplemental studies including blood gas analysis, end-tidal CO_2 measurements, or responses to CO_2 gas and NaCN can be conducted to gain after the ventilatory and lung mechanical findings have been evaluated. In general, these would be conducted as separate studies.

22.7.2.3 Capnography

The measurements of rates, volumes, and capacities provided by plethysmograph measurements have a limited ability to detect and evaluate some ventilatory disorders (Murphy 1994) that markedly affect blood gases.

Detection of hypo- or hyperventilation syndromes requires measurement of the partial pressure of arterial CO_2 ($PaCO_2$). In humans and large animal models, this can be accomplished by collecting arterial blood with a catheter or needle and analyzing for $PaCO_2$ using a blood gas analyzer. In conscious rodents, however, obtaining arterial blood samples by needle puncture or catheterization during ventilatory measurements is generally not practical. An alternative and noninvasive method for monitoring $PaCO_2$ is the measurement of peak-expired (end-tidal) CO_2 concentrations. This technique has been successfully used in humans (Nuzzo and Anton 1986) and recently has been adapted for use in conscious rats (Murphy 1994). Measuring end-tidal CO_2 in rats requires the use of a nasal mask and a microcapnometer (Columbus Instruments, Columbus, OH) for sampling air from the mask and calculating end-tidal CO_2 concentrations. End-tidal CO_2 values in rats are responsive to ventilatory changes and accurately reflect changes in $PaCO_2$ (Murphy 1994).

A noninvasive procedure in conscious rats has been developed for use in helping distinguish between the central and peripheral nervous system effects of drugs on ventilation. Exposure to CO_2 gas stimulates ventilation primarily through a central mechanism (Borison 1977). In contrast, a bolus injection of NaCN produces a transient stimulation of ventilation through a mechanism that involves selective stimulation of peripheral chemoreceptors (Heymans and Niel 1958). Thus, to distinguish central from peripheral nervous system effects, our procedure measures the change in ventilatory response (pretreatment versus posttreatment) to both a 5 min exposure to 8% CO_2 gas and a bolus intravenous (iv) injection of 300 ug/kg of NaCN. In this paradigm, a central depressant (e.g., morphine sulfate) inhibits the CO_2 response and has little effect on the NaCN response.

The species selected for use in safety pharmacology studies should be the same as those used in toxicology studies. The advantages of using these species (rat, dog, or monkey) is that (McNeill 1964) the pharmacokinetic data generated in these species can be used to define the test measurement intervals, and (Touvay and Le Mosquet 2000) acute toxicity data can be used to select the appropriate high dose. Further, the toxicologic/pathologic findings in these species can be used to help define the mechanism of functional change. The rat is the primary choice since rats are readily available, and techniques for measuring pulmonary function are well established in this species.

22.8 INHALATION EXPOSURE TECHNIQUES FOR THERAPEUTIC AGENTS

Many inhalation exposure techniques, such as the whole-body, nose-only, mouth-only, or head-only technique (Drew 1973; MacFarland 1976; Leong et al. 1981; Phalen 2009), the intranasal exposure technique (Elliott and De Young 1970), the endotracheal nebulization technique (Leong 1985, 1988; Schreck 1986), and the body plethysmographic techniques (Alarie 1966; Thorne and Karol 1989), have been developed for inhalation toxicity studies. Table 22.13 provides a summary of the advantages and disadvantages of each of the major inhalation exposure methodologies.

The main criteria for the design and operation of any dynamic (as opposed to static) inhalation exposure system are as follows:

- The concentration of the test atmosphere must be reasonably uniform throughout the chamber and should increase and decrease at a rate close to theoretical at the start or end of the exposure. Silver (1946) showed that the time taken for a chamber to reach a point of

TABLE 22.13

Advantages, Disadvantages, and Considerations Associated with Modes of Respiratory Exposure

Mode of Exposure	Advantages	Disadvantages	Design Considerations
Whole body	Easiest and only practical way of achieving longer-term (more than 4 h at a time) exposure	Uses large amounts of test substance Use with powder and liquid aerosols leads to mixed routes of exposure	Chamber mixing Animal heat loads
Head only	Good for repeated exposure Limited routes of entry into animal More efficient dose delivery	Stress to animal Losses can be large Seal around neck Labor in loading/unloading	Even distribution Pressure fluctuations Sampling and losses Air temperature, humidity Animal comfort Animal restraint
Nose/mouth only	Exposure limited to mouth and respiratory tract Uses less material (efficient) Containment of material Can pulse the exposure	Stress to animal Seal about face Effort to expose large number of animals	Pressure fluctuations Body temperature Sampling Airlocking Animals' comfort Losses in plumbing/masks
Lung only (tracheal administration)	Precision of dose One route of exposure Uses less material (efficient) Can pulse the exposure	Technically difficult Anesthesia or tracheostomy Limited to small numbers Bypasses nose Artifacts in deposition and response Technically more difficult	Air humidity/temperature Stress to the animal Physiologic support
Partial lung	Precision of total dose Localization of dose Can achieve very high local doses Unexposed control tissue from same animal	Anesthesia Placement of dose Difficulty in interpretation of results Technically difficult Possible redistribution of material within lung	Stress to animal Physiologic support

Source: Gad, S.C. and Chengelis, C.P., *Acute Toxicology Testing: Perspectives and Horizons*, 2nd edn., Academic Press, San Diego, CA, pp. 404–466, 1998.

equilibrium was proportional to the flow rate of atmosphere passing through the chamber and the chamber volume. From this, the concentration–time relationship during the *run-up* and *run-down* phase could be expressed by the equation

$$t_x = k\frac{V}{F}$$

where
t_x is the time required to reach $x\%$ of the equilibrium concentration
k is a constant of value determined by the value of x
V is the chamber volume
F is the chamber flow rate

The t_{99} value is frequently quoted for exposure chambers, representing the time required to reach 99% of the equilibrium concentration and providing an estimate of chamber efficiency. Thus, at maximum efficiency, the theoretical value of k at t_{99} is 4.605, and the closer to this the results of evaluation of actual chamber performance fall, the greater the efficiency and the better the design of the chamber.

- Flow rates must be controlled in such a way that they are not excessive, which might cause streaming effects within the chamber, but must be adequate to maintain normal oxygen levels, temperature, and humidity in relation to the number of animals being exposed. A minimum of 10 air changes per hour is frequently advocated and is appropriate in most cases. However, the chamber design and housing density also need to be taken into account, and some designs, such as that of Doe and Tinston (1981), function effectively at lower air change rates.
- The chamber or exposure manifold materials should not affect the chemical or physical nature of the test atmosphere.

For critical laboratory studies on inhaled drugs, a monodisperse aerosol of a specified range of MMAD should be used to increase the probability that the aerosol reaches the specified target area of the lungs. The Dautrebande aerosol generators (Dautrebande 1962c) and the DeVilbiss nebulizer (Drew and Lippmann 1978) are the classic single-reservoir generators for short-duration inhalation studies. For long-duration inhalation studies, the multiple-reservoir nebulizer (Miller et al. 1981) or the continuous syringe metering and elutriating atomizer (Leong et al. 1981) are frequently used. The nebulizers generate a polydisperse droplet aerosol either by the shearing force of a jet of air over a fine stream of liquid or by ultrasonic disintegration of the surface liquid in a reservoir (Drew and Lippmann 1978). The aerosols emerging from a jet nebulizer generally have MMADs ranging between 1.2 and 6.9 μm with GSDs of 1.7–2.2, and aerosols from an ultrasonic nebulizer have MMADs ranging between 3.7 and 10.5 μm with GSDs of 1.4–2.0 (Mercer 1981).

For testing therapeutic formulations, the liquid aerosols are usually generated by the pressurized metered-dose inhaler (MDI) (Newman 1984; Gad and Chengelis 1998; Newton 2000). The pressurized MDI generates a bolus of aerosols by atomizing a well-defined quantity of a drug that is solubilized in a propellant. Of concern in such formulations are the propellants (though these are generally inert gases) and excipients such as stabilizers (see Table 22.14). The aerosols, thus, consist of the drug particles with a coating of the propellant. As the aerosols emerge from the orifice, the mean particle size may be as large as 30 μm (Moren 1981). After traveling through a tubular or cone-shaped spacer, the propellant may evaporate, reducing the MMADs to a range of 2.8–5.5 μm with GSDs of 1.5–2.2 (Hiller et al. 1978; Sackner et al. 1981; Newman 1984) and making the aerosols more stable for inhalation studies. In a prolonged animal exposure study, multiple MDIs have to be actuated sequentially with an electromechanical gadget (Ulrich et al. 1984) to maintain a slightly pulsatile but relatively consistent chamber concentration.

For generating an aerosol from dry powders, various dust generators, such as the Wright dust feed, air elutriator or fluidized-bed dust generator, and air impact pulverizer, have been developed for acute and chronic animal inhalation studies and are described in many articles (Hinds 1980; Leong et al. 1981; Gardner and Kennedy, 1993; Gad and Chengelis 1998; Hext 2000; Valentine and Kennedy 2001; Phalen 2009). For generating powdery therapeutic agents, a metered-dose DPI, spinhaler, or a rotahaler is used (Newman 1984). The particle size of the drug powder is micronized to a specific size range during manufacture, and the spinhaler or the rotahaler only disperses the powders.

More recently, another approach for administering dry powders to both humans and test animals has arisen. Dry powders, while less frequently used in nasal drug delivery, are becoming

TABLE 22.14

Some Examples of Excipients Used for Dry Powder Aerosols

Active Ingredient	Excipient Carrier
Salbutamol sulfate	Lactose (63–90 μm): regular, spray-dried, and recrystallized
Budesonide	Lactose (α-monohydrate [≤32, 63–90, 125–180 μm])
rhDNase	Lactose (50 wt% < 42 and 115 mm)
	Mannitol (50 wt% < 43 mm)
	Sodium chloride (50 wt% ≤ 87 mm)
Bovine serum Albumin—Maltodextrin (50–50)	Lactose (a-monohydrate [63–90 mm])
	Fine-particle lactose (76 wt% < 10 mm)
	Micronized polyethylene glycol 6000 (97.5 wt% ≤ 10 mm)
Recombinant human Granulocyte colony stimulating factor—mannitol	Polyethylene glycol 8000 (38–75 mm, 90–125 mm)

more popular. Powders can be administered from several devices, the most common being the insufflator. Many insufflators work with predosed powder in gelatin capsules. To improve patient compliance, a multidose powder inhaler has been developed, which has been used to deliver budesonide. These devices can also be used for administration to test animals delivery, both in terms of amounts and aerodynamic size of the particles. While early DPIs such as the Rotohaler® used individual capsules of micronized drug, which were difficult to handle, modern devices use blister packs (e.g., Diskus®) or reservoirs (e.g., Turbuhaler®). The DPIs rely on inspiration to withdraw drug from the inhaler to the lung, and hence, the effect of inhalation flow rate through various devices has been extensively studied. The major problem to be overcome with these devices is to ensure that the finely micronized drug is thoroughly dispersed in the airstream. It has been recommended that patients inhale as rapidly as possible from these devices in order to provide the maximum force to disperse the powder. The quantity of drug and deposition pattern varies enormously depending on the device, for example, the Turbuhaler produces significantly greater lung delivery of salbutamol than the Diskus. Vidgren et al. (1987) demonstrated by gamma scintigraphy that a typical dry powder formulation of SCG suffers losses of 44% in the mouth and 40% in the actuator nozzle itself.

It must also be emphasized that the major mass of a heterodispersed aerosol may be contained in a few relatively large particles, since the mass of a particle is proportional to the cube of its diameter. Therefore, the particle-size distribution and the concentration of the drug particles in the exposure atmosphere should be sampled using a cascade impactor or membrane filter sampling technique, monitored using an optical or laser particle-size analyzer, and analyzed using optical or electron microscopy techniques.

In summary, many techniques have been developed for generating gas, vapor, and aerosol atmospheres for inhalation toxicology studies. By proper regulation of the operating conditions of the nebulizers and the formulation of MDIs, together with the use of spacer or reservoir attachments to MDIs, more particles within the respirable range can be generated for inhalation. An accurately controlled exposure concentration is essential to an accurate determination of the dose–response relationship in a safety assessment of an inhalant drug.

Finally, comparisons of various techniques for animal exposures indicate that the whole-body exposure technique is the most suitable for safety assessment of gases and vapors and permits simultaneous exposure of a large number of animals to the same concentration of a drug; however, this technique is not suitable for aerosol and powder exposures because the exposure condition represents the resultant effects from inhalation, ingestion, and dermal absorption of the drug (Phalen 2009; Gad and Chengelis 1998; Gad 2009).

22.9   REGULATORY GUIDELINES

There is very limited regulatory guidance for the safety evaluation of inhalation drugs. CDER (2002) has issued one piece of guidance, but this speaks more to CMC issues than safety evaluation. The operative guidance is limited to the following:

1. Water-soluble inhalation drugs shall be sterile.
2. Exposure of test animals will be in a manner and by a regimen as similar as possible to that to be employed clinically.

These can lead to unexpected issues because they may lead to conflicts with animal welfare guidance or technical limitations. For example, some therapeutic gases are administered 24 h a day to patients. Such is not possible at all by nose-only techniques in animals (too much stress and no access to food and water for lab animals) and not strictly possible by even whole-body regimens due to the requirements of animal husbandry.

22.10   UTILITY OF TOXICITY DATA

Regardless of the type of test and the parameters to be monitored, the ultimate goal is to interpolate or extrapolate from the dose–response data to find a no-observable-adverse-effect level (NOAEL) or a no-observable-effect level (NOEL). By applying a safety factor of 1–10 to the NOAEL, a safe single-exposure dose for a phase I clinical trial may be obtained. By applying a more stringent safety factor, a multiple-exposure dose for a clinical trial may also be obtained. After the drug candidate has successfully passed all the drug safety evaluations and entered in the production stage, more toxicity tests may be needed for the establishment of a threshold limit value-time-weighted average (TLV-TWA). A TLV-TWA is defined as "the time weighted average concentration for a normal 8 h workday and a 40 h workweek, to which nearly all workers may be repeatedly exposed, day after day, without adverse effect" (ACGIH 2006). Using TLVs as guides, long-term safe occupational exposures during production and industrial handling of a drug may be achieved. Appropriate safety assessments of pharmaceutical chemicals and drugs will ensure the creation and production of a safe drug for the benefit of humans and animals. Furthermore, inhalation toxicity data are needed for compliance with many regulatory requirements of the Food and Drug Administration, the Occupational Health and Safety Administration, and the Environmental Protection Agency (Gad and Chengelis 1998).

More comprehensive descriptions and discussions on inhalation toxicology and technology may be found in several monographs, reviews, and textbooks (Willeke 1980; Leong et al. 1981; Witschi and Netterheim 1982; Clarke and Pavin 1984; Witschi and Brain 1985; Barrow 1986; McFadden 1986; McClellan and Henderson 1989; Gardner and Kennedy 1993; Gad and Chengelis 1998; Hext 2000; Valentine and Kennedy 2001; Witschi 2001; Pauluhn 2002; Salem and Katz 2006; Phalen 2009).

QUESTIONS

1. Haber's rule relates effect to exposure concentration and what other variable?
   Answer: Time (duration) of exposure
2. Protein therapeutics can be effectively achieved delivery to what two respiratory regions?
   Answer: Nasal passage and deep lungs
3. BALF allows the evaluation of an inhaled dug on what tissue?
   Answer: Bronchial region of the lungs
4. Respiratory safety pharmacology evaluates what parameters?
   Answer: Respiratory rate and volume, blood gas concentration
5. Clearance of aerosol particles from the deep lungs is by means of?
   Answer: Mucociliary *elevator*

GLOSSARY

Acceptance Criteria: Numerical limits, ranges, or other criteria for the test described.

Batch: A specific quantity of a drug or other material that is intended to have uniform character and quality, within specified limits, and is produced according to a single manufacturing order during the same cycle of manufacture (21 CFR 210.3(b)(2)).

Container Closure System: The sum of packaging components that together contain, protect, and deliver the dosage form. This includes primary packaging components and secondary packaging components if the latter are intended to provide additional protection to the drug product (e.g., foil overwrap). The container closure system also includes the pump for nasal and inhalation sprays. For nasal spray and inhalation solution, suspension, and spray drug products, the critical components of the container closure system are those that contact either the patient or the formulation, components that affect the mechanics of the overall performance of the device, or any protective packaging.

CRF: Case Report Form.

CTM: Clinical Trials Material.

CTU: Clinical Trials Unit.

CYP: P450 isoenzymes.

Drug Product: The finished dosage form and the container closure system.

Drug Substance: An active ingredient that is intended to furnish pharmacological activity or other direct effect in the diagnosis, cure, mitigation, treatment, or prevention of disease or to affect the structure or any function of the human body (21 CFR 314.3(b)).

Excipients: Any intended formulation component other than the drug substance.

Extractables: Compounds that can be extracted from elastomeric or plastic components of the container closure system when in the presence of a solvent.

GLP: Good Laboratory Practice.

GMP: Good Manufacturing Practice.

Inhalation Solutions, Suspensions, and Sprays: Drug products that contain active ingredients dissolved or suspended in a formulation, typically aqueous based, which can contain other excipients and are intended for use by oral inhalation. Aqueous-based drug products for oral inhalation must be sterile (21 CFR 200.51). Inhalation solutions and suspensions are intended to be used with a specified nebulizer. Inhalation sprays are combination products where the components responsible for metering, atomization, and delivery of the formulation to the patient are a part of the container closure system.

Insufflator: Dry powder nasal inhaler used with Rynacrom cartridges. Each cartridge contains one dose; the inhaler opens the cartridge, allowing the powder to be blown into the nose by squeezing the bulb.

Leachables: Compounds that leach into the formulation from elastomeric or plastic components of the drug product container closure system.

MDI: Metered-dose inhaler, consisting of an aerosol unit and plastic mouthpiece. This is currently the most common type of inhaler and is widely available.

Nasal Sprays: Drug products that contain active ingredients dissolved or suspended in a formulation, typically aqueous based, which can contain other excipients and are intended for use by nasal inhalation. Container closure systems for nasal sprays include the container and all components that are responsible for metering, atomization, and delivery of the formulation to the patient.

Nociception: Perception of pain in the nose.

Placebo: A dosage form that is identical to the drug product except that the drug substance is absent or replaced by an inert ingredient.

Pump: All components of the container closure system that are responsible for metering, atomization, and delivery of the formulation to the patient.

Specification: The quality standard (i.e., test, analytical procedures, and acceptance criteria) provided in the approved application to confirm the quality of drug substances, drug products, intermediates, raw material reagents, components, in-process materials, container closure systems, and other materials used in the production of drug substances or drug products.

Specified Impurity: An identified or unidentified impurity that is selected for inclusion in the drug substance or drug product specification and is individually listed and limited to ensure the reproducibility of the quality of the drug substance and/or drug product.

Spinhaler: A DPI used with Intal capsules specifically designed for the spinhaler. Each capsule contains one dose; the inhaler opens the capsule such that the powder may be inhaled through the mouthpiece.

Syncroner: MDI with elongated mouthpiece, used as training device to see if medication is being inhaled properly.

Turbuhaler: A DPI. The drug is in the form of a pellet; when body of inhaler is rotated, prescribed amount of drug is ground off this pellet. The powder is then inhaled through a fluted aperture on top.

REFERENCES

ACGIH. (2006). *Documentation of the Threshold Limit Values and Biological Exposure Indices*, 7th edn. American Conference of Governmental Industrial Hygienists, Cincinnati, OH.

Adams, E.M., Spencer, H.C., Rowe, V.K., and Irish, D.D. (1950). Vapor toxicity of 1,1,1,-trichloroethane (methylchloroform) determined by experiments on laboratory animals. *Arch. Ind. Hyg. Occup. Med.* 1:225–236.

Adams, E.M., Spencer, H.C., Rowe, V.K., McCollister, D.D., and Irish, D.D. (1952). Vapor toxicity of carbon tetrachloride determined by experiments on laboratory animals. *Arch. Ind. Hyg. Occup. Med.* 6:50–66.

Agnew, J.E. (1984). Physical properties and mechanisms of deposition of aerosols. In: *Aerosols and the Lung, Clinical Aspects* (S.W. Clarke and D. Pavia, eds.). Butterworth, London, U.K., pp. 49–68.

Agu, R.U., Jorissen, M., Kinget, R., Verbeke, N., and Augustigns, P. (2002). Alternatives to in vivo nasal toxicological screening for nasally administered drugs. *STP Pharma Sci.* 12:13–22.

Akoun, G.M. (1989). Natural history of drug-induced pneumonitis. In: *Drug Induced Disorders*, Volume 3: *Treatment Induced Respiratory Disorders* (G.M. Akoun and J.P. White, eds.). Elsevier Scientific Publishers B.V., New York, pp. 3–9.

Alarie, Y. (1966). Irritating properties of airborne material to the upper respiratory tract. *Arch. Environ. Health* 13:433–449.

Alarie, Y. (1973). Sensory irritation by airborne chemicals. *CRC Crit. Rev. Toxicol.* 2:299–363.

Alarie, Y. (1981a). Toxicological evaluation of airborne chemical irritants and allergens using respiratory reflex reactions. In: *Inhalation Toxicology and Technology* (B.K.J. Leong, ed.). Ann Arbor Science, Ann Arbor, MI, pp. 207–231.

Alarie, Y. (1981b). Bioassay for evaluating the potency of airborne sensory irritants and predicting acceptable levels of exposure in man. *Food Cosmet. Toxicol.* 19:623–626.

Alarie, Y. and Luo, J.E. (1984). Sensory irritation by airborne chemicals: A basis to establish acceptable levels of exposure. In: *Toxicology of the Nasal Passages* (C.S. Barrow, ed.). Hemisphere, New York, pp. 91–100.

Alarie, Y., Kane, L., and Barrow, C. (1980). Sensory irritation: The use of an animal model to establish acceptable exposure to airborne chemical irritants. In: *Toxicology: Principles and Practice 1* (A.L. Reeves, ed.). John Wiley & Son, New York, pp. 48–92.

Aldridge, S. (2003). Inhaled antibodies work better for chronic sinusitis. *My Health and Age*, March 4, 2003, pp. 18–23.

Altman, P.L. and Dittmer, D.S. (1974). *Biological Data Book*, Vol. III. Federation of American Societies for Experimental Biology, Bethesda, MD.

Amdur, M.O. and Mead, J. (1958). Mechanics of respiration in unanesthetized guinea pigs. *Am. J. Physiol.* 192:364–368.

American Thoracic Society. (1991). Lung function testing: Selection of reference values and interpretative strategies. *Am. Rev. Respir. Dis.* 144:1202–1218.

Andersen, M.E., French, J.E., Gargas, M.L., Jones, R.A., and Jenkins, L.J. Jr. (1979). Saturable metabolism and the acute toxicity of 1,1-dichloroethylene. *Toxicol. Appl. Pharmacol.* 47:385–393.

Aso, Y., Yoneda, K., and Kikkawa, Y. (1976). Morphologic and biochemical study of pulmonary changes induced by bleomycin in mice. *Lab. Invest.* 35:558–568.

Aviado, D.M. (1981). Comparative cardiotoxicity of fluorocarbons. In: *Cardiac Toxicology*, Vol. II (T. Balazs, ed.). CRC Press, Boca Raton, FL, pp. 213–222.

Aviado, D.M. and Micozzi, M.S. (1981). Fluorine-containing organic compounds. In: *Patty's Industrial Hygiene and Toxicology*, Vol. 2B (G.D. Clayton and F.E. Clayton, eds.). Wiley, New York, pp. 3071–3115.

Balazs, T. (1981). Cardiotoxicity of adrenergic bronchodilator and vasodilating antihypertensive drugs. In: *Cardiac Toxicology*, Vol. II (T. Balazs, ed.). CRC Press, Boca Raton, FL, pp. 61–73.

Barrow, C.S. (1986). *Toxicology of the Nasal Passages*. Hemisphere, New York.

Bell, K.A. (1978). Local particle deposition in respiratory airway models. In: *Recent Developments in Aerosol Science* (D.T. Shaw, ed.). Wiley, New York, pp. 97–134.

Bice, D.E. (1985). Methods and approaches to assessing immunotoxicology of the lower respiratory tract. In: *Immunotoxicology and Immunopharmacology* (J.H. Dean, M.I. Luster, A.E. Munson, and H.A. Amos, eds.). Raven Press, New York, pp. 145–157.

Boggs, D.F. (1992). Comparative control of respiration. In: *Comparative Biology of the Normal Lung*, Vol. I (R.A. Parent, ed.). CRC Press, Boca Raton, FL, pp. 309–350.

Borison, H.L. (1977). Central nervous system depressants: Control-systems approach to respiratory depression. *Pharmacol. Ther. B* 3:211–226.

Brain, J.D. (1971). The effects of increased particles on the number of alveolar macrophages. In: *Inhaled Particles III Proceedings of ISOHS Symposium*, Unwin Brothers Ltd., London, U.K., pp. 220–223.

Brain, J.D., Knudson, D.E., Sorokin, S.P., and Davis, M.A. (1976). Pulmonary distribution of particles given by intratracheal instillation or be aerosol inhalation. *Environ. Res.* 11:13–33.

Brown, L.K. and Miller, A. (1987). Full lung volumes: Functional residual capacity, residual volume and total lung capacity. In: *Pulmonary Function Tests: A Guide for the Student and House Officer* (A. Miller, ed.). Grune & Stratton, Inc., New York, pp. 53–58.

Burham, S.R. (2002). The ideal nasal corticosteroid: Balancing efficacy, safety and patient preference. *Clin. Exp. Allergy Rev.* 2:32–37.

CDER. (2002). *Guidance for Industry: Nasal Spray and Inhalation Solution, Suspension, and Spray Drug Products-Chemistry, Manufacturing and Control Documentation*. Food and Drug Administration, Washington, DC.

Chan, T.L. and Lippmann, M. (1980). Experimental measurements and empirical modelling of the regional deposition of inhaled particles in humans. *Am. Ind. Hyg. Assoc. J.* 41:399–409.

Chenoweth, M.B., Leong, B.K.J., Sparschu, G.L., and Torkelson, T.R. (1972). Toxicities of methoxyflurane, halothane and diethyl ether in laboratory animals on repeated inhalation at subanesthetic concentrations. In: *Cellular Biology and Toxicity of Anesthetics* (B.R. Fink, ed.). Williams & Wilkins, Baltimore, MD, pp. 275–284.

Cherniack, N.S. (1988). Disorders in the control of breathing: Hyperventilation syndromes, In: *Textbook of Respiratory Medicine* (J.F. Murray and J.A. Nadal, eds.). W.B. Saunders Co., Philadelphia, PA, pp. 1861–1866.

Clarke, S.W. and Pavia, D. (eds.) (1984). *Aerosol and the Lung: Clinical and Experimental Aspects*. Butterworth, London, U.K.

Cox, J.S.G., Beach, J.E., Blair, A.M.J.N., Clarke, A.J., King, J., Lee, T.B., Loveday, D.E.E. et al. (1970). Disodium cromoglycate. *Adv. Drug. Res.* 5:115–196.

Curry, S.H., Taylor, A.J., Evans, S., Godfrey, S., and Zeidifard, E. (1975). Disposition of disodium cromoglycate administered in three particle sizes. *Br. J. Clin. Pharmacol.* 2:267–270.

Dahl, A.R. (1990). Dose concepts for inhaled vapors and gases. *Toxicol. Appl. Pharmacol.* 103:185–197.

Dallas, C.E., Bruckner, J.V., Maedgen, J.L., and Weir, F.W. (1986). A method for direct measurement of systemic uptake and elimination of volatile organics in small animals. *J. Pharmacol. Methods* 16:239–250.

Dallas, C.E., Ramanathan, R., Muralidhara, S., Gallo, J.M., and Bruckner, J.V. (1989). The uptake and elimination of 1,1,1-trichloroethane during and following inhalation exposure in rats. *Toxicol. App. Pharmacol.* 98:385–397.

Dautrebande, L. (1962a). Importance of particle size for therapeutic aerosol efficiency. In: *Microaerosols*. Academic Press, New York, pp. 37–57.

Dautrebande, L. (1962b). Practical recommendation for administering pharmacological aerosols. In: *Microaerosols*. Academic Press, New York, pp. 86–92.

Dautrebande, L. (1962c). Production of liquid and solid micromicellar aerosols. In: *Microaerosols*. Academic Press, New York, pp. 1–22.

Davies, R.J. and Pepys, J. (1975). Asthma due to inhaled chemical agents—The macrolide antibiotic spiramy-cin. *Clin. Allergy* 5:99–107.

Davis, B., Marin, M.G., Fischer, S., Graf, P., Widdicombe, J.G., and Nadel, J.A. (1976). New method for study of canine mucus gland secretion *in vivo*: Cholinergic regulation. *Am. Rev. Respir. Dis.* 113:257 (abstract).

Davis, C.N., Heyder, J., and Subba Ramv, M.C. (1972). Breathing of half micron aerosols. I. Experimental. *J. Appl. Physiol.* 32:591–600.

Dennis, W.L. (1961). The discussion of a paper by C.N. Davis: A formalized anatomy of the human. In: *Inhaled Particles and Vapours*, (C.N. Davis, ed.). Pergamon Press, London, U.K., p. 88.

Diamond, L. and O'Donnell, M. (1977). Pulmonary mechanics in normal rats. *J. Appl. Physiol.* 43:942–948.

Doe, J.E. and Tinston, D.J. (1981). Novel chamber for long-term inhalation studies. In: *Inhalation Toxicology and Technology* (K.J. Leong, ed.). Ann Arbor Science, Ann Arbor, MI, pp. 472–493.

Dorato, M.A. (1994). Toxicological evaluation of intranasal peptide and protein drugs. *Drugs Pharm. Sci.* 62:345–381.

Drazen, J.M. (1984). Physiological basis and interpretation of indices of pulmonary mechanics. *Environ. Health Perspect.* 56:3–9.

Drew, R.T. and Laskin, S. (1973). Environmental inhalation chambers. In: *Methods of Animal Experimentation*, Vol. IV. Academic Press, New York, pp. 1–41.

Drew, R.T. and Lippmann, M. (1978). Calibration of air sampling instruments. In: *Air Sampling Instruments for Evaluation of Atmospheric Contaminants*, 5th edn. American Conference of Governmental Industrial Hygienists, Cincinnati, OH, Section I, pp. 1–32.

Durham, S.R. (2002). The ideal nasal corticosteroid: Balancing efficacy, safety and patient preference. *Clin. Exp. Allergy Rev.* 2:32–37.

Eldridge, F.L. and Millhorn, D.E. (1981). Central regulation of respiration by endogenous neurotransmitters and neuromodulators. *Ann. Rev. Physiol.* 3:121–135.

Elliot, G.A. and DeYoung E.N. (1970). Intranasal toxicity testing of antiviral agents. *Am. N. Y. Acad. Sci.* 173:169–175.

Evans, M.J. (1982). Cell death and cell renewal in small airways and alveoli. In: *Mechanisms in Respiratory Toxicology*, Vol. 1 (H. Witschi and P. Nettesheim, eds.). CRC, Boca Raton, FL, pp. 189–218.

Fariba, A., Kellie, M., Stephen, M., Oune, O., Kafi, A., and Mary, H. (2002). Repeated doses of antenatal cor-ticosteroids in animals: A systematic review, *Am. J. Obstet. Gynecol.* 186:843–849.

Ferin, J. (1977). Effect of particle content of lung on clearance pathways. In: *Pulmonary Macrophage and Epithelial Cells* (C.L. Sanders, R.P. Schneider, G.E. Dagle, and H.A. Ragan, eds.). Technical Information Center, Energy Research & Development Administration, Springfield, VA, pp. 414–423.

Fieldner, A.C., Kazt, S.H., and Kinney, S.P. (1921). Gas masks for gases met in fighting fires. Technical paper no. 248. US Bureau of Mines, Pittsburgh, PA.

Gad, S.C. (2009). *Drug Safety Evaluation*, 2nd edn. Wiley, Hoboken, NJ.

Gad, S.C. (2012). *Safety Pharmacology*, 2nd edn. CRC Press, Boca Raton, FL.

Gad, S.C. and Chengelis, C.P. (1998). *Acute Toxicology Testing: Perspectives and Horizons*, 2nd edn. Academic Press, San Diego, CA, pp. 404–466.

Gamsu, G., Singer, M.M., Vincent, H.H., Berry, S., and Nadel, J.A. (1976). Postoperative impairment of mucous transport in the lung. *Am. Rev. Respir. Dis.* 114:673–679.

Gardner, D.E. and Kennedy Jr., G.L. (1993). Methodologies and technology for animal inhalation toxicology studies. In: *Toxicology of the Lung*, 2nd edn. (D.E. Gardner, J.D. Crapo, and R.O. McClellan, eds.). Raven Press, New York.

Godfrey, S., Zeidifard, E., Brown, K., and Bell, J.H. (1974). The possible site of action of sodium cromoglycate assessed by exercise challenge. *Clin. Sci. Mol. Med.* 46:265–272.

Haber, F.R. (1924). Funf Vortage aus den jahren 1920–1923, No. 3. *Die Chemie im Kriege*. Julius Springer, Berlin, Germany.

Hafner, R.E., Jr., Watanabe, P.G., and Gehring, P.J. (1975). Preliminary studies on the fate of inhaled vinyl chloride monomer in rats. *Am. N. Y. Acad. Sci.* 246:135–148.

Hatch, T.F. and Gross, P. (1964). *Pulmonary Deposition and Retention of Inhaled Aerosols*. Academic Press, New York, pp. 16–17, 51–52, 147–168.

Henderson, R. (1988). Use of bronchoalveolar lavage to detect lung damage. In: *Toxicology of the Lung* (D.E. Gardner, J.D. Crapo, and E.J. Massaro, eds.). Raven Press, New York, pp. 239–268.

Henderson, R. (1989). Bronchoalveolar lavage: A tool for assessing the health status of the lung. In: *Concepts in Inhalation Toxicology* (R.O. McClellan and R.F. Henderson, eds.). Hemisphere, Washington, DC, pp. 414–442.

Henderson, R.F. (1984). Use of Bronchoalveolar lavage to detect lung damage. *Environ. Health Perspect.* 56:115–129.

Henderson, R.F. and Loery, J.S. (1983). Effect of anesthetic agents on lavage fluid parameters used as indicators of pulmonary injury. *Lab. Anim. Sci.* 33:60–62.

Henderson, R.F., Barr, E.B., and Hotchkiss, J.A. (1991). Effect of exposure rate on response of the lung to inhaled particles. *Toxicologists* 11:234 (abstract).

Henderson, R.F., Benson, J.M., Hahn, F.F., Hobbs, C.H., Jones, R.K., Mauderly, J.L., McClellan, R.O., and Pickrell, J.A. (1985). New approaches for the evaluation of pulmonary toxicity: Bronchoalveolar lavage fluid analysis. *Fundam. Appl. Toxicol.* 5:451–458.

Henderson, R.F., Mauderly, J.L., Pickrell, J.A., Hahn, F.F., Muhle, H., and Rebar, A.H. (1987) Comparative study of bronchoalveolar lavage fluid: Effect of species, age, method of lavage. *Exp. Lung Res.* 1:329–342.

Hensley, M.J., O'Cain, C.F., McFadden, E.R. Jr., and Ingram, R.H. Jr. (1978). Distribution of bronchodilatation in normal subjects: Beta agonist versus atropine. *J. Appl. Physiol.* 45:778–782.

Heyder, J., Gebhart, J., and Stahlhofen, W. (1980). Inhalation of aerosols: Particle deposition and retention. In: *Generation of Aerosols and Facilities for Exposure Experiments* (K. Willeke, ed.). Ann Arbor Science, Ann Arbor, MI, pp. 80–99.

Heymans, C. (1955). Action of drugs on carotid body and sinus. *Pharmacol. Rev.* 7:119–142.

Heymans, C. and Niel, E. (1958). The effects of drugs on chemoreceptors. In: *Reflexogenic Areas of the Cardiovascular Systems* (C. Heymans and E. Neil, eds.). Churchill, Ltd., London, U.K., pp. 192–199.

Hiller, F.C., Mazunder, M.K., Wilson, J.D., and R.C. Bone, R.C. (1978). Aerodynamic size distribution of metered dose bronchodilator aerosols. *Am. Rev. Respir. Dis.* 118:311–317.

Hinds, W.C. (1980). Dry–dispersion aerosol generators. In: *Generation of Aerosols and Facilities for Exposure Experiments* (K. Willeke, ed.). Ann Arbor Science, Ann Arbor, MI, pp. 171–187.

Hirai, S., Yashiki, T., and Mima, H. (1981). Mechanisms for the enhancement of the nasal absorption of insulin by surfactants. *Int. J. Pharm.* 9:173–184.

Hocking, W.G. and Golde, D.W. (1979). The pulmonary alveolar macrophage. *N. Engl. J. Med.* 310:580–587, 639–645.

Hodson, M.E., Penketh, A.R., and Batten, J.C. (1981). Aerosol carbenicillin and gentamicin treatment of *Pseudomonas aeruginosa* infection I patients with cystic fibrosis. *Lancet* 2:1137–1139.

Holma, B. (1967). Lung clearance of mono- and di-disperse aerosols determined by profile scanning and whole body counting: A study on normal and SO_2 exposed rabbits. *Acta Med. Scand. Suppl.* 473:1–102.

Illum, L. and Davis, S.S. (1992). Intranasal insulin. Clinical pharmacokinetics. *Clin. Pharmacokinet.* 23:30–41.

Indans, I. (2002). Non-lethal end-points in inhalation toxicology. *Toxicol. Lett.* 135(1):53.

Ingram, R.H., Wellman, J.J., McFadden, E.R. Jr., and Mead, J. (1977). Relative contributions of large and small airways to flow limitation in normal subjects before and after atropine and isoproterenol. *J. Clin. Invest.* 59:696–703.

Irish, D.D. and Adams, E.M. (1940). Apparatus and methods for testing the toxicity of vapors. *Ind. Med. Surg.* 1:1–4.

Johnson, H.G., McNee, M.L., and Braughler, J.M. (1987). Inhibitors of metal catalyzed lipid peroxidation reactions inhibit mucus secretion and 15 HETE levels in canine trachea. *Prostaglandins Leukot. Med.* 30:123–132.

Johnson, H.G., McNee, M.L., Johnson, M.A., and Miller, M.D. (1983). Leukotriene C_4 and dimethylphenylpiperazinium-induced responses in canine airway tracheal muscle contraction and fluid secretion. *Int. Arch. Allergy Appl. Immun.* 71:214–218.

Kannisto, S., Voutilainen, R., Remes, K., and Korppi, M. (2002). Efficacy and safety of inhaled steroid and cromane treatment in school-age children: A randomized pragmatic pilot study. *Pediatr. Allergy Immunol.* 13:24–34.

Karol, M.H. (1988). Immunologic responses of the lung to inhaled toxicants. In: *Concepts in Inhalation Toxicology* (R.O. McClellan and R. Henderson, eds.). Hemisphere Publishing, Washington, DC, pp. 403–413.

Karol, M.H., Hillebrand, J.A., and Thorne, P.S. (1989). Characteristics of weekly pulmonary hypersensitivity responses elicited in the guinea pig by inhalation of ovalbumin aerosols. In: *Toxicology of the Lung* (D.E. Gardner, J.D. Crapo, and E.J. Massaro, eds.). Raven Press, New York, pp. 427–448.

Kavet, R.I., Brain, J.D., and Levens, D.J. (1978). Characteristics of weekly pulmonary macrophages lavaged from hamsters exposed to iron oxide aerosols. *Lab. Invest.* 38:312–319.

Keats, A.S. (1985). The effects of drugs on respiration in man. *Ann. Rev. Pharmacol. Toxicol.* 25:41–65.

King, T.K.C. (1966). Measurement of functional residual capacity in the rat. *J. Appl. Physiol.* 21:233–236.

Lalej-Bennis, D. (2001). Six month administration of gelified intranasal insulin in 16 type 1 diabetic patients under multiple injections: Efficacy vs subcutaneous injections and local tolerance. *Diabetes Metab.* 27(3):372–377.

Landy, T.D., Ramsey, J.C., and McKenna, M.J. (1983). Pulmonary physiology and inhalation dosimetry in rats: Development of a method and two examples. *Toxicol. Appl. Pharmacol.* 71:72–83.

Last, J.A. (1982). Mucus production and ciliary escalator. In: *Mechanisms in Respiratory Toxicology*, Vol. 1 (H. Witschi and P. Nettesheim, eds.). CRC Press, Boca Raton, FL, pp. 247–268.

Lauweryns, J.M. and Baert, J.H. (1977). Alveolar clearance and the role of the pulmonary lymphatics. *Am. Rev. Resp. Dis.* 115:625–683.

Lawrence, S. (2002). Intranasal delivery could be used to administer drugs directly to the brain. *Lancet* 359:1674.

Lee, W.C. and Wang, C.S. (1977). Particle deposition in systems of repeated bifurcating tubes. In: *Inhaled Particles IV* (W.H. Walton, ed.). Pergamon, Oxford, U.K., pp. 49–60.

Leigh, D.E. and Mead, J. (1974). *Principles of Body Plethysmography.* National Heart and Lung Institute, NIH, Bethesda, MD.

Leong, B.K.J. and MacFarland, H.N. (1965). Pulmonary dynamics and retention of toxic gases. *Arch. Environ. Health* 11:555–563.

Leong, B.K.J. and Rop, D.A. (1989). The combined effects of an inhalation anesthetic and an analgesic on the electrocardiograms of beagle dogs. *Abstract-475. International Congress of Toxicology.* Taylor & Francis, London, U.K., p. 159.

Leong, B.K.J., Coombs, J.K., Petzold, E.N., and Hanchar, A.J. (1988). A dosimetric endotracheal nebulization technique for pulmonary metabolic disposition studies in laboratory animals. *Inhal. Toxicol.* Premier issue:37–51.

Leong, B.K.J., Coombs, J.K., Petzold, E.N., Hanchar, A.H., and McNee, M.L. (1985). Endotracheal nebulization of drugs into the lungs of anesthetized animals. *Toxicologist* 5:31 (abstract).

Leong, B.K.J., Lund, J.E., Groehn, J.A., Coombs, J.K., Sabaitis, C.P., Weaver, R.J., and Griffin, L. (1987). Retinopathy from inhaling 4,4′-methylenedianiline aerosols. *Fundam. Appl. Toxicol.* 9:645–658.

Leong, B.K.J., Powell, D.J., and Pochyla, G.L. (1981). A new dust generator for inhalation toxicological studies. In: *Inhalation Toxicology and Technology* (B.K.J. Leong, ed.). Ann Arbor Science, Ann Arbor, MI, pp. 157–168.

Lichtiger, M., Landa, J.F., and Hirsch, J.A. (1975). Velocity of tracheal mucus in anesthetized women undergoing gynaecologic surgery. *Anesthesiology* 42:753–756.

Lippmann, M. (1970). "Respirable" dust sampling. *Am. Ind. Hyg. Assoc. J.* 31:138–159.

Lippmann, M. (1977). Regional deposition of particles in the human respiratory tract. In: *Handbook of Physiology*, Section 9, pp. 213–232.

Lippmann, M., Yeates, D.B., and Albert, R.E. (1980). Deposition, retention and clearance of inhaled particles. *Br. J. Ind. Med.* 37:337–362.

Logemaan, C.D. and Rankin, L.M. (2000). Newer intranasal migraine medications. *Am. Fam. Physician* 61:180–186.

Lourenco, R.V. and Cotromanes, E. (1982). Clinical aerosols. II. Therapeutic aerosols. *Arch. Intern. Med.* 142:2299–2308.

MacFarland, H.N. (1976). Respiratory toxicology. In: *Essays in Toxicology*, Vol. 7 (W.J. Hayes, ed.). Academic Press, New York, pp. 121–154.

Marple, V.A. and Rubow, K.L. (1980). Aerosol generation concepts and parameters. In: *Generation of Aerosols and Facilities for Exposure Experiments* (K. Willeke, ed.). Ann Arbor Science, Ann Arbor, MI, p. 6.

Mauderly, J.L. (1974). The influence of sex and age on the pulmonary function of the beagle dog. *J. Gerontol.* 29:282–289.

Mauderly, J.L. (1989). Effects of inhaled toxicants on pulmonary function. In: *Concepts in Inhalation Toxicology* (R.O. McClellan and R.F. Henderson, eds.). Hemisphere Publishing Corp, New York, pp. 347–401.

McClellan, R.O. and Henderson, R.F. (1989). *Concepts in Inhalation Toxicology.* Hemisphere, Washington, DC.

McFadden, E.R., Jr. (1986). *Inhaled Aerosol Bronchodilators.* Williams & Wilkins, Baltimore, MD, pp. 40–41.

McNeill, R.S. (1964). Effect of a β-adrenergic blocking agent, propranol, on asthmatics. *Lancet* 21:1101–1102.

Menon, M.P.S. and Das, A.K. (1977). Tetracycline asthma—A case report. *Clin. Allergy* 7:285–290.

Menzel, D.B. and Amdur, M.O. (1986). Toxic responses of the respiratory system. In: *Casarett and Doull's Toxicology*, 3rd edn. (C.D. Klaassen, M.O. Amdur, and J. Doull, eds.). Macmillan, New York, pp. 330–358.

Mercer, T.T. (1981). Production of therapeutic aerosols; principles and techniques. *Chest* 80(Suppl. 6):813–818.

Merlin, R.G. (1981). Cardiac toxicity of inhalation anesthetics. In: *Cardiac Toxicology*, Vol. II. CRC Press, Boca Raton, FL, pp. 4–10.

Milic-Emili, J. (1982). Recent advances in clinical assessment of control of breathing. *Lung* 160:1–17.

Miller, F.J., Gardner, D.E., Graham, J.A., Lee, R.E., Jr., Wilson, W.E., and Bachmann, J.D. (1979). Size considerations for establishing a standard for inhalable particles. *J. Air Pollut. Cont. Assoc.* 29:610–615.

Miller, J.L., Stuart, B.O., Deford, H.S., and Moss, O.R. (1981). Liquid aerosol generation for inhalation toxicology studies. In: *Inhalation Toxicology and Technology* (B.K.J. Leong, ed.). Ann Arbor Science, Ann Arbor, MI, pp. 121–207.

Moren, F. (1981). Pressurized aerosols for oral inhalation. *Int. J. Pharm.* 8:1–10.

Moren, F. (1993). Aerosol dosage forms and formulations. In: *Aerosols in Medicine*, 2nd edn. (S. Moren, M.B. Dolovich, M.T. Newhouse, and S.P. Newman, eds.). Elsevier, Amsterdam, the Netherlands, pp. 329–336.

Morris, J.B. and Shusterman, D.J. (2010) *Toxicology of the Nose and Upper Airway*. Informa, New York.

Mueller, R.A. (1982). The neuropharmacology of respiratory control. *Pharmacol. Rev.* 34:255–285.

Murphy, D.J. (1994) Safety pharmacology of the respiratory system: Techniques and study design. *Drug Dev. Res.* 32:237–246.

Murphy, D.J., Joran, M.E., and Grando, J.C. (1994). Microcapnometry: A non-invasive method for monitoring arterial CO_2 tension during ventilatory measurements in conscious rats. *Toxicol. Methods* 4:177–187.

Mussatto, D.J., Garrad, C.S., and Lourenco, R.V. (1988). The effect of inhaled histamine on human tracheal mucus velocity and bronchial mucociliary clearance. *Am. Rev. Respir. Dis.* 138:775–779.

National Academy of Sciences (NAS). (1958). *Handbook of Respiration*. NAS National Research Council/W.B. Saunders, Philadelphia, PA, p. 41.

Newman, S.P. (1984). Therapeutic aerosols. In: *Aerosols and the Lung, Clinical and Experimental Aspects* (S.W. Clarke and D. Pavia, eds.). Butterworth & Company, London, U.K., pp. 197–224.

Newton, P.E. (2000). Techniques for evaluating hazards of inhaled products. In: *Product Safety Evaluation Handbook*, 2nd edn. Marcel Dekker, Inc., New York, pp. 243–298.

Nuzzo, P.F. and Anton, W.R. (1986). Practical applications of capnography. *Respir. Ther.* 16:12–17.

Oberst, F.W. (1961). Factors affecting inhalation and retention of toxic vapours. In: *Inhaled Particles and Vapours* (C.N. Davis, ed.). Pergamon Press, New York, pp. 249–266.

O'Neil, J.J. and Raub, J.A. (1984). Pulmonary function testing in small laboratory mammals. *Environ. Health Perspect.* 53:11–22.

O'Neil, M.J. (2006). *The Merck Index*, 14th edn. Merck & Co., Inc., Whitehouse Station, NJ.

Palecek, F. (1969). Measurement of ventilatory mechanics in the rat. *J. Appl. Physiol.* 27:149–156.

Parent, R.A. (1991). *Comparative Biology of the Normal Lung*. CRC Press, Boca Raton, FL.

Paterson, J.W. (1977). Bronchodilators. In: *Asthma* (T.J.H. Clark and S. Godfrey, eds.). Chapman and Hall, London, U.K., pp. 251–271.

Paterson, J.W., Woocock, A.J., and Shenfield, G.M. (1979). Bronchodilator drugs. *Am. Rev. Respir. Dis.* 120:1149–1188.

Patterson, R. and J.F. Kelly. (1974). Animal models of the asthmatic state. *Ann. Rev. Med.* 25:53–68.

Patty, F.A. (1958). *Industrial Hygiene and Toxicology*, 2nd edn. Interscience, New York.

Pauluhn, J. (2002). Overview of testing methods used in inhalation toxicology: From facts to artifacts. *Eurotox 2002*, Budapest, Hungary, pp. 5–7.

Paustenbach, D.J., Carlson, G.P., Christian, J.E., Born, G.S., and Rausch, J.E. (1983). A dynamic closed-loop recirculating inhalation chamber for conducting pharmacokinetic and short-term toxicity studies. *Fundam. Appl. Toxicol.* 3:528–532.

Pavia, D. (1984). Lung mucociliary clearance. In: *Aerosols and the Lung, Clinical and Experimental Aspects* (S.W. Clarke and D. Pavia, eds.). Butterworth & Co, London, U.K., pp. 127–155.

Pavia, D., Bateman, J.R.M., Sheahan, N.F., Agnew, J.E., Newman, S.P., and Clarke, S.W. (1980). Techniques for measuring lung mucociliary clearance. *Eur. J. Respir. Dis.* 67(Suppl. 110):157–177.

Pavia, D., Sutton, P.P., Agnew, J.E., Lopez-Vidriero, M.T., Newman, S.P., and Clarke, S.W. (1983a). Measurement of bronchial mucociliary clearance. *Eur. J. Respir. Dis.* 64(Suppl. 127):41–56.

Pavia, D., Sutton, P.P., Lopez-Vidriero, M.T., Agnew, J.E., and Clarke, S.W. (1983b). Drug effects on mucociliary function. *Eur. J. Respir. Dis.* 64(Suppl. 128):304–317.

Phalen, R.F. (2009). *Inhalation Studies: Foundations and Techniques*, 2nd edn. Informa, Boca Raton, FL, pp. 35–46, 51–57.

Pigott, G.H. (2009) Inhalation toxicology. In: *General and Applied Toxicology*, 3rd edn. (B. Ballantyne, T.C. Marrs, and T. Syversen, eds.). Wiley, Hoboken, NJ.

Rinehart, W.E. and Hatch, T. (1964). Concentration-time product (Ct) as an expression of dose in sublethal exposures to phosgene. *Am. Ind. Hyg. Assoc. J.* 25:545–553.

Rosnow, E.C. (1992). Drug-induced pulmonary disease: An update. *Chest* 102:239–250.

Rylander, R. (1966). Current techniques to measure alterations in the ciliary activity of intact respiratory epithelium. *Am. Rev. Respir. Dis.* 93(Suppl.):67–85.

Sackner, M.A., Brown, L.K., and Kim, C.S. (1981). Basis of an improved metered aerosol delivery system. *Chest* 80(Suppl. 6):915–918.

Salem, H. and Katy, S.A. (2006). *Inhalation Toxicology*, 2nd edn. Tayler & Francis, Philadelphia, PA.

Schreck, R.W., Sekuterski, J.J., and Gross, K.B. (1986). Synchronized intratracheal aerosol generation for rodent studies. In: *Aerosols, Formation and Reactivity, Second International Aerosol Conference*, Berlin, Germany. Pergamon Press, New York, pp. 37–51.

Schwettmann, R.S. and Casterline, C.L. (1976). Delayed asthmatic response following occupational exposure to enflurane. *Anesthesiology* 44:166–169.

Shao, Z., Krishnamoorthy, R., and Mitra, A.K. (1992). Cyclodextrins as nasal absorption promoters of insulin-mechanistic evaluations. *Pharm. Res.* 9(9):1157–1163.

Shao, Z. and Mitra, A.K. (1992). Nasal membrane and intracellular protein and enzyme release by bile salts and bile salt fatty-acid mixed micelles-correlation with facilitated drug transport. *Pharm. Res.* 9:1184–1189.

Sharma, S., White, G., Imondi, A.R., Plaelse, M.E., Vail, D.M., and Dris, M.G. (2001). Development of inhalation agents for oncological use. *J. Clin. Oncol.* 19:1839–1847.

Sidorenko, G.I. and Pinigin, M.A. (1976). Concentration-time relationship for various regimens of inhalation of organic compounds. *Environ. Health Perspect.* 13:17–21.

Silver, S.D. (1946). Constant flow gassing chambers: Principles influencing design and operation. *J. Lab. Clin. Med.* 31:1153–1161.

Sinnet, E.E. (1981). Fast integrated flow plethysmograph for small mammals. *J. Appl. Physiol.* 50:1104–1110.

Smaldone, G.C. (1997). Determinants of dose and response to inhaled therapeutic agents in asthma. In: *Inhaled Glucocorticoids in Asthma: Mechanisms and Clinical Actions* (R.P. Schleimer, W.W. Busse, and P.M. O'Byrne, eds.). Marcel Dekker, New York, pp. 447–477.

Stahlhofen, W., Gebhart, J., and Heyder, J. (1980). Experimental determination of the regional deposition of aerosol particles in the human respiratory tract. *Am. Ind. Hyg. Assoc. J.* 41:385–398.

Stahlhofen, W., Gebhart, J., and Heyder, J. (1981). Biological variability of regional deposition of aerosol particles in the human respiratory tract. *Am. Ind. Hyg. Assoc. J.* 42:348–352.

Stott, W.T. and McKenna, M.J. (1984). The comparative absorption and excretion of chemical vapors by the upper, lower and intact respiratory tract of rats. *Fundam. Appl. Toxicol.* 4:594–602.

Swift, D.B. and Proctor, D.F. (1982). Human respiratory deposition of particles during oronasal breathing. *Atmos. Environ.* 16:2279–2282.

Tamulinas, C.B. and Leach, C.L. (2000). Routes of exposure: Inhalation and intranasal. In: *Excipient Toxicology and Safety* (M.L. Weiner and L.A. Kotkoskie, eds.). Marcel Dekker, New York, pp. 185–205.

Task Group on Lung Dynamics. (1966). Deposition and retention models for internal dosimetry of the human respiratory tract. *Health Phys.* 12:173–207.

Tattersfield, A.E. (1986). Beta adrenoreceptor antagonists and respiratory disease. *J. Cardiovasc. Pharmacol.* 8(Suppl. 4):535–539.

Taulbee, D.B., Yu, C.P., and Heyder, J. (1978). Aerosol transport in the human lung from analysis of single breaths. *J. Appl. Physiol.* 44:803–812.

Thorne, P.S. and Karol, M.H. (1989). Association of fever with late-onset pulmonary hypersensitivity responses in the guinea pig. *Toxicol. Appl. Pharmacol.* 100:247–258.

Touvay, C. and Le Mosquet, B. (2000). Systeme respiratoire et pharmacology de securite. *Therapie* 55:71–83.

Ts'o, T.O.T., Leong, B.K.J., and Chenoweth, M.B. (1975). Utilities and limitations of behavioral techniques in industrial toxicology. In: *Behavioral Toxicology* (B. Weiss and V.G. Laties, eds.). Plenum Press, New York, pp. 265–291.

Uemitsu, N., Minobe, Y., and Nakayoshi, H. (1985). Concentration-time-response relationship under conditions of single inhalation of carbon tetrachloride. *Toxicol. Appl. Pharmacol.* 77:260–266.

Ugwoke, M.I., Verbeke, N., and Kinget, R. (2001). The biopharmaceutical aspects of nasal mucoadhesive drug delivery. *J. Pharm. Pharmacol.* 53:3–21.

Ulrich, C.E., Klonne, D.R., and Church, S.V. (1984). Automated exposure system for metered-dose aerosol pharmaceuticals. *Toxicologist* 4:48.

Valentine, R. and Kennedy, G.L., Jr. (2001). Inhalation toxicology. In: *Principles and Methods of Toxicology*, 4th edn. (W. Hayes, ed.). Raven Press, New York, pp. 1085–1143.

Vidgren, M.T., Karkkainen, A., Paronen, T.P., and Karjalainen, P. (1987). Respiratory tract deposition of ^{99}Tc-labeled drug particles administered via a dry powder inhaler. *Int. J. Pharm.* 39:101–105.

Wanner, A. (1979). The role of mucociliary dysfunction in bronchial asthma. *Am. J. Med.* 67:477–485.

Wanner, A. (1981). Alteration of tracheal mucociliary transport in airway disease. Effect of pharmacologic agents. *Chest* 80(Suppl. 6):867–870.

Webber, S.E. and Widdicombe, J.G. (1987). The actions of methacholine, phenylephrine, salbutamol and histamine on mucus secretion from the ferret in-vitro trachea. *Agents Actions* 22:82–85.

Weibel, E.R. (1963). *Morphometry of the Human Lung.* Springer-Verlag, Heidelberg, Germany.

Weibel, E.R. (1983). How does lung structure affect gas exchange? *Chest* 83:657–665.

Westfall, T.C. and Westfall, D.P. (2011) Adrenergic agonists and antagonists. In: *Goodman & Gilman's: The Pharmacological Basis of Therapeutics*, 12th edn. (B.L. Brunton, B. Chabner, and B. Knollman, eds.). McGraw Hill Medical, New York, pp. 277–304.

Weston, R. and Karel, L. (1946). An application of the dosimetric method for biologically assaying inhaled substances. *J. Pharmacol. Exp. Ther.* 88:195–207.

Widdicombe, J.G. (1974). Reflex control of breathing. In: *Respiratory Physiology*, Vol. 2 (J.G. Widdicombe, ed.). University Park Press, Baltimore, MD.

Willeke, K. (1980). *Generation of Aerosols and Facilities for Exposure Experiments.* Ann Arbor Science, Ann Arbor, MI.

Williams, M.H. (1974). Steroids and antibiotic aerosols. *Am. Rev. Respir. Dis.* 110:122–127.

Wills, P. and Greenstone, M. (2001). Inhaled hyperosmolar agents for bronchiectasis. *Cochrane Database Syst. Rev.* CD002996.

Wilson, A.G.E. (1982). Toxicokinetics of uptake, accumulation, and metabolism of chemicals by the lung. In: *Mechanisms in Respiratory Toxicology*, Vol. 1 (H. Witschi and P. Nettesheim, eds.). CRC Press, Boca Raton, FL, pp. 162–178.

Witschi, H.P. and Brain, J.D. (1985). *Toxicology of Inhaled Materials.* Springer-Verlag, New York.

Witschi, H.P. and Nettesheim, P. (1982). *Mechanisms in Respiratory Toxicology*, Vol. 1. CRC Press, Boca Raton, FL.

Zhang, Y. and Jiang, X. (2001). Detoxification of nasal toxicity of nasal drug delivery system. *Zhongguo Yiyao Gongye Zazhi* 32:323–327.

Index

A

A549 cells
  oxidative stress, 431–453
  silica-induced cytotoxicity, 426–427
Acidity-related epithelial injury, phosgene,
    475–476
Acrolein, CFD model
  flux values and lesion incidence, 163–165
  I–IV levels of, 163–164
  LOAEL/NOAEL, 163–164
  nasal uptake of, 161, 163
  rat *vs.* human nasal passages, 161–162
  Reference Concentration (RfC), 163
  uncertainty factors, 164–165
Activator protein 1 (AP1), 442–443
Acute Exposure Guideline Levels (AEGLs),
    106, 247–248; *see also* Emergency
    response planning guidelines
    (ERPGs)
Acute lung injury (ALI)
  phosgene, 460
  statins, 380
  toxic inhalation injury, 362–363
Acute radiation syndrome (ARS), 508–511
Acute respiratory distress syndrome (ARDS)
  corticosteroids, 377–378
  phosgene, 460
  statins, 380
  toxic inhalation injury, 362–363
  ventilation injury, 373–374
Adamsite
  chemistry of, 218–219
  human toxicology of, 232–233
  toxicological effects, 231–232
Aerodynamic diameter (AD), 540–541
Aerosol
  definition, 73
  particle inhalability, 47
  therapeutic agents, safety assessment
    absorption and clearance, 541–542
    pharmacotoxicity of, 542–543
    pulmonary delivery of, 540–541
Aerosol deposition in nose
  experimental, 50–51
  measurement, 49–50
  prediction, 51
Airway-centered injury, 360
Akaike's information criterion (AIC)
    scores, 139, 141
α7 Nicotinic acetylcholine receptor (α7nAChR)
    agonists, 380–381
Alveolar damage, 360–361
Alveolar macrophages (AMs), 425

Alveoli
  microchannel model
    cell culture, 65
    device construction, 64–65
    linear stretching exacerbates nanoparticle and drug
      toxicity, 65
  microfluidic terminal sac model
    cell culture, 66
    combined solid and fluid mechanical stress
      exacerbates cell injury, 66–67
    device construction, 66
American-European Consensus Conference (AECC), 361
American Industrial Hygiene Association (AIHA), 526–527
Ammonia
  animal toxicity data
    acute lethality studies, 262, 277–278
    effects of, 281–283
    lethality studies, 281, 284
    mouse data, 262
    nonlethal effects, 263–264
    odor threshold, 258–260
    pulmonary/sensory irritation, 260–262, 276–277
    rat data, 262–263
  characteristics, 257–258
  environmental consideration
    Jack Rabbit Project, 28–288
    Minot, North Dakota NH_3 release, 284–287
    pipeline NH_3 release, 287
  pulmonary delivery of, 539
  toxicity in humans, 264–265
    accidental exposures, 265–267
    concentration levels, 278, 280
    dose–response data, 276
    effects for, 280
    ERPG levels, 275–276
    escape and lethality levels, 275
    Houston and Potchefstroom accidents, 272–275
    symptoms and toxicity end points, 278–279
    in volunteers, 267–272
Ampakines, 234
Amphibole asbestos; *see also* Fiber toxicology
  characteristics
    chemical composition, 300
    principal types, 301
    silica-based structure, 301
    structure of, 299–300
  epidemiology
    controlled use, 316
    evaluation of, 319
    fiber lung burden analyses, 317
    high density cement plants, 318
    potency of, 316–317
    short-term exposure, 318
    technical support document, 317